Energy, Transport, & the Environment

Oliver Inderwildi · Sir David King
Editors

Energy, Transport, & the Environment

Addressing the Sustainable Mobility
Paradigm

For James Cameron
with many thanks for all
your help on climate change
and its implications.

David King

Oliver Inderwildi

Oxford June 2012

🐎 Springer

Oliver Inderwildi
Smith School of Enterprise & the
 Environment
Oxford University
75 George Street
Oxford
OX1 2BQ
UK

Sir David King
Smith School of Enterprise & the
 Environment
Oxford University
75 George Street
Oxford
OX1 2BQ
UK

ISBN 978-1-4471-2716-1   e-ISBN 978-1-4471-2717-8
DOI 10.1007/978-1-4471-2717-8
Springer London Heidelberg New York Dordrecht

Library of Congress Control Number: 2012932742

Preface

I had the pleasure to be one of the speakers at the 2010 World Forum on Enterprise and the Environment, held by the Smith School of the University of Oxford. The participants, drawn from academia, industry, government and NGOs, were asked to explore the challenges and offer pathways to a sustainable, low carbon transportation future. To continue that spirited set of discussions and focus groups, a number of the participants have contributed to this volume on *Energy, Transport and the Environment*.

The topic of the forum is part of the larger energy and environmental challenge we face today. The Industrial Revolution and all that followed over the past 250 years liberated us from the constraints of human and animal power. In this brief instant in the history of human existence, we achieved remarkable advances in prosperity, but our success has brought challenges as well as opportunities. This revolution has sustained an approximately tenfold increase in the world population, from 700 million to an estimated 7 billion people by the end of 2011. It elevated the living conditions of an ever increasing fraction of humanity to heights that were unimaginable to the kings, queens, and emperors of earlier times. Our homes are warm in winter, cool in summer, and lit at night. The mobility of information, goods, and people has been fueled and transformed by our use of energy. We travel to a local market under the pull of more than a hundred horses to buy produce grown halfway around the world. We fly across continents and oceans using engines that have the power of a hundred thousand horses.

Our present path is not sustainable. We are becoming increasingly adept in our ability to find and extract fossil energy, but there is no debate that the supply of these fuels is finite. As discussed in the first section of this book, new discoveries of conventional oil, the dominant source of transportation energy, are peaking. There is a growing strain between projected demand and future supplies. The demand for oil will increase, driven in large part by rising prosperity in the developing world. On the supply side, more than 70% by volume of major discoveries since 2007 have been in deep offshore reservoirs, and oil companies are about to embark on greatly expanded exploration of northern arctic regions. "Unconventional" sources such as bituminous oils may play an increasingly important role as higher oil prices and improved technology make their extraction economically feasible.

Even if technology advances were able to accommodate increases in demand with only modest increases in the cost of extraction, we will still need dramatically to de-carbonize our use of energy. The scientific evidence that the climate is changing, largely due to human greenhouse gas emissions, grows more compelling with each passing year. In order to mitigate the most dire risks, we will need to make substantial and timely changes in an arena where established technologies are naturally favored and the timescale of change is often slow. Despite the inherent advantages of liquid fuel, it took several decades to transition to an oil-based transportation economy. Sustained and creative actions will be needed to replace the embedded infrastructure, especially when full current and future costs of business are not yet included.

Transport consumes roughly 19% of the world energy supply and produces 23% of the carbon dioxide emissions. An IEA report predicts that the quantity of transport emissions will double by 2050 in the absence of new government policies.[1] Furthermore, we know of no practical technology that can be used to capture greenhouse gas emissions from mobile platforms. Thus, to achieve economic prosperity, energy security, and environmental responsibility, we will need to make fundamental changes. These changes will require significant technological advances, serious commercial investments, and wise government policies to guide this transition. The articles on the various transportation sectors—road, aviation, sea, rail and cargo—indicate that solutions to sustainable mobility may be *the* most challenging aspect of our energy and climate challenge.

This volume considers transportation in the urban environment. The vast majority of the future infrastructure of the world will be built in locations where we have the greatest opportunity to transition to sustainable mobility. What set of acceptable, effective, and affordable government policies, tailored in each country, will most effectively assist this transition? A variety of policies are discussed: they span the gamut from financial incentives and disincentives to extensive and heavily supported public transportation systems to new businesses such as shared "personal" vehicles. All of these ideas are aimed at mitigating the growth of vehicles in cities. In this section, I would have liked to have seen more emphasis on urban planning in developing countries. By 2030, China and India alone are projected to expand or build cities for roughly 600 million people. Developing countries have an enormous opportunity to design urban infrastructure where thoughtful "system" approaches can be used to integrate living, working, and shopping areas with urban transport and recreational spaces. There are enormous opportunities here—the creation of sustainable cites and surrounding areas can greatly improve the quality of life of its inhabitants while reducing costs, congestion, and carbon emissions. One can also learn from innovative master plans of urban renewal projects and university campus expansions.

---

[1] *Transport, Energy and CO₂, Moving Towards Sustainablilty*, IEA Report, (2009).

Many of the articles in the book describe technical solutions. The de-carbonization of transportation would greatly benefit from technical breakthroughs in a number of areas. I personally believe that the fastest gains in reducing the cost and carbon footprint of road transport will come from fuel efficiency standards. More radical approaches such as improved batteries for plug-in hybrids and electric vehicles coupled with clean generation of electrical power, the development of affordable fuel cells and low-carbon sources of hydrogen, and alternative low-carbon fuels all have potential and should be explored. No one can predict which technologies will prevail, and it is therefore prudent to support a diverse research and development portfolio rather than down-select prematurely. Similarly, we should not deploy nascent technologies prematurely at great public cost. Technologies where substantial new infrastructure would have to be established can be field-tested in cost effective ways. For example, hydrogen fuel cell vehicles should be first piloted in situations where centralized refueling is possible and where hydrogen is readily available. What is clear is that the standard for technologies that can compete in price and performance with the internal combustion engine and liquid fossil fuel is high. We either must improve our low-carbon technology choices, change market demand, or both.

Finally, I want to return to the importance of finding solutions to sustainable mobility. Mobility is considered an integral part of the world's increasing standard of living. Ownership of a cell phone, a television, a refrigerator, and an automobile are often seen as important symbols of an improving standard of living. Furthermore, indigenous automobile production is seen as an important driver of the economic engine of developed and developing countries. When coupled with the fact that the population is projected to grow to approximately 9 billion by 2050 and 10 billion by the end of this century,[2] it is easy to see we face a stark challenge: meeting the world's rising energy needs with clean, sustainable sources. Countries with the highest per capita standard of living are beginning to decouple economic prosperity from carbon emissions, but we need to accelerate this trend. The developing world is following a proven path to prosperity; it is in the best interests of the developed *and* developing world to find a better path. The Red Queen in *Alice in Wonderland* describes our dilemma: "It takes all the running you can do to keep in the same place. If you want to get somewhere else, you must run at least twice as fast as that!"

Steven Chu
U.S. Secretary of Energy
Nobel laureat 1997

---

[2] See, for example, *Science* 333, Special Section on Population, pp 562–594, (2011).

Acknowledgments

The editors are grateful to the Smith School, Oxford University and the Smith Family Educational Foundation for their support of this work.

Contents

Part II   Road

Foreword
James Smith

Introduction

Oliver Inderwildi and Sir David King

Abstract In 1872 Phileas Fogg set off on his trip around the world from the Reform Club on Pall Mall in London, according to Jules Vernes [1]. The journey starts with a trip in a horse carriage to London's Charing Cross station, and from there Fogg and his valet travel by steam-boat, hot-air balloon, railway and even elephants. They manage the journey in just less than 80 days. The equivalent round-the-globe trip today might start with a walk to Piccadilly Circus and then an underground train trip to Heathrow airport on the Piccadilly Line. A three-legged flight via Tokyo Narita and San Francisco would get you back to Heathrow, and the London Underground would take you back to St. James. It is indeed now possible to do the previously tedious 80-day journey in less than 2 days, even with delays. The most tedious part of the modern version of Phileas Fogg's journey is in all likelihood passing through airport security at Heathrow. Not only would the modern trip be more than 40 times faster and significantly more convenient, it would also be virtually risk free.

In 1872 Phileas Fogg set off on his trip around the world from the Reform Club on Pall Mall in London, according to Jules Vernes [1]. The journey starts with a trip in a horse carriage to London's Charing Cross station, and from there Fogg and his valet travel by steam-boat, hot-air balloon, railway and even elephants. They manage the journey in just less than 80 days. The equivalent round-the-globe trip today might start with a walk to Piccadilly Circus and then an underground train trip to Heathrow airport on the Piccadilly Line. A three-legged flight via Tokyo Narita and San Francisco would get you back to Heathrow, and the London Underground would take you back to St. James. It is indeed now possible to do the

O. Inderwildi (✉) · Sir David King
Smith School of Enterprise and the Environment, University of Oxford,
Hayes House, 75 George Street, Oxford OX1 2BQ, UK
e-mail: oliver.inderwildi@smithschool.ox.ac.uk

O. Inderwildi and Sir David King (eds.), *Energy, Transport, & the Environment*,
DOI: 10.1007/978-1-4471-2717-8_1, © Springer-Verlag London 2012

Fig. 1 Horse-drawn street car in Dresden (Kingdom of Saxony, German Empire), pedestrians and carriages on Pall Mall, London (Kingdom of Great Britain)

previously tedious 80-day journey in less than 2 days, even with delays. The most tedious part of the modern version of Phileas Fogg's journey is in all likelihood passing through airport security at Heathrow. Not only would the modern trip be more than 40 times faster and significantly more convenient, it would also be virtually risk free.

In the process of the Industrial Revolution, the Victorians were instrumental in developing the first steam-powered transport infrastructure which Jules Vernes described in his epic novel. The steam engine, critically improved by the Scottish inventor James Watt, made it possible to utilise energy produced by the combustion of coal or wood for transport purposes in large vehicles such as boats or trains. Prior to this era there was an exclusive reliance on human and animal powered transport to move people and goods. During the nineteenth century intercity and international movements became dominated by steamers and steam railways. Urban transport, however, still relied heavily on horses for both individual and public transport (Fig. 1). At that time horse manure posed a significant problem for cities and the future of transport was intensely discussed by authorities [2]. The problem, however, was soon resolved. In the late nineteenth century Gottlieb Daimler and Karl Benz invented the internal combustion engine, an invention that transformed the world of transport. These engines provided the direct creation of motion from the combustion of fuels without steam, which, together with liquid fuel replacing coal, meant that the engine size was drastically reduced, and they could be installed in smaller vehicles such as carriages. This technological advancement rapidly spread and horses in cities were gradually replaced by vehicles powered by these combustion engines. Combustion-engine vehicles, cars, of course now dominate the townscapes in the developed and developing worlds.

Initially, the engines of these vehicles were fuelled with alcohol bought from pharmacies, but not for long.

Alongside the rapid uptake of the combustion-engine vehicle, a company in the United States, Standard Oil, established a market for kerosene produced from 'rock oil' that was initially marketed to replace expensive whale oil for lighting [3].

Rockefeller, Rothschild and Nobel were involved in this rapidly growing business that essentially lit up the night by providing cheap, quality oil for illumination to the mass market [3]. The demand for kerosene was expected to collapse due to the electrification of cities and the invention of the light bulb. This proved to be a massively incorrect forecast due to the new market for liquid fuel, the car [3].

The internal combustion engine and 'rock oil', now known as petroleum or crude oil, was a perfect match. As the demand for lighting oil decreased, the demand for oil for combustion engines rapidly increased; oil rapidly emerged as the big driver of the twentieth century economy. Road vehicles, diesel trains and later on larger buses were fuelled with different types of oil-derived fuels. On the eve of World War I Winston Churchill decided that Britain would have to base its 'naval supremacy on petroleum' due to the strategic benefits provided by oil, such as greater speed for ships and more efficient use of manpower on board [4]. This decision promoted oil as the unchallenged source of transport fuel, and access to oil became critical for military strategies in both the World Wars [5]. After World War II aviation entered the mobility market. Technological advancements, many originating from the arms race during the wars—novel materials, reliable propulsion technology and improved aircraft design made aviation safe. These safety improvements combined with cheap kerosene derived from petroleum have made commercial aviation the impressive industry it is nowadays.

Mass-produced vehicles became ubiquitous in the industrialised world and, reduced transport costs significantly as cheap, abundant petroleum became less available, globalising trade and integrating industrialised economies [6]. These advances have effectively shrunk the globe and built the fundamentals of the globalised society [7]. The transport revolution has not only changed our way of life, but has also massively transformed global economic development, human welfare and technological development [6]. At present, we consume 86 million barrels of crude oil everyday, enough to fill the Empire State Building 11 times over. Most refined products are used as transport fuels [8]. This enormous amount of crude oil is the main fuel of a highly complex multi-modal global transport system which has become the backbone of the global economy. Consequently, energy security concerns deeply influence geopolitics, as crude oil supply is inextricably tied to economic activity and development [9]. The objective of stable oil supplies is therefore a key part of foreign policy and diplomacy [10]. Enhancing the security of supply is a major aspect of the security agenda of oil-importing countries, while oil-exporting countries are often heavily dependent on stable oil markets [3]. Secure access to oil, especially during times of war, is a critical component of national security policies and military strategy; without access to oil military activity being incapacitated [5, 11]. Any threat to the continual availability of oil can trigger political crises and can ultimately provoke the use of military force [12].

More recently, concerns have been raised by the risk of potential oil shortages due to depleting oil reserves [13], which will eventually lead to economically damaging increases in oil prices, less secure supplies and consequently a greater risk of conflict [12].

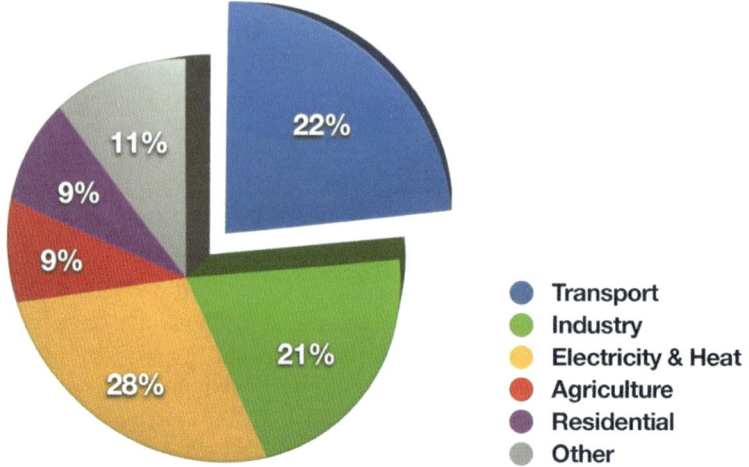

Fig. 2 Greenhouse gas emission in the European Union by sector (2007), European Environment Agency, 2009

Moreover, unease over looming climate change induced by anthropogenic greenhouse gas (GHG) emissions has intensified this discussion [14–16]; the transport sector is the second largest emitter of GHGs in the industrialised world (Fig. 2). The crude-oil addicted transportation culture hence has two main problems: declining fuel supply, the input problem, and increasing GHG emissions, the output problem. Both problems are elucidated below.

1 The Input Problem

In the transport sector the fuel mix has been dominated by fuels derived from so-called light crude oil, which is accessible and cheap to produce [3]. However in recent years, concerns have grown over the capacity for oil reserves to service rising demands [13, 17, 18]. These reserves can roughly be classified as conventional resources, such as light crude, and unconventional resources, such as tar sands, heavy oil and coal. The status of conventional oil reserves is obscured by a lack of binding international standards that define conventional oil (reserve volume and grade) [13], by intentional misreporting to suit political or financial agendas, and by inherent technical uncertainty [13]. Reporting and information agencies estimate that the remaining world oil inventory is between 1,184 and 1,342 Gb (giga barrels) [8, 19], while independent authors and academic institutions are more conservative and estimate conventional oil reserves at between 800 and 900 Gb [13, 18, 20]. At current demand, conventional oil reserves are forecast to run out by 2035. It should be noted, however, that reserve-production ratios are

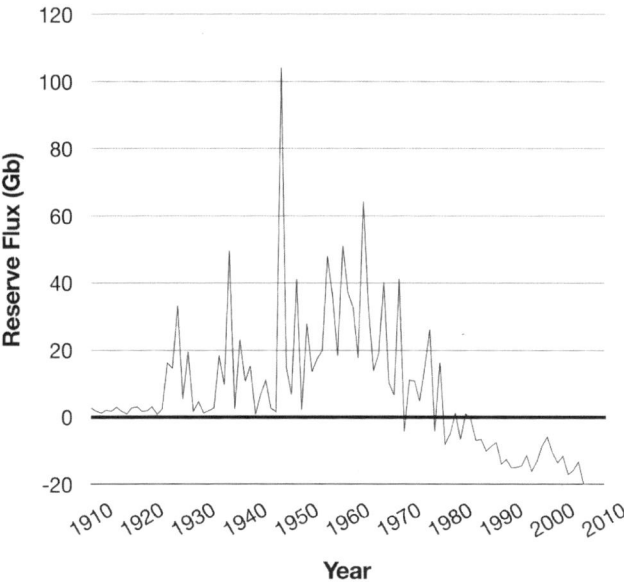

Fig. 3 Oil flux entering and exiting the global conventional oil reserves adapted from [13]

not sensitive to declining production rates, even if the net amount produced over an extended period remains the same.

Oil reserves are defined as the fraction of oil resources that can be commercially and technically recovered at current market prices [13]. Best practice assessment demands that estimated reserve volumes should be stated together with a 50% probability (P50) of achieving the specified volume to limit the inherent assessment uncertainty [13]. This uncertainty is increased by ambiguity over the point at which sub-commercial resources can be reclassified as commercially exploitable reserves—the price-reserve relationship [13]. Moreover, data from reporting agencies are inconsistent on the inclusion of Canadian tar sands in world reserve estimates and usually report sub-commercial reserves prematurely [13]. If data presented by information and reporting agencies is amended to reflect conventional P50 reserves, and to account for widely acknowledged false additions, figures become consistent with those quoted from independent institutions, discussed above.

In the context of rising liquid fuel demand, it is necessary to consider the effect that limited conventional oil resources may have on the liquid fuel mix. Figure 3 gives a history of the net flux of oil entering and exiting the conventional global oil reserve inventory, based on backdated P50 data.

Data below the zero flux axes refer to periods of net withdrawal from reserves. This first occurred in 1972 and has consistently occurred since 1980, indicating that conventional oil reserves have been in decline since then. This is in sharp contrast to figures published by reporting and information agencies that indicate oil reserves are continuously rising. Since records show that the peak of conventional oil discovery occurred in the early 1960s, it is unlikely that many significant and accessible conventional oil fields remain to be found. The World Energy Outlook 2008 estimates that the producing oil fields of the world are declining at such a rate that by 2020, only 50% of liquid fuel demand will be serviced by resources that are in production today [8]. Shortages in the supply of conventional oil will most likely be met by unconventional oil. But the major drawback of unconventional reserves is that extraction is much more energy consuming, as is the conversion to usable liquid fuels. Consequently, fuels derived from these sources have a higher carbon footprint, i.e. the same amount of fuel burnt would result in a larger amount of GHGs formed. They also have higher costs due to the energy consumed in production. Rising fuel demand and higher emissions per volume of fuel thus raises oil prices and multiplies overall emissions from the transport sector, and we conclude that while unconventional reserves could mitigate the input problem, they will necessarily exacerbate the output problem. The impact of a major shift to unconventional oil on oil prices will in turn slow down the global economy: it is estimated that a 10% rise in oil prices produces a 0.2% drop in global GDP. A doubling of oil prices, which should be anticipated over the coming decades, could reduce global GDP by 2% [21].

2 The Output Problem

The average temperature at the earth's surface has risen by $0.74 \pm 0.18°C$ over the last 100 years, according to the intergovernmental panel on climate change (IPCC) [22]. A further increase of between 1.1 and 6.4°C is likely this century. This acceleration in climate change is largely attributable to anthropogenic GHG emissions—carbon dioxide (CO_2), methane (CH_4) and nitrous oxides (NO_x) [23]. Since the Industrial Revolution the combustion of fossil fuels, together with ongoing deforestation, has caused atmospheric CO_2 concentrations to rise by 36% [15]. The atmosphere is heated through the absorption of infrared radiation by these GHGs [16]; consequently, raising the atmospheric GHG concentrations accelerates the rate of warming. The current level of atmospheric GHG concentrations is equivalent to 430 parts per million (ppm) of CO_2, so-called CO_2 equivalents ($CO_{2(eq)}$), compared to 280 ppm before the industrial revolution [24]. Even if the annual rate of GHG emissions were to stagnate at the current level, atmospheric GHG concentrations will still reach 550 ppm $CO_{2(eq)}$ by mid-century, double the pre-industrial levels. But a business-as-usual scenario, with rising rates of GHG emissions, will yield 550 ppm of $CO_{2(eq)}$ by 2035 [24].

The consequences of anthropogenic climate change are wide ranging. Warming oceans, glacial retreat and the melting of land-based ice, particularly in Greenland and the Antarctic, means that sea levels are likely to rise by between 0.18 and 0.59 metres by the end of the century, putting at risk those living near coasts. Rainfall patterns are also likely to change, extreme weather events could become more frequent, water scarcity will increase in many regions and crop yields will fall [16]. The spread of diseases, such as malaria and dengue fever, could accelerate, potentially causing turmoil in large parts of the developing world [6].

The Stern Review [24] warned that the costs of extreme weather alone could reach 0.5–1% of world gross domestic product (GDP) per annum by the middle of the century, and will keep rising if the world continues to warm. The same report says that "...climate change [under business as usual conditions, will] reduce welfare by an amount equivalent to a reduction in consumption per capita of between 5 and 20%" [24]. Moreover, studies by Barker have shown that the costs of mitigating and preventing the worst environmental effects of climate change will be insignificant compared to the risks and potential costs of an unhindered and unmitigated climate change [25].

This potential threat has led to intergovernmental and international environmental treaties such as the Kyoto Protocol to the United Nations Framework Convention on Climate Change, which aims at a stabilisation of GHG concentrations by reducing emissions [22]. In order to comply with the GHG emission targets set out by the EU and the 67 nations committed to action following the 2009 Copenhagen Accord [26], countries will have to reduce the emission intensity of all sectors, especially the emission intensive transport sector. However, reducing GHG emissions from the transport sector without decreasing human mobility remains a non-trivial challenge.

3 The Challenge

Economic activity and hence societal welfare are intrinsically linked to the mobility of humans and the transportation of goods. Reducing emissions from the transport sector by reducing mobility would have severe consequences for the global economy [27]. In 2009, we published a roadmap study that assessed technologies and policies that have the potential to reduce emissions from the transport sector while enhancing human mobility [24]. In 2010, the Smith School of Oxford University invited world-leading experts from academia, captains from industry as well as global decision makers to its annual World Forum on Enterprise and the Environment to speak and brainstorm on the topic in the debating hall of Oxford Union. The two-day conference focused on low-carbon transport and set out where the technological and political problems lie. Financial tools and economic measures were evaluated which could accelerate the transition.

The World Forum concluded that detailed futures analyses would be critical in order to plan the transformation of the global transportation system while avoiding further resource scarcity issues and other potential lock-ins.

As a follow-up to both the Roadmap study and the World Forum event, we therefore invited key thinkers to contribute to this volume in order to gauge issues and opportunities of the highly complex nexus of problems in energy, transport and environment in more depth, based on the critical issues determined during the brainstorming at the Oxford Union. In parallel the Smith School has also instigated an integrated, in-depth futures analysis of the transition of the global transport sector into a fossil-fuel free economy.

4 The Content of this Volume

The mobility challenge is here addressed from multiple perspectives by practitioners from the public, private and academic sectors. The compendium determines the challenges and opportunities embedded in a transformation of the transportation sector by taking an integrated approach.

Transport is intrinsically an energy problem. This compendium therefore introduces the complex challenges with an in-depth discussion on energy issues relevant to transport. Experts from all walks of life discuss energy security, oil reserves and the potential of alternative fuels produced from various feedstocks. Moreover, the political implications and potential consequences of resource scarcity are addressed in this first section.

Thereafter, a comprehensive section covers road transport, the biggest CO_2 emitter of all areas. This section is introduced by a detailed discussion of the mosaic of road transport fuels that are available and then analyses the technological challenges that have to be tackled in order to renew road transport. The section also covers policy options that allow us to reduce the impact of road transport using technologies that are available today.

Subsequently, urban mobility is addressed in detail. This is particularly important in the light of rapid current urbanisation in developing countries. Not only does this section cover a vision for a fully integrated urban transport system, but also discusses how we can develop and manage the existing system to make it more convenient and accessible while reducing its environmental impact. Last but not least, this section addresses the use of economic tools to assist this transition.

Aviation is an ever-growing sector of transport and, largely due to safety concerns, it is the most challenging sector for emissions reductions. The section therefore covers all areas that will enable us to reduce emissions and fuel consumption in commercial aviation without compromising on safety. Aviation governance as well as management of airspace, airports and airlines is included in the analysis. Moreover, technological issues such as propulsion technology, aircraft design, novel materials as well as alternative fuels, are all addressed by leading experts. The section concludes with an assessment of the potential of modal substitution from air to rail, which could significantly reduce the environmental impact of aviation.

Railways are of course not only important as alternatives to aviation, but are also critically important for cargo, including connections with freight ships and areas with high population densities. The chapters on sea, rail and cargo address both the potential and the challenges for this sector. The need to improve global logistics in order to reduce emissions from cargo is critically assessed. This section pays special attention to worldwide navigation and the potential to reduce environmental impact. Following this there are two contributions which critically discuss high speed rail as a potential solution for the constant growth in demand for transport in the United Kingdom.

Finally, a section on finance and economics assesses the role of these levers to assist the transition of the current transport infrastructure. This section starts by tackling the use of financial instruments and their ability to direct capital to where it is needed the most. Second, appraisal methodologies are investigated in order to determine whether these are still fit for purpose. Third, incentives are critically discussed in the context of the avoidance of perverse incentives.

5 The Benefits

Greenhouse gas emissions are not the only problem caused by transport; vehicles emit toxic, local pollutants, such as nitrous and sulphur oxides, volatile organic compounds, carbon monoxide and soot, which cause asthma and other respiratory diseases and produce acid rain that destroys forests [27]. Although the emission levels of these pollutants in developed countries have been reduced significantly through technologies such as advanced combustion and exhaust treatment systems as well as low sulphur fuels, those in the developing countries remain high [27]. Noise from busy streets as well as landing and departing aircraft can affect both people and wildlife [27]. In other words, improving our transport systems can reap many environmental benefits, while green technologies could create jobs in both developed and underdeveloped areas. Previous technologically driven revolutions, such as the transformation of our communication system in the 1990s, have had a tremendous impact on economic growth. Last but not least, the relief from the current crude oil addiction will ease geopolitical tensions.

There are many success stories. Aviation was for many years an aspiration of mankind, and over the past 100 years has become the safest mode of transport, due to advances in aircraft technology and air traffic management. Air quality in urban areas has dramatically improved in the developed world, key to this being the introduction of emission standards and regulations and the progressive development of exhaust-gas after treatment systems for cars.

Moreover, many cities, such as Paris and Bogota, have shown that a thought-through redesign can increase livability while preserving original flair. Optimising urban mass transit and integrating it into the wider transport system, for instance rail and aviation, can incentivise modal substitutions and reduce other impacts of mobility such as congestion and its detrimental economic impact.

However, achieving all these benefits will neither be straightforward nor cheap as modern mobility is an intrinsically complex problem. It is not only a resource scarcity problem, but includes technological and management issues, human psychology, financial markets and of course political will and economics play crucial roles.

This book tackles technical issues and policy options in the context of a holistic examination which includes human psychology, traffic management, finance and economics.

References

1. Verne J (2007) Around the world in eighty days. Penguin Classics, London
2. Burrows EG, Wallace M (1999) Gotham: a history of New York city to 1898. Oxford University Press, Oxford
3. Yergin D (2009) The price—the epic quest for oil, power and money. Free Press, New York
4. Jenkins R (2002) Churchill: a biography. PanMacmillan, London
5. Gray C (1999) Modern strategy. Oxford University Press, Oxford
6. Sachs J (2008) Common wealth: economics for a crowded planet. Penguin Press, London
7. Stiglitz JE (2003) Globalization and its discontents. W.W. Norton & Co., New York
8. IEA (2008) World energy outlook 2008. International Energy Agency, Paris
9. Moran D, Russell JAE (2008) Energy security and global politics: the militarization of resource management routledge global security studies. Routledge, Oxford
10. Kissinger H (1994) Diplomacy. Simon & Schuster, New York
11. Shaffer B (2009) Energy Politics. University of Pennsylvania Press, Philadelphia
12. Klare MT (2002) Resource Wars. Henry Holt & Co., New York
13. Owen NA, Inderwildi OR, King DA (2010) The status of conventional world oil reserves–hype or cause for concern? Energy Policy 38(8):4743–4749
14. Giddens A (2009) The politics of climate change. Polity, Cambridge
15. Gore A (2006) An inconvenient truth: the planetary emergency of global warming and what we can do about it rodale books. Emmaus, Pennsylvania
16. Walker G, King D (2008) The hot topic: how to tackle global warming and still keep the lights on. Bloomsbury, London
17. Leggett J (2005) Half gone: oil, gas, hot air and the global energy crisis. Portobello Books Ltd, London
18. Campbell CJ, Laherrere JH (1998) Preventing the next oil crunch—The end of cheap oil. Sci Am 278(3):77–83
19. EIA (2009) International energy outlook 2009. Energy Information Administration, Washington
20. Hook M, Hirsch R, Aleklett K (2009) Giant oil field decline rates and their influence on world oil production. Energy Policy 37(6):2262–2272
21. http://www.economist.com/blogs/freeexchange/2011/08/energy
22. IPCC (2007) Summary for policymakers climate change 2007: the physical science basis. Contribution of working group I to the fourth assessment report of the intergovernmental panel on climate change
23. Meinshausen M, Meinshausen N, Hare W, Raper SCB, Frieler K, Knutti R, Frame DJ, Allen MR (2009) Greenhouse-gas emission targets for limiting global warming to 2 degrees celsius. Nature 458(7242):1158–U1196. doi:10.1038/nature08017
24. Stern N (2006) Stern review: the economics of climate change. HM Treasury, London
25. Barker T (2008) The economics of avoiding dangerous climate change. An editorial essay on the stern review. Clim Change 89(3–4):173–194. doi:10.1007/s10584-008-9433-x
26. King DA, Richards K, Tyldesley S (2011) International climate change negotiations: key lessons and next steps. University of Oxford Smith School, Oxford
27. King DA (ed), Inderwildi OR et al (2009) The future of mobility roadmap. University of Oxford, Oxford

Part I
Energy

Foreword
Sir Chris Llewellyn Smith

The biggest challenge of the twenty-first century is to provide sufficient food, water and energy to allow everyone on the planet to live decent lives, in the face of rising population, the threat of climate change and the declining availability of fossil fuels. Energy is a means, not an end, but it is a necessary means: without adequate supplies of energy, it will not be possible to support the world's growing and increasingly urbanised population. This book, with its focus on transport—which accounts for some 25% of the world's use of primary energy—is a timely contribution to the debate on how to moderate energy use, and make it more efficient, and the search for alternative, sustainable and environmentally responsible sources of energy.

The challenge is enormous. Despite assuming the successful implementation of all agreed national policies and announced commitments designed to save energy and reduce the use of fossil fuels (whether or not any steps have been taken to implement the polices or meet the commitments), the International Energy Agency's 2010 'new policies scenario' shows energy use increasing 35% by 2035 (relative to a 2008 baseline), with consumption of fossil fuels up 24% (almost all of the increases being in developing countries, where additional energy is needed to raise living standards and populations are growing most rapidly). We should clearly aspire to do better.

Yet implementing the new policies scenario, which envisages a 70% increase in nuclear power, a 72% increase in hydro power (think of the civil engineering involved), and a 13-fold increase in wind power, would be extremely difficult. Indeed, most dispassionate observers expect energy use to increase by more than 35% by 2035. For example, as Iain Conn's excellent overview in the opening chapter of this book recalls, BP's recent energy outlook 2030, which—while not being a 'business as usual' extrapolation—analyses what BP thinks will happen and not what they would like to happen, foresees a 39% increase in energy use by the earlier date of 2030.

Decarbonising the economy, and moving away from oil-based transport, is frequently discussed in the context of reducing climate change. But the possibility, pointedly discussed by Colin Campbell in Chap. 2, that the anticipated peak in world oil production will soon be reached, or may already have been passed, and the certainty that eventually gas and coal will also become scarce, provide an equally if not more compelling reason to wean the world from fossil fuels. The urgency is underlined by the historical examples considered by Jörg Friedrichs in Chap. 3, which suggest that the political consequences of a rapid decline in the availability of fossil fuels could be extremely unpleasant.

The next four chapters deal with alternative fuels: unconventional oil (from the Canadian tar sands and other sources); synthetic fuels and the conversion of coal and gas to oil (which is likely to be done on a large scale if oil production declines rapidly before non-oil-based transport systems have been widely deployed, thereby in turn speeding up the exhaustion of coal and gas); biofuels; and hydrogen, as a possible means of storing energy generally as well as powering transport. There are useful discussions on the critical question of well-or field-to-wheels carbon emissions associated with these alternatives, and the challenges that must be overcome to roll them out on a large scale. In the penultimate chapter, Jacometti et al forcefully remind us that weaning the world from fossil fuels will require major changes in behaviour and lifestyles as well as new technologies.

These opening chapters provide an excellent introduction to the following discussions on different transport sectors, and the financial challenge of transforming the transport system.

Professor Sir Chris Llewellyn Smith FRS
Director of Energy Research Oxford University
President SESAME Council

Energy, Transport and the Environment: Providing Energy Security

Iain C. Conn

Abstract Growing global demand for energy as well as recent geopolitical and technical concerns and issues have served to move energy security up the policy agenda. At the same time the challenge of stabilising greenhouse gas emissions at "safe" levels remains a clear scientific imperative. It is time to stop 'polishing the 2050 diamond' and to take practical steps today towards a lower carbon, secure energy future. Moving to action is more important than ever. We must now take the practical steps available to us that can begin to make a material impact on meeting both climate and energy concerns affordably. Pathways for both transport and power are available to us today that can radically reduce emissions without threatening energy security and at a reasonable cost to consumers. The geopolitical relationships necessary to accelerate progress must also be leveraged to accelerate alignment and convergence towards a coherent set of global energy relationships and markets which enable economic progress and stability.

1 Introduction

Energy is at the heart of economic development and indispensible to our way of life. We need energy to be secure but we also need it to be affordable. Increasingly we also need lower-carbon energy as part of a sustainable lower-carbon economy. The balancing of these objectives in a robust policy framework is a truly demanding task and one of the main challenges facing us all in the twenty-first century.

I. C. Conn (✉)
BP p.l.c., 1 St. James's Square, London, SW1Y 4PD, UK
e-mail: connic@bp.com

O. Inderwildi and Sir David King (eds.), *Energy, Transport, & the Environment*,
DOI: 10.1007/978-1-4471-2717-8_2, © Springer-Verlag London 2012

Equally we cannot afford to be so taken up with the long-term challenge—what might be described as 'polishing the 2050 diamond' in the endless search for future perfection—that we fail to move into action. There are practical steps open to us now that can deliver material progress towards a secure and lower-carbon economy. We should not hesitate to get started on this long-term journey.

The framing of future energy policy would be a difficult enough task without allowing for the unexpected but recent events have been a reminder that the unexpected should always be factored into our thinking. Tragedies such as those in March 2011 in Japan can overtake industries and companies as well as countries. Fukushima could well change the immediate outlook for the nuclear industry, just as the Gulf of Mexico accident in 2010 asked fundamental questions of the oil industry about safeguards in deepwater drilling.

Any serious review of energy policy consequently needs to begin with an 'eyes wide open' look at the realities of global energy. This is what BP has tried to do in its Energy Outlook to 2030 [1] publication. This essay starts with a look at this analysis before reviewing some of the practical steps open to policy-makers and then looking at the international alignment that will be needed to deliver the desired outcomes. There are many policy choices to be made and they can have far-reaching consequences. It is hoped that this paper can provide some useful signposts for this task.

2 Energy Outlook 2030

BP's best judgement of the likely path of global energy markets to 2030 takes into account anticipated policy, technological and economic changes [1]. It is in many ways a 'reality check', a best view of the future energy world weighed on a balance of probabilities; it's not a 'business as usual' extrapolation (Fig. 1).

It recognises that global energy demand is fundamentally driven by population and GDP growth. By 2030, global population is projected to rise by around 20% but global income is projected to double. Energy consumption is likely to continue to grow as a result. This upward pressure will be mitigated to some degree by increased efficiency, which will be reflected through a reduction in energy intensity or the energy needed to produce one unit of GDP (Fig. 2).

Three key points emerge from this analysis. First is the projection of a 39% absolute increase in global energy consumption between now and 2030—a huge figure, but much lower than the expected doubling or more of global GDP.

Second is the changing balance of the energy world. In 2010 the total energy consumptions of the non-OECD and OECD were similar. However, by 2030 it is expected that the non-OECD will grow by a further 68% but OECD consumption will remain nearly static. This rebalancing has significant implications for the geopolitics of global energy and this is discussed in more detail later in this paper.

The third key point is on global energy mix. Energy evolution has never been quick, due to the scale of the "installed base" and the pace of technology.

Global Energy Consumption and Energy Mix

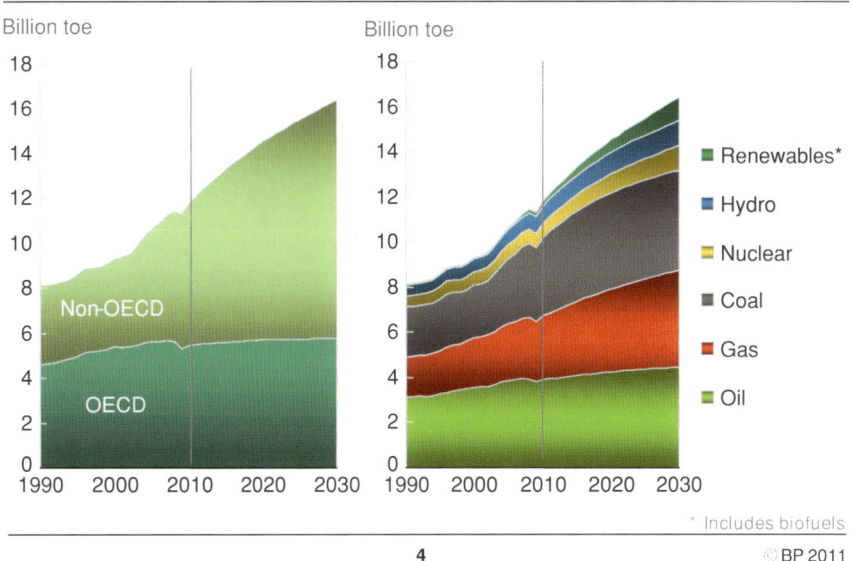

Fig. 1 Shows expected global energy consumption and energy mix from 1990 through to 2030. (Published with kind permission of © BP 2011. All Rights Reserved.)

However, over time important shifts in the global energy balance are clearly evident. Although oil use continues to grow gradually in absolute terms, the share accounted for by oil as a proportion of total world primary energy is declining steadily, while coal is maintaining its market share and growing significantly in absolute terms. However, by far the fastest growing source of energy to 2030 will be renewables. These include biofuels, produced and traded as a global commodity, which by 2030 could meet about 9% of transport fuel demand. Equally, despite these impressive growth rates, non-hydro renewables including biofuels will still account for only some 6% of global primary energy by 2030. Hydro and nuclear are projected to account for another 7% each of global primary energy over this period (Fig. 3).

The balance of the remaining 80% will be split almost equally among oil, coal and natural gas. Of these, natural gas will increasingly be the fuel of choice for power generation. It is a plentiful resource, flexible and economic and burns with half the CO_2 emissions of coal and half the capital cost per unit of generating capacity. It is also ideal for matching with intermittent renewable supplies. As a

Energy Intensity of Development

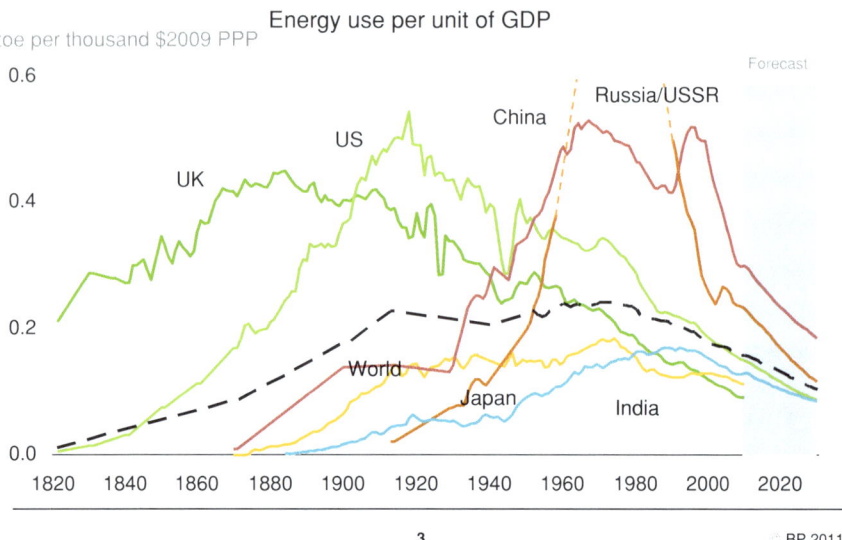

Fig. 2 Is compiled from recorded national economic data. The *left hand* vertical axis shows tons of oil equivalent per unit of GDP and the horizontal axis runs from 1820 to the present day. The series of curves shows the development of energy intensity for a range of countries over this period. The analysis shows that energy intensity peaks in all economies at the high point of heavy industrial development. Thereafter there is a general convergence towards a consistent and much lower intensity of energy use (Published with kind permission of BP 2011. All Rights Reserved.)

result, natural gas is expected to be the 'big winner' among the hydrocarbon fuels, gaining market share from both coal and oil.

Under much more aggressive climate policies, it is still unlikely that the world will get near to the Intergovernmental Panel on Climate Change (IPPC) Fourth Assessment Report's Working Group III stabilisation target of 450 parts per million of atmospheric CO_2eq. required to keep temperature increases within reach of 2°C. Ultimate stabilisation levels will not be known until well into the next century. However, the choices made in the next 20 years, particularly in power generation, will set the likely long-term path (Fig. 4).

This is a sobering reflection but the purpose here is not to step back from the challenge but to caution a sense of realism into the energy policy debate. It can be asserted that the promotion of renewables will create employment and substitute for oil and gas imports. However, the world will still be hugely dependent on fossil fuels in 2030 and there will be economic and competitive consequences for countries adopting this approach if the remainder of the world is using a more cost-competitive and carbon-intensive energy mix.

Global Energy Mix

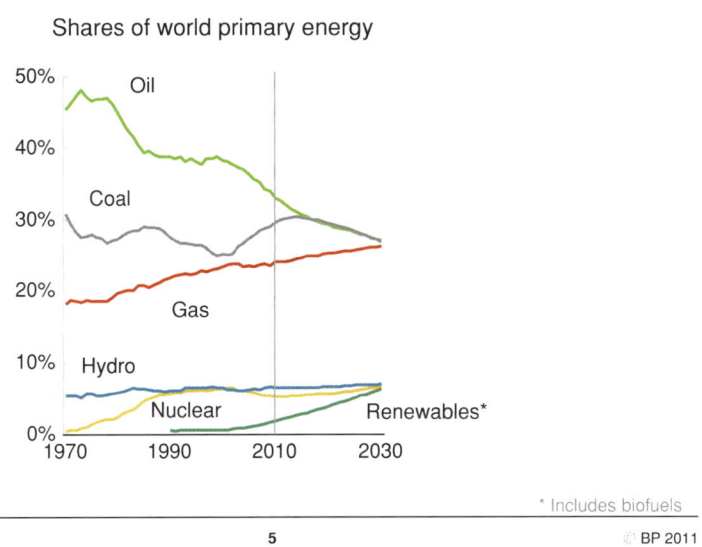

© BP 2011

Fig. 3 Shows the same data as the previous analysis but with fuel mix plotted as a percentage of total world primary energy. The category of renewables includes biofuels but excludes the small-scale dispersed use of biomass, particularly for heating and cooking, in parts of the world. (Published with kind permission of © BP 2011. All Rights Reserved.)

3 Practical Steps Forward

The challenge is significant but equally there are policy choices open in technology, energy mix, carbon pricing and effective markets that can help to meet the triple objectives of energy security, competitiveness and climate change.

Of the areas of potential public policy action that are open to legislators, four can be seen as relatively easier to achieve: encouraging competition; energy efficiency programmes; promoting energy research and development; and, education and communication. Others are much harder: developing long-term and economy-wide price signals for CO_2; the implementation of transitional incentives to speed up the deployment of near-commercial technologies which need a helping-hand down their cost curves; targeted regulatory action; and, international tax and trade mechanisms. Most governments find, in reality, that even that the 'easier' policy areas present significant challenges.

The first, and best, practical step to take—good for security, good for affordability, good for availability and good for lowering CO_2 emissions—is to increase

Global CO2 Emissions

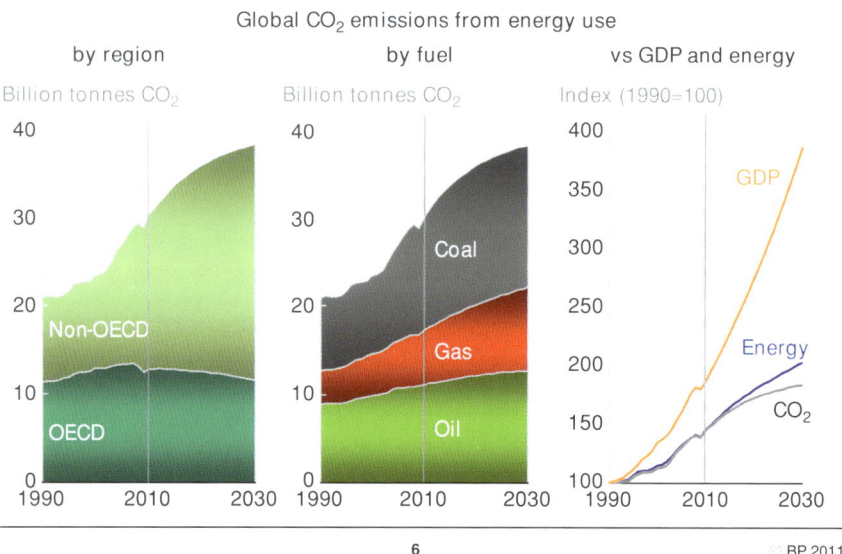

Fig. 4 Shows the projected trend of global emissions in line with these projections of energy mix. Even with reducing energy intensity and tightening national, regional and international climate change policies, global CO_2 emissions are forecast to rise by 27% by 2030. (Published with kind permission of © BP 2011. All Rights Reserved.)

the efficiency of energy use. Transport accounts for around 23% [2] of global CO_2 emissions and remains a fast growing sector. By far the most cost effective way to more efficient transport, without fundamentally changing patterns of consumption or existing infrastructure, is greater efficiency of existing internal combustion engines. Combining greater efficiency with progressive hybridisation and the correct biofuels can significantly reduce emissions from transport.

The blue line in Fig. 5 shows declining CO_2 per km driven, versus the additional purchase cost of vehicles with new technology, relative to conventional gasoline cars today. The key is to distinguish between near-term and longer-term options. In the longer-term—beyond 2030—battery electric vehicles and maybe even hydrogen fuel cells are likely to play a more material part in vehicle transport. It will be more rational to do this once the grid has been more fully decarbonised in order to deliver the full CO_2 reduction potential. Electricity storage technology also needs to evolve for the performance and cost of batteries to be competitive. Even just on environmental grounds, the current average emissions performance for battery electric vehicles in Europe remains above that which can be achieved with hybrids. Electric vehicles will have their time, but not just yet.

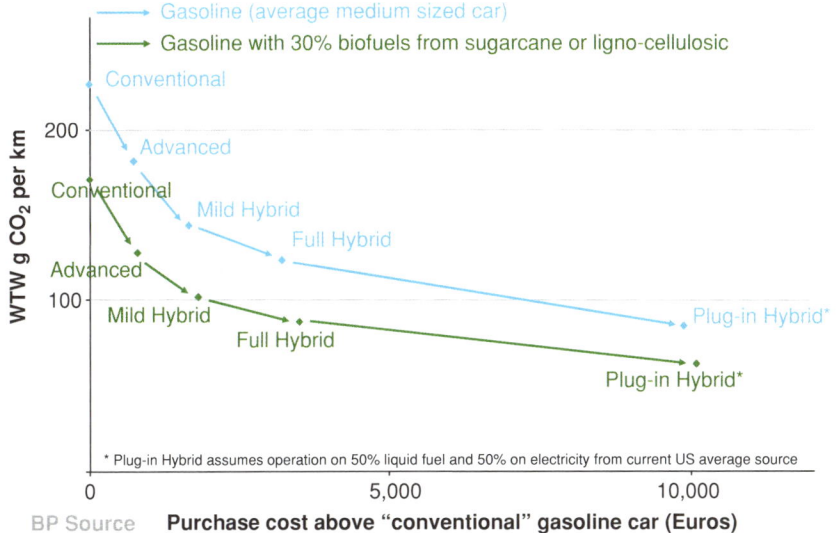

Fig. 5 Declining CO_2 per km driven, versus the additional purchase cost of vehicles with new technology, relative to conventional gasoline cars today (*blue*) and the additional benefit that can be gained from the use of sustainable biofuels (*green*) (Published with kind permission of © BP 2011. All Rights Reserved.)

In the shorter term, at least up to 2030, by far the most effective pathway to lower-carbon transport is to make existing vehicle engines more efficient. There are major gains to be obtained from advanced gasoline engine technology in particular. Combined with step-by-step hybridisation, we can see the potential for nearly halving CO_2 emissions per km at a much lower cost than for a battery electric vehicle.

The green line in Fig. 5 shows the additional benefit that can be gained from the use of sustainable biofuels.

When such a vehicle pathway is combined with the use of sustainable biofuels it becomes even more effective in reducing CO_2 from transport and can reduce emissions by about as much as electric vehicles run off a gas-powered grid—but at a fraction of the cost. For this reason, several companies are already investing heavily in the global supply of sustainable and CO_2-efficient sugarcane and ligno-cellulosic-based gasoline components. The biofuels also usefully add to the diversity of supply of future liquid transportation fuels.

A key point is that these technologies are either already available or experiencing valuable breakthroughs and build on existing deep industrial strengths and capabilities. We can be confident that such an approach can deliver progressive, achievable and material efficiency gains and CO_2 reductions in the transport sector. The pathway for transport is clear and can be pursued today, making a material impact using the existing energy infrastructure—a point not to be over-looked given the costs associated with a switch towards an electrified road transport system.

The other key use of energy is for electrical power generation, accounting for around 40% [3] of global CO_2 emissions. Economic growth in the developing economies will demand huge global additions of electrical power capacity by 2030. The types of capacity installed are likely to impact energy security and CO_2 emissions to 2050 and beyond. If we make the wrong decisions now, we are locked into the consequences for a long time to come. For power generation by far the most productive pathway is composed, once more, of energy efficiency plus greater use of natural gas in combined cycle gas turbine plant, growth in nuclear and ultimately renewables.

Some of the reasons why gas is an attractive and practical policy choice have already been mentioned. It is a plentiful resource, flexible and economic and burns with half the CO_2 emissions of coal, and half the capital cost per unit of generating capacity. In addition, the flexibility of gas fired plant can be very effective in complementing the natural intermittency of wind and solar power operations. On this basis, natural gas should be seen both as a preferred transition fuel to a lower-carbon economy and as a fundamentally advantaged energy supply option in its own right. The surprising and regrettable thing is that, instead of incentivising gas in power for these reasons, policymakers risk squeezing gas out—by promoting expensive, risky and relatively inefficient zero-carbon technologies that are not yet ready for deployment on a very large scale, and themselves risk being superseded by more advanced technologies fairly soon. Zero-carbon power technologies are evolving fast and, like electric vehicles, the time will come for them to be deployed at scale. In the meantime, natural gas can deliver low-carbon and security benefits for power generation on a large scale at low cost now, just as efficiency plus biofuels can deliver the same for transport. This is perhaps the most obvious area where governments need to 'stop polishing the 2050 diamond' and move ahead with these practical pathways as soon as possible.

One final but important point is on the physical availability of natural gas. Recent appraisals of global unconventional gas reserves [1], based on the huge expansion in the production of shale gas, tight gas and coal bed methane in the US, could add as much as 30 years of supply to proven global gas reserves. The transformation of the North American gas market has had the knock-on effect of depressed gas prices: US gas prices are now chasing parity with coal rather than fuel oil. This is leading to the displacement of coal in US power generation on price. Plentiful domestic gas supply has also freed up cargoes of liquefied natural gas (LNG), which had originally been targeted for the US market, to be attracted by prices offered in other regions. In Europe, this has already caused pressure on traditional oil-indexed contract prices as unprecedented amounts of cheaper spot LNG became available from 2009 and into early 2011. LNG that can be diverted from North America is now offering the possibility to Japan of alleviating power shortages thereby resulting from the devastating earthquake of 11 March 2011.

This increased availability of gas supply from diverse sources today, together with the anticipation of additional indigenous supplies of unconventional gas tomorrow, has very substantially reduced concerns over gas security of supply.

So, having looked at the practical steps which can be taken now, the focus in the rest of this paper is on three key external relationships and how they will shape the energy future in which the important energy choices have to be taken.

4 Europe and Russia

These policy choices are rightly a central preoccupation in European energy policy thinking. However, given the approach set out in the first part of this essay, it can be argued that the basic objectives and direction of travel are well understood and the main questions are around implementation and competitive impact. It can consequently be argued that the really big choices for Europe are not so much about internal energy policy as external energy relationships. If Europe does not get these external choices right, policy implementation will become disconnected from the global picture with damaging implications for both economic competitiveness and energy security.

Europe is a significant global energy power but other countries and regions have an increasing interest, and say, in energy too. Furthermore these interests are changing in quite fundamental ways. In 2010 China became the largest global energy consumer, accounting for around 20% of global energy consumption. India and Indonesia are following and Brazil is emerging as both a major producer and consumer. All of these shifts are real and substantive and cannot be wished away.

Geography and history ensure that Europe and Russia are bound together in the same political and economic space. A theoretical question can be asked about the need for a Europe/Russia relationship. In reality it is evident that there is no choice—Europe and Russia are obliged to live and work together and the only real choices are about how well or otherwise to go about this task.

This is true right across the energy sector. An extraordinary network of natural gas pipelines joins the Eurasian continent from the depths of Siberia to the core European energy markets. The pattern for oil pipelines is basically similar. Russia accounts for over 30% of European oil supplies and around 23% of natural gas [4]. In total, Russia supplies about one quarter of European energy requirements, without counting supplies from other countries which also transit via Russia. The infrastructure that makes this possible should be seen not as a liability but a valuable asset. It joins producers and customers and allows both to find competitive advantage in a global market.

Importantly for Europe, this infrastructure also provides a competitive basis for energy to continue to flow west, even as demand continues to grow in the east. The supply lines have also proved to be reliable and Russia kept energy supplies flowing to Europe throughout the course of the Cold War. Of course, this infrastructure is not and cannot be exclusive in terms of access into the EU market.

European Gas Supply Diversity

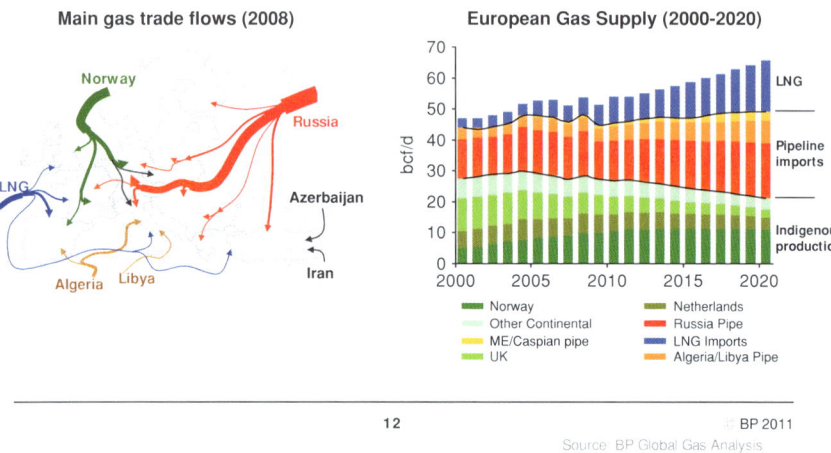

Fig. 6 Gives a view of the diversity of natural gas options open to Europe through to 2020 and beyond. To this can be added the potential for unconventional gas within Europe's own boundaries. There are many choices but a stable and mutually beneficial relationship with Russia will also strengthen the security of supplies from the Caspian, Central Asia and the Middle East. Unless Europe becomes self-sufficient in gas from unconventional sources such as shale gas, Russia will in all circumstances remain an indispensible partner in Europe's energy mix (Published with kind permission of BP 2011. All Rights Reserved.)

Europe will also inevitably encourage other gas infrastructure systems, such as the Southern Corridor, North Africa pipelines and LNG re-gasification to find their place in the market, while also making it easier for gas to cross internal member state borders in case of gas shortages in any part of the EU (Fig. 6).

So Europe and Russia have everything to gain from working together and much to lose from standing apart. Gains include energy security, co-investment opportunities, key sources of economic competitiveness and stability on a long border in an increasingly uncertain world.

5 China

Over the last 10 years, Chinese GDP has almost tripled and energy consumption has more than doubled. According to the IEA [5], this growth has made China the leading global emitter of energy-related CO_2 and the Chinese leadership recognises that this energy intensive pattern of growth is not sustainable into the longer-term, not least given China's concern for energy security given its growing import dependency for fossil fuels. In the last Five Year Plan ending 2010, China set a target to reduce energy intensity of GDP by 20% and official statements [6] show an achieved outcome of 19.1%. China also has one of the world's largest programmes to develop non-fossil fuels, with the intention that non-fossil fuels should account for 15% of total energy consumption by 2020 [7].

The next Five Year Plan up to 2015 focuses on the issues of energy mix and sustainability and sets out an intention to:

- increase the share of non-fossil fuels in total primary energy consumption from 8.3 to 11.4%
- increase the share of natural gas in the total energy mix from 4 to 8%
- reduce energy GDP intensity by 16%
- reduce CO_2 GDP intensity by 17%.

However, despite all these efforts, it needs to be recognised that China is still a developing country, with GDP per capita on a Purchasing Power Parity basis of around $6,000 per annum, compared to $30,000 for the EU27 and $45,000 for the US. As this gap closes, Chinese growth will continue to drive global energy demand. It is expected that Chinese total energy demand will increase by 80% over the next 20 years, accounting for over 40% of the global energy demand increase in this period [1] (Fig. 7).

However, there is no need to see Chinese energy consumption as a cause for global alarm, nor a threat to good energy policy. The reality is that China's success is increasingly at the heart of a prosperous globalised world. The global financial system has been largely stabilised by Chinese finance. Chinese imports underpin world export demand, while Chinese exports satisfy global consumer needs and support international competitiveness. This is also the case in the energy sector. China's success in energy diversification, energy efficiency and adoption of lower-carbon technology, is presenting major economic opportunities as well as being critical for global energy policy success. Indeed, China's ability to deal effectively with its environmental challenges will largely determine the global environmental outcome.

So in reality there is little choice—in terms of self-interest—except to engage with China at every stage of its journey. China's success is and will be shared by the rest of the world. For business there are worries—not without reason—about security of intellectual property and access to investment opportunity. The appropriate means must of course be found to protect these and other interests. Equally, other governments need to be clear that non-engagement is no choice at all, although the manner of engagement is a critical choice. The issue is whether countries and regions can make the crucial step to align around common interests, or will these be lost in the traditional but ultimately futile search for narrow commercial or national advantage.

On this basis there is a strong argument that engagement with China should stand at the very front of the foreign policy agenda, providing coherence on every aspect of analysis, assessment, representation and strategy. China is not only

China Energy Consumption and Imports

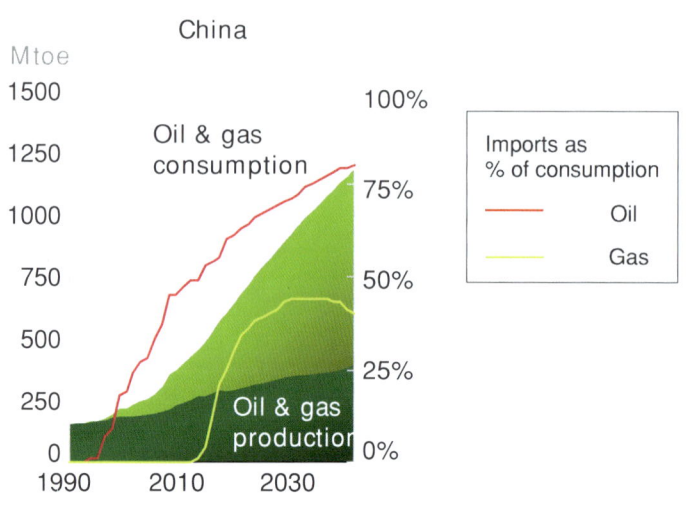

Fig. 7 Shows the impact of this growth on Chinese oil and natural gas consumption and imports. The darker green shows combined domestic oil and natural gas production. The lighter green shows total oil and gas consumption and illustrates the import 'gap'. It can be seen that Chinese oil demand is expected to more than double by 2030, accounting for 65% of global oil demand growth from today. By 2030, oil import dependency is projected to be at levels similar to Europe. However, gas import dependency may slow after 2020, as a result of the development of indigenous unconventional gas. (Published with kind permission of © BP 2011. All Rights Reserved.)

changing within its borders, it is also changing the world. This change will be an important consideration for future energy policy and governments need to understand and make sure they are part of this change. Working together, it should be possible to reach alignment on a coherent approach to energy policy, encourage partnership between energy companies and co-develop and deploy key research and technologies.

6 The United States

Finally it is important to turn to the United States, where the same challenges in energy policy need to be addressed at a time of economic recovery.

There are differences of course. The US will remain a major importer of oil but, in contrast to Europe and China, will also continue to be a significant oil producer. The

US will, in addition, be a major producer of natural gas, where unconventional sources give it a good chance of complete supply independence for the next 100 years or more. Given this reality, it might incidentally be thought a surprise that the coal industry can lobby so successful in resisting the promotion of natural gas.

However, the differences in energy balance should not obscure shared wider interests. In a multi-polar world, there is a common interest in aligning the key global relationships for the future—whether with Europe, China or the other major emerging economies. The intention should be to address these relationships as friends as well as competitors.

In energy policy there is much to gain from an aligned approach—whether on energy markets, financial regulation, carbon pricing, new energy technologies or key international relationships. It makes no sense at all to divide international markets by ill-matched policy and regulation, still less by fragmented international action. The US and EU economies could be viewed as a single energy market in the context of intense global competition and this points to the importance of aligned energy policy. This does not necessarily mean a treaty but it does require broad alignment on the pace and intensity of policy interventions, in order to avoid unintended and damaging dislocations on both sides.

A good example is tariff policy on biofuels. If both sides of the Atlantic are not aligned, this will create an unintended arbitrage resulting in biofuels flowing preferentially to either the US or Europe. This dislocation could also prevent the evolution of the Atlantic basin fuel pool as a commoditized source of cheap bio-components. However, perhaps the biggest issue and opportunity is in CO_2 pricing. Europe is pursuing cap and trade for CO_2. The US is still debating its own policy approach. In the end which mechanism is used is not important. What matters will be the timing and intensity of the application of such policy measures. If these are not coherent, then such policy interventions will result in dislocations in the markets between different regions. If coherent, policy interventions can crystallise alignment on global CO_2 pricing and accelerate the global process. No third country or region could afford to ignore an aligned EU and US in carbon markets and pricing.

Therefore, the most important reason for alignment is that it would result in more rapid international climate action and provide a critical accelerant to the necessary but slow UNFCCC process. It would probably also encourage alignment with China. If it were possible to achieve this coherence of the major trading blocs, the world would find that it had successfully laid the foundation for global energy policy in the twenty-first century.

7 Conclusions

This essay has looked at some of the challenges and choices facing the world of energy looking to 2030 and beyond.

The provision of energy security today is more complex than before, particularly if you embrace the challenges of the shifting mix of energy demand growth and that of climate change. For many decades, the solution to energy security has largely been one of secure supplies, largely of fossil fuels, combined with a drive for limited diversification. Energy demand continues to increase, and we will still be dependent on fossil fuels for decades to come. However today, "more energy" and "alternative energies" will not solve the equation on their own. We must add the hugely important ingredients of "less energy" (i.e., energy efficiency), "energy research and development", and, if we are to get anywhere quickly, more pro-active "foreign energy policy" along key axes. These of course must be bound up in sound energy policy within each major jurisdiction to ensure that competi-tiveness and economic growth are also maintained. This is a complex equation.

There is no one answer for energy security and different countries and blocs will take different pathways. The growth in global energy demand is such that all energy sources and technologies will be required. However, there are today practical pathways which make a material difference in transport and power and these should be pursued with urgency as a matter of pragmatic imperative. Both start with energy efficiency using today's technologies, and natural gas, biofuels, and indeed nuclear power can all play major roles in moving economies materially in the right direction while we also research and innovate around longer-term technologies and options. The important thing is to stop only "polishing the 2050 diamond" and start to make practical and material steps today through imple-mentation of the right policies.

Finally, it is also clear that the very necessary UNFCCC process could see some welcome acceleration if key economic blocs could strive towards energy policy alignment and coherence. This can be achieved without being bound up in a formal agreement or treaty. The US and Europe face very similar challenges, with industries in a similar stage of development. Transatlantic alignment on the intensity of policy changes could make a material contribution. Engagement and alignment around areas of mutual benefit with a number of key economic blocs such as Russia and China will also accelerate progress.

Today there are some really big choices which have to be made if we are to achieve progress while also delivering the provision of greater energy security. Although progress is slow, we have come a long way in the last 15 years, and there are some very pragmatic options available, both within our economies and between them, which provide more than a glimmer of hope.

References

1. BP p.l.c. (2011) BP energy outlook 2030. BP p.l.c., London, based on data from the International Energy Agency, The U.S. Energy Information Administration, the United Nations Statistics Divison, The World Bank et alia
2. King DA, Inderwildi OR (eds) (2010) The future of mobility roadmap, 2nd edn. Oxford University, Oxford
3. International Energy Agency (2010) World energy outlook. International Energy Agency, Paris. ISBN: 978-92-64-08624-1

4. BP p.l.c. (2011) BP Statistical review of world energy. BP p.l.c., London
5. International Energy Agency (2010) CO_2 emissions from fuel combustion. International Energy Agency, Paris. ISBN: 978-92-64-08027-8
6. U.S.-China Economic and Security Review Commission (2011) 12th five-year plan (2011–2015). http://www.uscc.gov/researchpapers/2011/12th-FiveYearPlan_062811.pdf
7. National Development & Reform Commission (NDRC) (2007) Medium and long-term development plan for renewable energy in China. National Development & Reform Commission (NDRC), Beijing. http://en.ndrc.gov.cn/

The Anomalous Age of Easy Energy

C. J. Campbell

Abstract Planet Earth goes back billions of years, and the first forms of life appeared long ago, with Modern Man arriving only 200,000 years ago. Agriculture changed his the way of life, as communities acquired land rights, leading to trade and the use of money. They faced the consequences of depleting their natural resources, prompting conflict. For most of history, energy needs were met mainly by Man's own muscles, but over the last two centuries he has tapped new sources from coal, followed by oil and gas, that changed the world radically, allowing the population to grow 10-fold. But these are finite natural resources, formed in the geological past, which means that they are subject to depletion. The *Second Half* of the Oil Age, which now dawns as production declines, may be marked by a contraction to match the expansion of the *First Half*. A debate rages as to the precise date of peak production, with much confusion from ambiguous resource definitions and lax reporting, but misses the point when what matters is the vision of the long decline on the other side of it. The transition is a turning point of historic magnitude, likely to be associated with much social, economic and political tension before people learn how to adapt. It is by all means an important subject.

1 Background

Scientific evidence suggests that the Universe came into being long ago with a *Big Bang*, and the Solar System followed 4.6 billion years ago. Planet Earth evolved differently from the other known planets because of its size and distance from the

C. J. Campbell (✉)
ASPO Ireland, Ballydehob, Co Cork, Ireland
e-mail: colin@aspo-ireland.org

O. Inderwildi and Sir David King (eds.), *Energy, Transport, & the Environment*,
DOI: 10.1007/978-1-4471-2717-8_3, © Springer-Verlag London 2012

Sun. It has a molten core, a viscous mantle and a hard crust, segments of which, forming continents, have moved around on the back of deep-seated convection currents [32]. Water too was formed as much as 3.8 billion years ago to give oceans and lakes and an atmosphere, a few kilometres thick, surrounded the Planet. Photosynthesis by algae changed its composition converting carbon dioxide into oxygen with nitrogen and other elements being added later. Moisture from the oceans was also transported by winds and deposited as rain and snow, giving the varied climate and vegetation enjoyed by the Planet.

Mountain ranges rose as the continents collided, and volcanoes erupted. The uplifted rocks were subjected to erosion under the fluctuating heat of the sun, rainfall and the forces of gravity. The debris slipped down the mountain slopes to be deposited by rivers in lakes and seas, which became home to early forms of life. Some species began to protect themselves with shells, whose remains were preserved as fossils, especially since the opening of the Cambrian period about 550 million years ago. The limpet (or *Patella*, to give it its scientific name) remains little changed, but other species evolved to provide the wide spectrum of living creatures that occupy the Planet today. It was not however plain sailing, for there were periodic cataclysmic events, caused by earthquakes, volcanic eruptions, meteor impacts and climate changes that adversely affected the living environment, leading to the extinction of especially the more sophisticated species. One such event occurred in Permian times, about 275 million years ago, when lava flows covered much of the Planet, causing a mass extinction. Another, resulting from a meteor impact about 65 million years ago at the end of the Cretaceous Period, put an end to the dinosaurs.

Modern Man, in an anatomical sense, arrived about 200,000 years ago, and spent his early years hunting and gathering edible plants, as had his antecedents. Settled agriculture followed about 12,000 years ago, especially in the *Fertile Crescent* of the Middle East, when people learned how to sow crops. The historical record reveals a pattern of how communities grew in size as they exploited the soils and forests at their disposal, only to decline when they exhausted the fertility of the soil and cut down their trees. They then often tried to conquer neighbouring territories, in some cases successfully building empires, which in due course also went into decline when they over-reached themselves. Conflict has been endemic, such that about one-quarter of adult males are thought to have lost their lives in battle over time [26].

The food had to be stored between harvests to feed the people, giving rise to the establishment of storehouses, which ran accounts to record how much was received and returned to the farmers. There was evidently scope for exploitation if a storehouse returned less than it received, or gave preference to privileged members of the community. Debt arose, sometimes to extreme levels, as people negotiated access to food before the crops could be gathered, or in the case of a bad harvest. For example, the Roman Emperor, Brutus, is reported to have charged interest at the rate of 48%, selling food to a starving town in Cyprus [42]. There were religious objections to the practice, with the priests, who played an influential role in society, moving to limit exploitation. Indeed, different religions may have arisen from the situation as groups of people invoked divine support for their particular positions. The objections to usury are easy to understand. The crop was the same, so if the owners of the storehouse charged interest, they could have bigger meals, doing no extra work in return, while the rest of the people had less to eat. Such a situation was a breeding ground for resentment and conflict.

Food storage was effectively a form of early money before gold and silver became a medium for exchange, its value being set by its natural scarcity in each region. It had political overtones, as the King or Emperor could engrave his effigy on the coins, strengthening the sense of nationalism among the people under his control. But it was heavy stuff to cart around, leading a storekeeper to offer to hold it, and charge a fee for doing so. The receipt, which in the days before paper was inscribed on a clay tablet, became a form of currency. Before long, an enterprising storekeeper realised that he could issue more receipts than he had gold on deposit, confident that not everyone would cash in simultaneously, thereby laying the foundations for fractional banking. Money fuelled investment, and has played a key role in history as economic expansion by the more successful countries triggered conflicts and political changes. For example, Government bonds had to be issued to fund the Napoleonic wars, and themselves became subject to speculation and manipulation. Its use has also given power to the financiers, and impacted politics in many hidden ways [11]. It is a very large and complex subject, as discussed for example by Gollwitzer [13] in his book *Europe in the Age of Imperialism.*

Although agriculture remained the main occupation for most people throughout history, there were important technological advances. Stone age man had fashioned flints into knives and arrowheads before people took to bronze, iron and steel for better tools and weapons, especially after the invention of gunpowder around 1250. At first, they used wood and charcoal for smelting but later turned to coal dug from surface outcrops as a better fuel. The pits were then deepened to become regular mines, and minerals were extracted in a similar way. But the mines were subject to flooding on hitting the water table, which prompted one of the most remarkable technological developments of all time. The bucket had given way to the hand pump before a way was found to inject steam under pressure into the cylinder of one pump to drive another. That evolved into the steam engine, which led to the rapid expansion of industry, transport, trade and agriculture, ushering in the *Industrial Revolution.*

2 The Coal Age

Coal is derived from plant material as found in peat bogs that was preserved and compacted under special conditions in the geological past, much having been formed in the Carboniferous period, about 300 million years ago. Britain was one of the pioneers of coal mining, and came to employ thousands of miners working underground in gruelling conditions.

Substantial coal deposits were also found in many other countries, including the United States. Estimates for global reserves range widely, being hard to make, because, like other minerals, coal mining is subject to the concentration of the resource in the ground, with thinner and deeper seams becoming viable if prices rise or costs fall. There are also several different categories from anthracite to lignite, each having its own characteristics. Extraction rates range widely, as do the net energy yields, but the evidence seems to suggest that overall production is set to decline in the not too distant future. Being a finite resource, coal is clearly subject to depletion.

3 The Oil Age

The Industrial Revolution gathered pace in the 1870s when a German engineer, by the name of Nikolaus Otto, perfected the *Internal Combustion Engine,* which had evolved from the steam engine, with the fuel being injected directly into the cylinder to make it more efficient. At first, it relied on benzene distilled from coal before turning to petroleum refined from crude oil.

Oil from surface seepages had been known since from the earliest days, having been used as mortar in the construction of Babylon, but the first wells deliberately drilled for it were sunk around 1850 in Pennsylvania, Romania and on the shores of the Caspian. It was at first used for lamps to replace whale oil, but soon became the prime fuel for transport. The first automobiles took to the road around 1880, and the first tractor ploughed its furrow in 1907, increasing food supply greatly.

The international oil industry was born, and it did not take long for the pioneering oil explorers to understand the basic geological requirements for finding an oilfield in places having the right combination of source, reservoir, trap and seal. It was relatively easy to identify the prime oil territories, most being endowed with natural seepages, and to find the major fields within them. In fact, the world's largest petroleum province around the Persian Gulf was identified when a well in the foothills of the Zagros Mountains of Iran blew out just over 100 year ago, sending a plume of oil into the sky with far-reaching consequences.

The modern world runs on oil and money, which are linked insofar as cheap, mainly oil-based, energy has fuelled economic prosperity and the related growth of money supply. Cities and highways are now choked with traffic, burning oil, and even in remote rural settings, the distant throb of an engine can often be heard, or vapour trails from aircraft seen in the sky above. Foresters in the Swiss Alps even use helicopters in place of mules to pull out the fallen trees. Oil and gas, like coal, are however finite natural resources formed but rarely under special and well-understood conditions in the geological past, which means that they are subject to depletion. In other words, for every gallon of oil used one less remains.

In considering this matter, it makes sense to concentrate on *Conventional* oil because it has supplied most to-date and will dominate all supply far into the future. Discovery peaked almost 50 years ago, as confirmed by an Exxon executive [28], and in 1981, we started using more than was found in new fields despite an

Fig. 1 The growing gap: regular conventional oil patterns

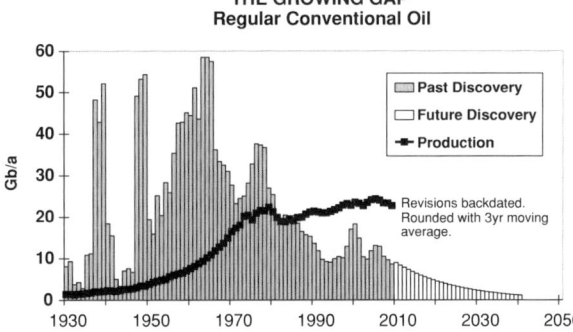

THE GROWING GAP
Regular Conventional Oil

intensive search applying advanced technology. Future discovery may be estimated within limits by projecting the long downward past trend. The peak of discovery must inevitably deliver a corresponding peak of production.

There is in fact a natural depletion pattern, whereby production starts and ends, passing a peak in between. The peak in an individual field tends to come early in its life, but for countries, regions and the world as a whole, the individual profiles average out over time to give an overall peak close to the so-called midpoint of depletion when half the total endowment has been extracted. It is worth noting in this connection that *Conventional Oil* is unlike coal and other minerals in that it is subject to a certain polarity, being normally either present in profitable amounts or not there at all, as even small oilfields can be very profitable. *Non-conventional* oils have very different characteristics and depletion profiles, as discussed in Fig. 1.

Industrialisation, driven by the new energy supplies, helped bring about the end of slavery but was not an unqualified blessing as it changed the structure of society, with the growth of cities inhabited by industrial workers, many living under abject conditions. They worked for wages becoming ever more aware of money and commerce. Furthermore, economic growth prompted more conflict as countries sought to expand their trade and spheres of influence by which to make money and feed their growing populations. Massive emigration to the New World over the last two centuries helped ease the population pressures of Europe, leading to the near extermination of the indigenous people. Economic factors and social reactions to the Industrial Age were largely responsible for two world wars of unparalleled severity in the twentieth century [41]. There had been local conflicts throughout history but the new technology, especially in the field of transport and munitions, radically changed the nature of war. The *Second World War* was basically a continuation of the *First*, being followed by the so-called *Cold War*, when the victorious former allies, having different economic principles, came to glower at each other across a divided world. But they somehow managed to avoid much direct conflict, having appreciated that weaponry had also overreached itself such that nuclear warfare would deny a meaningful victory.

The economic expansion was driven by financial capital as banks lent more than they had on deposit, confident that *Tomorrow's Economic Growth* was collateral for *Today's Debt*. Furthermore, there was a progressive change in the nature of investment. In earlier years, investors had clubbed together to support a specific project, such as building a railway or a canal. They had a detailed knowledge of the undertaking, and a commitment towards it because the receipt of regular long-term dividends was their reward. But gradually these investments were traded on Stock Exchanges by brokers, who naturally could have little detailed knowledge of the underlying projects, and still less loyalty towards them. Favourable stock market valuation, built in part on imagery, became the principal objective of the business manager and investor alike. The system has further evolved in recent years with complex derivatives and hedge funds, which have become ever more divorced from the underlying reality as banks started lending to each other. Making more money by whatever devious strategy became a primary motivation: there are even examples of banks profiteering by selling short their own investment funds.

Currency too played an important role. The principal benefit of the British Empire at its prime may have been the use of the pound sterling for world trade that delivered a massive and largely unseen tribute to the banks of London, dominated by a few well-known names. These self-same entities duly conquered the US market in 1913 when President Woodrow Wilson gave them control of what became the Federal Reserve Bank, amounting to a privately owned State bank, which was very profitable. Money buys political influence as Amschel Rothschild had already noted in the nineteenth century with his famous comment: *"Give me control of a nation's money and I care not who makes her laws"*.

The overall prosperity was however unevenly distributed, with the United States emerging supreme after the Second World War that extinguished the British Empire. Much of the rest of the world remained relatively undeveloped, in some cases facing the heavy burden of servicing foreign debt. Ecuador, for example, found itself having to dedicate its substantial oil revenues in their entirety to that end, with the lenders reaping a corresponding benefit. At the same time, manufacturing in the United States has declined in recent years in the face of cheaper overseas labour, such that its empire has become essentially a financial one, backed by military might. Parts of Detroit, once the heartland of the motor industry, are now described as almost a ghost town.

The evolving financial system has been generally successful in securing economic growth over the past two centuries, although there were reversals when it overstepped itself prompting periodic economic recessions. The most notable was that of 1930, which is termed the *Great Depression* although it may turn out to have been less severe than what now unfolds. It gave rise to far-reaching political reactions, with the growth of Fascism in several European countries, and the almost Socialist response of the so-called *New Deal* in America.

The foregoing discussion has touched on some of the evolving circumstances that have led to the present situation. It is a large and complex subject with many sensitive overtones. Population is one of them, and another is to ask if those controlling the financial world have anything in common in terms of background, affiliation and hidden agenda.

The world population at the time of Christ was about 300 million, and it no more than doubled over the ensuing seventeen centuries, as most people lived simple, rural lives, fed by whatever their particular region could provide. Their energy came mainly from their own muscles, although supplemented by that from slaves, draught animals, firewood, wind- and water-power. A new chapter opened when coal, followed by oil and gas, provided abundant new sources of energy allowing the world population to expand 10-fold. Today, 29 billion barrels of oil a year support 6.9 billion people with an energy supply equivalent to that of billions of slaves working around the clock (A barrel contains 42 US gallons, or 159 litres, delivering 5.7 gigajoules of energy).

The economic expansion brought social and psychological developments accentuated by television and the internet, using what a Pope might one day come to call *Satan's Antenna*. It tended to impose a consumeristic mindset on people, also breeding certain resentment as the less well-off compared their lot with affluent images on the screen. People came to believe that material wealth was the ultimate goal of their lives, and sought to achieve it by fair means or foul. In part, the change was driven by female emancipation in the more prosperous countries. In earlier years, women had been low consumers, managing the household for a family supported by male bread-winner, but now they form more than half of the workforce in developed countries, building successful careers and enjoying shopping, the latest fashion and the thrill of the bargain-hunt. In many cases, home-life has degenerated into screen-watching with a major, yet not necessarily positive, impact on the children. Divorce rates have increased in may countries, and increasing numbers of couples ignore the formality of marriage with its intrinsic commitments.

Migration too grew as the more successful countries admitted cheap labour from abroad, but the immigrants may soon find themselves increasingly isolated and resented as the affluence to support them declines. The changed circumstances seem to have raised the level of crime and violence over the past 50 years, with many cities being subject to riots and gang warfare, some having ethnic overtones. Even, once gentle, London now has violent demonstrations, and there are districts in which it might be dangerous to tread, with conditions in some other capitals being much worse. The invasions of Iraq and Afghanistan have stimulated much understandable resentment by Moslems, raising the threat of retaliatory actions.

4 The Status of Oil and Gas Depletion

As every beer-drinker knows: the glass starts full, and ends empty, and the quicker he drinks it, the sooner it is gone. It is a simple concept that applies equally to oil. The onset of the decline in oil production in any country, with its far-reaching economic consequences, could be readily identified and predicted where valid data on past production and the size of discoveries available to the public, which is far from the case.

Estimating the size of an oilfield early in its life poses no particular technical challenge, although naturally there is a certain legitimate range in the estimates. The difficulties arise primarily in relation to the reporting practices which call for some explanation. First, is the issue of defining what to measure: the terms *Conventional* and *Unconventional* are in wide usage but the boundary lacks a standard definition, which is a cause of confusion. For example, The Oil & Gas Journal reports Canada's reserves at 175 Gb (billion barrels), including the heavy oils from tarsands, whereas the equally reputable World Oil reports 27 Gb, excluding them. The boundary between *Light* and *Heavy* varies from place to place, with Canada having a cutoff at 25° API (a measure of density), Venezuela at 22° API and other countries sundry ill-defined limits.

Production data are relatively sound, although war-loss was not counted at all, despite being production in the sense that it depleted the reservoirs by like amount, and information in some OPEC countries remains almost a State secret. We may also note that *supply* is not the same as *production* at the wellhead because the refining process adds 2–3% to the volume. Furthermore the production profile of some countries is controlled by the capacity of long-distance pipelines, tending to give a plateau rather than a peak.

The position in regard to so-called *Reserves,* describing the estimates of the amount yet to be produced from known fields, is much worse, calling for further explanation. Only a certain percentage of the *oil-in-place* is recoverable, with much being held immovable by capillary forces and other physical factors in the pore space of the reservoir rocks. The percentage recoverable, known as the *Recovery Factor*, ranges widely, depending on the nature of the reservoir and the applied technology. On earlier years, a 30% recovery was considered normal, but in some cases it has now risen to as much as double that value, thanks to advanced engineering. Estimates of *oil-in-place* can however be no more than approximate, with the reservoirs being in some cases re-charged from deeper pools or flanking source-rocks.

As a starting point in discussing this issue, we may note that mineral rights in the United States belong to the landowner, meaning that the ownership of the old onshore fields was fragmented. Accordingly, it made good sense for the Stock Exchange to impose strict rules to prevent fraudulent exaggeration for financial purposes, while smiling on under-reporting as laudable commercial prudence. The rules recognised *Proved Producing Reserves* for the estimated future production of current wells, and *Proved Undeveloped Reserves* for that expected from low-risk infill wells between the existing ones before they had been drilled. One consequence was that the reported reserves of a field grew over time. A classic example is the Kern River Field of California which was found by a hand-dug well in 1899. It was drilled up slowly because it contains a heavy viscous oil, such that its reported reserves have been progressively increased under the rules, although its actual known size remained about the same. This has misled some economists of the *flat-earth school* to suggest that oil resources are near infinite to be progressively captured by technological advance and market forces [29]. The industry has indeed made amazing technological achievements. Modern geophysics and geochemistry mean that no stone need to be left unturned in the search for ever smaller and obscure prospects; drilling technology has developed such that it has become possible to drill to depths of more than 5,000 m below deep oceans, despite the occasional accident; and advanced petroleum engineering allows complex reservoirs to be drained efficiently. But there is a certain irony insofar as advances in technology also tend to accelerate depletion.

The so-called *Probable* and *Possible Reserves* were also recognised but had little financial significance. More weight is now being given to *probability* under new reserve reporting rules. In part, it reflects the use of *probability* values as a way of describing to management ever more sophisticated geophysical evaluations of the reservoir and nature of the trap. Strictly speaking, the *Mean Probability* value is the best estimate, which is roughly equivalent to *Proved Reserves* plus two-thirds *Probable Reserves* and one-third *Possible Reserves*. The probabilistic approach sounds more scientific but does not materially alter the assessments, because the degree of probability is itself a matter of judgment. For example, the US Geological Survey assessed undrilled East Greenland in 2000 as having a *95% Probability* of more than zero, namely at least one barrel, and a *5% Probability* of more than 112 Gb (billion barrels), giving a *Mean Probability* of 47 Gb, which was quoted to three decimal places, and duly added to the world assessment. The *Mean* value from such a wide range has little real meaning.

The international oil companies were subject to the Stock Exchange rules, and found that it made good commercial sense to report the minimum reserves needed for financial purposes. While the unreported amounts were useful to cover any temporary setback in operations around the world arising from political and other circumstances, the steady growth of what was reported also delivered a valuable, if slightly misleading, image to the Stock Market. Shell was one of the companies that routinely announced reserve growth for financial purposes, but in 2004 woke up to discover that it lacked the physical resources to support the claims, causing a financial scandal that cost the Chairman his job.

Those days are, however, now substantially over as the *giant fields* mature. They are defined as holding more than 500 Mb (million barrels), and offered the main scope for upward revision. This helps explain the behaviour of the major oil companies over the past decade or so. They evidently began to find it easier to secure reserves by acquisition than exploration, such that the seven major international companies, dubbed the *Seven Sisters*, are now reduced to four by merger. These transactions incidentally also provided good rewards for the financial institutions arranging them, with which some had close links—in one case even sharing a Chairman. The major companies are now selling off secondary refineries and marketing chains, as well as shedding staff, which suggests that they actually anticipate a shortfall in supply, although they cannot admit to such for fear of damaging their standing on the Stock Market. One stratagem by which to obscure the issue is to refer to the *Reserve to Production Ratio* quoted in years, as if it were remotely plausible for production to stay constant for 45.7 years (BP 2009), and then stop dead over night when production in every oilfield is observed to decline gradually to final exhaustion.

Table 1 Anomalous OPEC reserve reporting (Gb)

Date	Abu Dhabi	Iran	Iraq	Kuwait	Saudi Arabia	Venezuela
1980	28	58	31	65	163	18
1984	30	51	43	64	166	25
1985	31	49	45	**90**	169	26
1987	31	49	47	92	167	25
1988	**92**	**93**	**100**	92	167	**56**
1990	92	93	100	92	**258**	59
2000	92	90	113	94	259	78
2008	92	136	115	102	264	99
2009	92	138	115	102	260	99
2010	98	150	**143**	104	262	**211**

To step again back into history, we may note that the discovery of the major fields of East Texas in the 1930s led to a glut, which depressed the price of oil, adversely affecting the other producing areas in the country. The Government, abandoning free market principles, reacted by restricting production to a given number of days a month. Most oil in those days was moved by rail, and so it was convenient to entrust administration of the policy to the Texas Railroad Commission.

In 1960, the world's main producing countries followed this precedent with the formation of the *Organisation of Petroleum Exporting Countries* (*OPEC*), which agreed to support price by production quotas, based on production capacity, reserves, population and other factors. In 1974, some of them decided to restrict exports to the United States in reply to the latter's support for Israel, which had occupied Arab lands, prompting the *First Oil Shock* when prices rose fivefold. It was followed by the *Second Oil Shock* in 1979, occasioned by fears associated with the fall of the Shah of Iran. The high oil prices caused by these moves led to a mild economic recession which dampened demand and lowered prices, while the entry of new offshore oil from the North Sea and elsewhere further undermined the position of OPEC, putting pressure on its individual members. In 1985, Kuwait reacted by announcing a massive increase in its reported reserves, although nothing particular had changed in its oilfields. This had the effect of raising its OPEC quota, which allowed it to produce more oil, and thereby make more money (Table 1).

The basis for the increase is uncertain. It may have been arbitrary, or it might have reflected a change in the assumed *Recovery Factor* from the traditional 30% to a more realistic 40%. Alternatively, the new number might refer to total discovery, termed *Original Reserves* rather than *Remaining Reserves*, failing to deduct past production. To add to the confusion, the Oil Minister, perhaps inadvertently, has recently announced *Proved Reserves* of 24 Gb and *Proved & Probable* of 51 Gb. Another difficulty is the varying treatment of the Neutral Zone, a territory whose oil is shared between Kuwait and Saudi Arabia. It is in any case difficult to accept that reserves have risen progressively since 1985 despite production, given that most of its oil is in one field, the Burgan Field, which was found as much as 72 years ago.

The other OPEC countries found themselves having to match Kuwait in varying degrees to protect their quotas, as shown in the table, with the anomalous increases being highlighted. In 1988, Abu Dhabi decided to match Kuwait's 92 Gb exactly (up from 31 Gb); Iran went one better at 93 Gb (up from 49 Gb) and Iraq capped both at 100 Gb (up from 47 Gb). Saudi Arabia could not match Kuwait because it was already reporting more, but in 1990 announced a massive increase from 167 to 258 Gb to hold its own. Venezuela, for its part, went from 25 to 56 Gb by including its *non-conventional* heavy oils that had not previously been counted. It has also announced a massive increase in 2010 on the same basis.

In particular, it is clearly implausible that new discovery or genuine reserve revision in Abu Dhabi should have exactly matched production year on year to leave its reserves unchanged at 92 Gb since 1988, the date when it decided to match Kuwait's claim. In late 2010, Iran and Iraq announced further increases to respectively 150.53 and 143 Gb. Iran has made a major discovery of oil in a deeper reservoir beneath the Ferdowsi gasfield in the Persian Gulf, but the claims remain to be validated.

Speaking of OPEC, we may note in passing that Indonesia has now resigned, having passed its production peak in 1977. It evidently concluded that natural depletion removed the need for a production quota, suggesting indeed that the days of OPEC as a significant element controlling price on world markets may be numbered.

Kuwait's action offended its neighbour, Iraq, which was losing its rightful revenue under the OPEC rules as a result. It was also in dispute over the share of the oil in the South Rumaila Field that straddles the boundary between the two countries. It was then an ally of the United States and was encouraged to persuade the other OPEC countries to enforce their quotas better, as the low oil prices were adversely affecting the US domestic oil industry, which no doubt had political influence (see Schweizer [36], Klare [24]). But it overstepped its mandate in 1991 with a full-scale invasion of Kuwait, prompting a successful US counter-attack, known the First Gulf War. It was then subject to oil export restrictions, making it a so-called *swing producer* to control price, with the production restrictions being lifted from time to time for *humanitarian reasons* if prices rose uncomfortably. The United States and Britain subsequently launched the Second Gulf War. Although other reasons were proclaimed, it is noteworthy that President Bush later justified it with the words: *our energy supply was at risk.* The decision to go to a war that has caused so much civilian death and suffering, including the long-term health effects of depleted uranium, which was in the ammunition used, is now widely questioned.

In addition to OPEC's strategy for controlling production and price, there have been moves to outright nationalisation by the major producers: Russia in 1928, followed by Mexico in 1938, Iraq in 1972, Kuwait in 1975, Venezuela in 1976 and Saudi Arabia in 1979. National companies with privileged access were also established in many countries. As a result, the major international companies, which formerly controlled world supply, now own no more than a small fraction of its reserves (Fig. 2).

Fig. 2 Oil and gas
production profiles
2010 base case

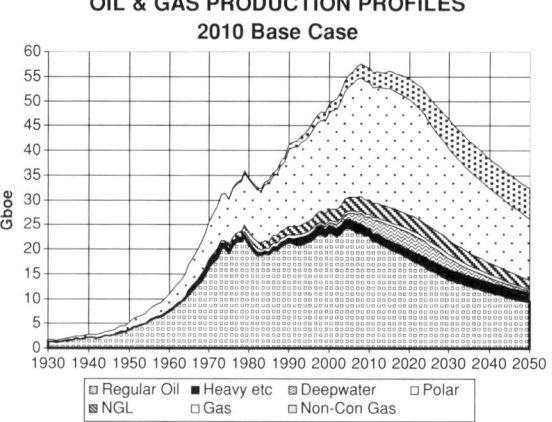

The skills of a detective are called upon to collect valid evidence on the issue of depletion. Ideally, the study should be based on confidential industry data for the individual fields so that the natural patterns of discovery and field-size distribution in the geological provinces can be identified, but analysts lacking that information have to be content with published national data. Certainly, it is important that the world model be built on a country-by-country analysis so that the details, including the anomalies and discrepancies, can be investigated (See Ref. [7]). Bland global statements lacking such support deserve little attention.

While no one can pretend to make accurate forecasts, given the appallingly weak public database, the general pattern of depletion, as illustrated in the graph, can be accepted with confidence. A debate rages as to the precise date of peak production, but really misses the point when what matters is the vision of the long decline on the other side of it. Like a mountain range, the profile has subsidiary peaks, plateaux and valleys in addition to an overall summit. The slope on the downside is unlikely to be as smooth as depicted.

Emphasis is here given to so-called *Regular Conventional Oil and Gas* that has supplied most to date and will dominate all production far into the future. Excluded by definition are: oil from coal and shale, bitumen, and heavy oil (with a density greater than 17.5°API); deepwater oil and gas (>500 m); Polar oil and gas; Natural Gas Liquids (NGL) from gasplants; and non-conventional gases (coalbed-methane, shale gas etc.). These *Non-Conventional* categories have little impact on *Peak* itself but are important in ameliorating the post-peak decline.

The discovery pattern of a country delivers a clear message. Any reserve revisions clearly have to be backdated to the discovery of the field concerned to obtain a valid trend. The larger fields tend to be found first for obvious reasons, being too big to miss. This is well illustrated by the example of Norway, which does publish reliable information. The Government there has let concessions from year to year since 1965, and some 800 exploration boreholes have now been drilled to test the best possibilities as progressively identified. The firm trend

Fig. 3 Norway oil discovery

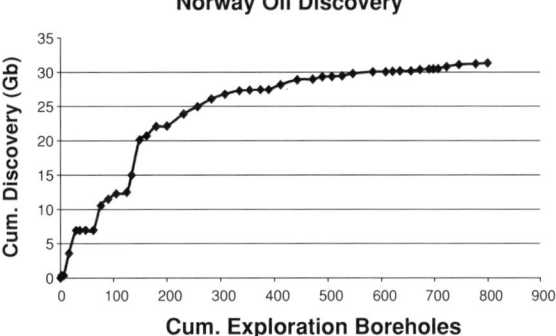

points to an eventual total of just over 30 Gb. It is evident that if the large old fields turn out to yield a little more than currently reported, it will simply deliver a flatter subsequent trend without materially changing the outcome. Admittedly, the country does have some additional *non-conventional* Polar and Deepwater potential that remains to be fully evaluated (Fig. 3).

The evidence presented in the appendices suggests that the world's production of *Regular Conventional* oil passed its peak in 2005. The growing shortages were made good by relatively expensive oil from tarsands and deepwater fields, prompting a rise in prices. Shrewd traders spotted the trend and bought profitable contracts on the Future's Market, while the industry built storage, watching it appreciate in value. It is worth stressing that oil price is set by much more than the physical transactions between producer and consumer, with the daily volume of traded contracts on the oil markets exceeding actual production by a factor of ten to thirty [40]. The rising prices also delivered a flood of petrodollars to the Middle East, where it still mainly costs less than $20 to produce a barrel, with much of the surplus being recycled to international financial institutions, undermining their stability.

By mid 2008, the traders correctly concluded that the price surge was reaching a limit at almost $150 a barrel, and started selling *short*, while the industry began draining its tanks to benefit from the high prices while they lasted. The high prices themselves triggered a serious economic recession which dampened demand, such that oil prices fell back to 2005 levels before edging up to around $90 at the close of 2010. Although many different estimates have been published, the evidence suggests that the production peak of all categories of oil coincided with the 2008 surge in prices, being partly influenced by the subsequent fall in demand. Some confirmation of this assessment is provided by BP which reports that world production did indeed fall by 2.6% in 2009. That is close to the current 2.4% *Depletion Rate* of *Regular Conventional*, namely annual as a percent of assessed future production. The post-peak decline may continue at about this rate, although by all means interrupted by anomalous spikes. The assessed decline of all categories of oil works out at about half that rate for the next decade or so before edging upwards. Although the decline is gentle, there is a fundamental difference between going up and coming down, which represents a turning point of historic magnitude, given the central role of oil-based energy in the modern world.

Table 2 Status of depletion (%)

Saudi Arabia	39
Iran	44
Iraq	29
Kuwait	37
UAE	35
Neutral Zone	64
Region	**38**

The oil price surge raises the issue of spare capacity, which has geopolitical overtones, especially in relation to the main suppliers in the Middle East. Table 2 shows an assessment of the status of oil depletion for that region. It is yet to reach the midpoint of depletion (50% depleted), which normally marks the onset of natural decline, suggesting that it does have spare capacity in terms of resources.

However, the actual spare capacity depends on the installed facilities, the level of new investment and operating environment. Iraq is relatively under-depleted but it is less than sure that the evolving post-war political situation will be such as to allow a rapid build up of capacity. In any case, it evidently makes eminent sense for these countries to hold production at current levels to postpone for as long as they can the onset of natural decline, given that oil revenue plays a central role in their economies. This seems to be well understood by King Abdullah of Saudi Arabia, who recently commented that he desired to leave some wealth in the ground for the sons and grandsons of his people. Most other countries are currently producing at capacity, although that may change as they too begin to see their limits and resolve to retain as much as possible for their future.

Another element worth mentioning is tax, or more particularly the allowances permitted against it, which represent an oblique subsidy. Norway, for example, had an 80% marginal tax rate on oil production, meaning that most of the cost of exploration drilling, not to mention the oil executives champagne, could be taken as a charge against taxable income. It did not stop there, for the rules even allowed an international company to treat as a charge part of the cost of its Head Office, Regional Office and research establishments that were deemed to be supporting its affiliate in Norway. National companies do not enjoy this benefit, and tend to take a more cautious line with exploration as they are paying for it in full. Since the role of national companies is likely to increase in the future, it follows that the global tax-driven incentive for exploration will diminish.

So far as future prices are concerned, it might be logical to expect an average of around $100 a barrel, as recent experience shows that higher prices impose recession dampening demand. Such a price limit would restrain the entry of expensive *Non-Conventional* oil, but that would carry the advantage of allowing it to last longer. It is important to stress that this suggested price limit is to be quoted in current dollars, as rampant inflation and dollar devaluation may arise as a consequence of the unfolding situation. It is worth noting in passing that oil is traded under various price categories, such as *Brent Crude* (based on North Sea production) or *West Texas Intermediate*, with the retail cost to the consumer ranging widely from country to country under differing tax regimes.

Table 3 Oil and gas production levels

Production regular conventional	Oil (Gb)	Gas (Tcf)	Total (Gboe)
Past	1,093	3,931	1,800
Future	908	6,319	2,096
From known fields	793	5,274	1,742
From new finds	115	1,045	303
Total to 2100	2,000	10,250	3,845
Discovered to 2010	1,885	9,205	3,542

Gas is still more difficult to analyse, as the data are even less reliable, with much uncertainty about the amounts flared, re-injected into the reservoir or used as an operating fuel. The depletion profile is also more influenced by the level of investment in facilities and pipelines than is the case for oil. It is here assumed that an idealised production profile for a country would show a plateau between 30 and 70% depletion, although there are many departures. Such a profile ends with a relatively steep cliff, as exemplified by the United States whose *Conventional* gas is now more than 80% depleted, with production falling at almost 10% a year. Countries with production far from market have to rely on costly liquefaction facilities to transport their supply, which restrains the development of their resources. World production is here expected to pass its maximum at the midpoint of depletion around 2015. *Non-conventional* gases such as coalbed methane or shale gas are important additional sources but subject to generally slow and costly extraction rates (Table 3).

In view of the unreliable and confused reserve reporting practices, it makes sense to describe everything in terms of *production* to a convenient cut-off at the end of the century, thereby avoiding having to worry about some insignificant tail-end. The present assessment of *Regular Conventional Oil and Gas* gives the estimates shown in the tables (to be generously rounded), with further details in the Appendix (Table 4).

As conventional production declines, increasing attention will no doubt turn to *non-conventional* sources, deserving a few brief comments. They currently provide about one-fifth of total liquid production and are here expected to rise to about one-third by 2030, passing their peak around 2025.

- *Heavy oils*: the resources of heavy oil and bitumen in the ground are enormous, especially in Canada and Venezuela, but the extraction rates are slow and costly, delivering a low, if not in some cases a negative net energy yield. The operations also carry some adverse environmental impacts. Oil Shale refers to immature source-rocks from which oil can be extracted by heating the rock in retorts, few such projects being viable. Shale oil refers to oil in impermeable shales that can be extracted by artificial fracturing with the help of highly deviated wells, and is now receiving attention in the United States and elsewhere. It can be conveniently classified with heavy oil, although not necessarily in terms of density.
- *Deepwater oil and gas*: there is no shortage of deep water in the world's oceans, but very few areas have the special geological conditions for oil and gas. The

Table 4 Estimated oil production to 2100

Amount			Gb	Annual production—regular oil							Total	Peak
Regular oil				Mb/d	2000	2010	2015	2020	2030		Gb	Date
Past	**Future**		**Total**	US-48	4.2	3.3	3.4	1.9	1.1		200	1970
Known fields		New		Europe	6.1	5.0	3.6	1.9	1.0		76	2000
1,094	788	118	2,000	Russia	6.5	8.4	8.7	5.9	4.0		230	1987
	906			ME Gulf	19	20	19	20	18		763	1974
All liquids				Other	29	30	27	20	14		731	2006
1,231	1,369		2,600	**World**	**64**	**67**	**62**	**50**	**38**		**2,000**	**2005**
2010 base scenario				**Non-conventional**								
Regular oil excludes heavy oils (inc.				Heavy etc.	3.8	4.1	4.2	4.4	5.5		198	2030
tarsands, oilshales); Polar, and				Deepwater	1.6	6.7	8.7	9.4	6.0		100	2021
Deepwater oil, and gasplant NGL				Polar	1.3	1.4	1.7	2.0	2.3		52	2030
and refinery gains of ~3%				Gas liquid	6.5	8.1	8.2	8.2	8.2		239	2020
Reference date: end 2010				*Rounding*			0	1			11	
Revised 16/03/2011				**All**	**77**	**87**	**85**	**75**	**60**		**2,600**	**2008**

prime tracts are related to the opening of the South Atlantic and Gulf of Mexico as the adjoining continents moved apart during the Cretaceous Period, which gave rise to prospective rifts. The Eastern Hemisphere is mainly characterised by convergent plate-tectonic situations giving less promising conditions, although some finds have been made. Operations test technology to the limit, being subject to the occasional serious accident.

- *Polar oil and gas*: the Polar Regions are now attracting new interest despite the harsh operating conditions, but carry some geological drawbacks. First, prime source-rocks were laid down in tropical areas, meaning that those found in Polar Regions depended on long range plate-tectonic movements. They are accordingly relatively old in geological terms, such that not all the petroleum systems have survived intact in the face of subsequent earth movements. Furthermore, the crust of the Earth has been subject to substantial vertical movements partly under the weight of fluctuating ice-caps, adversely affecting seal integrity. On balance, it appears to be a gas-prone domain, although some anomalous oil finds, including the giant Prudhoe Bay Field of Alaska, have been made.
- *Gasplant liquids (NGL)*: natural gas can be processed into liquids in specialised plants, currently producing about 8 Mb/d, although the product yields only about 64% of the energy provided by crude oil. The United States is by far the largest producer responsible for 22% of the world's output (The term NGL is easily confused with LNG, which refers to natural gas which is liquefied at low temperatures for transport).
- *Non-conventional gases*: there are several categories including *coalbed methane,* being the gas contained in coal measures; *shale gas*, being found in impermeable reservoirs; and *methane hydrates*, being methane in ice-like crystals, making up thin laminae, found in cold conditions in Polar Regions or the ocean depths.

5 The Second Half of the Oil Age

Taking a long historical perspective, it is evident that the *First Half of the Age of Oil* was an anomalous chapter of economic expansion, allowing the world population to increase sixfold. Average life expectancy of the increased population also rose to 69 years, with about half the people living in urban circumstances, which are dependent on imported energy. Agriculture, needed to feed the people, came to rely heavily on not only mechanised farming methods but also on the widespread provision of synthetic nutrients and pesticides made from petroleum. It is even said that agriculture has become a process that converts oil into food.

The easy energy from coal, oil and gas, changed the world, but they are finite resources subject to depletion, as discussed above. Their decline, as imposed by their natural limits, now dawns, and will have far-reaching consequence for Mankind, who has become so dependent upon them. But in fact, the turning point reflects no more than extreme example of a well-established pattern of growth followed by decline that has been repeated many times in history [16].

Scientific evidence suggests that the increased emissions and the reduction of forest cover during the *Oil Age* are giving rise to global warming, possibly with far-reaching consequences. Much attention is given to the role of carbon dioxide, although it makes up no more than 0.04% of the atmosphere, with the balance being mainly nitrogen (78.1%) and oxygen (20.9%). The climate has changed many times in the past due to fluctuations in solar radiation, volcanic activity and shifting oceanic currents as the continents moved under *Plate-tectonic* stresses. For example, the so-called *Little Ice Age,* which lasted from about 1650 to 1850, affected life in Europe significantly. In fact, much of the world's oil comes from two epochs of extreme global warming, 90 and 150 million years ago, when algae, from which it is derived, proliferated in the warm, sunlit waters, with the organic residue accumulating in stagnant rifts to be converted to oil by raised temperature when it was buried to a depth of about 2,000 m beneath younger sediments.

Not only is the climate vulnerable, but many other natural resources are under stress. The forests of Borneo are being decimated by logging companies to the detriment of the natives living in them. Fresh water supplies are seriously at risk as the aquifers deplete and deserts encroach in many areas, including populous China and much of the Middle East. Even the world population of bees, important for pollination, is declining, perhaps from the impact of pesticides. The destruction of farmland and natural habitats, together with energy supply, obviously reduces the level of supportable population.

The transition to the *Second Half of the Oil Age* threatens to be a time of great economic, political and geopolitical tension (see Postscript). As the economic powers from the *First Half* face eclipse, they may be inclined to fight for resources to delay the loss of their hegemony. The victors of the Iraq war are already moving to produce its oil, but the war in Afghanistan, which lies on a planned pipeline route from the Caspian, has been less successful, having now lasted 10 years, if anything strengthening the resolve of the indigenous tribesmen and poppy-farmers. The Caspian itself has proved less prolific than was hoped on the fall of the Soviets.

Iran, with its substantial oil and gas resources, is under threat but it remains to be seen if it will be attacked. Whatever the explanations for these wars, it is evident that foreign interest in the Middle East is heavily influenced by its substantial control of future world oil supply. Other countries with less satisfactory regimes have been left in peace. Wars prompt a sense of national loyalty. There is, indeed, an old political adage: *if you don't have an enemy, make one*. Certainly, real or perceived threats strengthen the hand of government in exercising control, which will be called for as the economic assumptions and practices of the past face major adaptation.

The response to the recent collapse of the banking system deserves comment. Governments have intervened under outdated economic principles by issuing yet more credit, lacking real collateral, to rescue failed banks and countries in difficulty, in the hope of stimulating consumerism to restore past prosperity. But logic suggests that any success will be brief, because economic recovery would lead to a rise in the demand for oil, an important driver of the economy, which would soon again pass the supply barrier leading to another price surge and consequential renewed recession. Indeed at the time of writing many fears about a so-called *double dip* recession are being aired. The financial system permeates the world in unseen ways, having acted as a hidden catalyst fuelling expansion. It has reacted to local recessions and changing political situations around the world, but has never had to face such a critical change as is now imposed by dwindling energy supplies affecting the entire Planet.

As mentioned above, one likely response to the recognition of *Peak Oil* is that producing countries will increasingly move to preserve what they have left for their own use, thereby reducing the amount available to importers. It makes eminent national sense although offending the principles of globalism, under which the resources of any country are deemed to belong to the highest bidder. The countries may learn from the example of Britain, whose oil policy was, with hindsight, not in the best national interest. It produced its oil at the maximum rate possible under free market principles, such that it exported at a time of low oil prices only to find itself now importing at high prices, having passed its production peak in 1999. Furthermore, the efficient exploitation of its resources results in relatively high current decline rates of 7% a year for oil and 8% for gas. But fortunately, a certain realism begins to manifest itself in the country with a new coalition government announcing a draconian budget to try to balance the books and to try to wean itself from oil dependency.

Norway, for its part, did use the revenues from its share of the North Sea to build an oil fund of over 500 billion dollars for the future, but it has lost some of its value in the financial crash having been partly entrusted to international financial institutions. The decline in Norway's oil production will undermine the extreme consumeristic affluence currently affecting the country, such that members of future generations may again have to live by catching fish from an open boat in midwinter, as did their antecedents. With hindsight, Norway too would have done better to have followed the advice of King Abdullah by leaving more of its wealth in the ground for the future. In fact, in earlier years it did have a cautious oil policy giving preference to its national companies. But it evidently came under pressure to open its doors. Speaking at an oil conference in Stavanger in 1980, the German Minister ended his speech with the words: *Norway—do not forget your history*, successfully conjuring up images of storm troopers on the frozen wartime streets. We may note in passing that Germany has drilled four times as many exploration wells as has Norway but has found less than one-tenth as much oil, emphasising that some countries are richly endowed with oil-bearing source-rocks whereas others are not.

Changes affecting the World of the magnitude discussed are likely to be accompanied by serious social unrest, signs of which are already being observed around the world. People naturally feel resentful of economic recessions unless they realise that this one is imposed by Nature (*see* Postscript).

The early years of the *Industrial Revolution* also saw new political responses to the changed economic circumstances with the emergence of Anarchist, Socialist, Communist, and Fascist movements aiming in different ways to provide a more egalitarian distribution of resources and protect national interests. New political pressures will no doubt emerge as the *Expansion* of the *First Half of the Oil Age* gives way to the *Contraction* of the *Second Half.* Many claims for *Democracy* are made. People who vote are encouraged to believe that they are responsible for the resulting government and its actions, when in reality it is the political parties and the financial and other powers behind them that exert a major influence on policy. It is perhaps significant that the business world is run on more authoritarian lines even if the employees do not exactly have to salute their bosses, yet the much more difficult challenge of running a country efficiently is assigned to democracy. Perhaps new forms of genuine democracy will arise, freed from financial influences. They are likely to be dominated by new moves to regionalism and local markets as people again try to find sustainable futures for themselves within co-operative communities to which they feel they truly belong.

The behaviour of the International Energy Agency [23], the OECD energy watchdog, is revealing in relation to political influence. It is significant because many governments base their policies and statements on its pronouncements.

A team within the organisation came to understand the issue of *Peak Oil* some years ago, and succeeded in publishing an oblique reference to it in the 1998 edition of the *World Energy Outlook* in the form of a table showing that oil demand would outpace supply by 2010 save for the entry of an item, termed *Unidentified Unconventional.* When it was decoded to imply shortage [12], the Agency evidently came under pressure from its masters, such that the *Unidentified Unconventional* became *Conventional Non-OPEC* in the subsequent issue of the *Outlook* without comment or explanation. This reaction is understandable because the organisation saw itself as representing the consumers in the face of OPEC, and realised that any admission of finite limits would strengthen the latter's hand. But now it begins to come clean with the slogan: *Let's leave oil before it leaves us.* In the current *World Energy Outlook,* it admits in a key graph to the decline of production from existing fields, but offsets that by unrealistic projections of new discovery and yet-to-be-developed fields, such that overall production continues to grow to 2035. It is difficult for politicians to present a serious problem without

Fig. 4 UK energy
production (Mtoe)

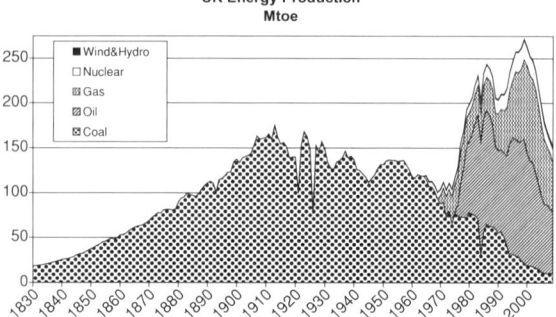

UK Energy Production
Mtoe

coming up with a palatable solution, but they begin to move, finding environmental justification easier to address than the raw reality of *Peak Oil* itself.

Nuclear energy has an important role to play, especially over the transition, although it too has its limits, as the production of prime-grade uranium has itself also passed its peak [10]. It is also subject to risks as illustrated by the earthquake in Japan in March 2011. Substantial coal reserves remain, as mentioned above, although they are evidently also mis-reported and subject to depletion, such that production may be approaching its peak [22]. There is, of course, great scope for bringing in renewable energy from tides, waves, winds, geothermal heat, sunlight, hydro-power and biofuels. Another contributor is *anaerobic digestion*, which is a system that processes urban and agricultural organic waste, of which there is plenty, into methane for heating or electricity generation, with the residue being a rich nutrient returned to the land. However, renewable energy is not cheap, with, for example, electricity from the growing number of Britain's wind farms costing about double that currently generated from gas. The wind is free but there is a lot of embedded fossil fuel energy and other costs in the turbines, structures and transmission lines. Lastly, it is also well to remember the issue of net energy return [15]. For example, a nuclear plant may take as much as 5 years to return the energy consumed in its construction (Fig. 4).

The example of Britain's energy supply, as illustrated in the graph (after Ref. [31]), is particularly noteworthy in illustrating the limits. Coal production, on which its *Industrial Revolution* was built, rose gradually to pass a peak in 1914 before declining to near exhaustion today. Oil and gas provided a replacement that surged to a peak around the end of the last century and are now in steep decline. Nuclear power offered comparatively little, and the contribution of wind and hydro is so small that it fails to show on the graph. Annual consumption is running at about 230 Mtoe, meaning that an increasingly severe shortfall will mark the years ahead. While improved efficiency and imports at rising cost can help make good the gap, it is hard to avoid the conclusion that the country will eventually be unable to support a population much above that at the beginning of the last century, namely around 40 million, if that. No doubt other industrial countries exhibit similar profiles.

Considering the world situation, it is noteworthy that by 2050, oil supply will have fallen to a level able to support less than half the current world population in

Fig. 5 Population and petroleum with a 1.5% annual decrease per capita consumption

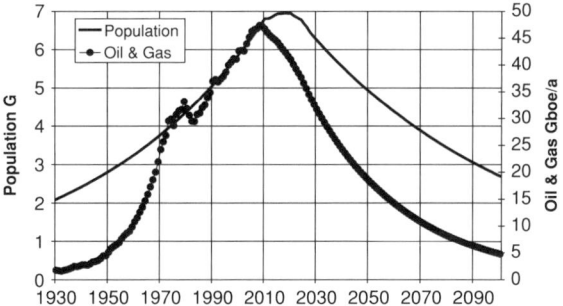

its present way of life. A failure to adapt means fewer people. The graph illustrates the relationship, assuming a modest reduction in per capita consumption of *Regular Conventional* oil and gas, the primary drivers of the economy (Fig. 5).

While the scale of change imposed by the new *post-peak* circumstances is daunting, marking a turning point in history, it is not necessarily a doomsday message, for the survivors may find a new benign age in which they build more respect for themselves, their neighbours and, above all, the environment in which Nature has ordained them to live. People have survived many radical changes in history. As an analogy, think for example of France in the summer of 1939. The war clouds were gathering as tensions built, but people went about their daily lives, reassured to some extent by the words of the British Prime Minister, proclaiming *Peace in our Time.* But then a few months later, the guns of an invading army began to be heard leading to years of harsh occupation. Even so, people soon came to terms with the new situation that had affected their lives so radically.

In investment terms, there may be merit in encouraging a return to the earlier system of investors concentrating on dedicated long-term dividends rather than speculating on superficial market fluctuations. Banks may again be required to back their loans by deposits with a shift of emphasis to encourage saving rather than spending.

As always, there will be winners and losers: the winners being those who better understand and adapt to the new circumstances in which they have to live. Some very positive moves have already made an appearance, as for example illustrated by the *Transition Town Movement*, which even introduces local currencies to foster local trade [21]. The idea of a *Depletion Protocol*, whereby countries agree to cut consumption to match world depletion rates, has been mooted [19]. It would prevent profiteering by the Middle East and allow the poorer countries to obtain their minimal needs at a reasonable price.

The ground-breaking study, entitled *The Limits to Growth* [30], has been followed in recent years by a proliferation of books and articles (some listed in the references below) that address the circumstances of the new world that opens. The book *Half Gone* by Jeremy Leggett [27] gives a penetrating and readable account of the situation. Even the German Military establishment has recently analysed the serious impact of peak oil in political and military terms [39]. Hallett and Wright [16] see the current situation as an extreme example of the pattern of growth and decline repeated many times on history. Another note-worthy study by Gretener [14] addresses the many anomalous aspects of the modern age with its rampant consumerism and failure to comprehend the underlying resource limits. It suggests that *Homo sapiens* may even soon find his place in the fossil record, which is of course his ultimate fate. The geological record shows how species that found a niche that suited them proliferated, only to die out when it closed for natural reasons. There are few, if any, precedents for a reversion to simplicity.

Academic research, led by Uppsala University [1], has been followed by no less than Oxford University where Sir David King, the British Government's former Chief Scientist, now addresses Peak Oil [34]. Prominent industrialists in Britain begin to use their influence to capture the ears of Government ministers on the topic [38].

There is evidently a very positive new awakening but much more needs to be done to prepare for what unfolds and postpone the inevitable demise of the species for as long as possible.

6 An Important Postscript

Revolutions broke out in Tunisia, Egypt, Libya and certain other Middle East countries during the early part of 2011. They were in part triggered by rising food prices that bred increased resentment against the governments, some of whom had enjoyed great oil wealth. It is too soon to evaluate the impact, but it seems likely that Libya's oil production, which approximates to consumption in the United Kingdom, will collapse, whatever the political outcome. If so, oil prices are likely to soar, prompting deeper recessions, which in turn trigger more social unrest and upheaval. If this circle widens to include more major producing countries, espe-cially Saudi Arabia, the consequence for the world would be devastating. The importers may attempt military intervention but experience suggests that that might make a bad situation worse. At root, the cause of this historical turning point is no less than Peak Oil.

Appendix

Table A.1 Oil

RESOURCE BASED PRODUCTION FORECAST — 2010

Regular Conventional Oil by Country to 2100

Region	Mb/d	2000	2005	2010	2020	2030	Past (Gb)	Future (Gb)	Disc %	Dep %	Expl	Disc	Prod
C	Russia	6.48	8.41	8.70	5.90	3.99	150	80	93	65	1988	1960	1987
F	Saudi Arabia	7.77	8.97	7.82	7.81	6.71	117	183	97	39	1967	1948	1980
F	Iran	3.70	4.14	4.12	3.80	3.12	66	84	92	44	1967	1964	1974
C	China	3.25	3.61	4.02	2.52	1.57	39	31	91	56	2003	1960	2010
H	US-48	4.21	3.34	3.01	1.81	1.09	179	21	99	89	1956	1931	1970
E	Mexico	3.01	3.33	2.59	1.29	0.64	25	13	97	73	2003	1977	2004
F	UAE	2.37	2.54	2.41	2.74	2.33	30	55	93	35	1952	1972	2016
F	Iraq	2.57	1.88	2.39	3.48	3.48	34	81	89	29	1978	1928	2025
F	Kuwait	1.76	2.24	2.09	2.08	1.96	42	58	97	37	1963	1938	1971
D	Norway	3.22	2.70	1.89	0.87	0.40	23	9	96	73	1997	1979	2001
A	Algeria	1.25	1.80	1.81	1.11	0.68	21	12	96	61	1970	1956	2007
A	Libya	1.41	1.63	1.65	1.33	1.07	27	25	95	50	1963	1961	1970
C	Kazakhstan	0.72	1.29	1.51	2.05	1.86	11	34	89	24	1988	2000	2025
D	UK	2.67	1.81	1.29	0.64	0.32	27	7	99	80	1990	1974	1999
A	Nigeria	2.17	2.59	1.21	1.21	1.01	28	24	95	51	1966	1968	2005
G	Qatar	0.74	0.84	1.08	1.10	1.10	9	26	96	27	1988	1940	2030
C	Azerbaijan	0.28	0.43	1.03	0.86	0.63	10	14	94	43	1953	1871	2014
B	Indonesia	1.43	1.07	0.95	0.63	0.41	24	8	97	75	1983	1944	2010
H	Canada	0.91	0.92	0.91	0.49	0.26	21	5	97	80	1980	1958	1974
G	Oman	0.97	0.77	0.86	0.50	0.29	9	6	96	62	1984	1962	2000
E	Venezuela	2.47	1.51	0.81	0.72	0.64	50	25	95	67	1981	1914	1970
E	Colombia	0.69	0.53	0.77	0.46	0.29	7	6	94	57	1988	1988	1999
B	India	0.65	0.66	0.73	0.44	0.26	8	5	98	61	1991	1974	2010
E	Argentina	0.76	0.70	0.65	0.35	0.18	10	4	96	74	1985	1962	1998
A	Angola	0.75	0.59	0.60	0.34	0.19	6.3	3.3	96	63	1968	1978	2000
B	Malaysia	0.69	0.63	0.55	0.37	0.25	7	5	98	59	1970	1971	2004
F	N.Zone	0.63	0.58	0.53	0.35	0.24	8.3	4.7	96	64	1962	1964	2003
A	Egypt	0.77	0.66	0.52	0.29	0.16	11	3	98	77	2006	1965	1996
A	Sudan	0.19	0.35	0.51	0.39	0.24	1	4	85	24	2002	2003	2013
E	Ecuador	0.39	0.53	0.48	0.29	0.17	4.7	3.3	96	59	1972	1960	2006
B	Australia	0.72	0.45	0.45	0.32	0.23	7.2	4.8	94	60	1985	1967	2000
G	Syria	0.52	0.43	0.37	0.21	0.12	4.9	2.3	95	68	1992	1966	1996
B	Vietnam	0.32	0.37	0.32	0.34	0.20	2.0	3.6	94	35	1994	1975	2016
A	Congo (B)	0.28	0.23	0.30	0.16	0.09	2.3	1.7	96	57	1992	1984	2009
A	Yemen	0.44	0.40	0.26	0.13	0.06	2.7	1.3	94	68	1975	1978	2001
B	Thailand	0.11	0.18	0.25	0.13	0.06	1.0	1.2	89	45	1983	1981	2011
D	Denmark	0.36	0.38	0.24	0.12	0.06	2.3	1.2	93	65	1985	1971	2004
A	Gabon	0.32	0.27	0.23	0.12	0.06	3.6	1.1	97	75	1991	1985	1997
E	Brasil	0.50	0.46	0.23	0.17	0.13	6.1	2.9	97	68	1982	1975	1990
C	Turkmenistan	0.14	0.18	0.18	0.13	0.10	3.5	2.0	93	63	1982	1956	1973
B	Brunei	0.19	0.19	0.13	0.08	0.05	3.5	1.0	98	77	1975	1929	1979
A	Chad	0.00	0.18	0.13	0.14	0.09	0.4	1.6	68	19	2002	1977	2015
E	Trinidad	0.12	0.14	0.10	0.07	0.05	3.5	1.0	99	78	1972	1959	1981
D	Italy	0.09	0.11	0.09	0.06	0.04	1.2	0.8	94	58	1962	1989	1997
C	Romania	0.12	0.11	0.09	0.07	0.06	5.5	1.5	98	80	1969	1890	1976
C	Ukraine	0.07	0.09	0.09	0.07	0.06	2.9	1.6	98	64	1990	1962	1970
A	Tunisia	0.08	0.08	0.08	0.06	0.04	1.5	0.8	95	65	1981	1964	1983
E	Peru	0.07	0.08	0.07	0.06	0.05	2.6	1.4	93	65	1975	1869	1982
A	Cameroon	0.08	0.08	0.07	0.04	0.02	1.3	0.5	93	73	1970	1977	1985
B	Pakistan	0.05	0.07	0.05	0.03	0.02	0.7	0.4	97	65	2003	1984	2006
G	Turkey	0.05	0.04	0.05	0.03	0.01	1.0	0.3	96	77	1992	1961	1991
C	Uzbekistan	0.09	0.07	0.04	0.06	0.09	1.0	2.0	80	35	1991	1985	2029
E	Bolivia	0.03	0.05	0.04	0.03	0.03	0.6	0.7	92	45	1962	1999	2005
B	Bahrain	0.04	0.04	0.04	0.02	0.01	1.2	0.2	99	83	1983	1932	1970
B	Papua	0.07	0.04	0.03	0.02	0.02	0.5	0.5	97	46	1990	1987	1993
D	Germany	0.05	0.04	0.03	0.02	0.02	2.0	0.5	96	78	1958	1949	1967
D	Netherlands	0.03	0.03	0.02	0.02	0.01	0.9	0.3	97	72	1985	1943	1986
D	France	0.03	0.02	0.02	0.01	0.01	0.8	0.2	97	83	1959	1954	1988
D	Austria	0.02	0.02	0.02	0.01	0.01	0.8	0.1	98	88	1975	1944	1955
C	Hungary	0.03	0.02	0.01	0.01	0.01	0.7	0.3	93	72	1964	1965	1979
C	Croatia	0.02	0.02	0.01	0.01	0.01	0.5	0.1	86	54	1985	1957	1988
C	Albania	0.01	0.01	0.01	0.01	0.01	0.5	0.3	93	67	2013	1928	1983
E	Chile	0.01	0.00	0.00	0.00	0.00	0.4	0.1	99	86	1972	1960	1982
A	Uganda	0.00	0.00	0.00	0.15	0.15	0.0	1.0	60	0	2010	2008	2030

Regular Conventional Oil by Region

Region	Mb/d	2000	2005	2010	2020	2030
F	ME.Gulf	18.8	20.3	19.4	20.3	17.8
C	Eurasia	11.2	14.2	15.7	11.7	8.4
H	N.America	5.1	4.3	3.9	2.3	1.4
E	L.America	8.1	7.3	5.7	3.4	2.2
A	Africa	7.3	8.4	7.1	5.3	3.8
D	Europe	6.5	5.1	3.6	1.8	0.9
B	Asia-Pacific	4.2	3.7	3.5	2.4	1.5
G	ME Minor	2.8	2.5	2.7	2.0	1.6
	Minor	0.5	1.0	0.8	0.5	0.3
	Non ME.Gulf	46	47	43	30	20
	MEGulf Share	29%	30%	31%	41%	47%
	Total	65	67	62	50	38

Non Conventional Oil

	2000	2005	2010	2020	2030
Heavy etc.	2.4	2.9	4.6	5.0	6.2
Deepwater	1.6	3.3	6.7	9.4	6.0
Polar	1.3	1.5	1.4	2.0	2.3
Gas Liquid	6.5	7.7	8.1	8.2	8.2
Total	12	15	21	25	23
WORLD	76	82	83	74	61

Regular Conventional World Summary

PRODUCTION to 2100	Gb	%
PAST	1093	55
FUTURE	908	45
Known	793	40
To be found	115	6
Discovered	1885	94
TOTAL	2000	

NOTES

Regular Conventional Oil includes condensate
ME-Gulf =UAE, Iran, Iraq, Kuwait, NZ, S.Arabia.
Eurasia = FSU, E.Europe & China.
N.America = USA & Canada.
Venezuela I = ordinary heavy.
Venezuela II = 4 Extra-Heavy oil projects.
The Production Forecast assumes decline at the
Current or Midpoint Depletion Rate, whichever
comes first.
Depletion Rate = annual production as % of Future.
Deepwater >500m WD.
The statistics refer to Production to a cutoff
at the end of the Century not Ultimate recovery.

Revised   10/03/2011

Table A.2 Gas

| | RESOURCE BASED PRODUCTION FORECAST | | | | | | | | | | | | 2010 |

Regular Conventional Gas by Country to 2100

Region		Tcf/a	2000	2005	2010	2020	2030	Past	Future	Disc	Dep	Expl	Disc	Prod
C	Russia		20.5	24.2	22.7	30.0	19.1	676	824	84	45	1988	1966	2014
H	US-48		23.7	23.0	23.3	8.5	3.1	1181	219	98	83	1956	1996	1979
A	Algeria		5.8	6.6	7.0	6.5	3.2	143	157	95	48	1961	1957	2011
F	Iran		3.9	5.4	6.5	6.5	6.5	116	1084	82	10	1967	1964	2030
H	Canada		7.7	7.7	5.6	1.5	0.4	211	39	99	82	1980	1993	2001
D	Norway		3.2	4.6	5.2	4.3	1.6	78	92	97	46	1997	1979	2018
G	Qatar		1.3	2.0	3.8	6.2	10.1	36	964	95	4	1991	1971	2030
F	Saudi Arabia		1.9	2.9	3.3	5.5	5.5	84	316	84	21	1967	1948	2030
C	China		1.0	1.8	3.2	5.0	5.0	43	157	92	22	2003	2000	2022
C	Turkmenistan		1.6	2.2	1.5	3.8	4.3	78	297	80	21	1986	1973	2030
B	Indonesia		2.9	3.0	3.0	3.0	3.0	78	142	90	35	1983	1973	2020
F	UAE		1.8	2.4	2.8	2.8	2.8	53	122	93	30	1952	1978	2023
B	Malaysia		1.7	2.7	2.6	2.6	2.1	42	83	93	34	1970	1970	2017
A	Nigeria		1.2	1.9	2.1	3.5	3.5	50	170	92	23	1966	1967	2029
D	Netherlands		2.6	2.8	2.6	1.5	0.9	117	48	99	71	1985	1959	1976
E	Venezuela		2.14	2.04	2.50	2.50	2.50	65	185	93	26	1981	1941	2030
A	Egypt		0.86	1.66	2.00	2.70	1.87	25	70	89	26	1985	1996	2019
C	Uzbekistan		1.99	2.11	2.20	2.20	1.29	68	62	93	52	1991	1974	2015
D	UK		4.12	3.38	2.10	0.96	0.44	94	26	98	79	1990	1966	2000
E	Mexico		1.51	1.58	1.70	1.45	0.84	58	42	96	58	2003	1977	2013
B	Australia		1.17	1.45	1.83	4.76	5.00	32	188	83	15	1985	1971	2030
E	Argentina		1.58	1.82	1.70	1.05	0.48	43	27	92	61	1985	1977	2004
B	Pakistan		0.86	1.21	1.50	1.50	0.89	26	39	88	40	2003	1952	2014
E	Trinidad		0.58	1.17	1.50	1.36	0.51	21	29	91	44	1971	1968	2012
C	Kazakhstan		0.16	0.78	1.26	3.26	3.60	14	111	91	11	1988	1979	2027
B	India		0.91	1.19	1.55	1.30	1.30	24	56	86	30	1991	1976	2023
A	Libya		0.36	0.70	1.00	1.28	1.60	21	74	84	20	1963	1965	2027
G	Oman		0.48	0.85	1.15	1.30	1.30	15	45	93	25	1991	1973	2021
B	Thailand		0.71	0.93	1.20	0.60	0.19	16	14	98	52	1983	1973	2008
C	Azerbaijan		0.49	0.21	0.91	1.30	1.30	16	54	85	23	1953	1999	2030
E	Colombia		0.51	0.53	1.00	0.27	0.06	15	7	98	67	1988	1973	2010
C	Ukraine		0.64	0.69	0.70	0.67	0.53	64	36	96	64	2000	1950	1975
E	Bolivia		0.20	0.52	0.50	0.80	0.80	10	22	87	15	1962	1999	2030
G	Yemen		0.67	0.73	0.50	0.50	0.34	10	15	97	42	1992	1989	2013
F	Iraq		0.15	0.40	0.61	1.00	1.57	14	136	73	10	1978	1953	2030
D	Germany		0.84	0.75	0.79	0.42	0.22	36	12	97	75	1958	1969	1987
B	Brunei		0.42	0.47	0.45	0.30	0.17	16	9	98	64	1975	1963	2006
E	Brasil		0.47	0.50	0.50	0.50	0.26	11	12	95	48	1982	2003	2011
G	Bahrain		0.41	0.47	0.50	0.19	0.07	16	5	99	77	2004	1932	2009
A	Angola		0.25	0.30	0.40	0.40	0.28	7	11	95	38	1968	1971	2016
F	Kuwait		0.40	0.53	0.40	0.40	0.40	21	49	97	31	1963	1938	2015
E	Peru		0.03	0.20	0.35	0.35	0.35	5	20	84	19	1975	1986	2025
C	Romania		0.49	0.41	0.31	0.09	0.03	45	2	99	95	1954	1985	1985
A	Congo (B)		0.13	0.25	0.30	0.11	0.02	3	3	93	54	1992	1984	2012
G	Syria		0.28	0.30	0.30	0.30	0.19	7	9	89	44	1992	1987	2013
D	Italy		0.59	0.43	0.27	0.18	0.12	26	6	98	80	1962	1968	1994
B	Vietnam		0.05	0.17	0.30	0.48	0.56	2	23	98	9	1996	1995	2030
D	Denmark		0.42	0.38	0.27	0.09	0.03	7	2	97	74	1985	1968	2000
A	Tunisia		0.08	0.11	0.15	0.12	0.05	2	3	94	45	1981	1974	2008
C	Hungary		0.11	0.11	0.09	0.07	0.05	7.7	3.3	97	70	1964	1965	1985
E	Chile		0.10	0.08	0.05	0.05	0.05	4.7	5.3	97	47	1972	1960	1992
A	Gabon		0.08	0.07	0.06	0.01	0.00	2.9	0.4	98	88	1990	1965	1995
D	Austria		0.06	0.06	0.06	0.04	0.03	3.4	4.9	98	67	1975	1949	1975
A	Cameroon		0.07	0.07	0.07	0.07	0.07	2	4.6	93	30	1977	1979	2025
E	Ecuador		0.04	0.04	0.05	0.04	0.02	1.0	1.0	93	52	1972	1969	2013
D	France		0.07	0.06	0.04	0.02	0.01	11.4	0.4	99	97	1959	1949	1978
F	N.Zone		0.05	0.06	0.04	0.05	0.05	2.7	7.3	96	27	1962	1967	2030+
G	Turkey		0.03	0.03	0.03	0.03	0.01	0.5	0.6	97	45	1975	1965	2012
A	Sudan		0.00	0.00	0.03	0.05	0.06	0.0	2.0	51	2	2002	2003	2030
C	Croatia		0.06	0.05	0.08	0.06	0.03	1.5	1.5	90	51	1985	1974	2014
A	Chad		0.00	0.00	0.01	0.01	0.01	0.0	0.5	66	2	2002	1975	2030
A	Uganda		0.00	0.00	0.01	0.02	0.02	0.0	0.5	41	2	2010	2015	2023
B	Papua-NG		0.00	0.00	0.01	0.30	0.30	0.1	19.9	90	0	1990	1990	2030+
C	Albania		0.00	0.00	0.00	0.01	0.01	0.5	0.8	91	39	1987	1977	1982

Regular Conventional Gas by Region

	Tcf/a	2000	2005	2010	2020	2030
C	Eurasia	27	33	33	47	35
H	N.America	31	31	29	10	3
A	Africa	8.8	11.7	13.1	15	11
F	ME.Gulf	8.1	11.7	13.6	16	17
B	Asia-Pacific	8.8	11.1	12.4	15	14
D	Europe	11.9	12.4	11.4	8	3
E	L.America	7.2	8.5	9.9	8	6
G	ME Minor	3.2	4.4	6.3	9	12
	Minor	1.2	1.6	3	2	1
	Rounding	0.0	0.0	0.0	0.1	0.3
	Total	108	125	131	129	102

Non Conventional Gas

	2000	2005	2010	2020	2030
US	9.6	12.7	15.4	20	21
Other	1.3	1.6	2.0	6	15
Total	10.9	14.3	17.4	26	36
WORLD	119	139	149	155	138

Regular Conventional World Summary

PRODUCTION to 2100	Tcf	%
PAST	3931	38
FUTURE		
Known	5274	51
To be found	1045	10
Discovered	9205	90
TOTAL	10250	

NOTES

Regular Conventional Gas excludes gas from coal and shale; and Deepwater (>500m) and Polar areas.

ME-Gulf =UAE, Iran, Iraq, Kuwait, NZ, S.Arabia.
Eurasia = FSU, E.Europe & China.
N.America = USA-48 & Sub Arctic Canada.
An idealised depletion profile assumes a production plateau from 30% to 70% depletion.
The statistics refer to Production to a cutoff at the end of the Century not Ultimate recovery.

Revised   10/03/2011

References

1. Aleklett K (2010) The peak of the oil age—analyzing the world oil production reference scenario in World Energy Outlook, 2008. Energy Policy
2. Bardi U (2003) La fine del Petroleo. ISBN 88-359-5425-8
3. Bentley RW, Mannan SA, Wheeler SJ (2007) Assessing the date of the global oil peak: the need to use 2P reserves, Energy Policy, vol 35, pp 6364–6382, Elsevier
4. BP, BP Statistical review of world energy, published annually
5. Campbell CJ, Laherrère JH (1998) The end of cheap oil. Scientific American, Camp Hill
6. Campbell CJ (2005) Oil crisis. ISBN 0906522-39-0
7. Campbell CJ, Heapes S (2009) An atlas of oil & gas depletion. ISBN 978-1-906600-42-6
8. Campbell CJ (2010) The second half of the oil age dawns. Swiss Derivatives Review. 43
9. Deffeyes KS (2005) The view from Hubbert's peak. Hill & Wang, New York
10. Dittmar M (2009) The future of nuclear energy: facts and fiction. www.theoildrum.com Aug 5–Nov 10
11. Ferguson N (2009) The ascent of money: a financial history of the world. ISBN 978-0-141-03548-2
12. Fleming D (2000) After oil, prospect. Nov 2000
13. Gollwitzer H (1969) Europe in the age of imperialism 1880–1914. Thames & Hudson, London
14. Gretener P (2010) The vanishing of a species? ISBN 978-1-897093-82-5
15. Hall C, Hallock J, Cleveland C, Jefferson M (2003) Hydrocarbons and the evolution of human culture. Nature 426:318–322
16. Hallett S, Wright J (2011) Life without oil. ISBN 978-1-61641-402-9
17. Heinberg R (2003) The party's over. ISBN 0-86571-482-7
18. Heinberg R (2006) The oil depletion protocol. ISBN 13-978-0-86571-563-9
19. Heinberg R (2007) Peak everything. ISBN 978-1-905570-13-3
20. Homer-Dixon T (2006) The upside of down. ISBN 1-59726-064-9
21. Hopkins R (2008) The transition handbook. ISBN 978-1-900322-18-8
22. Inman M (2010) Mining the truth on coal supplies. Nat Geogr News. Sept 8
23. International Energy Agency. World energy outlook (published annually)
24. Klare MT (2002) Resource wars. ISBN 0-8050-5576-2
25. Laherrère JH (1999) World oil supply: what goes up must come down—but when will it peak? Oil Gas J. Feb 1st
26. Le Blanc S, Register K (2003) Constant battles: the myth of the peaceful, noble savage. ISBN 312-31089-7
27. Leggett J (1999) Half gone. ISBN 1-84627-004-9
28. Longwell H (2002) The future of the oil and gas industry; past approaches, new challenges; World Energy 5/3
29. Maugeri L (2009) Squeezing more oil from the ground, Scientific American, Oct 2009
30. Meadows DH et al (1972) The limits to growth. ISBN 0855644-008-6
31. Mearns E (2010). Europe.theoildrum.com
32. Moorbath S (2009) Time and earth history, Oxford Magazine, 2nd Week, Trinity Term
33. Murphy P (2008) Plan C: community survival strategies for peak oil and climate change. ISBN 978-0-86571-607-0
34. Owen NA et al (2010) The status of conventional oil reserves—hype or cause for concern. J Energy Policy 2010(02):026
35. Petroleum Economist (2009) World energy Atlas 2009. ISBN 1.186186-273-3
36. Schweizer P (1994) Victory. ISBN 0-87113-567-1
37. Simmons MR (2005) Twilight in the desert. ISBN 0-474-73876-X0ccs
38. Skrebowski C (2010) In: The oil crunch—a wake-up call for the UK economy. Industry taskforce on peak oil and energy security. ISBN 978-0-9562121-1-5

39. Spiegel Online International (2010) Military study warns of a potentially drastic oil crisis. Sept 2
40. Tolub L, Erb MA (2010) Oil price band for the next decade: Utopia versus Reality; Swiss Derivatives Review, Issue 43, Summer 2010
41. Tuchman B (1966) The proud tower. ISBN 0-333-30645-5
42. Wright R (2004) A short history of progress. ISBN 978-0-88784-706.6
43. Ziegler WH, Campbell CJ, Zagar JJ (2009) Peak oil and gas. Swiss bulletin for applied geology v.14/1 + 2

Peak Oil Futures: Same Crisis, Different Responses

Jörg Friedrichs

Abstract Peak oil theory predicts that global oil production will soon start a terminal decline. Most proponents of the theory imply that no adequate alternate resource and technology will be available to replace oil as the backbone resource of industrial society. To understand what may happen if the proponents of peak oil theory are right, I analyze the historical experience of countries that have gone through a comparable experience. Japan (1918–1945), North Korea (1990s) and Cuba (1990s) have all been facing severe oil supply disruptions in the order of 20% or more. Despite the unique features of each case, it is possible to derive clues on how different parts of the world would react to a global energy crunch. The historical record suggests at least three possible peak oil trajectories: predatory militarism, totalitarian retrenchment, and socioeconomic adaptation.

1 Introduction

The Stone Age came to an end not for a shortage of stones. The Coal Age came to an end not for a shortage of coal. But, contra former Saudi Oil Minister Sheikh Yamani, the Oil Age may come to an end for a shortage of oil. This is what the proponents of "peak oil theory" suggest. Peak oil theory predicts that oil

This is a carefully revised version of my 2010 article 'Global energy crunch: how different parts of the world would react to a peak oil scenario', Energy Policy 38 (8): 4562–4569. Thanks to Elsevier for permission to reprint.

J. Friedrichs (✉)
Department of International Development, University of Oxford,
Queen Elizabeth House, 3 Mansfield Road, Oxford OX1 3TB, UK
e-mail: joerg.friedrichs@qeh.ox.ac.uk

production will soon start a terminal decline. Most authors imply, further, that no adequate alternate resource and technology will be available to replace oil as the backbone resource of industrial society.[1]

To be sure, the demise of oil has been predicted many times over. Oil shortages were predicted in the 1920s, 1930s, and 1940s. Peak oil theory was first introduced in 1956 by oil geologist Marion King Hubbert. During the oil crisis of 1973, US Ambassador to Saudi Arabia James Akins declared that "This time the wolf is here" [1]. Similar cries were heard in the second oil crisis of 1979.

Although the history of oil is the chronicle of a death foretold, oil is a finite resource. It is bound to first become scarce and then run out. The extrapolation of unfettered growth into the indefinite future is therefore misleading, and a peak of global oil production is only a matter of time. Peak oil theorists proffer serious arguments why, despite many false alarms in the past, Cassandra will turn out to be right this time. In this chapter I am not going to repeat their arguments. I am not personally committed to peak oil, and I will be more than happy if Cassandra is proven wrong this one more time.

However, given the momentous importance of oil for our industrial way of life the precautionary principle mandates to take warnings of peak oil seriously and assess possible consequences. In this spirit, I am not debating peak oil but asking the "what if" question: what is likely to happen if peak oil occurs? As a baseline for my assessment, I assume a decline of global oil production in the order of 2–5% per year for a couple of decades.[2] In line with most peak oil theorists, I further assume that no adequate alternate resource and technology will be available to replace oil as the backbone resource of industrial society.

While a global peak of oil production would per definition be a planetary event, reactions would vary in different parts of the world. Insofar as globalization has been fueled by cheap and abundant energy, traded as a commodity on a free market, increasing conflict over scarce energy resources would undermine the very foundations of the world-wide social, economic, and political normalization processes that have been observed over the past few centuries. As a consequence the world would once again become more diverse and, thus, less "global".

In this chapter, I focus on oil importing countries, which constitute the vast majority of states. Because an event comparable to peak oil has never happened at the planetary level, I resort to cases where severe oil supply disruptions in the order of 20% have occurred at the national level.[3] I believe that studying national

---

[1] For select readings on peak oil see Owen et al. [36], Aleklett et al. [2], Sorrell et al. [44], Hirsch [22], Brandt [7], United States Government Accountability Office [49], Hirsch et al. [24], Hubbert [26].

[2] The predictions of most peak oil theorists are in this band; see the overviews provided by Sorrell et al. ([45], pp. 4998–4999) and Hirsch [22].

[3] This is far above the formal threshold that the International Energy Agency stipulates for an international oil supply disruption (7%), and also higher than the shortfalls of global oil production during the oil crises of the 1970s (less than 7%).

analogs to peak oil as "proxy" cases is the best analytical strategy available to gain clarity about the effects peak oil would have on oil importing countries.

My first case is Japanese *predatory militarism* before and during the Pacific War. The specter of future resource shortages had played an important role in shaping Japan's imperialist strategy ever since the end of World War I. When an American oil embargo became imminent, in 1941, Japan preemptively attacked the US Naval Base at Pearl Harbor and radicalized its war of conquest in order to gain access to the rich oil supplies of the East Indies.

My second case is *totalitarian retrenchment* in North Korea after the end of the Cold War. When subsidized deliveries of oil and other vital resources from the Soviet Union were disrupted, the "Hermit Kingdom" reacted in a shockingly reckless way. Elite privileges were preserved in the face of hundreds of thousands of North Koreans dying from hunger. While this may be morally repugnant, it represents another possible peak oil scenario.

My third case is *socioeconomic adaptation* in Cuba, which was challenged by a similar disruption of subsidized deliveries from the Soviet Union. While this plunged Cuba into a deep crisis, there was no mass starvation comparable to North Korea. Instead, Cubans relied on social networks and non-industrial modes of production to cope with energy scarcity and the concomitant shortage of food. They were actively encouraged to do so by the regime in Havana.

My cases suggest three possible "peak oil futures", i.e. trajectories that different parts of the world may take in case of peak oil. This obviously does not imply that responses to a global peak of oil production would follow exactly the same lines as the national reactions to oil supply disruptions described in my case studies. Japan in the 1940s, as well as North Korea and Cuba in the 1990s, were unique places. It must make a difference that, in the 2000s, all oil importing countries are integrated in global market structures. Their circumstances vary. Therefore, we can easily imagine additional peak oil trajectories such as the mobilization of national sentiment by populist regimes. Nevertheless, my three case studies are sufficiently similar to a peak oil scenario to conjure up plausible conjectures on how different parts of the world would react to a global energy crunch.

So-called "techno-optimists" object to "Malthusians" that a global decline of oil production would not only lead to higher prices but also trigger a transition from oil to other energy sources, such as renewable energy or a new generation of nuclear reactors. But alas, this argument is countered by another piece of historical evidence. After the American War of Secession, the South of the United States was deprived of slaves as the backbone resource of its socioeconomic way of life. One would expect this to be the easiest case for a smooth energy transition. After the Civil War, Southerners only had to look to the North of their own country for investment and innovative technologies. Nevertheless, the modernization of "Dixieland" took at least a century. Insofar as a similar "upgrade" does not seem to be available in the event of peak oil, there is no reason to be particularly optimistic about a smooth transition to a post-oil (or post-carbon) society.

In the first three sections, I present my case studies. Each outlines the historical response of a country to an acute or (in the Japanese case) anticipated severe oil supply disruption. Next, I formulate generic hypotheses about the factors on which it would depend how different parts of the world would react to a global energy crunch. To counter the view that the transition to a post-oil society will be easy, I subsequently present my fourth case study on "Dixieland" after the American Civil War. I then formulate specific conjectures on how different parts of the world would react to a peak of global oil production. In the final section, I discuss possible factors that may mitigate the negative impact of peak oil.

2 Predatory Militarism: Japan, 1918–1945

In September 1945, defeated Japan was so fuel-starved that it was difficult to find an ambulance with sufficient fuel to transport Premier Tojo to a hospital after his attempted suicide. Pine roots had been dug out from mountainsides all over the country in a desperate attempt to find a resinous substitute to fossil fuel. Much of the Japanese air force and navy had been sacrificed in kamikaze raids, at least in part because there was not sufficient petrol to refuel planes and ships to return from their sorties and keep fighting ([58], pp. 362–367).

Ultimately, this is a dramatic case of a self-fulfilling prophecy. The main lesson the Japanese military had taken home from World War I was that a country cut off from access to raw materials was bound to lose in a military contest due to a trade embargo. In their view, Germany had lost the War because it did not muster the necessary industrial base or access to foreign markets to achieve wartime autarky. To be prepared for a similar war, resource-poor Japan would have to control access to strategic resources. Only a self-sufficient economic bloc in East Asia would sufficiently prop up Japanese industrial capacity to secure the desired status of a great power ([4], pp. 9–21, [5]).

It was precisely to prevent fuel starvation and external dependency on other strategic resources that Japan embarked on aggressive military campaigns. After a liberal interlude in the 1920s, the next decade saw the invasion of Manchuria (1931) followed by the invasion of China (1937). The paramount goal was to achieve self-sufficiency in an economic bloc that was later, in 1940, to be proclaimed as the "Greater East Asia Co-prosperity Sphere".

Even from the cynical viewpoint of Japanese military planners, however, it soon turned out that the targets had not been selected wisely. While Manchuria and the other occupied territories yielded significant quantities of food, coal, and iron ore, very little oil came from these areas. Instead of becoming more self-sufficient, Japan grew even more dependent on the importation of critical commodities— especially from the United States. The situation was particularly dramatic for petroleum, which was entirely indispensable as a military transportation fuel. Since the US was the dominant producer of petroleum at the time, Japan was heavily dependent on American deliveries. Japan imported 90% of its petroleum consumption, of which 75–80% was shipped from California. For the critically important gasoline, the dependence was even higher ([33], pp. 156–157).

With that in mind, it is easy to understand (not to condone) why the Japanese onslaught in East Asia degenerated into the total War in the Pacific when Japan felt threatened by the specter of a US trade embargo. The only alternative to importing oil from the US was looting it from Borneo and Sumatra in the East Indies. To reduce Japanese vulnerability to a US embargo, a southward advance was thus irresistibly appealing—especially to elements in the Japanese navy.

The idea of a southward advance became even more compelling after the start of the Second World War in Europe, when increasing demand for resources in the European theatre led to rising commodity prices. In the late 1930s the US, which had hitherto limited itself to token gestures, gradually began introducing real economic sanctions against Japan. Given the worsening fuel scarcity and in anticipation of a full-blown embargo, the Japanese army began its southward advance. Japan started an offensive in southern China in 1939, and occupied the northern part of French Indochina in September 1940 ([4], pp. 136–175).

When the full-blown American trade embargo finally came in July 1941, Tokyo took it as the ultimate confirmation that there was no other choice than to move further southwards and to tap the rich mineral resources available in the Dutch East Indies, and particularly the petroleum that was being extracted in the British part of Borneo. To secure its flank in the imminent military offensive, the Japanese navy famously endeavored a pre-emptive attack on the US Pacific Fleet stationed at Pearl Harbor. The intention was to roll over East Asia and create a geopolitical bloc while America was directing most of its attention toward the European theatre, and later to negotiate some settlement with the US from a position of relative strength [29, 33, 39].

None of this is to deny that Japanese imperialism reaches back to the late nineteenth century, that imperial Japan was a military aggressor, and that the war in East Asia started in 1931 rather than 1941. On the contrary, all of this is an important part of the story. During the 1930s, resource-starved Japan tried to build a regional economic bloc to prevent strangulation. Japan was prompted by the specter of fuel starvation to scrap the Open Door policy of free trade and to radicalize its strategy of predatory militarism to secure access to vital energy resources. The American trade embargo further radicalized this geopolitical bent.

3 Totalitarian Retrenchment: North Korea, 1990s

Whereas Japan in the 1930s and early 1940s went on conquest to assert its status as a great power and secure access to vital supplies, the totalitarian regime of North Korea in the 1990s retrenched in order to preserve elite privileges after the demise of the Soviet Union. As a consequence, a terrible famine between 1995 and 1998 led to the starvation of an estimated 600,000 to 1 million people, or 3–5% of the North Korean population ([17], p. 234).

This was in glaring contradiction to the self-proclaimed national ideology of self-reliance (*juche*). In line with that ideology, up until the 1980s the regime had heavily invested in coalmines and hydropower to satisfy the country's enormous energy needs. Furthermore, Pyongyang had developed a toxic industrial agriculture to feed the highly urbanized North Korean population. Farming was based on irrigation, mechanization, electrification, and the prodigal use of chemicals. In 1990, estimated per capita energy use was twice as large in North Korea as in China and more than half that of Japan ([54], p. 112).

All of this came to naught with the demise of the Soviet Union, when it turned out that access to oil was the Achilles heel of the North Korean economy. Since North Korea does not possess any proven reserves of petroleum, oil was mostly imported from the Soviet Union in exchange for political allegiance. In 1991, post-Soviet Russia stopped subsidized exports of oil and other vital goods to North Korea. Two years later, Russian exports to North Korea were down by 90% ([20], pp. 27–32).[4] This had dramatic effects. While the North Korean regime reserved most remaining fuel for the military, the rest of the industry nearly collapsed and agricultural production languished around subsistence level. Already in 1991, Pyongyang launched a "Let's Eat Two Meals a Day" campaign. In 1994, when Kim Il-sung bequeathed leadership to his son Kim Jong-il, a serious food crisis was looming. After a series of decent harvests due to favorable weather conditions in the early 1990s, severe floods and droughts led to the Great Famine between 1995 and 1998 ([20], pp. 73–76).

The Great Famine is a paradigm example of how the lack of an economic backbone resource such as oil can have momentous systemic ripple effects. To begin with, North Korean land machines depended on oil. Without fuel, tractors and other machines were not running. The next problem was transportation. Fuel was needed to bring fertilizer and other inputs to farms, and agricultural products to urban consumers. Fuel was also needed to ship coal from mines to fertilizer plants, where coal was converted into soil nutrients.[5] Fuel was further needed to get coal to power stations for electricity generation. As a consequence, electricity was yet another problem. Without sufficient electricity, irrigation pumping, and electrical railways became intermittent. The intermittency of electrical railways further affected transportation. Without reliable trains, it became even more difficult to bring coal to fertilizer plants or power stations, to transport fertilizer to farms, and to get agricultural products to urban consumers [54].

Thus, interlocking energy shortages combined with shortages of industrial inputs and a general decline of infrastructure to produce a dramatic decline of production, and thus an almost hopeless situation. While the entire economy was

---

[4] In 1993 China refused to step in for Russia, demanding hard currency for any further exports and radically cutting deliveries of "friendship grain".

[5] In North Korea, coal was used in the production of fertilizers both as an energy source and as a chemical feedstock ([54], pp. 117–119). Fertilizer use fell by more than 80% from 1989 to 1998 ([14], p. 14).

damaged, the consequences were most dramatic in agriculture where there was plummeting food production, considerable loss of arable land, and a rapid depletion of soil fertility. Restoring soil fertility would have required large amounts of lime, which however could not be transported without fuel. In a desperate attempt to replace land machines, draft oxen slowly became more numerous. But, unlike tractors, work animals compete with humans for food. The energy crisis also compelled many poor people to rely on biomass for cooking and heating. Unlike fossil fuel, however, the extraction of biomass reduces soil fertility, which in turn aggravated the agricultural crisis.

As a result of such interlocking vicious circles, the production of rice and maize fell by almost 50% between 1991 and 1998. North Korea was thus compelled to apply for international food aid. After a considerable time lag, the worst starvation was stopped in the late 1990s. But since North Korea's industrial agriculture cannot be restored without a viable energy regime, even today there is still a protracted food crisis with an ever-present risk of further starvation.[6]

To some extent it is of course true that the Great Famine was due to a malfunction of North Korea's Stalinist regime [35]. However, Pyongyang's performance is dysfunctional only when measured against Western humanitarian standards. On its own (considerably more cynical) terms, the regime has been incredibly successful. The crisis prompted North Korean elites to abandon the Stalinist path of wasteful industrialism, and administer systemic scarcity instead. This negative policy choice made it possible to avoid an economic and political opening, thus preserving cherished elite privileges. While the Soviet Union and most other communist regimes have disappeared, the Democratic People's Republic of Korea is still on the map. North Korea has even become a nuclear power, which sometimes enables Pyongyang to extort international concessions. While such brinkmanship may be morally repugnant, Korean-style totalitarian retrenchment is without doubt a possible response to a severe energy supply disruption.

4 Socioeconomic Adaptation: Cuba, 1990s

Cuba faced an energy supply disruption in the 1990s similar to the one experienced by North Korea. When taking into account subsidized oil deliveries from China to North Korea which lasted until 1993, the Cuban supply shock was even more dramatic. Subsidized energy supplies from the Soviet Bloc ceased to 100%. The CIA estimated the decline of Cuban fuel imports between 1989 and 1993 at a whopping 71% (quoted in Díaz Briquets and Pérez López [12], pp. 250).[7]

---

[6] See the Special Reports of the Crop and Food Security Assessment Mission to the DPRK (especially [13, 14, 53]).

[7] Official Cuban figures for the decline of imported raw materials and other vital inputs to industrial production and electricity generation were on a similar level (reported in Wright [57], p. 68). Even according to the most conservative estimate of the US Energy Information

In 1990, the Cuban leader Fidel Castro was forced to proclaim a national emergency called the "Special Period in Time of Peace". The crisis devastated the entire Cuban economy. Machines lay idle in the absence of fuel and spare parts. Public and private transportation were in shambles. Workers had difficulties getting to their jobs. Factories and households all over the island were struck by unpredictable electrical power outages ([37], pp. 138–140). As in North Korea, the most painful effects were felt in the food sector. The nutritional intake of the average Cuban—especially protein and fat—fell considerably below the level of basic human needs ([3], pp. 154–169). Consumers resorted to chopped-up grapefruit peel as a surrogate for beef, and some people started breeding chicken in their flats or raising livestock on their balconies ([37], p. 138).

Nevertheless, people in Cuba were not dying from malnutrition and starvation. Homeless people and gangs of street children, turned into scavengers, were not characteristic features of Cuban townscapes. Nor were violence, crime, desperation, and hopelessness characteristic features of Cuban neighborhood life [47]. This is in remarkable contrast to North Korea. Although reliable reports on the situation in North Korea are in short supply, reports from exiles indicate that during the 1990s everyday life in the "Hermit Kingdom" could be characterized as solitary, poor, nasty, brutish, and short [35]. As mentioned, there was a famine killing 3–5% of the North Korean population. While life was certainly hard during the "Special Period", nothing of that sort happened in Cuba.

Overall, the regime in Havana was more humane than its counterpart in Pyongyang. After some initial tinkering, it undertook cautious reforms. The country was opened for tourism, parts of the informal sector were legalized, and various forms of local self-help were encouraged [37]. To some extent, Cubans were also helped in their efforts to cope with the crisis by a benign climate, remittances, foreign investment, and international aid.

However the real miracle was done by the Cuban people. Against all odds, ordinary people managed to get by due to the remarkable cohesion of Cuban society at the level of local communities. Although Cuba is highly urbanized, the typical *barrio* is an urban village. Households are tightly embedded in neighborhood life. Most families have lived in the same home for generations. The typical Cuban household is shared by an extended family. Cuba's multi-generational family households include aunts, uncles, and cousins. People cultivate close relationships with friends and relatives inside and outside their *barrio* [47].[8]

One should not idealize this. Families were stuck in their homes because the regime had frozen the property structure after the revolution. Thus, people were cramped into narrow spaces because they had no other choice. The regime had invested in community cohesion not so much to create social glue but rather to

---

(Footnote 7 continued)

Administration, between 1989 and 1992 the consumption of petroleum in Cuba fell by 20% and the net consumption of electricity by 24% (http://www.eia.gov/countries/, viewed on 22 April 2011).

[8] In a survey, 86% of people from vulnerable neighborhoods in Havana declared that they could count on support from relatives, 97% from friends, and 89% from neighbors ([47], p. 142).

sustain political control. But be that as it may, the result was that most Cubans could rely on their families, friends, and neighbors. This local solidarity, or social capital, helped them to make ends meet during the "Special Period". As one inhabitant of a vulnerable neighborhood put it, the crisis brought people closer together because it forced them to rely on one another (quoted in Taylor [47], p. 140).

Traditional knowledge was another decisive factor in feeding the population. Although most land had been collectivized after the revolution of 1959, about 4% of Cuban farmers had kept their plots. Another 11% was organized in private cooperatives [8]. The survival of traditional family farms and private cooperatives alongside industrial agriculture turned out to be an important asset. Independent farms were more resilient to the crisis than state farms, because they operated with less fuel and agrochemical inputs. Cuba's surviving family farmers kept important traditional knowledge that could now be recovered. Other formerly independent farmers had moved to state farms or urban areas, where they could provide valuable know-how for self-supply and urban agriculture.

Urban agriculture was a local self-help movement, facilitated by the availability of traditional knowledge in combination with technologies of organic gardening and the Cuban-specific rustic ingenuity. Idle stretches of land between concrete blocks or in urban peripheries were turned into makeshift organic gardens. Vacant or abandoned plots in close vicinity to people's homes were transformed into plantation sites. People used whatever urban wastelands they could occupy to grow vegetables and other foodstuffs. By the mid-1990s, there were hundreds of registered horticultural clubs in Havana alone. An urban cultivator explained in an interview: "When the Special Period started, horticultural clubs were organized by farmers themselves (…). Special emphasis was made to involve the whole family in these activities (…). We wanted also to develop more collaboration and mutual help among ourselves; we exchanged seeds, varieties, and experiences. We achieved a sense and spirit of mutual help, solidarity, and we learned about agricultural production" (quoted in [9], p. 98).

As already mentioned, one should be careful not to idealize this. Environmentalists have exalted urban farming during the Special Period as a social "experiment", or even alternative "model" of organic agriculture (see for example Rosset and Benjamin [40], Cruz and Sánchez Medina [11]). In reality, Cuba's detour into low-input agriculture was of course not driven by ecological consciousness but by dire necessity. From the second half of the 1990s, when the economic situation improved and industrial inputs became more widely available, Cuba started drifting back to industrial farming. This was helped by the discovery of Cuban oil reserves and subsidized deliveries from Venezuela. Nevertheless it is encouraging to note that, during the early and mid-1990s, Cubans managed for a few years to mitigate an atrocious oil supply shock by their remarkable community ethos. The comparison with North Korea shows that this was not a minor achievement.

5 Peak Oil Trajectories

The historical cases of Japan, North Korea, and Cuba suggest three different patterns of how societies may respond to a severe energy supply disruption. Despite the fact that peak oil would initially be experienced as a global energy crunch rather than as a series of national crises, it seems reasonable to expect a comparable gamut of reactions. Countries prone to military solutions may follow a Japanese-style strategy of predatory militarism. Countries with a recent authoritarian tradition may follow a North Korean path of totalitarian retrenchment. Countries with a strong community ethos may embark on Cuban-style socioeconomic adaptation, relying on their people to mitigate the effects of energy scarcity.

As mentioned, it is of course possible to imagine further reactive patterns, such as the mobilization of national sentiment by populist regimes. But even so, the three peak oil scenarios identified can help us to derive plausible scenarios of how different parts of the world would be likely to react to a peak oil scenario.

Given its unrivaled military capabilities, the United States would be the most obvious candidates for a "Japanese" strategy of predatory militarism. In case of peak oil, the US may be tempted to use its unique power projection capacity to secure privileged access to oil. It has happened sometimes in the past, and may happen more often in the future, that US decision makers find military coercion more effective than trade. China is no match for the US, but it would be capable of using its military muscle to secure access to oil and gas in Central Asia and, possibly, the South China Sea. Elsewhere the PRC would be unlikely to use a predatory strategy because, for the foreseeable future, its maritime forces and air power are not strong enough. Countries like India or Israel have even more limited military clout, but may be tempted to engage in geopolitical operations in their regional neighborhood to secure access to important energy resources.

A "North Korean" solution of totalitarian retrenchment that "screws" the population to preserve elite privileges is most likely in countries with a strong authoritarian tradition. In consolidated democracies, totalitarian retrenchment is much harder to imagine. Nevertheless, the history of twentieth-century Europe shows that even liberal democracies can and do sometimes degenerate into tyranny. It is difficult to predict to what point even in consolidated liberal democracies political culture could deteriorate in a protracted and serious crisis. Political elites in less consolidated democracies might experience fewer constraints and scruples right from the start. For example, elites in the second-wave democracies of Latin America may have lesser qualms than their counterparts in Western Europe about "screwing" their own population to preserve elite privileges.

Compared to predatory militarism and totalitarian retrenchment, "Cuban-style" socioeconomic adaptation is far more desirable from a normative viewpoint. At the local level, people in many developing countries may be able to mitigate the effects of peak oil by reverting to community-based values and a subsistence lifestyle.[9] Such a regression would be relatively easy for people in societies where individualism, industrialism, and mass consumerism have not yet struck deep roots. Socioeconomic adaptation would be far more difficult for people in Western societies, where individualism, industrialism, and mass consumerism have held sway for such a long time that a smooth regression is hard to imagine. And yet, survival in many presently industrial Western societies may ultimately depend on support from local communities and a subsistence-based lifestyle.

In abstract terms, this leaves us with three causal propositions, or "hypotheses".

Hypothesis 1: The greater a country's military potential and the stronger the perception that force will be more effective than the free market to protect access to vital resources, the more likely there will be a strategy of predatory militarism.

Hypothesis 2: The shorter and the less a country or society has practiced humanism, pluralism, and liberal democracy, the more likely its elites will be willing and able to impose a policy of totalitarian retrenchment on their population.

Hypothesis 3: The shorter and the less a country or society has been exposed to individualism, industrialism, and mass consumerism, the more likely there will be an adaptive regression to community-based values and a subsistence lifestyle.

This is of course not to deny that oil exporting countries would be in a somewhat more comfortable position. Other things being equal, they could use the increased revenue from oil exportation to increase their power and wealth, while subsidizing domestic consumption and bolstering their economies—if, that is, they do not fall prey to predatory militarism; and if they evade the "resource curse" that has bedeviled so many developing countries in the twentieth century.

In the transition, large private Western oil companies such as Exxon and Shell would lose further ground to the state-controlled companies of oil exporting countries such as Saudi Arabia's Aramco or Nigeria's NNPC. As a consequence, even oil importing countries would increasingly rely on state-controlled companies such as China's CNPC [50]. Both in the realm of power politics and on the "marketplace of ideas", the ability of Western countries to impose liberal democracy through instruments such as development assistance and economic conditionality would further dwindle [31].

This can be formulated as yet another causal proposition, or "hypothesis".

Hypothesis 4: In the event of peak oil, there will be winners and losers. It seems reasonable to expect a redistribution of power and wealth from oil importers to oil exporters, and from private to state-controlled companies.

---

[9] Given the high population pressure in most developing countries, however, large segments of the population would fall victim to famine, disease and conflict.

6 Energy Transition?

So-called "techno-optimists" object to "peak-oil pessimists" that scarcity would not only lead to higher oil prices but would also trigger a transition to alternate energy sources, such as renewable energy or a new generation of improved nuclear reactors [43]. Some optimists are even as confident as to predict that revolutionary technologies such as solar energy or nuclear fusion will eventually make oil redundant [6, 52].

Could there not be a transition from oil to some alternate technology or resource? Since energy shifts have happened in the past, for example from coal to oil, is it not unimaginative and unnecessarily defeatist to discard such a possibility for the future [38]? Could there not be a revolutionary technological breakthrough, or some other positive surprise, around the corner that would catapult industrial society "beyond oil" (or even "beyond carbon")?

Alas, time is an issue. Developing and rolling out new technologies takes a tremendous amount of time. Moreover, while it is highly alluring to imagine the sudden appearance of a *deus ex machina*, such as the discovery of a new energy resource or a revolutionary technological breakthrough, past transitions such as the energy shift from coal to oil do not seem to be appropriate historical precedents [42]. After all, oil was a superior surrogate for coal. No such superior surrogate for oil seems to be available today [28].

Therefore, rather than looking at past energy shifts, we need to look at a situation where the challenge was to radically alter an entrenched socioeconomic way of life. This leads us to another case study: Dixieland after the American Civil War. What can be gleaned from this case is that the formation of the "new consciousness" necessary for radical social change is a painfully slow process.

The socioeconomic backbone resource of the Old South was neither wood nor coal nor oil, but human slaves. Precisely because the slave economy worked, white Southerners were willing to defend it in the bloody War of Secession of 1861–1865 [16, 55]. The abolition of slavery after the War plunged the South into a deep crisis. The War was followed by the Reconstruction Era (1865–1877), when the victorious North tried to enlist dissident elites and former slaves to impose its political and socio-economic institutions on a reluctant South. Despite the introduction of representation and suffrage for former slaves, reconstruction was mostly thwarted by the recalcitrance of traditionalist Southern elites. Heavy subsidization of railroads by Republican state governments in the South did not lead to the hoped-for modernization but rather to corruption, making a few investors rich and otherwise contributing to soaring public deficits. After the withdrawal of the last federal troops from the South, race inequality was re-established under the banner of white supremacy [15].

Later in the nineteenth century, Southern elites started to try and move on. As a matter of fact, they were not prevented by their conservative values from embracing industrial capitalism. Initially, this amounted to an uneasy compromise between cherished industrialization and dreaded modernization. On the one hand,

Southern elites became obsessed with the idea that an industrializing "New South" would rise like phoenix from the ashes of the "Old South". On the other hand, they remained loyal to time-honored values of agrarianism and patriarchal society. As Mark Twain put it in 1883, cultural life on the Mississippi was characterized by "practical, common-sense, progressive ideas, and progressive works, mixed up with the duel, the inflated speech, and the jejune romanticism of an absurd past that is dead, and out of charity ought to be buried" ([48], p. 264).

This was reflected in a quasi-colonial economy. While railroads were finally built on a massive scale, often with Northern capital, Southern industrialization was initially dominated by low-wage and labor-intensive manufacturing. Most industries were dedicated to the processing of agricultural goods (e.g. in cotton mills) or natural resources (e.g. in blast furnaces). The real industrial takeoff came much later, after several generations of socio-economic backwardness, and after the New Deal of the 1930s (electrification) and the war economy of World War II. In the mid-twentieth century, Dixieland finally became a growth region and came to be seen as part of the American "Sunbelt" [10, 56]. The Civil Rights Act of 1964 famously put an end to official race segregation in the South, although some race issues remain until the present day.

While there is a decently happy ending to the story, it took a century for the South to recover and catch up. This is remarkable because, to understand how a technological and socioeconomic upgrade might look like, Southerners only had to look to the North of their own country. There, industrial capitalism with its superior technologies and know-how was unfolding before their very eyes. With the right incentives in place, attracting Northern investment and technology transfers would not have been too difficult. But although conditions for an industrial upgrade were uniquely favorable, the 100 years from the Civil War to the Civil Rights Act are replete of unpleasant memories such as race riots and labor revolts, as well as the Ku Klux Klan and Jim Crow Laws.

Dixie is a cautionary tale for those who believe that, after peak oil, there will be an easy technological upgrade. If even in the US South, despite uniquely favorable circumstances, adaptation took a full century, then a technological upgrade will be much harder under the more challenging circumstances of a global energy crunch. The world would be struggling with an energetic downgrade, rather than an industrial upgrade as in the case of the American South. Developing new energy technologies is never fast and easy, and even less so in times of crisis. After peak oil, we should expect extremely slow and painful processes of social and technological adjustment that may last for a century or more.

This can be stated as my last and most general proposition, or "hypothesis".

Hypothesis 5: In the event of peak oil, we should not expect either immediate collapse or a smooth transition. People do not give up their lifestyle easily. We should expect painful adaptation processes that may last for a century or more.

7 Same Crisis, Different Responses

Based on the heuristic insights and causal hypotheses gleaned from the case studies, we are now in a position to second-guess how different parts of the world would react to a peak oil scenario. *Please note that my conjectures are limited to the first couple of decades after peak oil.*[10] Please do also note that the picture provided in this section is deliberately broad-brush. Nothing of what I am going to say must be understood in a deterministic way. All I can offer is a set of tentative and indicative conjectures, rather than scientifically exact point predictions.

As a baseline, I need to make some assumptions. First, I assume that, after a few years on a "bumpy plateau", oil production will fall by about 2–5% per year.[11] Second, I assume that no adequate alternate resource and technology will be available to replace oil as the backbone resource of industrial society. Third, I rely on knowledge about historical and institutional path dependencies in particular countries and world regions. While the long-term future is fundamentally open, in the short and medium term there are path dependencies that make some trajectories far more likely than others. For example we roughly know which countries have strong power projection capabilities, recent authoritarian traditions, and high levels of "social capital". We also know which regions possess significant reserves of energy resources, and how these resources have been managed so far.

In *North America*, the United States combines strong dependency on foreign oil deliveries with an unrivaled capacity to project military power. A predatory strategy will therefore be tempting. To be sure, America's free trade ideology militates against the open recourse to military coercion. The US will support the free market for oil as long as it is convenient. When the oil market comes under pressure because of tightening international supply, the US is likely to continue to defend it for a while. But when soaring oil prices start crippling the national economy, US leaders may find that military coercion is more effective. The US is then likely to put the blame on foreigners and pursue a geopolitical strategy of "energy security" to protect the American way of life [30]. Why keep negotiating with recalcitrant leaders such as Chavez if there is a military option? This is not to say that the military option is easy, as the Iraq war has shown. However, military coercion is likely to gain ascendancy relative to free market rhetoric as oil supplies become scarcer. The resource-rich neighbors of the US, Canada and Mexico, would become tied more closely to the American core.

In *Latin America*, medium-sized oil producing countries such as Venezuela and Ecuador may try to profiteer from soaring oil prices. If they engage in a strategy of brinkmanship and deny the US oil on favorable terms, their political regimes may be toppled. While this would further increase anti-American resentment in the region, political elites are likely to ultimately acquiesce to

---

[10] For the long-term perspective see Greer [18, 19].

[11] The predictions of most peak oil theorists are in this band; see the overviews provided by Sorrell et al. ([45], pp. 4998-4999) and Hirsch [22].

American hardball tactics. In the past, Latin American elites have often oppor-
tunistically colluded with the US. Eventually, resource-rich Brazil may be able to
escape intervention due to its larger size and geographical distance from the US. If
Brazil manages to offer sufficient benefits to neighboring countries, a regional state
complex around Brazil may be possible. Otherwise, energy-poor Latin American
countries would enter a serious crisis. We may then see how much Cuban-style
socioeconomic adaptation is possible in other Latin American societies.

After peak oil, *Western Europe* would enter a difficult quandary. Although
advanced industrial countries like Germany and France have the capacity to
rearm, predatory militarism is not a credible option for them. Since Europeans
have good historical reasons to dread militarism, the social consensus necessary
for this strategy would not be forthcoming at the decisive initial stages of
geopolitical positioning. For the same historical reasons, in most of Western
Europe the path of totalitarian retrenchment does not seem to be available either.
Concomitantly, Western European countries would be forced to strike opportu-
nistic "bargains" with Russia and the oil exporting countries across the Mediterranean. Due to their asymmetrical nature, however, such deals are inherently
fragile and subject to constant renegotiation. Investment in renewable energy and
innovative technologies might somewhat mitigate the transition, but ultimately
Europeans could hardly avoid a transition to a more community-based lifestyle.
Despite the present affluence of Western European societies (and, in part, precisely because of it), this would be extremely painful and last for several
generations.[12]

Ordinary Western Europeans would be forced to rely on local communities
for their welfare if not survival. However, a regression to community-based
values and a subsistence lifestyle would be difficult because the habits of
industrial society are deeply rooted. Western Europe's problems would be
compounded by social segregation along immigrant groups and/or religious fault
lines which, on the one hand, might enhance communal support for specific
communities but, on the other, would conjure up severe conflict in Europe's
multiethnic societies.

The situation in *Japan* would be largely comparable, although Japan is far less
multi-ethnic than Western Europe and people may be more willing to accept
disruptions to their taken-for-granted lifestyles.[13] Like in the Western European
case, the unavoidable transition to community-based values and a subsistence
lifestyle would be painful and last for generations.

The situation would be somewhat different in countries and regions that have
industrialized later and/or have a more recent authoritarian tradition that may be

---

[12] The UK might try to evade the quandary by stressing its special relationship with the US, but
it is debatable whether Britain could offer enough benefits to its North American ally to justify the
burden of provisioning another 60 million people with subsidized fuel.

[13] This has been confirmed in 2011, when the Japanese responded in a highly calm and
disciplined way to a Tsunami followed by serious social mayhem and a spectacular nuclear
meltdown at the facilities in Fukushima.

recovered. Thus, totalitarian retrenchment and socioeconomic adaptation are far more likely and easy to imagine in various countries of *Eastern Europe* and *Southeast Asia* than in Western Europe and Japan.

In *Least Developed Countries*, common people with limited exposure to industrial lifestyles would be forced to rely on the cohesion of social groups for their survival. Particularly but not exclusively in *Sub-Saharan Africa*, state failure and conflict over scarce resources would become endemic. The inevitable end of the oil-based "green revolution" in agriculture and the demise of international aid would wreak environmental havoc and human insecurity. The production of "biofuels" might mitigate the energy situation of wealthy strata, but would crowd out food production and thus exacerbate the plight of the poor. The ecological situation would be aggravated by vital biomass being removed from the soil as a combustible. In most places, the unavoidable consequence would be famine, disease, and mass exodus. In some places, however, a revival of community-based values and a return to a subsistence lifestyle may mitigate the effects.

The elites of oil exporting countries such as Nigeria, Angola, and Equatorial Guinea would keep selling their oil to the highest bidder, especially when the bid is backed by sufficient military clout, and when there are no onerous obligations with regard to democratization and human rights. If the US gives up its dysfunctional democratization agenda, it will have better access to African resources than Europe, China, or Japan. It is an open question how much ordinary people in the African petro-states would benefit from the revenues (but people in African countries that do not have fossil fuel reserves would certainly suffer more).

In Asia, *Russia* has enough energy to provide for its own needs. It would become a more important regional player due to its abundant energy resources. *China*, by contrast, heavily relies on imported oil. To preserve its industrial capacity, the country might be tempted to secure access to vital resources from Central Asia by military means. Authoritarian retrenchment may be lurking as an additional option.[14] *India* has more limited military clout and a less authoritarian state tradition, but may nevertheless be tempted to engage in limited geopolitical operations in its regional neighborhood. Small and resource-poor outposts of industrial civilization, such as Singapore, would struggle to survive.

The oil exporting countries of *Central Asia* and the *Middle East* would benefit more than in the past from their comparative advantage. Due to the effects of rising oil prices, their economies would continue to grow in relative and absolute terms. As a result, their oil consumption would be stable or even increase at a time when it would be declining in the rest of the world ([41], pp. 57–83). While the "resource curse" would persist in countries with particularly corrupt elites, in others political freedom may improve alongside the level of industrialization and

---

[14] Despite an increasing internal market, China still heavily relies on the exportation of industrial products. It may not yet have accumulated enough economic wealth to insulate itself against the demise of international free trade.

the standard of living. The Middle East would almost certainly replace Western Europe as the most attractive destination for Muslim migrants.

8 Final Considerations

So far it has been assumed that, in the event of peak oil, no obvious alternate resource and technology would be available to replace oil as the backbone resource of industrial society. To mitigate the impact of peak oil, there is a desperate need for a massive crash program to develop and implement a mix of surrogate resources and adequate new technologies. The program would have to start early, as it would come to fruition only after far more than a decade [24]. In the absence of such a crash program, and after the onset of the crisis, rather than grandiose designs we should expect haphazard moves to make the best of a difficult situation [23, 34].[15]

The most appealing energy resource to mitigate the impact of peak oil is *natural gas* as a transition fuel. Although reserves are limited, gas is more abundant than oil. Recently there have been encouraging developments in the extraction of unconventional gas, most notably shale gas. However, the "shale gas revolution" is uncertain and may have severe environmental side effects [46, 51]. Also, it is important to bear in mind that gas is not a liquid fuel. While oil is easily transported and traded, gas requires pipelines or needs to be liquefied. Most of the world's existing vehicle fleet runs on oil, not gas. Compensating a decline of 2–5% of oil production per year with gas would be a tall order.

Coal is still relatively abundant in the United States, Asia, and Australia [21, 25, 32]. At least for a few decades it would inevitably become a more important source of energy. Rising oil prices would make mining and transportation more expensive, but coal-rich countries are motivated to tackle such challenges as long as coal production makes financial and energetic sense. Related carbon emissions would lead to harmful consequences for the environment and global climate, not least because heavy investment in clean coal technologies is unlikely under crisis conditions.

To gain access to recoverable *oil reserves*, protected areas from the Arctic to Antarctica would be cleared for exploitation. Unconventional oil from tar sands and oil shale would to be exploited, regardless of the harmful environmental consequences. As in the case of clean coal technologies, heavy investment of scarce financial resources in environmentally friendly technologies is unlikely under crisis conditions. An obvious downside of such otherwise desirable technologies is that they tend to reduce the energy return on energy invested (EROEI), which would hardly be acceptable in a situation of increasing energy scarcity.

---

[15] Especially the decline of petrol-based transportation after peak oil may pose a challenge to the implementation of ambitious modernization programs.

In the unlikely event that massive financial resources and planning horizons of more than 15 years are still available after peak oil, *nuclear reactors* may be rushed through regardless of the risks involved. However, this can only make a limited contribution. The current share of nuclear power in world energy production is only about 6%, and it could hardly expand very much under crisis conditions. Moreover, uranium is as finite as any other energy resource.[16]

To the extent possible in times of turmoil, there would be further investment in *renewable energy*. But this could hardly make up for the losses. The share of renewable energy in the global mix is larger than that of nuclear power, but it is still relatively low.[17] It is debatable how much and how quickly it could be expanded under crisis conditions. Even the most appealing forms of renewable energy have downsides, such as intermittency and a low EROEI. Expanding the share of renewable energy beyond 20% is therefore likely to require a demanding energy infrastructure, and thus a shift from incremental to systemic change. The production of a wind turbine requires significant inputs of energy and raw materials, which are currently supplied from non-renewable resources. Biofuels displace food production and may crowd out important ecosystem services. Despite all these problems, the odds for renewable energy are better than for nuclear power.

From a technical viewpoint, the dependency of industrial society on any *specific* primary energy source, such as oil, could be reduced by relying more on electricity. In theory, this has the potential of mitigating the impact of peak oil and postponing the decline of overall world energy consumption. However, the low price elasticity of oil indicates that this would require considerable investment over a sustained period of time ([27], pp. 89–124). While the conversion of the existing vehicle fleet to electricity is a tall order, the *generic* dependency of industrial society on fossil fuels for electricity production would remain in place.

From an ecological viewpoint, electrification must therefore be supplemented by a combination of conservation, increased efficiency, and renewable energy. Even so, infinite growth on a finite planet is impossible. Industrial society as we know it cannot last forever, and at some point it is bound to become unviable. While my analysis covers only the first couple of decades after peak oil, even in this limited time-span the social and political prospects are daunting.

Most of us would certainly prefer industrial society to continue unabated, perhaps mitigated to avoid the worst effects of anthropogenic climate change. Even though we may not like the idea of a global energy crunch, however, it would be utterly imprudent not to take the prospects of peak oil very seriously.

---

[16] For a range of realistic estimates see Moriarty and Honnery ([34], p. 2472).

[17] Renewable energy accounts for about 20% of global energy consumption, but only if we include traditional biomass and hydropower. This is somewhat odd, as poor Ethiopian women collecting firewood, and gigantic installations such as the Chinese Three Gorges Dam, are hardly what we mean by the term. If we do not include traditional biomass and hydropower, the figure is much smaller.

Acknowledgments Thanks to Martin Kraus for stimulating discussions and amicable feedback. I would also like to express my gratitude to Jocelyn Alexander, Andreas Goldthau, Barbara Harriss-White, Eva Herschinger, David Von Hippel, Dan Hicks, Robert Hirsch, Peter Katzenstein, John Mathews, Rana Mitter, Avner Offer, Gianfranco Poggi, Jochen Prantl, Jörg Schindler, Mary Stokes White, Marisa Wilson, and the anonymous reviewers of Energy Policy, for helpful suggestions and comments.

References

1. Akins JE (1973) The oil crisis: this time the wolf is here. Foreign Aff 51(3):462–490
2. Aleklett K, Höök M, Jakobsson K, Lardelli M, Snowden S, Söderbergh B (2010) The peak of the oil age: analyzing the world oil production reference scenario in World Energy Outlook 2008. Energy Policy 38(3):1398–1414
3. Alvarez J (2004) Cuba's agricultural sector. University Press of Florida, Gainesville
4. Barnhart MA (1987) Japan prepares for total war: the search for economic security, 1919–1941. Cornell University Press, Ithaca
5. Beasley WG (1987) Japanese Imperialism, 1894–1945. Clarendon Press, Oxford
6. Bradford T (2006) Solar revolution: the economic transformation of the global energy industry. MIT Press, Cambridge
7. Brandt AR (2007) Testing Hubbert. Energy Policy 35(5):3074–3088
8. Burchard H-J (ed) (2000) La última reforma agraria del siglo: La agricultura cubana entre el cambio y el estancamiento. Nueva Sociedad, Caracas
9. Carrasco A, Acker D, Grieshop J (2003) Absorbing the shocks: the case of food security, extension and the agricultural knowledge and information system in Havana, Cuba. J Agric Education Ext 9(3):93–102
10. Cobb JC (1984) Industrialization and Southern Society, 1877–1984. University Press of Kentucky, Lexington
11. Cruz MC, Sánchez Medina R (2003) Agriculture in the city: a key to sustainability in Havana, Cuba. Ian Randle, Kingston
12. Díaz Briquets S, Pérez López J (2000) Conquering nature: the environmental legacy of socialism in Cuba. University of Pittsburgh Press, Pittsburgh
13. FAO/WFP (1999) Special Report: FAO/WFP crop and food supply assessment mission to the Democratic People's Republic of Korea, 8 Nov 1999
14. FAO/WFP (2008) Special Report: FAO/WFP crop and food security assessment mission to the Democratic People's Republic of Korea, 8 Dec 2008
15. Fitzgerald MW (2007) Splendid failure: postwar reconstruction in the American South. Dee, Chicago
16. Fogel RW (1989) Without consent or contract: the rise and fall of American Slavery. Norton, New York
17. Goodkind D, West D (2001) The North Korean famine and its demographic impact. Popul Dev Rev 27(2):219–238
18. Greer JM (2009) The ecotechnic future. New Society Publishers, Gabriola Island
19. Greer JM (2008) The long descent: a user's guide to the end of the Industrial Age. New Society Publishers, Gabriola Island
20. Haggard S, Noland M (2007) Famine in North Korea: markets, aid, and reform. Columbia University Press, New York
21. Heinberg R (2009) Blackout: coal, climate and the last energy crisis. New Society Publishers, Gabriola Island
22. Hirsch RL (2008) Mitigation of maximum world oil production: shortage scenarios. Energy Policy 36(2):881–889

23. Hirsch RL, Bezdek R, Wendling R (2010) The impending world energy mess. Griffin Media, Burlington
24. Hirsch RL, Bezdek R, Wendling R (2005) Peaking of world oil production: impacts, mitigation and risk management. Available online at http://www.netl.doe.gov/publications/others/pdf/Oil_Peaking_NETL.pdf. Downloaded 31 March 2010
25. Höök M, Aleklett K (2009) Historical trends in American coal production and a possible future outlook. Int J Coal Geol 78(3):201–216
26. Hubbert MK (1969) Energy resources, in: Committee on Resources and Man (ed.), Resources and man: a study and recommendations. Freeman, San Francisco, pp 157–242
27. IMF (2011) Tensions from the two-speed recovery: unemployment, commodities, and capital flows. International Monetary Fund. World Economic Outlook, Washington, DC April 2011
28. Kerr, RA (2010) Do we have the energy for the next transition? Science 329:780–781, 13 Aug 2010
29. Kershaw I (2007) Fateful choices: ten decisions that changed the world. Penguin, New York
30. Klare, MT (2008) Rising powers, shrinking planet: the new geopolitics of energy. Metropolitan Books, New York
31. Leder F, Shapiro, JN (2008) This time it's different: an inevitable decline in world petroleum production will keep oil product prices high, causing military conflicts and shifting wealth and power from democracies to authoritarian regimes. Energy Policy 36(8):2850–2852
32. Lin, B-Q, Liu, J-H (2010) Estimating coal production peak and trends of coal imports in China. Energy Policy 38(1):512–519
33. Miller, ES (2007) Bankrupting the enemy: the US financial siege of Japan before Pearl Harbor. Naval Institute Press, Annapolis
34. Moriarty P, Honnery D (2009) What energy levels can the earth sustain? Energy Policy 37(7):2469–2474
35. Natsios A (2001) The Great North Korean famine. United States Institute of Peace Press, Washington DC
36. Owen NA, Inderwildi OR, King DA (2010) The status of conventional world oil reserves: hype or cause for concern? Energy Policy 38(8):4743–4749
37. Pérez-López, JF (1995) Cuba's second economy: from behind the scenes to center stage. Transaction Publishers, New Brunswick
38. Podobnik B (2006) Global energy shifts: fostering sustainability in a turbulent age. Temple University Press, Philadelphia
39. Record J (2009) Japan's decision for war in 1941: some enduring lessons. Strategic studies institute, Carlisle
40. Rosset P, Benjamin M (eds) (1994) The greening of the revolution: Cuba's experiment with organic agriculture. Ocean Press, Melbourne
41. Rubin J (2009) Why your world is about to get a whole lot smaller: oil and the end of globalization. Random House, New York
42. Smil V (2010) Energy transitions: history, requirements, prospects. Praeger, Santa Barbara
43. Smil V (2008) Global catastrophes and trends: the next fifty years. MIT Press, Cambridge
44. Sorrell S, Speirs J, Bentley R, Miller R (2010) Global oil depletion: a review of the evidence. Energy Policy 38(9):5290–5295
45. Sorrell S, Miller R, Bentley R, Speirs J (2010) Oil futures: a comparison of global supply forecasts. Energy Policy 38(9):4990–5003
46. Stevens P (2010) The "shale gas revolution": hype and reality. Chatham House, London
47. Taylor, HL (2009) Inside el barrio: a bottom-up view of neighborhood life in Castro's Cuba. Kumarian Press, Sterling
48. Twain M (2006) Life on the Mississippi. Folio, London
49. United States Government Accountability Office (2007) Crude oil: uncertainty about future oil supply makes it important to develop a strategy for addressing a peak and decline in oil production. United States Government Accountability Office, Washington (GAO-07-283)
50. Vivoda V (2009) Resource nationalism, bargaining and international oil companies: challenges and change in the new millennium. New Political Econ 14(4):517–534

51. Walsh B (2011) Could shale gas power the world? Time Magazine, 31 March 2011
52. Weiss C, Bonvillian WB (2009) Structuring an energy technology revolution. MIT Press, Cambridge
53. WFP/FAO/UNICEF (2011) Special Report: WFP/FAO/UNICEF rapid food security assessment mission to the Democratic People's Republic of Korea, 24 March 2011
54. Williams JH, Hippel D Von, Nautilus Team (2002) Fuel and famine: rural energy crisis in the DPRK. Asian Perspective 26(1):111–140
55. Wright G (2006) Slavery and American economic development. Louisiana State University Press, Baton Rouge
56. Wright G (1986) Old South, New South: revolutions in the Southern Economy since the civil war. Basic Books, New York
57. Wright J (2009) Sustainable agriculture and food security in an era of oil scarcity: lessons from Cuba. Earthscan, London
58. Yergin D (1991) The prize: the epic quest for oil, money and power. Free Press, New York

Bioenergy Innovation and Sustainable Mobility: Deployment Feedstock Full Potentials

Weber Antonio Neves do Amaral, Guilherme Ary Plonski and Eduardo Giuliani

Abstract This chapter addresses how technological innovation in agribusiness should strengthen the synergies among different feedstock production systems, lower carbon footprint and create new business ventures based on bioenergy and bioproducts. We draw on examples from sugarcane from Brazil, where multiple feedstock production systems are being deployed for the production of bioenergy, biofuels and bioproducts, creating operational models for biorefineries. We make the case that sustainable mobility frameworks based on biofuels depend more on how agricultural landscapes are used, than on the feedstock conversion systems themselves. The production of multiple feedstock products at landscape level is crucial for sustainable mobility because it optimizes land use, diversifies the sources of income from farmers, reduces the food versus fuel competition and reduces biofuel carbon footprint. We also present an assessment of the biodiesel program in Brazil, which was launched in 2005 and is currently blended with diesel at a mandatory 5% level. By November 1, 2011, the world population will have reached 7 billion people and more likely by 2050, 9 billion, a demographic turning point as well as a major shift in global geopolitics and consumption patterns of food, energy, water and other resources. The challenges that this presents will continue to be the case for the foreseeable future, while climatic instability and increased severity of pests and disease will affect our capacity to produce goods and services. Agriculture must be at the center of supplying food in quantity and quality necessary to feed an aging population, with people living

W. A. N. do Amaral (✉)
USP—University of São Paulo, ESALQ, Av. Padua Dias 11, Piracicaba, SP, Brazil
e-mail: wamaral@esalq.usp.brwanamaral@gmail.com

G. A. Plonski
USP—University of São Paulo, Escola Politécnica da USP, Sao Paulo, Brazil

E. Giuliani
Venture Partners do Brasil—VPB, Sao Paulo, SP, Brazil

O. Inderwildi and Sir David King (eds.), *Energy, Transport, & the Environment*,   77
DOI: 10.1007/978-1-4471-2717-8_5, © Springer-Verlag London 2012

longer and healthier lives. But several looming challenges remain which must be addressed for agriculture to be successful in fulfilling these needs in a sustainable manner and requires a global coordinated effort. Feeding 9 billion people in 2050 will require the development of new agribusiness models, new technologies and innovative ways of delivering an array of bioproducts from multi-functional agricultural landscapes. Innovation at different feedstock chains can also mitigate price volatility associated with the uncertainties of single product delivery, and the development of competencies and distinctiveness of the bioenergy and bioproducts coming from these landscapes, and thus contributing to sustainable mobility frameworks based on biofuels.

1 Introduction

Low oil prices in the recent past have kept many agriculturally based chemicals and materials less profitable. Many now perceive that the era of cheap oil is over. Global oil production is nearing its maximum rate (so-called "peak oil") and rising demand for oil is occurring in the developing world, making projected oil prices higher in the foreseeable future [7]. The current oil price is $80/barrel which is considerably higher than the $20/barrel of the recent decades. Oil is projected to average $100/barrel from 2008 to 2015, and be $200/barrel by 2030 (nominally). This oil price trend strongly suggests that industrial agricultural products will be much more competitive in the near future.

Biotechnology is a central tool to increase the efficiency of chemical production from agricultural resources such as grains, novel crops and residues, and forestry. Sales of industrial-biotechnology products were ~ $140 billion in 2007, and 6% of all chemicals sales were generated with the help of biotechnology (The Economist, "Third time lucky", Jun 4, 2009). It was recently estimated that the global renewable chemicals market will be worth $59 billion by 2014 (www.marketsandmarkets.com). Industrial biotechnology companies have developed more quickly in Europe (e.g. Novozymes, DSM). Ethanol can be thought of as the first major non-food industrial biotechnological product; ethanol is produced using advanced enzymes and microbes to efficiently break down sugar and ferment it to ethanol. These biorefineries can further increase profitability by diversifying their operations with production of more high-value co-products from major feedstocks (sugarcane, corn, sweet sorghum, soy bean, etc.).

In addition, several *drivers at the global level are stimulating the production and use of biofuels*. These range from environmental concerns related to the need to *mitigate greenhouse gas emissions (GHGs), to energy security and diversification of oil-based fuels*. These have resulted in the development of different supporting policy frameworks. At regional and local levels, agricultural issues, social concerns and development programs are also drivers.

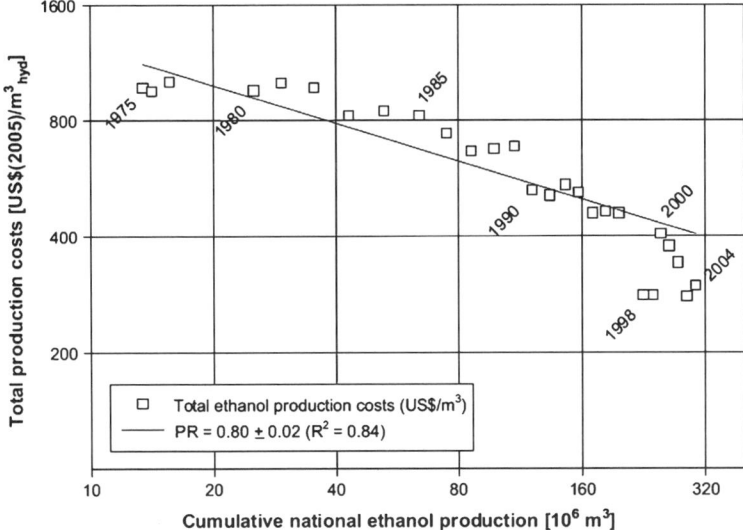

Fig. 1 Experience curve for total hydrated ethanol (1975–2004) including feedstock costs.
Source

Countries are currently implementing supporting policies to produce and use biofuels with different blending regimes in order to create demand for these fuels. While this promotes economic development and technological improvement throughout the entire biofuels chain, it also raises concerns about their sustainable growth, their impact on land use processes, food competition and their real contribution to the reduction of GHGs.

There has been recent scientific debate concerning the *sustainable production of biofuels*. Several initiatives are being developed in Europe and the United States relating to *the implementation of certification, mechanisms for traceability and definitions of criteria and indicators* for sustainable production. While the concerns are well-justified, some criticisms are motivated by protectionism and agricultural policies from developed countries, especially in Europe and the United States. If not properly developed and deployed, proposed certification schemes may become technical barriers rather than sound initiatives. Scientific and technological assessments of key criteria comparing different kinds of biofuels are needed to reduce the role of such interests and to establish the potential strengths of biofuels along with their limitations.

Bioethanol has a long history in Brazil. Its production started back in the 1930s, when a 5% blend with gasoline was compulsory, but only during the 1970s, with the oil crisis, the national ethanol program (Pro-álcool) was launched, and the production was boosted. In the last 30 years technological evolution in both agricultural and industrial systems led to important cost reductions (see Fig. 1). Today sugarcane ethanol is cost-competitive with fossil fuels, without any need for subsidies, and further improvement is foreseen in the near future.

The technology advances in ethanol production have also impacted its environmental performance. Considering the current conditions, ethanol can mitigate more than 80% of gasoline GHG emissions. For the future, emission reductions can be even higher, as mills adopt, for example, more efficient technologies to produce electricity (or even more ethanol) from cane's residual biomass. In this case, co-product credits are going to play a much more important role, as they can offset the emissions related to ethanol production. However, land use change effects must also be addressed, as well as other environmental aspects, for a better evaluation of sugarcane ethanol sustainability.

Brazil's commitment to sustainability in the agribusiness can be assessed by concrete examples such as the development and implementation of stringent legal environmental frameworks, agricultural zoning, massive investments in research and development and rural social policies, being the ethanol business a good example from which best practices could be disseminated. The long track record of Brazilian sugarcane ethanol proved its economic sustainability over time, while improving its social and environmental indicators, involving technology transfer from Europe, US and other regions and developing several innovations at the national level.

Increasing internal demands and the possibility of future exports will lead to considerably higher production levels. Several steps will be necessary to achieve these production targets, including sustainable planning of the sugarcane expansion into new areas, improving the logistics, the development of global markets and continuously developing new technological innovations, while at the same time improving the environmental performance of existing brown fields and especially from new green fields.

Data on important environmental sustainability of sugarcane production and processing are presented, including survey results about soil carbon stocks (below and above ground) for land uses involved in the recent sugarcane expansion. Land use change dynamics of sugarcane crop expansion was assessed for the current and future scenarios in Brazil, as well as ethanol life cycle GHG emissions, with a separated evaluation of LUC emissions considering initial collected data on soil carbon stocks and land use change.

2 The Brazilian Environmental Legal Framework Regulating Ethanol Production

The Brazilian environmental legal framework is complex and one of the most stringent and advanced in the world. As an agribusiness activity, the ethanol/sugar industry has several environmental restrictions that require appropriate legislation or general policies for its operation. Some of them are pioneers in the area which define principles in order to maintain the welfare of living beings and to provide resources for future generations: the first version of the Brazilian forest code dated from 1931 already addressed the need to combine forest cover with quality of life and livelihoods.

Brazil has a wide range of federal and state laws regarding environmental protection, aiming at combining the social economic development with environmental preservation. They also involve frameworks such as the Environmental Impact Assessment and Environmental Licensing, among others, especially for the implementation of new projects.

Volunteer adherence to Environmental Protocols represents also a major breakthrough for the sugar business. For example, the "Protocolo Agroambiental do Setor Sucroalcooleiro" (Agriculture and Environmental Protocol for the ethanol/sugar industry) signed by UNICA and the Government of the State of São Paulo in June 2007 deals with issues such as: conservation of soil and water resources, protection of forests, recovery of riparian corridors and watersheds, reduction of greenhouse emissions and improve the use of agrochemicals and fertilizers. But its main focus is anticipating the legal deadlines for ending sugarcane burning by 2014 from the previous deadline of 2021. In February 2008, the State Secretariat of Environment reported that 141 industries of sugar and alcohol had already signed the Protocol, receiving the "Certificado de Conformidade Agroambiental" (Agricultural and Environmental Certificate of Compliance). These adherences correspond to more than 90% of the total sugarcane production in São Paulo. A similar initiative is happening in the State of Minas Gerais with the "Protocolo de Intenções de Eliminação da Queima da Cana no Setor Sucroalcooleiro de Minas Gerais" from August 2008.

3 Environmental Indicators

3.1 Greenhouse Gases (GHG) Balance

One of the goals of using biofuels is to contribute with net reduction of GHG emissions and thus not affecting carbon stock negatively in different subsystems of production, below and above ground biomass (roots, branches and leaves) and in the soil (carbon fixed in clay, silt, sand and organic matter). Figure 2 shows that ethanol from sugarcane reduces 86% of the GHG emissions when compared to gasoline. It has also a leading performance when compared to other biofuels from other feedstocks.

Carbon stock changes in the soil due to land use change for biofuels production are now being accounted for in the GHG balance. In this case, it is necessary to know how much carbon would be fixed or released into the air under different land use regimes compared with the previous baseline of use. One limiting factor to perform an in-depth analysis of these balances is the lack of long-term monitoring plots assessing precisely these dynamics through time. The stock and flows of carbon for major crops like soybean, maize, cotton and sugarcane have been extensively studied, but in general using different methodologies. There are also other factors that affect the results: crop productivity and management, soil physical and chemical properties, climate and land use history for example.

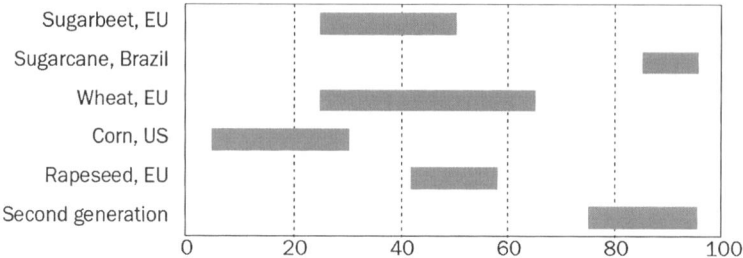

Fig. 2 GHG emissions avoided with ethanol or biodiesel replacing gasoline. *Source* International Energy Agency [11]

Table 1 Carbon stock in soil for selected crops and native vegetation

Biomass	Carbon stocks in soil (Mg/ha)
Campo Limpo—grassland savannah (a)	72
Sub-tropical forest (b)	72
Tropical forest (c)	71
Natural pasture (d)	56
Soybean (e)	53
Cerradão—woody savannah (a)	53
Managed pasture (f)	52
Cerrado—typical savannah (a)	46
Sugarcane without burn (g)	44
Degraded pasture (h)	41
Maize (h)	40
Cotton (i)	38
Sugarcane burned (g)	35

Sources (a) Lardy et al. [9]; (b) Cerri [2]; (c) Trumbore [16]; (d) Jantalia [8]; (e) Campos [1]; (f) Rangel and Silva [13]; (g) D'andrea et al. [3]; (i) Neves et al. [10]

In large countries such as Brazil, there are many different soils and climatic conditions. The different characteristics of each region will influence the potential for carbon storage. The land use history is also relevant when assessing and explaining current levels of carbon, because when land use changes do occur, soil carbon stocks take several years to achieve a new carbon balance. Table 1 presents the carbon stocks in soil for some selected Brazilian crops and in the native vegetation. For carbon stored in the biomass, crop productivity is of great importance as indicator carbon stored in the above ground biomass per unit of area (Table 2), which is a measure much easier to obtain and with a larger data set from multiple management and production systems in Brazil. From these numbers, the carbon balance resulting from land use change to sugarcane crop can be estimated for each previous use (Table 3).

Table 2 Carbon stocks in the above biomass of selected crops and native vegetation (fully grown crop)

Biomass	Carbon stocks in biomass (Mg/ha)
Tropical rain forest (a)	200.0
Cerradão—woody savannah (b)	33.5
Cerrado—typical savannah (b)	25.5
Sugarcane without burn (c)	17.5
Sugarcane burned (c)	17.0
Campo Limpo—grassland savannah (b)	8.4
Managed pasture (d)	6.5
Maize (e)	3.9
Cotton (f)	2.2
Soybean (g)	1.8
Degraded pasture (d)	1.3

Sources (a) INPE; (b) Ottmar et al. [12]; (c) VPB Estimative; (d) Estimated from Szakács [14]; (e) Estimated from Titon et al. [15]; (f) Adapted from Fornasieri and Domingos [5]; (g) Adapted from Campos [1]

Table 3 Carbon balance under different land uses replaced by sugarcane

Biomass	Total carbon stocks (Mg/ha)	Carbon balance due to sugarcane replacement (Mg/ha)
Cotton (d)	40.1	21.8
Degraded pasture (b)	42.0	19.8
Maize (h)	44.1	17.7
Sugarcane burned (g)	52.1	9.7
Soybean (e)	54.9	6.9
Managed pasture (f)	58.5	3.3
Cerrado—typical savannah (a)	71.5	−9.7
Campo Limpo—grassland savannah (a)	80.4	−18.6
Cerradão—woody savannah (a)	86.5	−24.7
Tropical forest (c)	271.0	−209.2
Total carbon stocks in sugarcane not burned = 61.8 Mg/ha		

Sources (a) Lardy et al. [9]/Ottmar et al. [12]; (b) D'andrea et al. [3]/Estimated from Szakács et al. [14]; (c) Trumbore et al. [16]/INPE; (d) Neves et al. [10]/Adapted from Fornasieri and Domingos [5]; (e) Campos [1]; (f) Rangel and Silva et al. [13]/Estimated from Szakács [14]; (g) VPB Estimative; (h) D'andrea et al. [3]/Estimated from Titon et al. [15]

3.2 Water

Practically all of the sugarcane produced in São Paulo State is grown without irrigation. The levels of water withdrawn and released for industrial use have substantially decreased over the past few years, from around 5 m^3/ton sugarcane collected in 1990 and 1997 to 1.83 m^3/ton sugarcane in 2004 (sampling in São Paulo). Mills with better water management practice replace only 500 l in the industrial system, with a recycling rate of 96.67%. Recent developments might lead to convert sugarcane mills from water consumers into the water exporters industry. Dedini, the largest Brazilian manufacturer of sugar mills and equipment supplies, has developed technologies (to be available in 2009) leading to zero water intake for the industrial mill (actually, a fraction of the 700 l water/t cane in the harvested sugar cane will be exported from the mill).

3.3 Soil and Fertilizers

The sustainability of the culture improves with the protection against soil erosion, compacting and moisture losses and correct fertilization. In Brazil, there are soils that have been producing sugarcane for more than 200 years, with ever-increasing yields and soil carbon content. Soil erosion in sugarcane fields is lower than in soybean and maize and other crops. It is expected also that the growing harvesting of cane without burning will further improve this condition, with the use of the remaining trash in the soil. Recent sugarcane expansion in Brazil has happened mostly in low fertility soils (pasture lands), and thus improving their organic matter and nutrient levels from previous land use patterns. Sugarcane uses lower inputs of fertilizers: ten, six and four times lower than maize respectively for nitrogen, phosphorous and potassium.

An important characteristic of the Brazilian sugarcane ethanol is the full recycling of industrial waste to the field. Vinasse, a waste of the distillation process, rich in nutrients (mainly potassium) and organic matters is a good example, which is being used extensively as a source of fertiirrigation (nutrients associated with water). Investments in infrastructure have enabled the use of water from the industrial process and the ashes from boilers. Filter cake recycling processes were also developed, thereby increasing the supply of nutrients to the field.

3.4 Management of Diseases, Insects and Weeds

Strategies for disease control involve the development of disease resistant varieties within large genetic improvement programs. This approach kept the major disease outbreak managed by replacing susceptible varieties. The soil pest monitoring method in reform areas enabled a 70% reduction of chemical control, thereby reducing costs and risks to operators and the environment.

Insecticide consumption in sugarcane crops is lower than in citrus, maize, coffee and soybean crops; the use of herbicides is also low, and fungicides use is virtually null. Among the main sugarcane pests, the sugarcane beetle, *Migdolus fryanus* (the most important pest) and the cigarrinha, *Mahanarva fimbriolata*, are biologically controlled. The control or management of weeds encompasses specific methods or combinations of mechanical, cultural, chemical and biological methods, making up an extremely dynamic process that is often reviewed. In Brazil, sugarcane uses more herbicides than coffee and maize crops, less herbicides than citrus and the same amount as soybean.

3.5 Conservation of Biodiversity

Brazil has significant biodiversity hotspots and contains more than 40% of all the tropical rain forests of the world. Brazilian biodiversity conservation priorities were set mainly between 1995 and 2000, with the contribution of hundreds of experts; protected areas were established for the six major biomes in the National Conservation Unit System.

The percentage of forest cover represents a good indicator of conservation of biodiversity in agricultural landscapes. In the state of São Paulo for example, the forest cover remaining is 11%, of which 8% is part of the original Atlantic Forest. Data from São Paulo State show that while the sugarcane area increased from 7 to 19% of the State territory in the last 10 years, native forests also increased from 5 to 11%, demonstrating that it is possible to recover biodiversity in intense agricultural systems, when proper environment policies and law enforcement are in place.

3.6 Air Quality

Burning sugarcane for harvesting is one of the most criticized issues in the sugarcane production system, causing local air pollution and affecting air quality, despite the benefits of using 100% ethanol running engines instead of gasoline, which decreases air pollution from 14 to 49%. In order to gradually eliminate sugarcane burning, several attempts are being made. The São Paulo Green Protocol is being considered the most important one, setting an example for other regions and states in Brazil. Voluntarily, 141 of the total of 170 sugar mills from the state of São Paulo signed this Protocol, and recently 13,000 sugarcane independent suppliers, members of the Organization of Sugarcane Farmers of the Center-South Region (Orplana) also signed this protocol.

4 Initiatives Towards Ethanol Certification and Compliance

Several initiatives are being developed in Europe and in the United States related to certification, traceability and definition of criteria and indicators for sustainable production of biofuels, mainly due to different supporting policies. Governmental and multiple stakeholder initiatives are supporting different assessment studies.

The main environmental issues addressed in such studies are related to greenhouse gas reduction compared with fossil fuels, competition with other land uses (especially food competition), impacts on the biodiversity and on the environment. In all cases, we can say that Brazilian sugarcane ethanols do comply with the targets for GHG emissions reduction and present outstanding performance for the other criteria.

While the above concerns are well-justified, some criticism of biofuels and their impacts are motivated by protectionism and interest in agricultural subsidies and agribusiness production chains in several developed countries, especially from EU countries. Certification schemes suggested may become non-tariff barriers, rather than environmentally and socially sound schemes.

Scientific and technological assessments comparing different kinds of biofuels are needed to reduce the play of such interests and to establish the strengths of best potential of biofuels along with their dangers and limitations. The last OECD report [11] stated that sugarcane ethanol in Brazil (and some other biofuels) can substantially reduce greenhouse gases compared with gasoline and mineral diesel. But the report also recognized that while still trade barriers would persist to the international market, it will be difficult for the world to take advantage of the environmental qualities of the use of some biofuels, mainly the ethanol from sugarcane and so forth, as international markets are not yet fully created for biofuels.

5 The Case of Biodiesel Production and Governmental Policies in Brazil

In 2004, Brazil launched its *National Program for the Production and Use of Biodiesel* (*NPPB*) in an attempt to further diversity the energy matrix and stimulate income generation and social development, with a focus on the Northeast region (the poorest region in Brazil) and small farmers. This Plan complies with the *mandatory blending of 2% of biodiesel in diesel from 2008* (*B2—equivalent to 800 million l*) *and 5% from 2013* (*B5—equivalent to 2 billion l*). Brazilian biodiesel production is already currently meeting the targets of this National Programme which were *revised in July 2008 to a 3% blending regime*.

Investors have recognized the potential and established in Brazil *a capacity above local market demands*, implementing and projecting new biodiesel plants in different regions (Fig. 3). Currently, there are 56 biodiesel plants authorized to operate by the Brazilian Oil, Gas and Biofuels Agency (ANP), but only 34 are producing biodiesel. These authorized plants have an annual production capacity of 2.8 billion l, well above the current 3% blending level imposed by the government, which equates to a production level of 1.2 billion l/annum.

The main reasons biodiesel production is operating far below the installed capacity are: (a) *higher costs* which limit the market (current production only fulfills legislation); (b) *uncertainties on the availability and price of feedstock* (some compete with the food and chemicals market); and (c) *need for improvement of production processes* at industrial plants to meet ANP's quality standards.

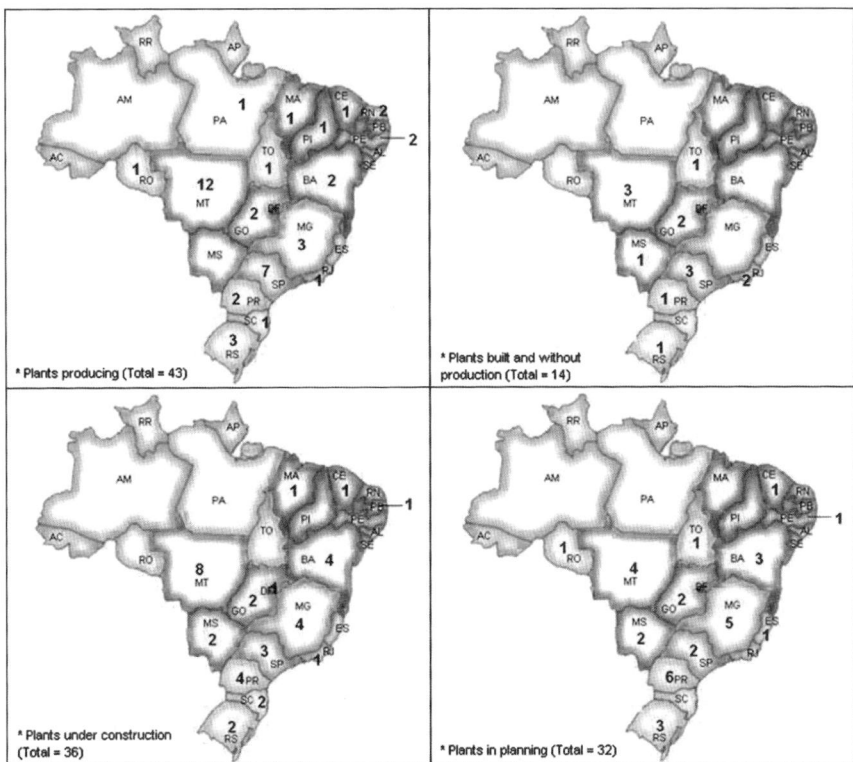

Fig. 3 Number of biodiesel plants by state (ANP; Dedini, mar-[4])

Additionally, policies created in developed countries have distorted the feed-stock market mainly because of the production of maize ethanol. This has caused a significant increase in agricultural commodity prices and damaged the profitability of investments in this industry in 2007/2008 [6].

Soybean has been used as the major feedstock (74% of total production) in the 34 biodiesel plants currently in operation in Brazil due to its wide availability. Other feedstocks, which have a smaller share of this market, include *sunflower* (10%), *animal fat* (7%), *castor bean* (6%), *palm oil* (1%), *peanut* (1%) *and jatropha* (1%). The Centre-South region is responsible for the production of about 70% of Brazilian biodiesel.

The study analyzed the current feedstock used for biodiesel production, including those with potential to replace soybean in the short and medium term, such as cotton and rapeseed. Recent research into other oil palms, especially from native species (buriti, macauba and babaçu), demonstrates there is no short- to medium-term prospect that these oils can be diverted to biodiesel production. They lack economies of scale and current uses (e.g. for cosmetics) have higher value (Fig. 4).

Fig. 4 Potential feedstock in Brazil versus time for deployment

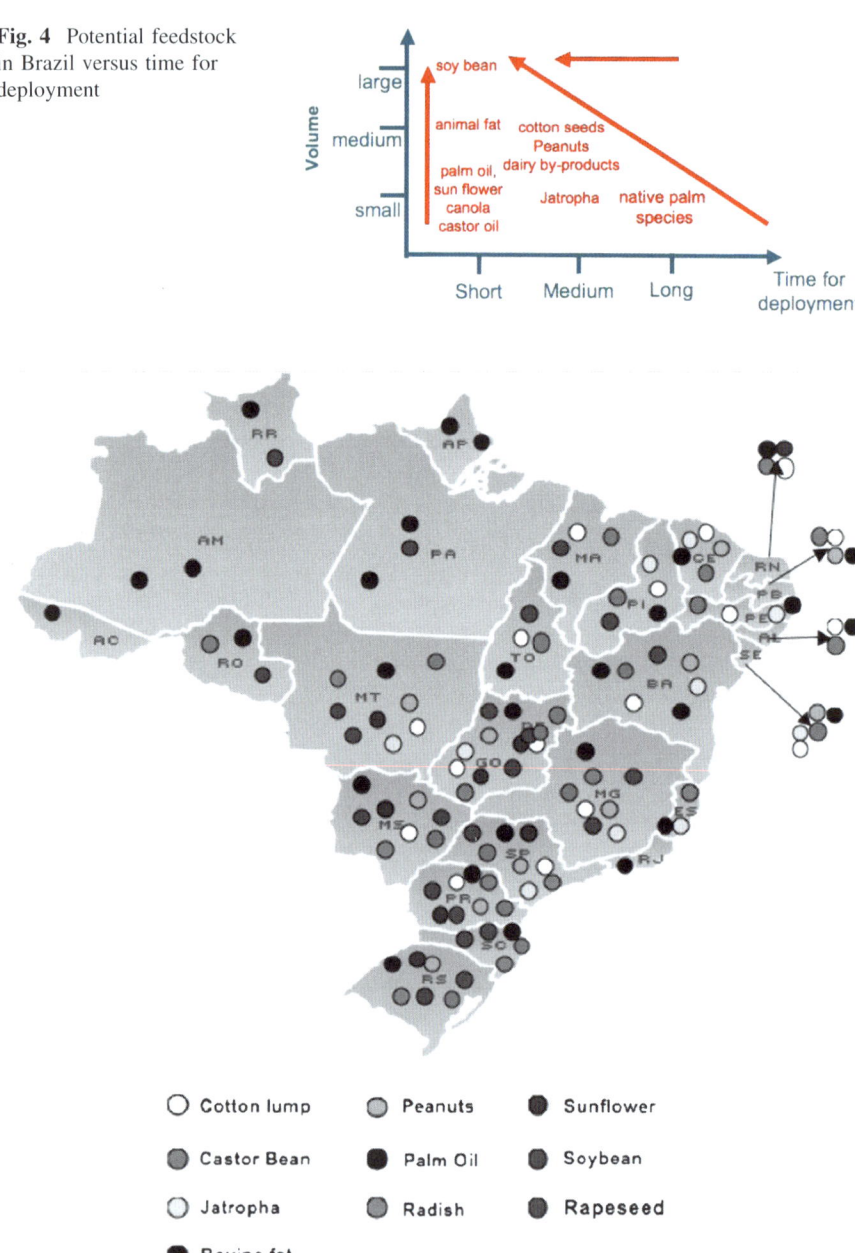

Fig. 5 Potential capacity of different crops in Brazil

A *positive impact* of the NPPB has been to *stimulate the production of multiple crops beyond traditional commodities* such as sunflower, castor oil, peanut and palm oil (Fig. 5). Several initiatives for the assessment of feedstock potential were carried

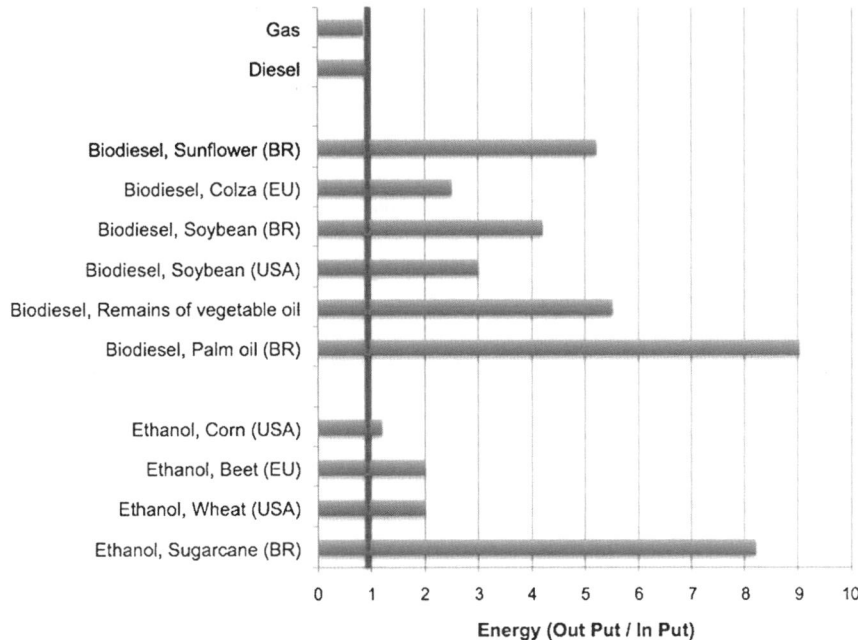

Fig. 6 Energy balances from feedstock for biodiesel and ethanol production (Author's analysis from multiple sources)

out. Agricultural technologies are pushing for higher productivity and eventually will determine the technical, environmental, economic and social viability of the production of biodiesel feedstock in each region of the country. Additionally, Brazil is developing the use of ethanol to replace methanol in the production process.

The NPPB is in its early stages when compared to the Brazilian ethanol. Proper quantification of the direct and indirect impacts on social and environmental indicators are not yet available for the entire country, but regionally is creating new opportunities for farmers and biodiesel producers.

5.1 Environmental Sustainability

5.1.1 Energy and Carbon Balance (GHG Emissions)

In-depth studies on LCA (Life Cycle Analysis) assessing *biodiesel energy and GHG balances* are still few in Brazil and do not cover all potential feedstock, technological routes and direct and indirect impacts of biodiesel production on land use change. Despite these uncertainties and the applicability of LCAs for biofuels, the production of biodiesel in Brazil *performs well above average when considering energy balances of other biodiesels* (Fig. 6).

These results are presented as the ratio between renewable energy produced to energy consumed in production from fossil fuels. The ratios for Brazilian biodiesels are 9.0 units for palm oil, 5.2 units for sunflower and 4.2 units for soybean. For biodiesel from rapeseed produced in the European Union, the ratio falls to 2.5 units. Overall, these *results demonstrate the comparative advantage of the energy performance of biodiesel produced in Brazil.*

In terms of GHG reduction balances, biodiesel performs better than diesel, as long as the feedstock used does not displace forest.

Biodiesel's carbon balance differs depending on the type of crops used, the land use change and the technology applied. *In order to produce the same amount of energy, soybean absorbs more carbon dioxide (CO_2) than palm.* However, *three times more land is required* to cultivate soybean in order to produce the same amount of energy.

6 Future Steps Towards Sustainable Production of Bioenergy and the Role of Innovation

Feeding 9 billion people in 2050 will require the development of new agribusiness models, new technologies and innovative ways of delivering an array of bioproducts from multi-functional agricultural landscapes. Therefore a great challenge facing policy makers, businesses, scientists and societies as a whole will be how to responsibly establish sustainable agricultural production systems and bioenergy supplies in sufficient volumes that meet the current and future demands globally.

Biofuels such as bioethanol are becoming viable alternatives to fossil fuels due to several technological advances and management improvements. Utilizing agricultural biomass for the production of biofuels has drawn much interest in many science and engineering disciplines, including molecular biologists and plant breeders. As a major crop for biofuels, sugarcane is to benefit from large research investments, from public and private sources.

Compared to other crops with biofuel potential, sugarcane and maize can provide both sugar/starch (seed) and cellulosic (bagasse/stover) material for bio-ethanol production. However, the combination of food, feed and fuel in one crop, although appealing, raises concerns related to the land delineation and distribution of these crops grown for energy versus food and feed. To avoid or reduce this dilemma and criticism, the conversion of this feedstock into bioethanol must be improved.

Conventional breeding, molecular marker assisted breeding and genetic engi-neering have already had, and will continue to have, important roles in sugarcane improvement. The rapidly expanding information from genomics and genetics combined with improved genetic engineering technologies offer a wide range of possibilities for enhanced bioethanol production from sugarcane.

It is now clear that multiple transgene strategies need to be developed (using synthetic biology and other approaches) to tackle complex traits, to engineer metabolic pathways and to combine the expression of different genes in biomass

Fig. 7 Market share of
Brazilian sugarcane business.
Source Cosan

**The industry is highly fragmented, top
10 producers account for less than 30% of
total production and the largest producer
for less than 10%.**

Market share in the Brazilian milling of sugar-cane

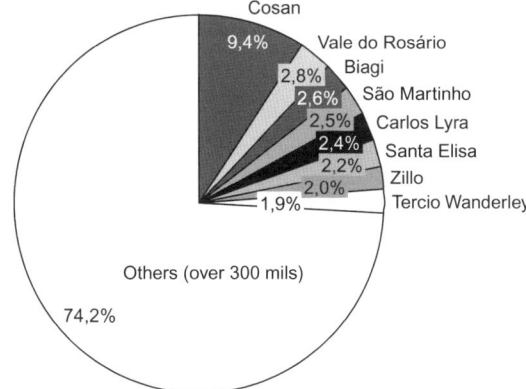

and in conversion technologies. Some studies have demonstrated the feasibility
of such technologies, but more effort is needed to make them both applicable to
bioethanol production and acceptable to the public. Indeed, the development of
genetically engineered crops raises issues of legislation relating to how these
technologies should be regulated and managed.

Brazil has its own legislation concerning plant biotechnology. However the
regulatory system often lags behind the advancement of these plant biotechnolo-
gies. An integrated agri-biotechnology system for food, feed and fuel production is
likely to be a challenge from the regulatory point of view, but will most certainly
be the future for sugarcane if it is to be bred for ethanol versus sugar production.

Therefore research programs addressing how to improve sustainable production
of ethanol using environmental indicators should assess the current and future
roles that biotechnology would have on productivity, and thus on environmental
performance of sugarcane and corn, and how legal and policy framework might
evolve to allow or not the deployment of multiple biotechnological applications
underlying these developments.

The examples provided here illustrate a sound baseline of sustainability compared with other current biofuels available in a large scale in the world, having the
smallest impact on food inflation, high levels of productivity (on average 7,000 l
of ethanol/ha and 6.1 MWh of energy/ha), with lower inputs of fertilizers and
agrochemicals, while reducing significantly the emissions of greenhouse gases.
The ending of sugarcane burning in 2014 is a good example of improving existing
practices. The proper planning of sugarcane expansion into new areas will be
another important step towards sustainable production of ethanol.

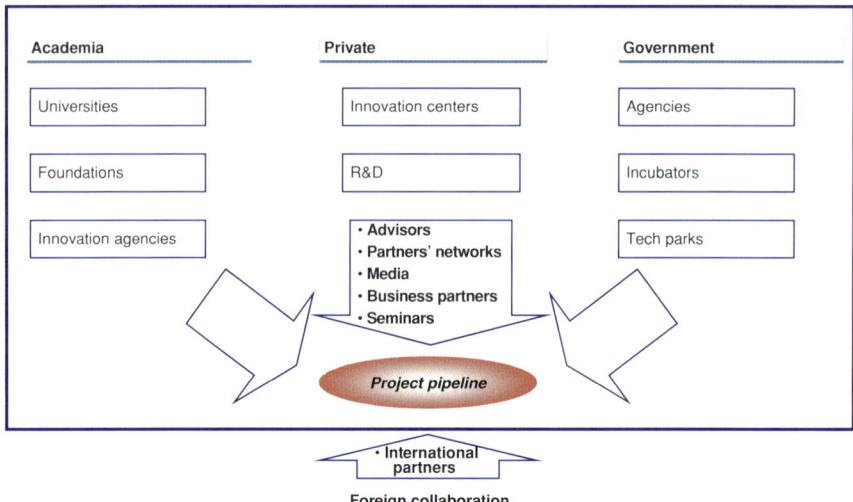

Fig. 8 Brazilian ethanol framework needed for innovation and deployment

Different initiatives in Brazil from the state of Sao Paulo Research Foundation (FAPESP), Ministry of Science and Education (MC&T–FINEP) and investments from the private sector are contributing to the deployment of new opportunities provided by the sugarcane biomass, and at the same time improving the environmental performances in agriculture and in the industry, particularly for the deployment of the full potential of sugarcane biomass, beyond the production of bioethanol.

The improvement of sugarcane production systems, especially at the agricultural level is a necessity as the amplitude of productivity among ca. 400 mills is very large, and jeopardizes the adoption rate of the best practices towards sustainability. Several technological innovations and management practices are already available, but not always implemented or adopted by the majority of mills.

The consolidation of the industry in large players (Fig. 7), and the presence of new investors in ethanol for national and foreign markets might accelerate the rate of adoption of best practices. Therefore, reducing the asymmetry of technological inputs and levels of productivity among sugarcane producers in Brazil will contribute significantly towards the optimization of use of fertilizers and for the elimination of sugarcane burning prior to harvesting.

Research programs should make the necessary bridges between demands from sugarcane producers and from suppliers of equipment and service providers, in the quest for technological innovation and thus for improving ethanol productivity and sustainability. A framework for ethanol innovation and deployment of pipeline technologies is necessary, complementing this transdiciplinary research agenda and making the links between academia, private sector and policy makers (Fig. 8).

The production of bioelectricity from sugarcane is already an important economic activity contributing to approximately 15% of Brazil's energy matrix and with the potential to reach 14,000 MW in the next decade [17].

Additional high value biochemical compounds are also making sugarcane an important feedstock asset for transnational and biotechnology enterprises, which are testing and deploying multiple biochemical and termochemical conversion platforms, therefore making the sugarcane business more attractive to investors and thus contributing to the diversification of the use of this important feedstock.

These new technologies and innovations are taking place in Brazil and elsewhere in the world, aiming at optimizing the use of feedstocks: using lignocellulosic materials (the second generation of biofuels); reducing waste; reducing the use of water and adding value to ethanol co-products and moving towards ethanol chemistry and biorefinaries full deployment.

References

1. Campos BC (2006) Dinâmica do carbono em latossolo vermelho sob sistemas de preparo de solo e de culturas. Santa Maria, RS. Tese (doutorado)—Universidade Federal de Santa Maria
2. Cerri CC (1986) Dinâmica da matéria orgânica do solo no agrossistema cana-de-açúcar. Tese (livre docência). Escola Superior de Agricultura "Luiz de Queiroz", Piracicaba
3. D'andrea AF, Silva MLN, Curi N, Guilherme LRG (2004) Estoque de carbono e nitrogênio e formas de nitrogênio mineral em um solo submetido a diferentes sistemas de manejo. In: Pesquisa agropecuária brasileira, Brasília, vol 39, no 2, p 179–186, fev 2004
4. Dedini (2008) Dedini lança usina de açúcar e etanol produtora de água. Press release. Dedini S.A. Indústria de Base, Piracicaba
5. Fornasieri F, Domingos VI (1978) Nutrição e adubação mineral do algodoeiro. In: Galdos MV (2007) Dinâmica do carbono do solo no agrossistema cana-de-açúcar. Tese (doutorado)— Escola Superior de Agricultura "Luiz de Queiroz", Piracicaba
6. International Food Policy Research Institute (IFPRI) (2010) Modeling the global trade and environmental impacts of biofuel policies. http://www.ifpri.org/publication/modeling-global-trade-and-environmental-impacts-biofuel-policies. Accessed 19 Aug 2011
7. International Energy Agency (2004–2006) www.iea.org. International Energy Agency, Paris
8. Jantalia CP (2005) Estudo de sistemas de uso do solo e rotações de culturas em sistemas agrícolas brasileiros: dinâmica de nitrogênio e carbono no sistema solo—planta—atmosfera. Tese (doutorado)—Universidade Federal Rural do Rio de Janeiro, Rio de Janeiro
9. Lardy LC, Brossard M, Assad MLL, Laurent JY (2002) Carbon and phosphorus stocks of clayey ferrassols in Cerrado native and agroecossystems, Brazil. Agric Ecosyst Environ 92:147–158. In: Aduan RE (2003) Respiração de solos e cilcagem de carbono em cerrado nativo e pastagem no Brasil central. Tese (doutorado)—Universidade de Brasília, Brasília
10. Neves CSVJ, Feller C, Larré-Larrouy M-C (2005) Matéria orgânica nas frações granulométricas de um latossolo vermelho distroférrico sob diferentes sistemas de uso e manejo. Ciências Agrárias, Londrina, vol 26, issue 1, pp 17–26
11. OECD (2007) Biofuels: is the cure worse than the disease? Paris. Available in http://www.oecd.org/dataoecd/40/25/39266869.pdf. Accessed 27 Sept 2007
12. Ottmar RD, Vihnanek RE, Miranda HS, Sato MN, Andrade SMA (2001) Séries de estéreofotografias para quantificar a biomassa da vegetação do cerrado no Brasil Central. Brasília: USDA, USAID, UnB. p 88. In: Ciclagem de Carbono em Ecossistemas Terrestres—O caso do Cerrado Brasileiro. Planaltina, DF—Embrapa Cerrados. Available in http://bbeletronica.cpac.embrapa.br/2003/doc/doc_105.pdf. Accessed 18 Set. 2008

13. Rangel OJP, Silva CA (2007) Estoque de carbono e nitrogênio e frações orgânicas de latossolo submetido a diferentes sistemas de uso e manejo. In: Revista Brasileira de Ciência do Solo, vol 31, pp 1609–1623

14. Szakács GGJ (2003) Avaliação das potencialidades dos solos arenosos sob pastagens, Anhembi—Piracicaba/SP. Piracicaba, 2003. Dissertação (mestrado)—Centro de Energia Nuclear na Agricultura

15. Titon M, Ros CO da, Aita C, Giacomini SJ, Amaral EBDo, Marques MG (2003) Produtividade e acúmulo de nitrogênio no milho com diferentes épocas de aplicação de N-uréia em sucessão a aveia preta. In: XXIX Congresso Brasileiro de Ciência do Solo, 2003, Ribeirão Preto—SP

16. Trumbore S (1993) Comparison of carbon dynamics in tropical and temperate soils using radiocarbon measurements. Global Biogeochemical Cycles 7:275–290. In: Silveira AM, Victoria RL, Ballester MV, de Camargo PB Martinelli LA, Piccolo MC 2000. Simulação dos efeitos das mudanças do uso da terra na dinâmica de carbono no solo na bacia do rio Piracicaba. Revista Pesquisa. agropecuária. brasileira, Brasília, vol 35, no 2, pp 389–399

17. Unica (2011) 3a Ethanol Summit, Sao Paulo, SP, Brazil. http://www.unica.com.br/ethanolsummit

Energy Security in an Emission-Constrained World: The Potential for Alternative Fuels

Tara Shirvani

Abstract The current transport fuel dilemma, which centres around the input problem of dwindling conventional crude oil reserves as well as the so-called output problem of increasing GHG emissions 1, has triggered an increased interest in alternative fuels from other hydrocarbon sources such as natural gas and coal as well as renewables. The increased consumption of unconventional oil resources would initially solve the input problem of falling conventional oil resources, but inevitably exacerbate the output problem of increased environmental pollution 1. Biofuels can be a viable substitute for fossil fuels, most notably when produced in a sustainable manner and from feedstock which is not in direct competition with food or animal feed.

1 Introduction

The current transport fuel dilemma, which centres around the input problem of dwindling conventional crude oil reserves as well as the so-called output problem of increasing GHG emissions [1], has triggered an increased interest in alternative fuels from other hydrocarbon sources such as natural gas and coal as well as renewables, see Fig. 1. The increased consumption of unconventional oil resources would initially solve the input problem of falling conventional oil resources, but inevitably exacerbate the output problem of increased environmental pollution [1]. Biofuels can be a viable substitute for fossil fuels, most notably when produced in

T. Shirvani (✉)
Smith School of Enterprise and the Environment, Hayes House,
University of Oxford, 75 George Street, Oxford OX1 2BQ, UK
e-mail: Tara.shirvani@smithschool.ox.ac.uk

O. Inderwildi and Sir David King (eds.), *Energy, Transport, & the Environment*,
DOI: 10.1007/978-1-4471-2717-8_6, © Springer-Verlag London 2012

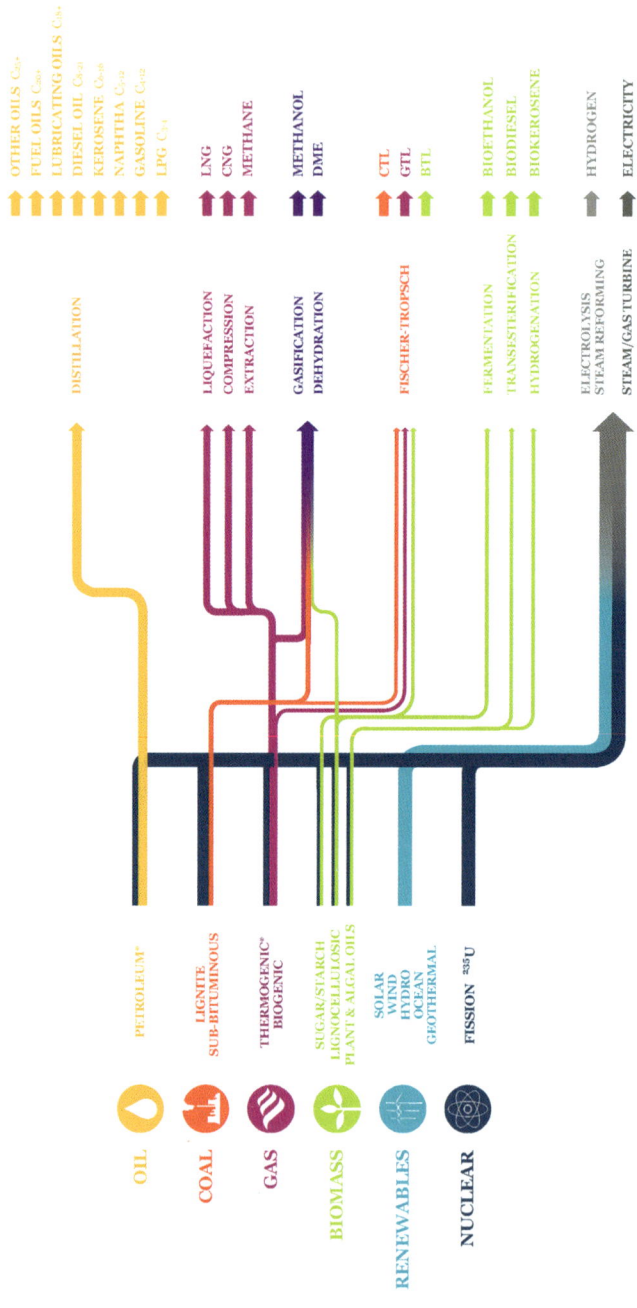

Fig. 1 Commercial fuels and primary energy [4]

Fig. 2 Fuel supply oil
derived versus alternative
(million barrels a day) [4]

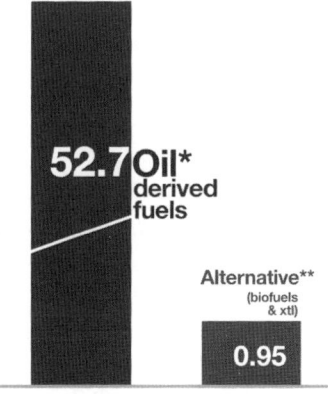

a sustainable manner and from feedstock which is not in direct competition with food or animal feed. The transition towards advanced biofuels may contribute towards a low carbon, sustainable fuel mix, but is unlikely to become the silver bullet for substituting the current energy demand of our global transport system. Currently, biofuels provide roughly only 1% of transport fuels and 0.2–0.3% of global energy supply [1], see Fig. 2. Such concerns have fuelled the interest in developing advanced biofuels from inedible biomass such as agave or microalgae, as these potentially help overcome problems faced by first-generation biofuels such as land-use change as well as trade-offs with food security [2, 3].

2 Unconventional Oil Resources

With the majority of conventional oil reserves located within the developing world and in particular under the management of the Organization of Petroleum Exporting Countries (OPEC) institution, the Western Hemisphere has fortified their search for new energy resources located in regions with less geopolitical tension. Unlike conventional oil, unconventional oil resources are mainly concentrated in developed countries. With more than 85% of global natural bitumen reservoirs located in South and North America, 81% of all explored tar sands concentrated in Canada [5] and 75% of world oil shale reservoirs situated in the US [6], in comparison to Saudi Arabia managing 20% of world's total conventional oil reserves, the unconventional oil industry can therefore successfully circumvent geopolitical issues related to oil supply. In the year 2000, unconventional oil already attributed to 12% of US crude oil production, particularly as most of the fuel feedstock has only just become economically feasible in light of oil prices around $100/barrel [6].

Unconventional oil products are commonly separated into natural bitumen and heavy oils which both are a precursor of crude oil. As the feedstock shares the same formation history as conventional oil, in theory it would develop over the turn of million years, into conventional crude as the reservoirs degrade through bacteria attack and erosion under light suspension [6]. Nowadays, to produce liquid fuels from unconventional oil resources, synthetic crude is first produced from tar sands, extra-heavy oil or oil shale before it is further refined into finished products. Bitumen is extracted from tar sands by open-pit mining method or in situ and is distilled, catalytically converted and hydro-treated to produce synthetic crude. Oil shale is a calcareous rock containing kerogen, a precursor of crude oil and can be extracted by underground mining or in situ method and is then heated to produce synthetic crude [5, 6]. As all unconventional oil resources are solid and have a high carbon-to-hydrogen ratio their conversion to liquid fuels is very energy intensive. For example, bitumen extraction in lower regions requires steam injection, whereas in deeper underground reservoirs process heat must be provided by either (i) controlled underground nuclear explosions, (ii) in situ combustion, (iii) electric heating elements immersed into bore holes [6]. Along with the energy intensity of the extraction process, the separation step for hydrocarbon compounds from associated sands, rocks and clay residues requires a considerable amount of additional feedstock which amounts to roughly 2,000 kg of oil sands/oil share processed per barrel of synthetic fuel produced [6]. Moreover, the high viscosity of unconventional fuel stock requires expensive transport methods, which further aggravate the financial feasibility. In addition to environmental degradation, the drawbacks associated with synfuel production from oil shale are production costs three times higher than conventional oil extraction; and considerable water usage of roughly three barrels of water per barrel of fuel produced [7, 8]. As conventional oil resources become depleted, the larger uptake of unconventional resources, here in the form of shale oil, is reflected in a higher fossil energy balance, i.e., Energy return on investment (EROI) ratio of 1.65 MJ/MJ and GHG emissions of around 182 g CO_2 eq/MJfuel [6]. In particular, unconventional oil resources have outlined higher volumes of nitrogen, oxygen, sulphur and heavy metal particles, which not only require special purification and separation methods, but regardless contribute to the fuel's atmospheric pollution level. For example, atmospheric emissions from oil shale processing can reach well-to-tank (WtT) emissions around 4–6 times the level of conventional fuels [6]. For liquid fuels produced from tar sands/extra-heavy oil and oil shale, life cycle GHG emissions are 5–40% [6, 9, 10] and 26–180% [6, 10] higher than those of conventional fuels, respectively. However, it should be noted that definitive conclusions about GHG emissions for tar sands and oil shale cannot be drawn with existing studies as additional research is needed to better understand these technologies [11].

3 Synthetic Fuels: Fischer–Tropsch Synthesis

In addition to the increased interest in unconventional oil reserves, there is renewed interest in producing liquid fuels from coal and natural gas. Production from both unconventional and renewable sources has been increasing rapidly in recent years and is projected in IEA's Reference Scenario to grow to 7.4 million barrels per day by 2030, or $\sim 10\%$ of global conventional oil supply [12]. Geopolitically, synthetic fuel production is an important process, as it allows the conversion of alternative fossil resources such as coal or gas into liquid fuels and has therefore been used to improve energy independence [13]. Nazi Germany and later the Apartheid regime in South Africa, both oil-poor and coal-rich, used the process to circumvent oil embargoes by the production of liquid fuels from coal [14]. These fuels indeed sustained these economies by ensuring transport fuel security. In the case of South Africa, a substantial amount of liquid fuel is still produced using the Fischer–Tropsch (FT) synthesis, vide infra [13].

Not only alternative fossil feedstock, but also renewable feedstock such as biomass can be used to produce Fischer–Tropsch fuels. One possible route to convert biomass to liquid fuels is via the gasification process: the feedstock reacts with oxygen or steam to produce a gaseous mixture of CO, CO_2, H_2, CH_4 and N_2. When gasifying the feedstock, the carbonaceous components are further broken down into a gaseous mixture of H_2 and CO which is commonly referred to as synthesis gas. The conversion process occurs under limited oxygen availability and as part of a thermal decomposition step. Syngas is currently used to generate transport fuels, by generating hydrogen via a water–gas-shift reaction [15] or hydrocarbons via the Fischer–Tropsch synthesis [16]. The chemists Franz Fischer and Hans Tropsch developed a chemical process that converts synthesis gas into liquid hydrocarbons that can be used as fuels [17, 18]. The Fischer–Tropsch process, catalytically converts syngas into liquid hydrocarbons ranging from C1 to C50 long chains [16, 19]. By varying the temperature and pressure level as well as the type of catalyst used, a distinct variety of fuel products can be generated [20], which for example in the case of biomass are of higher economic value than syngas produced from gasification. Synthesis gas is produced from fossil resources such as coal and natural gas, but it can also be produced from renewable biomass. Commonly, fuels made from natural gas are referred to as gas-to-liquid (GtL) fuels and analogously fuels manufactured from biomass or from coal are referred to as biomass-to-liquid (BtL) and coal-to-liquid (CtL), respectively. Hybrid approaches that use different fossil and renewable feedstocks are referred to as XtL processes. However, at the moment the state of technology for low sulphur FT-fuel from biomass is neither mature nor economically feasible.

One of the biggest advantages of liquid fuels from unconventional fossil sources over other alternative fuels is that they are compatible with existing vehicles and fuel infrastructure [12]. Another advantage for CTL and GTL is that they can achieve higher engine efficiencies and lower regulated air pollutant emissions mainly because of their several chemical and physical characteristics, including reduced density, ultra-low sulphur levels, low aromatic content and high cetane rating [21]. Although liquid fuels from unconventional fossil sources may help to address the energy security concerns and urban air pollution problems, they are likely to result in increases in GHG emissions from the transport sector. The life cycle GHG emissions for CTL without CCS are estimated to be more than 100% higher than those of conventional fuels [6, 19, 22, 23] and still 5–29% higher with CCS [6, 23]. One promising option to reduce CTL GHG emissions is to combine CTL with BTL as they both produce syngas from solid feedstocks through gasification.

Coal-to-liquid (CTL) is an old technology that was developed in the early twentieth century and includes three techniques, pyrolysis, direct coal liquefaction (DCL) and indirect coal liquefaction (ICL) [5, 24]. ICL is considered to be the most promising option out of the three for the following main reasons: (1) pyrolysis has low liquid fuel yields; (2) liquids produced via pyrolysis and DCL require further treatment to be used in existing vehicles; (3) ICL allows easier implementation of carbon capture and storage (CCS) and offers higher flexibility of final products; and (4) ICL enjoys stronger supporting experience as it has been in commercial operation in South Africa for over 50 years. The processes for ICL are rather similar to those for BTL described earlier, i.e., gasification of coal to produce syngas followed by FT synthesis generating synthetic gasoline and diesel or other desirable fuel products. Countries with large oil demands and domestic coal reserves such as the US and China are particularly interested in developing CTL [25, 26]. China's commercial CTL plants are already in the design and construction phase, with a projected annual production of 70 billion litres by 2030 in a high oil price scenario [8].

The main production processes for gas-to-liquid (GTL) can be described as syngas generation from natural gas followed by FT synthesis [21]. In order to convert natural gas into liquid hydrocarbons it first has to be produced, liquefied, and upgraded in a multistep process. After producing the natural gas, impurities such as sulphur components have to be removed after which the natural gas feedstock is cooled down to separate higher hydrocarbons from methane—the starting material for the catalytic liquefaction. Methane is further combined with oxygen for the production of synthesis gas via either catalytic partial oxidation or steam reforming, which is further used as feedstock for the production of FT-fuels. The synthetic crude is distilled to yield diesel, kerosene, naphtha, and oils. Commonly, the GTL process generates around 75% middle distillates and 25% non-fuel chemical products. In the case of syncrude, we assume that a barrel yields 31 gallons of middle distillates plus chemical by-products. Due to the higher energy density of the middle distillates a barrel of syncrude yields approximately 13% more energy in the form of liquid fuels that are conventionally crude. Fuels manufactured from natural gas using the Fischer–Tropsch process can replace conventional fuels at any ratio (0–100%) as they are miscible.

Just as with oil reserves, gas reserves are asymmetrically distributed, with three countries, Russia, Iran and Qatar in control of more than 50% of the conventional reserves [27]. The North Dome–South Pars complex located 3 km below the seabed of the Persian Gulf for instance accounts for 23% of proven conventional gas reserves. This complex is shared by Iran and Qatar and contains more than 50 trillion cubic metres of natural gas plus vast amounts of gas condensates [12]. The size of this gas field as well as geopolitical constraints is the reason why the two largest running GtL plants are located in the State of Qatar. Currently, there are four commercial GtL plants online: PetroSA's plant in Mossel Bay (SA) which

Table 1 Market entrance crude oil prices for liquid fuel production from different resources [5]

Resources	Market entrance oil price ($/barrel)
Tar sands	38
Extra-heavy oil	30
Oil shale	70 (short run); 30 (long run)
CTL	86
GTL	70
BTL	205
Corn ethanol	40

produces 35,000 bbl/d, Shell's Bintulu plant in Malaysia produces 12,500 bbl/d, in Qatar Sasol's Oryx plant produces 34,000 bbl/d and the first phase of Shell's Pearl plant produces 70,000 bbl/d. Consequently, the current global GtL capacity is 151,500 bbl/d, which is less than 0.2% of the global demand for liquid transport fuels [28]. In the near future, the most modern GtL plant in Qatar, Shell's Pearl will produce 140,000 barrels of GtL syncrude and 120,000 barrels of liquid petroleum gas (LPG), using 1.6 billions cubic feet (45 million m^3) of natural gas. Thus, this plant will convert approximately 1.6 trillion British Thermal Units (BTU) in the form of natural gas into approximately 612 billion BTU GtL middle distillates, 481 billion BTU LPG plus chemical products such as naphtha and lubricants.

Fuels manufactured using the Fischer–Tropsch process are high quality fuels with a cetane number of 75, virtually free of sulphur, metals and aromatics. This high fuel quality directly reflects in the emissions; according to a variety of experimental studies, CO, NOx, hydrocarbon and particulate emissions reduce drastically for both pure GtL diesel and various blends with conventional diesel [29, 30]. Experimental studies have furthermore shown that the thermal efficiency of a diesel engine is improved when GtL blends are used [31, 32]. While GtL fuels clearly reduce local pollutants, they are by no means environmentally friendly fuels as they emit significantly more greenhouse gases than fuels derived from conventional crude oil over their life cycle. There is a consensus in the scientific community that the carbon footprint of GtL fuels is approximately 10% higher than that of conventional, oil-derived fuels [19, 22]. GtL fuel are only advantageous from a GHG emission point of view when their synthesis is combined with CCS [22], which in turn lowers the efficiency of the process and hence reduces the energy gain.

With current technologies the surplus in conventional gas resources could be used to produce more than one trillion barrels of GtL syncrude, more than the projected demand over next two decades as projected by both IEA and EIA. Consequently, GtL fuel could theoretically mitigate oil supply shocks as there is no resource constraint. Nevertheless, mitigation using GtL fuels would come at a significant environmental cost as these fuels have an increased carbon footprint and hence, the growing contribution of the transport sector to GHG emissions would be exacerbated by large-scale deployment of GtL fuels. However, currently there is not enough GtL capacity online to absorb oil supply shocks. Due to significant lead times for additional capacity and exorbitant upfront investment, it is unlikely that a substantial percentage of global liquid fuel demand will be available. Therefore, a capacity constraint prevents GtL fuels from being produced on a scale sufficiently significant to have a major impact on energy security and global GHG emissions.

Table 1 shows the market entrance crude oil prices for liquid fuel production from different resources, i.e., oil prices that make the corresponding resource profitable. It can be seen that with current oil prices above 100 $/barrel, producing liquid fuels from all the unconventional fossil sources is already economically competitive.

4 Biofuels

Biofuels, in the form of ethanol or methanol from plants and woody biomass [13, 20, 33], synthetic oil products from agricultural or dedicated crops, and diesel fuel stock from vegetable oils, microalgae or animal fats [6], can act as a viable substitute for fossil fuels. Biofuels provide roughly 1% of transport fuels and 0.2–0.3% of global energy supply [33]. Currently, a total of 1.5 billion hectares of arable land mass (0.55%) is in use with a potential spare capacity of 250 million hectares in South America and 180 million hectares in Africa [34]. The size of occupied cropland is expected to rise between 17 and 44% by 2020 due to growing population figures, increased biofuels demand and changing diets [33].

An important prerequisite for biofuels to realise low life cycle GHG emissions is the consideration for careful land management [6]. For biofuels to provide truly sustainable transport solutions, the appropriate selection of cultivation sites and avoidance of land-use change is of paramount importance, see Fig. 3 [20]. When incorporating matters of land-use change, the environmental feasibility of biofuels from agricultural crops is considerably exacerbated and at times worse than that of conventional fossil fuels. By including the CO_2 emissions associated with the clearing of tropical forest areas or conversion of peat lands and savannas to new biomass cultivation sites [6, 35], the favourable GHG emissions balance for biofuels is rapidly negated through the increase of the associated biomass and soil carbon release [20]. In Southeast Asia where biomass feedstock is grown on carbon-rich rain forests a substantial biomass and soil carbon, i.e., 'carbon debt' [35], is released which would take up a payback period of 423 years [13]. Without the consideration of land-use change, the production of sugarcane-led ethanol in Brazil [34, 36], soybean-derived biodiesel in the United States [37, 38], rapeseed in Europe [39, 40] as well as palm oil in Southeast Asia [41, 42] results in considerable GHG emissions savings. Other fuel-lifecycles such as corn-ethanol production in the United States [38, 43] and China [44] yield marginal or no GHG emissions reductions when ignoring land-use change issues. The same effect may well occur indirectly (i.e. indirect land-use change), when existing arable land is

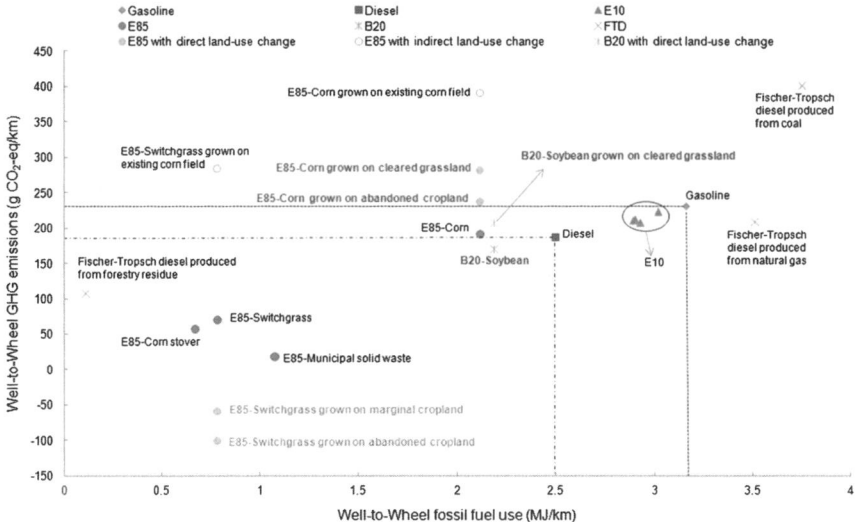

Fig. 3 Life cycle fossil fuel use and GHG emissions 'per vehicle km travelled' for biofuels and synthetic fuels in the US [51]

diverted to biofuels production at the expense of reduced food or feed production [34, 45]. For similar reasons, US corn-ethanol production has come under severe criticism by causing considerable soil erosion and requiring more nitrogen fertilizers, insecticides and herbicides than many other biofuel sources [1, 46]. The high emission rates of nitrogen and phosphorus released from crop cultivation areas and drainage water systems, strongly impact nearby freshwater and marine ecosystems [20, 47]. The application rate of artificial fertilizers must be carefully monitored and kept to a minimum to prevent emissions from nitrous oxide, a potent greenhouse gas [20, 47, 48]. For example, in the Gulf of Mexico US remnants of fertilizers are swept into the ocean and have created dead zones so devoid of oxygen that most sea life cannot exist [13, 49].

With first-generation biofuels mainly derived from agricultural crops such as sugar, starch, vegetable oils or animal fats and grain crops [13] most major production shifts have proven to trigger a direct competition with their use for food or animal feed [13, 20]. Parts of the developing world have already experienced the economic consequences of the 'food versua fuel' dilemma, particularly during the 2008 food price spike when rising energy prices and the weakening US dollar have been responsible for a 25–30% increase in soft commodity prices, and the remaining 70–75% was due to the expansion of biofuel production, low grain stocks, vast land-use shift, export bans and speculative trading [50].

Regardless of various contributors to the food crisis, we note that the rapid expansion of global ethanol production from Brazilian sugarcane did not contribute to the past food price surge. Triggered by the rising demand for grain, further price spikes in correlated soft commodity markets were recorded: with maize prices tripling, wheat prices rise by 127% and the price of rice increase by 170% [50].

Many have rightfully described the food versus fuel dilemma as the fuel hunger of the industrialised world opposing the demand for food in the undernourished developing world. The poor and undernourished population who are living on less than $2 per day will have to fight for their every day existence once being pushed into malnutrition, particularly the social stratum living on less that $1 per day is most likely to perish within the coming centuries [50, 52, 53]. The dramatic consequences for the human race are exacerbated as in 2008, 19% of global population was living in poverty and 22% in extreme poverty [54]. The continued diversion of grain away from food to fuel production, and the incentive to set aside agricultural land for biofuel production [55] continues to put pressure on the global food market as well as global land use. With food price spikes similar to the one in 2008 likely to become more frequent, a continuation of aggravated food security problems in the developing world is inevitable.

5 Carbohydrate Derived Fuels: Ethanol

Currently the alternative biofuels market is dominated by ethanol and biodiesel production [1], given their closely resembling chemical characteristics with conventional diesel and petrol [6] and 'blending flexibility'. With the exception of synthetic fuel products, high-level blends of biofuels require changes to the vehicle's engine infrastructure [20] as otherwise the fuel's vapour pressure levels, oxygen contents, lower volumetric energy content and hydrophilic nature increase the corrosion of the car engine [20, 33]. For the case of ethanol, the high octane number and oxygen level allow petrol fuel blends to operate on a higher compression ratio and is more thoroughly combusted than conventional gasoline [56, 57]. Ethanol is commonly blended in low proportions of up to 10% of volume (E10) with conventional gasoline-fuelled spark-ignition ICEV's, whereas in the United States flexible fuel vehicles (FFVs) operate on a modified engine/fuel system that is powered by higher blends of up to 85% volume (E85),and ethanol in its pure form (E100) is used in adapted vehicles in Brazil [1, 13].

The production of ethanol through the fermentation of sugars is a mature technological process. The fermentation technology is typically defined as a metabolic process in which an organic substrate undergoes chemical changes via enzyme activity of micro-organisms [15]. As part of the successful fermentation of starch and lignocellulosic crops, the sugar content of the feedstock must be released from the plant material in an initial pre-treatment step [58]. Starch crops are hydrolysed enzymatically to release the sugar solution after which they undergo the microbial fermentation step to generate bioethanol. Certain feedstock such as sugarcane is appropriate for fermentation without any pre-treatment of the biomass. However, for converting lignocellulosic material into ethanol fuel, the cellulosic and hemicellulosic biomass must be separated from the non-fermentable lignin content [59]. The covalent cross-links, which bind the cellulosic material together, are broken up mechanically and further undergo acid, alkali and/or steam treatment. The fermentable cellulosic and hemicellulosic biomass' component is hydrolysed enzymatically to release the sugar content, which is fermented to generate ethanol fuel. The residual lignin material is combusted on-site to provide additional heat volumes which can be used to offset the production cycle's energy requirements [59]. In comparison to the more straightforward conversion process for pure C6 sugars of starch or saccharose, the fermentation process for broken-down hemicelluloses requires the application of certain organisms capable of converting C5 sugars such as Xylose [16]. Current research efforts focus on the development of more efficient and robust micro-organisms that are more resistant to the high pressure and temperature levels of the fermentation process.

Ethanol is mainly derived from grain and sugar crops grown on dedicated agricultural land mass, with the majority being produced from maize in the United States and sugarcane in Brazil [34]. In the future, more environmentally friendly feedstock will centre around cassava [11], sweet sorghum [60], cellulosic biomass from crop residues, perennial grasses and municipal solid wastes (MSW) [48, 61]. In 2007, ethanol's global production volume of 46 billion litres replaced 4% of global petrol demand [62]. With present governmental production quotas in the United States, Europe and Asia on the verge of implementation, worldwide ethanol production is expected to reach 125 billion litres by 2020 [62]. The occupied land area used for ethanol production in 2006 stretched from 5.1 million hectares of US corn production to 2.9 million hectares of Brazilian sugarcane production [34].

By combining the synergies between the sugar market, a defossilized national heat and electricity grid and the initial assistance of a governmental subsidy programme [63], Brazil has become the world's second largest producer of ethanol fuel and the world's largest exporter [64]. The country benefits from high levels of solar radiation, year-round water supply and large untapped landmass used for ethanol plantations without diverting food and farmland [63]. The land mass occupied by sugarcane production only accounts for 10% of cultivated land and 1% of total arable land mass [34, 63] with plantations mainly expanded to degraded grassland situated far from the Amazonian tropical forests. Most importantly, the production of ethanol from sugarcane, benefitting from low nitrogen fertilizer rates and mature sugar fermentation technology, has been strongly promoted due to its positive net energy balance [1, 63, 65] consuming 1 MJ of fossil energy to generate approximately 8 MJ of biofuel. From a financial perspective, the Brazilian sugarcane production outcompetes all industry rivals by producing ethanol at US$ 30–35 per barrel of oil equivalent (boe), while currently being largely feasible without government subsidies [63, 66]. In comparison, the high costs of ethanol production in Europe of around US$ 80/boe and the United States of US$ 55/boe, support the investment case for Brazilian ethanol production. Coupled with the European Union running out of time to fulfil their target of supplying 10% of total transport fuel with biofuels by the 2020 [20, 34], the necessity for imports of crops and residues cultivated abroad has become undeniable. For the EU to meet its 10% biofuels directive domestically an area of between 20 and 30 million hectares of arable land will be required [67]. Reports by the European Commission [68] have estimated that the biofuels directive will trigger 12% of total rapeseed biodiesel production to be provided by outsourced palm oil plantation on peat lands with particularly high emission levels. However, for every litre of European rapeseed oil diverted to biodiesel, if only 2.4% of that rapeseed oil is replaced by palm oil grown on peat lands, the emissions from the oxidising peat cancel out the benefits from biodiesel indefinitely [45]. In this case, the European Union may well increase import levels of Brazilian ethanol up to a total of 3.9 billion gallons [20, 34] to meet the production target.

6 Lipid-Derived Fuels: Biodiesel

Biodiesel, is predominantly produced from edible oil crops such as rapeseed in Europe, palm oil in Asia and soybeans in Brazil [20, 34, 69], and have expanded to future feedstocks of microalgae [70–78] and jathropha [79]. Globally, with a production volume of 10 billion litres in 2007 [69], the biodiesel industry is a growing market expected to take up a larger share of the energy matrix in the future. In comparison to ethanol-fuelled vehicles, biodiesel blends can be used with compression-ignition ICEVs running on conventional diesel without altering the existing engine design and infrastructure. Similar to ethanol, biodiesel is able to achieve higher thermal efficiencies due to the combustion-enhancing oxygen level in the fuel, and results in lower criteria pollutant emissions, e.g. no sulphur oxides and particulates [34], without affecting the engine performance [1]. Biodiesel has proven a more favourable fuel efficiency than conventional diesel, as it achieves a higher thermal efficiency (MJ/km) and considerably lower criteria pollutant emissions (except NO_x) [1, 56, 80].

Biodiesel is a mixture of fatty acid alkyl esters (FAME) produced by the transesterification (ester exchange reaction) of a triglyceride (vegetable oil or animal fat) with methanol or ethanol in the presence of a base catalyst (usually sodium hydroxide or other advanced catalysts [73]). Lipid feedstocks are composed of 90–98% w/w of triglycerides, small amounts of mono- and diglycerides, free fatty acids (1–5%), residual amounts of phospholipids, phosphatides, carotenes, tocopherols, sulphur compounds, and water traces [73]. The transesterification process is a multiple step reaction involving three reversible reactions, whereby triglyceride molecules are converted into diglycerides which then are converted to monoglycerides and then completely cleaved into esters (biodiesel) and glycerol (co-product). Each step consumes one mole of alcohol and forms one mole of ester [81]. Thus, between 0.2 and 0.25 litre of methanol are required per litre of oil, which equates 4.8–6 mol of methanol per mole of oil [73, 81]. It is considered advisable to use methanol as the alcohol of choice, since the glycerol and ester products are almost immiscible and form two separate layers, thus simplifying the initial separation [81]. A homogenous or heterogeneous acid or basic catalyst is required to ensure a high transesterification reaction rate at moderate conditions of 60°C and atmospheric pressure. Two of the most commonly used catalysts include sodium hydroxide and potassium [82].

7 Advanced Biofuels

In order to avoid the characteristic limitations of first-generation biofuels, future biofuels must reduce fossil energy input, reduce GHG emissions, and be sourced from inedible biomass to avoid previously encountered environmental drawbacks [13]. Second-generation biofuels from the production of synthetic fuel and ethanol from inedible cellulosic materials such as stems, leaves, husks, etc. and have the potential to overcome the inherent problems of first-generation biofuels [13, 20]. Among the fuel products from advanced biomass are: ethanol produced from crop residues [83, 84], perennial grasses [43, 84, 85], MSW [86, 87], biodiesel derived from microalgae [75, 76, 78] and jathropha [88] as well as biomass-to-liquid (BTL) from perennial grasses and wood residues [1, 19]. Research efforts on genetic modification and breed for purpose of feedstocks with higher productivity levels, oil yields, lignocelluloses cell wall traits more amenable for processing, are ongoing [20]. Depending on where the biofuel feedstock is grown and whether it is sourced from dedicated plants, second-generation biofuels may have the potential to either induce a carbon release or lead to substantial carbon sequestration [35, 48]. Especially, biofuels derived from unwanted crop residues or MSW, have a more favourable GHG emissions balance than first-generation biofuels', as their use has the potential to reduce the pressure on food crops and decrease land use.

The conversion of lignocellulose materials into biofuels holds a promising potential, due to the amount of energy stored in the biomass as well as the volume of unused residues, co-products and wastes from various sectors at hand [13, 20]. However, on technical grounds, the technology for converting inedible plant residues is at the moment not as mature as the sugar/starch fermentation process [13, 43]. Both the chemical and biological routes for breaking down the protective shield of lignin and hemicellulose surrounding the cellulose have considerable shortcomings. While the chemical route of 'hydrolysation' is highly expensive due to large heat requirements, the biological 'enzymatic' conversion route merely results in limited yields and is further restrained by the high costs of enzymes used during the process [13]. Innovative new research on the wood-degrading abilities and action of enzymes from termites has proven to be of potential usage to degrade the cellulosic biomass into sugars. Still, the next conversion step from cellulose to sugars via cellulases, enzymes from bacteria or fungi, remains to be a slow and not yet mature process [89].

   The transition towards advanced biofuels synthesised from inedible cellulosic biomass may contribute towards a low carbon, sustainable fuel mix, but is unlikely to become the silver bullet for substituting the current energy demand of our global transport system [1]. Reasons for its marginal future application are driven by the technological drawbacks, limited availability of suitable land mass [90] as well as the amount of crop residues recoverable without damaging soil quality. Moreover, in an era of increasing concerns around the global 'water-food-energy nexus', the excessive water consumption volume of several biofuel production cycles, whether grown domestically or abroad, remains at the heart of the current debate [91]. Alongside conventional fossil fuels, many biofuels have water requirements that are of several magnitudes higher. Nevertheless, there are a number of social advantages attributed to the promotion of biofuels. When cultivating dedicated energy crops on marginal land in developing countries, new industries would emerge, that would create additional jobs in farming, transportation as well as fuel synthesis [13]. The concept of farming for both food and fuel [13], through the development of small-scale bio refineries would not only mitigate the urbanisation process by job creation in rural areas and simultaneously assist in regional poverty alleviation. By ideally producing transportation fuels on-site, the life cycle emissions and costs can be kept minimal. However, once again for a fuel cycle to be sustainable and environmentally viable, the simultaneous food, fuel and electricity production must be evaluated from a life cycle perspective which accounts for both the net energy consumed throughout the production process as well as the GHG emissions generated per unit of energy produced [13]. In addition, to guarantee the initial economic feasibility of local biofuel projects, appropriate policy levers such as tariffs and government subsidies must be incorporated. Other promising fuel paths include the conversion of biomass to synthetic oil products, such as jet fuel, synthetic diesel or gasoline [6], see next section.

8 Algae Biofuels

Such concerns surrounding the sustainable production of first and second generation biofuels have fuelled interest in producing fuel products from advanced biomass such as microalgae which potentially help overcome land-use issues as well as the food security dilemma [73, 78].

   With microalgae's high production yields, the required global land mass necessary to satisfy fossil fuel consumption could be considerably reduced. For algae-derived biodiesel with a yield of 850 GJ/ha/y, to replace the current total production of 1.1 billion tons of petroleum-derived diesel per year, a land mass of 57.3 million hectares would be required which approximates to an area somewhat larger than Spain and smaller than Texas [92]. The inherent potential advantage of biodiesel production from algae is lower life cycle GHG emissions as algae biomass converts atmospheric CO_2 through photosynthesis into bio plant material which are eventually released back to the atmosphere via micro-organisms when used as a fuel, via engine tail pipe emissions. In comparison, fossil fuel combustion releases additional carbon which took million of years to be removed from the atmosphere [93].

Microalgae can be grown in wastewater and/or in naturally resource limited environments, and have recognised superior biomass yields per hectare compared to other biofuel feedstocks [73, 74]. Relative to the particular cultivation requirements for other advanced biofuel sources, microalgae growth mainly requires solar radiation, carbon dioxide, water and nutrients in the form of inorganic salts [78]. Microalgae are defined as unique eukaryotic photosynthetic micro-organism whose cell walls are composed of saturated and unsaturated fatty acids with chain lengths of 12–22 carbon atoms [73]. From the existing 50,000 species of known microalgae, only a fraction are appropriate for biodiesel production, given the algae strains' varying lipid content and productivity levels [73, 94]. For most microalgae the average lipid content varies between 20 and 50% by weight of dry biomass, although some strains can under certain optimally induced conditions accumulate as much as 90% oil yield ratios [77, 78, 95]. Lipid content and growth rates follow an inverse relationship, with lipid accumulation occurring when nutrients are depleted and growth rates are lowered [96–98].

The two most common algae cultivation systems are centred upon open raceway ponds and closed photobioreactors (PBR) each of which have been designed in a variety of operating configurations [73]. The open pond cultivation system is limited by the few strains of microalgae that can be used, the lower efficiency of solar radiation utilisation, and finally lack of temperature control and the associated higher risk of a culture contamination or collapse [73, 94, 99, 100]. The alternative is to grow algae in PBRs, which are closed cultivation systems with tubular or flat-plate reactors consisting of an array of plastic or glass tubes through which the culture flows constantly. The closed cultivation system offers better control of growth parameters, the prevention of evaporation losses, minimised risk of invasion by competing micro-organisms and higher volumetric productivity levels [73]. Despite their advantages, the drawbacks of PBRs, which centre around the likelihood of overheating, scale-up difficulties, capital expenditures; still overshadow their large-scale development and have made their products uncompetitive with petroleum-derived fuels [73, 78]. By solely focussing on the cost of pumping ($0.22/kWh), the production cost of 1 ton of wet algal biomass is estimated to range from $9540.0 (horizontal tubular PBRs) to $227.0 (open raceway ponds) [2] and is therefore not competitive with hydrocarbon fuels [2, 78]. However, since this industry is not mature, there is ample space for optimisation. Increasing oil price further add to this effect.

Currently, algae biodiesel production is 2.5 times as energy intensive as conventional diesel from the United States and nearly equivalent to the high fuel-cycle energy use of oil shale diesel. The major disadvantage inherent in biodiesel production from microalgae is driven by the high energy input. Biodiesel from advanced biomass can only realise its inherent environmental advantages of GHG emissions reduction, once every step of the production chain is fully optimised and decarbonised, which includes the sourcing of all direct energy input in the form of heat and electricity, as well as indirect requirements for fertilizers, transport and building materials, from low-carbon energy sources.

As a priority, countries will need to defossilize primary energy sources used by their electricity grids, as only then can the transport sector move towards low GHG emissions. For example, China operating on a carbon-based electricity and heat grid, would eliminate the inherent environmental advantages of algal biodiesel, while Brazil and France which essentially operate on defossilized electricity grids, have the potential for biodiesel from algae to be a viable alternative to conventional diesel [92].

9 Conclusions

Fuels derived from unconventional oil such as tar sands, coal etc. could provide us with fuel security by mitigating potential shortages, though at a significant environmental cost due to the increased carbon footprint of these fuels. Combined with an increasing demand for transport fuels, these high-carbon fuels would drastically increase emissions from road transport [1]. However, emissions from ICEs can be reduced by using alternative low carbon fuels which emit less CO_2 per unit of energy retrieved than conventional fuels or only marginally more than consumed in their synthesis. Nevertheless, for biofuels to provide truly sustainable transport solutions, the appropriate selection of cultivation sites and the avoidance of problems faced by first-generation biofuels from edible biomass such as land-use change and food versus fuel tradeoffs, is of paramount importance. Second-generation biofuels, fuels synthesised from inedible cellulosic biomass, such as agricultural residues, municipal solid waste or dedicated energy crops, have the potential to be true low carbon fuels. Their effect on food security is minimal and they can be used in a variety of applications, such as cars and aircraft. Yet, second-generation biofuels will not be able to replace the current demand for fossil-based transport fuels given the limited availability of land [1]. Another set of advanced biofuels is derived from algae oil via hydrogenation or transesterification to yield kerosene or diesel-type fuels respectively. Algae are farmed in plastic bags or ponds filled with degraded water and minimise land-use change as well as freshwater issues whilst yielding large amounts of oil per unit area [92]. Nonetheless, biodiesel from advanced biomass can only realise its inherent environmental advantage of reduced GHG emissions, once every step of the production chain is fully optimised and decarbonised. Similarly, agave is attracting attention as potential ethanol feedstock due to its high productivities and sugar content as well as its ability to grow in naturally water-limited environments. Ethanol derived from agave is likely to be superior, or at least comparable, to that from corn, switchgrass and sugarcane in terms of energy and GHG balances, as well as in ethanol output and net GHG offset per unit land area [3]. In addition, biofuel productivity rates per area unit (MJ/ha) vary depending on the biomass feedstock and range from high-yield sources, such as sugarcane, microalgae and cassava to low-yield sources, such as corn, wheat and rapeseed. Ultimately, with the transport fuel industry remaining to be fuelled by hydrocarbons in the foreseeable future, biofuels will need to contribute as well as complement our current fuel mix as part of a larger mission to diversify our global transport system within the mid-to long-term future.

References

1. King DA, Inderwildi OR (2010) Future of mobility roadmaps. SSEE
2. Jorquera O, Kiperstok A, Sales EA, Embiruçu M, Ghirardi ML (2010) Bioresour Technol 101:1406–1413
3. Yan X, Tan DKY, Inderwildi OR, Smith J, King DA (2011) Energy Environ Sci 4:3110–3121
4. Williams AR (2011) Commercial fuels and primary energy graph. Smith School of Enterprise and the Environment, Oxford University, Oxford
5. Erturk M (2011) Renew Sustain Energy Rev 15:2766–2771
6. Schäfer A, Heywood JB, Jacoby HD (2009) Transportation in a climate-constrained world. The MIT Press, Cambridge
7. King CW, Webber ME (2008) Environ Sci Technol 42:7866–7872
8. King CW, Webber ME, Duncan IJ (2010) Energy Policy 38:1157–1167
9. Charpentier AD, Bergerson JA, MacLean HL (2009) Environ Res Lett 4:014005
10. McKellar JM, Charpentier AD, Bergerson JA, MacLean HL (2009) Int J Glob Warming 1:160–178
11. Jansson C, Westerbergh A, Zhang J, Hu X, Sun C (2009) Appl Energy 86:95–99
12. IEA (2009) OECD, Paris
13. Inderwildi OR, King DA (2009) Energy Environ Sci 2:343–346
14. Kalicki JH, Goldwyn DL (2005) Energy and security: toward a new foreign policy strategy. Woodrow Wilson Center Press, Washington
15. Naik S, Goud VV, Rout PK, Dalai AK (2010) Renew Sustain Energy Rev 14:578–597
16. King D (2010) World Economic Forum
17. Inderwildi OR, Jenkins SJ, King DA (2008) Angew Chem–Int Ed 47, 5253
18. Inderwildi OR, King DA, Jenkins SJ (2009) Phys Chem Chem Phys 11:11110–11112
19. van Vliet OPR, Faaij APC, Turkenburg WC (2009) Energy Convers Manag 50:855–876
20. Pickett J, Anderson D, Bowles D, Bridgwater T, Jarvis P, Mortimer N, Poliakoff M, Woods J (2008) The Royal Society, London
21. Gill SS, Tsolakis A, Dearn KD, Rodríguez-Fernández J (2011) Prog Energy Combust Sci 37:503–523
22. Jaramillo P, Griffin WM, Matthews HS (2008) Environ Sci Technol 42:7559–7565
23. Xie X, Wang M, Han J (2011) Environ Sci Technol
24. Höök M, Aleklett K (2010) Int J Energy Res 34:848–864
25. Vallentin D (2008) Energy Policy 36:3198–3211
26. Yan X, Crookes RJ (2010) Prog Energy Combust Sci 36:651–676
27. Lefevre N (2010) Energy Policy 38:1644
28. EIA (2009) US Department of Energy
29. Wu T, Huang Z, Zhang WG, Fang JH, Yin Q (2007) Energy Fuels 21:1908–1914
30. Wang HW, Hao H, Li XH, Zhang K, Ouyang MG (2009) Appl Energy 86:2257–2261
31. Abu-Jrai A, Tsolakis A, Theinnoi K, Cracknell R, Megaritis A, Wyszynski ML, Golunski SE (2006) Energy Fuels 20:2377–2384
32. Soltic P, Edenhauser D, Thurnheer T, Schreiber D, Sankowski A (2009) Fuel 88:1–8
33. Gallagher E (2008) Gallagher Review of the indirect effects of biofuels production

34. Goldemberg J (2007) Science 315:808
35. Fargione J, Hill J, Tilman D, Polasky S, Hawthorne P (2008) Science 319:1235
36. Macedo IC, Seabra JEA, Silva JEAR (2008) Biomass Bioenergy 32:582–595
37. Huo H, Wang M, Bloyd C, Putsche V (2008) Environ Sci Technol 43:750–756
38. Hill J, Nelson E, Tilman D, Polasky S, Tiffany D (2006) Proc Natl Acad Sci 103:11206
39. Kaltschmitt M, Reinhardt G, Stelzer T (1997) Biomass Bioenergy 12:121–134
40. Bernesson S, Nilsson D, Hansson PA (2004) Biomass Bioenergy 26:545–559
41. Thamsiriroj T, Murphy J (2009) Appl Energy 86:595–604
42. Yee KF, Tan KT, Abdullah AZ, Lee KT (2009) Appl Energy 86:S189–S196
43. Farrell AE, Plevin RJ, Turner BT, Jones AD, O'Hare M, Kammen DM (2006) Science 311:506
44. Yan X, Crookes RJ (2009) Renew Sustain Energy Rev 13:2505–2514
45. Searchinger T, Heimlich R, Houghton RA, Dong F, Elobeid A, Fabiosa J, Tokgoz S, Hayes D, Yu TH (2008) Science 319:1238
46. Pimentel D, Marklein A, Toth MA, Karpoff MN, Paul GS, McCormack R, Kyriazis J, Krueger T (2009) Hum Ecol 37:1–12
47. Tilman D (1999) Proc Natl Acad Sci U S A 96:5995
48. Tilman D, Socolow R, Foley JA, Hill J, Larson E, Lynd L, Pacala S, Reilly J, Searchinger T, Somerville C (2009) Science 325:270
49. Rabalais NN, Turner RE, Wiseman WJ (2002) Ann Rev Ecol Syst 33:235–263
50. Mitchell D (2008) Policy Res Work Pap 4682:20
51. King DA, Inderwildi OR Future of mobility roadmaps, SSEE
52. Dawe D (2008) Working papers
53. Ivanic MM, W (2008) World
54. Ravallion M (2009) Challenge 52:55–80
55. Kutas G, Lindberg C, Steenblik R I. I. f. S. D. G. S. (2007) Initiative, Biofuels–at what Cost?: Government support for ethanol and biodiesel in the European union, international institute for sustainable development
56. Agarwal AK (2007) Prog Energy Combust Sci 33:233–271
57. Graham LA, Belisle SL, Baas CL (2008) Atmos Environ 42:4498–4516
58. Corma A, Iborra S, Velty A (2007) Chem Rev 107:2411–2502
59. USDOE (2005) Breaking the biological barriers to cellulosic ethanol: a joint research agenda 7–9
60. Li SZ, Chan-Halbrendt C (2009) Appl Energy 86:S162–S169
61. Perrin R, Vogel K, Schmer M, Mitchell R (2008) BioEnergy Res 1:91–97
62. Balat M, Balat H (2009) Appl Energy 86:2273–2282
63. Nass LL, Pereira PAA, Ellis D (2007)
64. Hira A, De Oliveira LG (2009) Energy policy 37:2450–2456
65. Berg C (2004) Ediciones Le Monde diplomatique
66. Hazell PBR, Pachauri R (2006) I. F. P. R. Institute, A. Vision for Food, t. Environment, Energy and R. Institute, Bioenergy and agriculture: promises and challenges, International Food Policy Research Institute
67. Eickhout B, van den Born GJ, Notenboom J, Van Oorschot M, Ros J, Van Vuuren D, Westhoek H (2008) Netherlands Environmental Assessment Agency, MNP Report, 500143001
68. Edwards R, Szekeres S, Neuwahl F, Mahieu V (2008) Joint Research Centre of the European Commission, March 2008
69. Demirbas MF, Balat M, Balat H (2009) Energy Convers Manag 50:1746–1760
70. Schenk PM, Thomas-Hall SR, Stephens E, Marx UC, Mussgnug JH, Posten C, Kruse O, Hankamer B (2008) Bioenergy Res 1:20–43
71. Batan L, Quinn J, Willson B, Bradley T (2010) Environ Sci Technol 19:235–240
72. Becker EW (1994) Microalgae: biotechnology and microbiology, Cambridge University Press
73. Mata TM, Martins AA, Caetano N (2009) Renewable and Sustainable Energy Reviews

74. Sheehan J, Dunahay T, Benemann J, Roessler P (1998) *National Renewable Energy Laboratory*. Golden, CO 80401:580–24190
75. Lardon L, Helias A, Sialve B, Steyer JP, Bernard O (2009) Environ Sci Technol 43: 6475–6481
76. Campbell PK, Beer T, Batten D (2010) Bioresource Technol 102:50–56
77. Li Y, Horsman M, Wu N, Lan CQ, Dubois-Calero N (2008) Biotechnol Prog 24:815–820
78. Chisti Y (2007) Biotechnol Adv 25:294–306
79. Achten W, Verchot L, Franken YJ, Mathijs E, Singh VP, Aerts R, Muys B (2008) Biomass Bioenergy 32:1063–1084
80. Lapuerta M, Armas O, Rodriguez-Fernandez J (2008) Prog Energy Combust Sci 34: 198–223
81. Mittelbach M, Remschmidt C (2004) Biodiesel: the comprehensive handbook. Martin Mittelbach Graz, Austria
82. Freedman B, Pryde EH, Mounts TL (1984) J Am Oil Chemists' Soc 61:1638–1643
83. Sheehan J, Aden A, Paustian K, Killian K, Brenner J, Walsh M, Nelson R (2003) J Ind Ecol 7:117–146
84. Spatari S, Zhang Y, MacLean HL (2005) Environ Sci Technol 39:9750–9758
85. Schmer M, Vogel KP, Mitchell RB, Perrin RK (2008) Proc Natl Acad Sci 105:464
86. Chester M, Martin E (2009) Environ Sci Technol 43:5183
87. Kalogo Y, Habibi S, MacLean HL, Joshi SV (2007) Environ Sci Technol 41:35–41
88. Ou X, Zhang X, Chang S, Guo Q (2009) Appl Energy 86:S197–S208
89. Lin Y, Tanaka S (2006) Appl Microb Biotechnol 69:627–642
90. Campbell JE, Lobell DB, Genova RC, Field CB (2008) Environ Sci Technol 42:5791–5794
91. Gerbens-Leenes W, Hoekstra AY, Van der Meer TH (2009) Proc Natl Acad Sci 106:10219
92. Shirvani T, Yan X, Inderwildi OR, Edwards PP, King DA (2011) Energy Environ Sci 4:3773
93. Sheehan J, Camobreco V, Duffield J, Graboski M, Shapouri H (1998) Life cycle inventory of biodiesel and petroleum diesel for use in an urban busFinal report. National Renewable Energy Lab, Golden
94. Richmond A (2007) Handbook of microalgal culture: biotechnology and applied phycology. Blackwell Science, Oxford
95. Li Y, Horsman M, Wang B, Wu N, Lan CQ (2008) Appl Microb Biotechnol 81:629–636
96. Ratledge C (2002) Biochem Soc Trans 30:1047–1050
97. Wang B, Li Y, Wu N, Lan CQ (2008) Appl Microb Biotechnol 79:707–718
98. Khozin-Goldberg I, Cohen Z (2006) Phytochemistry 67:696–701
99. Carvalho A, Malcata FX (2006) Biotechnol Prog 22:1490–1506
100. Pulz O (2001) Appl Microb Biotechnol 57:287–293

Electricity, Mobility and the Neglected Indirect Emissions

Alexander R. Williams

Abstract Carbon dioxide (CO_2) emissions from passenger cars represent an important and growing contributor to climate change. Increasing the proportion of electric vehicles (EVs) in passenger car fleets could help to reduce these emissions, but their ability to do this depends on the fuel mix used in generating the electricity that energises EVs. This chapter analyses the indirect well-to-wheels CO_2 emissions from EVs when run in a range of countries and compares these to well-to-wheels emissions data for a selection of internal combustion engine vehicles (ICEVs) and hybrid electric vehicles (HEVs) (Holdway, Energy Environ Sci 3(12):1825−1832).

Carbon dioxide (CO_2) emissions from passenger cars represent an important and growing contributor to climate change. Increasing the proportion of electric vehicles (EVs) in passenger car fleets could help to reduce these emissions, but their ability to do this depends on the fuel mix used in generating the electricity that energises EVs. This chapter analyses the indirect well-to-wheels CO_2 emissions from EVs when run in a range of countries and compares these to well-to-wheels emissions data for a selection of internal combustion engine vehicles (ICEVs) and hybrid electric vehicles (HEVs) [1].

1 Background

The transport sector is the second largest emitter of greenhouse gases (GHGs) and its decarbonisation is key to achieving emissions reduction targets [2]. The dependence of the transport sector on fossil fuels, namely crude oil, has led to two

A. R. Williams (✉)
Smith School of Enterprise and the Environment, University of Oxford,
Hayes House, 75 George Street, Oxford, OX1 2BQ, UK
e-mail: alex.williams@smithschool.ox.ac.uk

O. Inderwildi and Sir David King (eds.), *Energy, Transport, & the Environment*,
DOI: 10.1007/978-1-4471-2717-8_7, © Springer-Verlag London 2012

main problems: conventional crude oil reserves are declining, and the increase in greenhouse gas emissions is destabilising our climate. The central challenge we are facing is to decrease our dependence on crude oil and control GHG concentrations without reducing human mobility [3].

As a global community, our response to this has been to develop synthetic hydrocarbon fuels and biofuels that can be used in conventional automobiles. Alternative drivetrains and propulsion technologies can also be a replacement for the internal combustion engine (ICE), most notably, electric motors powered by hydrogen fuel cells or batteries. It is the latter that this chapter analyses in more detail.

2 The Electric Vehicle

An electric vehicle is one that uses an electric motor as its primary method of propulsion. In existence since the nineteenth century, electric vehicles were the preferred method of transport before gasoline and the internal combustion engine started to dominate the motor car industry and became more common [4].

In 1899, 1,575 electric vehicles were sold in the US, compared with 936 gasoline cars. Over time their general-purpose commercial use reduced to specialist roles as platform trucks, forklift trucks, tow tractors and urban delivery vehicles, such as the iconic British milk float; for most of the twentieth century, the UK was the world's largest user of electric road vehicles [5].

There are many types of electric vehicle, spanning all modes of transport. Some are more common and promising than others. The main types are as follows:

- *Road* EVs include battery electric vehicles (BEVs), electric bicycles, and specialist commercial vehicles such as forklift trucks and milkfloats;
- *Hybrid* electric vehicles (HEVs) combine a conventional (ICE-based) powertrain with electric propulsion, usually as a passenger vehicle;
- *Railborne* EVs include electric locomotives, tramways, light rail systems, and rapid transit systems;
- *Airborne* EVs mainly include unmanned aerial vehicles;
- *Seaborne* EVs include submarines using electric motor driven propellers–where the batteries are often charged by diesel engines, although a growing number use a nuclear power source to drive the motor;
- *Spaceborne* EVs include a minor share of spacecraft such as the electrostatic ion thruster, the arcjet rocket, and the hall effect ion thruster. All of which are powered by either batteries, solar panels, or nuclear power (e.g. voyager spacecraft).

The variety of electric vehicles currently used across the world demonstrates their importance as a form of modern transport.

2.1 How do they Work?

The use of electricity to power transport hinges on providing a method for the vehicle to access power, either through a permanent connection to the transmission grid via powerlines, or by converting it into transportable forms of energy such as chemical or potential energy.

In the case of large electric transport vehicles such as trains and rapid transit, EVs are powered by static sources of electricity which are connected directly to the vehicle along power lines (overhead or tracks). This method also allows the use of regenerative braking which allows the motors to be used as brakes and themselves become generators of electricity which is then fed back into the power lines. These systems are provided by rotary electric motors. It is also possible to use linear motors (such as those used in Maglev trains) which allow the vehicle to float above rails by magnetic levitation.

In the case of small electric transport vehicles, EVs are powered by portable sources of electricity such as an internal battery which is recharged or replaced when discharged. These BEVs rely on a chemical reaction, which converts chemical energy into electrical charge. There are a number of different battery types with a range of chemical compositions, each with advantages and disadvantages. (See Table 1 or Chaps. 11 and 12 for discussion).

The batteries can be recharged via mains plug-in, replaced at a local replacement garage, or via an on-board ICE (as in a hybrid electric vehicle HEV). The HEV is an intermediate step between converting from a purely ICE transport system to a purely electrical transport system. HEV systems use an ICE to propel the vehicle, and to charge a BEV system when not required for drive (see Chap. 11). Other electricity storage technologies include static energy stored in electric double layer capacitors, and kinetic energy stored from flywheels, although these are much less common than battery methods.

At present the most common forms of electric road vehicles tend to be BEVs and HEVs, the environmental impact of which is further analysed later on in the chapter and in Chaps. 10 and 11 of this book. Large electric transport vehicles are used globally and their environmental effects are well studied. (See Chap. 29).

2.2 The BEV: Pros and Cons

Although progress has been made in recent years on some of the challenges facing BEVs, hurdles still remain. Their relatively low use of fuel and high operating efficiency means that BEVs are cheaper to operate than ICEVs and HEVs. The price of electricity versus that of petrol and diesel varies greatly by country, but for the UK in 2011,[1] the price of residential electricity was 3.19p/MJ compared to the

---

[1] Data taken on 20th June 2011.

Table 1 A comparison of chemical battery technologies for electric vehicles [3, 6–11]

Composition	Energy density (Wh/kg)	Peak power (W/kg)	Energy efficiency (%)	Life cycle	Cost (USD/kWh)	LCA score	Status	Notes
Lead–acid	35–50	150–400	>80	500–1000	100	503	Mature, proven technology	Little scope for improvement, low performance
Nickel–cadmium	50–60	80–150	75	800–1350	>550	544	Mature technology	Toxicity and safety issues
Nickel–metal hydride	70–95	200–300	70	750–1350	530	491	Proven technology	Safe, longevity in calendar and lifecycle
Sodium–nickel chloride	90–120	130–160	80–95	1000–1200	240	234	In development	High operating temperature, requires pre-heating to operate
Lithium-ion	80–130	200–300	>90	1000–3200	>800	278	In development	Safety concerns, complicated management, short calendar life

price of regular unleaded petrol at 3.91p/MJ.[2] Despite this, the upfront cost of purchase is still high due to the battery cost and the small scale of production[3] [12]. This high cost also includes a hidden carbon cost associated with manufacturing, shipping, and disposal. These are known as 'cradle-to-grave' emissions and may be higher for EVs than for ICEVs in some cases, mainly due to the energy required to manufacture the batteries, but also due to the lack of manufacturing process optimisation which often comes with new products [13].

Battery charging can often be done overnight, but BEVs take significantly longer to refuel than ICEVs (hours compared to minutes) [14–16], and for those charging during the daytime, the low number of charging points, lack of infrastructure to support EVs and current grid capacity, makes this difficult. Although there are notable exceptions and continuous improvements, to date most BEVs have had limited range compared to conventional ICEVs, due to restrictions in battery technology. This is shown in Table 1 and discussed further in Chap. 11.

Most significantly, however, while BEVs may help to decrease dependence on foreign oil, they have generally not decreased dependence on fossil fuels [17]. BEVs can become *de facto* zero emissions vehicles if they use electricity generated entirely from renewable sources such as hydro, solar, wind, or nuclear power [18]. The potential of BEVs to reduce greenhouse gas emissions, however, depends entirely on the fuel mix used in generating the electricity that charges the vehicles' batteries, which can vary widely from country to country, and even within countries. It is this issue that we investigate in detail in this chapter.

The environmental impact of running three selected battery electric vehicles in the fifteen largest economies is investigated as a response to growing emissions from passenger car fleets and their contribution to global GHG emissions as previously noted.

3 Indirect Emissions from Electric Vehicles

A number of recent studies have looked at emissions associated with plug-in hybrid electric vehicles (PHEVs) [19–21]. Most studies on the potential of EVs to reduce greenhouse gas emissions compared to conventional vehicle technologies are at least a decade old [22–25]. Two exceptions are Jacobson [26] and Arar [27].

In a review of options for powering EVs and fuel cell vehicles, Jacobson states that "The US could theoretically replace all 2007 on-road vehicles with BEVs powered by 73,000–144,000 5-MW wind turbines, less than the 300,000 airplanes the US produced during World War II." This study assumes that wind generated electricity is completely carbon neutral and that the plug-to-wheels efficiency

---

[2] At time of writing, regular unleaded gasoline price was 136.9 pence per litre; approximate electricity price from Scottish and Southern Energy was 11.50 pence per kWh.

[3] In the UK in 2011, the Nissan Leaf cost £25,990.

of BEVs (75–86%) is much greater than the average tank-to-wheels efficiency of fossil fuel vehicles (17%). Arar, using 2006 figures, finds that an instantaneous conversion of the US fleet of passenger cars and light-duty trucks to battery power could bring a CO_2 emission reduction of 58%. Arar also attempts to estimate fleet emissions for 2020 assuming a 10% uptake in EVs per year from 2011 to 2020 but using the present fuel mix. He finds that the 2020 fleet would emit 36% less CO_2 than if the conversion to EVs had not taken place.

Our previous study [1] on electricity usage for a selection of BEVs and the associated well-to-wheel CO_2 emissions showed that increasing the proportion of BEVs in passenger car fleets can help to greatly reduce the emissions associated with transport. The study looked specifically at three battery EVs which, at the time of writing (April 2010), were the only multi- passenger EVs available for sale in more than a single country: the Tesla Roadster, the TH!NK City, and the REVAi (marketed in the UK as the G-Wiz i).

For this chapter we have expanded on the previous study with an updated selection of EVs, HEVs, and ICEVs for twelve further countries. It investigates the correlation between the carbon intensity of electricty generation and the total emissions from the selected EVs.

Figures 1 and 2 show the well-to-wheels CO_2 emissions for three major commercial electric road vehicles; the Nissan Leaf, the TH!NK City and the Mitsubishi i MiEV—running in the world's fifteen largest economies ranked by GDP (PPP) according to the International Monetary Fund [28].

3.1 Methodology

Here we elaborate on the methodology used to calculate the results given in the figures. Data used are given in the tables compiled in Annexure 1. The average CO_2 emissions from electricity generation in each country (Table 2) were calculated as follows:

$$\text{Emissions (g/kWh)} = \frac{CO_2 \text{ produced (g)}}{\text{Net electricity generated (kWh)}}$$

Net electricity generation is defined here as electricity generated by major power producers—that is, companies whose main business is electricity generation—since national CO_2 emissions figures for electricity generation refer only to major power producers. The US Energy Information Administration's *Annual Energy Review* expressly excludes plants with a generator nameplate capacity less than 1 MW [29].

The well-to-wheels CO_2 emissions for the three EVs considered in this study are calculated by summing well-to-power-plant emissions and power-plant-to-wheels emissions.

Well-to-power-plant CO_2 emissions—analogous to well-to-tank for ICEVs— are defined here as emissions from primary fuel extraction (e.g. coal, oil, gas) and

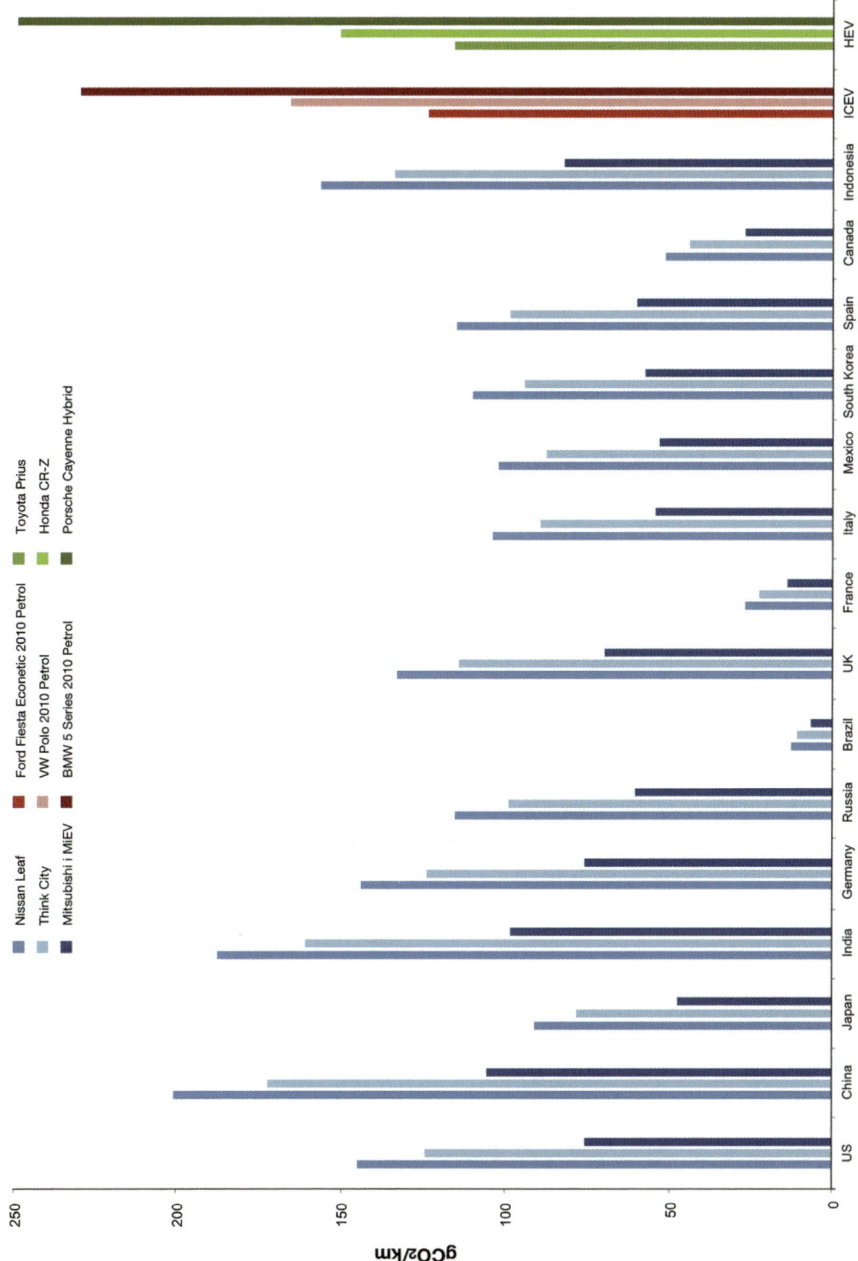

Fig. 1 Total CO_2 emissions for three commercially available electric vehicles in the world's fifteen largest economies according to the IMF. Ranked in order of GDP purchasing power parity. Data is given in Table 9

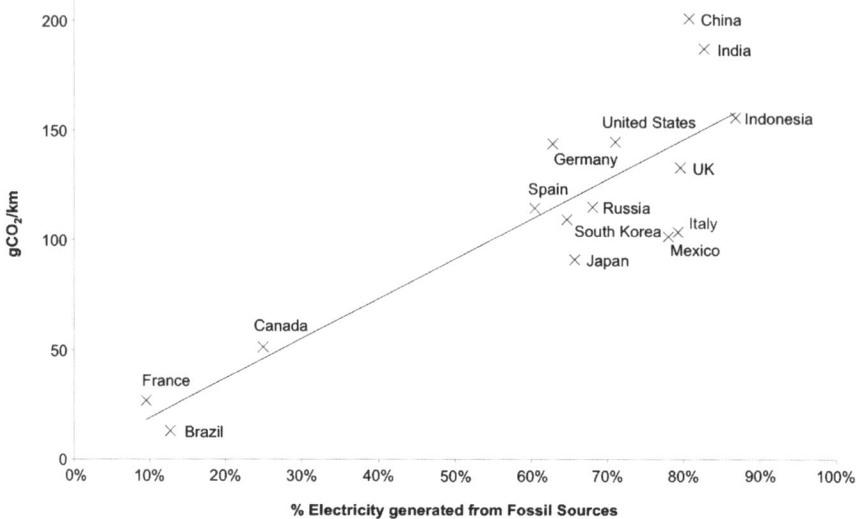

Fig. 2 Total CO$_2$ emissions versus the percentage of fossil-based electricity generation, by country

Table 2 Average CO$_2$ emissions from electricity generation (gCO$_2$/kWh) and the percentage of fossil fuel in electricity generation, by country

Country	Emissions[a]	Fossil fuel mix[b] (%)
US	611	71
China	868	81
Japan	365	66
India	805	83
Germany	612	63
Russia	484	68
Brazil	50	13
UK	557	80
France	88	10
Italy	429	79
Mexico	421	78
South Korea	444	65
Spain	487	60
Canada	213	25
Indonesia	662	87

[a] Average CO$_2$ emissions from electricity generation taken from CARMA database http://carma.org/
[b] Fossil fuel mix data taken from IEA World Statistics for 2010

all intermediate steps up to delivery to the power-plant for use in electricity generation (detailed in Table 3). Well-to-power-plant emissions for each EV in each country were calculated from fuel efficiency data (Table 4), the breakdown of total net electricity generation (taken from IEA world energy statistics, 2010), and the well-to-power-plant emissions for each type of fuel used in electricity

Table 3 Well-to-power-plant CO_2 emissions by type of fuel used in electricity generation (gCO_2/kWh)[a]

	Coal [48][d]	Oil [49][c]	Natural gas [48][e]	Nuclear [26][b]
Range	85–135	40–110	48–100	9–70
Mean	110	75	74	40

[a] The well-to-power-plant CO_2 emissions for hydro power (1.9 g CO_2/kWh) [50] are negligible and have been excluded. This figure does not include methane emissions, which may occur in a flooded reservoir from the anaerobic decomposition of biomass [49]

[b] For nuclear, this consists of fuel conversion, enrichment, and fabrication. Enrichment produces 95% of the CO_2 emissions from nuclear fuel processing. Emissions can vary greatly depending on the specific enrichment process employed

[c] For oil, this includes exploration, extraction, transportation, and refinement, and assumes residual fuel oil is used

[d] For coal, this includes mining and transport

[e] For natural gas, this includes gas processing, venting wells, pipeline operation (mainly compression), and system leakage in transportation and handling [49]

Table 4 Vehicle efficiency: amount of electricity required to operate EVs (kWh/km)

Electric vehicle (EV)	Power consumption[a]
Nissan Leaf	0.21
Mitsubishi i MiEV	0.11
Th!nk City	0.18

[a] All vehicle data are from www.nissan.com, www.think.no, www.mitsubishi.com

generation (Table 3). Power-plant-to-wheels CO_2 emissions—analogous to tank-to-wheels for ICEVs—are defined as emissions from generation of electricity at the power-plant to delivery to the EV. Power-plant-to-wheels emissions for each EV in each country were calculated from the fuel efficiency for each EV (Table 4) and the average direct CO_2 emissions from electricity generation (Table 2).

The sum of the well-to-power-plant (WtPP) and the power-plant-to-wheels (PPtW) CO_2 emissions for each EV in each country (C) is the well-to-wheels (WtW) CO_2 emissions (Table 5):

$$WtW_{EV}[\text{g/km}] = Eff_{EV}[\text{MJ/km}] \cdot (WtPP_C + PPtW_{EV})[\text{g/km}]$$

Analogously, well-to-wheels CO_2 emissions for the ICEVs and HEVs (Table 6) are the sum of well-to-tank emissions, defined as emissions from primary fuel extraction to delivery to vehicle fuel tank, and tank-to-wheels emissions, defined as tailpipe emissions from combustion of fuel.

The tank-to-wheels data are from the UK Government's Vehicle Certification Agency; the well-to-tank figures were calculated from fuel efficiency and the literature value for oil in Table 3. The range of well-to-tank emissions (25–62 gCO_2/kWh) are comparable to those from Silva et al. [30], who report 51.1 gCO_2/kWh for well-to-tank greenhouse gas emissions for diesel.

Table 5 Average well-to-wheels CO_2 emissions for EVs by country (gCO_2/km) (Tables 2–4)

Country	Nissan Leaf	Think City	Mitsubishi i MiEV
US	145	124	76
China	201	172	105
Japan	91	78	48
India	187	160	98
Germany	144	123	75
Russia	115	99	60
Brazil	13	11	7
UK	133	114	70
France	27	23	14
Italy	104	89	54
Mexico	102	87	53
South Korea	109	94	57
Spain	114	98	60
Canada	51	44	27
Indonesia	156	133	82
Average	*113*	*97*	*59*

Table 6 Data for well-to-wheels CO_2 emissions for a selection of commercial ICEVs and HEVs (gCO_2/km)

ICEVs[b]	Mpg[a]	Tailpipe Emissions[a]	Well-to-Tank[c]	Total
Fiesta Econetic 2010 Petrol	76.3	98	25	123
VW Polo 2010 Diesel	72.4	102	26	128
VW Polo 2010 Petrol	51.4	128	37	165
BMW 5 Series 2010 Petrol 2.3	37.2	178	51	229
BMW 5 Series 2010 Diesel 2.0	57.6	129	33	162
Porsche Boxter 2010 Petrol 2.8	30.1	221	62	283
HEVs[b]				
Toyota Prius	72.4	89	26	115
Honda CR-Z	56.5	117	33	150
Porsche Cayenne Hybrid	34.4	193	55	248

[a] For which fuel consumption (miles per gallon) and tailpipe emissions (gCO_2/km) values are taken from www.vcacarfueldata.org.uk (UK Government's Vehicle Certification Agency)
[b] The 2010 model year is used for all
[c] Well-to-Tank CO_2 emissions are calculated from the fuel efficiency (Mpg converted to km/L), the energy content of gasoline (32 MJ/L), and the literature value for emissions from oil refinement and extraction found in Table 6.4 (75 gCO_2/kWh converted to 20.8 gCO_2/MJ)

The well-to-wheels CO_2 emissions of the hypothetical EV fleet in each country (Table 7) is the average of each row in Table 5. The average well-to-wheels CO_2 emissions for the existing passenger car fleets in each country (Table 8) were calculated using annual country specific values for fuel consumption, the specific energy content of gasoline and diesel, and total distance driven.

Table 7 Well-to-wheels CO_2 emissions for hypothetical EV fleets, by country (gCO$_2$/km)

Country	Hypothetical EV fleet[a]
US	115
China	159
Japan	72
India	149
Germany	114
Russia	91
Brazil	10
UK	105
France	21
Italy	82
Mexico	81
South Korea	87
Spain	91
Canada	41
Indonesia	124
Average	*89*

[a] The figures for the hypothetical EV fleets are the average of the well-to-wheels CO_2 emissions of the Nissan Leaf, TH!NK City, and Mitsubishi i MiEV

Table 8 Average well-to-wheels CO_2 emissions for hypothetical EV fleets and existing passenger car fleets in the US, UK, and France (gCO$_2$/km)

	EV fleet[a]	Existing fleet[b]
US	115	361 [29, 36, 38, 39, 51]
UK	105	221 [32, 51–53]
France	21	228 [32, 37, 51, 54, 55]

[a] Taken from Table 7
[b] Existing fleet figures are calculated from annual values for fuel consumption, the specific energy content of gasoline and diesel, and total distance driven

The amount of electricity generation required to run each of the three EVs in each of the three countries considered in this analysis is shown in Table 4.[4] The figures are inclusive of electricity transmission and distribution losses[5] and battery charging inefficiency.

Figure 1 shows how the carbon intensity of electricity generation for the fifteen largest economies varies significantly, and demonstrates the dominance of fossil fuel generated power in these major economies. Figure 2 shows that there is a direct correlation between the well-to-wheels CO_2 emissions and the percentage of fossil fuel in the energy mix, with variations above and below the trend line due to difference between the carbon intensity of coal versus gas combustion. For example, at a fossil share of approximately 80%, there is a notable difference in the emissions in China and India compared with that of Italy and Mexico. The well-to-power-plant emissions from electricity generation are 110 and 74 gCO$_2$/km for coal and gas, respectively.

---

[4] All data in this study are for 2006, the last year for which all data were available.

[5] High voltage electricity transmission loss is approximately 7% [31].

3.2 EV Analysis

The difference in well-to-wheels CO_2 emissions between the three chosen electric vehicles is due to vehicle efficiency, or how much electricity is required to drive a unit of distance (kWh/km), given in Table 4.

As these figures demonstrate, the fuel mix of a nation's electricity grid is an overwhelming factor in determining the well-to-wheels CO_2 emissions from electric vehicles. In the US, China, India, Germany, UK, Spain, and Indonesia, the emissions for the Nissan Leaf are higher than those of the most efficient small ICEVs, with the Think City close behind. France, however, with 78% of its electricity coming from nuclear power, would find substantial emissions reductions from passenger cars if it were to replace a large part of its fleet with EVs (see *Fleets*). The well-to-power-plant CO_2 emissions associated with nuclear power generation are less than 50% of the average of coal, oil, and natural gas (Table 3), and the generation of electricity from nuclear fission produces no CO_2 emissions.

3.3 ICEVs and HEVs

The figures from Table 5 can be compared to well-to-wheels CO_2 emissions data for a selection of ICEVs and HEVs given in Table 6. This comparison underlines the importance of looking at vehicle CO_2 emissions on a well-to-wheels basis.

On a well-to-wheels basis, the lowest emitting ICEV or HEV model, the Toyota Prius (115 gCO_2/km), produces lower emissions than those associated with the Nissan Leaf and TH!NK City when run in the US, China, India, Germany, Indonesia, and UK (Nissan Leaf only), and the Ford Fiesta (123 gCO_2/km) is a close second. The EVs running in all countries are associated with lower emissions than the remaining ICEVs and HEVs shown in Table 6, except the Nissan Leaf running in China and India. While the Nissan Leaf's emissions are higher than those of the other two EVs, it fares well against comparable ICEVs such as the petrol BMW 5 Series and the Porsche Boxster, both of which have well-to-wheels emissions approximately twice that of the Nissan Leaf average of 113 gCO_2/km.

The well-to-tank CO_2 emissions for the ICEVs in Table 6 range from 25 to 60 gCO_2/km. The analogous well-to-power-plant emissions for the three EVs ranges from 11 to 88 gCO_2/km. The latter range is due to the variation in the share of fossil fuels in electricity generation in each country, which is analogous to the variation in fuel efficiency (mpg) seen in the selection of ICEVs.

Even though the average efficiency of electricity generation and supply to end users in the US, the UK, and France is only 30, 34, and 37%, respectively [29, 32, 33]—about the same as an efficient diesel engine [34]—these EVs lower use of fuel means their emissions are lower than comparable ICEVs or HEVs. The next section examines a hypothetical move to replace the current ICEV fleet in each country with EVs.

3.4 Fleets

The three EVs under analysis differ in size and power, and the average of their emissions can be used to represent those of a hypothetical EV fleet. If all passenger cars in each country were replaced with EVs with the average electricity generation requirements of the Nissan Leaf, Think City, and the Mitsubishi I MiEV (the three EVs considered in this analysis), the fleet well-to-wheels CO_2 emissions for each country would be as found in Table 7. (The results are based on the average CO_2 emissions from electricity generation from Table 2, which may change by the time electric vehicles reach significant fleet penetration.)

Table 8 also shows the average well-to-wheels CO_2 emissions for the existing passenger car fleets in three selected countries, the US, UK, and France. Compared to the existing fleets, the hypothetical EV fleets (Table 7) would produce 68% less CO_2 emissions in the US, 52% less in the UK, and 91% less in France.

The French fleet figures are lower than that of the US because of their greater use of diesel vehicles, whose engines are more efficient than gasoline engines; the higher use of manual transmissions compared to automatic; the smaller average car size; and the traditionally stricter fuel efficiency standards [35–37]. Even considering the 2006 model year only, the US well-to-wheels CO_2 emissions would still be higher, by 133 gCO_2/km, than those of the entire fleets in the UK and France [29, 38–41]. The difference would be even more marked if light-duty trucks were included. Light-duty trucks, many of which are used as passenger cars and include SUVs, pickup trucks, and minivans, make up 42% of all light-duty vehicles in the US, while accounting for only 10% in the UK and 16% in France [35, 37, 39].

3.5 The Challenges of Fleet Replacement

Moving a large part of the ICEV fleet to EVs would require a number of considerations. First, it may be difficult to get consumers to move to smaller cars such as the Th!nk City, particularly in the US, where, as mentioned above, SUVs, pickup trucks, and minivans make up two-fifths of the light-duty vehicle fleet. Second, EVs are still more expensive than comparable ICEVs, despite tax credits and other incentives (see Annexure 2). Third, the average vehicle lifetime in the US (for example) is about 15 years [42], meaning that major fleet penetration, under the best circumstances, would take many years [27]. Fourth, emissions reductions from taking ICEVs off the roads would be partially offset by increased emissions from power-plants, although controlling emissions from a few thousand power-plants may be easier than controlling emissions from millions of tailpipes. Fifth, a charging infrastructure and concomitant government policies would be required.

Finally, if the entire fleet in each country were replaced with EVs with the average electricity demand of the Nissan Leaf, Think City, and the Mitsubishi i MiEV, the net electricity consumption that would occur in each country would increase by approximately 10–20% [1]. However, since EVs are usually charged at night when demand is lower, and most EVs would not have to be charged, or charged fully, every night for average daily usage, the increase in demand could potentially be met in part by maintaining daytime electricity production levels overnight. In the UK, for example, where charging the hypothetical EV fleet would see a greater relative increase in electricity demand than in the US or France, half of the additional demand could potentially be met in this way, without the need for additional electricity generation infrastructure.

Furthermore, charging EVs in large numbers in this way, in off-peak hours, could lead to improved load shapes, reducing daily variability in electricity demand [19, 43]. The extent of this improvement may increase with the flexibility offered by smart electricity grids [44] and the introduction of EV batteries with shorter charging times. EVs could also potentially return to the grid excess electricity stored in their batteries during times of peak load and replace it with electricity in off-peak hours [45].

3.6 Overall Comparison

The well-to-wheels CO_2 emissions for EVs, a selection of ICEVs and HEVs, and the hypothetical EV fleets are compared by country, in Table 9. This table demonstrates that EVs emit less CO_2 than comparable ICEVs. Nevertheless, the data also underline the importance of the fuel mix in the electricity grid: the more electricity generation from nuclear and renewable sources, the lower the CO_2 emissions associated with EV use. In countries with particularly carbon intensive electricity grids, driving small, efficient ICEVs or HEVs is by far the best option in terms of emission reduction.

The data hide regional differences within each country, however. In the US, for example, in twelve states more than 70% of in-state electricity generation comes from the combustion of coal, while in four states more than 70% comes from nuclear and renewable sources.[6,7] At one extreme is North Dakota, where 94% of in-state electricity generation comes from coal, and emissions average 1012 gCO_2/kWh.[8] At the other extreme are states like Idaho, which generates 84% of its in-state electricity from hydropower, resulting in in-state electricity generation emissions of just 65 gCO_2/kWh. In the UK, 96% of electricity generated by major power producers in Northern Ireland comes from fossil fuels, while 37% in Scotland is from nuclear and renewable sources [32]. In France, 99% of electricity generation in the region of Île de France is from conventional thermal combustion, while in eight other regions more than 95% of generation is from nuclear and renewable sources [46].

---

[6] www.eia.doe.gov/cneaf/electricity/epa/generation_state.xls

[7] These values are indicative only, as electricity in much of the US is generated and transmitted on a regional basis, in regions that may span several states and may not follow state boundaries.

[8] www.eia.doe.gov/cneaf/electricity/epa/generation_state.xls and www.eia.doe.gov/cneaf/electricity/epa/emission_state.xls

Table 9 Comparison of well-to-wheels CO$_2$ emissions for EVs, hypothetical EV fleets and a selection of current ICEVs and HEVs, by country (gCO$_2$/km)

US	China	Japan	India	Germany	Russia	Brazil	UK	France	Italy	Mexico	South Korea	Spain	Canada	Indonesia	Average
76	105	48	98	75	60	7	70	14	54	53	57	60	27	82	59
115	115	72	115	114	91	10	105	21	82	81	87	91	41	115	89
115	123	78	123	115	99	11	114	23	89	87	94	98	44	123	97
123	128	91	128	123	115	13	115	27	104	102	109	114	51	124	113
124	150	115	149	123	115	115	115	115	115	115	115	115	115	128	115
128	159	123	150	128	123	123	128	123	123	123	123	123	123	133	123
145	162	128	160	144	128	128	133	128	128	128	128	128	128	150	128
150	165	150	162	150	150	150	150	150	150	150	150	150	150	156	150
162	172	162	165	162	162	162	162	162	162	162	162	162	162	162	162
165	201	165	187	165	165	165	165	165	165	165	165	165	165	165	165
229	229	229	229	229	229	229	229	229	229	229	229	229	229	229	229
248	248	248	248	248	248	248	248	248	248	248	248	248	248	248	248
283	283	283	283	283	283	283	283	283	283	283	283	283	283	283	283

EVs (blue)
HEVs (green)

Mitsubishi I MiEV
Toyota Prius (115)

Think City
Honda CR-Z (150)

Nissan Leaf
Porsche Cayenne Hybrid (248)

EV fleet

ICEVs: Fiesta Econetic Petrol (123), VW Polo 2010 Diesel (128), VW Polo 2010 Petrol (165), BMW 5 Series 2010 Petrol 2.3 (229), BMW 5 Series 2010 Diesel 2.0 (162), Porsche Boxter 2010 Petrol 2.8 (283)

Unshaded values correspond to ICEV well-to-wheels CO$_2$ emissions as given in brackets above.

Table 10 Average well-to-wheels CO_2 emissions for EVs (gCO_2/km) if using electricity generated from a single fuel in the UK. (Calculated from Tables 2, 3, 4)[a]

	Nissan Leaf	TH!NK City	Mitsubishi i MiEV	Hypothetical EV fleet
Coal	215	184	112	170
Oil	156	134	82	124
Natural gas	102	87	53	81
Nuclear	8	7	4	7

[a] Using CO_2 emissions estimates for electricity generation in the UK from [32]: 912 g/kWh for coal, 670 g/kWh for oil, and 412 g/kWh for natural gas

If the EVs in this study were to run in states or regions where one fuel predominates, the results would be similar to Table 10, which shows CO_2 emissions for the three EVs if running on just one fuel. These figures compare to well-to-wheels CO_2 emissions of 221 g/km for the existing UK passenger fleet in the 2006 model year (see Table 8). Thus, even though the EVs may be running on electricity generated entirely from coal, and even though the efficiency of electricity generation and supply to end users in the US, the UK, and France averages only 34% (see *ICEVs and HEVs*), the EVs' superior efficiency, means they have lower well-to-wheels emissions than the average ICEV.

Changing the fuel mix in the electricity sector, however, is difficult due to extremely long lead times from planning to operation, combined with plant lifetimes of up to 50 years. Jacobson suggests that nuclear, tidal, wave, and hydroelectric may not be the best options due to these long lead times. This delay produces "opportunity cost emissions" which can be significant when compared to technologies with the least delay: solar-photovoltaics, concentrated solar, wind, closely followed by geothermal [26].

Despite the environmental advantages mentioned above, the number of EVs on the road remains small, most likely due to cost and limited vehicle range.[9] In 2009, a mere 55 new EVs were sold in the UK according to the Society of Motor Manufacturing Traders (SMMT). Vehicle cost remains high at these low production volumes, although those who cannot yet afford an EV would do well to drive a fuel-efficient ICEV or HEV like those shown in Table 6. Continued improvement in ICEV technology is essential, as ICEVs will continue to dominate vehicle sales for the foreseeable future.

4 Conclusions

Analysing the indirect well-to-wheels CO_2 emissions from three production EVs has shown that EVs can reduce well-to-wheels CO_2 emissions over the existing fleet by more than 90%. This suggests that EVs should be particularly promoted

---

[9] The Nissan Leaf has a range of 175 km (109 mi) on the New European Driving Cycle. (www.nissan.co.uk).

where electricity generation is the least carbon intensive—as has been done, for example, in California and France [47]—as the potential for EVs to reduce CO_2 emissions depends on the fuel mix used in generating the electricity that powers them. The electricity grid must be decarbonised if EVs' full potential to reduce CO_2 emissions is to be realised, and should go hand-in-hand with the introduction of EVs. Decisions on building new electricity generation capacity in the coming years will affect the carbon intensity level of the grid for decades to come and affect the potential for EVs to greatly reduce CO_2 emissions.

4.1 The Wider Context

In the wider context of low carbon mobility, this chapter demonstrates the importance of lifecycle analysis as a tool to reveal the current environmental burden of large scale EV deployment. More importantly, however, the chapter shows that solving the mobility challenge needs to take the energy sector hand-in-hand with the transport sector. This hand should be further extended to other factors that have equal bearing on decarbonising transport, such as urban design and lifestyle choices. Mobility is a complex systemic challenge with multiple roots, and if electricity is to be one of our key approaches to reducing transport emissions, then we need to address the fact that the energy sector makes up a worrying 28% of global GHG emissions—6% more than transport! [3] If we neglect this, then electrifying any transport mode, not just passenger cars, will just be another techno-fix that just shifts the demand for energy to a different location. In the long run this will never reduce GHG emissions or our reliance on fossil fuels.

Annexure 1: Selected EV Descriptions

Nissan Leaf

The Nissan Leaf (also formatted "LEAF" as a backronym for leading, environmentally friendly, affordable, family car) is a five-door mid-size hatchback electric car manufactured by Nissan and introduced in Japan and the US in December 2010. The U.S. Environmental Protection Agency determined the range to be 117 kilometres (73 mi), with an energy consumption of 765 kJ/km (34 kWh per 100 miles) and rated the Leaf's combined fuel economy at 99 miles per gallon gasoline equivalent. The Leaf has a range of 175 km (109 mi) on the New European driving cycle.

Price (as of June 2011) £25,990*

Th!nk City

The Th!nk City is a small two-seater or 2 + 2-seater highway-capable electric car produced by Think Global and production partner Valmet Automotive, with a top speed of 110 kilometres per hour (68 mph) and an all-electric range of 160 kilometres (99 mi) on a full charge.

The Th!nk City is sold in Norway, the Netherlands, Spain, France, Austria, Switzerland, and Finland. As of December 2010 it is one of the four crash-tested, mass-produced, and highway-certified electric cars in the world.

Price (when released in UK, expected end-2011) £19,000*

Mitsubishi i MiEV

The Mitsubishi i MiEV (MiEV is an acronym for Mitsubishi innovative electric vehicle) is a five-door hatchback electric car produced by Mitsubishi Motors. According to the manufacturer, the i MiEV all-electric range is 100 miles (160 km) on the Japanese test cycle and 75 miles (121 km) on the US cycle. The i MiEV was launched for fleet customers in Japan in July 2009, and on April 1, 2010 for the wider public.

Price (as of June 2011) £23,990*

*All prices include £5,000 UK government plug-in car grant.

References

1. Holdway AR et al (2010) Indirect emissions from electric vehicles: emissions from electricity generation. Energy Environ Sci 3(12):1825–1832
2. Greenhouse Gas Emssion in the EU by sector, E.E. Agency, Editor 2007
3. Inderwildi OR (2009) Future of mobility roadmap. Smith School of Enterprise and the Environment, University of Oxford
4. Bellis, M (2010) Inventors—Electric cars (1890–1930). about.com2010
5. Cowan R, Hulten S (1996) Escaping lock-in: the case of the electric vehicle. University of Western Ontario, Canada
6. Yang C et al (2008) Appendix C: technical assessments of advanced vehicles for advanced energy pathways project (AEP). Public interest energy research (PIER) program, California Energy Commission
7. Thomas CE (2009) Fuel cell and battery electric vehicles compared. Int J Hydrogen Energy 34:6005–6020
8. Matheys J et al (2009) Comparison of the environmental impact of five electric vehicle battery technologies using LCA. Int J Sustain Manuf 1(3):318–329
9. Axsen J, Burke A, Kurani K (2008) Batteries for plug-in hybrid electric vehicles (PHEVs): goals and the state of technology circa 2008. Institute of Transportation Studies, University of California, Davis
10. Baker J (2008) New technology and possible advances in energy storage. Energy Policy 36(12):4368–4373

11. Kalhammer FR et al (2007) Status and prospects for zero emissions vehicle technology report of the ARB independent expert panel. State of California Air Resources Board Sacramento, California
12. Werber M, Fischer M, Schwarz PV (2009) Batteries: Higher energy density than gasoline? Energy Policy 37:2639–2641
13. Silva C, Ross M, Farias T (2009) Evaluation of energy consumption, emissions and cost of plug-in hybrid vehicles. Energy Convers Manag 50(7):1635–1643
14. Tesla Motors. http://www.teslamotors.com/roadster/specs. Accessed July 2011
15. Think Electric Vehicles. http://www.thinkev.com/. Accessed July 2011
16. Alex Williams (2012) REVA Electric Car Compant 15:54
17. Thomas CE (2009) Fuel cell and battery electric vehicles compared. Int J Hydrogen Energy 34:6005–6020
18. Campanari S, Manzolinia G, Garcia de la Iglesi F (2009) Energy analysis of electric vehicles using batteries or fuel cells through well-to-wheels driving cycle. J Power Sources 186: 464–477
19. Sioshansi R, Denholm P (2009) Emissions impacts and benefits of plug-in hybrid electric vehicles and vehicle-to-grid services. Environ Sci Technol 43:1199–1204
20. Samaras C, Meisterling K (2008) Life cycle assessment of greenhouse gas emissions from plug-in hybrid vehicles: implications for policy. Environ Sci Technol 42:3170–3176
21. Stephan CH, Sullivan J (2008) Environmental and energy implications of plug-in hybrid-electric vehicles. Environ Sci Technol 42:1185–1190
22. Lindly JK, Haskew TA (2002) Impact of electric vehicles on electric power generation and global environmental change. Adv Environ Res 6:291–302
23. Kobayashi O et al (1995) Environmental and economic evaluations of electric vehicles. In: SAE global vehicle development conference, Dearborn
24. Wang Q, Santini DL (1993) Magnitude and value of electric vehicle emissions reductions for six driving cycles in four U.S. cities with varying air quality problems. Argonne National Laboratory, Argonne
25. Bentley JM et al (1992) The impact of electric vehicles on CO_2 emissions: final report. Arthur D, Little: Cambridge, MA
26. Jacobson MZ (2009) Review of solutions to global warming, air pollution, and energy security. Energy Environ Sci 2(2):148
27. Arar JI (2010) New directions: the electric car and carbon emissions in the US. Atmospheric Environ 44:733
28. Nominal GDP list of countries, IMF, Editor 2010
29. Energy Information Administration (2008) Annual energy review 2007. US Department of Energy, Washington, DC
30. Silva CM et al (2006) A tank-to-wheel analysis tool for energy and emissions studies in road vehicles. Sci Total Environ 367:441–447
31. Hutchingson A (2011) The new energy fixes: 10 fixes. Pop Mech (73)
32. UK Department for Business, E., and Regulatory Reform (2009) Digest of United Kingdom energy statistics 2009. Norwich, UK: TSO
33. Ministère de l'écologie, d.l.é., du développement (2009) Repères: Chiffres clés de l'énergie, France
34. Pimentel D et al (2002) Renewable energy: current and potential issues. Bioscience 52: 1111–1120
35. Schäfer A et al (2009) Transportation in a climate-constrained world. The MIT Press, Cambridge
36. Code of Federal Regulations, Title 40, Protection of Environment (2008) US Office of the Federal Register, Washington
37. La Commission des comptes des transports de la Nation (2007) Les comptes des transports en 2006. Ministère de l'écologie, de l'énergie, du développement
38. Bureau of Transportation Statistics (2009) National transportation statistics 2009. US Department of Transportation, Washington

39. US Environmental Protection Agency (2008) Inventory of U.S. greenhouse gas emissions and sinks: 1990–2006. Washington

40. Heywood JB (2004) The performance of future ICE and fuel-cell-powered vehicles and their potential fleet impact. In: SAE World Congress & Exhibition, Detroit

41. Ford A (1995) The impacts of large scale use of electric vehicles in southern California. Energy Buildings 22:207–218

42. Coll-Mayor D, Paget M, Lightner E (2007) Future intelligent power grids: analysis of the vision in the European Union and the United States. Energy Policy 35:2453–2465

43. Tomić J, Kempton W (2007) Using fleets of electric-drive vehicles for grid support. J Power Sources 168(2):459–468

44. Energy Information Administration (2009) Alternatives to traditional transportation fuels 2007. US Department of Energy, Washington DC

45. European Parliament and Council (2002) Directive 2002/24/EC

46. Ministère de l'écologie, d.l.é., du développement durable et de l'aménagement du territoire (2008) Chiffres clés : L'énergie en France, France

47. Calef D, Goble R (2007) The allure of technology: how France and California promoted electric and hybrid vehicles to reduce urban air pollution. Policy Sci 40:1–34

48. Dones R, Heck T, Hirschberg S (2004) Greenhouse gas emissions from energy systems, comparison and overview. Encycl Energy 3:77–95

49. Weisser D (2007) A guide to life-cycle greenhouse gas (GHG) emissions from electric supply technologies. Energy 32:1543–1559

50. Hondo H (2004) Life cycle GHG emission analysis of power generation systems: Japanese case. Energy 30:2042–2056

51. Odeh NA, Cockerill TT (2008) Life cycle analysis of UK coal fired power plants. Energy Convers Manag 49:212–220

52. UK Department for Transport (2008) Transport statistics Great Britain. TSO, Norwich

53. UK Department for Environment, F., and Rural Affairs (Defra) (2008) UK Emissions of Air Pollutants 1970–2006. Didcot, UK: AEA

54. Centre Interprofessionnel Technique d'Études de la Pollution Atmosphérique (CITEPA) (2009) Inventaire des émissions de gaz à effet de serre en France au titre de la convention-cadre des Nations Unies sur les changements climatiques, Paris: CITEPA

55. Ministère de l'écologie, d.l.é., du développement durable et de l'aménagement du territoire (2009) Les chiffres du transport, France

The Future of Mobility
in a Resource-Constrained World:
An Overview and Solutions

Stéphanie Jacometti and Jack Jacometti

Abstract The previous chapters in this publication explored specific issues relating to the future of mobility, such as efficiency improvements, modular change and electrification. Rather than looking at technical and structural issues, this paper will concentrate instead on global resource constraints and how behavioural change will be a vital element of the solution.

1 Introduction

The previous chapters in this publication explored specific issues relating to the future of mobility, such as efficiency improvements, modular change and electrification. Rather than looking at technical and structural issues, this paper will concentrate instead on global resource constraints and how behavioural change will be a vital element of the solution.

There are increasingly strong indicators that global resource constraints, and the excessive resource consumption that has led to it, might require a much faster and fundamental response than currently anticipated.

This chapter addresses the nature of the challenges ahead and those factors that limit our in-depth understanding of the consequences of global resource constraints. Current global developments around resource security, combined with economic and political uncertainty, are currently only in military circles perceived as being included in the category of 'clear and present danger'.

S. Jacometti (✉) · J. Jacometti
Jacometti Associates, Cobham, Surrey, UK
e-mail: jacomettiassociates@gmail.com

O. Inderwildi and Sir David King (eds.), *Energy, Transport, & the Environment*,
DOI: 10.1007/978-1-4471-2717-8_8, © Springer-Verlag London 2012

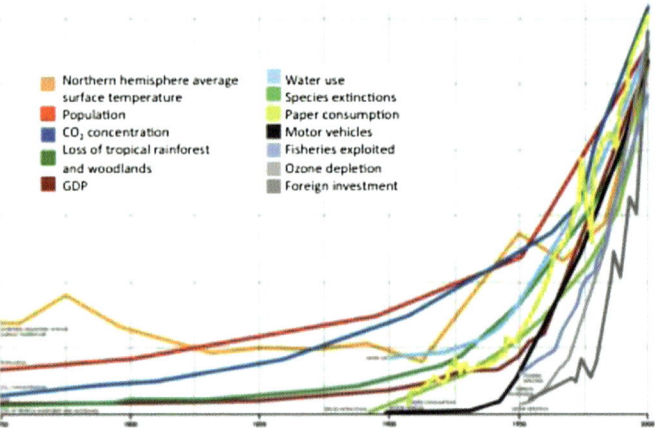

Fig. 1 Resource consumption

Since the invention of the motorcar, the automotive industry has developed into a cornerstone of the world's leading economies. The employment it provides, including secondary suppliers, is significant. Hence it is crucial to understand what forces are shaping its future and at what rate change could occur. This requires a clear vision combined with extraordinary leadership and political will, as well as an appreciation that purely technical solutions are unlikely to be sufficient. Emerging commercial opportunities need to be identified and captured, whilst at the same time legacies of the past must be disposed of. The shift from the Industrial to the Information Age is an exciting prospect, but it is likely to be disruptive, in particular for those sectors that are firmly anchored in the former.

There are promising signs that fundamental lifestyle and behavioural change, enabled by Information and Communications Technology (ICT) and the Future Internet, could play a prominent role in countering the global resource constraints challenge. Initiatives under development, such as those by the European Commission, will be briefly covered in this chapter.

2 Global Resource Constraints

The awareness of global resource constraints is gradually gaining momentum with governments and international institutions. Climate change remains an area of major focus and can be seen as a major symptom of resource constraints. Unfortunately, progress is slow, as is action. Signals of global resource constraints we are seeing include the rate of global resource consumption and alarming changes to a diverse number of factors as detailed below. Global resource consumption exceeds the Earth's regeneration capacity by 50% [1]. This means that the Earth takes 1.5 years to recover from 1 year of consumption. The diverse factors as mentioned above have increased at an alarming rate, particularly in the last 50 years. Population, water usage, number of motor vehicles, GDP and species extinction, amongst others, have increased dramatically, as change on all levels is accelerating. This is illustrated in Fig. 1 [2].

The aforementioned examples are only the tip of the iceberg. However, it is interesting and important to note that these warning signals are mostly ignored. The observation that humanity is at present consuming 50% more than the planet can regenerate should set alarm bells ringing. As this is clearly not happening on an adequate scale, it is important to understand why this is the case.

3 A Fragmented Global Perspective

The implications of the resource challenges are of a magnitude that is difficult to comprehend. One only needs to look at negotiations at the United Nations Climate Change Conference to see how leaders struggle to come to an agreement that will secure a sustainable future for everyone. This not only impairs the appreciation of the challenges, but also denies the exploration of commercial opportunities.

To this end, the book "The Master and His Emissary: The Divided Brain and the Making of the Western World" by McGilchrist [3], which provides remarkable insights into human psychology based on neuro-scientific evidence, is highly recommended. It provides fundamental insights into how the Western world tends to construct a worldview, which is not necessarily a reflection of reality.

McGilchrist concludes "the right hemisphere underwrites breadth and flexibility of attention, where the left hemisphere brings to bear focused attention. The right hemisphere sees things whole, and in their context, where the left hemisphere sees things abstracted from context, and broken in parts, from which it then reconstructs a whole, something very different."

In other words the Industrial Revolution has fostered left-brain focus leading to a lack of integral perspective and situational awareness, which reside in the right brain. Our consumption patterns reflect this as people find it difficult to keep an overview and take into account the long-term consequences of their choices.

This has led, in particular in the Western world, to the development of a fragmented global perspective, which fundamentally obstructs a clear and realistic view of the future and therefore appropriate responses.

Interestingly the East offers a different perspective. The left hemisphere of the brain appears less dominant in so far as they have not emulated the Western way of thinking. As such Asia Pacific might well play an increasingly prominent role in the shaping of a sustainable future.

Nair has outlined in "Consumptionomics" [4] how Asia has been encouraged by the Western world to consume more in order to dampen the effect of the recent financial crisis and how other solutions will be necessary to reshape capitalism as we know it. For example, China is looking at carbon and resource taxes as well as developing a network of smaller cities and towns that aim to discourage workers from moving to large cities. India is also taking a new approach and is challenging how development and the fair use of resources are linked [5].

4 Fundamental Societal Change

Global resource constraints have created challenges that amplify the need for a proactive approach. Being proactive will realise an optimal transition to a sustainable future. Contrary to popular belief, technology alone will NOT be able to quench our increasing thirst for resources and enable us to live on the Earth without environmental disaster. We cannot invent and innovate ourselves out of this situation in the short time that we have. Even if we had all the technology necessary, it would still take several decades to implement these technologies. In order to bridge the gap and to prevent disaster, we must make different choices and change our priorities. This means fundamental lifestyle change for the developed countries, while helping developing countries to avoid making the same mistakes as the Western world.

The economic system by which the world operates encourages growth. Infinite growth on a finite planet is a recipe for disaster. GDP is no longer an accurate way of measuring the wealth of a country, as it can be said that it measures how quickly a country uses up its natural resources. It does not measure the health or wellbeing of its citizens and it does not see these as a priority. For example, if someone is involved in a traffic accident, it is beneficial for the economy as that person needs to go to the hospital and needs to buy a new car. However, the person's health and wellbeing are both in disrepair.

Fundamental societal change, including lifestyle and behavioural change, is inevitable. It will either come as a shock and people will be forced to change or they can change, almost of their own accord, as outlined in "The Great Disruption—How the Climate Crisis Will Transform the Global Economy" by Gilding [6].

Governments tend to respond to this need to reduce resource demand by mounting awareness campaigns in the hope that this will make people change. Research has shown that information and knowledge alone is not enough [7, 8]. Feedback and community support are essential. One of the reasons why Weightwatchers and Alcoholics Anonymous are so successful is because they use the following three elements: information, feedback and community support.

If technology was used and the importance of financial savings stressed, people would be likely to drive more, as their cars are more efficient [9]. This is also known as the rebound effect. A real change can be made through giving people information, coupled with feedback and community support. Their mindset will shift and their values too.

When discussing behavioural change, the term "nudge" invariably comes up. Nudging is about choice architecture. It "alters people's behaviour in a predictable way without forbidding any options of significantly changing their economic incentives [10]." Nudging can be used in policy and works on the premise that people are inert. They are not likely to change and so this inertia can be harnessed. Unfortunately, nudging will not give us the changes we need in the period of time that we have.

Financial incentives are most often used when trying to change behaviour, but other incentives such as moral and social should be used too. People are very much influenced by their environment and, if they see or know about someone else's behaviour, they are likely to copy others' actions. The book "Nudge" by Thaler and Sunstein notes that "social influences come in two basic categories. The first involves information. If many people do something or think something, their actions and their thoughts convey information about what might be best for you to do or think. The second involves peer pressure. If you care about what other people think about you (…), then you might go along with the crowd to avoid their wrath or curry their favour [11]."

Nudging can be used to influence behaviour very effectively but unfortunately not everyone has the best interests of the environment at heart. Washington politicians encouraged the rise in demand for SUVs as they "avoided subjecting SUVs to the gas-guzzler tax and stringent emissions, safety, and energy regulations…" [12] Add to this rigorous marketing campaigns, and SUVs became a sign of status and wealth. So, in addition to nudging people in the right direction, a new kind of transportation must be developed to replace the old.

5 High Efficiency and Low Footprint Car Design and Beyond

High efficiency car designs have been developed by VW Group, BMW, Toyota and Mercedes Benz. Examples include:

- Highly efficient diesel engines in combination with BMW's Efficient Dynamics concept.
- Downscaled combined turbo-charging/super-charging petrol engines in a compact car design, such as the Audi A1 which achieves impressive fuel consumption figures.
- Plug-in hybrids such as Toyota's "plug-in" Prius, the GM Volt, Fisker Karma and the ultra efficient VW XL1 which show significant promise.
- The Nissan Leaf—a good example of a well-developed full electric vehicle. This type of vehicle will in the foreseeable future probably be primarily used in and around cities.

An interesting example of low footprint car design is Gordon Murray Design (GMD)'s T-25 city car. Rather than using an existing design and adjusting it so that it uses less fuel or adding an electric motor, GMD has taken the approach of putting every single nut and bolt of the vehicle through life cycle analysis. The car is lightweight and uses recycled materials. Rather than taking a cradle-to-grave approach (where the materials go to landfill after the car has come to the end of its useful life), GMD has focused on cradle-to-cradle and therefore analysed what can be done with the materials after the car has reached the end of its life as a vehicle, creating a closed loop and reusable materials. Such a car would help ease resource constraints significantly.

It has been shown that younger generations are more interested in mobility than transport. They aspire to live in cities, where having a car is more of a boon than an advantage. Politicians and governments have been slow to pick up on this. Recently, the European Commission published a Roadmap for Future Transport. Willy de Backer, former Head of Friends of Europe's Greening Europe Forum, noted that this White Paper is "not the transport paradigm change it pretends to be in its impact assessment [13]." De Backer also wrote that "the longer we wait to face the real hard choices (also in terms of changing lifestyles), the more 'tyrannical' and 'draconian' the solutions will have to be in the future [13]."

In other words, global resource constraints might well require even more drastic measures and potentially at a much earlier stage than currently envisaged. In order to conserve resources, public transport might become the main mode of transport, with car sharing, electric cars and bikes supplementing mobility, in combination with tele-commuting. There is no doubt that the transport systems of the future will look very different indeed.

6 Contextual or Situational Awareness Through Information Communication Technology and the Future Internet

An area of substantial promise is ICT and the Future Internet, as it has major potential—possibly also by acting as a 'collective consciousness'—to restore situational awareness on a global scale. This is particularly well understood in advanced military applications, where situational awareness is enabled by ICT. Situational awareness is a way of being aware of what is happening around you and how a change in circumstances can influence goals and objectives.

Since the role of ICT and the Internet has become so central in daily life, it has become one of the transformative powers that are reshaping the world. When deployed effectively, it could enable the socio-ecological transition required by an increasingly resource constrained planet. GDP growth as a measure of progress is becoming increasingly flawed, as supported by 'The Economics of Ecosystems and Biodiversity' [14] report, which was sponsored by the United Nations Environment Programme (UNEP), the European Commission and others.

An initiative by the Directorate General Information Systems and Media of the European Commission "A forward-looking analysis to identify new innovation paths for the Future Internet", identified changes in expectations and lifestyles such as:

- Post consumerism, rather than capitalism.
- A shift from a global to a more local economy.
- Growing environmental consciousness—doing more with less.
- From individualism to community—slower and simpler life.
- Wellbeing and quality of life as a personal priority.

The notion that societal change will be inevitable is also reflected in an increasing number of initiatives such as the Norfolk Island Carbon/Health Evaluation (NICHE) pilot project, which aims to test the effectiveness of a personal carbon-trading scheme in a closed system. Another initiative is the Japan Smart Community Alliance [15] which is unique in that it works with major industry players to develop smart grid initiatives.

Mobility is very much on the agenda of the FISITA World Automotive Summit, an annual strategic meeting of the leaders of the global automakers, technology suppliers, energy companies, government and NGOs. The November 2011 meeting asks "How can we shape personal mobility for the megacity? Previous meetings have tackled CO_2 reduction from road transportation and global traffic safety [16]."

7 Conclusion

It is becoming increasingly clear that we are facing global challenges on an unprecedented scale. The challenges we face are so daunting that the most common reaction appears to be denial. This observation is supported by neuroscientific insights regarding the left brain dominance of the Western world, which results in fragmented and distorted perspectives. In line with Einstein's profound observation that: "No problem can be resolved at the level of thinking that created it", we need to collectively raise our quality of thinking.

The great enabler is likely to be ICT and the Future Internet. The interesting parallel is that the Renaissance around 1450 in Europe coincided with the invention of the Gutenberg printing press. This might not be a coincidence as it enabled the dissemination of ideas. In today's world, cloud computing could be a metaphor for the collective consciousness.

A number of significant technological advances have been made to reduce the footprint and increase the efficiency of cars. It will be a challenge to disseminate these technologies at the necessary rate and we might have to accept that cars will become the exception rather than the rule. Mobility is about getting from A to B and this will most likely be public transport, rather than private transport.

In discussions with government representatives, international organisations and academia it was concluded that the development of scenario-led lifestyle or behavioural change programmes, focused on the community level, is crucial to shaping a sustainable future. A reduction in demand is necessary to allow time to build new infrastructures and develop renewable technologies and implement them. The development of these scenarios is required to identify business and therefore employment opportunities in the context of fundamentally different economic and business models.

This paper intends to provide essential food for thought—in the Greek tradition of Socratic dialogue—to develop an objective overview and the insights required to comprehend the challenges we are facing. These are a prerequisite to identify the commercial opportunities, which are likely to be of a similar magnitude as the challenges ahead. The transport system as we know it will have to adapt to an environment that is rapidly changing appears daunting but inevitable. However, human ingenuity appears to be at its best when facing major challenges, provided these are clearly understood.

References

1. Global Footprint Network (2010) Living planet report. http://www.footprintnetwork.org/press/LPR2010.pdf
2. Staff Writer (2008) How our economy is killing the earth. New Scientist, issue 2678
3. McGilchrist I (2010) The master and his emissary: the divided brain and the making of the western world. Yale University Press, New Haven, p 27
4. Nair C (2011) Consumptionomics: Asia's role in reshaping capitalism and saving the planet. Infinite Ideas Limited, Oxford
5. Nair C (2011) http://www.globalinstitutefortomorrow.com/article/ideas_for_tomorrow/itoys_wont_fix_asias_broken_growth_model/. Accessed 8 May 2011
6. Gilding P (2011) The great disruption. Bloomsbury Publishing, London
7. Strategy Unit (2004) Personal responsibility and changing behaviour: the state of knowledge and its implications for public policy. Strategy Unit, London, p 38
8. Global Action Plan (2006) Changing environmental behaviour: a review of evidence from global action plan. Global Action Plan, London, p 13
9. Wilkinson R, Pickett K (2009) The spirit level—why equality is better for everyone. Penguin Books, London
10. Thaler RH, Sunstein CR (2008) Nudge: improving decisions about health, wealth and happiness. Yale University Press, New Haven, p 6
11. Thaler RH, Sunstein CR (2008) Nudge: improving decisions about health, wealth and happiness. Yale University Press, New Haven, p 54
12. Sperling D, Gordon D (2009) Two billion cars—driving towards sustainability. Oxford University Press, New York, p 54
13. De Backer W (2011) http://www.friendsofeurope.org/Contentnavigation/Library/Libraryoverview/tabid/1186/articleType/ArticleView/articleId/2325/EU-Roadmap-for-future-transport-one-small-step-no-big-leap.aspx. Accessed 8 May 2011
14. The Economics of Ecosystems and Biodiversity (2011) http://www.teebweb.org/. Accessed 8 May 2011
15. Japan Smart Community Alliance (2011) http://www.smart-japan.org/english/tabid/103/Default.aspx. Accessed 8 May 2011
16. FISITA World Automotive Summit (2011) http://www.fisita-summit.com. Accessed 8 May 2011

Hydrogen: An End-State Solution for Transportation?

Asel Sartbaeva, Stephen A. Wells, Vladimir L. Kuznetsov and Peter P. Edwards

Abstract There is presently no global consensus on how our human society might ultimately transform from a hydrocarbon, fossil fuel-based energy economy to an alternative low-carbon, or zero-carbon economy. The same is true for alternative fuel options for transportation. Hydrogen fuel cell vehicles are highly promising in that their well-to-wheel carbon dioxide profile is very good and compare favorably against either battery electric vehicles or plug-in hybrid electric vehicles, since many national energy grids (e.g. China and the USA) are so dirty. These fuel cell vehicles have ranges similar to gasoline vehicles, e.g. the Honda Clarity has a range of some 250 miles. In that regard, many countries view the development and dissemination of hydrogen and fuel cell technologies as core technologies for a future sustainable economy which could contribute to environmental impact reduction, energy diversity and energy independence as well as new industry creation. However, the attraction of "competitor" electric vehicles is that much of the underlying infrastructure and technology already exists; national grids are in place with the right support infrastructure for scale-out across countries. When it comes to hydrogen, this is simply not the case. It is clear that this alternative technology is not yet ready for mass market and the infrastructure is not in place to support these vehicles. Thus the economics of a transition is difficult. These types of considerations led the US, for example, in 2009 to announce a significant reduction in research and development funding into automotive hydrogen fuel cells, arguing that a focus on areas such as plug-in vehicles has the potential to make the quickest impact on environmental issues. However, the long-term

A. Sartbaeva · V. L. Kuznetsov · P. P. Edwards (✉)
Department of Chemistry, Inorganic Chemistry Laboratory,
University of Oxford, South Parks Road, Oxford, OX1 3QR, UK
e-mail: peter.edwards@chem.ox.ac.uk

S. A. Wells
Department of Physics and Centre for Scientific Computing,
University of Warwick, Gibbet Hill Road, Coventry, CV4 7AL, UK

O. Inderwildi and Sir David King (eds.), *Energy, Transport, & the Environment*,
DOI: 10.1007/978-1-4471-2717-8_9, © Springer-Verlag London 2012

potential of hydrogen fuel cells is recognized while understanding the pressing scientific and technological and socio-economic challenges. As well as the vexing issue of the absence of any national infrastructure for hydrogen, these challenges center on the cost and durability of vehicle fuel cells; the current inability to store large volumes of hydrogen fuel onboard transport vehicles and the absence of large-scale processes to the manufacture of carbon-free hydrogen. In part of our contribution, we present a brief overview on the current states of hydrogen and fuel cell technologies, covering several of the key challenges of what one might see as a future hydrogen economy. But we stress that hydrogen can also fulfill an even broader, pivotal role in renewable energy capture and conversion. A variety of renewable or sustainable energy sources can be used to produce molecular hydrogen, which can then be used in multiple energy applications: *as* a fuel for personal transportation, in the conventional vision of a hydrogen transport economy; as an energy store in static applications, particularly as a buffer in energy generation; or *in* a fuel, using hydrogen as a feedstock in the synthesis of oxygenated fuels such as methanol or ethanol or even hydrocarbon fuels such as diesel. This complementary approach of using hydrogen for the synthesis of hydrocarbon-based liquid fuels at-a-stroke removes the burgeoning requirement noted earlier for large-scale infrastructure changes that are necessary in the use of molecular hydrogen *as* a fuel. Furthermore, the use of carbon dioxide (atmospheric or industrial by-product) as the source of carbon for the synthesis of liquid hydrocarbon fuels with hydrogen has the potential to reduce emission of fossil carbon into the atmosphere. Hydrogen generation as a buffer or energy store, using energy that would otherwise not be matched to load in the electricity grid, can be regarded as a potential key ally of renewable energy generation from intermittent natural sources including wind, wave, tide and solar power. Scientists and policy-makers should thus keep in mind a variety of possible "hydrogen economies".

Abbreviations

BEV	Battery-electric vehicle
CCS	Carbon capture and storage
CGH2	Compressed gas hydrogen (storage method)
CHP	Combined heat and power system
CSS	Carbon sequestration and storage
FC	Fuel cell
FCV	Fuel cell vehicle
FCEV	Fuel cell electric vehicle
H2FCEV	Hydrogen fuel cell electric vehicle
H2PEMFC	Hydrogen polymer electrode membrane (or Proton exchange membrane) fuel cell
ICE	Internal combustion engine
kWh	Kilowatt-hour measure of energy equal to 3.6 million joules
PEMFC	Polymer electrode membrane (or Proton exchange membrane) fuel cell
wt%	Weight per cent (composition)

1 The Hydrogen Economy

In many respects hydrogen is an ideal fuel. Although not naturally occurring as a primary fuel, it can be readily synthesized from natural gas, coal or oil. Perhaps most relevant to any sustainable energy vision, it can be produced by splitting molecules of water with input of electrical energy from renewable sources, e.g., solar, wind, wave etc. Of course perhaps the greatest attraction of hydrogen as a fuel source is that its only combustion product is water. In principle, then, one can imagine a future energy economy—the hydrogen economy—in which hydrogen is manufactured from water and electrical energy from renewable sources and is either stored until needed for use in either direct combustion or in fuel cells, or, as we shall illustrate (Sect. 3), combined with carbon dioxide to synthesize liquid hydrocarbon fuels. The term *hydrogen economy* was introduced during the energy crises of the 1970s to describe a national or international energy infrastructure based on hydrogen produced from non-fossil primary energy sources. It is now widely taken as a term for a future energy economy which relies on renewable rather than fossil energy sources, and in which hydrogen will play a major role as a suitable storage and transmission vector for energy [1–5]. We shall regard a mature hydrogen economy as one in which *the production and utilization of hydrogen has an economic significance comparable to the current production and use of fossil fuels*. This definition does not depend on the details of how hydrogen is used. We shall see that hydrogen for transport cannot sensibly be discussed without consideration of the entire energy economy.

One can distinguish between a *sustainable hydrogen economy*, in which fossil fuels can play no part, and a *transitional hydrogen economy*, in which hydrogen might be generated using fossil fuels (for example by the steam reforming of methane) while an infrastructure for hydrogen production, distribution and end utilization is established. To criticize a transitional economy for not yet being fully sustainable would be inappropriate, so long as the momentum of the transition is maintained and progress is being made toward sustainability.

The development of a future hydrogen energy economy is motivated by a desire to move away from the fossil fuel or carbon economy, because of the finite reserves of fossil fuels and the damaging environmental consequences of carbon dioxide emissions from their combustion, such as anthropogenic induced climate change and ocean acidification [6–10]. The vision is assisted by certain properties of hydrogen which make it attractive as future energy carrier.

One of the most attractive properties is that hydrogen can be produced from a variety of primary energy sources by water electrolysis using sustainable electricity, and the chemical potential energy of molecular hydrogen can be converted back into electricity with very high efficiency in a fuel cell. Invented in 1839, a fuel cell uses a catalyst to combine hydrogen and oxygen in a controlled electrochemical reaction to produce water and electricity. Their ability to generate carbon dioxide-free electricity—at least at the point of use within the fuel cell— has been the cornerstone of many visions of a carbon-free hydrogen economy.

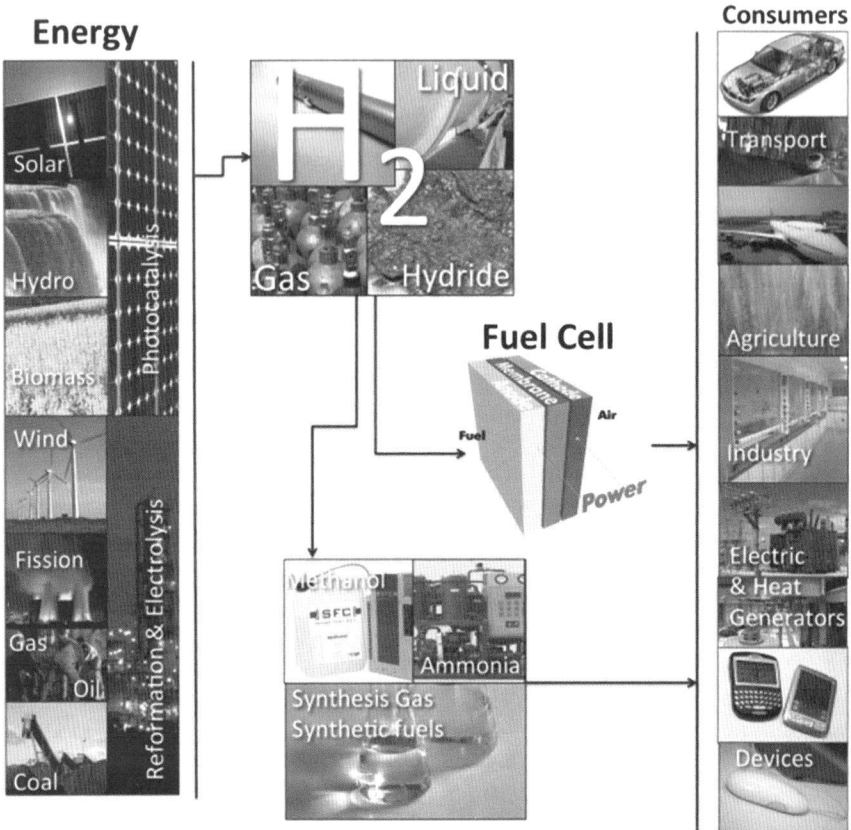

Fig. 1 Central role of hydrogen in a future energy economy

However, other applications of hydrogen—in particular as a chemical feedstock for synthesis of other fuels—may be equally important. A schematic representation of a possible scenario for a hydrogen economy [11], incorporating both applications, is illustrated in Fig. 1.

It is now clear, however, that significant scientific, technological and socio-economic challenges must be overcome before any such transition to a hydrogen economy is feasible. These challenges include the sustainable, high efficiency production of hydrogen, the effective, safe (and efficient) on-board storage of hydrogen in vehicles, the development of a hydrogen distribution structure and the development of fuel cells with a longer lifespan and much lower unit cost than is currently available [12, 13].

Various roadmaps have been proposed for the development of transitional hydrogen economies in, for example, Japan, USA and the EU, essentially setting out a time-line for the necessary technological and socio-economic developments to be achieved. These represent significant challenges—and also real opportunities for innovations in science and engineering.

2 Hydrogen in Fuel Production and Storage

An alternative and in many ways a complementary approach to the conventional view of the hydrogen economy is the following. First, renewable energy is stored in hydrogen by electrolysis of water but then for an additional energy penalty, the hydrogen is reacted, under appropriate catalytic conditions, with carbon dioxide to form an infrastructure-compatible liquid hydrocarbon fuel, such as methanol, dimethylether or indeed petrol or diesel. In this way, hydrogen is used in the fuel rather than as the fuel. Importantly, if all the processes, e.g. hydrogen production, are powered with carbon-free (renewable) energy, and the carbon dioxide used to make the fuel is captured directly from the atmosphere, then the subsequent combustion (utilization) of this fuel would result in a zero net carbon dioxide emissions. The synthesis and storage of such carbon-neutral liquid fuels offer the tantalizing prospect of decarbonizing transport without the paradigm shifts required by either a conversion to the hydrogen economy or the electrification of the vehicle fleet.

3 Visions and Scenarios

A transition to any kind of hydrogen-based economy is uncertain and certainly not inevitable. There are multiple possible routes to the use of hydrogen in a future energy economy. Much of the usual discussion is based on a perhaps narrow view of how a hydrogen economy should develop—the development of hydrogen and fuel cell electric vehicles for transportation. We broaden the view and outline a variety of visions for the role of hydrogen: these encompass "Hydrogen as Fuel", "Hydrogen: The Energy Store" and "Hydrogen in the Fuel".

"Hydrogen as Fuel". In this vision, hydrogen replaces hydrocarbon fuels for personal transportation, so that the consumer encounters hydrogen where once they encountered gasoline and diesel. We thus use the term "fuel" to mean an energy store in mobile applications. This application could be based on localized hydrogen generation, for example on-site electrolysis from renewable electricity at filling stations. However, it is more commonly assumed that hydrogen fuel will be produced by industrial-scale hydrogen generation infrastructure, with the management of hydrogen almost exactly paralleling the current use of petroleum. That is, hydrogen is produced in large quantities in industrial-scale "refineries" (for example, using nuclear heat in thermochemical cycles, or by steam reformation of methane with carbon capture and storage); it is transported in bulk over large distances by pipeline and trailer, kept at filling stations and filled into hydrogen fuel cell vehicles.

"Hydrogen: The Energy Store". Here the primary application of hydrogen is as a storage medium for renewable electricity; electrical energy is converted to the energy of the chemical bond. We use the term "energy store" rather than "fuel" to emphasize that this may be a static rather than a mobile application. The property exploited is the efficiency and relative ease of conversion from electrical power to hydrogen by electrolysis and back to power using fuel cells. In this vision, localized hydrogen production, storage and reconversion buffers many forms of renewable energy generation including wind, tide, wave and solar. This initial application is essentially static, and the first step away from fossil fuels in mass-market vehicular transportation is to BEVs rather than FCEVs. However, the resulting technological developments and improvements in electrolyzers, fuel cells and hydrogen storage may in turn lead to practical hydrogen fuel cell vehicles, with the hydrogen generated on-site at filling stations by electrolysis. The major infrastructure investment is in the renewable generation and efficient long-distance transmission of electricity, rather than the long-distance transport of bulk hydrogen.

"Hydrogen in the Fuel". In this vision we concentrate on the intrinsically high chemical energy of hydrogen. Hydrogen fuel cell vehicles may not be a major factor in this vision: rather, feedstock hydrogen is used for the synthesis of methanol for direct methanol fuel cells, and hydrocarbons usable in ICEs. The carbon component of these fuels can be obtained from carbon dioxide, captured either from the atmosphere or from the exhaust gases of chemical industry and power generation; thus even a hydrocarbon fuel could be made carbon-neutral, so long as the energy requirements of the fuel synthesis are met from renewable sources. Furthermore, at present very large quantities of hydrogen are used in industry, in particular in the synthesis of ammonia (Haber process) and hence of fertilizers. This hydrogen is currently produced mostly from fossil hydrocarbon sources-methane, oil and coal. This carbon cost of fertilizer is frequently neglected in discussion, leading to overly optimistic claims of the carbon neutrality of biomass-derived fuels. With a large increase in renewable electricity generation, and progressive improvements in the cost and efficiency of electrolysis, we could introduce renewably generated hydrogen into large-scale chemical industry. Indeed, a shift to renewable hydrogen in industry will eventually be absolutely necessary as fossil hydrocarbons become scarcer.

The visions of hydrogen as fuel, energy store and source of (carbon neutral) liquid hydrocarbon fuels are compatible and indeed complementary, as in every case a major emphasis on renewable energy generation, and the efficient production of hydrogen from renewable energy, is necessary in order to make hydrogen use possible and practical.

3.1 National and International Roadmaps

Tomei reviews multiple roadmaps from Canada, USA, UK, EU, Iceland, New York State and London [14]. She notes that the term roadmap has been used to cover many different kinds of document, many of which do not include time-lines or details but rather are technological reviews or aspirational policy guidelines. We will consider several cases for which timelines have been produced.

At present the hydrogen economy is at best embryonic [15]. Hydrogen vehicles exist in prototype form, with around 1,000 FCEVs in operation worldwide at the end of 2007; typically these development models use 350-bar or 700-bar CGH2 storage and have ranges of 200–300 km. In the public transport sector, around 100 hydrogen buses are in operation. Around 200 hydrogen fueling stations are active. Hydrogen transport projects are concentrated in cities, as the prototype vehicles are most suitable for short-range urban driving in the vicinity of hydrogen fueling station. In terms of global geography, hydrogen transport is concentrated in North America, Europe (especially Germany) and Japan. A small number of hydrogen buses were used as part of transportation service at the 2008 Beijing Olympic Games.

In Germany there appears to be a manufacturer push toward hydrogen transport [16] and considerable optimism about the prospects for H2FCEVs, with private vehicles and a hydrogen fueling infrastructure intended to be available from 2015.

In a recent report (2009), Japan's NEDO agency states that stationary fuel cell systems, FCVs, and hydrogen infrastructure have all been designated as priority technologies in the national "Cool Earth-Innovative Energy Technology Program" [17]. They provide parallel roadmaps, from the present to 2020–2030, for six technological fields: stationary PEMFCs, vehicle PEMFCs, solid-oxide stationary fuel cells, hydrogen storage, on-site hydrogen generation and off-site hydrogen generation [18]. Dissemination of these technologies in the commercial market is supposed to occur from around 2015. This comprehensive range of projects allows for stationary and mobile hydrogen applications and for localized or centralized hydrogen production; the use of parallel roadmaps illustrates that individual aspects of a hydrogen economy can be decoupled.

Iceland in particular, lying on the Mid-Atlantic Rift, has been a pioneer in both geothermal energy and in the introduction of hydrogen vehicles, with a policy aimed at phasing out fossil fuels by 2030. This small island nation is in a unique position to achieve this goal, as much of their domestic energy supply is already obtained from volcanic geothermal resources; use of fossil fuels is predominantly in land transport and in the country's large fishing fleets. A national timeline has been produced for the conversion to hydrogen, via an intermediate stage involving the use of methanol fuel cells in shipping [20, 21]. The effect of the recent collapse of the Icelandic economy, during the global financial crisis, is not yet clear.

In January 2004 the European Commission initiated the European Hydrogen and Fuel Cell Platform (HFP) with the expenditure of 2.8 billion € over a period of ten years, with the aim to prepare and direct an effective strategy for developing and exploiting a hydrogen-oriented economy for the period up to 2050. Key highlights and scenarios challenging European hydrogen vision are summarized in Fig. 2 [22].

This long-term vision regarding the potential of hydrogen and fuel cells illustrates the final goal of hydrogen economy that relies mainly on renewable energy production pathways. A target scenario for 2020 was developed for guiding the transition toward introducing hydrogen and fuel cells to the market. In March 2005, HFP published a Strategic Research Agenda and Deployment Strategy [23], followed by an Implementation Plan in January 2007 which combined these two documents into a long-term road map for Europe; a final version of this timeline

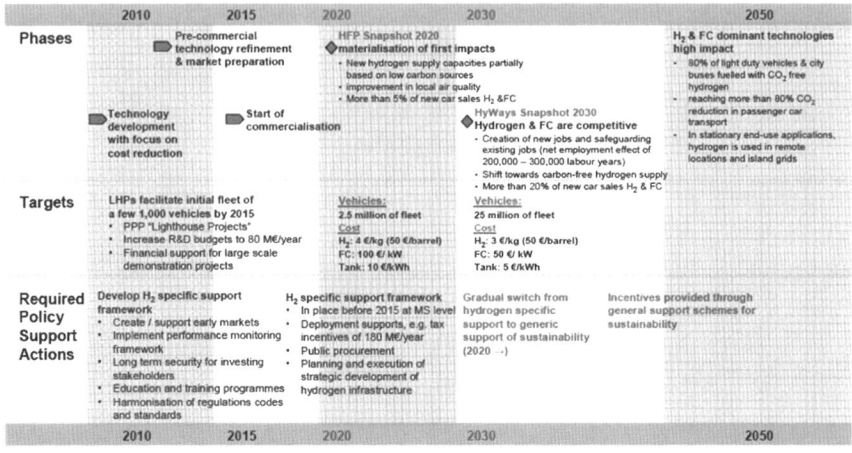

Fig. 2 European Union HyWays roadmap 2008 (from [19])

was published in 2008 [19]. The primary objective of this roadmap is to achieve by 2020 EU-wide availability of hydrogen and fuel cell vehicles with an appropriate coverage of refueling infrastructure. Key assumptions on hydrogen and fuel cell applications and the forecasts of several roadmaps for deployment status and targets for hydrogen are summarized in Ref. [23, 24].

In 2002 the US Department of Energy (DOE), in conjunction with the auto industry, established the US FreedomCAR Fuel Partnership. This partnership set technology goals and demonstration targets for 2015 that were considered sufficient to enable industry to move toward the commercialization of the hydrogen technologies [25].

In 2008 the US National Research Council's Committee has concluded that on the basis of the substantial financial commitments and technical progress the hydrogen production technologies and hydrogen and fuel cell vehicles could be ready for commercialization in the period 2015–2020. It was also estimated that, while fuel cell vehicles would not become competitive with gasoline-powered vehicles by 2020, they could account for more than 80 percent of new vehicles entering the fleet by 2050 [26].

These efforts will require considerable resources, especially government and private sector funding. The committee estimated that in addition to the current US government hydrogen R&D expenditures of about $300 million per year, additional $55 billion from 2008 to 2023 will be needed to support a transition to hydrogen fuel cell vehicles. This funding includes an extensive R&D program ($5 billion), support for the production of hydrogen ($10 billion) and support for the demonstration and deployment of the hydrogen vehicles at the earlier stages of commercialization ($40 billion). It is also estimated that private industry would need to invest about $145 billion for R&D, vehicle manufacturing and hydrogen infrastructure over the same period [26].

The commitment of the current US administration to hydrogen FCEVs is questionable, as an early action by the energy secretary Steven Chu was to focus funding on electric vehicle technology rather than hydrogen FCEVs. He cited several barriers, including infrastructure, development of long-lasting portable fuel cells and other problems [16, 27]. The budget proposal would have trimmed more than $100 million from the hydrogen program in the Office of Energy Efficiency and Renewable Energy, cutting it to $68 million for fuel cell research and development and steering the program away from areas related to transportation; the budget increased funding for other vehicle technology programs, including electric vehicles, lightweight materials and biofuels.

Funding for hydrogen transport projects was subsequently restored by Congress after protests from manufacturers [16]. The restructuring of the US auto industry in the aftermath of the economic collapse has involved a focus on battery electric vehicles and the associated charging infrastructure rather than H2FCEVs. Prospects for a US hydrogen economy seem uncertain: Lattin and Utgikar point out that in 2006, use of H2FCEVs was two orders of magnitude below expectations, and that hydrogen roadmaps from the 1970s had predicted a well-established hydrogen economy by the year 2000 [28], which clearly did not occur. A go/no-go decision on widespread commercial introduction of hydrogen vehicles is due to be made in the USA in 2015.

3.2 A UK Development Path?

Transitions have been investigated using various scenarios for future energy systems in the UK. Many early scenario exercises assumed that the use of hydrogen would require large-scale infrastructure development. Linked uncertainties in such scenarios resulted in infrastructure development being considered to be the primary barrier to the use of hydrogen, and so widespread use projected only for the long term, typically after 2050. However, several other studies consider that such a wholesale infrastructure change is not necessary. It is argued that development could originate from niche markets, with associated "islands" of hydrogen availability. However, it is only recently that scenario building has included the possibility of hydrogen being used to generate carbon-neutral liquid hydrocarbon fuels, with attendant small/no necessary changes to existing infrastructure. Although there is presently no official UK governmental roadmap, more wideranging scenarios have been suggested. The 'systems-approach' of looking at the future of hydrogen energy, outlined in the Hydrogen Strategic Framework for the UK [29] concludes that there is no one single route to a hydrogen economy, and indeed there is more than one kind of hydrogen economy. An analysis of how a transition to a hydrogen energy economy might occur in the UK energy system is the focus of an important new volume by Ekins [12]. Eames and McDowall have reported the results of an innovative foresighting study carried out by UKSHEC [30] and we review these. These are not predictions, but rather imaginative scenarios demonstrating possible routes to different kinds of hydrogen economy. The scenarios were developed for the UK economy but the storylines could be applicable across many countries.

The approach typical of existing governmental roadmaps, of hydrogen vehicles and their associated infrastructure being developed through public policy incentives, falls under Government Mission→Centralized Hydrogen for Transport. The major driver in this scenario is governmental concern over the issues of energy security and climate change. This scenario assumes strong governmental involvement, providing incentives for hydrogen technology development such as tax benefits on hydrogen vehicles, installation of hydrogen infrastructure and on production of renewable energy. Open questions for such a scenario include the democratic political feasibility of maintaining strong goal-directed government policy over several decades; the social acceptability of a substantial growth in nuclear power generation; and the technological issue of whether centralized production and large-scale long-distance distribution is really the best infrastructure model for hydrogen.

A technologically similar scenario with less emphasis on public policy and more on private-sector development falls under Corporate Race→Ubiquitous Hydrogen. This scenario explores a route to a hydrogen economy despite the failure of strong international and national government action over climate change and energy security. Rather, existing major private-sector entities, such as automobile manufacturers and energy companies, are driven to act, seeing the unsustainability of the current hydrocarbon economy as a threat to the corporate bottom line.

The scenario of Disruptive Innovation→Synthetic Liquid Fuel envisions high levels of hydrogen production, but not for direct use in hydrogen fuel cells. Rather, hydrogen and captured carbon (e.g. from CSS) are used to synthesize methanol for fuel cells, for electronics applications and small-scale personal mobility (scooters) and synthetic hydrocarbons for internal combustion engines. This scenario takes place in the context of a failure to make the H2FCEV commercially viable. It might emerge that the improvements in mobile hydrogen storage and in H2PEMFCs that are required to make a hydrogen car practical for consumers cannot be achieved at a cost that is practical for manufacturers without larger financial subsidies than governments are willing to provide.

The major fuel cell application in this scenario is for mobile power—for increasingly power-hungry high-performance portable electronics, in auxiliary power units for engine-off power in vehicles, and in scooters. For such small-scale applications, hydrogen storage is impractical, and so methanol and the dilute methanol fuel cell become the technology of choice. Synthetic fuels thus become increasingly widespread, and eventually fuels synthesized from renewable hydrogen and CO_2 (atmospheric or CCS) come to replace fossil fuels; methanol being used in fuel cells and synthetic hydrocarbons in ICEs.

Another scenario not dominated by hydrogen for transport is that of Structural Shift-Electricity Store. Here the primary role of hydrogen is to buffer and store the energy output of intermittent and/or highly distributed renewables. The major driver of the scenario is a wholesale shift to renewable energy, with government and regulators, corporations and consumers all feeling strong concerns over the risks of climate change and energy insecurity. This will require the exploitation of the widest possible range of renewable resources. Growth in localized and small-scale energy projects—microgeneration—encourages hydrogen generation by electrolysis, for buffering, and increasing use of stationary fuel cells, both to reconvert stored hydrogen to grid electricity and also in CHP applications. Steady improvements in fuel cell and electrolyzer technologies will allow hydrogen production and use to be as localized as possible, avoiding the need for a long-range hydrogen distribution or bulk storage infrastructure. Hydrogen storage technology will also improve as small-scale, temporary storage of buffer hydrogen becomes a major application. In the automotive industry, the initial shift is not toward H2FCEVs but rather toward BEVs. Fuel cells enter the automotive market subsequently, as range-extenders for electric vehicles.

4 Challenges

4.1 Background

Technologies for the production, distribution and use of hydrogen have been the subject of several recent reviews [5, 31, 32].The electrolytic dissociation of water to form hydrogen and oxygen can be achieved with high thermodynamic efficiency, up to 85% in a conventional alkaline water electrolyzer. Given sufficient electrical power, therefore, hydrogen could be produced in abundance; of course, as noted earlier, such power would have to be produced from sustainable sources in order for hydrogen itself to be correctly labeled as a sustainable or renewable energy carrier.

Recent advances in the catalytic production of hydrogen from renewable sources are also highlighted in a special issue of the journal Catalysis Today [33]; here, renewable sources such as biomass and biomass-derived oxygenates including methanol, ethanol and glycerol are outlined, as well as the photo-catalytic decomposition of water. This direct water dissociation is regarded by many as a potential "show-stopper" if conversion efficiencies could be increased by a factor of 2–3.

Hydrogen has the highest specific energy of any chemical substance at 33.3 kWh/kg, oxidizes to form only water and is readily applicable in fuel cells, which convert its latent chemical energy into electrical energy with very high efficiency.

4.2 Fuel Cells

Fuel cell vehicles are highly promising because their well-to-wheel carbon dioxide profile is very good and compares favorably against battery electric vehicles and plug-in hybrid electric vehicles, since most energy grids are so dirty (particularly China and the US, where coal usage is dominant). Comprehensive well-to-wheel analyses for a wide variety of transport technologies have recently been reviewed by Weindorf and Bunger [34].

The theoretical maximum efficiency of a fuel cell is given by the ratio of the free-energy change to the enthalpy change during the reaction; for a hydrogen fuel cell operating at temperatures around 100°C, this efficiency is around 90%. The theoretical voltage of a hydrogen cell is 1.23 V, though in practice cells operate at around 0.7 V, and multiple cells must be stacked together, connected by bipolar plates, to achieve higher voltages. Even so a hydrogen FCEV can operate with a net efficiency including the drive train of up to 45%. This greatly exceeds the performance of ICEs, which at partial load (typical of driving conditions) routinely operates at around 20% efficiency; the theoretical maximum efficiency of ICEs is limited by the Carnot cycle. Although the efficiency advantages of fuel cells have been appreciated since the nineteenth century, they have repeatedly failed to break into the mass market, largely due to the materials problems involved in developing effective, durable and inexpensive materials for the anode and cathode catalysts and for the electrolyte.

The main types of fuel cells can be described in terms of their operating temperatures and the nature of the electrolyte. Solid-oxide and molten carbonate fuel cells operate at temperatures above 600°C, are relatively tolerant of impurities in the fuel stream, and are considered more suitable for stationary electricity generation applications than for transport due to their slow start-up from cold. These high-temperature cells may have applications in combined heat and power (CHP) systems.

Lower temperature fuel cells fuelled by hydrogen include the alkaline fuel cell, the phosphoric acid fuel cell and the polymer electrolyte membrane or proton exchange membrane fuel cell (PEMFC), operating at around 80°C. The PEMFC is considered the leading candidate for the power source in hydrogen FCEVs. The low-temperature fuel cells are less tolerant of fuel variations than the high-temperature cells, as carbon monoxide is a catalyst poison, and must be provided with high-purity hydrogen as fuel. Fortunately, a pure oxygen supply is not required and fuel cells for transport are expected to be air-breathing.

A variant of the PEMFC is the direct methanol fuel cell, powered not by hydrogen but by a dilute methanol solution. This cell is proposed as a power supply for portable applications such as laptop computers, where it would offer much longer operating times than batteries can; the ease of handling of the methanol solution makes it more suitable than hydrogen for such small-scale applications. The methanol fuel cell does of course emit carbon dioxide in use; thus the cell would only be carbon-neutral if the methanol was generated from carbon dioxide using renewable energy and renewable feedstock hydrogen. The alkaline fuel cell can also be fueled by ammonia.

4.3 Production

As noted, hydrogen is not a primary energy source, but a secondary energy vector-an energy store. There is no reservoir of "fossil hydrogen" comparable to the fossil carbon reserves which have fuelled the global economy since the Industrial Revolution; hydrogen must be generated before it is used.

Total global hydrogen production is currently around 60 Mt per year, which could in principle fuel at least 600 million fuel cell vehicles-around 80% of the world vehicle fleet. However, hydrogen is presently used almost exclusively as an industrial chemical for a wide variety of processes, including ammonia production for fertilizers, use in refineries, in the food industry and methanol production. An enormous increase in hydrogen production will be required in order to realize a hydrogen economy.

At present, the overwhelming majority of hydrogen production, about 96%, comes from fossil fuels. The single largest method of production is the steam reforming of natural gas, producing about 48% of the total. Electrolysis, which is assumed to be a dominant production method in a sustainable hydrogen economy, currently accounts for only 4% of total production. Electrolysis is only economically viable where cheap hydroelectric power is available, for example in Canada, or Norway. Currently production of hydrogen by electrolysis (using grid electricity) is typically about three times more expensive than chemical production from methane [35, 36].

At present, renewable resources including biomass collectively contribute about 13% of world total primary energy supply, whereas coal, oil and gas collectively contribute about 80%. In electricity production, renewables contribute 18% of global production, almost all of which comes from large hydroelectric projects [5].

Hydroelectric power is currently the largest renewable energy contribution to electricity generation, though it still amounts to only a few per cent of global energy use. However, large-scale hydroelectric generation is only practical in a few locations with appropriate geography, such as Scandinavia or Kyrgyzstan. Ocean energy covers the harvesting of energy from the tides or from the waves and is not yet a major player in electricity generation, although pilot plants are in operation. Geothermal energy is in theory available almost anywhere on Earth if deep drilling is possible, but in practice is most accessible in regions of young volcanism where very hot rock lies close to the surface. Biomass is a potential source of renewable energy, but suffers from the drawbacks of low energy density, competition with food production and the carbon cost of fertilizer [5].

Nuclear fission is currently the largest non-fossil fuel contributor to electricity generation, and an expansion of nuclear power is one proposal for meeting the world's growing energy needs without increased combustion of fossil fuels and the attendant emissions of carbon dioxide. Fission power could contribute to the transitional hydrogen economy both through electricity production and, in theory, through thermochemical cycles of water splitting such as the sulfur-iodine process, powered by waste heat from fission; this, however, is still a subject of basic research and development. The nuclear industry has always had an uneasy relationship with the environmental movement, due to its links with the military production of plutonium, and due to the still unsolved problem of the disposal of highly radioactive nuclear waste from spent reactor fuel, which requires safe storage, sequestered from the environment, over geological timescales.

The ongoing high-profile nuclear disaster at Fukushima-1, Japan, appears already to have influenced the political acceptability of an expansion of nuclear power, in Europe at least.

The generation of energy from hydrogen fusion is, ironically, not usually considered an aspect of the "hydrogen economy" per se. Despite decades of intensive research [37], practical energy generation from hydrogen fusion is still at least decades away and thus cannot contribute to a transitional hydrogen economy in the near future.

The harvesting of wind energy using large turbines is one of the fastest growing forms of renewable energy generation. Solar power generation options include solar photoelectric, in which electrical power is generated using solar panels; solar thermal, where the heat of the sun is concentrated using mirrors and used to run a turbine generator; and solar photochemical, where sunlight incident on a photocatalyst splits water into hydrogen and oxygen directly. There is a natural geographical synergy between wind-energy generation at higher latitudes, e.g. northern Europe, and solar generation in well-insolated regions at lower latitudes, e.g. the Sahara.

The total solar energy flux incident on the Earth's surface exceeds our global energy consumption by a factor of around 10,000; therefore our civilization's energy needs could in theory be met by intercepting only a small fraction of the incident solar radiation [38]. In practice, however, this is a considerable scientific, technological and socio-economic challenge. Ambitious plans have been proposed for large-scale solar power projects in well-insolated regions. Abbott has proposed the term "solar hydrogen economy" to describe the near-exclusive use of solar energy for electricity and hydrogen generation [39].

A potentially important application for hydrogen as an energy vector, in the context of sustainable energy generation, is in energy load balancing, or buffering, and storage. The process of harvesting energy from the environment is susceptible to the variations in the natural environment, both predictable and unpredictable: solar power will not work at night or under heavy cloud, wind power cannot be generated during calm. Some form of load balancing will therefore be necessary in order for renewable energy to deliver reliable, continuous power to the electricity distribution grid.

The production of hydrogen from water by electrolysis is a process that copes very well with variations in load. Therefore a potential load balancing mechanism is to apply the variable or intermittent power obtained from renewables to the production of hydrogen by electrolysis, and either to make use of the hydrogen in vehicles or to reconvert the stored hydrogen to electrical power using fuel cells at a steady rate suitable for delivery to the electricity grid. A project of this nature, using an electrolysis facility to buffer and store wind-generated electricity and to fuel hydrogen buses, is under construction in Hamburg by Vattenfall [16]. Thus two problems—the need for energy to produce hydrogen, and the tendency of renewable energy generation to be intermittent and poorly matched to demand—may in fact be each other's solution.

4.4 Storage and Distribution

There are a number of key issues which have so far prevented the widespread use of hydrogen. One of the chief problems is, of course, that gaseous hydrogen at room temperature and pressure takes up an impractically large amount of space. Hydrogen is typically stored as highly compressed gas or as cryogenic liquid. The work done in compressing or liquefying the hydrogen is a significant fraction of its energy content. The weight and volume of a suitable high-pressure vessel or insulated cryogenic tank can be considerable, posing a serious difficulty for hydrogen fuel cell vehicles.

It is widely believed that a really practical hydrogen FCEV will require a different form of hydrogen storage, either by physisorption in a highly porous material or chemisorption in some form of hydride; it must be easily and repeatedly reversible, rapidly taking up hydrogen at a fueling station and releasing it to the fuel cell when needed, with high enough mass and volumetric densities when fully loaded that the mass and volume of the fuel storage system are practical. The achievement of such a hydrogen store is widely regarded as one of the greatest challenges in the transition to a hydrogen economy involving hydrogen fuel cell vehicles.

Figure 3 summarizes the principle methods of hydrogen storage, drawing attention to the system's materials weight percent hydrogen, the operating temperature and the energy required for hydrogen release [5, 40]. Although liquid and compressed gaseous hydrogen are of course 100 wt% hydrogen in themselves, this figure does not account for the weight of the cryogenic or high-pressure storage system; liquid H_2 and compressed gas H_2 storage systems achieve 5–10 wt% hydrogen overall. It is worth noting that hydrocarbons such as octane, considered as hydrogen storage materials, achieve better than 15 wt% hydrogen. It is therefore a major challenge for new hydrogen storage methods to compete with current fossil fuels in terms of energy density.

Arguably the best-developed methods for solid-state hydrogen storage are based on metal alloy or interstitial hydrides [41–43]. Several types of interstitial hydrides exist which operate at different temperatures and pressures from below zero up to 300°C and above. Transition-metal hydrides are too heavy for vehicular applications but have potential applications in stationary hydrogen storage (e.g. for hydrogen generated from intermittent sources) or in maritime applications, where heavy hydrogen storage materials can be used in ballast. Attractive, new-generation hybrid storage systems offer a combination of chemisorption and storage under pressure; essentially a pressurized storage tank is filled with an interstitial metal hydride, resulting in a system in which working pressures are lowered considerably. In conclusion, we note, however, that currently no storage material meets the US Department of Energy targets for storing and releasing hydrogen fuel on demand.

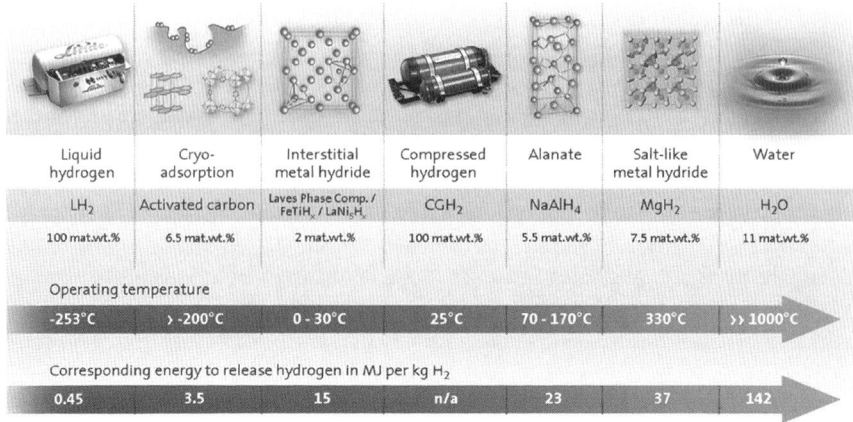

Liquid hydrogen	Cryo-adsorption	Interstitial metal hydride	Compressed hydrogen	Alanate	Salt-like metal hydride	Water
LH$_2$	Activated carbon	Laves Phase Comp. / FeTiH$_x$ / LaNi$_5$H$_x$	CGH$_2$	NaAlH$_4$	MgH$_2$	H$_2$O
100 mat.wt.%	6.5 mat.wt.%	2 mat.wt.%	100 mat.wt.%	5.5 mat.wt.%	7.5 mat.wt.%	11 mat.wt.%

Operating temperature

-253°C	> -200°C	0 - 30°C	25°C	70 - 170°C	330°C	>> 1000°C

Corresponding energy to release hydrogen in MJ per kg H$_2$

0.45	3.5	15	n/a	23	37	142

Fig. 3 Overview of hydrogen storage methods and materials (from [40])

4.5 Infrastructure

The safety of hydrogen production, storage and use is not only a technological issue but it is also a psychological and sociological issue facing the adoption of the hydrogen economy. To be accepted by the public, hydrogen must be considered safe; moreover, hydrogen technology must come to be seen as normal and mainstream rather than as a fringe interest. The risks of hydrogen use, and public attitudes to hydrogen technology, have recently been reviewed by Ricci et al. [44, 45].

The establishment of a hydrogen infrastructure also has its own challenges. Compared to natural gas (for which pipeline distribution over thousands of kilometers is quite practical), hydrogen is more difficult to work with as the small, light molecule diffuses through metals, causing not only loss of hydrogen but also dangerous embrittlement of the pipeline structure. Dedicated hydrogen pipelines must therefore be made from stainless steel and cost approximately six times as much as an equivalent natural gas pipeline. Some tens of percent of hydrogen can be mingled with methane for delivery, at least locally, through the natural gas network. This could be a viable delivery mechanism in a transitional hydrogen economy if appropriate gas separation technologies are also developed.

Distribution of liquid hydrogen or compressed gas by tanker trailers suffers from the issue of the energy input to achieve liquefaction or high compression, and is only practical over short distances and small amounts. We must therefore question whether the centralized production and long-range distribution of hydrogen is viable, compared to long-range distribution of electrical power with local generation of hydrogen by electrolysis where it is needed.

Another fundamental barrier to the widespread use of hydrogen is cost. Hydrogen production, storage, distribution and use have not had the century of large-scale, mass-market investment and development that has made petroleum so dominant in the transport sector. Perhaps the biggest challenge facing the utilization of the PEMFC for hydrogen vehicles are lifetime and cost. At present, the electrodes are made with a nanostructured platinum catalyst, an expensive material and one which is sensitive to poisoning by carbon monoxide, ammonia and by sulfur compounds. Materials costs, and the costs of manual production of components, give a current price of $2,000–4,000/kW, and the lifetime of the cell is below 2,000 h [25]. For hydrogen FCEVs to be economically competitive with gasoline vehicles, and thus be an attractive option for consumers in the transitional hydrogen economy, costs must be reduced below $100/kW and lifetime must be increased to above 5,000 h. While mass production can help in cost reduction, cheaper and more durable catalyst and electrolyte materials are urgently needed. Ekins et al. mention a company, ITM Power, which seems to have made some interesting developments in membrane technologies for electrolysis and fuel cells, making hydrogen production significantly cheaper [31].

4.6 Challenger/Complementary Technologies for Hydrogen

Electric vehicles already exist in the form of both battery electric vehicles (BEVs) and hybrid ICE/battery vehicles such as the well-known Toyota Prius. BEVs can be practical for short-range commuter driving, but suffer from limited operational range (100–150 km). The energy density of even a modern Li-ion battery (around 120 W h/kg) is only around 1% of the energy density of gasoline, so unreasonable large battery weights (on the order of 1000 kg) would be required to give BEVs the ~ 500 km range of a typical modern small car. Battery recharging is also very slow compared to the rapid refueling of a gasoline vehicle.

Batteries do possess attractive features for electric vehicles, especially their energy efficiency; about 90% of a battery's stored energy can be extracted as useful electricity, and electric motors are about 95% efficient, giving BEVs a net efficiency of around 85%. This is three to four times the efficiency of an ICE and at least twice the efficiency of a hydrogen FCEV. Well-to-wheel analyses indicate [34] that it would be environmentally and energetically advantageous to operate BEVs from the current EU grid electricity mix rather than conventional ICEs; by contrast it would be disadvantageous to generate hydrogen from grid electricity and operate hydrogen FCEVs.

In principle, a dramatic improvement in battery technology could make the range, power and charging/discharging kinetics of an electric vehicle fully competitive with conventional ICEs. In this scenario the sustainable energy economy would be based on electricity generated from renewable resources, distributed by an improved electricity grid, and used in fully electric vehicles. The principal contribution of hydrogen technology to this scenario would be providing buffering and load-leveling for intermittent renewables.

Another approach centers on ICEs running on "biofuels" generated from biomass or with synthetic fuels produced by CO_2 conversion. The use of biofuels is often regarded as carbon-neutral as the carbon in the fuels was fixed from the atmosphere by vegetation. Bioethanol, from the fermentation of plant sugars, is currently the mostly widely used biofuel. Biodiesel, produced by processing of vegetable oils, is potentially a direct replacement for diesel in ICEs. The environmental benefits of current approaches to biofuels are questionable, however, as their production competes strongly with agricultural food production and planting for biofuels is associated with deforestation. Moreover, any use of fertilisers in crop production carries with it a significant carbon penalty; hydrogen is generated from methane for ammonia synthesis in fertiliser production.

As noted earlier, an attractive potential alternative to biodiesel is using hydrogen to manufacture synthetic hydrocarbon fuels (methane, methanol or diesel) by chemical means, e.g. the Fischer–Tropsch process, from water, sustainable hydrogen and carbon dioxide. The development of efficient and cost-competitive methods of converting carbon dioxide into fuels could address the issue of both carbon sequestration and sustainable and economic production of carbon-neutral fuels. Such carbon-neutral or sustainable organic fuels offer the attractive possibility of decarbonising our transport system without the undoubted paradigm shifts required by conversion to a hydrogen energy economy, or indeed by the electrification of the vehicle fleet. Synthetic fuels are of course only carbon-neutral if both the energy and the hydrogen required for their synthesis are renewably generated.

We can thus visualize a form of hydrogen economy based, not on hydrogen fuel cell vehicles, but on the large-scale renewable production of hydrogen as a chemical feedstock for the synthesis of fuels, and to replace hydrocarbon-derived hydrogen in chemical synthesis, particularly of ammonia for fertilizers. The synthesis of fuels could be significant in keeping the enormous current base of internal-combustion vehicles in action while escaping from our current dependence on fossil fuels. We should also note that hydrogen can be used as a fuel in internal combustion engines with minimal modification, allowing the use of bivalent ICEs capable of employing hydrocarbon or hydrogen fuels according to availability. Although this application of hydrogen lacks the efficiency benefits of fuel cell use, it could be a valuable part of the transition to a hydrogen economy, allowing for the gradual development of a widespread hydrogen fueling infrastructure and of large-scale renewable hydrogen generation, and thus making any later transition to hydrogen fuel cell vehicles less challenging. The synthesis of fertilizer from renewable hydrogen is significant not only for the carbon economics of biofuels, but also for global agriculture generally.

5 Conclusions

There are many forms of "hydrogen economy" and the conventional vision, with hydrogen fuel cell vehicles becoming dominant in personal transport, is only one approach. Other forms of a hydrogen economy include the use of hydrogen essentially as an energy store, making hydrogen technologies a key ally of renewable electricity generation technologies and thus of electric vehicles. We have also highlighted the utility of hydrogen as a chemical feedstock, with renewably generated hydrogen taking over from "black" hydrogen for ammonia (fertilizer) production, and being a key feedstock for the conversion of carbon dioxide to synthetic fuels including methanol for fuel cells and synthetic hydrocarbons for diesel ICEs. Certain technologies, most importantly for the efficient generation of hydrogen using renewable energy, are common to any vision of a hydrogen economy and thus must be a very high research and development priority. Parallel technologies, including battery electric vehicles, should be considered as allies rather than competitors.

It is most important that policy-makers are kept aware of the full breadth of possibilities for hydrogen economies, so that difficulties in one area of hydrogen technology—in particular, the development of a commercially practical on-board hydrogen store for vehicles—are not regarded as a disincentive for investment, research and development in other areas such as electrolytic or photochemical water-splitting, static hydrogen storage and static fuel cells.

A new programme, "UKH2Mobility", has been announced on 18 January 2012 by the Department for Business, Innovation and Skills. The programme brings together public and private participants: three government departments (DBIS, the Department for Transport, and the Department for Energy and Climate Change) and thirteen industry participants (Air Liquide, SA; Air Products PLC; Daimler AG; Hyundai Motor Company; Intelligent Energy Limited; ITM Power PLC; Johnson Matthey PLC; Nissan Motor Manufacturing (UK) Limited; Scottish and Southern Energy plc; Tata Motors European Technical Centre plc; The BOC Group Limited; Toyota Motor Corporation; Vauxhall Motors). UKH2Mobility will evaluate the potential for hydrogen as a fuel for Ultra Low Carbon Vehicles in the UK before developing an action plan for an anticipated roll-out to consumers in 2014/15. This programme indicates a recognition by both public and private sectors in the UK of the potential role of hydrogen in transport.

Acknowledgments AS thanks Royal Society and SAW thanks Leverhulme Trust for funding. The authors are grateful to an anonymous reviewer for the helpful comments.

References

1. Crabtree GW, Dresselhaus MS, Buchanan MV (2004) The hydrogen economy. Phys Today 57(12):39–44
2. Edwards PP, Kuznetsov VL, David WIF (2007) Sustainable hydrogen energy, In: Armstrong F, Blundell K (eds) Energy...beyond oil. Oxford University Press, Oxford, pp 156–168
3. Dresselhaus MS (2009) Progress and challenges of a hydrogen economy. In: Materials issues in a hydrogen economy, pp 3–12
4. Young S (2001) Tomorrow's energy: hydrogen, fuel cells and the prospects for a cleaner planet. Nature 414(6863):487–488

5. Wells SA et al (2010) Hydrogen economy, in energy production and storage. In: Crabtree RH (ed) Inorganic chemical strategies for a warming world. Wiley, Chichester
6. Stern N (2006) The economics of climate change. The stern review. Cambridge University Press, New York
7. IPCC (2007) Climate Change 2007: impacts, adaptation and vulnerability. In: Parry ML et al (eds) Contribution of working group II to the fourth assessment report of the intergovernmental panel on climate change. Cambridge University Press, Cambridge
8. Vincent D (2007) Arresting carbon dioxide emissions: why and how? In: Armstrong F, Blundell K (eds) Energy…beyond oil. Oxford University Press, Oxford, pp 9–34
9. Rokke NA (2006) CO2 capture, transport and storage for coal, oil and gas: technology overview. In: Jamasb T et al (eds) Future electricity technologies and systems. Cambridge University Press, Cambridge, pp 179–194
10. IPCC (2005) In: Metz B et al (eds) Special report on carbon dioxide capture and storage. Prepared by working group III of the intergovernmental panel on climate change. Cambridge University Press, Cambridge
11. Sartbaeva A et al (2008) Hydrogen nexus in a sustainable energy future. Energy Environ Sci 1(1):79–85
12. Ekins P (2010) In: Ekins P (ed) Hydrogen energy: economic and social challenges. Earthscan Publications Ltd, Milton, p 272
13. Edwards PP et al (2008) Hydrogen and fuel cells: towards a sustainable energy future. Energy Policy 36:4356–4362
14. Tomei J (2009) Planning for a transition to a hydrogen economy: a review of roadmaps. In: UKSHEC social scince working paper no. 37. Policy Studies Institute, London
15. Wietschel M, Ball M, Seydel P (2009) Hydrogen today. In: Ball M, Wietschel M (eds) The hydrogen economy. Cambridge University Press, Cambridge, pp 254–270
16. Tollefson J (2010) Fuel of the future? In: Nature. Macmillan Publishers Limited, London. pp 1262–1264
17. Development of fuel cell and hydrogen technologies (2009–2010) New energy and industrial technology development organization (NEDO), fuel cell and hydrogen technology development department. http://www.nedo.go.jp/content/100079670.pdf
18. Roadmap for fuel cell and hydrogen technology development by new energy and industrial technology development organization (NEDO) (2008) https://app3.infoc.nedo.go.jp/informations/koubo/events/FA/nedoeventpage.2008-06-18.1414722325/
19. HyWays (2008) The European hydrogen roadmap 2008. http://www.hyways.de/docs/Brochures_and_Flyers/HyWays_Roadmap_FINAL_22FEB2008.pdf
20. Icelandic Ministry for Foreign Affairs—Towards the hydrogen economy (2003) http://www.mfa.is/media/MFA_pdf/Hydrogen.ppt
21. Icelandic New Energy (2008) Promoting hydrogen in Iceland 2008. http://www.newenergy.is/en/
22. European Commission (2003) Hydrogen energy and fuel cells, a vision of our future. European Commission, Westminster
23. European Hydrogen & Fuel Cell Technology Platform. Deployment Strategy (2005) https://www.hfpeurope.org/hfp/keydocs
24. International Energy Agency (2005) Prospects for hydrogen and fuel cells. http://www.iea.org/textbase/nppdf/free/2005/hydrogen2005.pdf
25. U.S. Department of Energy (2006) Hydrogen posture plan: an integrated research, development and demonstration plan. http://www.hydrogen.energy.gov/pdfs/hydrogen_posture_plan_dec06.pdf
26. US National Research Council, National Academy Press (2008) Transitions to alternative transportation technologies-a focus on hydrogen. US National Research Council, National Academy Press, Washington, p 142
27. Wald M (2009) U.S drops research into fuel cells for cars. The New York Times, New York
28. Lattin WC, Utgikar VP (2007) Transition to hydrogen economy in the United States: a 2006 status report. Int J Hydrogen Energy 32(15):3230–3237
29. E4tech, Element Energy, Eoin Lees Energy (2004) A strategic framework for hydrogen energy in the UK. Final report to Department of Trade and Industry, UK

30. Eames M, McDowall W (2010) Sustainability, foresight and contested futures: exploring visions and pathways in the transition to a hydrogen economy. Technol Anal Strategic Manag 22(6):671–692
31. Ekins P, Hawkins S, Hughes N (2010) Hydrogen technologies and costs. In: Ekins P (ed) Hydrogen energy: economic and social challenges. Earthscan Publications Ltd, London, pp 29–58
32. Ball M, Wietschel M, (2009) In: Ball M, Wietschel M (ed) The hydrogen economy: opportunities and challanges. Cambridge University Press, Oxford
33. Subramani V et al (2007) Recent advances in catalytic production of hydrogen from renewable sources. Catal Today 129(3–4):275–286
34. Weindorf W, Bunger U (2009) Energy-chain analysis of hydrogen and its competing alternative fuels for transport. In: Ball M, Wietschel M (eds) The hydrogen economy. Cambridge University Press, Cambridge, pp 199–253
35. Ball M, Weindorf W, Bunger U (2009) Hydrogen production. In: Ball M, Weindorf W (eds) The hydrogen economy. Cambridge University Press, Cambridge, pp 277–308
36. IEA (International Energy Association) (2007) Hydrogen production & distribution. IEA energy technology essentials. OECD/IEA, Paris
37. ITER (2007) www.iter.org
38. Zweibel K, Mason J, Fthenakis V (2008) A solar grand plan. Scientific American, pp 64–73, Jan 2008
39. Abbott D (2010) Keeping the energy debate clean: how do we supply the world's energy needs? Proc IEEE 98(1):42–66
40. Eberle U, Felderhoff M, Schuth F (2009) Chemical and physical solutions for hydrogen storage. Angew Chem Int Ed 48(36):6608–6630
41. Grochala W, Edwards PP (2004) Thermal decomposition of the non-interstitial hydrides for the storage and production of hydrogen. Chem Rev 104(3):1283–1315
42. Harris IR et al (2004) Hydrogen storage: the grand challenge fuel. Cell Rev 1(1):17–23
43. Orimo SI et al (2007) Complex hydrides for hydrogen storage. Chem Rev 107(10):4111–4132
44. Ricci M, Bellaby P, Flynn R (2010) Hydrogen risks: a critical analysis of expert knowledge and expectations. In: Ekins P (ed) Hydrogen energy: economic and social challenges. Earthscan Publications Ltd, London, pp 217–240
45. Ricci M et al (2010) Public attitudes to hydrogen energy: evidence from six case studies in the UK. In: Ekins P (ed) Hydrogen energy: economic and social challenges. Earthscan Publications Ltd, London, pp 241–264

Part II
Road

Foreword
James Smith

We are hooked on travel. It is part of the human condition. Moreover, the development of the global economy over the last 150 years has increasingly been underpinned by massive transportation of raw materials, work in progress, finished goods, food and people. In the first half of this century the global economy is likely to increase by a factor of five. Hundreds of millions of people will be lifted out of poverty. Huge growth in transport is an essential enabler.

The number of vehicles on global roads will have doubled to about 2 billion by the middle of the century. They will travel the equivalent of 80,000 trips to the sun and back each year. Confronted with the surging demand for transport, three vital questions emerge. How do we contain energy demand for transport? Where do we get the increased energy? How do we significantly reduce CO_2 emissions?

The modern industrial world has been enabled by the phenomenal energy density of fossil fuels. Energy flows out of a petrol pump over 2,000 times faster than out of a 13 amp plug. This means 2 min at the petrol pump gives a car a range of about 600 miles. But 5 min with a 13 amp plug gives about 1 mile. This disparity is the essence of the challenge of creating a low carbon future for road transport. Replicating the energy density of fossil fuels is a challenge to fundamental science and will demand engineering on an enormous scale.

The solutions to the energy and environmental challenges of road transport will come from many sources and emerge over a series of time horizons. We have to work on all the solutions and all the time horizons. Major reduction in the carbon intensity of the installed technological base for road transport is a crucial starting point. We need to combine ambition, determination and realism. We need collaboration among a wide range of interests. Deploying the power of the market will be vital.

The need for action is urgent. A huge array of technological and engineering skills will be required. But that will not be enough. We need also to harness the social and political dynamics of change. And one of the most important sciences will be in understanding our behaviour, so we, as individuals, make low energy and low carbon transport choices.

We ought to be daunted by this challenge. But more importantly we should be invigorated by the worth of the cause and by the opportunities for individuals, companies and countries.

James Smith
Chairman of the Carbon Trust UK
Former Chairman Shell UK

Pathways to Sustainable Road Transport: A Mosaic of Solutions

Mark Gainsborough

Mobility means access—access to markets, to jobs, to education, to economic opportunity, to extended communities. Road transport is therefore fundamental to economic growth and improved living standards, enabling trade and social interaction through the movement of goods and people. It is also fundamental to Shell's business. Most of our products enable some form of mobility. This includes the fuels that power vehicles, such as petrol, diesel and biofuels. It also includes less obvious products, such as lubricants to help engines and drive trains run smoothly, bitumen that paves roads and petrochemical-based plastics that make up more than 10% of the average new car.

It is therefore critical that as a commercial business, Shell anticipates, leads and evolves with developments in the road transport sector. To succeed, we will have to meet the demands of our shareholders and the wants and needs of our customers. However, we will have to balance this with the role we must play in developing a more sustainable road transport system.

Transport is a significant consumer of energy and a source of greenhouse gases, including carbon dioxide (CO_2), which lead to climate change. Today, road transport accounts for 17% of global energy use and energy-related CO_2 emissions [5]. Demand is expected to grow rapidly as emerging nations continue through their most intensive phase of economic development. By 2050, the global car park could triple to 2 billion and the number of trucks could double [5]. Unrestricted, this surging demand could see fuel consumption and emissions of CO_2 from the world's cars double. This is the challenge for road transport.

Mark Gainsborough—Executive Vice President of Strategy, Portfolio and Alternative Energies, Shell.

M. Gainsborough (✉)
Shell Centre, London, SE1 7NA, UK
e-mail: alex.burnett@shell.com

O. Inderwildi and Sir David King (eds.), *Energy, Transport, & the Environment*,   167
DOI: 10.1007/978-1-4471-2717-8_10, © Springer-Verlag London 2012

In response, governments around the world are introducing ambitious CO_2 reduction targets. But there is no single solution. Alternative, low-carbon fuel and energy sources and new drive train technologies will be essential. But barriers to achieving global mitigation targets in transport are significant. They include the challenge of scaling-up low-carbon alternatives and the dependence of promising low-carbon vehicle technologies on the evolution of a decarbonised energy supply and associated infrastructure.

The internal combustion engine and liquid fuels, which have fuelled mobility for over 100 years, will continue to be indispensible. Vehicle manufacturers will continue to improve the efficiency of vehicles and fuel providers will make further improvements to conventional hydrocarbon fuels that will be blended with increasing volumes of low-carbon biofuels and other fuel components such as gas to liquids (GTL) gasoil.

Any response will also have to take into consideration significant demographic shifts. By 2050, three-quarters of the world's 9 billion people will live in cities, which already account for 80% of CO_2 emissions [12]. Smarter infrastructure and behaviour change brought about by smarter policy will be required to keep people and goods moving and avoid the worst impacts of climate change and local air pollution.

There is no 'silver bullet' solution. Political, industrial and individual choices will determine our progress towards sustainable mobility.

1 Energy Supply and Demand: A Zone of Uncertainty

Despite recent economic turbulence, growth in global population and prosperity continues to drive energy demand as emerging nations enter their most energy-intensive phase of economic development. In 2010, oil demand increased by 3% and underlying global demand for energy by 2050 could triple from its 2000 level if emerging economies follow historical patterns of development [2, 3]. Up to 3 billion energy consumers will be added to the world's population and these people will require access to commercial energy and transport.

The energy system will struggle to match this surging demand for easily accessible energy. In broad-brush terms, natural innovation and competition could spur improvements in energy efficiency to moderate underlying demand by about 20% over this time. Ordinary rates of supply growth—taking into account technological, geological, competitive, financial and political realities—could naturally boost energy production by about 50%. But this still leaves a gap between business-as-usual supply and demand of around 400 Ej/a—the size of the energy industry in 2000 [10].

The Shell scenario planning team describes this as a 'zone of uncertainty' looming over the global economy. To bridge this gap we will need to see an enormous expansion in energy supplies, coupled with extraordinary measures to moderate demand.

2 Towards a More Sustainable Energy System

Satisfying this surging demand for energy is not the only challenge. We are also witnessing a historic shift as global efforts to develop an environmentally sustainable energy system gain momentum. At the Copenhagen Climate Conference in 2009, there was consensus that we needed to significantly reduce CO_2 emissions to keep the global temperature rise to below 2°C in order to mitigate the worst impacts of climate change. We cannot afford to wait.

Delays in reducing CO_2 will result in much higher likelihood of global temperature rises and later 'catch up' will be extremely difficult because of the reduction rates that will be required.

In response, governments around the world have introduced legislation to curb their CO_2 emissions. In Europe, the European Union has a CO_2 emissions reduction target of 20% on 1990 levels by 2020 and the U.S. has committed to make an economy-wide greenhouse gas emissions reduction of 17% on 2005 levels by 2020. Importantly the Chinese Government also aims to deliver a 17% reduction in CO_2 emissions per unit of GDP as part of its new Five Year Plan.

These targets are driving demand for greater efficiency and for renewable sources of energy. The Shell scenarios teams estimate that renewable energy sources, such as wind and solar, could supply as much as 30% of global energy by 2050, compared with 13% today [11]. This would be a big achievement given the financial and technical hurdles facing new energy sources. Yet, even with renewables supplying 30%, fossil fuels will continue to supply around 60% of global energy, with nuclear accounting for the remaining 10% [11]. A sustainable energy system will, therefore, be one in which cleaner fossil fuels, as well as renewable energy sources, meet a growing share of demand.

3 The Challenge for Road Transport

Road transport already accounts for approximately 17% of energy use and energy-related CO_2 emissions [5] and demand for mobility is expected to surge in the coming decades. The IEA estimates that by 2050 the global car park could triple to 2 billion and the number of trucks, double. As a result, fuel consumption and emissions of CO_2 from the world's cars could double [5].

Eighty per cent of this growth is expected to occur in developing economies that will experience a sharp rise in vehicle numbers as their economic development continues. China's aggressive motorway building programme and rising prosperity is key, as is demand growth in the Middle East and other rapidly developing economies. Without significant Policy-led developments in the road transport sector, this surging demand could see transport CO_2 emissions increase by up to 80% by 2050 [5].

As part of their wider efforts to tackle CO_2 emissions, governments around the world are introducing road transport policies. For example, in the fuels sector the

European Union Renewable Energy Directive requires 10% (energy basis) of road vehicle fuel from renewable sources by 2020, the USA Energy Independence and Security Act 2007 requires 36 billion gallons of renewable road transport fuels by 2022 and in California, the Low Carbon Fuel Standard calls for a reduction of 10% in the carbon intensity of California's transport fuels by 2020.

Most developed countries also have legislation designed to improve the fuel efficiency of passenger cars. In the U.S., the Energy Independence and Security Act (EISA) requires a 40% improvement in miles per gallon offered by cars and light trucks by 2020 (compared to 2007). In Europe, the EU requires a reduction in tailpipe emissions from their average of 160 g CO_2/km in 2006 to 130 g CO_2/km by 2015. Japan also has fuel economy legislation and so, significantly, does China. However, many non-OECD countries do not and this is a cause for concern as well as an opportunity.

4 A Mosaic of Solutions

Meeting policy targets and reducing the impact of road transport is a challenge for fuel providers and vehicle manufacturers. Despite media hype around specific technologies, the reality is that there is no single technology solution. All sustainable transport solutions will be needed and countries and regions will pick portfolios of solutions based on a number of factors—cost, energy security, infrastructure, demographics, job creation and consumer behaviour. This will lead to a global 'mosaic' of transport solutions, in which all vehicle technologies and fuels will exist alongside each other.

Demand for traditional hydrocarbon fuel components, biofuels, hydrogen, electricity and natural gas—including GTL, CNG and LNG—will increase, resulting in multiple combinations of fuel and drive train pathways optimised to local circumstances. The choice of fuel and drive train options will also vary within countries. Some drive trains will be more suited to short journeys in urban areas, while others will be more appropriate for longer journeys or require power and torque for haulage, agriculture or construction.

This is the starting point for smarter road transport. There is no 'silver bullet' solution. Meeting the challenge will require coordination and cooperation between governments, fuel suppliers, vehicle manufacturers and road transport consumers.

5 Smarter Products

Automotive manufacturers are making significant progress, developing efficient drive trains that emit less CO_2 per unit of distance travelled. This is being achieved at relatively low cost by downsizing engines, light-weighting and reducing vehicle size. Manufacturers are also successfully developing new drive trains, powered by alternative technologies including electricity and hydrogen. These technologies have great potential, but can take a long time to achieve significant market penetration in a complex energy sector. This is not always an easy message to deliver, because people want to see similar rates of change in the energy system as have been seen in other sectors, such as information technology.

Because the scale of the energy system is so large, it takes time to build the human and industrial capacity to achieve substantial deployment. A paper for Nature Magazine by Gert Jan Kramer and Martin Haigh, showed that it takes 30 years to span the 10,000-fold growth needed to get from pilot-plant scale up to 1–2% of the world's total primary energy—a sustained growth rate of 26% each year [6].

To meet the increasing demand for mobility, the internal combustion engine will therefore continue to play a critical role in the coming decades. By extension, demand for liquid fuels, which have fuelled mobility for over 100 years, will continue to grow. At Shell, we expect consumption to rise by around 20% between 2010 and 2030. In response, fuel suppliers will continue to introduce smarter, more efficient conventional hydrocarbon fuel components that are more efficient and emit lower CO_2 emissions. But this is not easy. Smarter fuel means different things to different people. Customers demand fuels that are widely available, high performance, have good fuel economy and are compatible with today's infrastructure—at an affordable price. Vehicle manufacturers demand fuels that are high performance, have good fuel economy, reduced emissions and ease of re-fuelling. Governments and regulators demand reduced emissions, good fuel economy and improved energy security.

Progress is being made. New fuel formulations, such as Shell's FuelSave efficient fuel formula, can achieve significant efficiency gains today. Conventional hydrocarbon fuel components will also increasingly be mixed with other fuel components to diversify supply and reduce local emissions. For example, the rapid emergence of abundant and affordable gas is expected to play an increasing role in the diversification of transport fuels. For example, gas to liquids (GTL) gasoil is a versatile, cost-effective alternative diesel fuel that can be blended with conventional diesel and biodiesel and used in the same vehicles and infrastructure. It is also cleaner burning that conventional diesel and trials in London, Berlin and Shanghai have shown that buses, taxis and trucks running on high concentrations of GTL gasoil can contribute to improved local air quality—important given the growing global urban population.

6 Biofuels Offer a Low-Carbon Alternative Today

To keep concentrations of CO_2 within the acknowledged 'safe' limit we need to significantly reduce CO_2 emissions today. Despite their future potential, we cannot afford to wait for alternative transport technologies such as electric mobility and hydrogen fuel cell to achieve significant market penetration. Any delay will make it extremely difficult to achieve the emissions reduction rates required to stay within the safe limit. It is essential that we make use of the options available to us today.

In the fuel sector, liquid hydrocarbon fuel components blended with sustainable biofuels represent the most realistic commercial solution to reduce CO_2 emissions in the road transport sector over the next 20 years. Biofuels can significantly reduce CO_2 emissions, are compatible with the existing liquid fuel infrastructure and are available in commercial quantities today. Used in vehicles, biofuels emit similar amounts of CO_2 as conventional fuels, but unlike crude oil, the biomass they are made from has recently absorbed CO_2 from the air during growth. In theory, this leaves the carbon balance neutral. The actual CO_2 reduction of biofuels depends on a wide range of factors, including the feedstock used, how it is processed, distributed and used in vehicles. For example, Brazilian sugar cane ethanol produces around 70% lower CO_2 than conventional fuels and typically has lower CO_2 emissions than ethanol produced from other feedstocks such as corn (maize) grown in the US [1].

Biofuels also help diversify the liquid road transport fuel pool and reduce dependence on oil based transport fuels. This offers the prospect of improved energy security, particularly when domestic feedstocks are used. For some countries, biofuels also offer economic and rural development opportunities.

Recognising these benefits, policy makers in more than 50 countries around the world have developed or are developing renewable fuels policies that contain provisions for biofuels. Legislative drivers differ—cost, energy security, support for domestic agriculture, job creation and the environment. The result is an international market that represents 3% of the global road transport fuel supply today and could increase to as much as 27% of all global transport fuels by 2050 [4].

To support this growing market and further reduce CO_2 emissions, biofuels will need to be blended with conventional fuel components in higher concentrations than they are in most markets today. This is a challenge for fuel providers and vehicle manufacturers, who must simultaneously introduce modifications to car fuel systems and to fuels so that mandates are met and motorists driving new and older vehicles can find the right fuel for their car. Because governments have generally regulated the fuel and vehicle industries separately, there is no defined legislative pathway for the introduction of new fuels and compatible vehicles. This issue is referred to as the 'blend wall'. To overcome the 'blend wall' will require cooperation and coordination between vehicle manufacturers, fuel providers and governments.

There are also some difficult issues linked with increased production of bio-fuels. Strict environmental and social safeguards are needed. We know that the 'well-to-wheel' CO_2 performance of biofuels can vary widely depending on the feedstocks and production and processing techniques used. There are also concerns about working conditions, competition for agricultural land, land use change, impacts on local communities and use of water. It is essential that measures are taken to ensure that biofuels are produced in a more sustainable way—safeguarding the environment and delivering benefits to society.

This will be achieved through encouraging and rewarding biofuels that demonstrate good social and environmental performance through international sustainability certification schemes. These include bodies such as Bonsucro for sugarcane, the Roundtable on Sustainable Palm Oil for palm oil and the Round Table on Responsible Soy Association for Soy.

7 Advanced Biofuels will be Needed

To meet the increasing demand for sustainable biofuels, advanced biofuels using new feedstocks such as crop wastes or inedible crops and new conversion processes offer the potential for improved CO_2 reductions and improved fuel characteristics. These biofuels will reduce the likelihood of competition with other industries for commodities, including food. They could also allow for higher blends of biofuels to be used, further reducing the overall emissions of liquid fuels.

Despite their promise, from a practical and commercial standpoint, it is unlikely that advanced biofuels will emerge in commercial quantities until the 2020s. This is because like other technologies, it will take time to overcome technical challenges and reduce costs. The industry is making good progress, but breaking down and converting new biofuel feedstock options such as straw into fuel is far more complex than converting the crops used to produce some of today's biofuels. Processing them efficiently at scale, in terms of cost and CO_2 emissions is challenging. So it will take time and considerable investment for these technologies to progress from a lab-based process to demonstration plants and towards commercial roll-out.

8 Hydrogen Fuel Cell and Electrification: Compelling Opportunities for the Future

Electric mobility and hydrogen fuel cell vehicles (FCVs) will make a significant contribution to emissions reductions and diversification in the road transport sector. However, in the near-term their offering is limited by the CO_2 profile of the source energy, the challenge of developing new vehicles and refuelling infrastructure and consumer acceptance.

When driven, FCVs convert hydrogen into electricity and produce only heat and water. The overall CO_2 footprint, however, depends on how the hydrogen has been produced and its journey to the vehicle. Currently, most (around 95%) hydrogen is produced at refineries and gas production sites from natural gas or by gasifying coal, for example at refineries or fertilizer plants. Hydrogen produced in this way can reduce local air emissions at point of use, but well-to-wheels CO_2 emissions are unlikely to be significantly reduced.

FCVs also require new drive trains and new dispensing systems. This presents both technical and commercial challenges. Good technical progress has been made at hydrogen demonstration refuelling stations, which have allowed suppliers, energy companies and the automotive industry to evaluate a range of different technologies and learn valuable lessons about costs, consumer behaviour and how to safely store and dispense hydrogen at different pressures. However, hydrogen continues to face significant challenges to commercial roll-out. To reduce costs and achieve its potential, hydrogen will require considerable cooperation between car makers, fuel and equipment suppliers and governments.

The industry will not be able to fund the commercialisation of hydrogen alone. Policy and funding mechanisms will be needed to establish public–private coalitions to accelerate development and provide incentives to ensure that companies making early investments are not disadvantaged. Without this political support, hydrogen will take longer to reach its commercial potential as a road transport fuel. A promising example in development is the German government sponsored H_2 Mobility programme set up to accelerate commercialisation. Policy makers around the world will be watching how the H_2 Mobility programme evolves.

Like hydrogen powered vehicles electric vehicles produce no emissions in use, but the CO_2 savings of electric vehicles and plug-in hybrids are strongly dependent on decarbonisation of the electricity grid. To significantly reduce CO_2 per unit of distance travelled will require substantial investment in low carbon electricity generation, such as renewable, nuclear or fossil fuels combined with carbon capture and storage (CCS). In the foreseeable future, coal is likely to be the main source of (marginal) electricity for cars with clear consequences for the well-to-wheel CO_2 profile of electric vehicles.

Electrification also faces technology and infrastructure challenges. The uptake of electric vehicles is critically dependent on the cost of batteries and range limitations. Two types of electric vehicles that take electricity from the grid to charge their onboard batteries are emerging. Battery electric vehicles (BEVs) are powered only by a battery and are most suited to short-distance travel in urban areas. This is because their range is limited by battery life and dependency on the availability of recharging infrastructure. Globally, 60% of daily drives are less than 40 miles, but motorists have become accustomed to a vehicle range of at least 200 miles. The cost, mass and volume of batteries to meet this requirement remain considerable.

Plug-in hybrid electric vehicles (PHEVs) attempt to address these limitations. They are powered primarily by the battery, but switch to liquid fuel when needed. This provides users with greater flexibility since they can be used for unplanned journeys or for longer distance travel where there is insufficient electricity charging infrastructure or time to recharge the vehicle. But PHEVs incur the extra costs of dual power sources.

Even after these challenges have been overcome cost reduction and consumer acceptance of a new technology, infrastructure and driving experience remain unknown.

9 Governments can Accelerate Technology Development

Fuel providers and automotive manufacturers are making significant investments in new transport technologies and good progress is being made. But the reality is that new technologies are only economically viable at the commercial deployment stage of development. The 'discover', 'develop' and even 'commercial demonstration' stages of new transport technologies are unlikely to generate a profit.

Governments can help accelerate the scale-up of these technologies. In the longer term they can create and support the market conditions required for development of the market. Mechanisms include rewarding sustainable low carbon fuels, creating regulatory frameworks that stimulate market driven innovation, providing regulatory certainty for periods that will encourage long-term investment and setting challenging but achievable goals—with realistic timescales for implementation.

For example in Brazil, tax breaks and mandated infrastructure encouraged the introduction of flex fuel vehicles and stimulated the demand for sustainable low carbon sugarcane ethanol. This transformed an energy infrastructure that was fossil fuel based in the late 1960s into one that used substational amounts of ethanol by the mid 1980s.

In the near-term, governments can accelerate the pace of development with specific and time-bound support to assist new technologies in the demonstration phase. This is not unique for new fuels. It is the same, for example, for carbon capture and storage and for wind.

For example, in the United States, the Department of Energy has set aside up to $36 million to fund small-scale projects that will advance the technology improvements and process integration needed to produce 'drop-in' advanced biofuels. This funding is provided to projects in the develop and demonstrate stages of technology development to help improve the economics and efficiency of biological and chemical processes that convert non-food biomass feedstocks into replacements for petroleum-based feedstocks, products and fuels.

10 Smarter Use

Much attention is focused on smarter products and technologies being developed by fuel providers and vehicle manufacturers. But to meet the mobility challenge, smarter products will have to be complemented by smarter use. Our behaviour as transport consumers and the decisions we make about why, when and how we travel will have a significant impact on energy consumption and the impact that we have on the environment. This applies to a customer making a trip to a retailer for groceries as much as to large corporations making deliveries to their customers around the world.

Increasingly, we will ask why we are travelling. Limiting use of road transport is politically and socially contentious because of the economic and social opportunities that road transport provides. But embracing technological advances increasingly allows us to make use of alternatives to travel. The internet, conference calling or video conferencing can have a significant impact on energy consumption and emissions.

When we travel is also important. Avoiding peak times of the day or year will often mean there is less traffic, journeys take less time, are more economical and reduce congestion for others during peak traffic times. Increasingly, broad-minded employers will support more flexible working to enable people to change their travelling patterns and reduce their impact on the environment.

Perhaps most important in terms of our behaviour as transport consumers is how we travel. Good quality public transport and innovative alternatives, such as car clubs and bike hire schemes are critical to enabling choice.

Driving the right-sized vehicle for the task in hand also has a significant impact, as does our behaviour as drivers. 'Eco-driving'—driving at a moderate, steady pace and avoiding heavy breaking and acceleration—can have a significant impact on fuel consumption.

11 Where is the Incentive?

As transport consumers, our willingness to make smarter choices about the way we travel and transport goods is determined by a number of factors. Economics, information and politics play a significant role.

Cost is often the most significant determinant of choice. Simply put, if the alternative costs more, consumers are unlikely to take it up. Smarter options also need to be at least as appealing as their alternatives. Surprisingly, the opposite is not always true. For example, high fuel prices do not necessarily equate to smarter behaviour or technology choices.

This is often because consumers are, understandably, risk averse. Consumers are unlikely to take up an untested technology in any volume until there is evidence that the cost is not prohibitive, it is as reliable and easy to use as conventional technologies and the infrastructure is in place to support it. Without these guarantees, consumer demand will be limited. In turn, this has a knock-on effect on the willingness of fuel providers and vehicle manufacturers to invest significantly in new fuels and vehicles.

To make smarter choices, consumers need clear, accessible and reliable information they can trust. For example, clearer and more consistent standards and labelling of fuels and vehicles will help consumers to better understand the impacts of their transport choices and make more informed choices. Likewise, information about the impact of driver behaviour on fuel consumption and associated costs is likely to change the approach that consumers and organisations take to driving.

Fuel providers can also work directly with customers, partners and governments to help moderate their energy usage. Products, including more energy efficient fuels, as well as services, such as driver education initiatives, will have a significant impact on energy consumption.

Governments and policy makers also play a vital role in encouraging consumers to make smarter choices. They have the incentive and the means to provide clear information about the choices available to consumers. For example, in many countries, car showrooms are obliged to display the results of fuel economy testing with standard windscreen labels.

Governments can also incentivise behavioural change through financial mechanisms such as vehicle taxation, road tolling and congestion charging. Often this direct intervention is the most effective way to bring about changes in consumer behaviour.

12 Smarter Infrastructure

By 2050, three-quarters of the world's 9 billion people will live in cities, which already account for 80% of CO_2 emissions [7]. That is a rate of development equivalent to a new city of one million people every week for the next 30 years [12].

How cities develop will be the difference between chaotic infrastructure, with sprawling mobility needs and highly inefficient energy consumption and smarter infrastructure with high population densities and efficient mass-transit infrastructures.

Smarter infrastructure will allow consumers to make smarter mobility choices, facilitate the introduction of new vehicle and fuel technologies and can make a significant contribution to energy efficiency and emissions reductions. But as populations grow and urbanise, the pressure on transport infrastructure is increasing.

This is a significant challenge for mass-transport networks that have grown organically in cities such as Los Angeles or Sydney. Retrofitting smart transport infrastructures is costly. Conversely, cities in the rapidly expanding economies of the East have an opportunity to engineer sophisticated infrastructure strategies that will facilitate smarter mobility choices and keep cities moving efficiently into the future.

An integrated systems approach, linking urban planning and land use policies that directly influence transport demand with support for public and other transport systems will be critical to mitigate the expected rapid growth in transport emissions from rapidly industrialising economies such as China and India. The quality of urban mobility infrastructure development can hard-wire either energy profligacy or energy efficiency into the system for decades.

In the private sector, energy, resources and infrastructure companies face a significant opportunity to innovate in new supply chains for an increasingly urban world. But they also face major up-front investment costs. Already, multinational companies like IBM and Siemens have reshaped their organisations for the urban world in an effort to capture market demands for urban mass-transit systems, electric mobility, energy-efficient heating and power schemes and high-speed information technologies.

If policy makers enable better management of future city development—and this is a bold assumption on the evidence of history—then the challenge to business will be whether rewards for innovation offset the risks of diversifying early.

Infrastructure developed through integrated partnerships between policy makers, regulators, manufacturers and fuel providers will achieve results. These partnerships have the potential to instil the confidence and trust of consumers who will ultimately determine the viability of infrastructure.

13 Conclusions

There are scores of theoretical solutions to the problem of keeping people and goods moving in a world constrained by environmental concerns, population density, energy security and economic constraints. There are technical solutions, policy solutions and solutions based on the way we plan our cities and shape the communities of the future. The mobility challenge is being worked and debated by experts in laboratories, think tanks, design centres and legislative bodies.

But solutions will have to work in the market. The choices that customers—big and small—make when they buy a car; select a fuel; step on the accelerator; select a route; even when they choose to drive, walk or take a train are shaped only in part by experts with great ideas. Just as often they are shaped by individual customer needs and wants—from low cost, to performance, to status, to a sense of security, to simply the convenience of driving to the corner shop to pick up a litre of milk.

So the mobility challenge demands solutions that customers will embrace and pay for—that are, as we in business put it, commercially viable. And that means that all of us who are committed to shaping a more sustainable mobility future—fuel companies, car companies, policy makers, governments, academics, inventors, city planners—need to be working not only with each other but also with our customers and investors to solve the problem.

At heart, Shell is a mobility company and we look forward to a future of exciting mobility possibilities. But we also know that we have to make changes today, one customer at a time.

References

1. European Union Renewable Energy Directive Annex IV
2. International Energy Agency Energy Technology Perspectives (2010)
3. International Energy Agency Oil Market Reports (2010)
4. International Energy Agency Technology Roadmap: Biofuels for Transport (2011)
5. International Energy Agency Transport, Energy and CO_2: Moving Toward Sustainability (2009)

6. Kramer GJ, Haigh M (2009) No quick switch to low-carbon energy. Nature 462:568–569
7. London School of Economics, The Urban Age (2007)
8. Meinshausen N, Hare W, Raper SCB, Frieler K, Knutti R, Frame DJ, Allen MR (2009) Greenhouse gas emission targets for limiting global warming to 2°C. Nature 458:1158–1163
9. Renewables 2010 Global Status Report, REN21
10. Shell World Energy Model—not publicly available
11. Signals and Signposts: Shell Energy Scenarios to 2050
12. UN Habitat
13. UN Population Division

Meeting 2050 CO_2 Emissions Reduction Targets: The Potential for Electric Vehicles

Julia King and Eric Ling

Abstract This chapter explores the potential for electric vehicles to contribute to decarbonising surface transport. Decarbonising transport is a major global challenge—meeting CO_2 emissions reduction targets for 2050, with a rapidly growing, and urbanising global population.

1 Introduction

This chapter explores the potential for electric vehicles to contribute to decarbonising surface transport. Decarbonising transport is a major global challenge—meeting CO_2 emissions reduction targets for 2050, with a rapidly growing, and urbanising global population.

While the analysis in this chapter focuses on the picture in the UK, the global context is included as it is this that will drive the large-scale change needed in the automotive industry. The scale and pace of change required are significant: building an electric vehicle industry which, within the next 30 years, needs to be at least the size of the current global car industry. To start this process the electric vehicles being developed across the industry today need to be taken up in significant numbers by 'early adopter' consumers, in countries with strong commitments to emissions reduction. The Climate Change Act (2008) [1] commits the UK

J. King (✉)
Aston University, Aston Triangle, Birmingham, B4 7ET, UK
e-mail: julia.king@aston.ac.uk

E. Ling
Committee on Climate Change, 7 Holbein Place, London, SW1W 8NR, UK
e-mail: eric.ling@theccc.gsi.gov.uk

O. Inderwildi and Sir David King (eds.), *Energy, Transport, & the Environment*,
DOI: 10.1007/978-1-4471-2717-8_11, © Springer-Verlag London 2012

Government to an 80% reduction in greenhouse gas emissions by 2050, putting the UK at the forefront of delivering a low-carbon economy, including the critical area of surface transport. In this chapter we look at the scope for decarbonising surface transport through the deployment of electric vehicles: the types and development of electric vehicles, and their recharging; the limitations of today's vehicles, and the potential for early adoption, assessed through an analysis of how cars are actually used. High battery costs are expected to continue to fall with mass production and technology progress, and we use economic modelling to indicate that within the next 10 years electric cars are likely to become a more cost-effective choice for consumers in the UK than petrol and diesel powered cars, as fossil fuel and carbon prices increase.

The conclusion for the UK is that there is a strong case for government to support the early development of the electric vehicle market at this time. While the technology is still maturing and electric vehicles do not have the same perceived level of utility, specifically in terms of range and refuelling time, as conventional vehicles, there is a large pool of potential users for whom these factors would not be a constraint. Appropriate policy measures are required to support both demand and supply at this early stage, for a technology which has a strong likelihood of being a major contributor to decarbonising land transport during the 2030s and beyond.

This chapter is structured in the following way.

1. *Context*: the UK's 80% target for reduction in greenhouse gas, GHG, emissions by 2050; the requirement for a 90% reduction in surface transport emissions; the importance of light duty vehicles.
2. The history of electric vehicles and previous attempts to introduce mass market electric vehicle technology.
3. Electric vehicle, battery and charging technology solutions.
4. The impact on the electricity system.
5. Future prospects for technology improvement and consumer behaviour change.
6. Cost effectiveness of electric vehicles.
7. Conclusions and implications for policy.

2 The Emissions Reduction Context

2.1 CO₂ Emissions Reduction Targets

Mounting evidence for the impacts of CO_2 emissions on climate has led to international agreement that global emissions must be reduced by at least 50% by 2050 to have an acceptable probability of keeping average temperature increase by the end of the century to around 2°, and hence avoid the onset of climate impacts which would be severely damaging to life on the planet [2]. This requires total

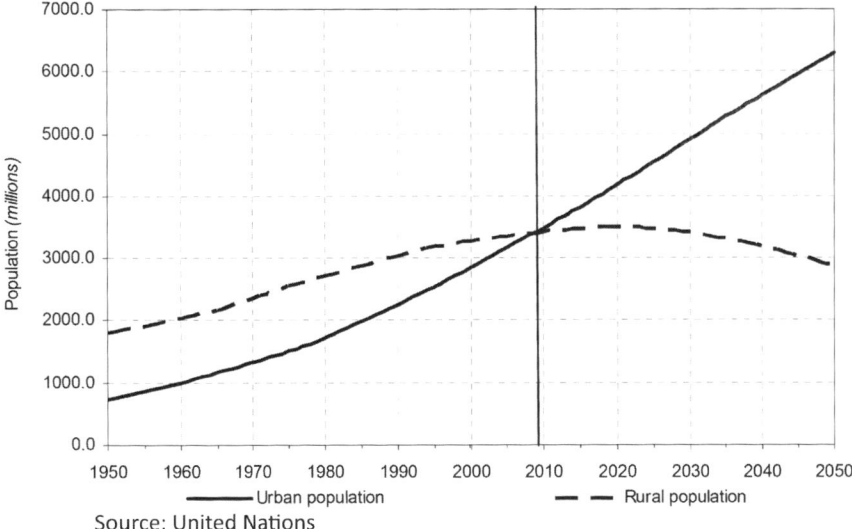

Source: United Nations

Fig. 1 Urbanisation

global emissions of greenhouse gases (GHGs, primarily CO$_2$ but also including the other GHGs covered by the Kyoto agreement) of no more than 20–24 Gt in 2050. With the global population predicted to exceed 9 billion by 2050, this implies per capita emissions of only 2–2.5 tonnes. In the UK today per capita emissions are around 12.5 tonnes, so an 'equal shares' approach to allowable emissions by mid century requires at least an 80% overall reduction. This is typical of the scale of reduction required in developed nations—around 80–90% by 2050 [2].

The global population is also changing as it grows, with two key factors for emissions being urbanisation and increasing affluence. The United Nations predicted that 2010 would be the year when the percentage of the world's population living in cities exceeded that in the countryside [3], and by 2050, 70% of us are likely to be city dwellers, Fig. 1. Megacities, cities with more than 10 million inhabitants, are becoming increasingly common, especially in Asia and other rapidly urbanising parts of the world. Urban populations are typically more affluent, more 'middle class'—increasing electricity demand and exchanging bicycles for cars. Between 2010 and 2050, the urban population will have grown from around 3.3–6.5 billion, with the rich and middle classes making up a rapidly increasing proportion of the population [4]. Reducing global emissions, while addressing the increasing demand for flexible personal mobility, will be one of the major challenges of the twenty-first century.

This chapter will focus on the UK, but the picture is similar across the developed countries. However, the UK has taken an innovative approach to driving the development and delivery of policy to reduce emissions. In 2008, the UK Government legislated the Climate Change Act [1], setting a legally binding

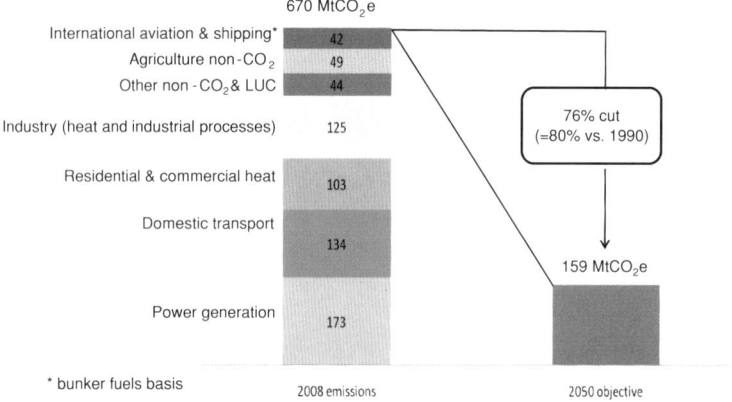

Fig. 2 The scale of the UK challenge

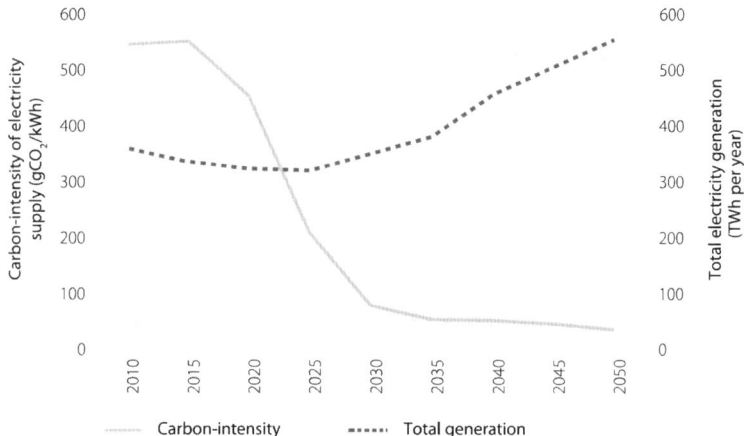

Committee on Climate Change October 2009: AEA (2008) MARKAL-MED modelling

Fig. 3 Required power sector emissions reduction

target for a reduction in GHG emissions of at least 80% by 2050, relative to 1990 levels, about 70% compared to 2009 emissions [2]. Figure 2 [5] shows the scale of the challenge and the relative importance of different sectors in terms of their contribution to emissions. In the UK, as is the case globally, power generation is largest contributor to emissions. Reducing the carbon intensity of electricity generation is the first priority in cutting emissions. The path that the UK needs to take, if it is to meet its carbon budgets, is shown in Fig. 3 [6], reducing from approximately 500 g/kWh in 2010 to around 300 g by 2020 and below 90 g by 2030. Having committed to these investments, it is then a logical step to use decarbonised electricity to support the decarbonisation of two further major areas for emissions: transport and heat (domestic and commercial).

It is clear that, if the UK is to meet the challenge of over 80% reduction by the middle of the century, transport will have to at least deliver this share. However, Fig. 2 includes some particularly difficult to reduce sectors: agriculture, accounting for the majority of the non-CO_2 GHGs; aviation and shipping; and some sectors of industry. Given the challenge of reducing emissions in these sectors, it is likely that land transport (along with electricity generation and the provision of heating in commercial and residential buildings) will have to deliver a reduction of more than 80%. Indeed in Sweden the target is for a fossil-fuel-free vehicle fleet by 2030 [7] and the UK has identified a requirement at around a 90% emissions cut by 2050 [5].

2.2 Surface Transport Emissions

Transport is the third largest contributor to global CO_2 emissions, at around 14%, after power generation and land use change [8]. In developed countries transport is typically the second largest source of emissions, the cause of around 20% of emissions in many countries, but rising to a high of 33% in the USA [9].

Surface transport emissions in developed countries are predominantly generated by the use of light duty vehicles: cars and vans. This is illustrated in Fig. 4 [8] for the UK in 2008, where such vehicles contribute 73% of emissions. The focus for both policy and technology development, in the early stages of implementation of transport emissions reduction plans, therefore falls on addressing the usage (distance driven) and the energy consumption of these vehicles.

The number of cars in the UK—the UK car parc—is currently approximately 31 million [10]. This is a small proportion of the global total of almost a billion vehicles, a figure which is predicted to rise to around 2 billion between 2020 and 2030 [9] and potentially to 3 billion or more by 2050 as the world population grows and becomes increasingly affluent. Today road transport already consumes around 53% of total oil production [11], so as vehicle numbers grow, the drive to improve vehicle efficiency and to reduce oil dependence will come from the pressure on oil resources as well as climate change considerations.

Globally, both distance driven and car ownership are increasing, especially in urban areas [12, 13]. Developed countries exhibit similar trends, for example, Fig. 5 shows the increase in distance driven in the UK over the past 50 years [14]. A step change in vehicle efficiency and emissions will be required to deliver the challenge of both meeting mobility expectations and mitigating climate change. The scale of the challenge is illustrated by the following simple calculation. The average new car purchased in the UK in 2010 had an emissions level of 144 g/km [15]. This car, driven 14,000 km per year, would emit 2 tonnes of CO_2—the total per capita allowance in 2050. Globally, if we are to reduce emissions by 50% with a trebling in car ownership/use, per km emissions would need to reduce to 1/6 of present levels, from 144 to 24 g/km. The UK target of a 90% reduction, if it were to be achieved by vehicle technology alone, implies per km emissions of less than 14 g. While neither of these figures is 'correct' or 'accurate', they indicate that, if

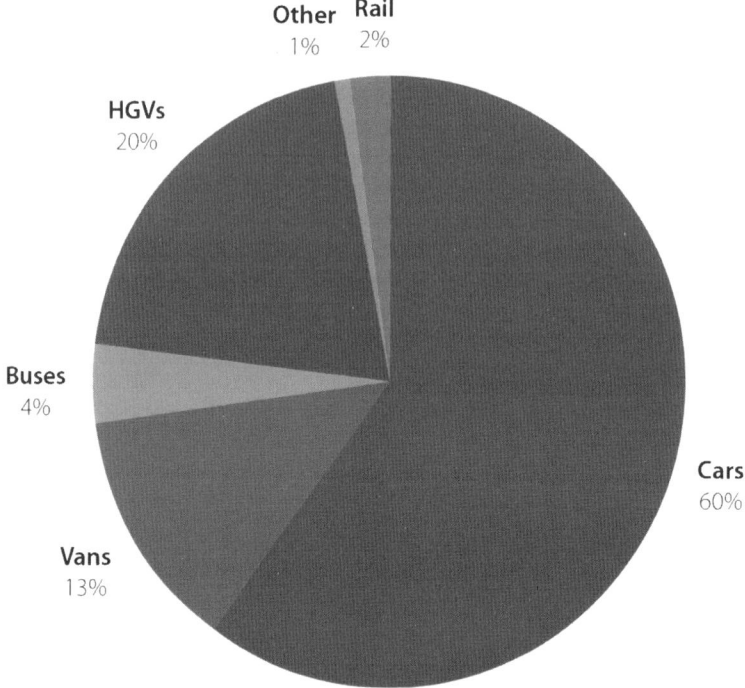

(DECC 2009, *2008 final UK greenhouse gas emissions: data tables*)

Fig. 4 UK surface transport emissions 2008

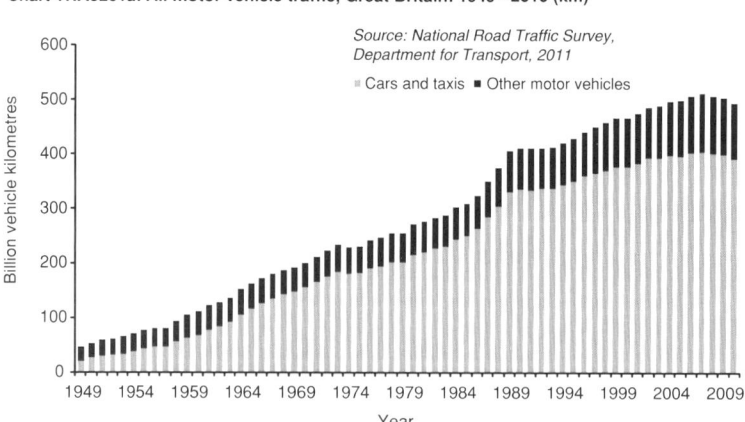

Fig. 5 Increase in UK distance driven

we are not to constrain people's freedom to drive their own vehicles very significantly in the future, per km emissions need to fall below a level of about 20 g by 2050. This is unlikely to be achieved with the internal combustion engine.

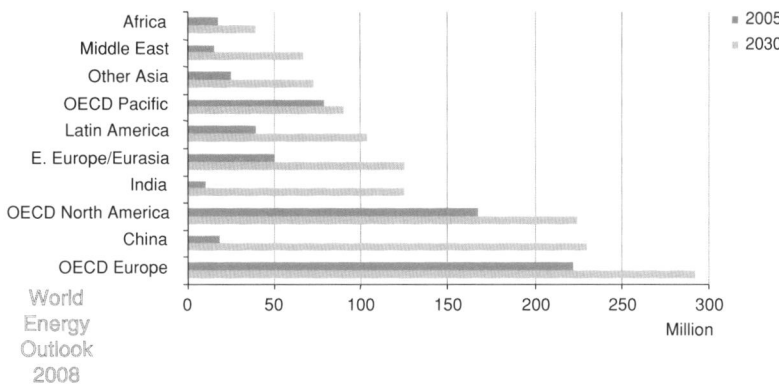

Fig. 6 Increase in global vehicle ownership

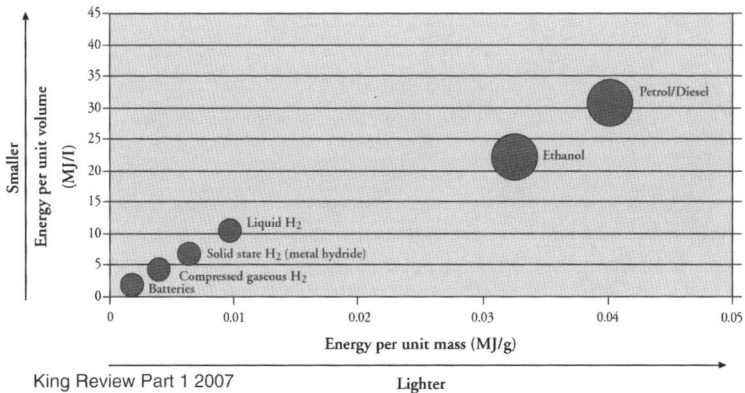

Fig. 7 Energy densities—oil, biofuels, hydrogen, batteries

In a recent report, the Institution of Mechanical Engineers reviewed the range of possible future advances for the internal combustion engine and concluded that the very best efficiency achievable would correspond to an emissions level of around 80 g/km, about 50% of the values we see today [16]. Similar conclusions have been reported elsewhere [17].

The vehicle options to achieve per km levels of 20 g (which must be implemented alongside modal shift to public transport combined with improved city design and planning to ensure that such services are economically viable) are, potentially, electric and electric hybrid vehicles, hydrogen-fuelled vehicles and biofuels. In a world of three billion vehicles, all three solutions will be needed—for different applications and because a diversity of solutions will be critical. We must avoid repeating the mistake by moving from one dependence—fossil fuels—to another, such as lithium, platinum or miscanthus (Fig. 6).

However, the biggest challenge of replacing oil for transport is that it is such a good fuel: a high energy density, in terms of both volume and weight, Fig. 7 [17];

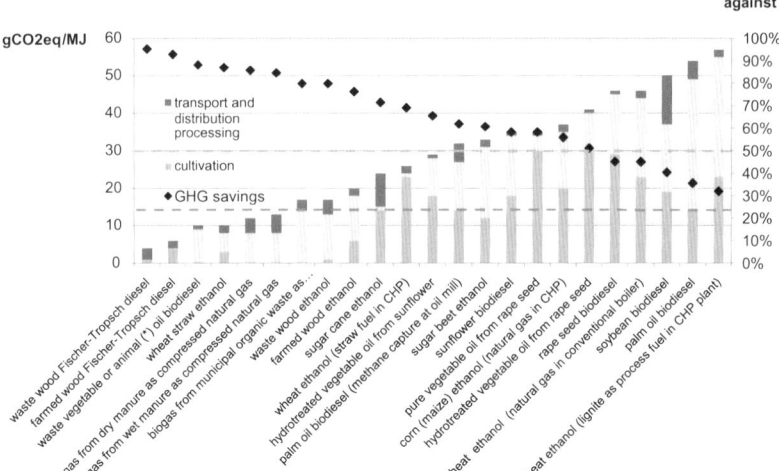

Fig. 8 Biofuels: GHG savings using RED typical values

easy to handle and store at room temperature; plentiful, readily available and cheap—at least up to now. These features have shaped the transport system we have today.

Biofuels are an appealing solution because they work with existing vehicle technologies and distribution infrastructure, with energy densities approaching those of fossil fuels. However the concerns over land use change, the impact on food prices, and the availability of land and water in a world with over 9 billion inhabitants, indicate that biofuels are always likely to be a precious and limited resource. In combination with the best internal combustion engine technology a biofuel would need to deliver a 75–80% reduction in emissions compared to conventional petrol or diesel. The actual reduction in emissions from current biofuels, if they were to form a 100% replacement for oil-based fuels, is still an area of some controversy, but the figures used in the EU's Renewable Energy Directive [18] indicate that only sugarcane ethanol, wheat straw ethanol and a small number of products made from waste wood and oil meet this level of reduction, Fig. 8. However, these figures exclude the impact of land use change—when this is included, the list will be even shorter.

Research and development work continues on second and third generation biofuels, to improve the production of fuels from wastes and woody crops and to introduce new routes which do not compete with land for food, such as the growth of algae. Progress in developing improved liquid biofuels seems certain, but, given the potential impacts of biofuel production at the scale required for the world's growing car fleet, this can only be part of the solution. Indeed, it may be necessary to concentrate the use of biofuels where there are no alternatives, for example, in aviation and in heavy vehicles where electric solutions are not feasible [19].

Hydrogen as a fuel (or an energy carrier) is another technology which has the potential to deliver a clean fuel for land transport. While hydrogen can be stored at better energy densities than batteries, it is still lower than oil, Fig. 7, and storage either involves high pressures/very low temperatures or solid storage technologies which are still at the R and D stage. Plentiful hydrogen is not available in nature, it has to be produced, as a by-product of other chemical processing or by hydrolysis of water. In the latter case, the high energy input to turn water into oxygen and hydrogen comes from electricity. Hydrogen can be burned in a modified internal combustion engine, at similar efficiency levels to a fossil-fuelled vehicle, less than 25%, or used in a fuel cell, at efficiencies of closer to 70%. However even in a fuel cell vehicle, assuming the hydrogen is produced using grid average electricity, the effective g/km will be significantly higher than a battery electric vehicle. Indeed, with current UK electricity generation, MacKay [20] calculates that hydrogen fuel cell vehicles would be significantly worse than conventional fossil-fuelled vehicles. Further R and D is needed to overcome some of the remaining challenges to hydrogen as a transport fuel, including fuel storage and fuel cell costs, as well as the provision of refuelling infrastructure. This is particularly relevant to the electric vehicle picture as fuel cells could, in the longer term, offer the final step in replacing the ICE in plug-in hybrids.

Electric and electric hybrid vehicles represent a low-carbon solution which we can start to implement today in countries such as the UK which are aiming to take a leading role in reducing transport emissions. The electricity supply runs along city streets and is in peoples' homes. The technology has been demonstrated, and, despite some limitations, offers a practical and efficient solution to many of our personal mobility requirements.

3 History and Near Term Outlook

3.1 History

Battery electric vehicles were first developed at the end of the nineteenth century [8]. However, the cost, size and weight of the batteries placed the electric drivetrain at a strong disadvantage to the rapidly developing internal combustion engine (ICE) combined with its high energy density fuel. In the USA, between 1900 and 1920 the proportion of electric vehicles being produced fell from 60 to 4% of the total as the ICE swiftly came to dominate the car and van markets [21].

The energy crisis of the 1970s spurred a renewed interest in electric vehicles. The Electric and Hybrid Vehicles Program was established in the United States to encourage R and D on electric and hybrid vehicle technologies. However, the first major efforts to promote the *deployment* of electric vehicles started in September 1990 with California Air Resources Board's (CARB) proposed new regulations to require the phased introduction of low-emission vehicles to address air quality standards in the state of California. These included the zero-emission vehicles

(ZEV) mandate that required a specified proportion of all new passenger cars and light trucks by major car manufacturers to be ZEVs, beginning with 2% of new vehicles in 1998 and increasing to 5% in 2001 and to 10% in 2003. The dominant battery technologies at the time (lead-acid and nickel–cadmium), provided insufficient performance for cars. Recognising that advances in battery technology were insufficient to achieve the ZEV mandate, in 1996 CARB began to reduce the ZEV mandate, initially to a more modest number of up to 3,750 EVs (around 0.02% of new cars), then to allow 6 of the 10% ZEV requirement in 2003 to be met by "partial zero emission vehicles" (PZEVs), and in 2001 awarding multiple credits to pure ZEVs, such that less than 2% battery electric vehicles were required to meet the 10% ZEV requirement. Following still further changes, only a few thousand electric cars were deployed in California by 2002.

As a result of the mandate, a range of models including electric cars (General Motors EV1, Nissan Altra EV, Honda EV Plus), sports utility vehicles (the Toyota RAV4 EV), light commercial vehicles (Chrysler TEVan) and pickup trucks (Ford Ranger EV pickup truck, GM S10 EV pickup) were developed. Initially these used early stage battery technology (the General Motors EV1 used lead-acid batteries), picking up more advanced battery technologies in later models (the Nissan Altra EV used cobalt-based, then manganese-based Li-ion batteries, while General Motors moved on to nickel-metal hybrid batteries in the EV1 which extended its range from 112 to 220 km per charge). However, the costs of even the latest generation of electric vehicle battery packs remained prohibitive: at over $1,000/ kWh [22] a 35 KWh battery pack (the typical capacity in 2000 [23]) would have cost in excess of $35,000 (£22,000).

As mainstream car manufacturers postponed plans to bring electric vehicles to market, a number of smaller manufacturers developed vehicles in niche market segments. In 2001, the REVAi Electric Car Company (then an India-US joint venture) launched the REVAi electric vehicle, a very small car with a top speed of 65 km/h and a range of 80 km/50 mi per charge. The REVAi was launched in India, arriving in London in 2004 under the G-Wiz brand. At the other end of the market Californian firm Tesla Motors introduced the Tesla Roadster, an electric sports car based on the Lotus Elise, in 2008.

Despite the apparent failure of electric cars to present a credible choice for consumers in the 1990s and early 2000s, by the late 2000s they were again in the spotlight. This period has seen the highest oil prices since the energy crises of the 1970s, as well as increased policy interest in mitigating climate change. Electric vehicles, which in the 1990s were seen primarily as a solution to the local environmental problem of poor urban air quality, are now seen an important option to mitigate both macroeconomic risks associated with high and volatile oil prices, and the global environmental risks associated climate change. Meanwhile, decreases in battery costs since the early 2000s have brought the prospects of affordable electric vehicles. By 2009, a number of major car manufacturers had announced development of electric vehicles for the mass market: General Motors announced the release of the Chevrolet Volt, a plug-in hybrid car with a 40 mile electric range and a petrol engine for longer distance travel; Mitsubishi announced

Table 1 Electric car models currently under development [46]

Brand	Model	Type	Likely UK launch date
Hyundai	Ix-metro	BEV	2011
Westfield (race car)	iRacer	BEV	2011
Lightning Car Company	The lightning GT	BEV	2012
Renault	Fluence Z.E.	BEV	2012
Vauxhall	Ampera	PHEV (RE)	2012
Renault	ZOE	BEV	2012
Audi	E-Tron	BEV	2012
Chevrolet	Volt	PHEV (RE)	2012
Ford	Focus	BEV	2012
Morgan	Lifecar2	BEV	2012
Smart	ED	BEV	2012
Tesla	Model S	BEV	2012
Westfield	Sport-E	BEV	2012
Axon	E-PHEV	PHEV	2012
Toyota	Prius PHV	PHEV	2012
BMW (sub-brand)	Megacity Vehicle	BEV	2013
Ford	C-MAX	PHEV	2013
Westfield	GMT Electric	BEV	2013
Volkswagen	Up! blue-e-motion	BEV	2013
Volkswagen	Golf blue-e-motion	BEV	2013/2014
Fisker	Karma	PHEV	TBC
Ginetta	G50	BEV	TBC
Porsche	918 Spyder	PHEV	TBC
Th!nk	City EV	BEV	TBC
Th!nk	Ox	BEV	TBC

the i-MiEV, a small car with an advertised range of 90 miles; and Nissan announced the Leaf, a medium-sized car with an advertised range of 110 miles. Concurrently, in several countries, a number of electric vehicle field trials were launched to test the performance of the new generation of electric cars, and learn about consumers' requirements for range and charging infrastructure. In 2009, the UK Government, through the Technology Strategy Board, initiated a number of collaborative demonstrator programmes with industry, involving over 240 cars from Ford, BMW, Jaguar Land Rover, Allied Vehicles, Mercedes-Benz, Toyota, Mitsubishi and Nissan in eight locations around the UK (including three in London). The UK Government has awarded funding for pilot projects to roll out electric vehicle charging infrastructure in eight cities under the "Plugged-in Places" programme, and has introduced a consumer incentive scheme funded at around £300 m, offering up to £5,000 per car to reduce the up-front cost of electric and plug-in hybrid cars. At the time of writing (2011), six electric car models were available on the UK market (Mitsubishi i-MiEV, Smart fortwo ED, Peugeot iOn, Nissan Leaf, Tata Vista, Citroen CZero), and car manufacturers have announced release of a significantly expanded range of models over the next few years (Table 1).

4 Current Technology

4.1 Electric and Plug-in Hybrid Cars

The term 'electric vehicles' can include a wide range of vehicle types, including many types of hybrid, as well as fuel cell vehicles. In this chapter we use it only for battery electric vehicles (BEVs) and range-extended and plug-in hybrid vehicles (PHEVs). The architectures and operation of these vehicles are covered elsewhere at both the detailed technical and more general levels [8, 16, 17, 24], so a high level overview is offered here.

A BEV, or pure electric vehicle, is the simplest form of electric vehicle (EV)—one in which a battery driving one or more electric motors provides the fuel source and powertrain of the car. There is no additional internal combustion engine (ICE), fuel tank, or on board generator (except that the electric motor, typically, can be used to charge the battery on braking, in kinetic energy recovery mode.) The car is 'fuelled' by plugging into an electric socket to charge the battery. The simplest types of BEV have been in daily use for a long time—e.g. fork lift trucks and golf carts—and practical BEV cars, such as the Leaf, the i-MiEV and others mentioned above are now becoming generally available. With current battery technology—variants on the Li-ion battery chemistries—BEVs are suitable for small city cars, with typical ranges of 100–190 km (60–120 mi), used for short trips and daily commuting. BEVs have significant advantages in cities, the small size makes parking easier and reduces congestion, and the lack of tailpipe emissions and the quiet operation improve air quality and reduce noise pollution. Another characteristic of battery-powered cars is outstanding acceleration, power can be delivered much more rapidly by discharging a battery than from a combination of combustion and mechanical action in an ICE, so high performance sports cars are another interesting application typified by the Tesla. Interest is also now growing in BEVs for large chauffeur driven cars, where space for the batteries is not an issue, silent operation is part of the luxury experience and short trips to deliver distinguished occupants are the norm. Electric vehicles are fundamentally simpler systems than ICE vehicles, so apart from the battery costs, they are not expensive, and have lower maintenance requirements.

Current plug-in hybrid electric vehicles (PHEV) combine a smaller battery (again recharged by plugging into the mains) typically with a range of about 16 mi/20 km, as well a fuel tank and an internal combustion engine. The motor and ICE are arranged in a 'parallel hybrid' configuration, so both the ICE and the electric motor can drive the wheels. The vehicle will run in full electric mode when the battery is charged, with the ICE coming on when battery power levels drop. Limited battery recharging takes place during use through kinetic energy recovery on braking. The plug-in version of the Toyota Prius is an example of this configuration.

Extended range electric vehicles are another form of EV, with a greater battery range and a smaller ICE with a 'range extender' role. In this series hybrid configuration, the vehicle is more like a BEV, with the electric motor driving the

Fig. 9 Volume occupied by the lead-acid batteries in the 1970 BMW 1602 test vehicle

wheels and the role of the range extender engine being to recharge the battery during longer journeys. An example of this type of vehicle is the Chevrolet Volt, with a battery electric range of around 40 mi/70 km.

The advantage of PHEVs is that they do not have the range limitations of BEVs, but they can make a significant contribution to reducing emissions if the batteries are kept charged. The average daily driving distance in the US and Europe is in the range 30–69 km/20–40 mi and 93% of UK car trips (accounting for 62% of CO$_2$ emissions) are less than 25 miles [25]. While the smaller batteries reduce the costs compared with BEVs, the more complex systems with additional components (ICE, fuel tanks, more complex controls) increase system costs and maintenance requirements.

4.2 Batteries

Battery energy density, Fig. 7, and cost are the two major challenges for electric vehicles. Early prototype electric cars used lead-acid batteries, for example, in this 1970s demonstrator from BMW, Fig. 9 [26]. Such battery technology was unable to deliver a practical car and electric vehicle development came to a halt.

Modest improvements in the performance of lead-acid and nickel–cadmium batteries were driven by the Californian ZEV mandate in the 1990s. These were overtaken by the emergence of advanced battery technologies, including nickel-metal hydride, lithium-ion and sodium-nickel chloride, promising better performance. Today there is a developing consensus that Li-ion batteries, which combine both high power and energy densities, are likely to be the solution for vehicle applications in the medium term. Li-ion covers a range of electrode materials and electrolyte chemistries: cathode systems currently being used and evaluated include being Li–nickel cobalt aluminium; Li-manganese spinel; and Li-iron phosphate [22]. Different chemistries are being developed to optimise performance to meet the demands of different vehicle types: BEVs need the highest energy densities, maximising energy storage, whereas PHEVs require a combination of both power and energy density, Fig. 10 [47].

Research into batteries to meet the needs of the next generation of electric vehicles is a rapidly growing field. Beyond today's Li-ion batteries, researchers are looking at

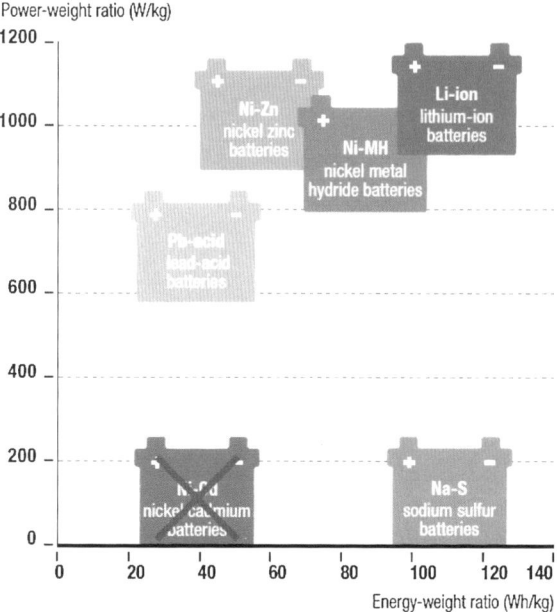

Fig. 10 Battery power and energy densities. This figure is from reference [47]

nanostructured Li-ion cathodes, Li-organic and Li-air batteries, as well as magnesium—sulphur and Al-CF$_x$ chemistries and proton batteries [21]. This is picked up in more detail in Sect. 6.1 on prospects for improvements in battery technology.

Batteries for electric vehicles are made up of large numbers of cells. Different applications use batteries of different sizes. A BEV might have 200 cells with 20–30 kWh of energy storage, giving a range of around 100 mi, PHEVs might range from 40 to 90 cells for a series hybrid with 15 mi range to a range-extended model with 30–40mi electric range.

Battery performance deteriorates with increasing numbers of charge/discharge cycles, and is influenced by the way the vehicle is used and the battery changed and discharged. Battery control and management systems are a critical part of the design of an EV, impacting vehicle performance and the life of the most expensive component of the vehicle. To continue to be useful in the vehicle, the battery needs to reach a maximum state of charge of at least 80% of the maximum level when new. This is driving an interest in the second use of car batteries for local energy storage, where batteries removed from cars could be used, for example, for storing cheap, night time electricity in homes for use during the day; storing energy from intermittent generators such as wind turbines or solar cells; or for other types of grid and network balancing. Batteries in some new EVs models are being guaranteed for a 5-year life, indicating confidence that the maximum state of charge will be at least 80% after 2,000 or more full charge/discharge cycles. It is probable that batteries in most cars will last much longer than 5 years, but until we have the data from large numbers of the new cars currently coming to market, manufacturers will remain cautious.

Fig. 11 Electric vehicle battery costs

Quoted costs for current Li-ion batteries vary widely, in a range from about $1,000 to 500 per kWh, Fig. 11 [22, 23, 28–30]. This makes the cost of a 20 kWh BEV battery between $10,000 and $20,000—the dominant reason for the high current prices of EVs. When EVs become a genuinely cost-effective solution for drivers depends on battery cost reduction, Fig. 11, and the effects of the carbon price and the cost of oil. While there are many uncertainties in carbon, and indeed oil, price predictions, with the pressures of both climate change mitigation and reducing oil supplies, within the next 20 years, EVs are likely to become one of the cost-effective solutions for personal mobility, as discussed in Sect. 7.

There have been some early concerns about battery safety, arising from isolated incidents in early technology vehicles and the incidence of overheating leading to fires in laptop batteries. The cause has been identified as 'overcharging' leading to chemical changes and Li plating at the anode in some battery chemistries. Modified battery chemistries and battery management systems are now in place to address this issue [22].

4.3 Vehicle Charging and Charging Infrastructure

One reason why electric cars are candidates for early adoption among the potential low-carbon vehicle technologies is the relatively simple infrastructure requirements. In developed countries electricity runs along every city street, and in the UK 70% of suburban households [31] have off street parking.

4.3.1 Slow Charging

The simplest approach to charging an EV is to plug it into a standard domestic socket (240 V/13 amp in the UK). A full charge typically takes 6–8 h, but 80% charge can be achieved in as little as 2 h. The obvious advantage of this approach is that limited infrastructure is required, especially if drivers charge at home, overnight, using a smart meter to access low night time rates and helping to balance the grid, see Sect. 5, below. This is fine for suburban commuters, using their cars mainly to drive to work, distances typically less than 25 miles, Sect. 6. The next stage of infrastructure development for this potential group of early adopters is workplace charging, the average parking time at work in the UK is 6.2 h [31].

However, BEVs are, in many ways, ideal small city cars, so the challenge is to provide an infrastructure to give city dwellers, living in flats with no access to off street parking, the confidence that they can charge their vehicles at locations convenient for home and or work. A number of demonstrations of city charging infrastructure have recently been undertaken or are underway, e.g. in the UK (Plugged in Places), Japan (EV/PHV Towns), Germany and the USA (Electric Mini trials). The Mayor of London has announced the ambition that by 2015 there will be 25,000 public and private vehicle charging facilities in London and that no Londoner will be more than one mile from a charging post [32]. Studies of consumer preferences in relation to charging are currently underway—results from the Mini-E trial in Berlin, the Technology Strategy Board low-carbon vehicle demonstrator across the UK and the Energy Technologies Institute study [33–35] indicate that drivers tend not to use public charging points where they have alternatives. It is important to understand driver behaviour in relation to charging, because although the technology is simple, the cost of installing large numbers of on street charging posts is significant [6].

4.3.2 Fast Charging

Fast charging uses three phase electricity to fully charge a vehicle in less than 30 min. Having such facilities available appears to be important to consumers, whether they use them or not: in an early study in Japan of EV users at a TEPCO facility [6, 31], the installation of a fast charging post had a dramatic impact on user confidence in terms of using the full range of the vehicles. Again the technology is simple, and would probably need to be provided in a supervised location or for use by trained personnel, such as a filling station or a vehicle depot.

However, fast charging is not an ideal development either from an electricity system point of view, or from a battery life perspective. Managing large numbers of short duration/high load events is challenging for the grid, the benefits of grid balancing from slow charging vehicles are lost, and battery life is shortened by frequent fast charging.

4.3.3 Inductive Charging

An elegant solution for cities could come in the form of inductive charging, where coils under the road surface could be used to inductively charge vehicle batteries. The advantages of non-contact charging come in safety in exposed locations and reducing the amount of street furniture, but the vehicle requires a coil which can be lowered to be in close proximity to the buried coil. Significant improvements in charging efficiency will be needed to make this a relevant technology for cars, developments are ongoing and a better understanding of the potential will emerge over the next few years.

4.3.4 Battery Exchange

Potentially the fastest solution for 'recharging' is battery exchange, pioneered by Better Place [36]. The is driven into a battery exchange station and positioned such that a robot removes the drained battery from below the vehicle and replaces it with a fully charged one, in the space of a few minutes, perhaps faster than filling a tank with fuel. This could be an effective solution for vehicle fleets with identical or a small number of battery types—for example city taxi fleets—but is unlikely to be cost-effective for general application with a wide range of vehicles with different batteries.

Each of these charging solutions has particular advantages and disadvantages, and it will be very interesting to see them demonstrated over the coming years as EV ownership grows. They would all benefit from urgent international agreement on standardisation—as a minimum sockets and charging conditions, in the case of battery exchange and inductive charging, standardisation of the battery itself, its location and, for exchange, fixing.

Battery life, costs and the implications of some of the solutions to the time taken to charge an EV battery also highlight the issue of battery ownership. Consumers may be reluctant to buy a vehicle with an uncertain battery life, especially one which has a significant probability of being shorter than the vehicle. Recharging solutions such as battery exchange are likely to be unacceptable to vehicle owners who could find they have swapped a new battery for an older one. Battery ownership by utilities, who are likely to be selling the 'fuel' to the electric motorist, would fit with a second life for batteries as a form of static energy storage; vehicle leasing, or alternative models of battery leasing, are all potential routes to the introduction of this new technology.

5 Electric Cars and the Electricity System

Electrification of transport is an effective route to reducing carbon emissions where the electricity generation system has a low or relatively low carbon intensity. In the UK, the current electricity generation system has operated for the past two years

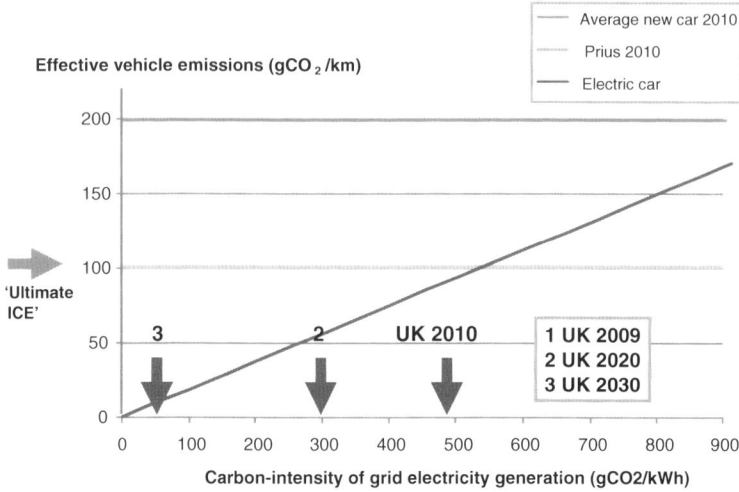

Fig. 12 Electric cars: potential for ultra-low emissions

(2009 and 2010) at an average carbon intensity of 500 g/kWh. The path to meet the UK's carbon budgets requires reductions to levels below 90 g/kWh by 2030. Current electricity generation emissions vary widely in developed countries, from close to zero in Sweden where generation is dominated by nuclear and hydroelectric plants to over 1,000 g/kWh in Australia with a predominantly coal fired system. However, in many countries power generation is the largest contributor to GHG emissions, as it is globally, so decarbonising power is central to most national emission reduction plans and commitments.

5.1 Effective Emissions of Electric Vehicles

It is important to look at the effective emissions of electric vehicles, i.e. the emissions from generating the electricity used to charge the battery, in the context of their contribution to mitigating climate change. These emissions should be compared to the 'well-to-wheel' emissions of ICE vehicles, adding about 15% [17] to quoted tailpipe emissions to account for the fuel production and transport processes. Figure 12 shows the effective emissions of a 25 kWhr BEV with a 100 m/160 km range (the Nissan Leaf has a 24 kWhr battery and a 110 m/176 km range) as a function of the carbon intensity of electricity generation, compared to the emissions from a new Toyota Prius (89 g/km, increased by 12.5% to account for the well-to-wheel emissions of the petrol used) and the average emissions of the UK fleet in 2010 (173 g/km increased by 15% to give effective well-to-wheel emissions). The diagram indicates that the BEV is a low emissions vehicle today,

using the current electricity system, and with the generating system decarbonisation targeted by 2030, could provide around a 90% reduction in per km emissions when compared with today's average car.

5.2 Impact on the Generating System: Capacity

Concern is often raised about the impact of electric vehicles on existing generating
systems. Some simple calculations based on the UK system indicate that this is a
relatively minor issue, reducing to a very small impact if motorists can be
encouraged to charge at night through the use of preferential tariffs and smart
meters.

Current UK generating capacity is approximately 70 GW, or 600 TWh per
year, with peak demand reaching 60 GW, and annual usage of around 350 TWh.

Making some simple assumptions we can estimate the impact of a fully electric
UK car fleet on the current system:

- 30 million vehicles;
- electric vehicle batteries storing 25 kWh of energy;
- fully charged every three days, the average in the BMW MiniE trial in Berlin in
  2009, or assuming an average daily driving distance of up to 50 km.

This uses 91 TWh per year, approximately 15% of current generating capacity
or 26% of current usage. The Committee on Climate Change has recommended a
challenging target which the UK should aim to achieve of 1.7 million EVs, about
5% of the UK car parc, by 2020 [31]. This would need less than 1% of current
generating capacity or 1.5% if current usage levels.

The conclusion is that, even with ambitious rates of introduction of electric
vehicles, the load on the electricity generating system will grow slowly and this
can be readily built into requirements as new capacity is planned.

The impact on generating capacity is even lower if owners can be incentivised
to charge their vehicles at night. Figure 13 shows the average daily electricity
consumption through a 24 h period in the winter and the summer of 2010 [37].
The available energy during the winter night time dip is approximately
8 h × 16 GW ×365 days = 47 TWh per year. This is enough energy to fuel over
15 million BEVs, about half the UK car parc, with no additional generating
capacity.

5.3 Distribution System Considerations

While the impact on generating capacity will be readily manageable, the impact on
the distribution system is also important. Modelling by Element Energy for the
Committee on Climate Change has looked at a simple model for the UK

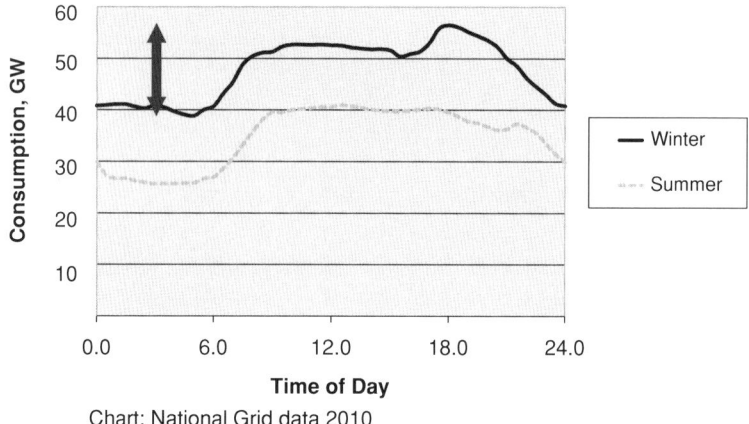

Chart: National Grid data 2010

Fig. 13 Average electricity consumption over 24 h periods

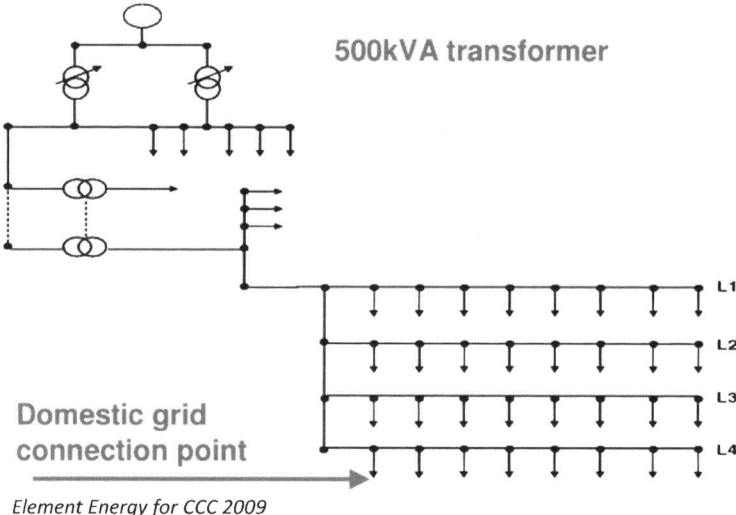

Element Energy for CCC 2009

Fig. 14 Model of the electricity distribution system

distribution system, illustrated in Fig. 14 [31]. With a complement of 15 million EVs, plugged into the system as and when drivers returned from work, current transformer ratings would be likely to be exceeded between 5 and 8 pm on most evenings, Fig. 15. However, it is unlikely that electric vehicle numbers approaching 15 million could be achieved before the 2030s, giving ample time to upgrade the system.

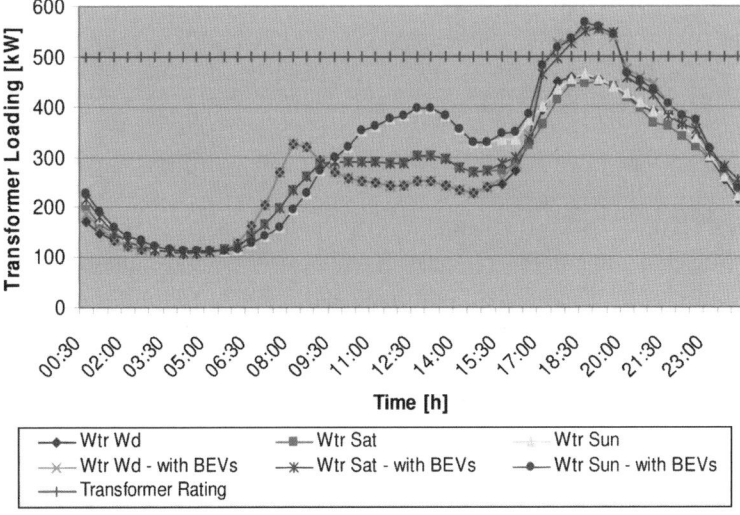

Element Energy for CCC 2009

Fig. 15 Potential impact of uncontrolled slow charging of 15 million EVs on the distribution system

5.4 Benefits to the Electricity System from Electric Vehicles

The implementation of smart meters would remove the problem of unconstrained charging, and overloading of distribution system transformer capacity, by providing the means to turn charging on and off remotely for short periods. But it offers a more important benefit, the possibility of using the electric vehicle fleet for balancing the grid—effectively managing peak loads—something which will become increasingly important with more use of intermittent renewable generation as the electricity system is decarbonised.

In a recent study by Ricardo for the National Grid [38] the potential for electric vehicle charging to be used for balancing the operation of the grid was examined in two modes: Demand Side Management, DSM, enabled by smart meters; and Vehicle-to-Grid, V2G, which needs a smart grid system enable two way flow of current. DSM simply involves the interruption of charging for short periods at times of high electrical demand. Assuming a BEV/PHEV fleet of just 600,000 vehicles in 2020—just over a third of the number recommended as a target by the Committee on Climate Change—the research suggests that the vehicles could provide an average of 6% of the daily network balancing requirements, rising to 10% in the evening and overnight, simply using standard domestic 3 kW chargers. The vehicle owner could earn a small return for this service, potentially £50 per annum, or around an 18% saving on vehicle recharging costs.

V2G would allow two way flow of energy between the battery and the grid. The smart grid and smart meter control environment would enable the use of the battery as an energy buffer for balancing the loads on the grid, while ensuring that the vehicle owner retained an appropriate level of control over the state of charge of the vehicle. Given the typical daily driving distances discussed above, the study assumed that, on average, 8.1 kWh of charge remained in each battery at the end of the day. The modelling looked at a range of charger power levels from the standard domestic 3 kW up to 50 kW 3-phase. At a charger power level of 22 kW for all vehicles, an average on 79% of daily grid balancing requirements could be met, and 100% during the evening and night time period. There would be a significant increase in revenue for the vehicle owner, offset by the investment cost for a 3-phase charging installation, making this a potentially interesting opportunity for specialist fleet operators, rather than domestic users.

V2G operation raises a potential concern over the impact on battery life of the increased number of charge and discharge cycles. The modelling shows that the additional energy transferred to and from the battery for balancing purposes is less than 1% of that associated with charging and discharging in normal use, suggesting that the impact on battery life would be minimal.

5.5 *Electricity System Impacts: Conclusions*

Much of the discussion above assumes that electric vehicles are slow charged, ideally during the night when electricity usage is at a minimum. Assuming users can be appropriately incentivised to charge their vehicles in this way, very large numbers of vehicles could be accommodated with little impact on generating capacity. Even at quite low levels of EV penetration the grid balancing opportunity is a significant one, and with no additional investment requirements beyond the current planned roll out of smart meters in the UK.

A significant move to fast charging, however, would increase the costs and impact of electric vehicles, and should not be encouraged if electric vehicles are to deliver a cost-effective route to decarbonisation of surface transport.

6 Prospects: Future Technology and Consumer Behaviour

Section 4, above, provided an overview of the technology for a generation of EVs that is now beginning to enter the market with comparable performance to conventional vehicles. Large-scale adoption of EVs will now require a combination of:

- advances in technology to address drawbacks to the current generation of electric vehicles, particularly in the areas of range and cost;
- behaviour change to adapt to electric vehicles and to address perceived constraints, where EV performance and capability differs from traditional ICEs.

This section discusses prospects for both improvements in technology and behaviour change.

6.1 Prospects for Improvements in Technology

As indicated above, the principal drawbacks to the current generation of electric vehicles are their high cost (in the case of both battery electric and plug-in hybrid vehicles) and limited range (in the case of battery electric vehicles); for example, at the time of writing the Nissan Leaf had a retail price in the UK (before Government subsidy) of £30,990, and an advertised range of 110 miles. The high cost of electric vehicle batteries currently limits the maximum battery capacity that can be installed before the cost of the vehicle becomes prohibitive. An understanding of the appropriate role for electric vehicles in reducing transport CO_2 emissions therefore requires an assessment of the potential for future cost reductions of electric vehicle batteries.

A report commissioned by CARB, by Kalhammer et al. [29], to assess the current status of sustainable zero-emission vehicle technologies and the prospects for near-and long-term technology advancement concluded that costs of currently available (i.e. in low volume production or tested in prototype vehicles) lithium-ion battery technologies of $342–475/kWh (BEVs) and $505–816 (PHEVs) could be achieved at limited production volumes (around 500 MWh per year, equivalent to production of 12,500 large battery electric cars or 20,000 small battery electric cars), while costs as low as $232–326/kWh (BEVs) and $346–560 (PHEVs) could be achieved in mass production (around 2,500 MWh per year), with the lower end of the ranges relating to larger battery capacities, Fig. 11.

Kalhammer et al. also identify prospects for further cost reductions of lithium-ion battery technologies, through development of new electrode materials with lower costs and/or higher capacities, cost reductions of inactive cell materials (especially separators and electrolyte salts) through improved and larger scale production processes, lower costs of balance-of-system components and advanced large-scale manufacturing processes. Boston Consulting Group [28] estimates a similar reduction in battery costs to $360–440/kWh by 2020 due to learning, economies of scale and automation occurring as production volumes increase.

Finally, a number of advanced battery chemistries offer additional prospects for cost reductions through improvements in energy density. While the energy density of the current generation of battery technologies (around 100 Wh/kg [29]) is approaching the U.S. Advanced Battery Consortium's "Minimum Goal for Long Term Commercialization" of 150 Wh/kg, research is underway to develop battery chemistries with significantly higher energy densities. For example, nickel–cobalt manganese (NCM) lithium-ion battery chemistries have theoretical maximum energy densities of 230–260 Wh/kg; alloys such as transition metal oxide (TMO)-silicon alloy lithium-ion battery chemistries have theoretical energy densities of three to four times the levels of current batteries, at 310–410 Wh/kg. A number of battery chemistries at a much earlier stage of research and development promise even greater energy densities: Zinc-Air batteries have a theoretical maximum energy density of 1,100 Wh/kg, Lithium−Sulphur batteries of 2,600 Wh/kg and Lithium−Air batteries of 5,200 Wh/kg. As higher energy density is achieved by producing a given energy storage capacity with a reduced quantity of materials, improvements in energy density should translate into improvements in cost, provided the reduction in materials required is not offset by significantly more expensive materials or production processes.

These prospects for longer term cost reductions are reflected in a number of industry forecasts and targets. The United States Advanced Battery Consortium (USABC) [39], a subsidiary of the United States Council for Automotive Research (an organisation comprising Chrysler Group, Ford Motor Company and General Motors Corporation) has defined a 'minimum goal for commercialization' of $150/kWh and a long-term goal of $100/kWh. EUROBAT [40] (the trade organisation for European manufacturers of storage batteries) has a current objective for cost reduction in Li-ion battery packs of €200 ($246)/kWhr. The Japanese Ministry of Economy, Trade and Industry (METI) [41] has announced a target cost of 5% of 2010 levels, which equates to around $50/kWh. Independent analysts also forecast significant cost reductions. In November 2009, Deutsche Bank forecast that battery prices would decrease from $650/kWh in 2009 to around $325/kWh by 2020, a 50% reduction [42]; in December 2010, Deutsche Bank, noting the steep decline in current battery prices and revision of industry expectations since their 2009 estimate, forecast that battery prices would decrease from $450/kWh in 2010 to around $250/kWh by 2020 [43].

The prospects for improvements in both energy density and cost look good. There is a consistent message from recent assessments that a halving in cost is achievable by 2020 with the benefits of 10 years progress up the manufacturing learner curve and the onset of volume manufacturing. A significant increase—perhaps as high as 50%—in energy density seems a realistic expectation, contributing to cost reduction. Such improvements will deliver the option of significantly improved range or reduced cost, with more to come with increasing volume and as new battery technologies reach maturity.

6.2 Prospects for Behaviour Change

Despite good prospects for reductions in the cost of EV batteries, it is unlikely that a combination of reduced costs and increased energy density will be achieved that will enable BEVs to have the full range of an ICE vehicle today. (However, PHEVs have no practical range restrictions, but for maximum benefit in terms of emissions reduction, both a longer electric range and drivers who keep their batteries charged are needed.) However, even with significantly improved range, recharging time is likely to remain a constraint. It is not clear that a battery charging infrastructure that is sufficiently widespread (i.e. with nationwide coverage), local (ensuring certainty over ability to recharge at any location), and fast (so that charging did not significantly increase journey times) is viable, at least in the medium term, for both technology and cost reasons. So a conservative assumption is that BEVs will, at least in early adoption, be restricted to uses where they can do a full day's driving on a single charge.

Fig. 16 Proportion of cars travelling a given distance at least one day per week

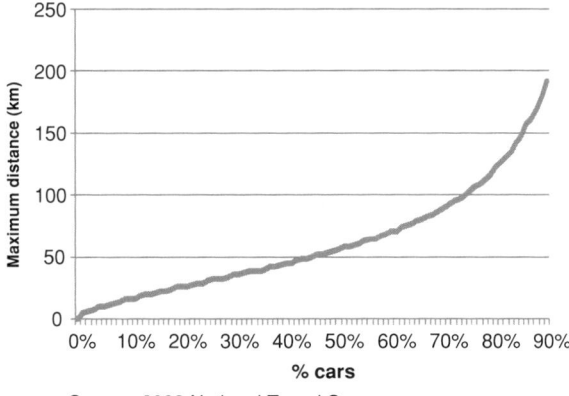

Source: 2008 National Travel Survey

Taking this conservative assumption, we explore the nature and scale of behaviour change that would be required for BEVs to replace conventional ICE vehicles in the UK. Potential behavioural changes are:

- switching to public transport for long journeys;
- selecting an alternative to travel such as teleconferencing or internet shopping;
- renting a conventional car—which might be part of a BEV ownership package including the occasional provision of a PHEV or conventional vehicle;
- transferring the trip to a conventional car in a multi-car household.

Whether a driver needs to or would be prepared to make these changes, and hence the potential for BEVs to replace conventional vehicles, is likely to depend on the frequency with which the conventional vehicle is currently used for longer range travel:

- some drivers would not require longer range and switching to a battery electric car would not require any behavioural change;
- some drivers would require longer range infrequently enough that a battery electric car would not necessitate an unacceptable level of behavioural change;
- some drivers would require longer range frequently enough that a battery electric car would require an unacceptable degree of behavioural change.

Evidence on the proportion of drivers who do not require longer range is limited. However, Fig. 16 is based on data from the National Travel Survey [44] and shows the proportion of cars which travel a given distance at least one day per week. The majority of cars do not travel significant distances in the average seven day period (the duration of the National Travel Survey). Fifty percent of cars travel no more than 56 km on any day of the average week; 86% of cars travel no more than the range of a 160 km electric car. Similar results are observed in a Department for Transport survey [45], which indicates that in 2009 20% of car drivers made long-distance journeys about once a month, 40% made long-distance journeys less than once a month, and 33% never made long-distance journeys.

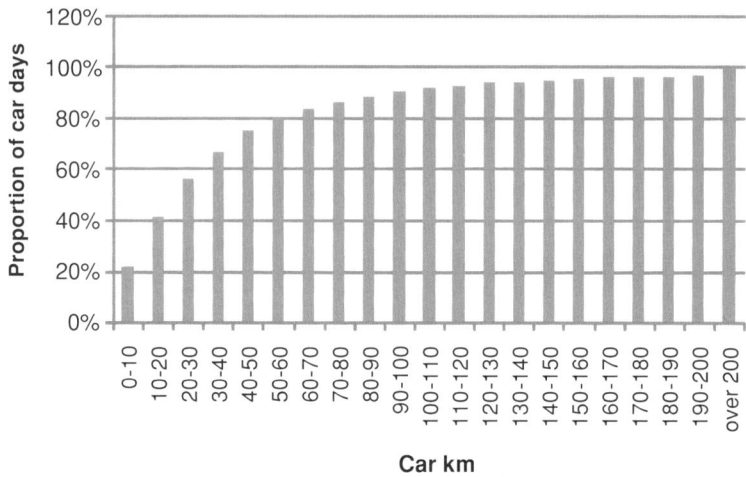

Car km

Source: 2008 National Travel Survey

Fig. 17 Cumulative distribution of daily car travel

So far we have assumed that drivers can and will make use of the full stated range of an EV. Early survey evidence in 2009 [31] indicated that users of electric vehicles are generally unwilling to use more than a third to a half of the vehicle's technical range. However, more recent data from trials show that, as user confidence grows, more of the range is exploited [31, 33, 34]. We have therefore assumed that a usable range ratio of 1.5 (the ratio of technical range to maximum distance driven) could be achieved with increased familiarity with electric vehicles and sufficient availability of public charging infrastructure. Taking this into account, Fig. 16 indicates that 75% of cars travel no more than the 106 km effective range, of a nominal 160 km range EV, in a week.

Although these results do not allow us to give a confident estimate the proportion of cars that could be replaced by a battery electric vehicle with no behaviour change, they suggest that the potential market share of battery electric vehicles could be substantial, perhaps a third to one half of vehicles, with today's technology and limited charging infrastructure.

For the remainder of drivers, replacement of a conventional car with a BEV would require some degree of behaviour change. To give a feel for the level of change required, Fig. 17 shows the proportion of "car days" (a day of motoring for an individual car) where total cumulative travel falls within a given distance.

On over 90% of total car days, individual vehicles are not driven further than 100 km. On 95.9% of total car days, cars do not travel further than 160 km, and on 97.4% of total car days cars do not travel further than 200 km. These data suggest that battery electric cars with a range of 160 km (an effective range of 106km) could fulfil the same requirements as conventional cars on the majority (92%) of days, i.e. a battery electric car replacing a conventional car would require some kind of behavioural change on only 8% of days, i.e. 30 days per year for the average driver.

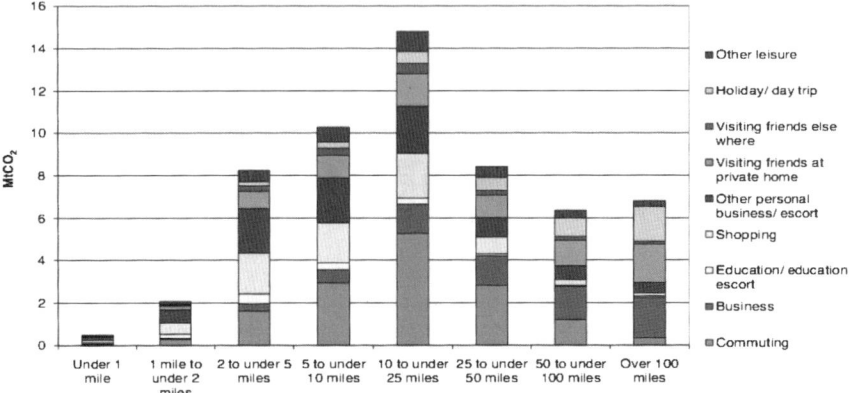

Fig. 18 Trip length and purpose and CO_2 emissions

(Clearly this concept of an 'average driver' does not take account of the distribution in lifestyles, which should be included in an ideal analysis.) The degree of behavioural change required would be lower for cars with greater range, or for segments of the car market with lower than average frequency of long-distance trips. Given our conservative assumption that cars are charged only at night, other small changes such as charging provision at the workplace and, for example, restaurants and shopping malls—locations where a car will be parked for periods of more than about an hour—would reduce the number of days on which a BEV would not be suitable.

The potential benefit of workplace charging is indicated by the dominance of commuting as a primary use for cars in the UK today, Fig. 18, and the observation that the average parking time at work is 6.2 h [31].

A relatively minor behavioural change is for multi-car households to replace a conventional car with a BEV, and rebalance the distribution of trips between the household cars such that the BEV does predominantly shorter distance trips and the conventional car(s) is used for longer journeys. In the UK 34.2% of households have more than one car [44].

This section has established that the potential early market share of battery electric vehicles could be substantial, as:

- there is a large group of drivers who do not drive outside the range of a current BEV on any day, so no behavioural change would be needed;
- while some drivers do currently require long range, for a significant proportion of these, this occurs on a minority of days, on average, 8% of days, or around 30 days per year. For this group, a battery electric car would not require an unacceptable level of behavioural change, particularly given the options available to multi-car households, which make up over a third of UK households.

This analysis has focussed on opportunities for BEVs. Those drivers who would require longer range frequently enough that a BEV would necessitate an unacceptable degree of behavioural change could opt for a PHEV.

7 Cost Effectiveness

In the sections above we have set out the current status of electric vehicle technology, the contribution of a low-carbon electricity system to emissions reductions delivered by electric vehicles, the feasibility of charging infrastructure to facilitate long-distance travel, and prospects for improvements in electric vehicle technology, performance and cost. We have also argued that even in the absence of sufficiently developed charging infrastructure, the potential market share for BEVs among drivers who do not require long range, or who would be prepared to adapt to their limited range, could be substantial. Over two-thirds of drivers are in multi-car households and/or never drive long distances. A key consideration in determining whether these people will purchase EVs in the future is therefore cost.

This section discusses the cost effectiveness of both BEVs and PHEVs, both in terms of the vehicles' economic cost to society (the social basis), and from the perspective of consumers (the private basis). This analysis looks at the situation for cars, but the factors affecting the cost effectiveness of electric vans, and hence the conclusions, are very similar.

The cost differential between electric vehicles and conventional vehicles is determined by differences in capital and operating costs. For simplicity, we estimate the cost differential for a medium size car with a petrol car as the conventional vehicle.

We estimate the capital cost differential using the approach taken by AEA [46]:

- The capital cost differential between a BEV and a conventional vehicle is taken as the cost of the electric motor and Li-ion batteries minus the cost of the engine and transmission of the equivalent conventional vehicle.
- The capital cost differential between a PHEV and a conventional vehicle is taken as the additional cost of the electric motor and Li-ion batteries (i.e. retaining the cost of the engine and transmission of the equivalent conventional vehicle).
- We use an electric motor (including the associated control system) cost of £2K, and a conventional vehicle engine and transmission cost of £2.2K for a medium-sized car.

The costs of electric vehicle batteries are a critical determinant of BEV and PHEV capital costs, and of their lifetime cost effectiveness relative to a conventional vehicle. We assume a conservative value for the current costs of an electric vehicle battery pack of $1,000 per kWh, although some car manufacturers have stated that they are sourcing batteries today at costs as low as €400($507)/kWh [30]. Given the prospects for future cost reductions of EV batteries described in Sect. 6.1, we assume in our central case that battery pack costs decrease to $500/kWh by 2015 (on the basis that this figure is already being quoted by manufacturers today), $285/kWh by 2020 and $200/kWh by 2030. For sensitivity, we also consider a more pessimistic scenario in which battery pack costs decrease to \$500/kWh by 2015, but then only to \$300/kWh by 2030.

The operating cost differential between electric and conventional vehicles is predominantly determined by differences in: maintenance costs, fuel costs and insurance costs.

There is little consensus on what the insurance and maintenance cost differentials are likely to be in 2030, but the expectation is that maintenance for a BEV will be less costly than for a conventional vehicle, given the simplicity of the system. In the absence of good data, we assume the operating cost differential to be the difference in fuel costs only. The fuel cost differential arises from the difference in the efficiency with which the electric motor and the internal combustion engine use energy to move the vehicle, and the relative costs of petrol or diesel vs. electricity. For a PHEV the fuel cost differential is lower because only a proportion of PHEV vehicle km is covered in electric mode. The fuel cost differential is therefore determined by:

- average annual distance travelled;
- the relative efficiency of the electric and conventional vehicles;
- petrol and diesel costs;
- electricity costs;
- vehicle lifetime;
- the appropriate discount rate.

We assume:

- average annual distance travelled as 13,000 km;
- BEVs have a 160 km range requiring a 32 kWh battery pack; PHEVs have a 65 km range with a 13 kWh battery pack;
- vehicle 'efficiencies' (in terms of energy content of fuel required to move the vehicle a set distance) of 0.7 MJ/km (approximately 0.2 kWh/km) for electric drive versus 2.4 MJ/km (0.7 kWh/km) for petrol ICE drive in a medium-sized car [46] (these figures are not specific to particular EVs or ICE models, but correspond to a 32 kWh BEV with a 160 km range and, for example, a BMW 318i at 30 mph [20]);
- vehicle lifetime of 12 years for all vehicles;
- PHEVs travel 70% of total distance in electric drive, and 30% in ICE drive, with the efficiencies set out above. Some sources estimate greater efficiency in ICE drive due to hybridisation of the PHEV (e.g. series rather than parallel hybrid). However, this greater efficiency typically incurs some additional cost, which our analysis suggests may not be justified given the limited emissions savings from improving efficiency for only 30% of distance driven;
- different petrol costs, electricity costs and discount rates are used to estimate social and private cost effectiveness, and are set out within the relevant subsection below.

7.1 Social Cost Effectiveness

This section sets out our estimates of the costs of BEV and PHEVs, and their cost effectiveness on a social basis (i.e. cost effectiveness as a CO_2 abatement option, £/tCO_2 abated). In Sect. 7.2 we examine cost effectiveness on a private basis (i.e. net present value, NPV, of lifetime costs of a BEV or PHEV compared with a conventional car.).

While major elements of social cost effectiveness (e.g. fuel costs in relation to world energy prices and EV capital costs driven by global battery prices) are relatively country independent (although electricity costs will depend on the country's demand profile, and hence off-peak spare capacity, as well as the decarbonisation scenario), the analysis on the private basis is specific to the UK scenario of high levels of tax on petrol and diesel fuels.

To estimate social costs, we assume:

- A petrol cost of 44.0 p/l (consistent with oil at $90/bbl) between 2010 and 2030 for our central scenario, and, for sensitivity, a high fuel price scenario with petrol cost of 44.0 p/l in 2010 rising to 90.7 p/l (consistent with oil at $200/bbl) in 2030. The UK Department for Energy and Climate Change (DECC) forecast an oil price trajectory with oil at $70/bbl in 2010, rising to $90/bbl in 2030. However, current oil prices have been significantly higher than DECC's 2010 price (ranging between $90 and over $120/bbl in 2011), suggesting that DECC's 2010 petrol cost is too low for the central scenario, and that the possibility of a significantly higher oil price should also be considered;
- Electricity cost of 2.72 p/kWh (see below);
- HMT Green Book recommended discount rate of 3.5%.

Our assumption on electricity costs is a weighted average of the short run marginal costs of nuclear generation and gas generation with carbon capture and storage. We use the short run marginal costs because, as discussed in Sect. 5.2, if sufficient low-carbon generating capacity is deployed to meet future levels of electricity demand for sectors excluding transport, there would be sufficient spare generating capacity during the night time off-peak period to fuel all the electric vehicles in our scenario.

Figure 19a shows the relationship between the abatement cost (£/tCO_2 abated) and battery costs. At higher battery costs (greater than around $350/kWh), the PHEV has the lower abatement cost, because the extra cost of a PHEV is lower than the cost of the BEV's extra battery capacity. BEV and PHEV abatement costs are roughly equal at battery costs of around $350/kWh, and at lower battery costs, the BEV has the lower abatement cost. (The results are of course sensitive to the assumption about the range and battery size). In our high fuel price sensitivity scenario (Fig. 19b), abatement costs are much lower.

As set out above, we assume, in our central case, that battery pack costs decrease from a high assumption of current levels of around $1,000/kWh to $500/kWh by 2015, $285/kWh by 2020, and to $200/kWh by 2030 and for our sensitivity analysis to $500/kWh by 2015, but only down to $300/kWh by 2030.

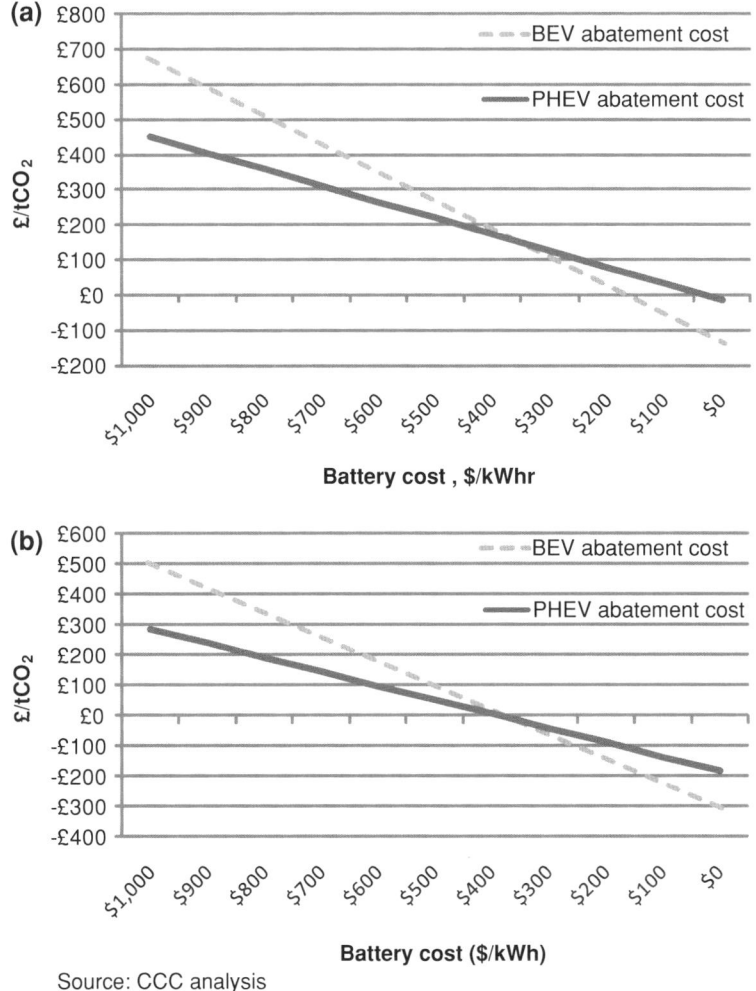

Fig. 19 The relationship between the abatement cost (£/tCO₂ abated) and battery costs, **a** central scenario, **b** high fuel price scenario

Figure 20a shows the corresponding reduction in abatement costs over time, in the central case, and compares abatement costs with the average carbon price over a twelve year vehicle life, indicating that BEVs are likely to become cost-effective when battery costs decrease to around \$250/kWh, by the mid-2020s on these assumptions, with an the abatement cost of around £60/tonne; it will take longer for PHEVs to become a cost-effective abatement solution.

The sensitivity analysis shows that under our high battery cost scenario, electric vehicles are not yet cost-effective by 2030 (Fig. 20b). This underscores the need for effective policy to reduce electric vehicle battery costs to 2030, through R&D

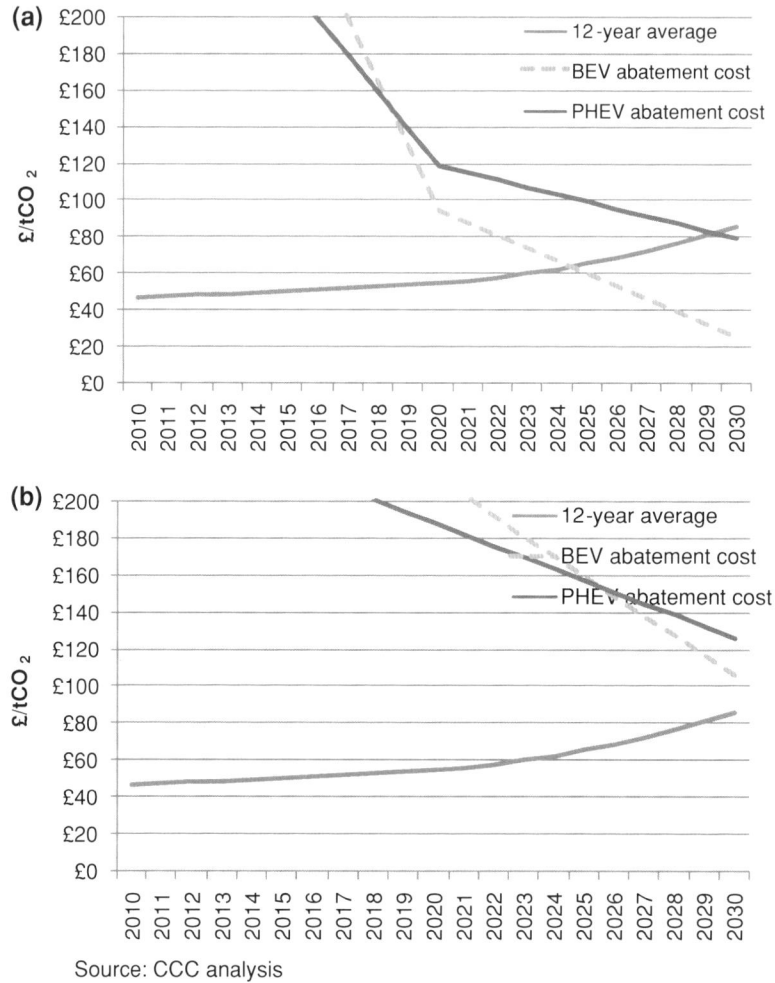

Source: CCC analysis

Fig. 20 Carbon price and abatement costs for BEVs and PHEVs **a** central scenario, **b** high battery cost scenario, **c** high fuel price scenario, **d** high fuel price and battery cost

investment and experience/learning gained in producing sufficient numbers of vehicles. The approach now being taken in a number of countries, including, for example, France, of targets for EV deployment in the range 1–2 million by 2020 is beneficial in recognising the manufacturing scale that will enable significant product cost reduction.

In the high fuel price scenario with oil rising to \$200/bbl (and a petrol cost of 90.7 p/l) by 2030, electric vehicles become cost-effective between 2019 and 2020, and actually become negative-cost measures by 2021–2023, with a BEV delivering a saving of £142/tCO$_2$ and a PHEV a saving of £88/tCO$_2$ by 2030 with battery costs at \$200/kWh (Fig. 20c). Even with higher battery costs of \$300/kWh,

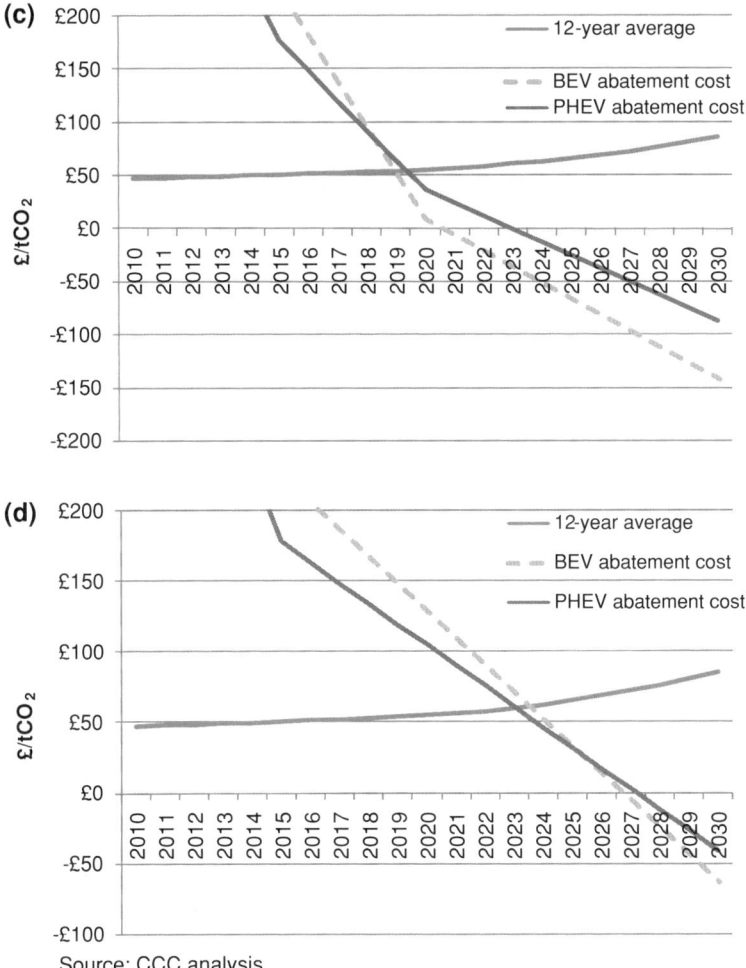

Fig. 20 (continued)

electric vehicles become cost-effective by 2023–2024, with a BEV delivering a saving of £61/tCO$_2$ and a PHEV a saving of £41/tCO$_2$ by 2030 (Fig. 20d). Once electric vehicles become cost-effective at the prevailing carbon price, the optimum solution from a cost-effective CO$_2$ reduction perspective would be for longer range BEVs to displace PHEVs.

7.2 Private Cost Effectiveness

Whereas we evaluate social cost effectiveness according to the cost of CO$_2$ abated (£/tCO$_2$), we evaluate private cost effectiveness according to the NPV of the lifetime costs of a BEV or PHEV, in comparison with that of a conventional vehicle.

To reflect the incentives facing the private consumer in the UK, we use prices inclusive of tax (e.g. VAT, fuel duty) rather than costs, and a private sector discount rate rather than the HMT Green Book recommended discount rate of 3.5%.

To estimate private prices, we assume:

- A petrol price of 127.8 p/l (consistent with oil at $90/bbl) between 2010 and 2030 for our central scenario, and, for sensitivity, a high fuel price scenario with petrol price of 127.8 p/l in 2010 rising to 178.4 p/l (consistent with oil at $200/bbl) in 2030. As discussed in Sect. 7.1, current oil prices suggest that DECC's 2010 petrol cost of $70/bbl in 2010 is too low for the central scenario, and that the possibility of a significantly higher oil price should also be considered;
- electricity price of 8.9 p/kWh (Committee on Climate Change estimate of the retail price of electricity associated with the cost of generation used in Sect. 7.1, of 2.72 p/kWh [5]);
- discount rate of 7%. The choice of discount rate for the private perspective is always controversial. Our reason for choosing 7% here is that individuals typically require relatively low rates of return (<7%) on their savings. Therefore, the assumption that the higher rates of return typically required for energy efficient technologies (e.g. low-carbon cars) represent genuine private discount rates is problematic. It is likely that these high observed required rates of return are due to market failures: bias against new and unfamiliar technologies, failure to recognise extent of operating cost savings, psychological factors relating to high capital costs. While these market failures certainly exist, they need not persist over the longer term: unfamiliar technologies become familiar through continued take up, policy should address information deficiencies (uncertainty over viability of the technology, total cost of ownership, etc.), private sector agents could arbitrage away the discount rate differential (e.g. through battery leasing), etc. Furthermore widespread availability by 2020 of low-carbon vehicles that offer significant operating cost savings is likely to challenge the established tendency to give undue weight to capital costs in making purchase decisions.

Figure 21a shows that, from a private perspective the initial lifetime cost premium of a BEV car is around £12,400, and that of a PHEV car around £4,800. This lifetime cost premium is likely to decline to zero by around 2017–2018 for both BEVs and PHEVs, when we expect battery costs to decrease to around $390/kWh. Therefore, electric cars are cost-effective on a private basis earlier, and at higher battery costs than on a social basis. In our central case, the calculations suggest that EVs will become a cost-effective purchase for (rational) consumers before the end of the 2010s.

Our sensitivity analysis shows that even under a high battery cost scenario, the initial lifetime cost premia are likely to decline to zero by around 2021–2022 for a PHEV, and by 2022–2023 for a BEV, with battery costs decreasing to around $400/kWh (Fig. 21b).

In the high fuel price scenario with oil rising to $200/bbl (and a petrol price of 178.4 p/l) by 2030, PHEVs become negative cost from 2015 from the private perspective, and BEVs become negative cost from 2016 to 2017. Under this

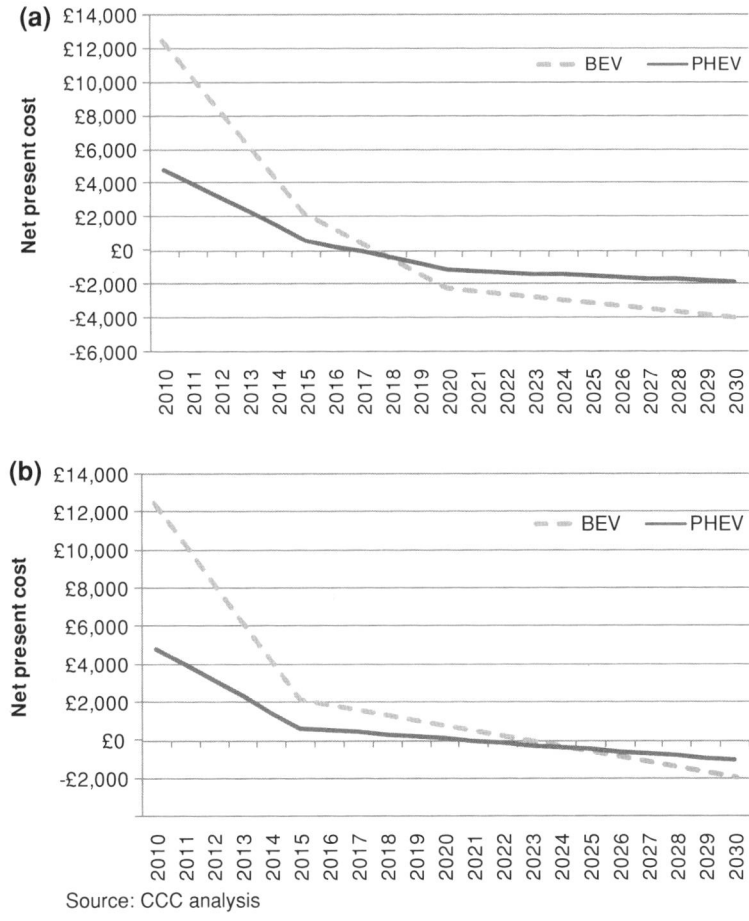

Source: CCC analysis

Fig. 21 The private perspective: reduction in the NPV of EV lifetime costs over time **a** central scenario, **b** high battery cost scenario, **c** high fuel price scenario, **d** high fuel price and battery cost

scenario, a BEV delivers a lifetime cost saving of £7,886 and a PHEV a saving of £4,582 by 2030 with battery costs at \$200/kWh (Fig. 21c). Even with higher battery costs of \$300/kWh, electric vehicles become negative cost from 2015 to 2018 from the private perspective, with BEV delivering a lifetime cost saving of £5,844 and a PHEV a saving of £3,752 by 2030 (Fig. 21d).

8 Conclusions and Implications for Policy

A generation of electric vehicles is reaching the UK and global car markets for the first time. Early releases have been BEVs with relatively limited range; forthcoming releases will include PHEVs with no practical range restrictions. Both

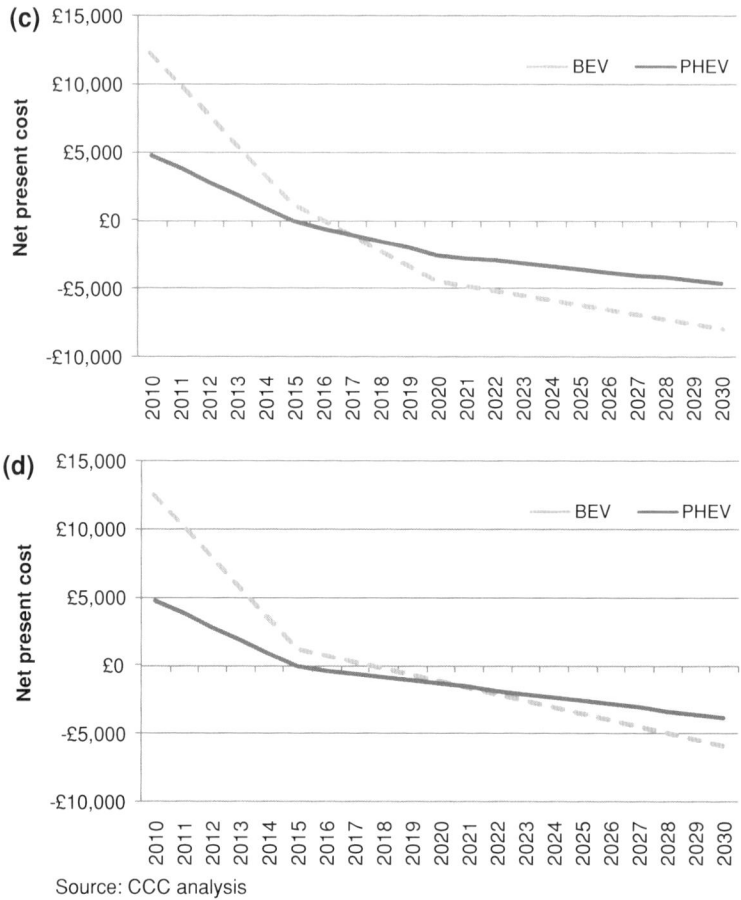

Source: CCC analysis

Fig. 21 (continued)

BEV and PHEV will carry a cost premium compared with conventional vehicles for some years, but this premium will decline over time. In previous sections we have argued that:

- in the UK the decarbonisation of the power sector required to supply low-carbon electricity to the residential, commercial and industrial sectors is critical to meeting our emissions reduction targets enshrined in the Climate Change Act 2008;
- without additional generating capacity our current system, with the roll out of smart meters, could provide sufficient 'spare' electricity, mainly at night, to power over one-third of the car and van fleet of 2030. Subsequent electrification of transport may require investment in additional low-carbon generation capacity;

- by 2030 such a low-carbon generating system would deliver effective BEV emissions of below 20 g/km;
- even with the assumption that the battery charging infrastructure is not sufficiently widespread, locally available, and fast (to accommodate significant mid-journey charging), such that battery electric vehicles would be largely limited to doing a full day's driving on a single charge, they could meet a large proportion of motoring needs with limited behavioural change, potentially around one-third of drivers today and an increasing proportion as battery energy density increases;
- with expected decreases in battery costs, BEVs are likely to become cost-effective abatement options by around 2020 and PHEVs a few years later. From a private perspective, the initial lifetime cost premium of both BEV and PHEV cars is likely to decline to zero by around 2017–2018, and possibly even earlier given our conservative assumption on cost of maintenance. Initially PHEVs are likely to be lower cost than BEVs because of the dominant effect of battery prices on BEVs, so it seems likely that consumers will see PHEVs as a better option, for both price and range considerations;
- with a lower than expected decrease in battery costs, electric vehicles do not become cost-effective as early, underscoring the need for effective policy to reduce electric vehicle battery costs to 2030. However, with higher than expected fuel prices, electric vehicles emerge as a very cost-effective option from both the social and private perspective, even with relatively high battery costs.

Our overall conclusion from this assessment of the state of the technology, and its potential to provide both a cost-effective solution for the motorist as well as an important route to decarbonising road transport, is that now is the right time for governments to be supporting the development of an important new electric vehicle industry, through appropriate policy interventions on both the supply and demand side. The next ten years will see a period of rapid change.

The optimal policy for reducing transport emissions is one that delivers the most cost-effective low-carbon technologies, and catalyses manufacturing cost reduction to accelerate affordability and the delivery of benefits, incentivising the deployment of these technologies over time on a trajectory which minimises total economic costs. The straightforward comparison of abatement costs with prices would imply widespread deployment of BEV vehicles from 2024 to 2025, and PHEV vehicles from 2029 to 2030.

However, new technologies take time to achieve significant market shares. Automotive companies will require time to develop sufficient electric vehicle production capacity to supply the large and growing global market. Strong indications that early demand for electric vehicles exists and will grow is needed to give them confidence to invest in production capacity as well as new technology and model development. While a certain proportion of consumers (early adopters) may purchase an electric vehicle as soon as they are on the market, the majority require evidence that the vehicles are a suitable, reliable and mainstream option.

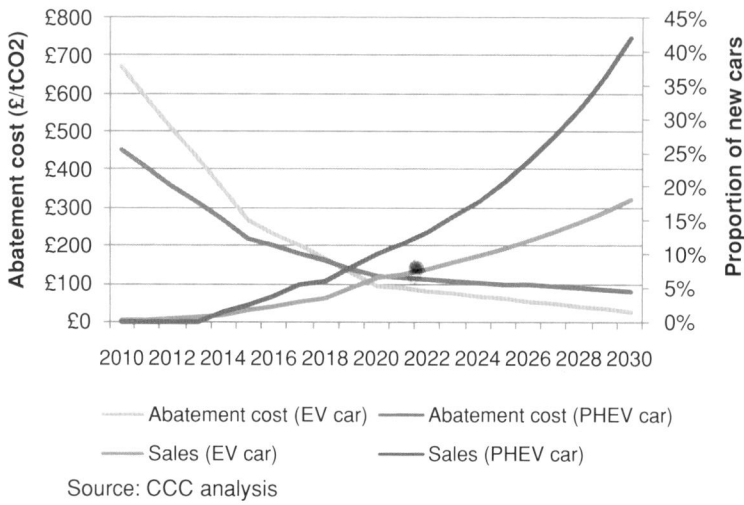

Source: CCC analysis

Fig. 22 Electric vehicle deployment

Moderate, and growing, production volumes of electric vehicles, over a critical period—the next 5–7 years—are required to realise the cost reductions that will make the vehicles a cost-effective low-carbon technology and to develop consumer confidence in a new product. The appropriate trajectory for deployment of electric vehicles is a one of gradual take up, delivering the benefits of early deployment while limiting total expenditure whist EVs are still a costly solution.

The Committee on Climate Change recommend a trajectory of take up beginning with very low volumes of electric cars and vans (BEVs and PHEVs) in 2010, rising to:

- a UK market share of 16% of new cars and van sales in 2020, resulting in a share of 5% of the total car and van fleets by this date; and
- a UK market share of 60% of new car and van sales in 2030, resulting in a share of 30% of the total car and van fleets by this date.

Under this trajectory, numbers of electric vehicles deployed increase in roughly inverse proportion to the costs of the vehicles, Fig. 22, to strike an appropriate balance between the costs and benefits of deployment.

As electric vehicles incur an initial lifetime cost premium, a price support package is required to achieve this trajectory. This price support would be required until this cost premium declines to zero, in around 2018 based on the assumptions in the central scenario calculations above. An indication of the levels of price support required every year to deliver the Committee on Climate Change recommended UK trajectory is set out in Fig. 23, with a total net present cost of around £900 million. Our sensitivity analysis shows that even under a high battery cost scenario, in which battery pack costs decrease only down to around $430/kWh

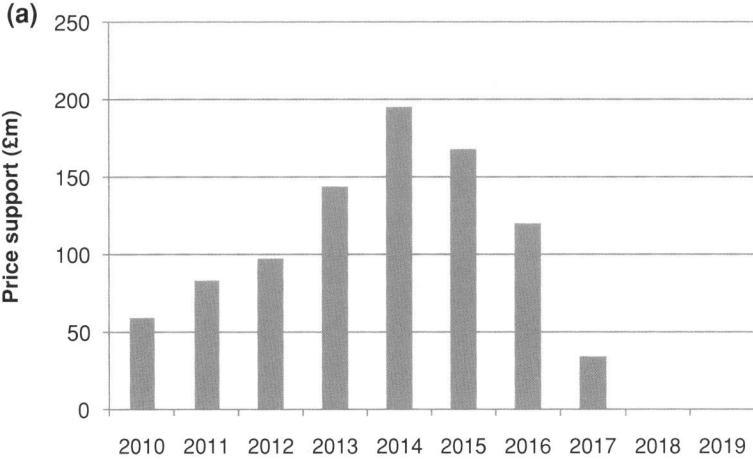

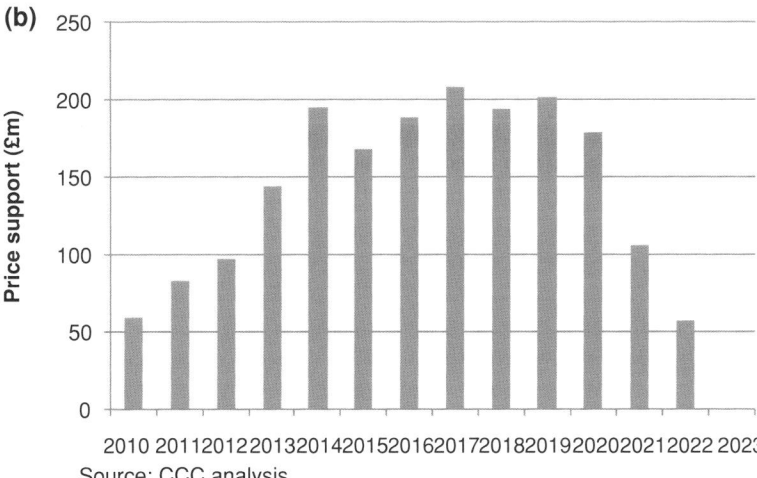

Source: CCC analysis

Fig. 23 Annual price support required to deliver the Committee on Climate Change's recommended **a** EV trajectory, **b** high battery cost scenario

by 2030, price support would be required until around 2023, with a total net present cost of around £1.9 billion (Fig. 23b).

Price support alone is not likely to be enough to stimulate this early market in electric vehicles, but it is a major and expensive element of the package. A range of support will be needed to stimulate demand, including for example: zero rating for VED; exemption from congestion charges; free parking at key locations; use of bus and multi-occupancy vehicle lanes [6, 17]. Measures to provide confidence for early adopters, such as a visible public charging infrastructure, will be very important. Emerging findings from EV trials indicate that a visible public charging infrastructure is needed to give comfort to early adopters of EVs, but these drivers do not,

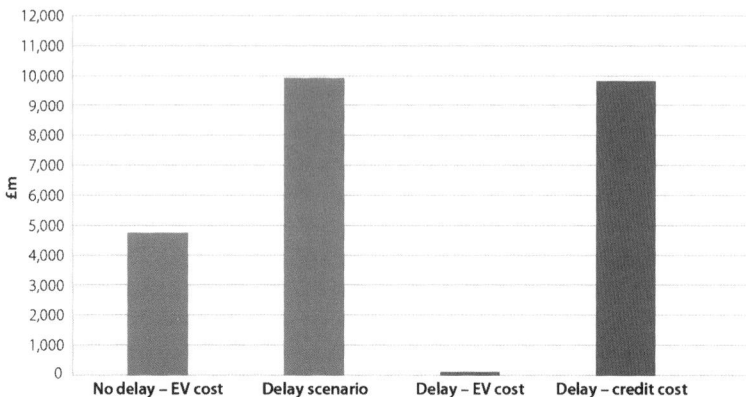

Fig. 24 The case for early action

in fact, make significant use of this infrastructure, preferring to charge at home or at work. Avoiding over-investment in infrastructure will be a particular challenge. (In the UK, London is an exception to this comment about infrastructure, and other large cities may be similar, where many people live in flats with no access to off street parking, and will need to rely on public charging infrastructure.) Stimulation of supply, including support for R and D, demonstration and capital investment will help to ensure development of local capability, while national targets for EV fleets, combined with long-term regulatory emissions standards, such as those adopted by the EU, will give industry the assurance of a stable long-term environment for new product and facility investment, and enable cost reduction.

In this area of relatively long product lives and new product introduction time-scales, early investment in emissions reduction measures avoids longer term costs. In the 2040s EVs will be making a large contribution to reducing road transport emissions if they are present in the car fleet at the levels recommended by the Committee on Climate Change [5]. If not, the UK will be paying for the excess emissions from our road transport system at the carbon price of the day. Our final figure, Fig. 24, shows the cost of investment now in support of developing the EV market, compared to the purchase of carbon credits in the 2040s, assuming the DECC forecast carbon price of £135—200/tCO$_2$e in 2040–2050. This figure highlights substantial cost saving from abatement associated with extensive penetration of electric vehicles in the 2040s, based on the high level of take up by 2030.

There have been several 'false starts' in relation to the widespread introduction of electric vehicles. The conditions today: urbanisation and the growth of a new global middle class, concerns about air quality, fuel security and peak oil, combined with a compelling need to reduce global CO$_2$ emissions; indicate that this will be seen as the age of the electric car.

The rate of change needed to meet the Committee on Climate Change's recommendations for the proportion of EVs in UK new car sales in 2030 is high, but it is simply a reflection of the scale of change that will be needed to meet our GHG reduction targets.

References

1. Climate Change Act (2008) UK legislation: www.legislation.gov.uk/ukpga/2008/27/contents
2. Committee on Climate Change (2008) Building a low-carbon economy—the UK's contribution to tackling climate change. www.theccc.org.uk
3. United Nations Secretariat (2011) World population prospects: the 2006 revision and world urbanisation prospects; the 2007. http://esa.un.org.unup. Accessed Tuesday 26 July 2011, 7:18:21am
4. Wilson D, Dragusanu R (2008) The expanding middle: the exploding world middle class and falling global inequality. Goldman Sachs Global Economics Paper No:170 July 2008
5. Committee on Climate Change (2010) The fourth carbon budget: reducing emissions through the 2020s. www.theccc.org.uk
6. Committee on Climate Change (2009) Meeting carbon budgets—the need for a step change. www.theccc.org.uk
7. Swedish Government (2009) An integrated climate and energy policy. Information sheet about the Government bills 2008/09: 162 and 163 March 2009. www.sweden.gov.se
8. Sperling D, Gordon D (2009) Two billion cars: driving towards sustainability, Oxford University Press 2009. ISBN 978-0-19-537664-7
9. Department for Energy and Climate Change (2011) UK emissions statistics: 2009 final UK figures
10. Society of Motor Manufacturers and Traders (2011) SMMT data for UK car parc in 2009. www.smmt.co.uk
11. International Energy Agency (2011) Oil in the world 2008. http://iea.org/stats/oildata.asp? COUNTRY_CODE=29
12. Mitchell WJ, Borroni-Bird CE, Burns LD (2010) Reinventing the automobile. The MIT Press Cambridge, Massachusetts
13. International Energy Agency (2008) World Energy Outlook 2008. OECD/IEA, Paris
14. Department for Transport (2011) Traffic volume—road TRA 0210. www2.dft.gov.uk/pgr/ statistics/datatablespublications/roads/traffic/index.html
15. Society of Motor Manufacturers and Traders (2011) New Car CO$_2$ Report 2011. www. smmt.co.uk
16. Institution of Mechanical Engineers (2009) Low carbon vehicles driving the UK's transport revolution. IMechE
17. King J (2007) The king review of low-carbon cars part 1: the potential for CO$_2$ reduction HM Treasury 2007. ISBN-13 978-1-84532-335-6
18. EC (2009) Directive 2009/28/EC on the promotion of the use of energy from renewable sources (renewable energy directive). Official Journal of the European Union, p 52
19. Committee on Climate Change (2009) Meeting the UK aviation target—options for reductions to 2050. CCC December 2009 www.theccc.org.uk
20. MacKay DJC (2009) Sustainable energy—without the hot air. UIT, Cambridge Ltd. UK
21. Armand M, Tarascon J-M (2008) Building better batteries. Nature 451:652–657
22. Arup/Cenex (2008) Investigation into the scope for the transport sector to switch to electric vehicles and plug-in hybrid vehicles. BERR & DfT, UK
23. Gaines L, Cuenca R (2000) Costs of Lithium-ion batteries for vehicles. Argonne National Laboratory Center for Transportation Research, Illinois, p 6
24. Royal Academy of Engineering (2010) Electric vehicles: charged with potential. ISBN 1-903496-56-X RAEng, May 2010
25. Department for Transport (2008) Carbon pathways analysis: informing development of a carbon reduction strategy for the transport sector DfT, UK July 2008 www2.dft.gov.uk
26. BMW (1971) BMW Versuchswargen mit Elektroantrieb aus Basis 02er Reine, BMW Werkfoto BMW AF 4267/1; 4270/1, BMW Werkszeichnung BMW AF 9202/1, used with the permission of BMW

27. Boston Consulting group (2010) Batteries for electric cars: challenges, opportunities, and the outlook to 2020. pp 5–7
28. Kalhammer FR, Kopf BM, Swan DH, Roan VP, Walsh M (2007) Status and prospects for zero emissions vehicle technology: report of the ARB independent expert panel 2007 prepared for State of California Air Resources Board. Sacramento, California
29. McKinsey & Company (2010) A portfolio of power-trains for Europe: a fact-based analysis: the role of battery electric vehicles, Plug-in Hybrids and Fuel Cell Electric Vehicles
30. Element Energy (2009) Strategies for the uptake of electric vehicles and associated infrastructure. Final report for CCC, October 2009 www.ccc.org.uk/element_energy_- _EV_infrastructure_report_for_ccc_2009_final.pdf
31. Mayor of London (2009, 2011) Electric vehicles for London/Electric vehicle delivery plan. www.london.gov.uk/priorities/transport/green-transport/electric-vehicles Accessed 26 July 2011
32. BMW (2011) BMW internal report (final report of the Berlin field trial) Klimaentlastung durch den Einsatz erneuerbarer Energien im Zusammenwirken mit emissionsfreien Elektrofahrzeugen MINI E 1.0, 28 Feb 2011
33. TSB Low ultra low carbon vehicle demonstrator (2011) CABLED project outcomes 2011. Posted 18 July 2011 http://cabled.org.uk/news/first-years-findings Accessed 26 July 2011
34. Turrentine TS, Garas D, Lentz A, Woodjack J (2011) The UC Davis MINI E consumer study. Institute of Transportation Studies, University of California, David, Research Report UCD-ITS-RR-11-05
35. Better Place (2011) Battery switch technology demonstration. www.betterplace.com/the-solution-switch-stations Accessed 26 July 2011
36. National Grid (2010) Average daily demand profiles. www.nationalgrid.com/uk/electricity/data/demand+data
37. Ricardo and National Grid (2011) Bucks for balancing: can plug-in vehicles of the future extract cash—and carbon—from the power grid? Ricardo/National Grid White Paper www.ricardo.com
38. United States Advanced Battery Consortium USABC Goals for Advanced Batteries for EVs. http://www.uscar.org/commands/files_download.php?files_id=27 Accessed 18 July 2011
39. EUROBAT (2005) Battery systems for electric energy storage issues: battery industry RTD position paper, Brussels, p 36. Available http://www.eurobat.org/sites/default/files/documents/pdf/pr_pp_bat-rtd-0705.pdf, Accessed 18 July 2011
40. Kawaguchi Y (2010) Japan's policy for electric vehicle (Electric Vehicle and Advanced Technology Office, Japan Ministry of Economy, Trade and Industry (METI) Presentation to Michelin Challenge Bibendum conference, Rio de Janeiro 2010 www.challenge.bibendum.com
41. Deutsche Bank (2009) Electric cars: plugged. In: 2: a mega-theme gains momentum, p 45
42. Deutsche Bank (2010) The end of the oil age: 2011 and beyond: a reality check, p 19
43. Department for Transport (2010) National Travel Survey 20008. http://www2.dft.gov.uk/pgr/statistics/datatablespublications/nts/
44. Department for Transport (2009) Public experiences of and attitudes towards rail travel: 2006 and 2009. http://www2.dft.gov.uk/adobepdf/162469/221513/336278-/railtravel2009.pdf
45. AEA (2009) Review of cost assumptions and technology uptake scenarios in the CCC transport MACC model. www.theccc.org.uk
46. Society of Motor Manufacturers and Traders (2011) Electric car guide 2011. p 30 www.smmt.co.uk
47. Michelin 'Let's drive electric! Electric and hybrid vehicles' Challenge Bibendum Booklets (2011) www.challenge.bibendum.com

Using Electric Vehicles
for Road Transport

Malcolm D. McCulloch, Justin D. K. Bishop
and Reed T. Doucette

Abstract Road vehicles account for almost half of the energy used in all transport modes globally. Reducing energy use in vehicles is key to meeting the forecast increase in demand for transport, while improving energy security and mitigating climate change. Non-powertrain vehicle options may reduce fuel consumption by at least 15%. Electric motors are the significant powertrain option to reduce energy use in vehicles because they are more efficient than the internal combustion engine and can recover a portion of the vehicle kinetic energy during braking. Conventionally, batteries are used to meet both the power and energy demands of electric vehicles and their variants. However, batteries are well-suited to store energy, while ultra-capacitors and high-speed flywheels are better placed to meet the bidirectional, high power requirements of real-world driving. Combining technologies with complementary strengths can yield a lower cost and more efficient energy storage system. While pure and hybrid electric vehicles use less energy than internal combustion engine vehicles, their ability to mitigate climate change is a function of the emissions intensity of the processes used to generate their electricity.

M. D. McCulloch (✉) · R. T. Doucette
Department of Engineering Science, Energy and Power Group,
University of Oxford, Parks Road, Oxford OX1 3PJ, UK
e-mail: malcolm.mcculloch@eng.ox.ac.uk

R. T. Doucette
e-mail: reed.douette@eng.ox.ac.uk

J. D. K. Bishop
Department of Engineering Science, Institute for Carbon and Energy Reduction in
Transport, Oxford Martin School, c/o University of Oxford, Parks Road,
Oxford OX1 3PJ, UK
e-mail: justin.bishop@eng.ox.ac.uk

O. Inderwildi and Sir David King (eds.), *Energy, Transport, & the Environment*,
DOI: 10.1007/978-1-4471-2717-8_12, © Springer-Verlag London 2012

1 Introduction

This chapter considers the design and impact of the advanced powertrains of electric vehicles (EV) and hybrid electric vehicles (HEV) along with their plug-in hybrid electric vehicle (PHEV) variants. All these powertrains have the potential to reduce transport's dependence on oil and the environmental impact.

In 2008, fossil fuels accounted for 97% of transport's energy use [1]; light-duty passenger vehicles used 39 EJ [2], equivalent to 40% of the total energy consumption of transport [3]. Organization for economic co-operation and development (OECD) countries have the highest per capita transport demand and are locked into using foreign fossil fuel resources to supplement their domestic production [1].

Concerns arise, not only because of the security and stability of the energy supply, but also due to the increasing carbon dioxide-equivalent greenhouse gas (GHG) emissions from vehicle use. The emissions associated with the energy used by light-duty vehicles were 2.7 Gt GHG in 2008[1] [4]. The population of the developing world also has aspirations to use motorised transport. Consequently, the total energy used in road transport is forecast to increase to 73 EJ by mid-century [2]. Associated GHG emissions are expected to nearly double to 5.1 Gt over the same period [4].

New vehicle technologies must be investigated, developed and commercialised to meet the growing demand for affordable transport within these resource and emissions constraints. This can be accomplished in three ways. First, by reducing the energy used in vehicles by improving their efficiency. Second, by expanding the types of energy vectors so that the vehicles can make use of energy derived from an array of sources. Third, by reducing both the number of vehicles on the roads and the distances that they travel. EVs, HEVs and PHEVs deliver on the first two ways through more efficient powertrains and by using electricity as an energy vector.

2 Energy and Power

Satisfying real-world driving demands requires periods of high power during acceleration and deceleration and high energy, when travelling at high velocity and over long distances. The new European drive cycle (NEDC), shown in Fig. 1, is the standardised velocity–time profile used to determine vehicle fuel consumption and emissions in Europe. The combined NEDC consists of four low speed, urban subcycles and a higher speed, exurban or motorway subcycle. *The standard drive cycles used in the United States and Japan are the FTP and 10–15*

---

[1] The 2008 value was determined by a linear interpolation on the 2000–2050 dataset of energy and emissions, yielding correlation coefficients of $R^2 = 0.997$ and $R^2 = 0.997$, respectively.

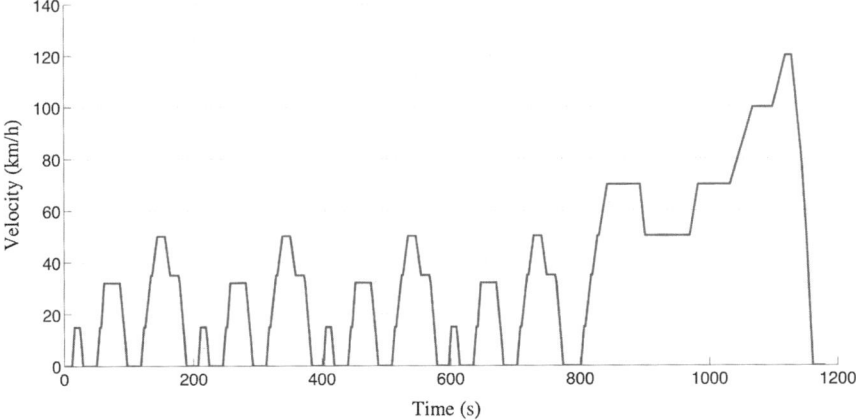

Fig. 1 The combined urban and extra-urban new European drive cycle (NEDC) used to determine light-duty passenger vehicle fuel use and CO_2 emissions in Europe

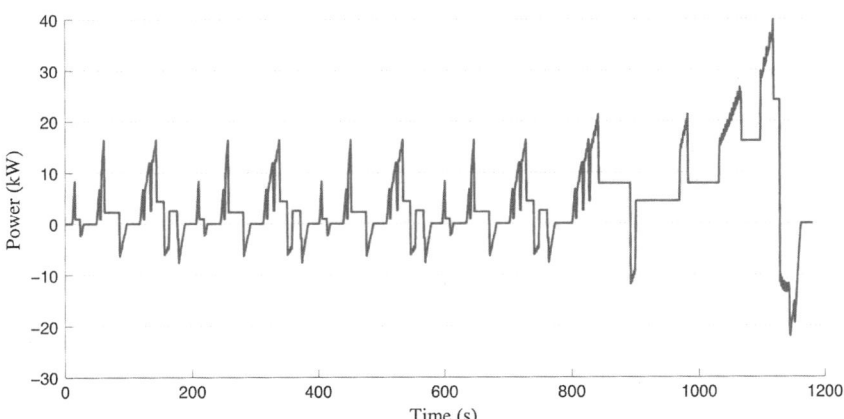

Fig. 2 Power required by a mid-size EV to complete the NEDC

cycles, respectively. Generally, such standard drive cycles underestimate energy use and associated emissions because they can fail to reflect factors which are important parts of real-world driving, including the presence of auxiliary loads (Fig. 3) *and changes in road gradient. For example, the NEDC can underestimate real-world cold and hot emissions by 30 and 50%, respectively* [5, 6].

Figure 2 presents the power demands for a mid-size EV. It shows that such a mid-size car requires approximately 5 MJ to complete the 11 km combined NEDC, leading to an energy use of about 45 MJ/100 km. Accelerating to the NEDC's top speed of 120 km/h from 100 km/h in 20 s requires a peak power of

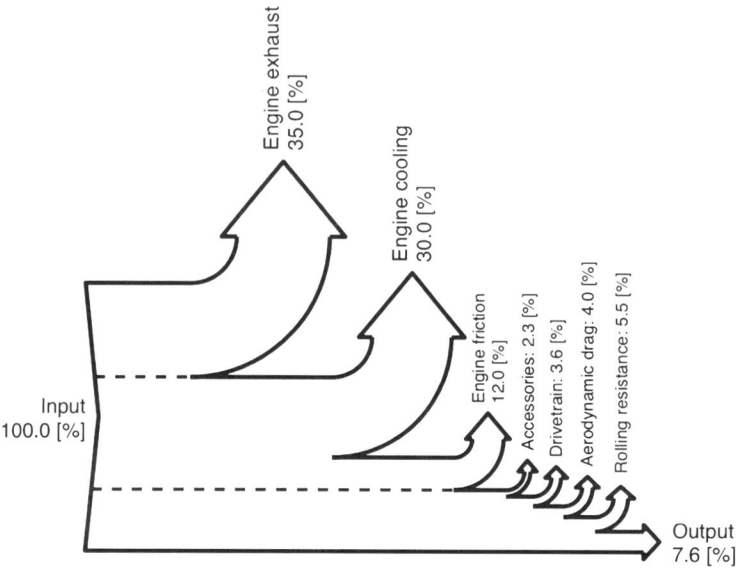

Fig. 3 Sankey diagram highlighting the magnitude of energy lost in the powertrain and non-powertrain components of a passenger vehicle on the US environment protection agency (EPA) combined city/highway drive cycle. Adapted from [8]

40 kW. Therefore, such a medium-sized vehicle requires an electric motor of at least a 40 kW to complete the NEDC. Note that the NEDC is not a very aggressive drive cycle and that more realistic cycles would likely have higher power demands. To achieve a 300 km range, 135 MJ (38 kWh) of energy storage capacity is required. This energy is present in fewer than four litres of petrol or diesel when used in a conventional vehicle. The battery pack required to yield an all-electric range similar to that achievable with an equivalent conventional vehicle is both large and expensive (Table 2).

Over the entire cycle, the average (root mean squared) power is only 9 kW. Moreover, the power required of the internal combustion engine (ICE) only exceeds the average power for 18% of the time and for generally fewer than ten seconds in each instance. The ICE in a conventional vehicle spends much of its time in the NEDC operating under inefficient, part-load conditions. Moreover, the battery pack in a pure EV must be sized to supply the full energy and power demands, though it spends relatively little time supplying power over the average value. Hence, it is important to note that the nature of the energy and the power demands of a vehicle can lead to interesting technological solutions, such as hybridizing different types of energy storage and conversion technologies. In this way, components that are better suited to handling high power can be matched with those optimised to store energy.

3 Understanding Energy Losses

To understand how the efficiency of vehicles can be improved, it is useful to describe the factors that affect vehicle losses. These factors can be disaggregated usefully into (1) the forces impeding motion and (2) the losses incurred by vehicles' powertrain components. Losses associated with these mechanisms are illustrated using the best-selling medium vehicle in the UK in 2010 [7].

However, the manner in which a vehicle is driven will directly affect most of the loss mechanisms. Driver behaviour can be represented by drive cycles which express the speed of the vehicle as a function of time. Many types of drive cycles exist. The losses which arise in a passenger vehicle following the United States environmental protection agency (EPA) combined city/highway drive cycle are illustrated in Fig. 3. The remainder of this section will examine the potential for future vehicles to reduce the losses they experience in both of these areas.

4 Losses Due to Travelling

When a vehicle is travelling on the road, it needs to overcome the force associated with displacing the air along its path—termed aerodynamic drag loss. As the tyres roll, they change shape—flatter when they are in contact with the road—which requires energy. This is termed as the rolling resistance loss. When the vehicle is brought to a halt, the kinetic energy is lost as heat in the brakes. This is termed the kinetic energy loss.

4.1 Aerodynamic Drag Loss

Aerodynamic drag comprises normal (pressure) and tangential (friction) forces acting on a vehicle shape as it moves through the air. The size of the drag force is dependent on the density of the air, the vehicle size, shape and square of the speed.

Slower travelling vehicles experience smaller aerodynamic losses but, at least in the short term, it is unlikely that lower speed limits will be introduced. Typically at speeds above 55 km/h, aerodynamic drag becomes the dominant force opposing vehicle motion. Vehicle design can focus on reducing aerodynamic losses by reducing a vehicle's drag coefficient and frontal area which are primarily determined by a vehicle's shape. For instance, the presence of a car boot leads to flow separation at the rear edge of the roof, spreading downwards. The horizontal distance from the rear edge of the roof to the rear edge of the boot, the height of the boot [9] and the resulting angle impacts the flow separation. Thus, fastbacks (coupés) generally outperform notch-backs (saloon cars), which in turn are better than square-backs (estate cars) [10–12]. Other sources of pressure drag arise due to wheel housings, external vehicle features and engine ventilation [13].

The extent to which vehicle aerodynamic drag may be reduced is constrained by functional, economic and aesthetic demands [10]. The desire for vehicles in different size classes constrains their external dimensions and associated drag. Moreover, the aesthetics of low wind noise at high velocity due to the interaction of the airflow with external vehicle features, such as side mirrors and radio antennae, constrains the design of the vehicle shape. *The effect of reducing the drag coefficient was simulated on the best-selling medium-sized vehicle in the UK using the oxford vehicle model (OVEM). Reducing its drag coefficient 10% reduces its total energy use per kilometre by 4% on the NEDC and US EPA drive cycles. These findings are consistent with the estimates in the literature* [14–17]. Aerodynamic drag may be reduced by: using vortex generators [11]; covering the vehicle underbody [16]; using body coatings to reduce friction drag; and reducing the frontal area.

4.2 Rolling Resistance Losses

Rolling resistance acts in parallel to vehicle motion and along the road surface. It is dependent on the weight of the vehicle and the coefficient of rolling resistance of the tyre. At speeds below 55 km/h, rolling resistance is the dominant drag force and typically accounts for about a third of the energy at the wheels [18].

Reducing rolling resistance *of the best-selling medium-sized vehicle in the UK reduces its energy use per kilometre by 4% on the NEDC and US EPA drive cycles. These findings are consistent with the estimates in the literature* [14, 15, 18]. The tyre rolling resistance is a function of a number of factors, including: vehicle loading; inflation pressure; wheel diameter; speed rating and the road conditions for which the tyre is designed. The rolling resistance coefficients of modern tyres have decreased over the last 30 years by approximately 10% to a median value of 0.0099 [18]. There is certainly room for tyres to achieve lower rolling resistance coefficients in the future. However, that will be challenged by consumer desire for tyres which perform safely in all weather conditions and at high speeds.

4.3 Kinetic Energy Losses

Every time a vehicle accelerates, kinetic energy is transferred to the vehicle. This energy is largely lost as heat in the brakes as the vehicle slows down. The kinetic energy is dependent on the mass of the vehicle and the square of the speed of the vehicle. Mass is the largest contributor to the intermittent power requirements of a vehicle and is the most important factor in determining the power rating of powertrain components and the amount of energy a vehicle must store on board [19].

Light weighting strategies often focus on material substitution and vehicle component redesign [20]. Every 10% of primary mass saved can reduce fuel consumption by up to 7% [21]. Since "mass begets mass," saving primary mass often allows other components to be downsized to yield secondary mass savings [22], introducing a mass decompounding effect. For example, by reducing the mass of the vehicle body, the suspension, brakes and tyres may be downsized. Two-thirds of the total vehicle curb mass is found to be gross vehicle mass-dependent, indicating the extent of the decompounding potential [20]. The vehicle subsystems which are affected by light weighting are the suspension, engine, tyres and wheels, transmission, structure, steering and brakes, electrical and the exterior. Moreover, a lighter vehicle requires less power to achieve a given acceleration. Consequently, the vehicle can achieve similar performance with a smaller engine, which requires less massive structural supports. In general, mass decompounding yields 1.04 kg of secondary mass savings for each 1 kg of primary mass avoided [20], reducing the a vehicle's energy needs even further. Not all components of a vehicle are typically affected by such light weighting strategies: the interior, information and controls, fuel, exhaust, closures and heating, ventilation and cooling are largely gross vehicle weight-independent [20].

5 Losses in Power Conversion Components

Powertrain losses occur as a result of a powertrain's components working to overcome the resistive forces impeding the motion of a vehicle. Powertrains consist of components that store energy and convert that energy to kinetic energy. In the case of vehicular applications in EVs and HEVs, some of these components may be rather novel. These components can be arranged and the energy flows between them can be managed in ways that can produce significant improvements in vehicle efficiency. This section will present the currently available energy storage and conversion technologies that will likely shape future vehicle development.

5.1 Internal Combustion Engine

The ICE is the conventional device used to convert chemical energy in transport fuels to shaft work. ICEs have good specific power (W/kg) characteristics and use energy dense fossil-fuels (MJ/l). Therefore, conventional vehicles have normally used ICEs as their only primary energy conversion device to meet their driving demands.

The efficiency of the ICE is a function of its power output which is the product of its torque and speed. Losses arise due to thermodynamic limits, inefficient combustion of the fuel, friction, heat transfer to the cylinder walls during the

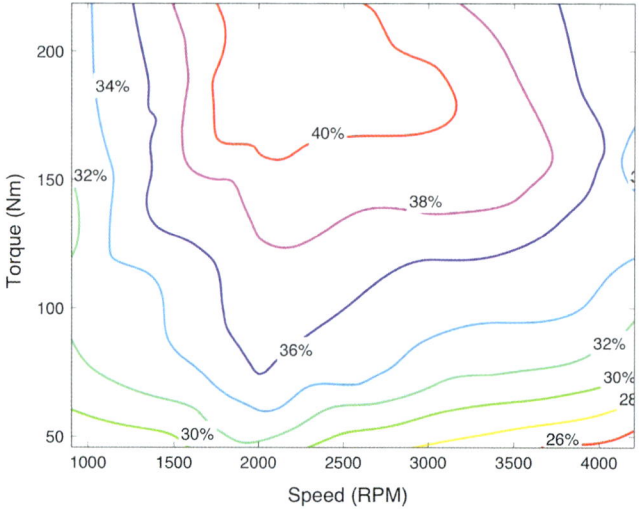

Fig. 4 Efficiency map of a diesel internal combustion engine, with efficiency shown as iso-contours. Adapted from [24]

piston linear motion and in the exhaust gas. There is a combination of torque and engine speed, corresponding to the peak output power, which yields maximum efficiency, as shown in Fig. 4. Under these full-load conditions, the peak efficiency is typically less than 45%. At normal driving speeds, whether in an urban setting or on the motorway, the ICE operates at part-load, where the losses are more significant [23].

A transmission is an essential auxiliary component to the ICE and is used to match the engine speed to that of the wheels. Typically, every gear of the transmission has a fixed ratio. For a given output power desired, the gear should be chosen to minimise energy losses in the engine.

There are a number of measures available to improve the efficiency of ICEs under real-world driving conditions. Downsizing the ICE brings its peak power closer to that required when driving normally. Therefore, it can operate closer to its peak efficiency. In urban settings, an engine start/stop function avoids the inefficiency associated with idling. Increasing the number of gears in a transmission gearbox (ultimately to a continuous variable transmission) allows the engine to operate closer to its most efficient point, for a given power output. Turbo charging allows some of the work potential in the exhaust gas to be recovered and improves the performance of vehicles with downsized engines. In some HEVs, ICEs can either be completely decoupled from the road load, or only operated at certain points in their efficiency map. This allows them to produce power much more efficiently than in a conventional vehicle where they must respond to meet all the transient power demands of normal driving.

5.2 Fuel Cells

Fuel cells (FC) are another energy conversion technology with the potential to find an application in HEVs. Fuel cells work by converting energy stored in chemical bonds into electrical energy that can go onto be used by a vehicle's energy storage media or electric motors. Many types of fuel cells exist, each with their own chemical processes and traits that come with distinct advantages and disadvantages. The fuel cell technology that has historically been used most frequently in light-duty fuel cell vehicles is the proton exchange membrane (PEM) fuel cell. PEM fuel cells use hydrogen to generate electricity and they are commonly used in passenger vehicles because of their relatively low operating temperatures, low mass and high durability in dynamic environments. The adoption of fuel cells has been hindered to a large degree on account of unresolved issues related to their cost, resource constraints, refuelling infrastructure and the storage of hydrogen. Please refer to chapter on batteries for more detail.

5.3 Electric Motors

Electric motors are a mature technology, with extensive use in industrial, domestic and power generation applications. However, they have not been widely deployed in mainstream, light-duty road passenger vehicles to provide tractive force. Electrification of a vehicle powertrain involves decoupling the shaft work output of the ICE from the rotational work required at the wheel. The three main reasons for considering the use of electric motors are that: they have a higher peak efficiency of converting potential energy to shaft work than ICEs when motoring; their torque profiles (100% torque from zero speed) are better suited to meet the demands of realistic driving without the need of a multi-speed transmission; and they can be used as generators to convert a portion of a vehicle's kinetic energy to electrical current during braking [25–27].

In contrast with the ICE, the rotor is the only moving part in an electric motor. Losses are dominated by copper and iron losses. The copper losses occur because of the electrical resistance of the windings and is proportional to the square of the torque. Iron losses arise due to the interaction of the magnetic fields with the iron components, and is proportional to the speed [28, 29]. An electric motor may not require a transmission to match its speed to that of the wheels. Moreover, the electric motor has a broad region of high conversion efficiency, when compared to an ICE of equal power rating (Fig. 5). Finally, electric motors generally have a high power density (kW/l). This permits flexible packaging within the vehicle, such as in or near to the vehicle's wheels.

The most common electric machines used for traction are induction machines (IM) and permanent magnet machines (PMM), see Table 1. IM are reliable and low cost, but have a limited constant-power range and suffer from

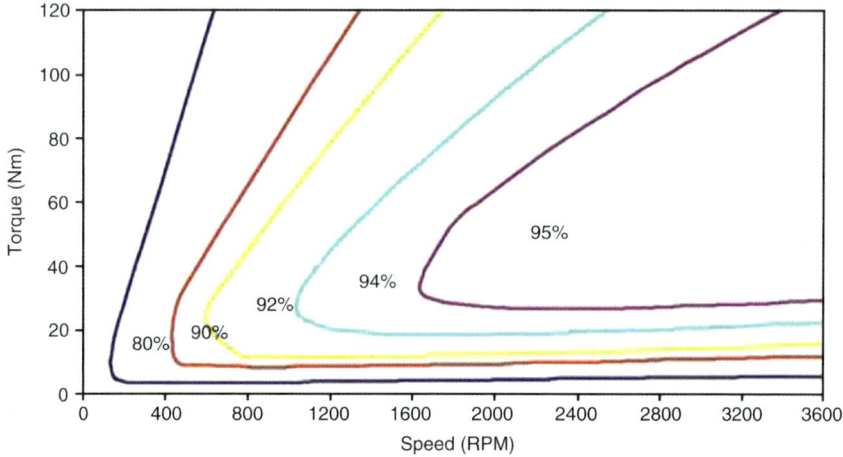

Fig. 5 Efficiency map of a permanent magnet electric motor, with efficiency shown as iso-contours [30]

Table 1 Comparison of electric motor technologies for energy conversion in EVs [13, 30, 33–36]

Electric motor technology	Specific power (W/kg)	Specific torque (Nm/kg)	Peak efficiency (%)
Permanent magnet	660–1,120	22	92.5–97
Induction	563–2,128	1.43	90–90.5

high weights and low efficiency, particularly at high speeds [31]. PMM are more compact in size, lighter in weight and are efficient. However, the magnets are expensive, leading to a high cost per unit power [25, 32]. Despite these drawbacks, PMM have been identified as the most promising motor technology for modern EVs [32]. New topologies of PMM have been specifically developed to meet the needs of the EV market in a very high efficiency and an extremely lightweight solution [30].

Power electronic converters use high power, efficient, reliable and fast-acting semiconductor switches to convert the output from the electrical system to a controlled input for use in the machine. A key advantage of the machine/converter sub-system is that bi-directional energy flow is possible. This implies that when the vehicle is slowed down, the kinetic energy of the vehicle can be recovered to an on board energy store, thus mitigating the kinetic energy loss. However, a bi-directional energy store is needed.

6 Energy Storage Technologies

Energy has traditionally been stored on board vehicles in the form of chemical energy of petrol. However, the conversion process is not (readily) reversible. Therefore, an alternative reversible energy store needs to be on board the vehicle to exploit the recovery of the kinetic energy. This section will address the various technologies capable of storing energy on board a vehicle: batteries, ultra-capacitors and high-speed flywheels.

The ideal characteristics of device energy storage in EVs and HEVs include high specific energy and power, long calendar life and cycle life and low cost. However, no single energy storage device currently satisfies all of these requirements to such an extent that efficient EVs and HEVs have been widely adopted [25, 37]. Thus, energy storage remains the weak link in the electrified powertrain [37], and the cost and performance of energy storage devices seems likely to remain the key determinant in the future development and successful marketing of EVs and their variants [38].

6.1 Batteries

Batteries convert chemical energy directly to electrical energy. The key technologies of interest here are those types that can do so reversibly. Not only can they then accept the recovered kinetic energy, but also provide the opportunity of using the energy from the electrical grid, with a route to a diversified energy mix.

Battery storage has traditionally comprised rechargeable lead–acid (Pb–A), nickel–cadmium (Ni–Cd) and nickel–metal hydride (Ni–MH) cell technologies, with lithium-ion (Li-ion) and lithium–polymer now being the dominant technology. Their characteristics are found in Table 2. There is energy embodied in batteries which is present in their constituent materials and added during the manufacturing process [39]. This energy is small in comparison to the total energy used during the vehicle operation over its lifetime.

The Pb–A battery is a mature technology with low costs that has been a commonly-used energy storage medium in previous EVs [25]. Despite their low specific energy and associated mass penalty, Pb–A batteries are appropriate for use in micro and mild hybrids [37] and may be paired with DC motors in low power applications [51]. However, the low specific energy of the Pb–A renders them unsuitable as an energy storage technology for pure EVs and HEVs where the electric powertrain satisfies more of the driving load.

Ni–MH and Ni–Cd batteries are usually thought of as occupying the next rung in the progression of batteries from Pb–A. The Ni–MH battery is both less expensive than the Ni–Cd technology and avoids the environmental issues associated with cadmium disposal [25, 37]. Moreover, Ni–MH batteries have high

Table 2 Comparison of battery technologies for energy storage in EVs [25, 37, 38, 40–50]

Battery technology	Specific power (W/kg)	Specific energy (Wh/kg)	Roundtrip efficiency (%)	Lifetime (cycles)	Cost (€/kWh)
Lead-acid (flooded)	157–300	20–50	72–82.5	100–2,000	38–150
Nickel–cadmium	150–300	20–80	72–78	5–1,500	110–600
Nickel metal hydride	154–1,500	46–120	70	300–3,000	766
Lithium-ion	300–3,000	35–190	90–95	300–3,000	422–1,000

specific energy, are tolerant of abusive overcharge and over discharge and display excellent thermal properties [37].

Lithium-based batteries are currently regarded as the state of the art in battery storage technology. There are many different forms of lithium-based batteries, each with a distinct chemical composition offering its own advantages and disadvantages. The high electrochemical reduction potential and low atomic mass of lithium allow batteries based on it to achieve a high specific energy.

Li-ion batteries comprise a number of cells. Cells are wired in series strings to satisfy the voltage requirements at the battery terminals. Parallel wiring of series strings allows the total battery energy capacity to be satisfied [25]. *The higher the voltage of the battery pack, the lower the transmission current (and losses) along the powertrain to the motor. Operating voltage of existing EVs and HEVs are on the order 200 V* [52, 53]. *This will increase to over 500 V, as with the new Toyota Hybrid Synergy Drive* [54]. *Li-ion batteries require a management system to ensure efficient pack operation and provide under voltage, overvoltage, short-circuit and thermal protection* [25, 37]. Therefore, a Li-ion battery is expensive on account of its management system and the elements used in its electrodes and electrolyte [25, 55]. These batteries also do not suffer the memory effects associated with Ni–MH chemistry [37]. The ability for Li-ion batteries to achieve lower costs in the future will be a function of the cost of the natural resources on which they are based, the potential to substitute one expensive natural resource for a less expensive one and the ability to achieve mass production.

Supplying high power requires the battery to discharge at high current, which reduces the battery capacity faster than a slow current discharge. Moreover, the battery life is shortened with the number of fast discharges. Batteries are suited for storing energy but are an expensive, heavy and inefficient medium to satisfy the frequent, large bi-directional current flows associated with real-world driving. Most commercial EVs and their variants use batteries to satisfy both the energy and power demands. Consequently, the battery capacity is sized to meet both the maximum power demanded to satisfy real-world driving requirements and the maximum energy required to enable the vehicle to achieve a suitable range [38].

Table 3 Comparison of high specific power technologies to complement energy storage using secondary batteries in EVs [25, 36, 38, 47–49, 58–60]

Technology	Specific power (W/kg)	Specific energy (Wh/kg)	Roundtrip efficiency (%)	Cost (€/kWh)
Ultra-capacitors	700–5,000	2.4–5.6	95	713–4,900
Composite flywheels	2,000	200–350	98	253

Use £1.00 = 1.15e; £1.00 = US$1.50

6.2 Ultra-Capacitors

Ultra-capacitors store charge on two plates separated by an insulating dielectric. See Table 3 for their key characteristics. The main benefit of using an ultra-capacitor is the large specific pulse power,[2] due to a short discharge time and low equivalent series resistance [49]. Ultra-capacitors have long calendar lives and up to 500,000 cycles [38]. Yet, ultra-capacitors have relatively low specific energies and are an expensive way to store large amounts of energy. Therefore, rather than a substitute or alternative, ultra-capacitors, as they currently exist, are commonly thought of as a natural partner to be hybridized with another form of energy storage or conversion. For instance, a battery and ultra-capacitor hybrid has the potential to yield an energy storage system which meets a vehicle's energy and power needs at lower cost than batteries alone [47]. *Some of the net cost savings from using ultra-capacitors may be reduced on account of the additional power electronics required to ensure correct operation of the more complex, hybrid energy storage system.*

6.3 High-Speed Flywheels

High-speed flywheels store energy mechanically. See Table 3 for there key characteristics. It is a technology that, through recent advances in high-strength, low density materials, has emerged as a low cost, robust and high specific power energy storage option [56].

In stationary applications, the energy capacity of a flywheel is typically increased by adding to its inertia. However, in mobile applications, where increased weight and inertia are detrimental, the energy capacity of flywheels can be increased by increasing their speed—hence, high-speed flywheels.

High-speed flywheels can be integrated into EV and HEV powertrains in one of two ways: electrically or mechanically. In an electrical integration, the rotational potential energy is stored in the flywheel and converted to shaft work to drive an electrical generator. The electrical energy output is directed onto the vehicle's

---

[2] Not all ultra-capacitors outperform all batteries. Some high power battery models have specific power performance that is comparable with ultra-capacitors [48].

electric bus where it can be used to power the motor or recharge another energy storage device, such as a battery. This type of flywheel integration is generally less efficient; however, it allows for greater packaging flexibility. Alternatively, high-speed flywheels may be mechanically linked directly to the drivetrain, usually via a continuously variable transmission. In this configuration, flywheels avoid the losses associated with the additional, roundtrip conversion to and from electrical energy [57]. However, high-speed flywheels can have relatively high self-discharge characteristics and can be inefficient at storing charge for extended periods, unlike batteries [26] and ultra-capacitors. As with ultra-capacitors, high-speed flywheels are well-suited to handling high power loads, and they may be hybridized with additional components which are better suited to meeting high energy demands.

7 Powertrain Topologies

The energy storage and conversion components that have been discussed can be arranged and managed in a number of ways once they are actually placed in a vehicle. Generally, these components are assembled in a manner based on one of the four powertrain topologies discussed below: pure EV, series HEV, parallel HEV and series–parallel HEV. Each of these topologies has its own strengths and weaknesses and, as the case studies in this chapter illustrate, there is usually no single answer as to which one is optimal.

7.1 Electric Vehicle Powertrain

A pure EV uses an electric motor for traction and that motor is powered by an energy storage system (ESS), which is usually a battery. Figure 6 shows a schematic of an example of a common powertrain topology for an EV. In 2011, the energy storage system onboard EVs most commonly takes the form of a battery array—though the battery array has the potential to be hybridized with other technologies, such as ultra-capacitors or high-speed flywheels. EVs have a tank-to-wheel efficiency that is typically higher than conventional vehicles, which rely solely on ICEs. Moreover, EVs produce no tailpipe emissions and use electrical energy which can be derived from any means—not just liquid fossil-fuels. One of the weaknesses of EVs is that in order to achieve a range comparable to a conventional vehicle, they need an energy storage system with a high energy storage capacity. Given the state of energy storage technology in 2011, such an energy storage system is typically heavy and expensive, especially if it is made large enough to achieve a range approaching that of a conventional vehicle.

The Tesla Roadster from Tesla Motors is an example of a pure EV. It uses a Li-ion battery pack of mass 450 kg, which accounts for 36% of the curb mass of the vehicle, to achieve a range of 354 km on the US EPA combined city/highway drive cycle.

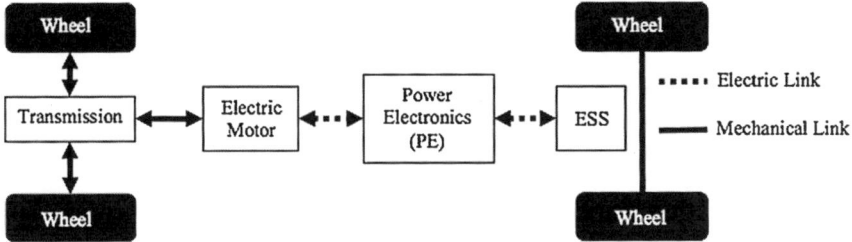

Fig. 6 Schematic of *an example of* an electric vehicle powertrain [61]

7.2 Series Hybrid Electric Vehicle Powertrain

A series HEV is similar to an EV in that an electric motor provides all of its tractive force, and that motor is powered by an energy storage system. However, a series HEV also has an onboard power generator that can be used to recharge the energy storage system or supplement its power while the vehicle is driving. The onboard power generator, also known as an alternative power unit (APU),
typically takes the form of an ICE connected to a generator or a fuel cell. Figure 7 shows a schematic of one example of a powertrain for a series HEV. The alternative power unit in a series HEV can often be downsized and operated at its most efficient point since it only has to recharge the energy storage system and it does not have to provide traction. The energy storage system can also be designed to meet the transient driving loads while the alternative power unit supplies the average power. This means the energy storage system can often be downsized since it shares the energy storage and power production load with the alternative power unit. The presence of the electric traction motor and energy storage system can improve the efficiency of the vehicle by enabling regenerative braking. The energy storage system of a series HEV can also be charged from the electric grid. Such a vehicle is known as a plug-in hybrid electric vehicle (PHEV). The Vauxhall Ampera (or Chevrolet Volt, as sold in the United States) is an example of a series PHEV that uses an ICE as its alternative power unit.

The Volt has a reported all-electric range of 64 km before it has to resort to using its alternative power unit, a four-cylinder ICE. According to EPA testing on the combined city/highway drive cycle, the Volt will achieve an equivalent of 2.4 1/100 km in electric-only mode and 6.4 1/100 km once its alternative power unit is engaged.

7.3 Parallel Hybrid Electric Vehicle Powertrain

In a parallel HEV, the tractive force is split between an ICE and an electric motor. Figure 8 shows a schematic of an example *of a parallel HEV powertrain. This topology represents only one possible parallel configuration, but it illustrates the*

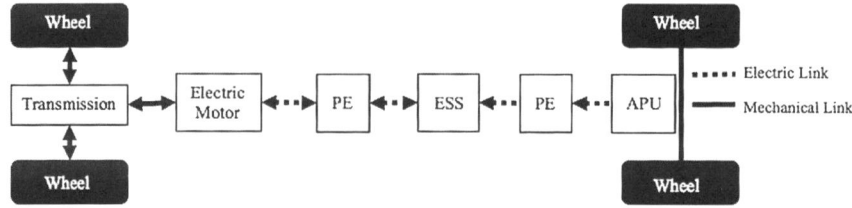

Fig. 7 Schematic of *an example of* a series hybrid electric vehicle powertrain [61]

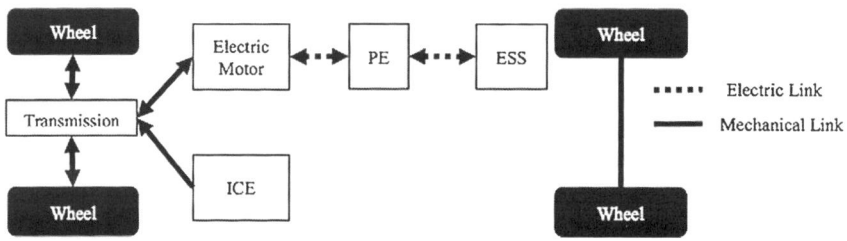

Fig. 8 Schematic of an example of a parallel hybrid electric vehicle powertrain [61]

primary points of parallel HEV operation. In this configuration the electric motor typically propels the vehicle at low speeds since it has a higher efficiency than an ICE during the stop-and-go pattern characteristic of urban driving. The ICE is typically used to power the vehicle during high speed driving where there is less speed fluctuation and the engine has to deviate less frequently from operating at its most efficient point. Parallel hybrids, such as the Ford Escape Hybrid, are capable of an extended range by utilising an ICE. Also, the ICE, motor and energy storage system can be downsized since neither one of them is solely responsible for the total power output of the vehicle. Parallel HEVs can also be PHEVs, but this is less common since their energy storage system generally has lower energy capacities. This is because since the energy storage system is only used to power low speed driving, generally in start and stop conditions.

7.4 Series–Parallel Hybrid Powertrain

A series–parallel, also known as a dual or power split, HEV combines elements from both the series and parallel HEV configurations. Like a parallel HEV, the road load in a series–parallel HEV can be split between the motor and the ICE. Like a series HEV, the ICE can also function as an alternative power unit to generate additional power to recharge the energy storage system while the vehicle is driving. A series–parallel HEV can also be a PHEV.

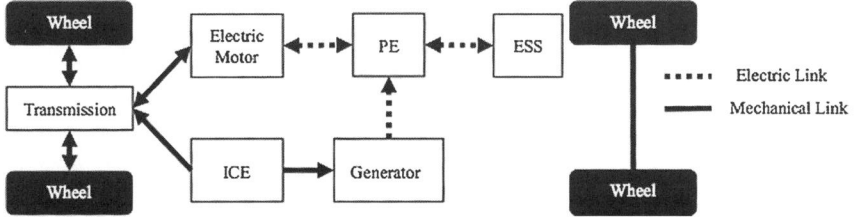

Fig. 9 Schematic of an example of a series–parallel hybrid electric vehicle powertrain [61]

The Toyota Prius implements this powertrain topology. Figure 9 shows a schematic of a common layout for the powertrain of a series–parallel hybrid.

8 Case Studies

How much energy will vehicles require? How will they store and convert that energy? Where will that energy come from? The previous section provided a look at the technologies and topologies that will likely play a prominent role in answering these questions and shaping the future of automobile transport. This section looks at two studies that provide practical examples for how the answers to these questions can intersect, sometimes to produce surprising results.

8.1 Case Study 1: Comparison of Batteries, Ultra-Capacitors and High-Speed Flywheels Functioning as the Energy Storage Medium in a Fuel Cell-Based Series HEV

Batteries are currently the most common form of energy storage onboard EVs and HEVs. However, their ubiquity belies the fact that they still have characteristics that make them less than ideal as a vehicle energy storage medium. For instance, their low power and energy density, low cycle life and high cost could all be improved in order to increase the adoption of EVs and HEVs. This study compares high-speed flywheels and ultra-capacitors to the more conventional form of energy storage—batteries—on the bases of cost and fuel economy in a series HEV [62]. *The results* show that batteries may have some legitimate competitors, at least in certain applications.

8.1.1 Methods

This study was conducted by using *the OVEM* computer simulation tool to model high-speed flywheels, ultra-capacitors, and Li-ion batteries each as the

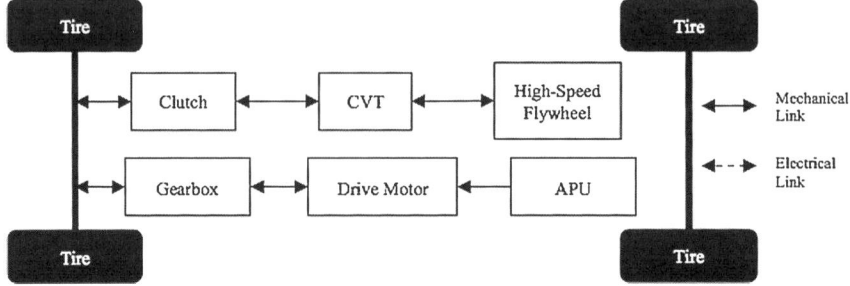

Fig. 10 Schematic of a mechanically integrated high-speed flywheel in a series HEV [61]

sole energy storage medium onboard a series HEV. The series HEV was simulated as using a fuel cell as its main power unit, so the fuel economy for the vehicle was expressed in km/kg H_2. It should be noted that the fuel economy figures in this study would be qualitatively similar for an ICE. The batteries and ultra-capacitor arrays were integrated into the powertrain as shown in Fig. 6 by connecting them to the electric bus. The high-speed flywheels were mechanically integrated into the powertrain via a continuously variable transmission, as seen in Fig. 10. These three different technologies were simulated in a range of sizes in the standard series HEV powertrain. As the size of the energy storage media increased, so did their cost, mass and energy storage capacity. The energy storage media were sized such that they would be able to meet the transient power demands of the drive cycle. The fuel cell was sized to meet the average power (load).

The simulation was run over two different drive cycles: the NEDC (Fig. 1) and the assessment and reliability of transport emission models and inventory systems (ARTEMIS) combined drive cycle [5] (ACDC, Fig. 11). The NEDC was chosen because it is used in Europe as the basis for the fuel economy ratings of commercially-available vehicles. The ACDC was used because it is a more aggressive drive cycle better representing actual urban and highway driving.

8.1.2 Results

The results from this study are presented in Figs. 12 and 13. Over both of the drive cycles, the batteries were shown to achieve the highest fuel economy. However, over both drive cycles, the ultra-capacitor and high-speed fly-wheel systems that achieved the highest fuel economy were within several percentage points of the best-performing battery array. Additionally, powertrains using the high-speed flywheel and ultra-capacitors have the potential to be cheaper than similarly fuel efficient powertrains using batteries, especially on the more realistic ACDC.

Such a result may seem counterintuitive on account of the lower specific energy of both the high-speed flywheels and ultra-capacitors relative to the batteries, as one

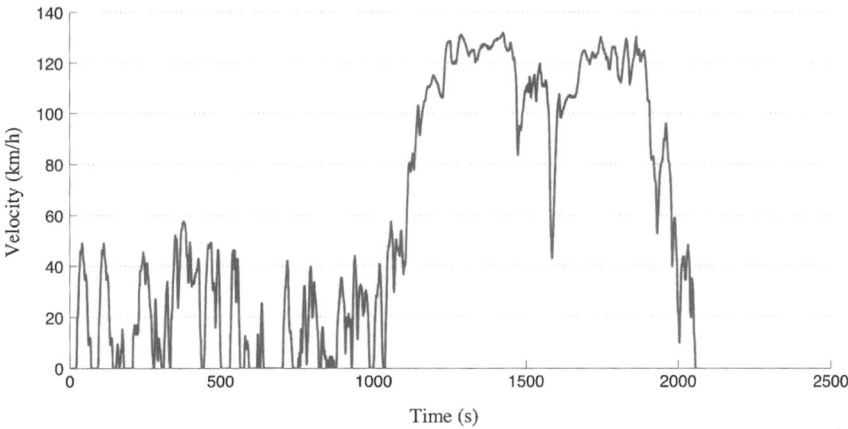

Fig. 11 The artemis combined driving cycle (ACDC) [5]

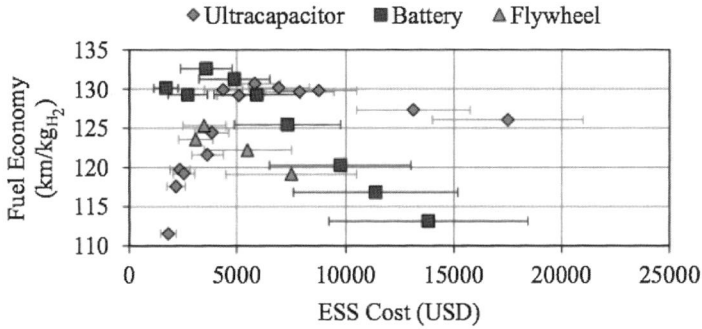

Fig. 12 Fuel economy as a function of ESS cost on the NEDC [62]

may envision that the alternative power unit would have to operate under a greater load to maintain the flywheel and ultra-capacitors at a useful state of charge. However, the results illustrate that the energy stored in both the flywheel and ultra-capacitor arrays is generally enough to power the vehicle through the transient spikes in road load enabling, the alternative power unit to operate at a high efficiency.

The flywheels, ultra-capacitors and batteries used in this case study are commercially-available and the comparison represents the state of the technology at present. The flywheel system costs range from the current value to the future estimates by the manufacturer based on mainstream use. The cost range used for both ultra-capacitors and batteries are based on a survey of the literature and represent mainstream production for non-automotive applications primarily [62].

It is also interesting to note that there are regions of optimal sizing for the energy storage media. As the size of the energy storage increased there was a point

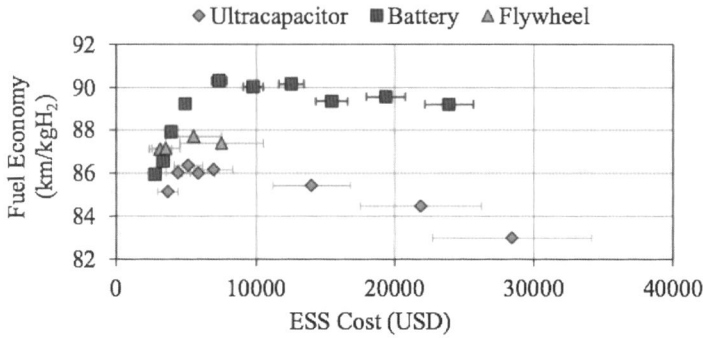

Fig. 13 Fuel economy as a function of ESS cost on the ACDC [62]

where they reached their maximum fuel economy. As they continued to expand, their fuel economy decreased due to the effects of the increased mass of the vehicle. Therefore, the optimal size for each ESS will vary depending on the drive cycle, the ESS technology and the characteristics of the rest of the vehicle.

This study did not consider life cycle costs, but both high-speed flywheels and ultra-capacitors have long calendar lives that could result in significant cost savings over their operating lives. For these reasons, it appears that high-speed flywheels and ultra-capacitors may be more legitimate competitors to batteries than is widely perceived.

8.2 Case Study 2: Comparison of Tank to Wheel CO_2 Emissions from a Conventional Vehicle, an EV, and PHEVs

One of the key attributes of pure EVs is that they produce no emissions at the point of use. However, this does not mean that EVs cannot still be responsible for emissions. *Complete well-to-tank accounting includes emissions resulting from the extraction, processing and transportation of the raw materials and their conversion to the final fuel. However, in this case study, the scope of well-to-tank emissions is limited those emitted during the electric power generation process alone.* If they obtain their electrical energy from a process that involves fossil-fuel combustion, then their emissions are simply shifted from the vehicle to the power station.

This fact is often overlooked, and it is also widely assumed that EVs produce significantly fewer tank-to-wheel CO_2 emissions than any other vehicle powertrain topology—fewer than PHEVs, and certainly fewer than conventional vehicles. However, this notion may not always be true. Efficient ICE-based vehicles may actually have lower tank-to-wheel CO_2 emissions than EVs if those EVs derive their electrical energy from CO_2-intensive sources. This study examines the relationship between the CO_2 emissions of ICE-based vehicles, PHEVs, and EVs given the varying CO_2 intensity of different regions' power generation mixes [63].

8.2.1 Methods

In order to make equitable comparisons across the three different types of vehicle topologies—conventional vehicle, EV, and PHEV—a single vehicle platform was needed. Since there is currently no single vehicle platform that uses all three of these topologies, this study compared the CO_2 emissions from an actual ICE-based vehicle to those from similar EVs and PHEVs via the use of OVEM [64]. The first step was to select a widely used ICE-based vehicle to serve as the template for all three vehicle topologies. For this purpose, the Ford Focus was selected because it is a popular mid-size car in many markets around the world.[3] All of the relevant manufacturer supplied figures from the ICE-based Focus were recorded, chief among them, the vehicle's tank-to-wheel energy use and CO_2 emissions.

Then the ICE-based version of the Focus was converted to a simulated EV and PHEV. This was done by replacing any components unique to the ICE-based vehicles with those components necessary to make the vehicle either an EV or PHEV. For instance, the ICE in each of the conventional vehicles was exchanged for an electric motor, and the multi-speed transmission was exchanged for a fixed speed gearbox. In the case of the EV, Li-ion batteries replaced the fuel tank. In the case of the series PHEV, a Li-ion battery pack was added along with a downsized ICE. The modelled EVs and PHEVs reflected the mass and efficiency characteristics of their new components, but retained all of the other relevant properties of the conventional vehicle upon which they were based. The EV's battery pack was sized such that it would be able to achieve a range of 300 km over the NEDC. The battery packs of the series PHEVs came in four different sizes such that the PHEVs would be able to achieve an all-electric range of 20, 40, 60 and 80 km. The PHEVs operated by having their ICEs switch on to recharge their batteries when the state of charge of the batteries fell to 20% and the ICE switched off again once the state of charge reached 80%. *It was assumed that the EVs and PHEVs were able to utilise regenerative braking to recover energy from the wheels when decelerating. Such energy was still subjected to the powertrain loss mechanisms. For braking powers within the motor capacity, regenerative braking was used. Standard friction brakes were employed for braking powers in excess of the motor capacity.* Table 4 shows the pertinent specifications for the ICE-based vehicles as well as the modelled EVs and PHEVs used in this study. The engine of the conventional vehicle operates the entire time the vehicle is driving, and the PHEV has two distinct modes: its electric-only mode where it is powered exclusively by

---

[3] In this study, the version of the Focus that was used was the 1.6 TDCi ECOnetic because it was the lowest CO_2 emitting Focus at the time this study was conducted.

Table 4 Specifications for the actual ICE and simulated EV and PHEVs used in this study

Powertrain type	Battery pack mass (kg)	APU mass (kg)	Curb mass (kg)	Energy consumption in electric-only mode (MJ/100 km)	Emissions when ICE is operating (g CO_2/km)
Conventional vehicle	–	–	1,340	–	11.4
EV	340	–	1,380	455	–
PHEV-20	30	104	1,168	424	10.8
PHEV-40	50	107	1,191	427	10.9
PHEV-60	80	110	1,224	431	10.9
PHEV-80	110	113	1,257	437	10.9

its batteries and the other where its ICE is operating to power the car and recharge its batteries. It is interesting to note that the simulations showed all of the PHEVs use less energy than the EV in electric-only mode. This is primarily due to the fact that the PHEVs have lower mass, since the mass of their alternative power unit is offset by the fact that it reduces the vehicles' need for a heavy battery pack. It is also worth noting that the CO_2 emissions from the PHEVs during the phase when their ICEs are operating are all largely the same. This can be explained by the fact that the ICE in the PHEVs is decoupled from the road load, and hence is not as sensitive to changes in vehicle mass.

The average CO_2 intensity of a nation's power generation mix is a measure of the amount of CO_2 produced per unit of generated electricity, weighted by the amount of power obtained from each process—expressed in this study as g CO_2/ MJ. This study used the current average CO_2 intensity of the national power generation mix for different key countries around the world—France, USA and China—to see how the CO_2 emissions from the EV and PHEVs varied depending on where they were recharging. France was selected because the CO_2 intensity of its grid is low due to its widespread use of nuclear power. The USA was selected because the average CO_2 intensity of its national grid is about in the mid-point of the range seen internationally and it is the second largest market for new vehicles in the world. China was selected because the CO_2 intensity of its grid is high due to its heavy reliance on coal and it represents the world's largest new vehicle market. It has even been projected that by 2050 China alone will have nearly as many cars on its roads than the approximately 600 million that existed worldwide in 2005 [65]. Table 5 contains the data used for the CO_2 intensity of these nations power generation mixes.[4]

The tank-to-wheel CO_2 emissions for the conventional vehicle, EV and PHEVs were all derived in different ways. The operating emissions from conventional vehicles can be measured by what comes out of their tailpipe. In Europe, automobile manufacturers publish the grams of CO_2 per kilometre (g CO_2/km) emitted by conventional vehicles. Therefore, in this study, that manufacturer reported

---

[4] Data from carbon monitoring for action (CARMA), available online at www.carma.org.

Table 5 CO_2 intensity of the power generation mix of the countries examined in this study

Country	CO_2 intensity of power generation mix (g CO_2/MJ)
France	24
USA	169
China	241

value is used as the constant CO_2 emissions from the conventional vehicle over its entire range. The emissions from EVs depend on their own energy use and on the CO_2 intensity of the power generation mix from which the EVs energy was obtained. The CO_2 intensity varies considerably depending on the composition of the power generation mix. By combining the energy use of the EV with the CO_2 intensity of its power generation mix, its emissions in g CO_2/km, EEV can be determined. The CO_2 emissions from the PHEVs depend on the CO_2 intensity of the power generation mix, the efficiency of the vehicle in its electric-only and its ICE-operating modes, and the distance travelled in both modes. *Note that the energy consumption and hence CO_2 emissions figures for the PHEVs were obtained by simulating the vehicle over multiple discharge–charge cycles as they were run over repeated iterations of the drive cycle. This enabled the study to obtain a reliable average value for their CO_2 emissions that was not unfairly biased by a particular state of charge profile over a specific drive cycle (or segment thereof).*

8.2.2 Results

Figures 14, 15, 16 present the tank-to-wheel CO_2 emissions from all three topologies as though they were operating in France, the USA and China. Note that the conventional vehicle produces the same tank-to-wheel emissions regardless of where it is operating. The tank-to-wheel CO_2 emissions from the EV depend directly on the CO_2 intensity of the power generation mix from which it recharges. The CO_2 emissions from the PHEVs in electric-only mode also depend directly on the CO_2 intensity of the power generation mix. However, once the PHEVs enter their ICE-operating mode, their cumulative CO_2 emissions averaged over their distance travelled pick up at the level they were at during electric-only mode before asymptotically approaching their ICE's CO_2 emissions value.

Figure 14 shows how the average emissions profiles compare for the conventional vehicle, EV and PHEVs when the EV and PHEVs charge in France. While the PHEVs were more efficient in their electric-only mode than the EVs, which led to them having lower emissions than the EVs in their all-electric range, the difference was almost negligible. Figure 14 shows that given France's low CO_2 intensity, there was less than a 1 g CO_2/km difference between the CO_2 emissions from the most efficient PHEV and EV. What was only a small difference in emissions in electric-only mode becomes much larger once the PHEVs ICE turns on. As the average CO_2 emissions from the PHEVs asymptotically approach their

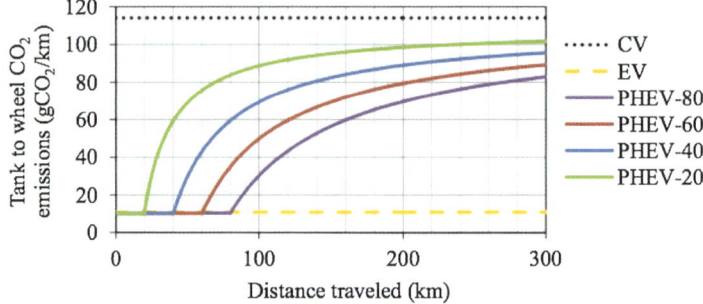

Fig. 14 Tank-to-wheel CO_2 emissions of various vehicle powertrain topologies in France

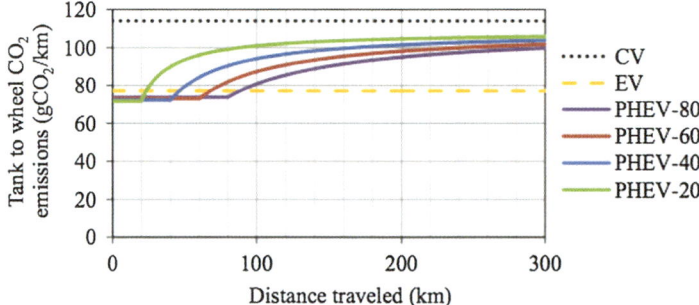

Fig. 15 Tank to wheel CO_2 emissions of various vehicle powertrain topologies in the USA

CO_2 emissions from their ICEs, they move farther away from the low, constant CO_2 emissions of the EV. Since the PHEVs still produce fewer g CO_2/km than the conventional vehicle when their ICE is operating, and since in France their CO_2 emissions in electric-only mode begin at a point below those from the conventional vehicle, their average emissions performance is lower at all points in their driving range than the conventional vehicle. From these results, the generalisation can be made that if the CO_2 intensity of the power generation mix is low, then EVs provide a more consistent option than PHEVs or conventional vehicles for lowering CO_2 emissions in automobile transport.

Figure 15 shows the average emissions for the three topologies over the course of their driving range as if they were charging in the USA. Given the USA's CO_2 intensity, the PHEVs emit slightly fewer g CO_2/km than the EV during their electric-only mode. This was expected because of the lower curb mass and higher efficiency of the PHEVs. The PHEVs even continue to emit less than the EV into their ICE-operating mode. For instance, the PHEV-80 continues to have lower emissions than the EV for several kilometres into its ICE-operating mode. There is clearly a range of distances over which the PHEVs emit less than the EV. Thus, in a country with a mid-range CO_2 intensity, PHEVs offer the possibility of reducing

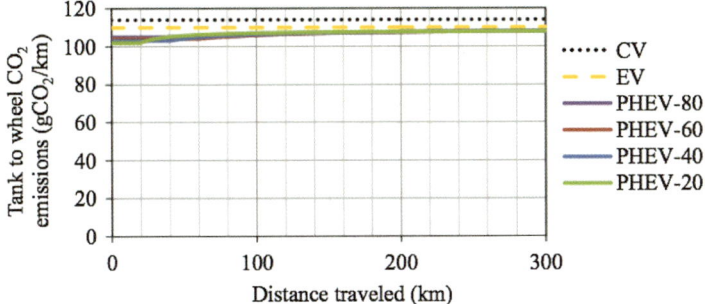

Fig. 16 Tank to wheel CO_2 emissions of various vehicle powertrain topologies in China

CO_2 emissions relative to EVs when the PHEV begins with a full charge and is not driven too extensively in its ICE-operating mode. Yet, once the PHEVs travel further in their ICE-operating mode, they begin to emit significantly more g CO_2/km than the EV. Both the EV and PHEVs emit less CO_2 at all points in their range than the conventional vehicle. Thus, with power generation mixes of mid-range CO_2 intensity, PHEVs may provide lower CO_2 emissions if the driving distances between charges are short relative to the maximum range of the vehicle (as they typically are for many drivers). However, the emissions reductions from PHEVs in electric-only mode compared to those from EVs are not too substantial and should not be overstated. The EV produced low CO_2 emissions over its entire range and it achieved over 90% lower CO_2 emissions than the conventional vehicle. Ultimately, EVs appear to be a better option than PHEVs for lowering CO_2 emissions in automotive transport given a power generation mix similar to the average one found in the USA.

When the CO_2 intensity of the power generation mix is high, a different picture begins to emerge as shown in Fig. 16. Given China's power generation mix, the simulated PHEVs achieve lower tank-to-wheel CO_2 emissions at all points over their range than the EV and conventional vehicle. This fact has been largely ignored in the literature and it demonstrates that PHEVs provide a stronger alternative than previously thought to lower the CO_2 emissions from transport in countries with a highly CO_2 intensive power generation mix. In these countries, PHEVs have the potential to be more than a mere stopgap in the transition from conventional vehicles to EVs—compensating for the current range deficiencies of EVs until battery technology improves. Depending on how the CO_2 intensity of the power generation mix evolves, PHEVs may be able to play a more prominent role in a long-term solution to lower CO_2 emissions in automobile transport. However, the reality of China's power generation mix also shows that even EVs and PHEVs will not be able to significantly lower the CO_2 emissions from automobile transport in the short to medium term. The EV and PHEV-20 in electric-only mode achieved less than a 5 and 11% CO_2 emissions reduction, respectively, compared to the conventional vehicle. This demonstrates that countries with highly CO_2

intensive power generation mixes clearly have an extra incentive to reduce their CO_2 intensity in order to maximise the potential of EVs and PHEVs to reduce CO_2 emissions in automotive transport.

The specific results on display in these studies are a product of the assumptions and modelling techniques outlined earlier. As such, the above predictions should not necessarily be expected to hold true for all possible EVs and PHEVs. Nevertheless, these results illustrate the complex nature of the questions posited at the beginning of this section. Clearly, there is no simple answer as to what vehicle topologies and technologies are optimal at all times and for all places. A thorough understanding of the characteristics and limits of various automotive solutions will be crucial if we are to make informed decisions about the future of road transport.

9 Conclusions

Road transport accounted for 40% of the total final consumption of energy across all transport modes globally in 2008. Energy use is expected to double by 2050 to satisfy growth in road transport demands. Much of the growth in demand for transport will be due to more extensive vehicle use in developed countries and the transition to motorised transport in developing countries. The consequence will be approximately 5 Gt GHG emitted in 2050 due to automobile transport alone. In order to ameliorate many of the environmental and economic issues related to automobile transport, vehicles' energy use must be reduced and the sources from which they can derive their energy should be diversified. The primary way that vehicle design can tackle these issues is through EV and HEV powertrains. EVs and HEVs are typically more efficient than conventional vehicles based on an ICE and they have the potential to make use of energy derived from an array of sources addressing the energy security issue.

The sizing of vehicle powertrain components is strongly dependent on the vehicle mass. Reducing vehicle component primary mass allows additional, secondary mass savings. Consequently, a 10% reduction in primary vehicle mass can yield at least 7% reduction in energy use. Reducing other non-powertrain resistive forces, such as aerodynamic drag and tyre rolling resistance, can lower energy use by up to 5% each.

Most conventional vehicles use the mature and application-flexible ICE as the main energy conversion technology. However, many modern vehicles have ICEs with peak power outputs far in excess of what is required under normal driving conditions. The consequence is that the ICE generally operates under part-load which is when it is less efficient. This is compounded by many idling periods when driving in urban or congested environments. Most notably, all of the energy used to achieve a particular vehicle velocity is converted to heat by the brakes when the conventional vehicle eventually slows to a halt.

Electric motors are also a mature technology, with widespread use in industrial, domestic and power generation applications. Motors have significantly higher peak efficiencies than ICEs and power profiles which favour part-load operation of the kind normally seen during realistic driving conditions. Moreover, an electric motor can be operated as a generator to recapture the vehicle's kinetic energy during braking. Thus, electric motors via electrified powertrains are a more efficient alternative for delivering road transport.

Many EVs, HEVs and their variants use batteries to meet the energy and power needs under normal driving conditions. In particular, batteries are used to satisfy the high power demands associated with sharp accelerations and the lower energy demanded when cruising. Batteries are better suited to storing energy than ultra-capacitors and high-speed flywheels on account of their high specific energy. However, batteries are generally ill-suited to the high current bi-directional flows associated with large power demands. The result is lower roundtrip efficiency, reduced cycle capacity and shorter battery lifetime. Using high specific energy lithium-based batteries in this way incurs a large cost penalty. Ultra-capacitors and flywheels are alternative technologies that may be able to address some of the shortcomings of batteries. Both technologies are well-suited to frequent charge–discharge cycling, have high specific power, and can be used to provide a blended energy storage system which is more efficient and less expensive.

This chapter included two case studies that examined the complex nature of just some of the economic and environmental issues surrounding the future of automobile transport. The first study compared ultra-capacitors and high-speed flywheels to batteries as the energy storage medium in a series HEV. The results showed that though batteries allowed the vehicle to achieve the highest fuel economy, ultra-capacitors and high-speed flywheels have the potential to be nearly as efficient at less cost. The second study compared the tank-to-wheel CO_2 emissions stemming from conventional vehicle, EV and HEV use. The results showed that PHEVs actually have the potential, in regions where electricity generation is CO_2 intensive, to produce fewer tank-to-wheel CO_2 emissions than similar conventional vehicles and EVs. Therefore, although EVs and their variants have more efficient powertrains than conventional vehicles, their tank-to-wheel emissions, and ultimate ability to mitigate climate change, is a function of the sources and processes from which they derive their electrical energy.

References

1. IEA (2010) Key World Energy Statistics 2010. International Energy Agency, Paris. http://www.iea.org/textbase/nppdf/free/2010/key_stats_2010.pdf
2. Fulton L, Eads G (2004) IEA/SMP Model documentation and reference case projection. International Energy Agency/World Business Council for Sustainable Development, Paris. http://www.wbcsd.org/web/publications/mobility/smp-model-document.pdf
3. IEA (2009) Key World Energy Statistics 2009. International Energy Agency, Paris
4. Fulton L (2004) The IEA/SMP Transportation model. International Energy Agency/World Business Council for Sustainable Development, Paris. http://www.wbcsd.org/web/publications/mobility/smp-model-spreadsheet.xls

5. André M (2004) Real-world driving cycles for measuring cars pollutant emissions—Part A: the ARTEMIS European driving cycles. INRETS-LTE 0411. Institut National de Recherche sur les Transports et leur Securite (INRETS), Bron
6. Joumard R, André M, Vidon R, Tassel P, Pruvost C (2000) Influence of driving cycles on unit emissions from passenger cars. Atmos Environ 34(27):4621
7. SMMT (2010) Motor Industry Facts 2010. London. http://lib.smmt.co.uk/articles/sharedfolder/Publications/SMMT_WEB_2010_100dpi.pdf
8. Fenske G, Erck R, Ajayi L, Erdemir A, Eryilmaz O (2006) Parasitic energy loss mechanisms impact on vehicle system efficiency. Argonne National Laboratory, 2006, Illinois. http://www1.eere.energy.gov/vehiclesandfuels/pdfs/hvso_2006/07_fenske.pdf
9. Hucho WH (ed) (1987) Aerodynamics of road vehicles: from fluid mechanics to vehicle engineering. Cambridge University Press, Cambridge
10. Hucho WH, Sovran G (1993) Aerodynamics of road vehicles. Annu Rev Fluid Mech 25:485. doi: 10.1146/annurev.fl.25.010193.002413
11. Koike M, Nagayoshi T, Hamamoto N (2004) Research on aerodynamic drag reduction by Vortex Generators. Mitsubishi Motors Tech Rev 16:11. http://www.mitsubishi-motors.com/corporate/aboutus/technology/review/e/pdf/2004/16E_03.pdf
12. Nunney MJ (2007) Light and heavy vehicle technology, 4th edn. Elsevier, Amsterdam
13. Guzzella L, Sciarretta A (2007) Vehicle propulsion systems: introduction to modeling and optimization, 2nd edn. Springer, Berlin
14. IEA (2008) Energy Technology Perspectives 2008—Scenarios and Strategies to 2050. International Energy Agency, Paris
15. IEA (2009) Transport, energy and CO_2. International Energy Agency, Paris. http://www.iea.org/textbase/nppdf/free/2009/transport2009.pdf
16. Kobayashi S, Plotkin S, Ribeiro SK (2009) Energy efficiency technologies for road vehicles. Energy Effic 2(2):125. doi: 10.1007/s12053-008-9037-3
17. Beaudoin JF, Cadot O, Aider JL, Gosse K, Paranthoën P, Hamelin B, Tissier M, Allano D, Mutabazi I, Gonzales M et al (2004) Cavitation as a complementary tool for automotive aerodynamics. Exp Fluids 37(5):763. doi: 10.1007/s00348-004-0879-y
18. TRB (2006) Tires and passenger vehicle fuel economy: informing consumers, improving performance—special report 286. The National Academy of Sciences, USA. http://onlinepubs.trb.org/onlinepubs/sr/sr286.pdf
19. Goldberg LH (2000) Green electronics, green bottom line: environmentally responsible engineering. Butterworth-Heinemann, Oxford, Chap. 24
20. Bjelkengren C (2008) The impact of mass decompounding on assessing the value of vehicle lightweighting. PhD thesis, Massachusetts Institute of Technology, Massachusetts. http://msl.mit.edu/students/msl_theses/Bjelkengren_C-thesis.pdf
21. Cheah LW (2010) Cars on a diet: the material and energy impacts of passenger vehicle weight reduction in the U.S. PhD thesis, Massachusetts Institute of Technology, Massachusetts. doi: http://web.mit.edu/sloan-auto-lab/research/beforeh2/files/LCheah_PhD_thesis_2010.pdf
22. Malen DE, Reddy K (2007) Preliminary vehicle mass estimation using empirical subsystem influence coefficients. Auto/Steel Partnership. Ann Arbor. http://www.asp.org/database/custom/Mass%20Compounding%20%20Final%20Report.pdf
23. Rakopoulos CD, Glakoumis EG (2006) Second-law analyses applied to internal combustion engines operation. Prog Energy Combust Sci 32(1):2. doi: 10.1016/j.pecs.2005.10.001
24. Brusstar M, Stuhldreher M, Swain D, Pidgeon W (2002) High efficiency and low emissions from a port-injected engine with neat alcohol fuels. 2002-01-2743. Society of Automotive Engineers, Byron Bay. http://www.stonis-world.net/docs/engine_with_neat_alcohol_fuels.pdf
25. Husain I (2003) Electric and hybrid vehicles. CRC Press, Florida
26. Boretti A (2010) Comparison of fuel economies of high efficiency diesel and hydrogen engines powering a compact car with a flywheel based kinetic energy recovery systems. Int J Hydrogen Energy 35(16):8417. doi: 10.1016/j.ijhydene.2010.05.031

27. Williamson SS, Emadi A, Rajashekara K (2007) Comprehensive efficiency modeling of electric traction motor drives for hybrid electric vehicle propulsion applications. IEEE Trans Veh Technol 56(4):1651. doi:10.1109/TVT.2007.896967
28. Hughes E (1995) Electrical technology, 7th edn. Longman, London
29. Kenjo T, Nagamori S (2003) Brushless motors: advanced theory and modern applications. Sogo Electronics Press, Tokyo
30. Woolmer TJ, McCulloch MD (2007) Analysis of the yokeless and segmented armature machine. In: Electric machines and drives conference (IEMDC). Antalya, Turkey, pp 704–708. doi: 10.1109/IEMDC.2007.382753
31. Zeraoulia M, Benbouzid MEH, Diallo D (2006) Electric motor drive selection issues for HEV propulsion systems: a comparative study. IEEE Trans Veh Technol 55(6):1756. doi:10.1109/TVT.2006.878719
32. Chau KT, Chan CC, Liu C (2008) Overview of permanent-magnet brushless drives for electric and hybrid electric vehicles. IEEE Trans Ind Electron 55(6):2246. doi:10.1109/TIE.2008.918403
33. Westbrook MH (2001) The electric car: development and future of battery, hybrid, and fuel-cell cars. The Institution of Electrical Engineers, London
34. West JGW (1994) DC, induction, reluctance and PM motors for electric vehicles. Power Eng J 8(2):77. doi:10.1049/pej:19930203
35. Uematsu T, Wallace RS (1995) Design of a 100 kW switched reluctance motor for electric vehicle propulsion. In: Tenth annual applied power electronics conference and exposition (APEC), vol 1. Dallas, USA, pp 411–415. doi: 10.1109/APEC.1995.468981
36. Miller JM (2004) Propulsion systems for hybrid vehicles. The Institution of Electrical Engineers, London
37. Lukic SM, Cao J, Bansal RC, Rodriguez F, Emadi A (2008) Energy storage systems for automotive applications. IEEE Trans Ind Electron 55(6):2258. doi:10.1109/TIE.2008.918390
38. Burke A, Zhao H (2010) Simulations of plug-in hybrid vehicles using advances lithium batteries and ultracapacitors on various driving cycles. UCD-ITS- RR-10-02. University of California-Davis Institute of Transportation Studies, Davis
39. Rydh CJ, Sandén BA (2005) Energy analysis of batteries in photovoltaic systems. Part I: performance and energy requirements. Energy Convers Manag 46(11–12):1957. doi: 10.1016/j.enconman.2004.10.003
40. Divya KC, Østergaard J (2009) Battery energy storage technology for power systems—an overview. Electr Power Syst Res 79(4):511. doi:10.1016/j.epsr.2008.09.017
41. Palacín MR (2009) Recent advances in rechargeable battery materials: a chemist's perspective. Chem Soc Rev 38:2565. doi: 10.1039/b820555h
42. Berndt D (2003) Battery technology handbook, 2nd edn. Expert Verlag, Renningen-Malsheim, Chap. 1, pp 1, 5, 6
43. Cross D, Hilton J (2008) High speed flywheel based hybrid systems for low carbon vehicles. In: IET hybrid and eco-friendly vehicle conference (HEVC). Coventry, UK
44. Schoenung SM, Hassenzahl WV (2003) Long- vs. short-term energy storage technologies analysis a life-cycle cost study a study for the doe energy storage systems program. SAND2003-2783. Sandia National Laboratories, Washington. http://prod.sandia.gov/techlib/access-control.cgi/2003/032783.pdf
45. Lipman TE, Delucchi MA (2003) Hybrid-electric vehicle design retail and life-cycle cost analysis. UCD-ITS-RR-03-01. Energy and Resources Group, University of California, Berkeley
46. Dixon J, Nakashima I, Arcos EF, Ortuzar M (2010) Electric vehicle using a combination of ultracapacitors and ZEBRA battery. IEEE Trans Ind Electr 57(3):943. doi:10.1109/TIE.2009.2027920
47. Frenzel B, Kurzweil P, Roönnebeck H (2011) Electromobility concept for racing cars based on lithium-ion batteries and supercapacitors. J Power Sour 196(12):5364. doi: 10.1016/j.jpowsour.2010.10.057

48. Burke A, Miller M (2011) The power capability of ultracapacitors and lithium batteries for electric and hybrid vehicle applications. J Power Sour 196(1):514. doi:10.1016/j.jpowsour.2010.06.092
49. Auer J, Sartorelli G, Miller J (2006) In: IET hybrid vehicle conference. Coventry, UK, pp 79–90
50. Van den Bossche P, Vergels F, Van Mierlo J, Matheys J, Van Autenboer W (2006) SUBAT: an assessment of sustainable battery technology. J Power Sour 162(2):913. doi:10.1016/j.jpowsour.2005.07.039
51. Chan CC (2007) The state of the art of electric, hybrid, and fuel cell vehicles. Proc IEEE 95(4):704. doi:10.1109/JPROC.2007.892489
52. Karden E, Ploumen S, Fricke B, Miller T, Snyder K (2007) Energy storage devices for future hybrid electric vehicles. J Power Sour 168(1):2. doi:10.1016/j.jpowsour.2006.10.090
53. Khaligh A, Li Z (2010) Battery, ultracapacitor, fuel cell, and hybrid energy storage systems for electric, hybrid electric, fuel cell, and plug-in hybrid electric vehicles: state of the art. IEEE Trans Veh Technol 59(6):2806. doi:10.1109/TVT.2010.2047877
54. Toyota Prius Technical Specifications. http://www.toyotagb-press.co.uk/protected/vehicles/current/press_packs/prius/tech_spec.pdf. Accessed 21 July 2011
55. Gaines L, Cuenca R (2000) Costs of lithium–ion batteries for vehicles. ESD-42. Argonne National Laboratory, Illinois
56. Gao Y, Gay SE, Ehsani M, Thelen RF, Hebner RE (2003) Flywheel electric motor/generator characterization for hybrid vehicles. In: IEEE 58th vehicular technology conference-fall, vol. 5. Orlando, USA, pp 3321–3325. doi: 10.1109/VETECF.2003.1286291
57. Diego-Ayala U, Martinez-Gonzalez P, McGlashan N, Pullen KR (2008) The mechanical hybrid vehicle: an investigation of a flywheel-based vehicular regenerative energy capture system. Proc Inst Mech Eng, Part D: J Automob Eng 222(11):2087. doi:10.1243/09544070JAUTO677
58. Bolund B, Bernhoff H, Leijon M (2007) Flywheel energy and power storage systems. Renew Sustain Energy Rev 11(2):235. doi:10.1016/j.rser.2005.01.004
59. Van Mierlo J, Maggetto G (2004) Innovative iteration algorithm for a vehicle simulation program. IEEE Trans Veh Technol 53(2):401. doi:10.1109/TVT.2004.823534
60. Solero L, Lidozzi A, Serrao V, Martellucci L, Rossi E (2011) Ultracapacitors for fuel saving in small size hybrid vehicles. J Power Sour 196(1):587. doi:10.1016/j.jpowsour.2009.07.041
61. Doucette RT (2010) The oxford vehicle model manual: a tool for modeling and simulating the powertrains of electric and hybrid electric vehicles. Energy and Power Group, Department of Engineering Science, University of Oxford, Oxford. http://epg.eng.ox.ac.uk/sites/default/files/OVEM%20Manual%20pdf%20July%2019%202010.pdf
62. Doucette RT, McCulloch MD (2011) A comparison of high-speed flywheels, batteries, and ultracapacitors on the bases of cost and fuel economy as the energy storage system in a fuel cell based hybrid electric vehicle. J Power Sour 196(3):1163. doi:10.1016/j.jpowsour.2010.08.100
63. Doucette RT, McCulloch MD (2011) Modeling the CO_2 emissions from battery electric vehicles given the power generation mixes of different countries. Energy Policy 39(2):803. doi:10.1016/j.enpol.2010.10.054
64. Doucette RT, McCulloch MD (2011) Modeling the prospects of plug-in hybrid electric vehicles to reduce CO_2 emissions. Appl Energy 88(7):2315. doi:10.1016/j.apenergy.2011.01.045
65. Chamon M, Mauro P, Okawa Y (2008) Mass car ownership in the emerging market giants. Econ Policy 23(54):243. doi:10.1111/j.1468-0327.2008.00201.x

Rechargeable Batteries for Transport and Grid Applications: Current Status and Challenges

Clare P. Grey

Abstract Lithium-ion batteries are poised to make a significant impact on the electrification of transport and may also play a role for some power regulation/ storage applications on the electric grid. In this article, we describe some of the applications where these batteries are either already being or are about to be used. The components that make up a lithium-ion battery are outlined along with the causes of capacity fade and safety issues. Current battery chemistries are then surveyed along with the factors that control possible scenarios to increase energy densities on both a volumetric and mass basis.

1 Introduction

The development of light, long-lasting rechargeable batteries has been an integral part of the portable electronics revolution. This revolution has transformed the way in which we communicate and transfer and access data globally, and has impacted developing nations as much as industrialized societies. The invention of the lithium-ion (Li-ion) battery, a rechargeable battery in which lithium ions (Li^+) shuttle between two materials ($LiCoO_2$ and graphitic carbon) has been an integral

C. P. Grey (✉)
Departments of Chemistry,
Cambridge University, Cambridge, CB2 1EW, UK
e-mail: cpg27@cam.ac.uk

C. P. Grey
Departments of Chemistry, Stony Brook University,
Stony Brook, NY 11794-3400, USA

O. Inderwildi and Sir David King (eds.), *Energy, Transport, & the Environment*,
DOI: 10.1007/978-1-4471-2717-8_13, © Springer-Verlag London 2012

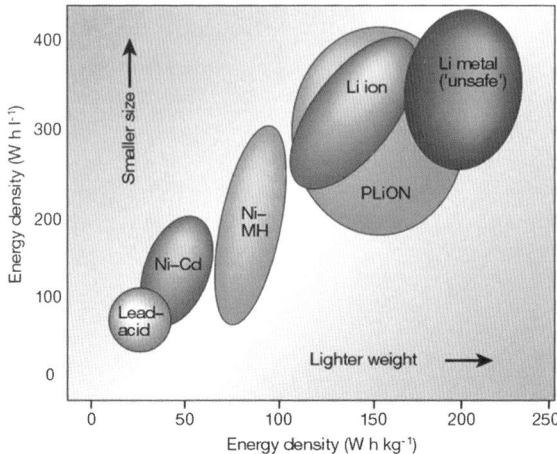

Fig. 1 A comparison of the energy densities of different types of batteries on both a volumetric (Whl^{-1}) and mass basis ($Whkg^{-1}$). In many applications where space is at a premium (e.g., in a passenger car) the volumetric energy density is as, if not more important, than the mass-based energy density. The figure shows the status in 2001 [1]. Higher energy densities are now possible for Li-ion batteries, as discussed in the text. Reprinted by permission from Macmillan Publishers Ltd: Nature, Ref. [1], copyright (2001)

part of the many advances. The first generation of such batteries allowed two to three times more the energy to be stored, as compared to nickel- or lead-based batteries of the same size and mass (Fig. 1). Furthermore, they do not suffer from problems such as "memory effects" (a problem for nickel–cadmium [Ni–Cd] batteries) or hydrogen evolution (lead-acid, PbA).

Rechargeable batteries are now poised to play an increasingly important role in transport and grid applications, but the introduction of these devices comes with different sets of challenges. This article aims to briefly outline arenas where these batteries will have an impact, some of the available battery technologies, and some of the areas where improvements must be achieved to increase market penetration and consumer acceptance. Since Li-ion batteries offer the highest energy density among the currently available technologies, with ever improving power capabilities [1], the major focus of this article is, therefore, on Li-ion. Mature as the technology may appear, both fundamental research and development are still needed to enable further technological breakthroughs and market penetration of lithium-ion batteries [2, 3]. Competitors to Li-ion technology include sodium and redox flow batteries for grid storage, and PbA and nickel metal hydride (NiMH) batteries and supercapacitors for transport applications; where appropriate, applications where these technologies are preferred are outlined. Li-air, lithium-sulfur and Li-(metal) batteries have also received considerable attention and these technologies, and these batteries, which although much further from commercial realization, are briefly discussed in Sect. 7.

2 Applications

Current concerns about limited fossil fuel resources, coupled with the need to decrease greenhouse gas emissions significantly, has resulted in a push to use renewable energies at a much larger scale. Energy security issues, particularly in countries with limited natural resources, are also playing an important role in motivating this change. The widespread development and use of hybrid and electric vehicles forms part of this strategy, since these vehicles both increase the efficiency of fuel use and allow for fuel diversification (since many types of fuels can be used to produce the electricity). In the automotive sector, the different technologies are divided into hybrid electric vehicles (HEVs), plug-in hybrids (PHEVs) and all electric vehicles (EVs) (which are often also called battery electric vehicles or BEVs). These different technologies come with different demands on the size and type of battery pack. HEVs, having started as smaller to mid-sized cars such as the Toyota Prius, are rapidly increasing their market share, particularly in higher end vehicles. They can increase fuel efficiency by allowing a car to operate in a "stop–start" mode (i.e., switching on and off the engine during idling) and via regenerative breaking, (i.e., the use of the kinetic energy generated from braking to charge the battery). The largest efficiency gain actually comes from the ability to use the battery as a mechanism to ensure that the on-board internal combustion engine (ICE) operates in its most efficient range (i.e., with the most efficient load and engine speed). These systems require only small battery packs, and depending on the type of hybrid [e.g., stop–start only, through to a full hybrid such as the Toyota Prius (with 1.5–2.5 kWh batteries)] make use of more established automotive battery technologies such as nickel metal hydride (NiMH) or the cheaper PbA batteries. In particular, hybrid (PbA–supercapacitor carbon) batteries, such as those made by Axion power, offer high rate capabilities coupled with longer lifetimes than conventional PbA technologies, and are being considered for stop–start applications. Note that the term "hybrid" is used here because this technology combines both battery and capacitor technologies. There are two types of PHEVs, both requiring much larger batteries. A range-extended HEV such as the new Prius, is similar to the older Prius, but it contains a larger battery, that can be charged on the grid, and is capable of a larger electric range (currently up to 12.5 miles) [4]. In a range-extended EV such as the Chevy Volt, (and in contrast to the range-extended HEV), only the electric motor is connected to the wheels. A much smaller ICE is present to allow for more extended range. The 5.5 ft T-shaped 16 kWh battery of the 2011 Volt weighs 198 kg and has a range of 35 miles [5].

Only electric engines are present in a BEV, the driving range of the vehicle being directly related to the size of the battery pack. Current battery packs range from 24 to 53 kWh in the Nissan Leap and the Tesla Roadstar, respectively, and result in ranges, under *ideal* driving conditions (and no use of other sources of electric drain such as heating and air conditioning) of 100 and 230 miles, respectively. The price and size of the battery pack are directly related to the kWh it can deliver, restricting the range that these vehicles will have, if price is the primary concern. Range can be increased noticeably by decreasing weight and production and sales of EV micro cars, scooters and bicycles have grown significantly, particularly in the developing nations where air pollution and traffic congestion is a concern.

Fuel savings vary depending on technology and how the vehicles are driven, but estimates of 25–45% gains in fuel economy have been made by using Argonne National Laboratory's "Greenhouse gases, Regulated Emissions, and Energy use in Transportation" (GREET) model for full HEV technologies used with gasoline and diesel ICEs [6]. Higher numbers are obtained by PHEVs and BEVs but now total CO_2 emissions now depend dramatically on the CO_2 emissions associated with electricity generation and the mix of fuel used to generate the electricity [7]. Use of either nuclear energy and/or renewable sources such as solar and wind result in the lowest CO_2 emissions. Despite concerns about the range of all-electric vehicles, safety is still perhaps the most important factor, with cost, energy density and power also being important concerns. The requirements and systems (and cost) goals, along with the different testing protocols, for the different batteries required for the different applications are summarized in the US Council for Automotive Research (USCAR) web site [8] [USCAR is a consortium involving the US automotive manufacturers, in partnership with the US (Government) National Laboratories].

Batteries will play a role in grid storage, a need that is becoming increasingly important as intermittent and/or diffuse renewable sources (e.g., solar and wind) are coupled to the grid. The requirements for the battery remain lifetime cost, scalability, power/energy density and safety, but now lifetime cost and scalability are key factors. Lithium-ion batteries, with their relatively high costs and modest storage ability may play a role in applications where rapid response is needed, such as in grid (voltage and frequency) regulation and load leveling. In particular, batteries can respond fast enough to the constantly fluctuating loads that occur over periods of minutes to hours [9]. Modeling has suggested that they become financially viable investments, if the savings associated with the delay in the building of new power plants (needed to take into account increased electricity demand and the handling of intermittent electricity sources) is factored in. Demonstration studies are underway including for example an 8 MW/4 h battery plant for wind integration in Tehachapi, CA (Southern California Edison) that makes use of A123 Li-ion batteries. Sodium-based technologies such as the sodium–sulfur and sodium–nickel chloride cells are more scalable, but are not *currently* significantly cheaper. They may be more practical in larger applications and pilot studies are underway, worldwide. Other technologies being considered in demonstration projects include redox-flow batteries and the PbA–carbon hybrids discussed above [9].

3 What is a Battery?

A battery is made up of one or more electrochemical cells, which are often wired in series or parallel to generate the desired voltage and power. The open circuit voltage (OCV) of the cell is determined by the materials that are used as positive and negative electrodes. On charging a lithium-ion battery, lithium ions are removed from the positive electrode, the positive electrode material being oxidized (by removal of electrons via the external circuit) (Fig. 2). The lithium ions diffuse through a non-aqueous electrolyte where they are inserted into the negative electrode. The negative electrode is reduced via the electrons from the external circuit. On discharge, i.e., when the battery is doing work, the lithium ions (or more generally cations), move towards the positive electrode and the process is reversed. This movement of the cations to the positive electrode on discharge, and the anions to the negative electrode in primary (i.e., non-rechargeable) batteries led to the use of the terms "cathode" and "anode" for the positive and negative electrodes, respectively. Although the ions reverse directions on charge and discharge, these terms continue to be used interchangeably in the secondary (rechargeable) battery field, and both sets of terms are used throughout this article.

Lithium-ion batteries were first commercialized by Sony Corp. in 1990 [10], and contain $LiCoO_2$ and graphite as cathodes and anodes, respectively (Fig. 2). The use of the strongly electropositive element lithium results in a cell that operates at approximately 3.75 V, an almost 2 V increase in cell voltage over the PbA cell which uses Pb^{2+}/Pb^{4+} and Pb/Pb^{2+} redox couples (OCV = 2.1 V). It has a greater than 2 V improvement over the Ni–Cd and NiMH batteries which use Ni^{2+}/Ni^{3+} and either Cd^{2+}/Cd^0 (Ni–Cd) or $(H^+ + M)$/metal hydride (NiMH) redox couples (the OCV = 1.2–1.3 V for both cells, where M in the NiMH cell is an alloy of lanthanum and nickel) [2, 11]. Electrolytes comprising a salt such as $LiPF_6$ dissolved in organic carbonates (typically a mixture of dimethyl carbonate, ethylene carbonate and sometimes ethylmethylcarbonate) are used. The blend represents a compromise between lithium ion conductivity, viscosity and the ability of these electrolytes to decompose in the initial charge to form a stable passivating (and protective) coating on the surface of the electrodes (in particular the carbon) called the surface electrode interphase (SEI). The electrolytes need to be stable over a large voltage window and the formation of a stable SEI coating on the graphite is critical since it protects the lithiated graphite from further reaction with the electrolyte. Additives are generally added to improve the stability of the SEI.

As the lithium ions are removed or inserted into the active materials, the active materials typically swell or contract ($LiCoO_2$ expands by 2% on extracting 50% of the Li) [12] and thus the electrodes are constructed by mixing together the active electrode materials with a highly conducting form of carbon (e.g., a carbon black such as "super-P" carbon) and a flexible polymer binder such as polyvinyldifluoride (PVDF). The carbon serves to electrically wire the particles together, while the binder holds the whole composite together. This rubbery composite is then spread on either a copper (at the anode) or an aluminum (at the cathode) current

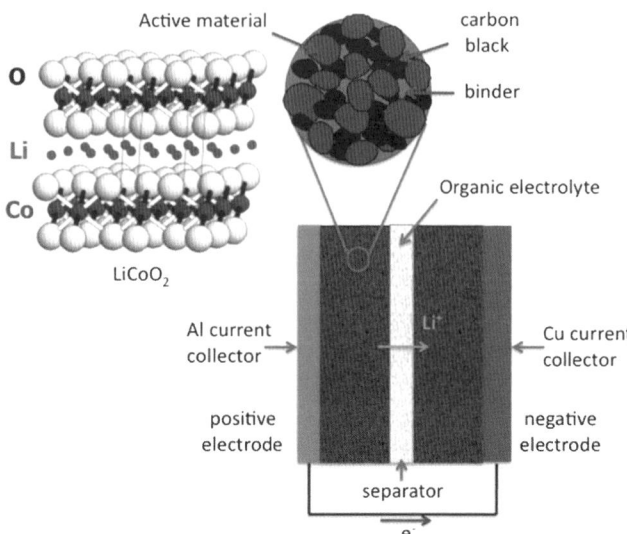

Fig. 2 A schematic of a lithium-ion battery. The directions of the flow of electrons and lithium ions correspond to those found on charging the battery. Both the positive and negative electrodes comprise porous composite structures made up of the active material, conductive carbon and a polymer binder coated onto the current collectors. In the Sony cell, the active materials on the positive and negative electrodes are $LiCoO_2$ and carbon, respectively. The layered structure of $LiCoO_2$ is shown on the *top left-hand side*

collector. (Since the lighter weight aluminum reacts with lithium, it cannot be used as the anode). The composite must have sufficient porosity to allow the electrolyte to penetrate through the electrode film so that the lithium ions in the electrolyte can reach all the particles of the composite. A porous separator (typically polymer based), which allows the electrolyte to pass through it, is used to physically separate the two electrodes and prevent short-circuiting. The layered structure is then wound to form cylindrical cells or packed in flat prismatic or "pouch-cell" batteries, (often called "coffee bag", or "plastic batteries", the terms referring to the Al-coated plastic bags used to seal the batteries). The packing density of prismatic cells is higher than cylindrical cells and they are widely being used in mobile phones and are being introduced into car batteries such as that of the Nissan Leaf. The cheaper-to-manufacture pouch cells are also increasing in their usage.

Pivotal to the success of the $LiCoO_2$ cell was the use of materials that are built up from layers of CoO_6 [in $LiCoO_2$ (Fig. 2)] and carbon (graphite) that are able to incorporate lithium reversibly in between their layers without destroying the host framework (i.e., the integrity of the layers). The second important development was the use of carbon as the anode instead of a lithium metal anode. Lithium metal anodes suffer from the formation of lithium metal dendrites on repeated cycling that can penetrate through the separator, leading to short-circuiting of the cells, rapid discharging and hence serious safety concerns. Lithium reacts with carbon at

Fig. 3 A schematic comparing the different potentials associated with lithium extraction from lithiated graphite (Li_xC_6), layered/spinel structured titanium sulfides (Li_xTiS_2/ $Li_x[Ti_2]S_4$ respectively), Li_xCoO_2 and olivine $LiCoPO_4$. The figure is adapted, with permission, from Ref. [14]. Copyright 2010 American Chemical Society

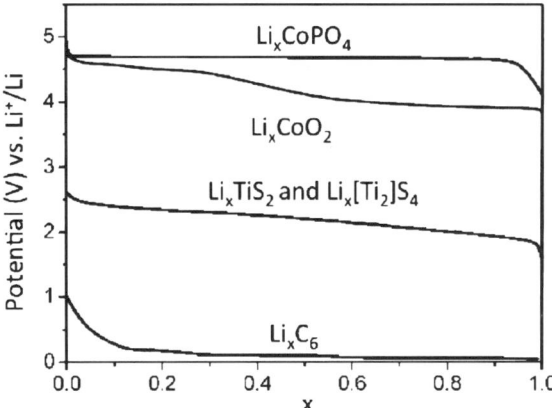

a potential between 40–250 mV higher than lithium metal [13] (Fig. 3) and so long as the battery is charged at a sufficiently slow rate, the lithium ions are safely incorporated within the carbon host material. The third important development was the movement away from the lithium titanium sulfide layered materials ($LiTiS_2$) used in the first lithium batteries, the use of an oxide over a sulfide, and cobalt versus titanium, leading to an increase in the potential of the cell by almost 2 V, as illustrated in Fig. 3.

The overall capacity of the cell is determined by the capacity of the two electrode materials, which is in turn dictated by (i) how much lithium can be reversibly removed from the materials at a rate that is compatible with the needs of the battery and (ii) the mass of the host material. For example, carbon can be cycled between C and LiC_6 resulting in a capacity of 372 mAhg^{-1}. Although in principle all the lithium ions can be removed from the heavier $LiCoO_2$ structure, an irreversible phase transition occurs on removal of the last few remaining lithium ions resulting in a shearing of the CoO_2 layers (i.e., they slip relative to each other) [15]. This structural transformation makes it difficult to reinsert lithium into the structure. In practice only 0.5–0.6 Li are removed from the cell to (i) avoid this phase transformation, (ii) avoid the formation of Li_xCoO_2 structures in which the lithium ions are ordered, and thus have lower Li$^+$ conductivities and (iii) maintain the voltage of the cell in the region that minimizes electrolyte decomposition on the cathode (generally considered to be about 4.3 V versus an Li metal anode) and oxygen loss from the cathode. This leads to a practical capacity of only 120–140 mAhg^1, as opposed to the theoretical capacity of this material of 273 mAhg^{-1}, i.e., the value that is obtained if all the Li is removed.

Cells are constructed by "balancing" the masses of the two electrode materials by taking into account their different capacities. The energy density is then calculated based on the capacity and the voltage at which the cell operates and is generally measured in Whkg^{-1} or Whl^{-1}. Clearly, the practical energy density of a cell will be much lower than the theoretical one (410 Whkg^{-1} for the Sony Cell)

[11], once all the packaging and masses of all the inactive components (e.g., separator, electrolyte, binder, conductive carbon and packaging) are taken into account. The practical energy densities were compared to those of PbA and NiCad and NiMH in Fig. 1. Improvements in cell design have, for example, resulted in energy densities of 188 $Whkg^{-1}$ and 520 Whl^{-1} for a $LiCoO_2$ 18650 MoliCell [2], the term "18650" representing a standard cylindrical battery type.

The power that can be extracted from a battery depends strongly on the rate at which the ions and electrons can be removed from electrode materials. Particularly when comparing different electrode chemistries or modifications the C-rate is often quoted, where 1C indicates that the battery has been fully charged or discharged in 1 h, while C/10 indicates that it has been charged/discharged in 10 h. Both rate and energy density are structure dependent as will be discussed in more detail below. Particularly in transport applications it is often the rate at which a battery can be charged that is more critical, because this will determine how much energy can be stored via regenerative breaking and how long a consumer will need to wait to recharge their battery. Furthermore, overcharging, which will occur in materials that cannot handle high rates, can result in materials that have been oxidized/ reduced beyond their safe/stable compositions and so care must be taken in terms of the electronic circuitry to prevent this happening. In a battery with multiple cells wired in series, this is particularly important. Chemists have added "redox shuttles" into the batteries as a mechanism for soaking up excess charge when the potential of the cell reaches a regime that is unsafe. These shuttles are molecules that are added to the electrolyte that are oxidized at the cathode at certain potentials (above the working, safe region) and then migrate to the anode where they are then reduced. However, it is not clear that this approach will be cost effective in large batteries.

4 Capacity Fade Mechanisms

Some of the mechanisms by which batteries fail are materials and chemistry specific while some are quite general. For example, the formation of an SEI coating removes lithium ions from the cell since this coating comprises a mixture of lithium salts (carbonates, fluorides, etc. and oligomeric/polymeric organics which can also be lithiated). With time, this coating can thicken, removing increasingly more Li from the cell. There is no excess lithium in the battery— every mechanism that removes lithium from the cell directly translates into a loss of capacity. Dissolution of metal ions from the cathode material can occur, leading to deposition of the same element as a metal on the anode. This can lead to structural changes in the cathode and surface metal sites that might aid in the catalytic decomposition of the electrolyte. The charged materials are rarely in their thermodynamic state and so with time, reactions to form either another more stable crystal structure, or structural defects can occur, both altering the ability of the material to reversibly incorporate lithium. The constant swelling and contraction of the active materials can lead to loss of electrical contact between particles forming so-called "dead lithium", i.e., lithium ions that cannot be accessed during the charge/discharge cycles. The considerable stresses on the particles can lead to cracking and "electrochemical grinding" again resulting in loss of electrical contact, and also to the formation of fresh surfaces which react to form more SEI (removing more Li from the cell). Great care is taken to assemble these batteries in dry rooms to minimize the moisture content of the cells, small concentrations of water leading to reaction with the PF_6^- anion in the electrolyte and further degradation mechanisms (to form, for example, HF and oxygenated fluorophosphates salts, and fluorinated organics). These failure mechanisms are greatly enhanced if the batteries overheat or are left at high temperatures, particularly when they are held in the charged state, where the electrode materials are the most reactive. As a result, the large batteries used in automotive applications have to be actively cooled. Good temperature control is probably one of the most important factors in ensuring long calendar life.

Other serious failure mechanisms include the plating of lithium metal on the anode on charge, particularly if the battery is overcharged (i.e., charged to too high potential) or is charged too fast. Even in batteries with overcharge protection, failure of parts of the electrode, can sometimes lead to distributions of potentials across the electrode due to different resistances, or to faster effective charge rates for different parts of the electrode. In the worst-case scenario, lithium dendrites can result in short-circuiting, but often the lithium detaches from the carbon electrode and thus results in more dead lithium in the cell, and in safety issues if the cells are disassembled.

5 Safety

As mentioned above, short-circuiting of the battery and rapid (and over-) charging represents a serious safety concern. Short circuits may result from lithium dendrites, penetration of the battery by a foreign object (e.g., in an accident), or as reported for some incidents, the presence of metal filings that found their way into the cell during the production of the battery. Short circuits result in rapid discharge of a charged battery and thus rapid heating. As the battery heats up, the metastable charged products will decompose, typically releasing oxygen gas. For example $Li_{1-x}CoO_2$ reacts to form Co_3O_4, $LiCoO_2$ and oxygen [16]. This is an exothermic reaction (the materials are reacting to form more stable products) and so more heat is released, speeding up the reaction and the battery can go into "thermal runaway". The presence of heat and oxygen can result in the ignition of the organic electrolytes and the other plastic components in the cell, which can result in serious fires and explosions. The melting of the separator and the consequent short circuiting of the cell is often a trigger for this. The need to pack multiple cells together in close proximity to make a battery with low volumetric energy density magnifies this problem. Although attempts have been made to replace the liquid electrolytes with solid electrolytes or less flammable solvents, to date, these electrolytes do not have sufficiently high lithium-ion conductivities to work in large-scale practical devices. Solid-state electrolytes come with other problems such as significantly reduced interfacial areas between the electrodes and electrolyte (again reducing rate) and difficulties in managing volume expansions of the different components, although the progress has been made in their implementation in microbatteries. New, safer electrode chemistries have been developed such as the $LiFePO_4$ cathode, and the $Li_4Ti_5O_{12}$ anode as will be described in more detail below. In general, these safer chemistries operate at lower voltages, and thus in regions where the electrolyte is more stable. Thus their usage represents a compromise between energy density and safety.

6 Electrode Chemistries

The energy density of a battery is a direct product of the electrode chemistry and we now review some of the materials currently used in batteries and some of the possible electrode materials that are potentially available to use in future technologies. The development of the Sony cell was made possible by the fundamental studies in the field of intercalation (solid state) chemistry in the 1970s, and in particular the work of John Goodenough and co-workers in Oxford University on $LiCoO_2$ [17–19]. In the almost 20 years of life of the lithium-ion battery, there has been continuous progress in the discovery and optimization of new intercalation materials for this application, as discussed in recent reviews [11, 20], and alternatives to $LiCoO_2$ have reached the market to different degrees, bringing about incremental—but significant—improvements in performance. The capacity is currently limited by the cathode and our major focus in on these chemistries. Increases in anode capacity, while needed, will only be *fully* utilized if cathodes with increased capacity are found, or methods for constructing thicker cathodes films can be engineered.

Cathodes (positives). $LiCoO_2$ is formed from a close-packed array of oxygen anions, with lithium and cobalt ions ordering in alternate layers, and adopting the so-called α-$NaFeO_2$-type framework (Fig. 2). On removal of only 4–5% of the Li^+ ions from the material, the material undergoes a transition from a semiconductor to a metal [21]. In a practical battery, as this material cycles within the range $Li_{1-x}CoO_2$ $0.05 < x < 0.6$, the material remains metallic throughout, which aids the insertion and removal of electrons on discharge and charge. Even on full discharge, the material does not return to stoichiometric $LiCoO_2$ since some of the lithium has been lost to the cell via the formation of the SEI on the graphitic negative electrode.

Rapid charging and discharging requires pathways for high lithium mobility through the electrode material. In $LiCoO_2$ (and graphite), this Li^+ transport occurs in the two dimensions within the lithium layers. Insertion and extraction is also anisotropic and can only occur on the faces of the $LiCoO_2$ (and graphite) crystal that slice through the layers. In principle, reducing the size of the particles of the active materials improves the rate, reducing the distance over which the lithium ions (and electrons) need to travel. However, the use of nanoparticles does not come without problems. They are difficult to pack (their so-called "tap density" is low), meaning that the volumetric energy density of the electrode is low. The particles also tend to be more mobile within the composite so that it is difficult to ensure that they remain electrically wired (connected) after multiple charge–discharge cycles. The most serious problem, at least for $LiCoO_2$, is that their higher surface areas result in more surfaces where side reactions can occur. This can result in more loss of Li^+ to the SEI and also mechanisms for self-discharging of the battery, if a stable SEI is not formed. Safety is the most important issue, as there are more surfaces where oxygen loss can occur (see below).

The major limitations of $LiCoO_2$ for use in transport and grid applications are (i) cost (ii) toxicity (iii) rate and (iv) stability. (i) Cost: The materials cost of $LiCoO_2$ represents a significant fraction of the cost of the battery and attempts have been made to replace or reduce the amount of Co in the cell via the use of different cathode chemistries. (ii) Toxicity: Although Co is toxic, alternatives that contain Ni and V are being considered where this issue will be more severe. Unlike batteries for portable electronics, recycling of the much larger batteries must occur, to recover the transition metal and lithium-ions. Indeed, some models of the cost and viability of lithium ion batteries, argue that a movement away from expensive cobalt battery technologies to cheaper manganese-based systems, changes the economic viability of recycling, also altering the overall cost of the technology. Thus, although the toxicity issue is real, given the success with widespread lead-acid recycling, it is in principle manageable. (iii) Rate: $LiCoO_2$ is not considered to be a very high rate material, because of its moderate lithium-ion transport combined with the relatively large (micron-size) particles used in the electrodes, and the 2D nature of the transport and insertion processes. (iv) Stability of the charged material Li_xCoO_2. Stability is important because it is directly related to safety and also to long-term degradation of materials, particularly if the battery is left in a charged state at higher temperatures. Differential scanning calorimetry (DSC) experiments of partially charged samples of $LiCoO_2$ show heat generation above 200°C, indicative of decomposition and then oxygen-loss, and the reaction of the electrolyte with the released oxygen (Fig. 4) [22]. Overcharging (caused by the use of too high rates) can also result in decomposition of the material, oxygen loss and dissolution of Co^{4+} in the electrolyte [16]. This is a serious concern for $LiCoO_2$.

A widespread approach to attempt to increase performance over that of $LiCoO_2$ has consisted in the partial or complete substitution of Co in the octahedral (transition metal) sites of the layered, α-$NaFeO_2$-type framework (Fig. 2) by other redox active metals such as Mn and Ni and dopants such as Al, leading to a vast array of candidates with different compositions and, in some cases, enhanced storage capacity (energy) and cycle life [20]. Unfortunately, the obvious replacement of the Co, by the cheap Mn and Fe ions, to form layered $LiMnO_2$ and $LiFeO_2$, results in materials that are not stable, the lithium and transition metal ions interchanging on cycling, resulting in the formation of a spinel structure

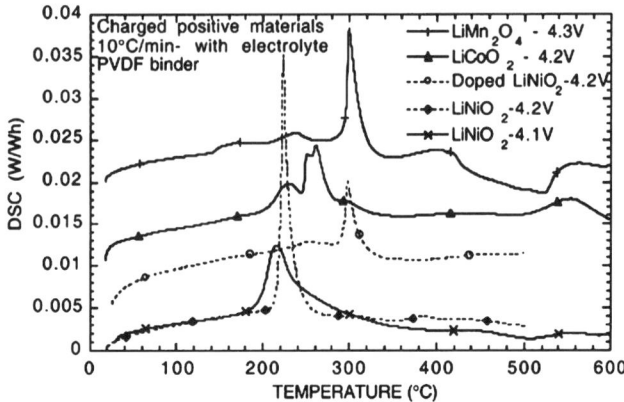

Fig. 4 Differential scanning calorimetry (DSC) measurements of various partially charged cathode materials that are used or have been considered for commercial applications. (Li_xNiO_2 charged to 4.1 and 4.2 V, doped Li_xNiO_2 (4.2 V), Li_xCoO_2 (4.2 V) and spinel $Li_xMn_2O_4$ (4.3 V), in the presence of binder and electrolyte. Reprinted from Ref [22], with permission from Elsevier

($LiMn_2O_4$) in the case of $LiMnO_2$ and a disordered rock salt for $LiFeO_2$. Substitutions that represent significant advances and that have received a great deal of attention are $LiNi_{0.8}Co_{0.2}O_2$ [23] and its Al^{3+}-doped variant (NCA), and $LiNi_{1/3}Mn_{1/3}Co_{1/3}O_2$ (NMC) [24, 25], the doping in $LiNi_{0.8}Co_{0.2}O_2$ [22] and NCA, serving to reduce the tendency of the nickel ions to disorder between the different metal layers [20]. $LiNi_{0.8}Co_{0.2}O_2$ and NCA largely operate via the Ni^{3+}/Ni^{4+} couple (the Co is oxidized at higher potentials) and show noticeably higher rate (power) performances over $LiCoO_2$. In their charged states, they are also more stable than undoped $LiNiO_2$ (see Fig. 4) [22]. SAFT (for example) has developed batteries with NCA and has demonstrated long-term cycling. However, Ni^{4+} on the surfaces of the charged particles is not stable with respect to decomposition of the electrolyte, and coatings are required if these materials are to be stable enough for applications requiring multiple charge–discharge cycles.

The discovery of NMC [24, 25] in 2001, was important because it is noticeably more stable than $LiCoO_2$ in the charged state, and it shows a higher capacity because more (almost 80% of the) Li^+ can be extracted reversibly from this material, leading to capacities of up to 200 mAhg^{-1} if the material is charged to 4.6 V. When optimized, it shows good rate performance, with almost 90% theoretical capacity achievable at a 1C rate [26]. It also operates via a two-electron ($Ni^{2+}-Ni^{4+}$) couple, suggesting new strategies to increase capacity. Attempts have now been made to combine some of the benefits of all these materials, by for example coating $LiCoO_2$ particles with NMC, to improve stability. Core–shell particles with NCA cores and manganese-rich shells have been prepared to combine rate with stability [27].

Another family of materials being considered that has the same basic layered structure is $Li[Ni_xLi_{(1/3-2x/3)}Mn_{(2/3-x/3)}]O_2$ [28, 29]. This class of materials can nominally be thought of as being formed from a solid solution between Li_2MnO_3

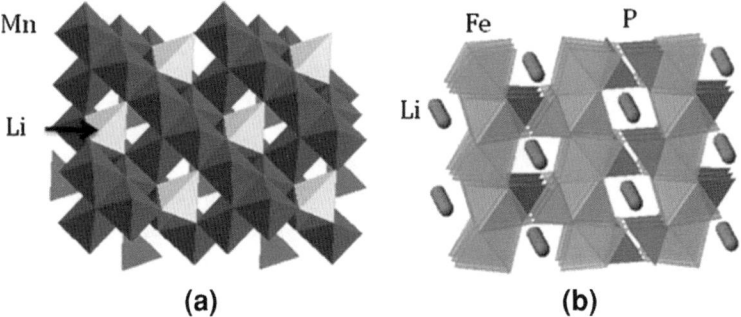

Fig. 5 The **a** spinel and **b** LiFePO$_4$ (olivine) crystal structures. The lithium ions (*light grey tetrahedra*) in spinel are contained within 3D tunnel structures, while in the olivine structure, they (*grey balls*) are contained within 1D tunnels

(or, as written in the layered notation Li[Li$_{1/3}$Mn$_{2/3}$]O$_2$) and the layered material Li[Ni$_{1/2}$Mn$_{1/2}$]O$_2$. This system is complex structurally, with local ordering of the cations in the predominantly transition metal layers occurring that is similar in nature, but shorter in coherence length, to that found in the end member Li[Li$_{1/3}$Mn$_{2/3}$]O$_2$ [30, 31]. The mechanism of lithium extraction is also complex, involving significant structural rearrangement in the first cycle [31], but leading to notable capacity gains with respect to LiCoO$_2$, with capacities of up to 200–220 mAhg^{-1} becoming feasible. However, the rearrangements are associated with oxygen-loss and the formation of surface coatings [31], the latter negatively impacting the rate performance of this material. Again, most of the capacity comes from the Ni^{2+}–Ni^{4+} redox couple, and as in NMC, the materials must be charged to at least 4.5 V versus Li to access capacities close to the full capacity. Despite the concerns associated with operation above 4.3 V, both NMC and this class of materials are being seriously evaluated for use in batteries for automotive applications.

Spinels. Historically, the spinel materials, LiMn$_{2-x}$M'$_x$O$_4$, represent the second most important insertion host for lithium-ion battery applications and they have been extensively investigated [20, 32].These materials can be cycled at approximately 4 V (versus Li) from LiMn$_2$O$_4$ to MnO$_2$, the structure retaining the spinel host framework throughout. The spinel structure is a three dimensional (3D) framework formed of MnO$_6$ octahedra, with the lithium ions in the 3D tunnels (Fig. 5a). The 3D connectivity and good electronic conductivity of this mixed valent (Mn^{3+}/Mn^{4+}) material is associated with high lithium transport and very high rates making it suitable for applications where power and cost are a primary criterion. The charged spinel structure is also more stable than LiCoO$_2$ (Fig. 4). Capacities of 148 mAhg^{-1} are, in theory, possible for the stoichiometric spinel, but in practice no more than 100–120 mAhg^{-1} has been achieved over multiple charge–discharge cycles because of the formulations used for practical applications [33]. Unfortunately, many of the spinels suffer from rapid capacity fade following extended storage or on cycling, particularly at high temperatures, which has been ascribed to the dissolution of Mn^{2+} ions [20], with a resulting oxidation of the spinel. The Jahn–Teller distortion that occurs on intercalation of $LiMn_2O_4$ to form $Li_2Mn_2O_4$ is associated with a large asymmetric change in cell volume [20] and has been implicated as another possible source of capacity fade, if the average oxidation state of the manganese drops below +3.5 during discharge (which can occur if the voltage drops to 3.5 V). Doping, which represents a method for controlling the oxidation state of the manganese at the end of discharge (but is also responsible for the decrease in theoretical capacity), along with surface coatings/treatments are used to prevent capacity fade. Li-excess (i.e., Li-doped) spinels $Li_{1+x}Mn_{2-x}O_4$, x = 0.05–0.1 are now being produced as cheap cathode materials, particularly for use in high power applications.

Phosphates. The search for new electrode materials was revolutionized by the discovery in 1999 that the application of a thin carbon-coating to particles of $LiFePO_4$ (LFP) enabled the extraction of essentially all of the Li ions, resulting in a capacity of 170 mAhg^{-1} [34]. This phospho-olivine (Fig. 5b) was first proposed for use as a positive electrode by Padhi et al. [35] but in this initial work, only 0.6 Li could be extracted, largely due to the low electronic conductivity of this phase. It has now been shown that by synthesizing nanoparticles, and ensuring good contact between the particles and the conductive carbon matrix (by carbon coating or intimate mixing between LFP and C) outstanding performance can be achieved [34, 36, 37]. As a consequence, $LiFePO_4$ is a strong candidate as a positive electrode in the next generation of batteries designed for transportation, because of the low cost of the raw materials, low toxicity, and excellent capacity retention and thermal stability [37]. Furthermore, the voltage at which this material operates (3.4 V, versus Li) means that it is considered to be a "safe" material, since it is well below the decomposition potential of the electrolyte (approximately 4.3–4.5 V, versus Li). This should be balanced with the low energy density of this material, because of its lower operating voltage, the presence of inactive phosphate ions and the poor packing density of the nanoparticles. However, the success achieved with LFP has led to considerable increased interest in other lithium transition metal phosphate phases as possible electrode materials, and to the exploration of other insulating oxyanion cathode phases such as lithium metal silicates and fluorosulfates [37].

7 Looking Forward

Cost. Cost is an extremely important consideration in battery design that can be measured in terms of an upfront cost but also in terms of the cost over the lifetime of the device. Clearly, if batteries can be made to last longer, while utilizing a larger amount of the total energy density (typically only about 80% usage of total energy density is assumed in calculations) this goes some way to mitigating the cost of the initial purchase. Materials represent 60–70% of the pack cost with cost of the raw materials for the cathode accounting for between 15–30% of the cost [38]. Separators are also extremely expensive representing approximately 9% of the cost. Surprisingly, the use of the various cathode chemistries has not yet translated into large differences in the costs of the different battery packs, despite some chemistries using cheaper raw materials such as Mn and Fe. This is partly because the cathodes made from Mn and Fe are associated with either lower capacities (spinels) and/or operate at lower voltages (olivines). Higher energy densities lead to lower costs since costs are typically measured in dollars per kWh. To date, the form of LFP used in batteries is still expensive to produce (in comparison to $LiCoO_2$) so that the cheaper cost of the raw materials has not yet translated into a cheaper electrode. NCA, NMC and the cobalt-free phases Li_2MnO_3–$Li(Ni_{0.5}Mn_{0.5})O_2$ are also more expensive to produce, the latter two systems requiring the synthesis of a precursor phase via solution routes, and then a second high temperature heating step. Since the anode capacity is larger than that of the cathode, one approach, which could lead to a noticeable reduction in pack cost is to build cells with even thicker cathodes (than assumed in the numbers given above). The integrity of the cathode after multiple cycles must still be maintained and these thicker electrodes must still be capable of withstanding high rates. This is non-trivial because the volume expansion issues still need to be managed, and there must still be pathways for the lithium ions in the electrolyte to access all the particles. However, this is conceptually a straightforward approach to reduce cost. The US Advanced Battery Consortium (USABC) has set an extremely tough cost goal of \$150/kWh (selling price) for EV batteries [39], given that current cost estimates vary from between about \$300–\$650/kWh [38]. The lowest estimated costs quoted here are only obtained by assuming no capacity fade and thicker electrodes (e.g., 3 mAhcm^{-2}) and the cheapest chemistry (spinel); the highest costs (which are more typical) are obtained with NCA or NMC, thinner electrodes and by taking into account capacity fade. Other reports find smaller differences in costs between the different cathode chemistries.

Prospects for Improved Anodes and Cathodes. Materials that react via intercalation mechanisms upon lithium extraction/insertion have intrinsic limitations in terms of capacity and rate, which are largely derived from structural aspects. The ability to reversibly intercalate lithium ions is limited by the change that a crystal structure is able to withstand before transforming into another structure or becoming disordered. This can vary dramatically, even within the same structure type, since different cations can be associated with quite different changes in size and coordination preference on change of oxidation state. Critically, there must also be sufficient numbers of Li ions that can be removed or vacancies into which Li ions can be inserted.

Incremental increases in energy density can be achieved by increasing the operating voltage of the cell. For this reason, the olivine $LiMnPO_4$ (4.1 V), the vanadate $LiVPO_4F$ (4.2 V) and the high-voltage spinel $Li(Ni_{0.5}Mn_{1.5})O_4$ (4.8 V), continue to be investigated [20, 37]. $LiMnPO_4$ is difficult to activate unless it is nanosized and/or doped with other transition metal elements. $LiCoPO_4$ is also difficult to activate and it operates in a regime where the electrolyte decomposition is a serious concern (Fig. 3). High power cells can be constructed from micron-sized particles of $Li(Ni_{0.5}Mn_{1.5})O_4$, which will help somewhat in electrolyte decomposition, but this problem is still not fully solved.

In order to increase the capacity of either the anode or cathode significantly, processes or redox reactions are required that involve multiple changes in (formal) oxidation state of the redox active element that are (i) reversible, (ii) occur in the appropriate voltage windows, (iii) involve minimal changes in volume and (iv) minimal changes in the host structure. Ideally, the materials should also be good electronic and ionic conductors. Since it is difficult, in practice, to identify materials that simultaneously fulfill all these criteria, research has focused on materials that meet some of these criteria, to design strategies to overcome or minimize some of their other limitations.

The first example of this strategy that brings about substantial increases in storage capacity is the use of metals and semimetals that can electrochemically and reversibly form alloys with lithium. The capacities that can be achieved from these alloying reactions can reach extremely high values, both on a mass and volumetric basis (e.g., 8,322 mAh/cm^3 and 3,572 mAh/g for silicon compared to 975 mAh/cm^3 and 372 mAh/g for graphite, where the volumetric/gravimetric capacity has been calculated based on the original volume/mass of graphite and silicon), because phases with a large lithium content (e.g., $Li_{3.75}Si$ and $Li_{4.4}Sn$) can be formed. However, the practical utilization of these reactions has been severely handicapped by the huge volume changes associated to the (de)alloying process, which introduce large strains in the particles of active material, leading to pulverization and, subsequently, to capacity loss due to formation of dead or inactive regions [40]. The strategies used to circumvent these issues involve (i) the dilution of the active material in an inert matrix that can absorb the stresses, (ii) nanosizing, (iii) limiting the range over which the materials are cycled and (iv) the development of new binders [41]. Efforts in this direction have proved successful, and have resulted in the commercialization of batteries that contain Sn or Si-based negative electrodes, for now aimed at the portable electronics markets. Challenges and questions remain in terms of the capacity retention of these systems for the number of cycles required for transport and grid applications. When fresh surface is exposed to the electrolyte, perhaps due to particle cracking, new SEI grows, gradually consuming more lithium. Very thick SEIs have been seen on these materials, and new additives/coatings will need to be found that are able to form stable SEIs/coatings that are able to withstand the large volume expansions of these systems.

A second advance in the field came with the discovery of materials that operate via what is now referred to as a "conversion reaction". Here, a metal salt, for example, MO reacts with lithium to form a composite comprising a metal particle surrounded by the salt Li_2O [42, 43]. While most of these compounds react at low voltages, in principle, making them suitable alternatives for the negative electrodes, some of the more ionic fluoride phases are active at potentials above 2 V, leading to a whole new array of candidates for the positive electrode with very high energy densities [44]. However, several issues remain that hinder the commercial realization of conversion-based electrodes: (i) unsatisfactory cycling performance, (ii) an unacceptably large, in terms of energy loss, voltage hysteresis between discharge and charge, (iii) the virtually ubiquitous large coulombic inefficiency observed in the first cycle. It is thus clear that more fundamental knowledge needs to be acquired before technological solutions to these three barriers can be proposed.

On the cathode side, silicates with the composition Li_2MSiO_4 have been widely investigated [37] because they are potentially very cheap, and they can, if both lithium ions can be extracted reversibly, yield very high capacities. To date, the iron silicate has shown promise, but it has not been possible to extract more than one electron/lithium from either the iron or manganese material. However, there is considerably activity in this area involving a wide range of oxyanion compounds (e.g., new phosphates, borates and sulfates), and the area holds much promise.

Finally, there are a number of other important areas where much research is being performed. First, a number of anode materials that operate at higher voltages and are therefore less likely to suffer from safety issues due to the formation of Li dendrites, have been investigated. Of particular note is the spinel $Li_4Ti_5O_{12}$, which operates at 1.5 V and has essentially no volume change on lithium extraction [45]. However, its lower capacity and the lower cell voltage result in a noticeable loss in energy density over cells containing graphite. It has, however, been coupled with higher voltage cathodes and may find a use in cells where safety is the major concern. Li-air and Li-sulfur batteries have generated much interest because of their extremely high energy densities (estimated practical energy densities of 370 and 1,700 Wh/kg, respectively) and potentially low costs [37, 46]. However, their volumetric energy densities is not as high, their power performance is poor and considerable technological challenges remain to be addressed. It is important to note that the quoted energy densities are calculated assuming that a Li metal anode is used; given the problems with Li dendrite formation, this is extremely problematic. Some current strategies include devising solid-state coatings or membranes on the Li anode to help mitigate this problem, but this is still an unsolved issue. The Li-air battery utilizes the reaction of oxygen in the air with Li^+ ions to form Li_2O_2, while in the lower voltage Li–S battery the Li^+ ions react with sulfur (which is both cheap and abundant) to form Li_2S. Other challenges in lithium-air technology include reducing the large hysteresis in the potential on charge and discharge, finding an electrolyte that does react with the superoxide generated during the reaction, and improving rate performance. The lithium sulfur cell has significant problems associated with the solubility of the partially reduced sulfur species in the electrolyte, which can act as redox shuttles migrating to the negative electrode where they can be reduced (resulting in self-discharge and loss of sulfur). Although in theory these technologies approach the storage capabilities of gasoline, they are currently far from commercialization.

8 Conclusions

Batteries are made up of a number of different materials that make use of a complex range of chemistries that must all work in concert over a wide range of temporal and spatial timescales. The materials requirements for transport and grid applications are severe, with stringent requirements in terms of cost, safety and efficiency (capacity retention). Despite this, a wide range of lithium-ion battery technologies already exist that are currently ready for use in transport applications and for grid applications. New chemistries and materials are being developed that hold promise for increasing energy density and safety and for reducing cost. The already-scheduled building of large-scale production plants to make batteries containing current technologies, but that can readily be modified to make use of new materials and modifications, will drive down costs, particularly if there is a glut of batteries on the market. This, in combination with strong legislation to reduce CO_2 emissions, will lead to their widespread use in a number of new electricity-based technologies.

Acknowledgments Support for the author's research in this field has come from the Assistant Secretary for Energy Efficiency and Renewable Energy, Office of FreedomCAR and Vehicle Technologies of the U.S. Department of Energy (DOE) via subcontract No. 6517749 with the Lawrence Berkeley National Laboratory and from the DOE office of Basic Energy Sciences, via support of the North Eastern Center for Chemical Energy Sciences, an Energy Frontier research Center. Discussions with Gerbrand Ceder, M. Stanley Whittingham, Jordi Cabana and Roger Thornton are gratefully acknowledged.

References

1. Tarascon JM, Armand M (2001) Nature 414:359
2. Department of Energy (2007) Basic energy sciences (BES) report "Basic research needs for electrical energy storage". April, http://science.energy.gov/ ~ /media/bes/pdf/reports/files/ees_rpt.pdf
3. Armand M, Tarascon JM (2008) Nature 451:652–657
4. http://www.toyota.co.uk
5. http://gm-volt.com/full-specifications/
6. Argonne National Laboratory (2005) Well-to-wheels analysis of advanced fuel/vehicle systems—A North American study of energy use, greenhouse gas emissions, and criteria pollutant emissions. http://www.transportation.anl.gov/pdfs/TA/339.pdf
7. Argonne National Laboratory (2009) Well-to-wheels energy use and greenhouse gas emissions analysis of plug-in hybrid electric vehicles. http://www.transportation.anl.gov/pdfs/TA/559.pdf
8. http://www.uscar.org/guest/article_view.php?articles_id=74
9. http://www.ge.com/battery/resources/pdf/ImreGyuk.pdf
10. Nagaura T (1990) 4th international rechargeable battery seminar, Deerfield Beach
11. Palacin MR (2009) Chem Soc Rev 38:2565
12. Reimers JN, Dahn, JR (1994) J Electrochem Soc 139:2091
13. Dahn JR (1991) Phys Rev B 44:9170
14. Goodenough JB, Kim Y (2010) Chem Mater 22:587–603
15. Amatucci GG, Tarascon JM, Klein LC (1996) J Electrochem Soc 143:1114
16. Doh C-H, Kim D-H, Kim H-S, Shin Jeon Y-D, Moon S-I, Jin B-S, Eom SW, Kim K-S, Kim K-W, Oh D-H, Veluchamya A (2008) J Power Sources 75:881
17. Guerard D, Herold A (1975) Carbon 13:337–345
18. Whittingham MS (1978) Prog Solid State Chem 12:41–99
19. Mizushima K, Jones PC, Wiseman PJ, Goodenough JB (1980) Mater Res Bull 15:783–789

20. Whittingham MS (2004) Chem Rev 104:4271–4301
21. Menetrier M, Saadoune I, Levasseur S, Delmas C (1999) J Mater Chem 9:1135
22. Biensan Ph, Simon B, Peres JP, de Guibert A, Broussely M, Bodet JM, Perton F (1999) J Power Sources 81:906
23. Delmas C, Saadoune I, Rougier A (1993) J Power Sources 44:595–602
24. Lu ZH, MacNeil DD, Dahn JR (2001) Electrochem Solid State Lett 4:A200–A203
25. Ohzuku T, Makimura Y (2001) Chem Lett (7):642–643
26. Park CW, Kang SH, Belharouak I, Sun YK, Amine K (2008) J Power Sources 177:177
27. Sun YK, Myung ST, Park BC, Prakash J, Belharouk I, Amine K (2009) Nat Mater 8:330
28. Lu Z, MacNeil DD, Dahn JR (2001) Electrochem Solid State Lett 4:A191–A194
29. Thackeray MM, Johnson CS, Vaughey JT, Li N, Hackney SA (2005) J Mater Chem 15:2257–2267
30. Yoon W-S, Iannopollo S, Grey CP, Carlier D, Gorman J, Reed J, Ceder G (2004) Electrochem. Solid St. Lett. 7:A167
31. Jiang M, Key B, Meng YS, Grey CP (2009) Chem Mater 21:2733–2745
32. Thackeray MM (1997) Prog Solid St Chem 25:1
33. Liu W, Kowal K, Farrington GC (1998) J Electrochem Soc 145:459
34. Ravet N, Goodenough JB, Besner S, Simoneau M, Hovington M, Armand M (1999) Abstract #127, 196th ECS meeting, Honolulu, 17–22 Oct
35. Padhi AK, Nanjundaswamy KS, Goodenough JB (1997) J Electrochem Soc 144:1188–1194
36. Chung SY, Bloking JT, Chiang YM (2002) Nat Mater 1:123–128
37. Ellis BL, Lee KT, Nazar LF (2010) Chem Mater 22:691–714
38. Barnett B, Rempel J, McCoy C, Dalton-Castor S, Sriramulu S (2011) Department of Energy Merit Review http://www1.eere.energy.gov/vehiclesandfuels/pdfs/merit_review_2011/electrochemical_storage/es001_barnett_2011_o.pdf
39. http://www1.eere.energy.gov/vehiclesandfuels/pdfs/program/electrochemical_energy_storage_roadmap.pdf
40. Timmons A, Dahn JR (2006) J Electrochem Soc 153:A1206–A1210
41. Larcher D, Beattie S, Morcrette M, Edstroem K, Jumas JC, Tarascon JM (2007) J Mater Chem 17:3759–3772
42. Idota Y, Kubota T, Matsufuji A, Maekawa Y, Miyasaka T (1997) Science 276:1395–1397
43. Poizot P, Laruelle S, Grugeon S, Dupont L, Tarascon JM (2000) Nature 407:496–499
44. Badway F, Cosandey F, Pereira N, Amatucci GG (2003) J Electrochem Soc 150:A1318–A1327
45. Ferg E, Gummow RJ, Kock AD, Thackeray MM (1994) J Electrochem Soc 141:L147
46. Girishkumar G, McCloskey B, Luntz AC, Swanson S, Wilcke W (2010) Phys Chem Lett 1:2193

Fuel Cell Technology

Aaron Holdway and Oliver Inderwildi

1 Introduction

Fuel cells offer the transport sector the promise of decreased dependence on fossil fuels, low or zero emissions, and high efficiency.[1] Unlike internal combustion engines, fuel cells convert chemical energy directly into electrical energy, producing much less waste heat and offering a much higher theoretical efficiency. Unlike batteries, fuel cells can run continuously with continuous input of reactants (fuel and oxidant). Fuel cells run best on pure or reformed hydrogen [1, 2] but some can operate directly on alternative fuels such as methanol or hydrocarbons [3].

Many prototype fuel cell automobiles have been produced [4–7]. The existing technical challenges, however, make it more likely that the initial market uptake may be for fleet vehicles such as buses, which often use a single refuelling station and can store larger quantities of hydrogen onboard. Limited trials of hydrogen buses have been carried out in cities such as Chicago, Vancouver, Beijing, Aichi (Japan), Perth (Australia), and ten European cities in a European Union-funded

---

[1] Fuel cells also offer promise in stationary power generation on a variety of scales, as well as in portable electronics such as mobile phones and laptops.

A. Holdway · O. Inderwildi (✉)
Smith School of Enterprise and the Environment, University of Oxford,
Hayes House, 75 George Street, Oxford OX1 2BQ, UK
e-mail: oliver.inderwildi@smithschool.ox.ac.uk

O. Inderwildi and Sir David King (eds.), *Energy, Transport, & the Environment*,   273
DOI: 10.1007/978-1-4471-2717-8_14, © Springer-Verlag London 2012

project [8]. Additional early introduction may be through niche markets such as forklifts and airport ground-support vehicles.

Nevertheless, fuel cells currently remain a niche technology. Designs, materials, components, and fabrication methods are continually evolving, but technological breakthroughs will be needed before fuel cells can become commercially viable in mobile applications. Significant improvements in cost, durability, reliability, and power density will be needed. The transition to a hydrogen economy for transport, if made at all, will be over the long term and will most likely be challenging and investment intensive.

2 General Advantages of Fuel Cells

Fuel cells offer the following general advantages over internal combustion engines:

- Potential to decrease dependence on fossil fuels and increase energy security
- Less emission of greenhouse gases (low or zero at point of use, but the method of hydrogen production must also be taken into account)
- Emissions of sulphur oxides (SO_x), nitrogen oxides (NO_x), and particulates is virtually zero
- Choice of many fuels (e.g., hydrogen, hydrocarbons, alcohols)
- Scalability of power: scale well from the watt range (e.g., mobile phones) to the megawatt range (power plants)
- High efficiency (40–60%, with the potential of reaching 80% in future [9]) compared to internal combustion engines (20–30% [9])
- Most efficient at partial load (unlike internal combustion engines), which is the typical running condition for automobiles
- Quiet due to noiseless electrochemical combustion and fewer moving parts.

3 General Limitations to Commercial Viability

Despite the above advantages, fuel cells face a number of limitations to commercialisation in automotive applications:

- Fuel cell system cost is high compared to the cost of an internal combustion engine, mainly due to the cost of the required precious-metal catalyst
- Durability and reliability, especially at low temperature
- Power density (see Table 1)
- fuel production, distribution, and storage
- some fuels require reforming, which is detrimental to fuel cell efficiency.

Table 1 Comparison of principal fuel cell types

	PEMFC	AFC	PAFC	MCFC	SOFC
Electrolyte	Flexible polymer membrane	$KOH_{(aq)}$	$H_3PO_{4(aq)}$ in porous silicon carbide matrix	Molten alkali metal carbonate in porous matrix	Ceramic (yttria-stabilised zirconia)
Operating temperature (°C)	80	60–220 (typically 70)	200	650	500–1,000
Catalyst	Pt	Pt (anode), Ni (anode and cathode)	Pt	Ni (Ni alloy at anode, NiO at cathode)	Perovskites (ceramic) (Ni cermet at anode, La-based compounds at cathode)
Fuel compatibility	H_2, methanol (see direct methanol fuel cell, below)	H_2	H_2	H_2, hydrocarbons (e.g., CH_4)	H_2, hydrocarbons (e.g., CH_4), CO
Reformer	External reformer required if H_2 unavailable	No (reformate fuels cannot be used because of presence of CO_2)	External reformer required if H_2 unavailable	Internal (thermally integrated)	Internal (thermally integrated)
Main poison	CO, S	CO_2, CO	CO (tolerance 1–2%), S	S	S
Electrical efficiency (%) [10]	40–60	45–60	35–45	45–60	45–55
Power density (mW/cm²) [11]	100–1,000 [S. Ding (2009)]	150–400	150–300	100–300	250–350
Power range (kW) [11]	0.001–1,000	1–100	50–1000	100–100,000	10–100,000
Advantages	Low operating temperature; best power density; rapid start-up (best start-stop cycling)	Extremely low-cost electrolyte; often do not need Pt catalyst at cathode	Relatively low-cost electrolyte; technology is mature; high reliability	Do not need Pt catalyst; fuel flexibility (external reforming not required); high-quality waste heat	Do not need Pt catalyst; fuel flexibility; relatively high power density; high-quality waste heat

(continued)

Table 1 (continued)

	PEMFC	AFC	PAFC	MCFC	SOFC
Disadvantages	Catalyst, membrane, and ancillary components are expensive [12]; waste heat rejection is difficult [13]; catalyst is susceptible to poisoning [14]	Requires pure H_2 and O_2; waste heat rejection is difficult; electrolyte is corrosive	Catalyst is expensive; catalyst susceptible to poisoning; electrolyte is corrosive; electrolyte must be replenished occasionally	Materials relatively expensive because of high operating temperature; electrolyte is corrosive; lower durability; very long start-up time	Materials relatively expensive because of high operating temperature; lower durability; long start-up time
Most promising applications	Portable electronics, automotive, stationary power	Suited only for auxiliary power in space applications	Stationary power	Stationary power	Stationary power
Suitable for commercial automotive applications?	Yes	Not at present, particularly because of need for pure H_2 and O_2; development has stagnated	Not at present, because of low power density and because electrolyte freezes at 42°C; further development has focused on stationary applications	No, because of very long start-up time (tens of hours); further development has been only for stationary applications	Possibly for automotive auxiliary power units [15]

4 Principal Fuel Cell Types

Fuel cells are typically classified according to the electrolyte they use:

- Polymer electrolyte membrane fuel cell, or proton exchange membrane fuel cell (PEMFC) [16–18]—includes direct alcohol fuel cells (e.g., direct methanol fuel cell [19])
- Alkaline fuel cell (AFC) [20, 21]
- Phosphoric acid fuel cell (PAFC) [22, 23]
- Solid oxide fuel cell (SOFC) [3, 24, 25]
- Molten carbonate fuel cell (MCFC) [26, 27].

PEMFCs, AFCs, and PAFCs operate at lower temperatures than MCFCs and SOFCs and use pure hydrogen as fuel. PEMFCs and PAFCs require an external reformer unless pure hydrogen is available.[2] If running on pure hydrogen, fuel cells produce zero emissions at point of use. If running on a hydrocarbon fuel, however, carbon dioxide (CO_2) will be produced, though less than for an internal combustion engine because of the greater efficiency of fuel cells. AFCs cannot use reformate fuels because of the presence of CO_2, which severely degrades AFC performance. The high operating temperature of MCFCs and SOFCs allows them to run on hydrocarbons such as methane and eliminates the need for an expensive platinum catalyst. The high temperatures, however, also contribute to lower durability and a long start-up time. PEMFCs generally have the highest power density. Table 1 provides a comparison of the above fuel cell types.

Examples of other types of fuel cells include biological fuel cells, or biofuel cells, such as enzymatic fuel cells [28–30] and microbial fuel cells [31–33]. Biofuel cells remain in the early stages of development and currently lack sufficient power density for practical applications, but the potential to one day replace expensive noble-metal catalysts with abundant biological catalysts in fuel cells operating at near-ambient temperatures and pH continues to drive research forward.

A further type of fuel cell is direct alcohol fuel cells (DAFCs), which are a type of PEMFC because they use a polymer electrolyte membrane. Fuels include methanol [19], ethanol [34], and ethylene glycol [35]. The most developed DAFC is the direct methanol fuel cell (DMFC), which offers the following advantages and disadvantages:

5 Advantages of DMFCs

- Low operating temperature
- Fuels more readily available than hydrogen

---

[2] This is not true of direct alcohol fuel cells, a class of PEMFCs which run on alcohols such as methanol.

- Fuel does not require external reforming (the fuel is used directly)
- Fuel storage more compact because fuel is denser than hydrogen
- Do not need ancillary systems such as humidifiers and coolant recirculators because fuel is liquid.

6 Disadvantages of DMFCs

- Technology relatively immature
- Higher Precious-Metal catalyst loadings than for hydrogen PEMFCs, which contributes to higher cost and increased pressure on platinum availability
- Lower operating efficiency than for hydrogen PEMFCs (35–40% [10])
- Lower power density and slower power response than for hydrogen PEMFCs
- Reduced performance (because reaction kinetics more complex)
- Electrolyte is toxic.

DAFCs are suited for portable electronics applications but not for automobiles, except in some cases as auxiliary power units. Hydrogen PEMFCs avoid the disadvantages of DAFCs [19].

7 Assessment

The widely differing materials, cell design, electrochemistry, and operating temperature of the different fuel cells bring about important advantages and disadvantages (summarised in Table 1, with an assessment of their promise for automotive applications). PEMFCs and SOFCs offer the best prospects for eventual wide commercial implementation, although for automotive applications, PEMFCs appear best positioned for commercialisation and currently receive the most attention from major automobile manufacturers [36–46]. Indeed, more than 90% of fuel cell vehicles on the road since 2000 have used PEMFCs [47]. Crucially, only PEMFCs currently meet the power density target of 800–900 mW/cm^2 required for automotive applications [48] (see Table 1).

PEMFCs' low operating temperature, relatively high power density (and consequent compactness), and rapid start-up make them superior to SOFCs for automotive and other mobile applications. The PEMFC operating temperature (80°C) is limited by the need for water to be present in the electrolyte membrane, although research is being done on novel polymer electrolytes for PEMFCs that do not require water, raising the operating temperature to as high as 200°C [3, 49, 50]. This mitigates some of the disadvantages of current PEMFCs—namely, reaction kinetics improve, CO poisoning is reduced, and heat rejection improves. The lack of low-cost CO-tolerant catalysts remains a major challenge for PEMFC commercialisation, so high-temperature PEMFCs may offer some promise.

SOFCs offer two major advantages over PEMFCs: no need for a platinum catalyst and greater fuel flexibility without the need for external reforming. SOFCs' high operating temperature and consequent lower durability and long start-up time, however, pose significant hurdles to use in automotive applications [25]. Advances in materials, however, have made possible so-called intermediate-temperature (IT) SOFCs, which operate at 500–750°C [3]. The lower operating temperature of IT-SOFCs, and consequently the faster start-up time, greater durability, and lower cost materials, makes this technology a possible candidate for automotive applications [3]. IT-SOFCs offer greater fuel flexibility, efficiency, and tolerance of impurities than PEMFCs, and unlike PEMFCs, do not require external fuel reforming. They face the challenge, however, of reduced activity for oxygen reduction at the cathode compared to high-temperature SOFCs [51]. Because of their still relatively long start-up time and bulkiness compared to PEMFCs, however, IT-SOFCs in the automotive sector might be suitable only as auxiliary power units, such as in heavy-duty lorries [15].

8 Market Penetration

Despite steady progress in fuel cell research and development, no fuel cell vehicles are yet ready for wide commercial application. As of 2009, only a few thousand light-duty fuel cell vehicles were deployed worldwide, with a larger number of niche vehicles such as forklifts [52]. A number of manufacturers have produced prototype fuel cell vehicles, and some, such as Honda, have begun leasing vehicles in limited numbers [53]. A 2008 report by the National Research Council in the US estimated that in a "maximum practical scenario," where mass production begins by 2015, technology goals are met, government support is given, and consumers accept the new technology even with an initially limited number of refuelling stations, fuel cell vehicles could become commercially competitive in the US by about 2023 [54]. Although the time taken to overcome the technological hurdles is unknown, the National Research Council estimates that producing commercially viable fuel cell vehicles by 2023 would require about US $55 billion in government investment in support of research and development, production, and refuelling infrastructure. Without such investment, it is unlikely there will be a compelling business case for automakers or a concomitant creation of the required infrastructure [55]. As of early 2011, Daimler, Ford, GM, Honda, Hyundai, Nissan, PSA Peugeot Citroën, and Toyota had announced plans to begin mass producing fuel cell vehicles anywhere from 2015 to 2030 [44, 46, 56–61]. The latest of these dates is from Ford. Frenette and Daniel, engineers at Ford, wrote in 2009 that Ford estimates fuel cells vehicles might not appear on the market in "any meaningful percentage" of total industry volume until at least 2030 [62]. Due to long fleet lifetimes of modern road vehicles, however, major fleet penetration could take *until* 2055 *or later* [62, 63].

To get to this stage, however, breakthroughs in materials and fabrication methods, rather than incremental change, will be needed. For PEMFCs, a particular challenge is catalyst cost reduction. The search for non-platinum catalysts has been ongoing for more than 40 years. At low temperatures and using hydrogen, reformate, or methanol fuel, platinum-based catalysts remain the most active of any of the pure metals [64]. One could potentially use a greater amount of a cheaper (non-noble), less active catalyst, but the acidic environment of PEMFCs rules out non-noble metals [48]. Researchers have demonstrated that it may be possible to reduce platinum loading by using bimetallic catalysts of platinum and another noble metal, such as rubidium, in nanostructures that increase catalyst surface area, such as carbon aerogels [65], nanofibers [66], and nanotubes [67]. The use of such catalysts, however, has been limited by a lack of procedures for controlled large-scale production. Lim et al. [68] suggest that seeded growth of Pd–Pt nanodendrites could be used as a model for reducing platinum loading through better control of nanostructure morphology.

The need for alternative catalysts is also driven by limitations on the availability of platinum. Platinum catalysts in automobiles already consume approximately half of the platinum sold globally each year [69]. Gordon et al. [70] conclude that even if the world's entire platinum supply were devoted to fuel cell vehicles, ignoring its many competing uses, a fuel cell fleet the size of the current world's light-duty fleet could be sustained for *less than* 20 years. Borgwardt [71] concludes that under the most favourable circumstances, and with the US auto industry consuming as much platinum as is currently consumed by the *global* auto industry, a complete transition of the US vehicle fleet to fuel cell vehicles would take *until the* 2060s. The effect this would have on the price of platinum would exacerbate the existing problem of making fuel cells more cost-competitive with other technologies. Unless breakthroughs are made, however, low-temperature fuel cells will continue to require platinum catalysts for the foreseeable future.

For automotive fuel cells to realise their potential, five hurdles must be overcome: they must become competitive on (1) cost and (2) convenience to internal combustion engines, and until significant advancement is made in (3) hydrogen production, and (4) hydrogen storage and massive investment made in (5) distribution infrastructure [72, 73] (see Chapter by Sartbaeva et al. Hydrogen). Until these obstacles can be surmounted, automotive fuel cells will remain a niche technology.

References

1. Ahmed S, Krumpelt M (2001) Hydrogen from hydrocarbon fuels for fuel cells. Int J Hydrogen Energy 26:291–301
2. Qi AD, Peppley B, Karan K (2007) Integrated fuel processors for fuel cell application: a review. Fuel Process Technol 88:3–22
3. Brett DJL et al (2008) Intermediate temperature solid oxide fuel cells. Chem Soc Rev 37:1568–1578

4. Lloyd AC (2000) The California fuel cell partnership: an avenue to clean air. J Power Sources 86:57–60

5. Johnston B, Mayo MC, Khare A (2005) Hydrogen: the energy source for the 21st century. Technovation 25:569–585

6. Van den Hoed R (2005) Commitment to fuel cell technology? How to interpret carmakers' efforts in this radical technology. J Power Sources 141:265–271

7. Dixon RK (2007) Advancing towards a hydrogen energy economy: status, opportunities and barriers. Mitig Adapt Strat Glob Change 12:325–341

8. Fuel Cell Bus Club (no date) http://www.fuel-cell-bus-club.com. Accessed 10 February 2011

9. Grant PM (2003) Hydrogen lifts off—with a heavy load. Nature 424:129–130

10. Rand DAJ, Dell RM (2008) Hydrogen energy: challenges and prospects. RSC Publishing, Camrbidge

11. O'Hayre R et al (2006) Fuel cell fundamentals. Wiley, New York

12. Tsuchiya H, Kobayashi O (2004) Mass production cost of PEM fuel cell by learning curve. Int J Hydrogen Energy 29:985–990

13. Van den Oosterkamp PF (2006) Critical issues in heat transfer for fuel cell systems. Energy Convers Manag 47:3552–3561

14. Cheng X et al (2007) A review of PEM hydrogen fuel cell contamination: impacts, mechanisms, and mitigation. J Power Sources 165:739–756

15. Steele BCH, Heinzel A (2001) Materials for fuel-cell technologies. Nature 414:345–352

16. Sopian K, Daud WRW (2006) Challenges and future developments in proton exchange membrane fuel cells. Renew Energy 31:719–727

17. Mehta V, Cooper JS (2003) Review and analysis of PEM fuel cell design and manufacturing. J Power Sources 114:32–53

18. Wu JX, Liu QY, Fang HB (2006) Toward the optimization of operating conditions for hydrogen polymer electrolyte fuel cells. J Power Sources 156:388–399

19. Kamarudin SK et al (2007) Overview on the challenges and developments of micro-direct methanol fuel cells (DMFC). J Power Sources 163:743–754

20. McLean GF et al (2002) An assessment of alkaline fuel cell technology. Int J Hydrogen Energy 27:507–526

21. Verma A, Basu S (2005) Direct use of alcohols and sodium borohydride as fuel in an alkaline fuel cell. J Power Sources 145:282–285

22. Sammes N, Bove R, Stahl K (2004) Phosphoric acid fuel cells: fundamentals and applications. Curr Opin Solid St M 8:372–378

23. Neergat M, Shukla AK (2001) A high-performance phosphoric acid fuel cell. J Power Sources 102:317–321

24. Yano M et al (2007) Recent advances in single-chamber solid oxide fuel cells: a review. Solid State Ion 177:3351–3359

25. Ormerod RM (2003) Solid oxide fuel cells. Chem Soc Rev 32:17–28

26. Bischoff M (2006) Molten carbonate fuel cells: a high temperature fuel cell on the edge to commercialization. J Power Sources 160:842–845

27. Dicks AL (2004) Molten carbonate fuel cells. Curr Opin Solid St M 8:379–383

28. Cracknell JA, Vincent KA, Armstrong FA (2008) Enzymes as working or inspirational electrocatalysts for fuel cells and electrolysis. Chem Rev 108:2439–2461

29. Willner I et al (2009) Integrated enzyme-based biofuel cells—a review. Fuel Cells 9:7–24

30. Ivanov I, Vidaković-Koch T, Sundmacher K (2010) Recent advances in enzymatic fuel cells: experiments and modeling. Energies 3:803–846

31. Franks AE, Nevin KP (2010) Microbial fuel cells, a current review. Energies 3:899–919

32. Logan BE (2010) Scaling up microbial fuel cells and other bioelectrochemical systems. Appl Microbiol Biot 85:1665–1671

33. Pant D et al (2010) A review of the substrates used in microbial fuel cells (MFCs) for sustainable energy production. Bioresour Technol 101:1533–1543

34. Wang Q et al (2008) High performance direct ethanol fuel cell with double-layered anode catalyst layer. J Power Sources 177:142–147

35. Livshits V, Peled E (2006) Progress in the development of a high-power, direct ethylene glycol fuel cell (DEGFC). J Power Sources 161:1187–1191
36. Honda Motor Company (no date) FCX Clarity. http://automobiles.honda.com/fcx-clarity. Accessed 1 April 2011
37. Kirubakaran A, Shailendra J, Nema RK (2009) A review on fuel cell technologies and power electronic interface. Renew Sust Energ Rev 13:2430–2440
38. Daimler (no date) Fuel cell drive technology. http://www.daimler.com/technology-and-innovation/drive-technologies/fuel-cell. Accessed 1 April 2011
39. Ford motor company (no date) How a fuel cell works. http://media.ford.com/article_display.cfm?article_id=1908. Accessed 1 April 2011
40. PSA Peugeot Citroën (no date) How does it work? The fuel cell. http://www.psa-peugeot-citroen.com/modules/pac/anglais/index.html. Accessed 1 April 2011
41. Matsunaga M, Fukushima T, Ojima K (2009) Advances in the power train system of Honda FCX clarity fuel cell vehicle. SAE World Congress & Exhibition, Detroit. http://www.sae.org/technical/papers/2009-01-1012. Accessed 10 February 2011
42. Hyundai Motor Company (no date) Fuel cell vehicles. http://www.hyundai.com/in/en/CompanyInfomation/Technology/FuelCellVehicles/FuelCellVehicles.htm. Accessed 1 April 2011
43. Shimoi R, Aoyama T, Iiyama A (2009) Development of fuel cell stack durability based on actual vehicle test data: current status and future work. SAE World Congress & Exhibition, Detroit. http://www.sae.org/technical/papers/2009-01-1014. Accessed 10 February 2011
44. Nissan Motor Company (no date) Next-generation fuel cell stack. http://www.nissan-global.com/EN/TECHNOLOGY/OVERVIEW/fcv_stack.html. Accessed 1 April 2011
45. Noto H et al (2009) Development of fuel cell hybrid vehicle by Toyota—durability. SAE World Congress & Exhibition, Detroit. http://www.sae.org/technical/papers/2009-01-1002. Accessed 10 February 2011
46. Toyota Motor Sales, USA (2010) Fuel cell hybrid vehicle-advanced. http://www.toyota.com/esq/articles/2010/FCHV_ADV.html. Accessed 1 April 2011
47. De Bruijn F (2005) The current status of fuel cell technology for mobile and stationary applications. Green Chem 7:132–150
48. Gasteiger HA et al (2005) Activity benchmarks and requirements for Pt, Pt-alloy, and non-Pt oxygen reduction catalysts for PEMFCs. Appl Catal B Environ 56:9–35
49. Zhang JL et al (2006) High temperature PEM fuel cells. J Power Sources 160:872–891
50. Shao YY et al (2007) Proton exchange membrane fuel cell from low temperature to high temperature: material challenges. J Power Sources 167:235–242
51. Shao ZP, Haile SM (2004) A high-performance cathode for the next generation of solid-oxide fuel cells. Nature 431:170–173
52. Fuel Cell Today (2010) Fuel cells: sustainability. Fuel cell today industry review 2010. Fuel Cell Today, Royston
53. Honda Motor Company (no date) Drive FCX Clarity FCEV. http://automobiles.honda.com/fcx-clarity/drive-fcx-clarity.aspx. Accessed 1 April 2011
54. National Research Council (2008) Transition to alternative transportation technologies—a focus on hydrogen. National Academies Press, Washington
55. Greene DL et al (2008) Hydrogen scenario analysis summary report: analysis of the transition to hydrogen fuel cell vehicles and the potential hydrogen energy infrastructure requirements. Oak Ridge National Laboratory, Oak Ridge
56. Daimler (2010) Annual report 2009. http://www.daimler.com/Projects/c2c/channel/documents/1813321_DAI_2009_Annual_Report.pdf. Accessed 1 April 2011
57. Ford Motor Company (2009) 2009/10 Blueprint for sustainability. http://corporate.ford.com/doc/sr09-blueprint-summary.pdf. Accessed 1 April 2011
58. General Motors (2010) GM's fuel cell system shrinks in size, weight, cost. Testing under way on production-intent system for 2015 commercialization. http://media.gm.com/content/media/us/en/news/news_detail.html/content/Pages/news/us/en/2010/Mar/0316_fuelcell. Accessed 1 April 2011

59. Reuters (2010) Honda drives toward home solar hydrogen refueling. http://www.reuters.com/article/2010/03/13/honda-hydrogen-idUSN1212479020100313?type=marketsNews. Accessed 1 April 2011
60. Hyundai Motor Company (2010) Hyundai completes development of Tucson ix hydrogen fuel-cell electric vehicle. http://worldwide.hyundai.com/company-overview/news-view.aspx?WT.ac=PressRelease&idx=324. Accessed 1 April 2011
61. PSA Peugeot Citroën (no date) Technology and the environment. http://mediacenter.psa-peugeot-citroen.com/eng/technologie_environnement.html. Accessed 1 April 2011
62. Frenette G, Forthoffer D (2009) Economic & commercial viability of hydrogen fuel cell vehicles from an automotive manufacturer perspective. Int J Hydrogen Energy 34:3578–3588
63. Schafer A, Heywood JB, Weiss MA (2006) Future fuel cell and internal combustion engine automobile technologies: a 25-year life cycle and fleet impact assessment. Energy 31:2064–2087
64. Arico AS et al (2005) Nanostructured materials for advanced energy conversion and storage devices. Nat Mater 4:366–377
65. Du H et al (2007) Carbon aerogel supported Pt-Ru catalysts for using as the anode of direct methanol fuel cells. Carbon 45:429–435
66. Guo J et al (2006) Carbon nanofibers supported Pt-Ru electrocatalysts for direct methanol fuel cells. Carbon 44:152–157
67. Prabhuram J et al (2006) Multiwalled carbon nanotube supported PtRu for the anode of direct methanol fuel cells. J Phys Chem B 110:5245–5252
68. Lim B et al (2009) Pd-Pt bimetallic nanodendrites with high activity for oxygen reduction. Science 324:1302–1305
69. Loferski PJ (2008) Platinum-group metals. In: 2007 Minerals Yearbook. U.S. Geological Survey
70. Gordon RB, Bertram M, Graedel TE (2006) Metal stocks and sustainability. Proc Natl Acad Sci USA 103:1209–1214
71. Borgwardt RH (2001) Platinum, fuel cells, and future US road transport. Transport Res D-Tr E 6:199–207
72. Edwards PP et al (2008) Hydrogen and fuel cells: towards a sustainable energy future. Energy Policy 36:4356–4362
73. Sartbaeva A et al (2008) Hydrogen nexus in a sustainable energy future. Energy Environ Sci 1:79–85

Fuel Taxes, Fuel Economy of Vehicles and Costs of Conserved Energy: The Case of the European Union

David Bonilla

Abstract This chapter is an overview of the changes in real-world fuel economy in key countries and of recent developments in fuel taxes imposed on all fuels across the EU-Member States. Coal and gas are undertaxed but diesel and gasoline are overtaxed; however, fuel economy is directly affected by fuel taxes (prices) and not by taxes on coal. Standards on fuel economy can be interpreted as taxes on fuel and both standards and fuel taxes can be triggers for investment in alternative energy technology. Costs of conserved energy show that hybrid trucks are cost-effective to buy for freight transport operators so long as fuel costs are high. Trends in the cost of conserved energy are likely to favour investment in fuel saving technologies so long as fuel and oil prices remain high as is currently the case. Taxes on fossil fuels are one way to save fossil fuels and EU Governments are aware of the need to save fossil fuels and to reduce dependency on them. EU fuel taxes have led to improved fuel economy on EU roads.

1 Introduction

In this chapter, we examine the annual change in fuel economy (L/100 km) and in the demand for motor fuels for the US, the UK, the EU and Japan. Energy use for transport (mainly gasoline, diesel and electricity but increasingly biofuels) consumption has increased by almost 2% a year in both Japan and Europe and by 1% in the US from 1973 to 2007 [24]. This increase in fuel consumption calls for policies that can regulate the

D. Bonilla (✉)
Transport Studies Unit, School of Geography and Environment, University of Oxford, Oxford, UK
e-mail: david.bonilla@ouce.ox.ac.uk

O. Inderwildi and Sir David King (eds.), *Energy, Transport, & the Environment*, DOI: 10.1007/978-1-4471-2717-8_15, © Springer-Verlag London 2012

consumption of motor fuels. First, we investigate the historical evolution of fuel economy over the period 1949–2004.[1] For the three key regions we compare the historical evolution of motor fuel use and briefly discuss the regulatory frameworks for controlling motor fuels consumption. Second, the paper should inform about how different countries, with different manufacturing capacities, and technological mixes, have tried to reduce motor fuel use and, in some cases, to reduce CO_2 emissions produced by land transport (cars and trucks). The EU holds market power in the world energy market which brings opportunities for the diffusion of energy-efficient cars and low-carbon fuels. The impact on fuel economy (of passenger cars) of EU-wide fuel taxes is also discussed since the evidence points that higher fuel taxes (in the EU), as opposed to the US case, do encourage sustained improvements in real-world fuel economy. Finally a case study of hybrid trucks is discussed.

The literature on motor fuel use (and how to save that fuel) has been wide. It is not our purpose to discuss the entire literature on the topic and only a few articles are discussed. For the UK case Bonilla and Foxon [1] summarise the major studies for several OECD countries on fuel economy of new cars, which use a range of econometric methods. Typically, such studies use, as the dependent variable, new car fuel economy in miles per gallon (mpg) and tend to focus on gasoline fuels. Some studies measure on-road fuel economy rather than that of new cars. Three studies [1, 26, 37, 46] focusing on the OECD region report widely different elasticities ranging from −0.01 to −0.6. Two studies [3, 5] review hundreds of studies too. The role of diesel fuel economy is examined in Schipper [35, 36] and in Bonilla [2]. Bonilla [2] undermines the case for further dieselisation of the UK vehicle stock. Studies tend to give wide variations of price elasticity of fuel economy because of wide differences in functional form, period of estimation and estimation technique. Studies also give inadequate attention to fuel economy and to models of fuel economy explicitly [20]. Third, unlike Witt [45] and Greene [21] who examine selected car makes, our work includes data on aggregate fuel economy (entire national fleet of cars). Because price elasticities are so low fuel prices and taxes would need to rise considerably to have a clear negative impact on the ratio of L/100 km (on-road fuel economy).

2 The Problems of Real-World Vehicle Fuel Economy: US and the EU Member States

The US. experience regarding on-road fuel economy has received the most attention from transport economists. In general differences in fuel economy (US and EU) can be explained by the dieselisation of the EU vehicle fleet, by the market share of efficient and lighter cars and by mileage per car, as well as by the

---

[1] The European measure of fuel economy in litres per 100 km is preferred to the US MPG metric. A reduction in that ratio shows an improvement in fuel economy (fewer litres per 100 km). This is the inverse of the US measure of fuel economy in miles per gallon.

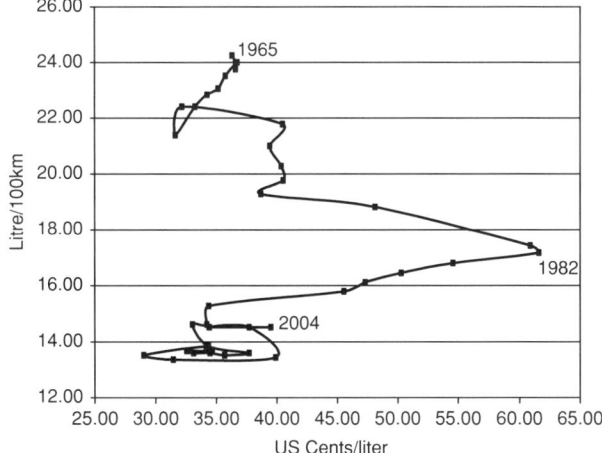

Fig. 1 Real-world fuel economy (vans, pick up trucks and sport utility vehicles, two wheel drives) 1965–2004. The US elaborated using data from US energy information administration website

high fuel tax regime in the EU. In all of these parameters the EU holds an advantage over the US. The EU has always had an advantage over the US in on-road fuel economy (and in new car fuel economy) since the car fleet has been historically lighter than that of the US. EU's average fuel economy (key nations only) is about 7.24 L/km in 2008 and the OECD average is 10.2 L/100 km [19]. US fuel economy (cars only) reaches 10.2 (in 2004) and above 14 (L/100 km; Light trucks). These levels lie above (and worse) the EU average. Figure 1 shows the time path (1965–2004) of fuel economy of on-road vehicles for large vehicles (vans, pick up trucks and SUV's) and Fig. 2 plots fuel economy of passenger cars only. Figures 1 and 2 plot-on-road fuel economy (with a year lag) on the y-axis. Fuel economy is plotted as a function of cost (US$ Cents/litre): in 1978–1982 the higher the fuel cost the lower the ratio of litres per Kilometer.

This negative relationship (gasoline price and fuel economy) varies year on year (Fig. 1 US only, Fig. 2 UK included) and it is strongest in the 1970s energy crisis when gasoline costs shoot up from 31 to 61 (US$ Cents per litre) in the 1973–1982 period. In the final years (2000–2005), increases in gasoline costs do not lead to improved (on-road) fuel economy because the CAFE standard was not tightened at the time and the lead time required for car purchases to react to higher fuel costs. The arrows show the direction of change of fuel economy in Fig. 2.

Commenting on fuel economy the US NRC argues "there is a marked inconsistency between pressing automotive manufacturers from improved fuel economy from new vehicles on the one hand and on insisting on low real gasoline prices on the other. Higher real gasoline prices, through taxes, would create both a demand for fuel efficient vehicles and an incentive for owners of existing vehicles to drive them less" [31]. The EU fuel taxes, in contrast to US ones, are high enough to create a large demand for fuel efficient cars. The role of tax is assessed in latter sections.

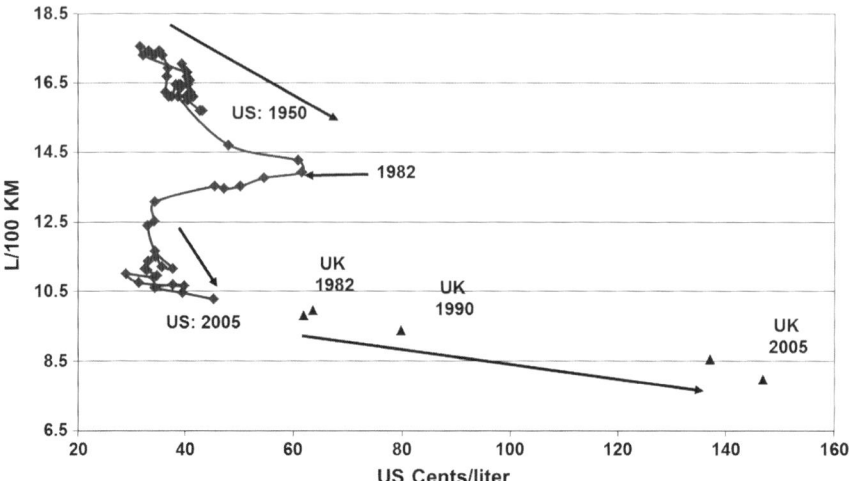

Fig. 2 The US on-road fuel economy (passenger cars) 1950–2005. Elaborated using data from US energy information administration website and from the UK (DFT) TSGB, (various years)

As is the case of light trucks vehicles (larger vehicles vans, SUV's) the passenger car sector did react to the fuel economy standard and to higher prices in the years of high gasoline costs. It is unknown the exact amount of fuel consumption that was saved because of the effect of the standard or of higher gasoline prices. In 1978–1982 large gains in on-road fuel economy are registered (Fig. 2) but the gains are cancelled in latter years.

Besides fuel prices the demand for autos and higher mileage in the last decades have contributed to the upward trend in fuel consumption. In 1970s 120 million passenger cars used US roads and these were driven by 1.3 trillion miles. In 2005, 237 million cars (passenger and light-duty vehicles) were driven 2.8 trillion miles [6]. In addition the "truck fraction" of total passenger car sales rose from 15% in the 1970s to 50% in 2006. By 2005, in the UK, 26 million cars used UK roads and these were driven 397 billion miles. In the same year, 54 million Japanese cars were driven 263 billion miles (423 billion km) on Japanese roads [30]. These large increases in distance travelled more than offset improvements at the engine level and produce even higher national consumption of fuels.

3 Comparisons Among Countries and Impacts on On-Road Fuel Economy (US, Japan, and the United Kingdom)

In 1990–2005 the rate of improvement declined (less litres per km driven) in on-road fuel economy in the US (Figs. 1 and 2). On-road fuel economy improvements, however, in the UK and Japan have been more pronounced than in the US. Out of the three countries Japan is the leader with the best fuel economy performance (of Litres per km travelled). In the period 1975–2005 US on-road fuel economy improved from 22.41 (L/100 km) to 10.6 by 2005 (EIA and Davis and Diegel 2006). For Japan the improvement from 10.38 (L/100 km) to 7.16 by 2005 (Japan data from: MLIT) is remarkable since the vehicle mix was already lighter and of a smaller size (in comparison to the US vehicle fleet) and efficient in 1975. The UK shows improvements of 10.34–7.97 (L/100 km) in 1978–2005 [7, 8] in the same period and it too had a vehicle mix which was smaller and lighter than the US one. The UK and Japan have increasingly registered large sales of heavier vehicles (with more horsepower) thus worsening their real-world fuel economy (before the financial crisis of 2008).

To sum, there are three key changes since 1970s shaping the growth of motor fuels. First, the increase in the volume of vehicles on the road, in the car distance driven and in vehicle weight have all outweighed the improvements in fuel economy for all three countries. Second, in the key three regions CO_2 emissions (passenger car only) continue to grow beyond the growth in CO_2 emissions of other sectors (residential-commercial, electricity generation). Third, Japanese CO_2 emissions (cars only) have tripled since 1975 [24] which surpasses the growth rate of CO_2 emissions in the UK (cars only). The US passenger car emissions (using data on gasoline consumption but excluding biofuels) have increased more slowly reaching 898 Mt-CO_2 from 1975 to 1243 Mt-CO_2 in 2006 [15].

4 Linking Car Sales to Fuel Economy: The US Experience

Considering all classes of LDVs (light-duty vehicles), the biggest overall increase in market share has been registered by SUVs since 1976 (Fig. 3).[2] Sales (domestic cars, non light trucks) have fallen by over 65% since 1975, and this is the biggest overall decrease among the LDVs. In contrast to domestic cars sales SUVs have registered a steady increase in sales since 1975. As of 2006, SUVs comprised around 53% of all LDV sales. Light truck sales have tripled in 1980–2005 but sales have fallen considerably during the great recession of 2008.

We see very little improvement (a decrease in the ratio of L/100 km) in the average fuel economy of cars and light trucks sold in the US since 1976 (Fig. 4). The combined average fuel economy for all new LDVs hardly changed (from 9.6–9.0 L/100 km) 1980–2004 (Fig. 4) because light trucks took a large market share and are 40% more fuel intensive than passenger cars. In recent years the financial crisis and high gasoline prices led to an income (falling household incomes) and price effect that, in turn, led consumers to chose smaller vehicles and new car fuel economy did improve to 7.9 (L/100 km, combined value in Fig. 4) from 9.06 (L/100 km) in 5 years. The following technical factors also partly

---

[2] Here, SUVs are used as light trucks less than 10,000 lbs.

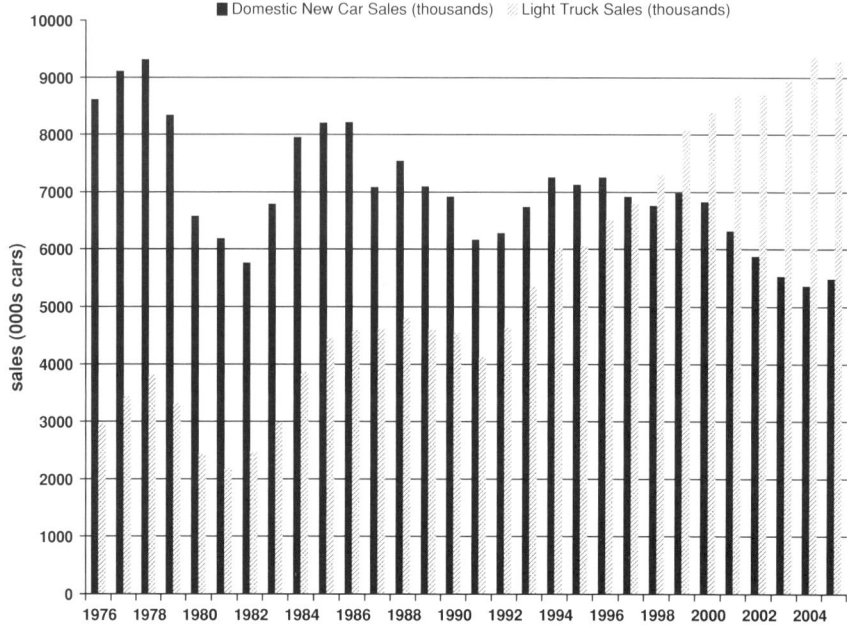

Fig. 3 Sales of domestic new car sales and light truck sales (ooo's units). *Source* US EPA, [16], [4]

explained the lack of improvement in fuel economy, relative to passenger cars, of light trucks and sport utility vehicles (SUVs): they have a relatively (1) higher vehicle mass leading to higher energy demand (2) a higher cross-sectional area causing high drag losses and (3) a higher rolling resistance due to all-terrain tyres.[3]

The debate over SUVs has kept policymakers, manufacturers and the public divided for the purposes of CAFE regulations. SUVs are high-performance four-wheel cars built on a truck chassis. The Corporate Average Fuel Economy regulations are intended to improve the average fuel economy of cars and light trucks sold in the US, has strong implications for SUVs. The National Highway Traffic Safety Administration (NHTSA) defines CAFE as the sales weighted average fuel economy expressed in miles per gallon, of a manufacturer's fleet of passenger cars or light trucks (with a gross vehicle weight rating of 3.8 tons or less) manufactured for sale in the US. In the late 1970s when CAFE was introduced the market share of light trucks (and SUVs) was almost nonexistent of the passenger car market and light trucks were allowed to meet lower fuel economy standards [40].

However, in the 1990s SUVs sales rose highlighting the need to regulate their fuel economy performance. Light-duty vehicles are those vehicles that the Environmental Protection Agency (EPA) classifies as cars and light-duty trucks.

---

[3] http://www.suv-qed.com

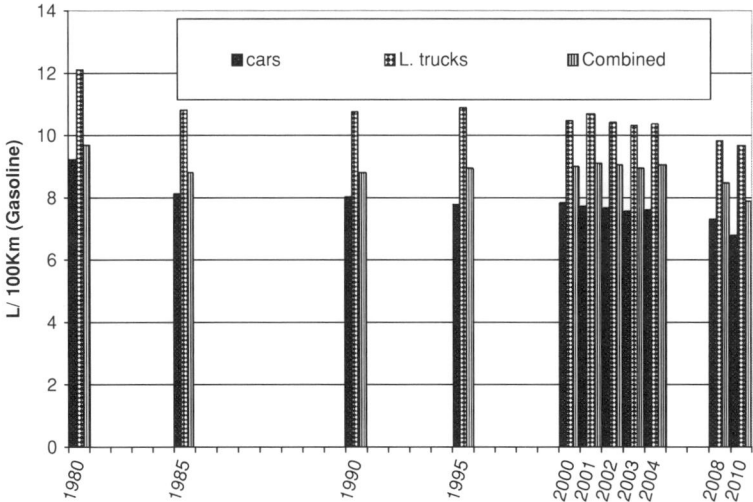

Fig. 4 Average fuel economy of new light-duty vehicles (l/100 km). *Source* US energy information administration, official energy statistics from the US government, 2010 (the official source used mpg. Data is converted to L/100 km)

Specifically, LDVs include SUVs, vans and trucks with less than 3.85 tonnes. Therefore, SUVs are classified as light trucks by the US. government, and thus are subject to the less strict light truck standards under the CAFE regulations. Currently, the CAFE requirement for light trucks is 9.8 L/100 km, versus 7.7 L/100 km for passenger cars. These figures for cars and light trucks were last changed in 2009. The new target is 6.7 L/100 km (light trucks and standard cars) to be achieved by 2016.

The EU equivalent of CAFE is the EU Voluntary Agreement (with car manufacturers) that was enforced in 1995 [9, 12]. Japan has adopted the top runner programme for its own fuel economy standard. These two agreements have been partially successful in controlling CO_2 emissions of new vehicles.

5 The Solution: The Role of Taxes for Initiating Technical Change in Vehicle Fleets

The focus of this section is on taxes on European motor fuels (gasoline and diesel) since the latter are low in the US as said above.[4] Standards on fuel economy can be taken as taxes on fuel since the implementing the standards require additional costs

---

[4] This section is partly based on findings of Freightvision [18].

to the car maker. Standards, however, spur investment in alternative power trains. A possible mechanism to reduce fuel use (and to raise average fuel economy for the national car fleet) is by imposing a tax or by pricing carbon in open trading system, see for example [11]. Taxation of fossil fuel is defined as "excise" tax (tax on producers), which is a direct tax affecting motor fuels and all freight companies whether these are small or large ones. Excise taxes are levied on fuel producers, although ultimately, the tax will be passed to consumers or freight hauliers and to passenger cars. Fuel taxes raise the nominal price of fuels and fuel taxes are mostly oil taxes. All EU-27 economies levy taxes on fuels and some (Denmark, Sweden, Finland and France) levy taxes on CO_2 emissions.

In this chapter, fuel taxes (mostly fossil fuels taxes) include:

- Traditional excise taxes set on oil products (diesel, gasoline, heavy fuel oil, coal, bio-fuels (taxes on producers and on consumers);
- tax exemptions i.e. biodiesel tax is lower than that of diesel or gasoline;
- taxes on consumers as final excise taxes (excluding VAT);
- CO_2 taxes (prices).

Taxes impact on fuel prices sending the right signals to consumers and producers and to freight companies to act and to internalise the externality. Taxes can also be used to reflect their social cost of CO_2 and other damage to the environment. Fuel tax regimes do not currently reflect the carbon content of fuels: diesel taxes in the EU are on average much higher than those of coal or gas.

This section describes taxation policy of fossil fuels (and pricing since the two are intimately linked) that can be applied, in the coming decades to 2050, to reduce (1) consumption of fossil fuels (thus improving fuel economy and reducing fossil fuel dependency), (2) GHG emissions (mainly CO_2), of passenger cars, light trucks and medium-sized ones.

5.1 Fuel Tax and Oil Price: The EU Case

Fuel taxes generally increase as oil price falls, and fall when oil price rises; this is called the fiscal drag. Fuel taxes can also work as revenue raising devices, pollution abatement ones and in general as a way to internalise externalities. Diesel taxes are around 70% of final (tax inclusive) diesel price. This final price is the retail price to car drivers and freight companies or consumers. According to Maugeri [29], crude oil prices comprise around 20–25% of the final retail price of a petroleum product (diesel). If this is so, diesel prices are mostly composed of taxes. The EU has set minimum excise tax on diesel and other products [13]. This minimum tax is set to rise to 330 £/1,000 L by 2010 [17]. The main fiscal component of diesel taxes is an excise tax, which is a fixed tax on each litre produced of diesel.

Fuel taxes can influence motor fuel consumption and fuel choice (and what is more vital vehicle size through, vehicle choice) by changing the nominal price of fuels and by making it financially attractive to consume greener (low carbon) electricity. Tax reform is vital to ensure that brown electricity does not replace green electricity, this is so, if the large-scale deployment of electric cars takes place in the EU. Taxation should be designed with the assumption that fuel taxes should reflect CO_2 emissions produced by electric cars.

Traditionally most car drivers and freight companies are unwilling to spend extra money on cleaner fuels (and vehicles) when cheaper options, such as diesel fuel, are available. All alternative fuels, however, are more expensive than conventional petrol and diesel to the user on a pre-tax basis. Hence the EU and UK Governments have ensured that environmentally friendly fuels compete equally in the market by introducing excise tax differentials in which the (lower CO_2 intensive fuel) alternative fuel is taxed at a lower rate (in the UK for example, an excise duty incentive, of 3 UK pence/L, is given to ultra low sulphur diesel). Therefore, tax instruments can influence choices for low-CO_2 fuels and thereby for fuel-efficient cars and trucks.

5.2 Interaction of Fossil Fuel Tax and CO_2 Tax

According to the OECD, a carbon tax "is an instrument of environmental cost internalisation. It is an excise tax on the producers of raw fossil fuels based on the relative carbon content of those fuels". OECD [33] this means that pricing CO_2 is equivalent to a fuel (energy) tax. According to Stern [38], there are three ways to price CO_2: through a traditional tax, through trading of emissions permits, through regulations or technical requirements which need more costly equipment, i.e. hybrid cars, trucks or biofuels. If environmental taxes (CO_2 taxes, for instance) are levied on motor fuels (mainly diesel, but some coal and gas are also taxed), the price of fuels will be higher by the increase in these taxes. The increase in the price of fuels, for instance, will be important for the decisions of car drivers, freight companies and oil production. Environmental taxes are of key importance when setting the level of fuel taxes because fossil fuels contain CO_2 and a tax on CO_2 will indirectly raise the price of a fuel and thereby the final tax level. Of course, higher CO_2 taxes and prices will induce changes in consumption patterns, leading to a shift in fuel use towards less CO_2-intensive fuels i.e. away from coal towards gas or away from diesel towards biofuels. Deciding a "sensible" CO_2 tax is no easy task but one way is to set the CO_2 tax equal to the abatement cost of a low-CO_2 fuel.

Taxes on fossil fuels are one way to save fossil fuels and EU Governments are aware of the need to save fossil fuels and to reduce dependency on them. The recent French proposal [34] for HGV in France is a case in point. Such fuels will be limited as the world's energy resources deplete which puts upward pressure on fuel prices and thereby on fuel taxes. Taxes can improve energy efficiency and they are often the most economic and readily available means for achieving superior fuel economy (less L/km travelled), lower fossil fuel dependency, for reducing greenhouse gas emissions and for reducing congestion.

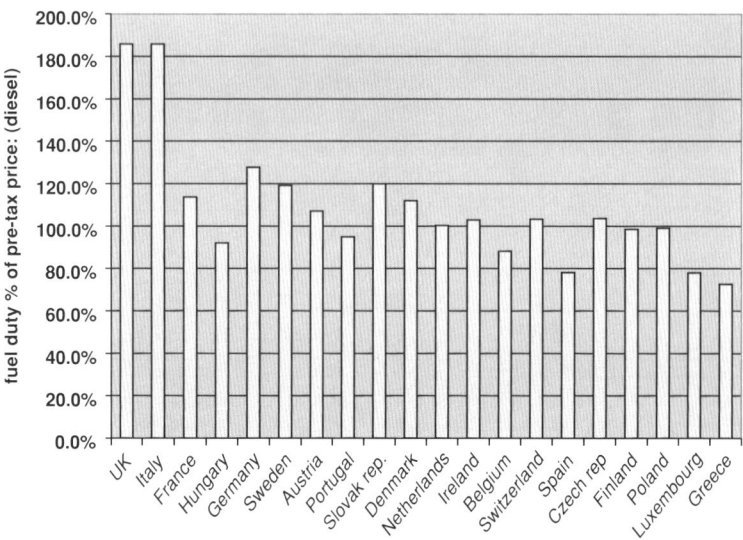

Fig. 5 Fuel duty as a fraction of diesel price (pre-tax). *Source* [23]

Taxes on fossil fuels may also be influenced by the stringency of the CO2 reduction target (Kyoto Protocol, or post Kyoto agreements), which recommended pricing CO_2 through a CO_2 trading scheme in the coming decades [38]. The IPCC 4th assessment report, and the EU [25] and the EU [12] recommend CO_2 trading. Despite this conventional taxes on coal, with a high rate of CO_2 per unit of energy, are lower than that of diesel or gasoline.

Because, collectively, the EU has market power in the world energy market (and in the car market) raising fuel taxes can reduce the demand for fossil fuels, in turn, depressing fossil fuel prices. Hence tax changes within the EU-27 Member States impact on the international price of oil, gas and coal, as well other fossil fuel inputs.

Figure 5 summarises the level of excise taxes (fraction of diesel price) in 2008 on diesel fuel, the main fuel used in private car and in trucks. Diesel taxes can be seen, Fig. 5, to vary widely within the EU-27, within those countries and overtime.

Taxes on the same fuel (say, diesel) vary also widely within the EU-27 reflecting a divergence in political opinion (Figs. 5 and 6). For example, diesel excise taxes (EU average) are about 109% of pre-tax price [23] or 130% according to Newbery [32] who uses Eurostat sources. Diesel excise duties vary from about 50% (of pre-tax price) in Greece to about 180% in Italy and the UK. The larger EU economies tend to tax diesel above the EU average (Fig. 6).

Tax rates on final diesel prices (Fig. 6) are 65% (UK), 56% (Germany) and 53% (France) (not shown in Fig. 5, [22]). The lowest tax shares exist in Greece (42%) and Spain (43.7%). In other words EU economies are highly dependent on fuel tax revenues. And reducing this dependence could imply a fall in tax revenues for EU economies.

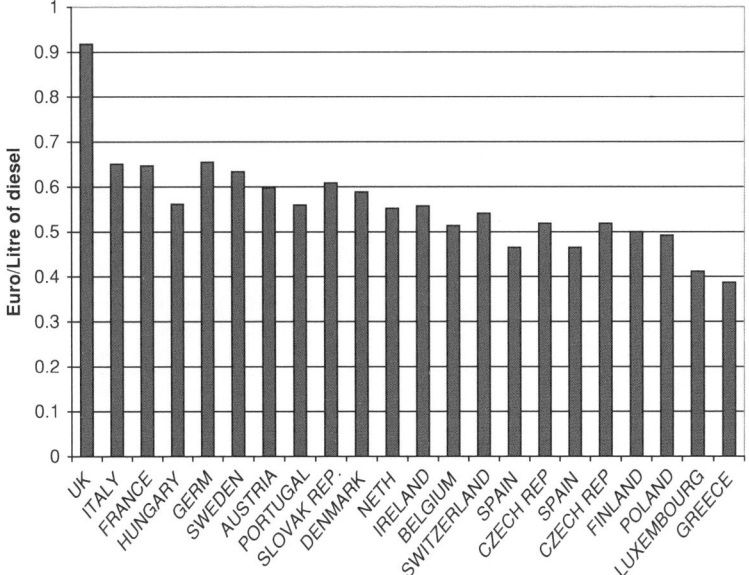

Fig. 6 Fuel taxes on diesel in selected countries. *Source* [23]

Figure 6 completes the tax picture by showing the actual level of taxes across the EU zone. Again the UK stands out and so does Greece. Both countries have different fiscal stands regarding their diesel taxation regimes.

The divergent level of fuel tax regimes (Fig. 6) distorts rates of dieselisation of national car fleets and road freight markets within EU-27 economies. In some countries the operational savings, of driving a diesel car, will be larger since the tax per litre is lower than in other EU economies.

In 2004, the EU Commission introduced a Directive targeting taxes on fossil fuels for the EU-27 in 2004 aiming to[5]:

- Reduce distortions of competition emanating from divergent rates of tax on energy products in Member states.
- Reduce distortions of competition between mineral oils and the other energy products that have not been subject to Community tax legislation up to now.
- Increase incentives for energy efficiency (to reduce dependency on imported energy and to cut CO_2 emissions); and
- Allow Member States to offer companies tax incentives in return for specific undertakings to reduce emissions.

Although taxation of fossil fuels traditionally falls under the jurisdiction of national legislation of Member States, the EU Directive (see above) has widened

---

[5] EU Commission (2003).

the scope of energy taxation across EU states. The directive in question also contains specific provisions for freight companies. For instance, it states:

> "As far as commercial diesel is concerned the Commission considers that it is necessary for member States to continue working on the Commission proposal for a directive for the harmonisation of taxation of commercial diesel fuel. The energy tax directive only provides for minimum rates of taxation and minimum rates do not remedy the problem of distortion of competition on-road haulage markets, which stems from the significant differences…" ([13], p. 2)

Ideally taxes should be set according to the level of environmental damage that each fossil fuel generates. This, however, is not the case in the EU-27. For example, taxes on coal are lower than taxes on oil and coal is the least heavily taxed in the EU as Newbery [32] finds. A tax regime that truly reflects CO_2 intensity of fuels is relevant if the objective is to reduce CO_2 emissions of energy use of transport from all sources of CO_2 emissions not just gasoline and diesel. This is specially so if car fleets are replaced by electric-powered ones. By providing the right price incentives it is possible to avoid an increase in brown electricity production and to favour green electricity for electric vehicle stocks.

In the main there are six effects of imposing a tax on fossil fuel use (and to a limited extent on fossil fuel dependency), namely:

1. Reduces vehicle km and truck km driven more than less fuel intensive modes of transport i.e. rail freight.
2. Reduces indirectly gasoline, diesel and oil prices by curtailing oil use but there is a time lag.
3. Provides incentive for limited modal shift (from road to electric rail reducing fossil fuel use).
4. Provides signals to car drivers and freight companies to choose fuel-efficient vehicles—fuel economy improvement.
5. Provides a price signal for freight companies s to adopt energy efficiency programmes.
6. Leads to competitive (dis) advantage of some EU-member States.

The EU currently promotes the 'fair and efficient pricing' policy of the European Commission [10, 12] which bears implications for fuel taxes. The EU intends to ensure that all external damage caused by road traffic is fully internalised in the price of transport. The EU Commission argues that pricing should be fair, implying that 'polluters' are "obliged to pay the marginal social cost of their activities, and efficient, giving them an economic incentive to reduce the negative effects of these activities" [17].

5.3 Setting up Fuel Taxes

How high should fuel (CO_2) taxes be set? It is usually argued that fuel taxes (or prices of fuel) should be high enough to match, or exceed, the current price of alternative technologies so that alternative fuels/technologies can compete equally. Thus one way is to raise the after tax price of diesel so that diesel no longer has a cost advantage. The current market price of 15 Euros a tonne of CO_2 is too low to provide an incentive for freight companies to adopt fuel saving measures, and costs of alternatives lie above this level.[6]

Abatement costs for alternative technologies and fuels are as follows. The UKPIA [42] study finds a wide range of cost estimates for CO_2 abatement options. Prices per tonne of CO_2 saved range from 115 (£/t CO_2) to 1,025 (£/t CO_2). Biodiesel fuels will cost 185 (£/t CO_2, and hybrid technology 115 £ tonne of CO_2. Clearly at 15 £/t CO_2 (market price) diesel excise duty is below the costs of alternative fuels in 2009 giving a cost advantage to diesel. For the transport sector the IPCC 4 AR [27] finds that most fuel savings are cost-effective at oil prices of 100 US$/bbl. (this is equivalent to a fuel tax of 0.31 £/L). In fact Greene and Schafer [22] forecast a decline in GHG of 6% after imposing a carbon price in transport.

A CO_2 abatement cost should be equivalent more or less to an increase in fuel tax (Table 1) thus persuading freight companies to switch into hybrid/electric trucks. In this way, fuel taxes, by raising final fuel prices, justify a shift into alternative options. At those prices of CO_2 such technologies can allow a reduction in fossil fuel use but not of fossil fuel dependency. Should there be a shift towards electric cars or trucks a reduction in dependency is feasible. Once the tax is established the fossil fuel saving can be obtained. For example a hybrid truck is profitable investment under high fuel taxes. Table 1 shows the equivalent increase in taxes due to a sensitivity analysis of cost per tonne saved given a technology option (hybrid trucks). Table 1 shows the cost per litre of alternative fuels.

Another example should help to clarify how costs per tonne CO_2 are calculated and how taxes are set. In one case for hybrid trucks the following is assumed from published sources. Assume the hybrid truck costs 30,000 US$ extra, emits 24 tonnes CO_2/year. Based on: truck MPG (miles per gallon) of 8.3 ($\times 0.5$; fuel economy is 50% superior to conventional engine) and truck mileage of 51,000 km a year [41]. The truck has a natural lifespan of 8 years (Davies and Diegel 2005). A 10% discount rate is added for freight company to work out present value of the investment.[7]

Equation 1 provides the cost of conserved energy (CO_2)

$$CC_k = \left[\frac{C_k}{CS_k}\right] \times (d) \times (1 - (1 + r)^{-n}) \qquad (1)$$

where,

---

[6] Another value that can be taken as a benchmark (the minimum excise duty set by the EU) is equivalent to £/t CO_2 (119 £ per tonne of CO_2 (0.33 £/L). The social cost of carbon can also be compared to the market price of carbon avoided.

[7] To determine cost of conserved CO_2 a year we compare extra investment cost as well as fuel economy levels of the options.

Table 1 Fuel taxes and cost of reducing Greenhouse gas; various options

Options	A £ per tonne CO_2 removed	B £ p/L (pre-tax price)	C Cost of conserved CO_2 (£/L) Per year (for 30 k US extra cost)	D Increase in fuel duty £/L
Conventional diesel		12		
Hybrids (UKPIA) [17]	115 (total lifetime)	31	0.03	0.03
Hybrid truck (Author calculation; Eq. 1)	360 (over lifetime)		0.12	0.12

Diesel price (inclusive of tax) in 2005 was 0.66 £/L (using 2009 exchange rates of £ to US) retail price of diesel = to 200 £/t- CO_2. Using Eq. 1 a sensitivity analysis shows that if extra costs of hybrid truck exceed 30,000 US then that technology is no longer cost competitive.

k = type of vehicle technology i.e. hybrid engines;
CC = Cost of conserved of CO_2 per year (fuel) (US\$/t CO_2);
C = extra investment cost compared to conventional diesel truck (US\$);
d = Discount rate per year to the freight company;
n = lifetime of truck (years);
CS = CO_2 or energy saved per year (tonnes).

Following Eq. 1 the cost of conserving CO_2 is valued at 39 £/t CO_2 at an extra investment of 30,000 US (hybrid truck) compared to a conventional truck.[8] The rise in the fuel tax on diesel would then close the cost advantage of diesel (other things equal final diesel prices would also increase by 36%) compared to greener fuels i.e. biofuels or low-carbon electricity.

How expensive is the cost of saving CO_2 for this technology? There are two methods to compare the cost of saving CO_2, (using a hybrid truck). One way is to compare the extra cost of the hybrid truck with the conventional truck. In other words the fiscal authorities would have to impose a negative tax (fee bate) to subsidise in some way that extra investment cost.[9] This feebate could be financed by a hike in fuel tax. A less precise way is to compare the cost of the measure with that of the minimum fuel duty on energy products in the EU [14] which is 0.33 £/L of diesel fuel (330 £/1,000 L).

Once we have simulated the tax increase (in absolute terms) it is possible to argue that, say, hybrid vehicle technologies will enter the market and reduce fuel consumption (and GHG of which CO_2 is the main gas) by at least 50% by 2050. This rests on the argument that vehicle stocks are replaced totally by 2050.

---

[8] (0.12 UK pence/L which is equivalent to almost 36% increase in fuel duty (0.12/0.33 × 100).

[9] A fee bate is a financial incentive added to or subtracted from the purchase price of a vehicles. Feebates are designed to encourage the manufacture and purchase of a fuel efficient truck. See (U. S DOE [44]) for further details.

Reductions in GHG emissions result directly from reduced fuel use. Those emission reductions would exceed the downward trend in GHG emissions resulting in declines in fossil fuel use. (The effect, however, on fossil fuel dependency is less clear.)

6 Conclusions

Energy use for transport (mainly gasoline, diesel and electricity but increasingly biofuels) consumption has increased by almost 2% a year in both Japan and in Europe and by 1% in the US. In this chapter an overview has been given on the interaction of fuel consumption (vehicle fuel economy, vehicle use) and fuel taxes. A cross country perspective of tax levels in the EU has been given. New car fuel economy in the EU and the US and Japan has improved; however, the improvements have been greatest in Japan and in the EU. The differences in fuel economy improvement rates is explained partly by differences in fuel taxes, dieselization, market penetration of compact cars and the level of mileage and consumer taste for larger vehicles.

The EU, US. and Japan have all instituted fuel economy standards incorporating their own economic philosophy and fuel tax regimes. Fuel economy is directly influenced by fuel price increases as signals are given to drivers and the tax regime (fuel taxes) in these countries affects fuel prices. EU-wide taxes are higher on gasoline and diesel than on other fuels. The evidence is clear: higher taxes have led consumers to be much more aware of the fuel economy of their cars. Taxes on coal and gas are lower than on gasoline but the electrification of the vehicle stock in the EU will require tax reform to control motor (fossil) fuel consumption and CO_2 emissions. Increasing taxes can lead to lower costs of adoption of power train technologies (trucks) such as the hybrid engine. The case study demonstrates that it is cost-effective to adopt these hybrid vehicles under higher fuel prices following the imposition of a fuel tax.

Acknowledgments ICERT—Institute of Carbon and Energy reduction in Transport, Oxford Martin School, University of Oxford, has provided funding for this work.

References

1. Bonilla D, Foxon T (2007) Estimating demand for new car fuel economy in the UK 1970–2004 using a two-stage error correction model. Working paper, No. 23, environmental economy and policy research. Department of land economy, University of Cambridge
2. Bonilla D (2009) Fuel demand in UK roads and dieselisation of fuel economy. Energy policy 37:3769–3778 (Special Issue of Carbon in Motion)
3. Brons M, Nijkamp P, Pels E, Rietveld P (2007) A meta analysis of the price elasticity of gasoline demand. A SUR approach. Energy Econ 30:2105–2122

4. CAFÉ Overview, Frequently asked Questions, NHTSA, available online at www.nhtsa.dot. gov/cars/rules/cafe/overview/htm
5. Dahl C (1995) Demand for transportation fuels: a survey of demand elasticities and their components. J Energy Lit 1(2):3–27
6. Davis SC, Diegel SW (2006): Transportation energy data book, 24th Edn ORNL-6973. Oak Ridge 24 National Laboratory, Oak Ridge
7. Department of Trade and Industry (now BERR), United Kingdom. Digest of United Kingdom energy statistics (London: HMSO, Various Years)
8. Department of Transport, United Kingdom. Transport Statistics Great Britain (London: HMSO, Various Years)
9. EC (European Commission) (1999) Official Journal of the European Communities L, 350 28 Commission recommendation of 5 February 1999 on the reduction of CO_2 from passenger cars (notified under document number c 107. (Text with EEA relevance). 1999/125/EC
10. EC (1995), Green paper on towards fair and efficient pricing in transport, 1995
11. EC (2000) Green paper on greenhouse gas emissions trading within the European Union, 2000
12. EC (2001, 2006 revised) White Paper—European transport policy for 2010: time to decide, 2001
13. EC (2003) Energy taxation: Commission welcomes Council adoption of new EU rules. Brussels, 27 Oct 2003. Press release. Available at: http://ec.europa.eu/taxation_customs/taxation/excise_duties/energy_products/legislation/index_en.html
14. EC (2009) Eurostat. www.eurostat.eu
15. EIA (2009) Energy Information Administration (US.) website at: http://www.eia.gov
16. EPA (2011) US Environmental Protection Agency. Website at www.epa.gov
17. EU (2008) Strategy for an internalisation of external costs, SEC. COMM: 2209
18. Freightvision (2009) Forecasts, Preliminary Vision, Conflict and Measures. Management Summary III, 1–64. Vienna, Austria. DG-Energy and Transport-7th framework programme
19. Fulton L, Cazzola P, Cuenot F (2009) Pricing and taxation related policies to save oil in the transport sector. Energy policy 37:3758–3768
20. Graham D, Glaister S (2002) The demand for automobile fuel: a survey of elasticity. J Transp Econ Policy 36:1–26
21. Greene DL (1991) Short run pricing strategy to increase corporate average fuel economy. Econ Inq 29:101–114
22. Greene DL, Schafer A (2003) Reducing greenhouse gas emissions from US. transportation. PEW Center, Arlington, p 68
23. IEA (2008) Fuel prices and transport indicators workshop on new energy indicators for transport: the way forward 28–29 January 2008. Energy Statistics Division Paris, 29th January 2008. Presentation by Lavagne d'Ortigue, O. (IEA official)
24. Institute of Energy Economics (2010) Japan handbook of energy economics and statistics. The Energy Conservation Center, Tokyo
25. IPCC 4th assessment report. 2007 Summary of policy Makers: Synthesis Report
26. Johansson O, Schipper L (1997) Measuring the long run fuel demand of cars: separate estimations of vehicle stock, mean fuel intensity, and mean annual driving distance. J Transp Econ Policy 31:277–292
27. Kahn Ribeiro S, kobayashi S, Beuthe M, Gasca J, Greene D, Lee D, Muromachi Y, Newton PJ, Plotkin S, Sperling D, Zhou PJ. In: Metz B, Davidson OR, Bosch PR, Dave R, Mayer LA (eds) climate change 2007 Mitigation, contribution of working group III to the Fourth assessment report of the intergovermental panel on climate change. Cambridge university Press, Cambridge
28. Maples J (2005) Fuel economy of the light-duty vehicle fleet. energy information administration. Official Energy Statistics from the US Government
29. Maugeri L (2006) The age of oil the mythology and history, and future of the world's most controversial resource. Praeger London 1–340
30. Ministry of land infrastructure and transport (Japan). http://www.mlit.go.jp/en/index.html
31. National Research Council (2002) Effectiveness and impacts of corporate average fuel economy standards. Committee on effectiveness and impact of corporate average average fuel economy standards. 1–184

32. Newbery D (2005) Why tax energy towards a more rational energy policy. Energy J 26(3):1–39 (Special issue of European electricity liberalisation)
33. OECD (2009) Glossary of Statistical Terms. Available at: http://stats.oecd.org/glossary/search.asp
34. Rocard M (2009) Rapport de la conference des experts et de la table ronde sur la contribution Climat et Energie. Report, 1–83
35. Schipper L, Marie Lilliu C, Fulton L (2002) Diesels in Europe:analysis of characteristics, usage patters, enegy savings and CO_2 emissions implications. J Transp Econ Policy 36:305–340
36. Schiper L (2009) Fuel economy, vehicle use and other factors affecting CO_2 emissions from transport. Energy Policy 37:3711–3713
37. Small KA, Van Dender K (2006) Fuel efficiency and motor vehicle travel: the declining rebound effect, UCI department of economics. Working Paper 05 06 03
38. Stern N (2009) A blueprint for a safer planet: how to manage climate change and create a new Era of progress and prosperity. Bodley Head, London
39. The Hypercar Concept (2001), http://www.suv-qed.com (author not known)
40. Tietenberg T, Lewis L (2009) Environmental and Natural Resource Economics, 8th edn. Pearson international Edition, Boston
41. UK DfT (2005) Freight transport statistics, London, UK
42. UKPIA (UK Petroleum Institute Association) (2008). Future Road Fuels. Report prepared for the UKPIA by Bishop N, Watson M 1–52 London
43. US Department of transportation, Highway Statistics, available online at http://www.fhwa.dot.gov/policy/ohpi/hss/hsspubsarc.cfm
44. U.S DOE (Department of Energy) (1995) Effects of Feebates on vehicle fuel economy, carbon dioxide emissions and consumer surplus. Technical report two energy efficiency in the us economy. Prepared by the energy analysis programme at Lawrence Berkeley laboratory. Principal authors area Davis WB, Levine M, Train K
45. Witt R (1997) The demand for car fuel economy: some evidence for the UK. Appl Econ 29:1249–1254
46. Zachariadis T, Clerides S (2006) The impact of standards on vehicle fuel economy: an international panel analysis. Working paper. University of Cyprus

Toward a Global Low Carbon Fuel Standard for Road Transport

Daniel Sperling and Sonia Yeh

Abstract A new policy instrument, known as a low carbon fuel standard (LCFS), is a promising approach to decarbonize transportation fuels. An LCFS has several important features: it applies a life cycle carbon intensity standard, incorporates market mechanisms by allowing credit trading, and targets all transport fuels. A harmonized international framework is needed that builds on newly enacted LCFS policies adopted in California and the European Union.

Keywords Performance standard · Life cycle emissions · Biofuels · Alternative fuels

1 Introduction

Vehicles, planes, and ships remain almost entirely dependent on petroleum and account for almost one-fourth of all greenhouse gas (GHG) emissions in the world. A central strategy to reduce GHG emissions, as well as to reduce oil use, is to decarbonize transport fuels. No country other than Brazil has been successful at replacing petroleum fuels in the transport sector. Many countries, especially the US, have jumped from one alternative fuel to another. The *fuels du jour* in the 1980s and 1990s were coal liquids, methanol, compressed natural gas, and electricity for battery vehicles. Early in this decade it was hydrogen, followed by corn ethanol, and now electricity for plug-in hybrid electric vehicles. The *fuel du jour*

D. Sperling (✉) · S. Yeh
Institute of Transportation Studies, University of California,
2028 Academic Surge Building, One Shields Avenue, Davis, CA 95616, USA
e-mail: dsperling@ucdavis.edu

O. Inderwildi and Sir David King (eds.), *Energy, Transport, & the Environment*,
DOI: 10.1007/978-1-4471-2717-8_16, © Springer-Verlag London 2012

phenomenon has much to do with oil market failures, overblown promises, the power of incumbents, and the short attention spans of governments, the mass media, and the public [10]. The ad hoc approach of the past needs to be replaced by durable policies that do not depend on government picking winners. A new approach is needed that would ideally be fuel neutral, performance-based, and harness market forces. Such an approach has emerged in Europe and the United States. It is furthest along in California, where it was adopted on April 23, 2009.

1.1 Greenhouse Gas Performance Standard for Low Carbon Fuels

The California Low Carbon Fuel Standard (LCFS) is imposed on all transport fuel providers, including refiners, blenders, producers, and importers. Aviation and certain maritime fuels are excluded either because the state does not have authority over them, or because including them presents logistical challenges.

California's LCFS requires a 10% reduction in the greenhouse gas intensity of transport fuels by 2020. The LCFS metric is total carbon and other greenhouse gases emitted per unit of fuel energy. The standard captures all GHGs emitted in the life cycle, from extraction, cultivation, land use conversion, processing, transport and distribution, and fuel use. Although upstream emissions account for only about 20% of total GHG emissions from petroleum, they represent almost the total life cycle emissions for biofuels, electricity, and hydrogen. Upstream emissions from extraction, production, and refining also comprise a large percentage of total emissions for the very heavy oils and tar sands that oil companies are increasingly embracing to supplement limited supplies of conventional crude oil. The LCFS is the first major public initiative to codify life cycle concepts into law, an innovation that will become more widespread as climate policies are pursued more aggressively.

To implement the LCFS, each fuel supplier must meet the GHG intensity standard that declines each year (reaching a 10% reduction in 2020, in the case of California). To maximize flexibility and innovation throughout the energy sector, the LCFS allows for the trading and banking of emission credits. An oil refiner could, for instance, buy credits (or the fuels themselves) from biofuel producers. Or they could buy credits from an electric utility that sells power to electric vehicles. Those companies that are most innovative and best able to produce low-cost, low-carbon alternative fuels would thrive. The combination of regulatory and market mechanisms makes the LCFS more robust and durable than a pure regulatory approach and, as indicated below, more acceptable and effective than a pure market approach. Companies failing to meet the standard, could face monetary penalties and/or legal action (in California, the State might take action via its administering agency, the Air Resources Board).

The European Union first unveiled an LCFS proposal at about the same time as California in early 2007. In December 2008, the European Parliament adopted a revised Fuel Quality Directive (FQD) that incorporated a low carbon fuel standard [3].

The FQD requires fuel suppliers to reduce life cycle GHG emissions by up to 10% by 2020. The 10% reduction is broader than the California LCFS by allowing credit from upstream reductions in gas flaring and venting and for the use of carbon capture and storage (CCS) technologies. It also allows the purchase of credits under the Clean Development Mechanism (CDM) of the Kyoto Protocol. Upstream emission reductions, CCS, and the CDM can be used to meet up to 4% of the 10% requirement.

2 The Shortcomings of Volumetric Mandates

In the past, volumetric mandates have been the preferred policy to reduce the use of petroleum fuels. The U.S. adopted a volumetric mandate for biofuels in 2005 and strengthened it in December 2007 as part of the Energy Independence and Security Act (EISA). This renewable fuels standard (RFS) requires that 36 billion gallons of biofuels be sold annually by 2022, of which 21 billion gallons must be "advanced" biofuels and the other 15 billion gallons can be corn ethanol. The advanced biofuels are required to achieve at least 50% reduction from baseline life cycle GHG emissions, with a subcategory of cellulosic biofuels required to meet a 60% reduction target. These reduction targets are based on life cycle emissions, including emissions from indirect land use.

Similarly, the UK's Renewable Transport Fuel Obligation (RTFO) targets 3.25% of all transport fuel sold in the UK to come from a renewable source by 2009–2010, and to reach 5% in 2013–2014. The European Union's Biofuel Directive (BD) initially set a target of 5.75% of biofuels by 2010 and 10% biofuels by 2020, but has since broadened it to include all renewable fuels and renamed as the Renewable Energy Directive (RED).

Volumetric biomass mandates have various shortcomings: first, they target only biofuels and not other alternatives. Second, setting GHG reduction targets within the volumetric mandates, as the US does with its RFS program, is a positive but clumsy effort to reduce GHGs. It forces biofuels into a small number of fixed categories and thereby stifles innovation. Once the regulatory agency concludes that certain biofuel pathways meet the specified GHG reduction target, there is little incentive for further improvement. As a result, there is less incentive to use very low-carbon materials, such as waste biomass, or adopt sustainable farming and management practices that reduce direct and indirect land use emissions [12].

The LCFS is a more robust and ultimately efficient approach than volumetric mandates. Unlike the RFS and other biofuel programs, the LCFS will encourage oil companies to pursue a fuller set of low-carbon fuel options. It will encourage companies to integrate their R&D portfolios across all energy options, including wind, solar, hydrogen, and natural gas, along with carbon capture and sequestration technologies. Compared to biofuel mandates, an LCFS has three key advantages: it inspires industry to pursue innovation aggressively; it is flexible and performance-based so that industry, not government, picks the winners; and it directly targets *actual* life cycle GHG emissions associated with the production, distribution, and use of the fuel from the source to the vehicle.

3 Why Not Carbon Taxes or Cap-and-Trade?

Carbon taxes, or closely related cap and trade programs, should be central to any regional or national initiative to reduce GHG emissions. Indeed, even a modest carbon tax works well with electricity generation. Electricity suppliers can choose among a wide variety of commercially available low-carbon energy sources, including nuclear power, wind, natural gas, and even coal with carbon capture and sequestration. A tax of as little as $25 per ton of carbon dioxide would increase the retail price of electricity made from coal by about 17% (in the US), which would be enough to motivate electricity producers to seek lower carbon alternatives. The result would be innovation, change, and decarbonization. Politically plausible carbon taxes promise to be effective in transforming the electricity industry [1, 11].

But transportation is a different story. A $50-a-ton tax, which would raise gasoline prices about 45 cents per gallon. This is well above what U.S. politicians have been considering but, judging by European experience, would produce very little response from consumers or producers (many European countries have had transport fuel taxes equivalent to $4 per gallon for many years, with virtually no effect in decarbonizing fuels–though the taxes are not based on carbon content). Recent studies suggest California's LCFS can be met at costs lower than or comparable to oil priced at $60–100 per barrel [2, 15] and that "alternative liquid fuel technology can be deployable and supply a substantial volume of clean fuels for U.S. transportation at a reasonable cost" [8]. However, because of market failures, uncertain oil prices, and risk aversion [6], companies are unlikely to invest in new fuel technologies and infrastructure for alternative fuels. More direct, performance-based policy instruments are needed to overcome carbon lock-in [13].

While some economists characterize the LCFS approach as second best because it is not as efficient as a carbon tax or cap and trade [7], given the huge barriers to alternative fuels and the limited impact of increased taxes and prices on transportation fuel demand, the LCFS appears to be the most practical way to begin the transition to alternative fuels. It is conceivable that in the long run when advanced biofuels and electric and hydrogen vehicles are commercially viable and overcome the infrastructure hurdle, cap and trade, and carbon taxes will become effective policies with the transport sector. But until then, more direct forcing mechanisms, such as a low carbon fuel standard for refiners, will likely be far more effective at stimulating innovation and overcoming the many barriers to change.

4 Scientific Uncertainty and Indirect Effects

Perhaps the most controversial and challenging issue in calculating life cycle emissions is indirect land use changes. When biofuel production increases, land is diverted from agriculture to energy production. The displaced agricultural production is replaced elsewhere, bringing new land into intensive agricultural production. By definition, this newly farmed land was previously used for less-intensive purposes. It might have been pasture land, wetlands, or perhaps even rainforest. Because a vast amount of carbon is sequestered in the roots and vegetation below and above ground—effectively storing more than twice the carbon contained in the entire atmosphere—any change in land usage can have a large effect on carbon releases.

If biofuel production does not result in land-use changes—for instance when fuel is made from crop and forestry residues or municipal waste—then the indirect land-use effects are small or even zero. But if rainforests are destroyed or peat is burned, then the carbon releases are huge [4, 5]. In the more extreme cases, these land-use shifts can result in each new gallon of biofuel releasing several times as much carbon as the petroleum fuel it is replacing. In the case of corn ethanol, preliminary analyses suggest that indirect land use changes may add 40% or more GHG emissions per unit of ethanol [2, 14]. Cellulosic fuels would have a much smaller effect (mostly because of much higher yields per unit of land).

The problem is that scientific studies have not yet adequately quantified the indirect land-use effect. It is a classic challenge of how to handle scientific uncertainty in a policy context. For biofuel producers, the prudent approach is to focus on biofuels with minimal indirect land-use effects: fuels created from wastes and residues or from degraded land and produced from algae and renewable hydrocarbons.

A broader concern is the environmental and social sustainability of biofuels. Unlike the biofuel program in the US, the European renewable energy mandates are met in large part through imports. In the UK, as of December 2008, 97% of the renewable fuels were imports—biodiesel made from American soy, rapeseed from Germany, and palm oil from Malaysia and Indonesia, and ethanol made from Brazilian sugarcane. In the EU, most of biofuel imports are ethanol from Brazil and palm oil from Malaysia and Indonesia [9]. Environmental groups have raised concerns about the local environmental and social impacts of these imported fuels. As a result, the Netherlands, UK, and Germany are adopting sustainability standards for biofuels. These sustainability standards typically address issues of biodiversity, and soil, air, and water quality, as well as social and economic conditions of local communities and workers. They require reporting and documentation but lack real enforcement. The effectiveness of these standards remains uncertain. More science-based research and technical analysis are needed to better quantify these diverse market-mediated effects and cumulative environmental damages. These issues about indirect land use change and environmental and social sustainability are equally relevant to debates over cap and trade and volumetric mandates, however, and thus are not an argument to prefer them over LCFS.

5 Energy Security and Climate Policy

Those more concerned with energy security than with climate change might be skeptical of the LCFS. They might fear that the LCFS disadvantages high-carbon alternatives such as tar sands and coal liquids. That concern is valid, but disadvantaging does not mean banning. Tar sands and coal liquids could still be introduced on a large scale with an LCFS. The LCFS would require producers of high-carbon alternatives to be more energy efficient and to reduce carbon emissions associated with production and refining. Producers could do so by using low-carbon energy sources for processing energy and could capture and sequester carbon emissions. They could also opt for ways of converting tar sands and coal resources into fuels that facilitate carbon capture and sequestration. For instance, gasifying coal to produce hydrogen allows for the capture of almost all the carbon, since none remains in the fuel itself. In this way, coal could be a nearly zero-carbon option.

6 Conclusions

The ad hoc policy approach to alternative fuels has largely failed. A more durable and comprehensive approach is needed that encourages innovation and lets industry and consumers pick winners. The LCFS does that. It provides a single GHG performance standard for all transport fuel providers and all transport fuels, and it uses credit trading to ensure that the transition is accomplished in a more economically efficient manner.

Although one might prefer more pure market instruments, such as carbon taxes and cap and trade, those instruments are not likely to be effective in the foreseeable future with transport fuels. At the envisioned (and politically plausible) price and cap levels, large investments in electric vehicles, plug-in hybrids, hydrogen fuel cell vehicles, and advanced biofuels would not be induced. More direct policies, such as the LCFS, are needed to stimulate innovations in low-GHG alternative fuels.

While the LCFS would be highly effective on its own, to be most effective it must be coupled with other policies—those that address the amount of fuel consumed (since the LCFS is an intensity standard), accelerate the initial provision of infrastructure to supply low-carbon fuels, and assure vehicles are available to use the low-carbon fuels.

Acknowledgments Sperling is a board member of the California Air Resources Board (CARB), the government agency that enacted California's LCFS on April 23, 2009. Yeh received financial support from CARB to lead research effort to support implementation of California's LCFS. The view and opinions herein are those of the authors alone and do not necessarily represent the views of the sponsors or any other organization.

References

1. Burtraw D (ed) (2007) Cap and trade policy to achieve greenhouse gas emission targets. Civil Society Institute, Newton
2. CARB (2009) Staff report: proposed regulation to implement the low carbon fuel standard—initial statement of reasons. Staff report, vol 1. California Air Resources Board. http://www.arb.ca.gov/fuels/lcfs/030409lcfs_isor_vol1.pdf

3. EC (2008) European Parliament legislative resolution of 17 December 2008 on the proposal for a directive of the European Parliament and of the Council amending Directive 98/70/EC as regards the specification of petrol, diesel and gas-oil and introducing a mechanism to monitor and reduce greenhouse gas emissions from the use of road transport fuels and amending Council Directive 1999/32/EC, as regards the specification of fuel used by inland waterway vessels and repealing Directive 93/12/EEC (COM(2007)0018–C6-0061/2007–2007/0019(COD))

4. Fargione J, Hill J, Tilman D, Polasky S, Hawthorne P (2008) Land clearing and the biofuel carbon debt. Science 319(5867):1235–1238

5. Gibbs HK, Johnston M, Foley JA, Holloway T, Monfreda C, Ramankutty N, Zaks D (2008) Carbon payback times for crop-based biofuel expansion in the tropics: the effects of changing yield and technology. Environ Res Lett 3:034001

6. Greene DL, German J, Delucchi MA (2009) Fuel economy: the case for market failure. In: Sperling D, Cannon JS (eds) Reducing climate impacts in the transportation sector. Springer Science + Business Media, Singapore

7. Holland S, Hughes J, Knittel C (2009) Greenhouse gas reductions under low carbon fuel standards? Am Econ J Econ Policy 1(1):106–146

8. NRC (2009) Liquid transportation fuels from coal and biomass: technological status, costs, and environmental impacts. America's Energy Future Panel on Alternative Liquid Transportation Fuels; National Academy of Sciences; National Academy of Engineering; National Research Council. http://www.nap.edu/catalog.php?record_id=12620

9. OECD (2008) Economic assessment of biofuel support policies. Organisation for Economic Co-operation and Development, Paris

10. Sperling D, Gordon D (2009) Two billion cars: driving toward sustainability. Oxford University Press, Oxford

11. Stavins R (2008) Addressing climate change with a comprehensive U.S. Cap-and-Trade System. Oxf Rev Econ Policy 24(2):298–321

12. Tilman D, Socolow R, Foley JA, Hill J, Larson E, Lynd L, Pacala S, Reilly J, Searchinger T, Somerville C, Williams R (2009) Beneficial biofuels—the food, energy, and environment trilemma. Science 325:270–271

13. Unruh GC (2002) Escaping carbon lock-in. Energy Policy 30(4):317–325

14. US EPA (2009) Regulation of fuels and fuel additives: changes to renewable fuel standard program. 40 CFR Part 80 U.S. Environmental Protection Agency

15. Yeh S, Lutsey N, Parker NC (2009) Assessment of technologies to meet a low carbon fuel standard. Environ Sci Technol 43(18):6907–6914. doi:10.1021/es900262w

Part III
Urban Mobility

Foreword
David Banister

The city provides great opportunities for clean, low energy and carbon-free transport for both passengers and freight. Such thinking is important, as cities also provide the living space for over half of the global population and they are likely to continue to provide the main sources of global employment, and act as the centres for education, innovation, knowledge and government. Their success is central to the continued globalisation of the world economy, for the wealth of individuals and for reductions in social inequalities. The environmental issues now form a much stronger element to complement the economic and social factors that together constitute sustainable development.

This group of chapters summarises the thinking from four different perspectives. Chapter 17 examines the role of new technologies in providing a well connected city through the introduction of flexible, light, short range electric vehicles that are affordable and autonomous. This is the optimistic view of urban mobility and perhaps provides the answer to the need to reinvent the car for cities in developed countries. Chapter 18 takes a more realistic view that technology needs to be supported by complementary actions that encourage high quality urban design, strong governance and leadership, with a clear move away from the dominant role that the car has played in structuring the city. The car is still seen to be important, but stronger integration with urban planning and design is proposed, together with a more sharing culture of car clubs, so that the number of cars in cities is reduced.

It is in Chap. 19 that the more pragmatic view of transport in cities is presented through an extensive case study that outlines the Mayor's Transport Strategy in London. The ambitious aim here is to deliver high quality public transport that is both accessible and affordable to all, and will help in regeneration and achieve substantial carbon reductions. The perspective given is integrative, as it covers all types of transport and urban form, and it links in with the wider social, economic and environmental issues. The quality of transport in London has been transformed over the last 10 years. The final chapter is more theoretical, and it demonstrates that efficiency gains can be made from transit through increased patronage, and if these gains are passed onto passengers then a road user charge should lead to a reduction in the overall equilibrium cost of travel. But such a reduction would not take place if the operators took the efficiency gains as additional profits. The conclusion reached is that transit fares need to be regulated when a road user charge is introduced so that benefits are distributed to users as well as operators.

Taken as a whole, these papers give a wide ranging perspective on the opportunities for more efficient, clean and low energy transport in cities, covering technology, pricing, design, governance and leadership, as well as addressing many of the actions that have already been introduced in cities.

David Banister
Director of the Transport Studies Unit
University of Oxford

Personal Urban Mobility
for the Twenty-First Century

Christopher E. Borroni-Bird

Abstract Mobility enhances our lives by allowing contact with others, offering new experiences and supporting the exchange of ideas and goods. Cities facilitate connecting people and goods but their roads are becoming increasingly crowded. This is particularly acute in emerging markets, with increased migration from rural to urban areas and with increasingly wealthy people aspiring to automobile ownership. Despite heavily congested roads, there is still a desire for personal mobility in urban centers because no other means of transportation has so far offered the same mix of freedom, comfort, utility and security as the automobile. However, a new kind of vehicle, based on having a small footprint, being electrically powered and being able to connect with other vehicles will be required in the future to provide personal mobility while addressing the societal challenges of energy, environment, safety, congestion and land use for parking. This new DNA for the automobile, based on electrification and connectivity, will also improve the integration of personal and public transport so that a new mobility system, offering the best features of both, can be realized.

1 Introduction

A typical automobile weighs about 20 times as much as its driver, can travel over 500 km without refueling and attain speeds above 150 km per hour, requires more than 15 m^2 of space for parking and is parked more than 90% of the time [1]. They are designed to meet almost all conceivable needs for moving people and

C. E. Borroni-Bird (✉)
General Motors, Detroit, MI, USA
e-mail: christopher.borroni-bird@gm.com

O. Inderwildi and Sir David King (eds.), *Energy, Transport, & the Environment,*
DOI: 10.1007/978-1-4471-2717-8_17, © Springer-Verlag London 2012

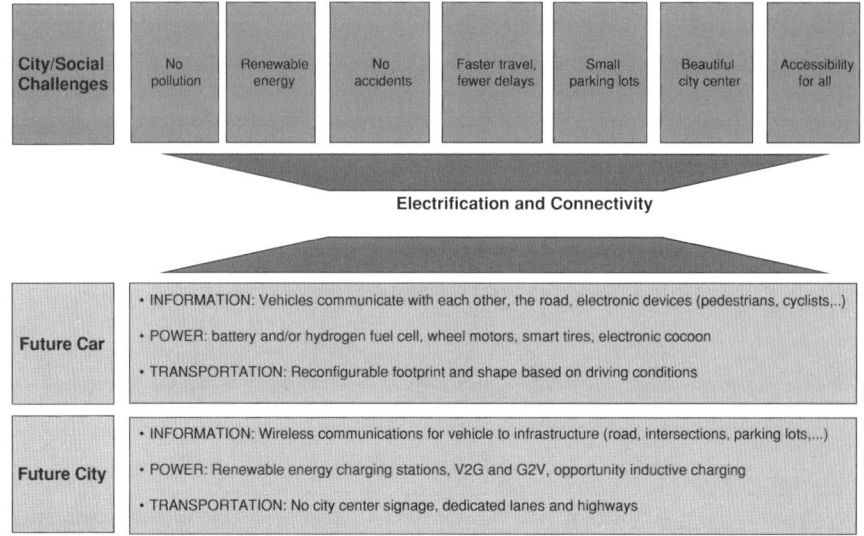

Fig. 1 Preserving personal urban mobility in a sustainable manner

cargo over long distances. No other means of transportation offers the same valued combination of safety, comfort, convenience, utility and choice of route and schedule. However, this flexibility can drive cost, mass, energy and space inefficiency especially when there is only one occupant in the vehicle. There may be situations where the vehicle user is willing to trade some of this flexibility for a smaller and lighter vehicle, especially when primarily used in an urban environment, which is where most of the world's people now live and where most of the driving is concentrated.

The societal challenges posed by the widespread use of automobiles in an urban environment include energy consumption and greenhouse gas emissions, air and noise pollution, crashes, traffic congestion, limited availability of land for parking, and limited accessibility for all city dwellers to personal transportation. A proposed solution that preserves personal urban mobility in a sustainable manner is described in Fig. 1.

This possible future solution proposes the use of lightweight, battery-powered, adaptive and connected vehicles that can operate within a wirelessly networked urban mobility system. In order to exploit the emerging technologies of electrification and connectivity, this solution will require reinventing not only the automobile but also the supporting transportation infrastructure.

This transition to green, smart, and connected vehicles is desirable because amid growing concerns regarding energy security, global climate change, and traffic congestion, it is clear that without transformative improvements the 120-year-old foundational DNA of the conventional automobile is not sustainable for dense urban environments where an increasing majority of the world's population lives.

2 Mobility Needs for Tomorrow's Cities

In the late nineteenth century, when the horseless carriage began to replace the horse, most of the world's people lived in rural areas, villages, and small towns so cars were designed for that environment. However, the majority of people now live in cities, many of which are very densely populated, and the percentage will increase—dramatically in the case of the developing world. From 1950 to 1990 the number of Asians living in urban areas increased from 234 million to 1 billion and is projected to reach 3.4 billion by 2025 [2].

According to the United Nations, it is expected that all of the world's projected population increase in the future will occur in urban areas and that 60% of the world's population will live in urban areas in 2030, up from 50% in 2007 (and from 29% in 1950) [3]. Because urban residents tend to be wealthier than rural dwellers, it is projected that they will consume a disproportionate amount of the world's energy and materials resources and will be responsible for over 75% of the world's greenhouse gas emissions in 2030 [2]. Since greenhouse gas emissions and the consumption of materials and energy will, increasingly, be driven by urban populations this means it will not be possible to address these global concerns without creating urban solutions.

Rapidly developing cities in countries such as China and India, which represent a vast emerging market for vehicles and mobility systems, have extremely high population densities when compared with Western cities. As presented in Fig. 2, Mumbai's 50,000 people per square kilometer population density is several times higher than in, say, London or New York City [4] and data from New York clearly shows a negative correlation between population density and vehicle ownership. Even though wealth is evenly spread among the different boroughs of New York City the residents of Manhattan are less likely to own a vehicle than those living in the other, less densely populated boroughs [5]. Generalizing, research does show that vehicle ownership drops by about a third as residential density doubles [6]. Reasons for this are easy to find: access to an effective public transport system and difficulties in driving a car, such as congestion that causes unpredictable travel times and parking that is both expensive and difficult to find, generally make it less likely for city center residents to own and use an automobile. London and some other cities have further deterred automobile usage in the city center by introducing congestion charging.

Higher population density also translates into shorter trips and lower driving speeds than those occurring within more dispersed settlement patterns. More than half of commuters in cities all around the world travel less than 50 km a day [7]. A vehicle that can reliably provide 50 km range, recharged at work to give up to a 100-km daily range, should be sufficient for over 75% of commuters. Traffic speeds in some city centers can be as low as 15 km/h and in most city suburbs the average travel speeds remain below 60 km/h [8]. Recent data, collected by General Motors, from several thousand drivers from six cities around the world confirms this and is shown in Fig. 3. A vehicle that can achieve 60 km/h can actually

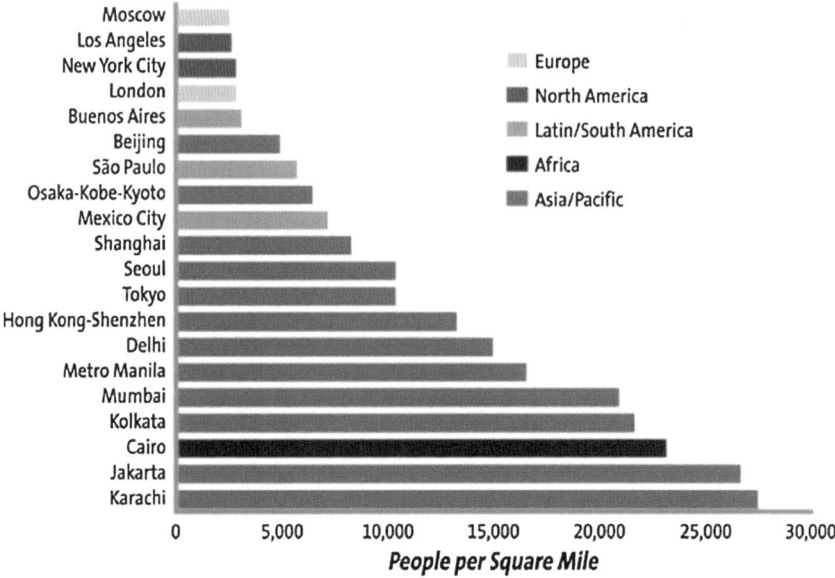

Fig. 2 Living density world wide

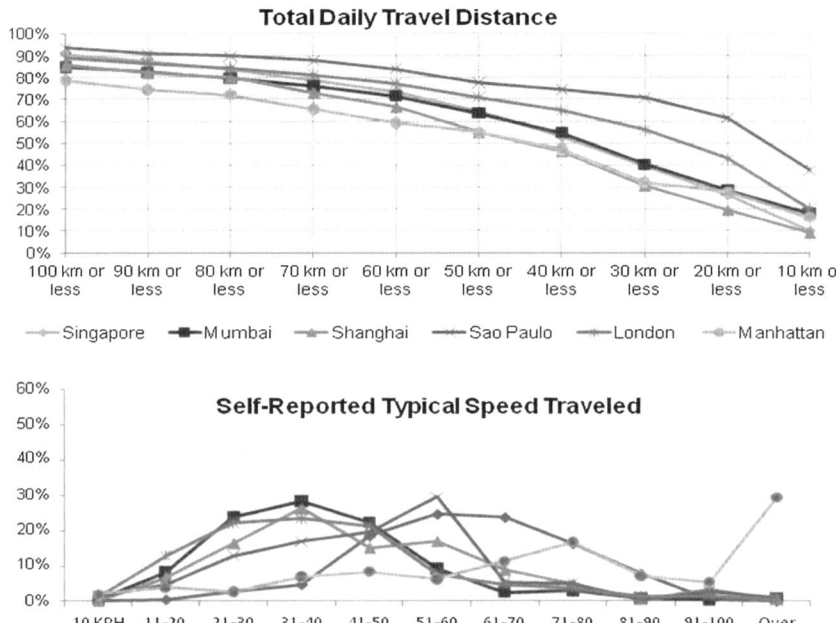

Fig. 3 Travel distances and speed in different agglomerations

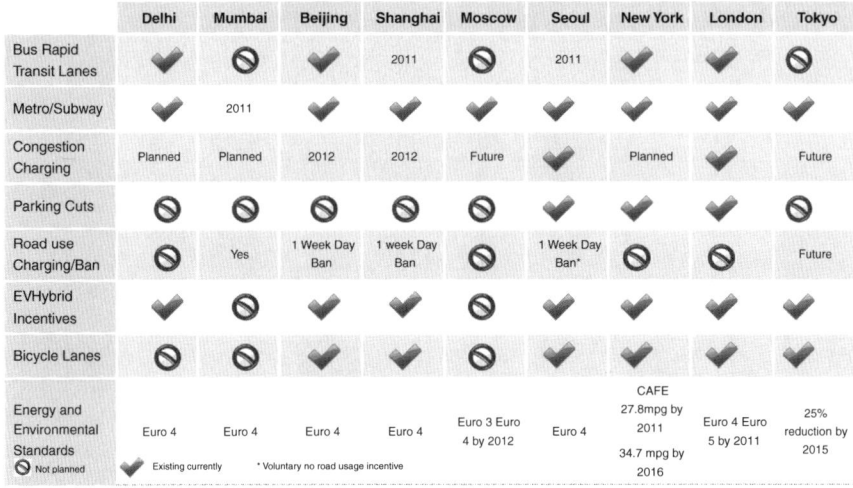

	Delhi	Mumbai	Beijing	Shanghai	Moscow	Seoul	New York	London	Tokyo
Bus Rapid Transit Lanes	✔	⊘	✔	2011	⊘	2011	✔	✔	⊘
Metro/Subway	✔	2011	✔	✔	✔	✔	✔	✔	✔
Congestion Charging	Planned	Planned	2012	2012	Future	✔	Planned	✔	Future
Parking Cuts	⊘	⊘	⊘	⊘	⊘	✔	✔	✔	⊘
Road use Charging/Ban	⊘	Yes	1 Week Day Ban	1 week Day Ban	⊘	1 Week Day Ban*	⊘	⊘	Future
EVHybrid Incentives	✔	⊘	✔	✔	⊘	✔	✔	✔	✔
Bicycle Lanes	⊘	⊘	✔	✔	⊘	✔	✔	✔	✔
Energy and Environmental Standards ⊘ Not planned ✔ Existing currently * Voluntary no road usage incentive	Euro 4	Euro 4	Euro 4	Euro 4	Euro 3 Euro 4 by 2012	Euro 4	CAFE 27.8mpg by 2011 34.7 mpg by 2016	Euro 4 Euro 5 by 2011	25% reduction by 2015

Source: Frost and Sullivan

Fig. 4 Different approaches in transport management

provide faster mobility than conventional automobiles having 150 km/h top speeds if it can be made smart enough to avoid traffic jams and find parking more easily. If it can park and retrieve itself autonomously then its average, or door-to-door, travel speed can almost be doubled, especially for short trips [9].

In essence, densely populated cities require a different type of vehicle that does not need to go as far or as fast as a conventional automobile but should be clean, smart, safe and easy to park. Failure to develop these types of vehicles may result in declining automobile ownership as cities may take further actions to promote bicycle and public transport usage and to deter usage of conventional automobiles, as outlined in Fig. 4 [10].

One potential solution to maintaining personal mobility even in the most challenging environment of densely crowded urban centers is to transition towards smaller, electrically-driven and wirelessly networked vehicles. Figure 5 is by no means exhaustive but, rather, illustrates a variety of different types of vehicles that are used for moving people around today in urban environments. For vehicles larger than a bicycle, reducing their size will make them easier to park and could lower their upfront and operating costs. Connecting these vehicles together could help to avoid crashes and assist in improving traffic flows, in reducing congestion, and in finding conveniently located available parking spaces more easily. Incorporating electric-drive will eliminate tailpipe emissions locally and expand the market for renewable energy sources, in contrast with petroleum-fueled, mechanically-driven automobiles. Each of these three ideas will, in turn, be discussed in the following sections.

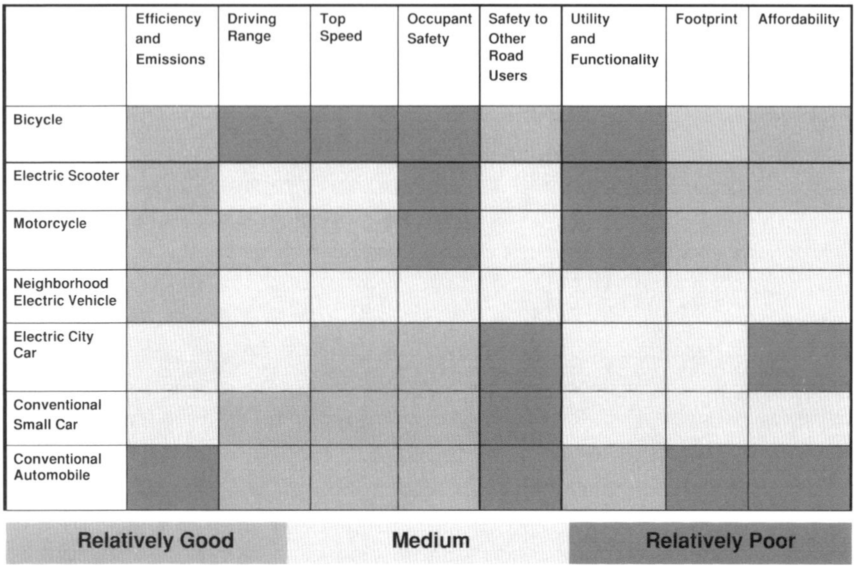

	Efficiency and Emissions	Driving Range	Top Speed	Occupant Safety	Safety to Other Road Users	Utility and Functionality	Footprint	Affordability
Bicycle								
Electric Scooter								
Motorcycle								
Neighborhood Electric Vehicle								
Electric City Car								
Conventional Small Car								
Conventional Automobile								

Relatively Good	Medium	Relatively Poor

Fig. 5 Modes of transport in urban environments

3 Electrification

Electricity and hydrogen are forms of energy that can be harnessed efficiently but are difficult to store, in contrast with petroleum-based fuels that can be stored conveniently but require complex controls to ensure clean and efficient consumption. For the last 100 years, this trade-off between energy density and efficiency for mobile propulsion has been settled in favor of the internal combustion engine; for a given cost, the industry has been able to develop technology to treat the exhaust emissions while the development of long-range electric vehicles continues. Despite the attention being given to electric vehicles, it is not expected that the traditional internal combustion engine-powered vehicle will disappear. Figure 6 shows that it will continually improve and play a role for the foreseeable future even as it is complemented by a variety of electric-drive vehicles.

For vehicles that are driven with electric motors, the extended-range and hydrogen fuel cell versions may be better suited to power larger vehicles that need to operate over longer distances. However, with improvements in battery technology and with emerging city consumer and city government needs there will be a growing demand for affordable, short range battery-electric vehicles for urban centers, as presented in Fig. 7. The short distances driven each day within cities and the generally slow urban traffic conditions reduce the energy required from the battery and the power demanded by the electric motor. By taking advantage of these reductions, it should be possible to reduce the vehicle's cost, weight, complexity, and energy consumption. This will make electric vehicles more

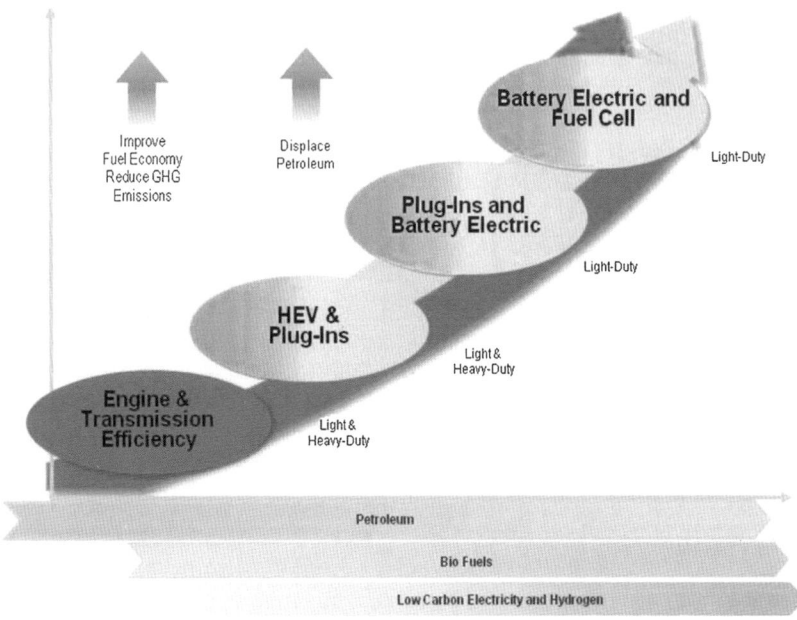

Fig. 6 Transitioning towards novel vehicles

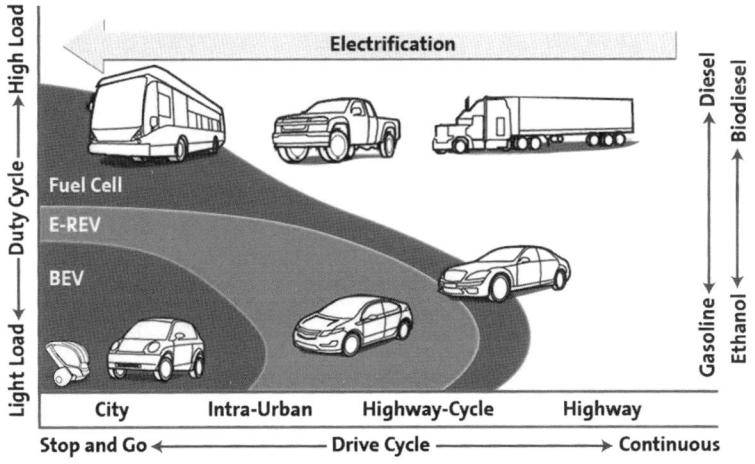

Fig. 7 Different drivetrains for different duties

commercially attractive and increase the positive energy and environmental impact associated with their adoption.

Because of the much higher efficiency of a ultra-lightweight electric vehicle, the net well-to-wheels greenhouse gas emissions are significantly lower than when using petroleum to fuel conventional automobiles. This should be true even if the source of electricity is coal, which produces more carbon dioxide emissions per unit of energy than the other common energy sources but which is the most widely available fuel in countries such as China and USA. Whereas the difference in well-to-wheels greenhouse gas emissions can be debated for the same automobile using either batteries or internal combustion engines [11] there is no doubt that using a battery in a vehicle several times lighter and with less demanding performance will realize significantly lower greenhouse gas emissions than a conventional petroleum-fueled automobile. Taking this to extreme, electric bicycles achieve 500 miles per gallon of gasoline energy equivalent which is 10–20 times more than a typical automobile [12].

A lightweight electric vehicle could also make it economically attractive to incorporate solar panels onto the vehicle's roof, perhaps, to further improve the vehicle's environmental performance. 1 m^2 of solar cells on the roof could generate as much as 1 kWh of electricity while the vehicle is parked for 10 h on a sunny day and this might be enough to support 10 km of "free" driving range for a lightweight EV. At the very least, it can significantly lower the amount of energy consumed by vehicle electrical systems, such as the air conditioning system on a hot day, and so extend the vehicle's range for a given size battery.

Lightweight electric vehicles can even be environmentally-friendly when compared with a bus, especially when the bus is not highly occupied for many journeys. Diesel buses realize 4 mpg fuel economy in the central business districts where they typically operate [13] and this is about 30 times less efficient than a four-wheeled lightweight electric vehicle on an equivalent energy basis. In other words, when a bus is full of passengers (60) it has about the same vehicle efficiency as a lightweight EV transporting two passengers in terms of energy consumed per passenger-mile basis; when the bus is not full it can have a substantially lower energy efficiency on the same basis. The lightweight EVs also produce no tailpipe emissions and offer more potential to increase energy diversity and reduce greenhouse gas emissions than the diesel bus, especially if clean, renewable power sources, such as sunlight and wind, are used. Finally, the cost of a diesel bus is around $300,000 which translates into around $5,000 per passenger and this could be more expensive on a per-passenger basis than the lightweight EV.

4 Connectivity

A lightweight and small electric vehicle can address several urban challenges (energy, environment, parking, affordability) but it will not solve safety and congestion challenges. One possible way to address these particular issues will probably require that vehicles operate as a network and communicate with each other to avoid collisions and optimize the use of road space to improve traffic flow. In essence, the vehicles need to become wirelessly networked computers on wheels, enabling the transportation equivalent of the Internet or "Mobility Internet".

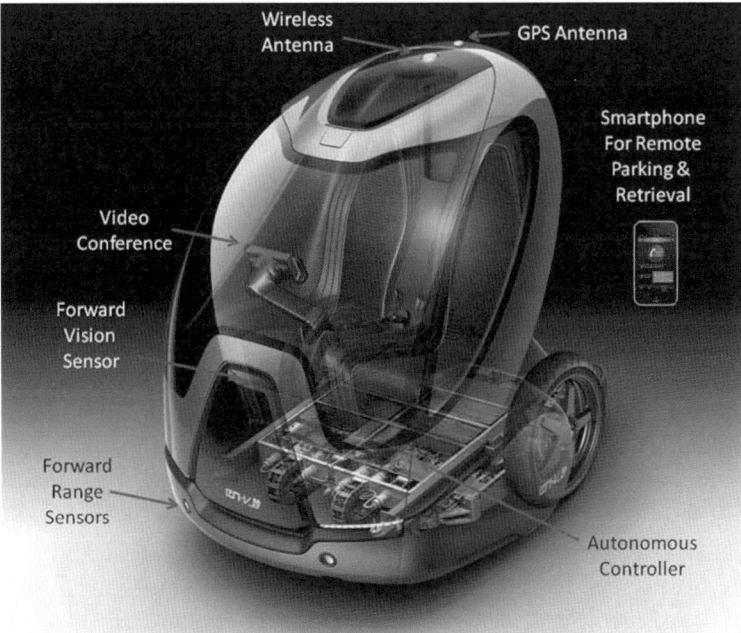

Fig. 8 An example EN-V developed by GM

A prototype physical embodiment of this type of vehicle is shown in Fig. 8. This is one of the EN-Vs that were developed by GM and demonstrated at the 2010 Shanghai World Expo. These Electric-Networked Vehicles are accurately located using GPS technology, can sense objects using vision and ultrasonic systems, and can wirelessly communicate with other vehicles and with the roadside infrastructure (using the 5.9 GHz portion of the spectrum for dedicated short range communications). The EN-Vs have demonstrated autonomous driving and platooning, automated parking and retrieval, and automated collision avoidance [14].

Several vehicles in the 2007 DARPA Urban Challenge demonstrated impressive autonomous driving over a distance of 100 km without incident in a mock urban setting while obeying all rules of the road, negotiating intersections and parking in designated spots [15]. This performance was achieved using GPS and sensors alone but a combination of sensing and connectivity (or "sixth" sense) could provide an even more reliable approach towards collision avoidance because there are scenarios where connectivity will be more useful. Sensors cannot "see" around corners or obstacles, for example, but vehicles can communicate with each other and with the roadside infrastructure this way. A combination of sensing and communication (non-verbal) is used in Nature when pedestrian flow occurs at a traffic intersection without collisions.

Vehicles that are aware of their surroundings can act to avoid a crash or decelerate to a low enough speed to potentially mitigate harm to pedestrians, cyclists, and vehicle occupants. This capability could potentially help to eliminate vehicle collisions at speeds and forces that cause injury or property damage. Vehicles crossing intersections autonomously may someday behave like swarms of bees or flocks of birds turning in flight without colliding or like people rubbing shoulders when "colliding". Much work remains before autonomous driving performance can reach these levels but recent developments are impressive.

A shift towards autonomous operation could lead to lighter, more energy efficient vehicles that in theory may not need as much crash structure as today's vehicles. In this scenario, other elements of the vehicle's design can be reinvented, leading the way to using new materials and developing new vehicle proportions, new ways of entering/exiting the vehicle, and new ways of controlling the vehicle. These possibilities will be described in more detail in the next section.

In addition to reducing the frequency and severity of vehicle to vehicle collisions and collisions between vehicles and vulnerable road users (such as pedestrians and cyclists), the co-ordination of vehicle movements and flows through networked intelligence could produce very high passenger throughput in streets and on roads. Wirelessly connected operation can potentially decrease traffic congestion and enhance energy efficiency by significantly reducing the variation in travel times and speeds. Digitally controlled, wirelessly connected vehicles can platoon, making use of communication links to maintain a constant separation and speed, and yet such a system would still allow individual vehicles the freedom to join and separate whenever they want. Intersections can also be controlled, in response to the dynamics of traffic flow, to minimize interruptions and to reduce the delays caused by successive vehicles accelerating from a stop sign. Studies in Japan indicate that raising traffic speeds from 10 to 20 km/h could reduce carbon dioxide emissions by nearly 40% for all the affected vehicles [16].

Two major sources of congestion in urban areas are crashes and searching for parking, and both can be potentially reduced through connectivity. Incidents are responsible for causing approximately 25% of non-recurring congestion [17] and many of these incidents could be avoided through vehicles communicating with each other and with the infrastructure. The search for parking can be responsible for nearly 30% of gasoline consumed by vehicles in dense, urban driving [18]. The parking infrastructure could be developed to communicate the location (and even price) of available parking which would reduce the search time and, effectively, take vehicles off the road which reduces congestion for the other driving vehicles.

These strategies can improve the throughput of roads and streets—in effect, adding "virtual lanes" without requiring more space. Even at low, but relatively constant, speeds below 30 km/h, intelligent, wirelessly connected vehicles may achieve higher average speeds than today's cars, particularly when traveling in downtown areas as traffic speeds in the center of Tokyo and London are below 20 km/h [19]. In Beijing, a 5 km journey cross-town can take between 30 and 90 min, which is slower than cycling and can be slower than walking. When the possibility of automated parking and retrieval is considered then travel speeds during autonomous operation could be reduced even further while still providing the same door-to-door time as today!

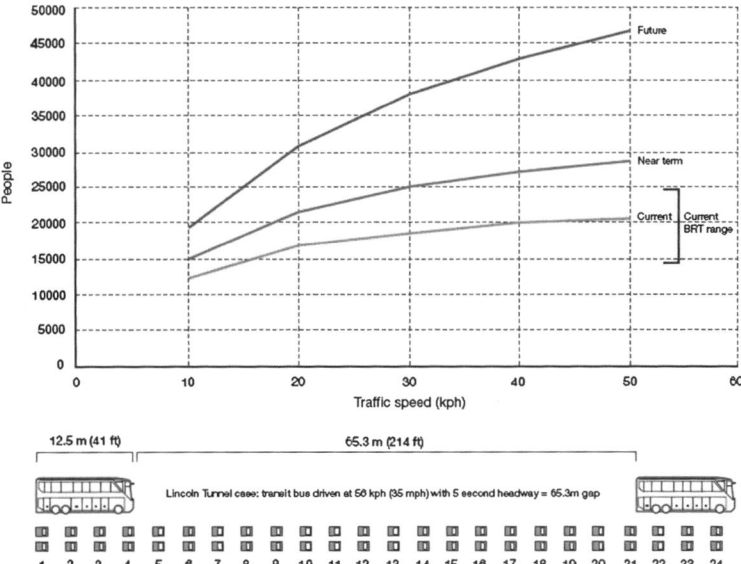

Fig. 9 Using novel concepts to optimize people movements

It may not be obvious but platooned vehicles might even match or exceed the passenger throughput of rapid transit bus systems. In the best case in the US (New York City Lincoln Tunnel buses accommodating 35 people each and operating at 56 km/h in a dedicated bus lane with a 5 s headway between them) a bus system can move about 25,000 passengers per hour along a street [20]. Figure 9 illustrates a scenario where two passenger EN-Vs, traveling at 40 km/h with a platooning gap of 3.5 m, enabled by future advances in wireless communications and sensing technologies, could be competitive with state-of-the-art BRT systems and be capable of moving as many as 30,000 people per hour [1].

In general, it is wasting energy and road space to allow crudely coordinated movement of relatively large and heavy vehicles along streets and roads and there is, clearly, a great deal of room for improvement. The intelligent management of swarms of small vehicles can provide much higher throughput on urban road networks, and may even surpass that of public transportation when passenger loading and unloading is taken into account. High throughput depends as much on overall system flexibility, coordinated speeds, and intersection management as it does on high top speed.

5 Vehicle Design

Vehicles for personal mobility take many forms and there is unlikely to be a single best design solution. Attention will be given here to reinventing the automobile by providing a future alternative that preserves the essential benefits of an automobile

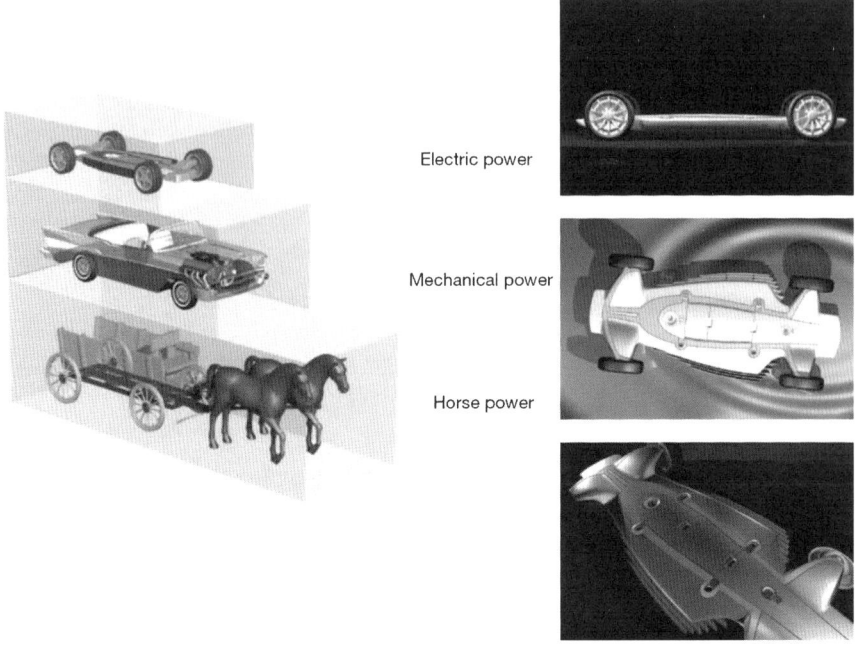

Electric power

Mechanical power

Horse power

Fig. 10 GM's Hy-wire electric-powered "skateboard" compared to other concepts

(weather protection, security and safety, privacy, cargo capacity and so on) while reducing the societal externalities in an urban environment.

The horseless carriage replaced the horse with an internal combustion engine that provided mechanical power and, not surprisingly, was located in the front of the vehicle. Today's hybrid electric vehicles retain this same layout as do battery-electric cars, even though there is no engine. The electric motor is normally placed where the engine is mounted and the batteries are packaged wherever there is available space on the vehicle but outside the passenger compartment.

A new generation of lightweight, compact electric vehicles can abandon this century-old architecture and place all the powertrain components under the floor, instead of at the front of the vehicle. This will raise the height and frontal area of the vehicle but it will also reduce the vehicle's footprint and so shrink the precious space required for parking and maneuvering. GM's Hy-wire electric-powered "skateboard", shown in Fig. 10, provided the pioneering demonstration of this concept (hydrogen fuel cells were the power source used in the Hy-wire but the concept may be even more valid with batteries since they have more shape flexibility than hydrogen storage systems) [21].

By moving the electric motors to inside the wheels rather than in the traditional space between the wheels, more space can be made available for batteries and the vehicle's footprint can be reduced even more, to less than half the size of a typical automobile. As shown in Fig. 11 each wheel can also be independently and

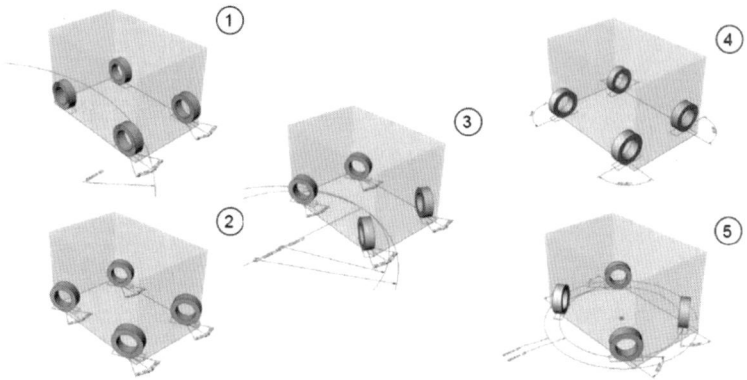

Fig. 11 Enhanced maneuvering with independently and digitally controlled wheels

digitally controlled, which allows for useful maneuvers like the O-turn in place of the U-turn, and even sideways motion for parking [22].

In most parking lots today, only one-third of the land area is actually occupied by vehicles when it is full because large aisle spaces are required for backing and turning and clearances are needed between the parked cars. With wheel motor-enabled maneuverability, the actual occupied space could be doubled. And if the vehicles themselves are substantially smaller, then the total improvement in parking density could be four-fold, as shown with MIT's folding stacked CityCar concept in Fig. 12 [1]. Such improvements promise a major change in the ratio of "car space" to "people space" in dense urban centers. This will be particularly crucial in the rapidly growing cities of Asia, with their rapidly expanding automobile ownership rates.

In addition to the omni-directional skateboard, the other potential design game-changer is to aim for achieving "zero crash" and autonomous operation. If vehicles do not crash then the safety standards and regulations relating to occupant protection in crashes could potentially be changed. Potentially, as much as 50% of the vehicle's mass could be eliminated as a result of this paradigm shift because the vehicle's structure would require less strength if there were no risk of vehicle impacts and other vehicle systems, such as propulsion, brakes and steering could be smaller and lighter if the rest of the vehicle were lighter [23]. In theory, content such as airbags and energy absorbing foam could also potentially be eliminated from the vehicle's interior in this future state.

In addition to these direct changes there could be new opportunities to re-design the interior with thinner seats so that the cabin is not only lighter but has lower thermal mass. Similarly, new glazing materials could be lighter and designed to reduce solar loading since light transmissibility requirements might change if vehicles do not crash. When combined, this could significantly reduce heating and air conditioning loads and yield further mass savings from the HVAC system. It could also improve passenger comfort, increase energy efficiency and allow a less expensive propulsion system to provide the same driving range and performance.

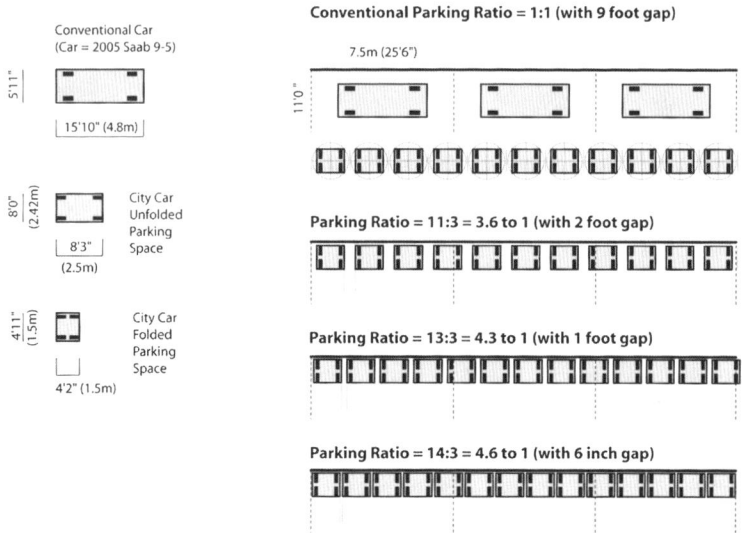

Fig. 12 MIT's folding stacked CityCar concept

Although frontal area may increase with a skateboard design, overall aerodynamic drag might be reduced because the vehicle may have the ability to "see" everything and avoid crashes so that side mirrors and windshield angles could be further optimized for reducing aerodynamic forces. For example, the driver could still see the outside world but with small cameras mounted all around the vehicle that relay images back into the cabin. In such a "virtual" environment it is conceivable that someday the driver could even have a driving experience completely decoupled from what the vehicle is actually doing!

If vehicles of the future could avoid crashes then new materials could be used to make the vehicle more affordable or, conversely, to accentuate it as a status symbol. For emerging markets, open top vehicles with inexpensive structures could be desirable while affluent buyers might want exotic materials and features not currently allows in automobiles. In other words, autonomous operation could increase the diversity of vehicle styling and usher in a "design renaissance" where regulations for fuel economy and crash safety are modified and new opportunities open up to make the vehicles more fashionable and more highly personalized.

As people age (another megatrend) we will need to find easier ways to interact with our vehicles. Autonomous operation is expected to be a huge enabler for providing mobility but if crash requirements were modified it could also be easier to optimally accommodate all types of passengers and enable universal design principles to be more easily executed. For example, it could be easier to enter and exit the vehicle, through the type of front door opening that was demonstrated with the EN-Vs. Autonomous operation could possibly enable personal mobility for nearly everyone, including children who could be driven to school in the future while they are able to videoconference with their parents or find some more time for studying!

Fig. 13 Three design themes, virtual gaming, fashion personalization and mobility

These 3 ideas—virtual gaming, fashion personalization and mobility for all age groups—were demonstrated with three design themes, shown in Fig. 13, for the EN-Vs [15]. Although each EN-V was technically identical in having the same propulsion system and autonomous operation technology platform they demonstrated unique design solutions and customer experiences. It is important to realize that the attraction for personal mobility is not simply defined by utility and flexibility; there is an emotional attachment that should not be dismissed.

6 Pricing Markets for Road and Parking Space

Smart, networked vehicles can help to optimize available road space and parking space because their movements can be coordinated in the most efficient way since traffic volumes can be monitored precisely, based on GPS location, and prices can be adjusted in real time [1]. If these prices are wirelessly communicated to the GPS navigation systems then the vehicles could choose the least expensive route, subject to a time constraint or the quickest route, within a budget constraint. Alternative choices could include the most energy-efficient route or even a randomly selected route, subject to time and cost constraints [24].

This creates a road usage economy which has precise, real-time dynamic pricing. There is feedback to this pricing because it rises when the demand, expressed by individual route selections, increases but the behavior of individual vehicles will also respond to any pricing changes. Although the dynamic behavior of this type of system is complex, appropriate rules and control algorithms should be better at allocating road space than is typically achieved today.

The same approach can also be applied to parking. At present, the price of parking is often not adjusted in response to demand and drivers are usually not given advance information about the availability and price of relevant parking spaces. If the occupancy of parking spaces is electronically monitored or sensed, and if this information is wirelessly communicated to GPS navigation systems then available parking spaces can become dynamically priced. If the driver is keen to find a parking space quickly then they might pay more to obtain a convenient

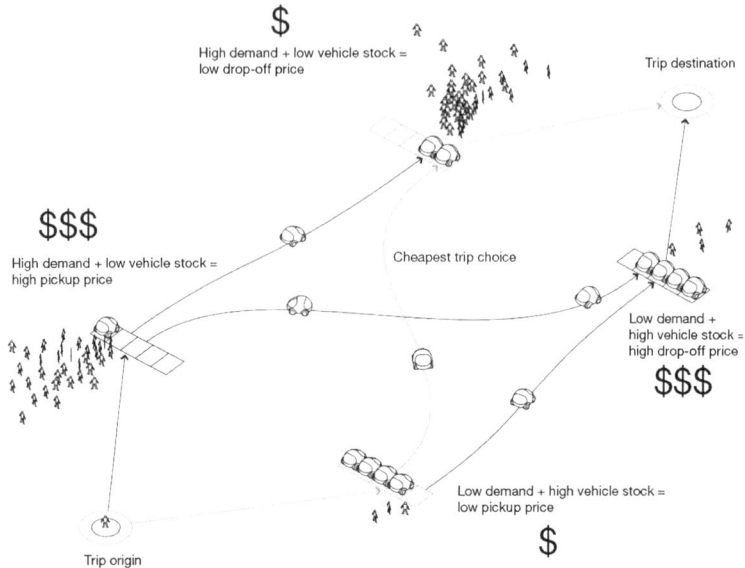

Fig. 14 Demand based one-way vehicle sharing

space whereas if they have more time and less money they could be satisfied with a cheaper space that is farther away and might require more walking.

7 Integration of Personal and Public Transport

Vehicles that spend more than 90% of their time parked can be managed in suburbs and low-density urban areas even though it is not the best use of space. However, in downtown areas and densely populated cities it is becoming increasingly difficult and expensive to provide sufficient parking for large numbers of privately owned vehicles. A large fraction of the mobility demand in affluent cities must be satisfied by taxis while in many Asian cities other shared-use vehicles, such as motorized rickshaws, are popular.

Taxi systems, in effect, use wireless communications and the intelligence of their drivers to match the supply of a mobility service to a dynamically varying demand. An alternative solution that provides one-way vehicle sharing and is based on price on demand is shown in Fig. 14 [1]. It uses racks of small vehicles, such as CityCars or EN-Vs, that are provided at conveniently spaced locations inside an urban area [24].

These small electric vehicles could be recharged while they are parked in their racks. When you want to make a trip, you walk to the nearest rack, swipe a credit card or otherwise electronically identify yourself, pick up a vehicle and drive it to a rack near your destination, and drop it off. These one-way rental systems are

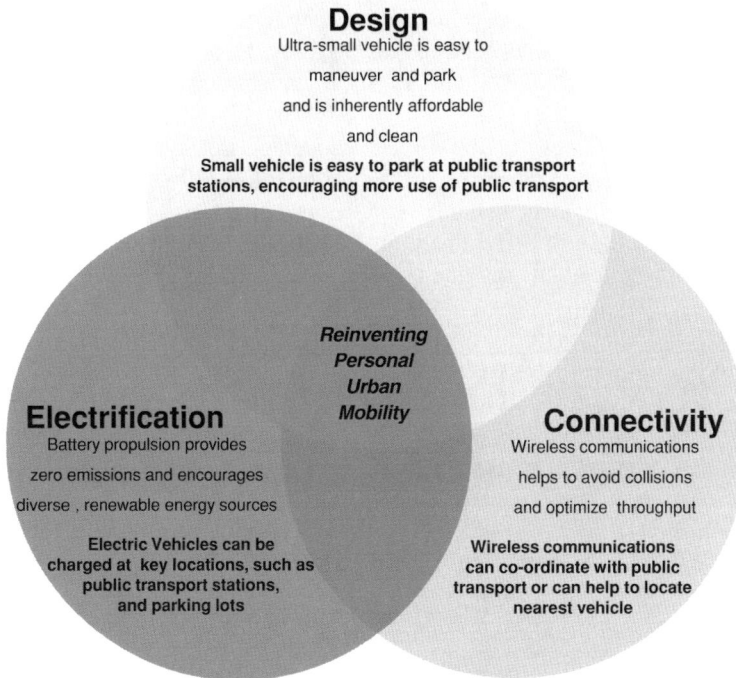

Design

Ultra-small vehicle is easy to

maneuver and park

and is inherently affordable

and clean

**Small vehicle is easy to park at public transport
stations, encouraging more use of public transport**

*Reinventing
Personal
Urban
Mobility*

Electrification

Battery propulsion provides

zero emissions and encourages

diverse , renewable energy sources

**Electric Vehicles can be
charged at key locations, such as
public transport stations,
and parking lots**

Connectivity

Wireless communications

helps to avoid collisions

and optimize throughput

**Wireless communications
can co-ordinate with public
transport or can help to locate
nearest vehicle**

Fig. 15 Reinventing personal urban mobility

easier to use than the better known two-way rental systems but they also make it harder to match the supply of vehicles and parking spaces, at locations throughout the city, in response to the dynamically varying demand.

For this system to be efficient and profitable there should not be too many surplus vehicles in the fleet or too many empty parking spaces. To provide a reliable service to users, the wait times for picking up and dropping off the vehicle should be short and predictable. This requires GPS-based vehicle tracking and real-time management of vehicle distribution throughout the system. These are more easily achieved if the vehicles are networked together. Ensuring that vehicles are optimally located for the users might require pricing incentives so that an inconveniently located vehicle is less expensive. Alternatively, if the vehicles can be driven autonomously to their desired location for use by the next customer then the cost of re-balancing the fleet is reduced as someone does not need to be paid to move them between the drop-off and pick-up locations [1].

Although commuters can park conventional automobiles at train stations it is clear that small, electric and connected vehicles are particularly well suited to seamless integration with public transport systems, as outlined in Fig. 15's Venn Diagram. Smaller vehicles, that are easier to park, can encourage more public transport usage because parking space limitations in the suburban train stations today can deter people from commuting by train into the city center. Alternatively, smaller vehicles would be easier to park in the city center and could be located conveniently at the train or bus station to provide last-mile mobility. Small vehicles could also be shared and might be owned by the city as part of the Public Transport System.

One of the incentives for the city to encourage this type of system is the significant value of the real estate that could be released by the combination of small footprint, highly maneuverable vehicles and shared use operation. Since land prices in the city center can be many times higher than in the suburbs there is a greater incentive to encouraging the use of smaller vehicles in the city center. It is possible that the land space required to support personal mobility could be decreased by as much as 95% if vehicles themselves occupy four times less space and if, say, five times fewer vehicles are used because they are shared. As an example, for a city of 5 million people and 2 million automobiles the land area that could be recovered might reach 50 km^2 since each automobile presently occupies 10–15 m^2 of space and actually consumes at least two spaces (one at home and one, on average, somewhere else in the city—at work, for example, or at the supermarket). In city centers the released parking space could be highly valuable and may help finance electrification and connectivity infrastructure enablers.

Affordable housing for city residents, valuable high rise commercial offices and green parks are a few of the options available to cities if this land could be released. Figure 16 illustrates that parking (shaded areas) accounts for a significant fraction of the land use in the center of Albuquerque [25] and the same is true for other cities. Although multi-level parking structures can reduce the land area needed there are limits to the height above ground that can be achieved and subterranean development is expensive and can be technically challenging. Complex, robotized parking structures can approximately double the packing density of vehicles over conventional multi-storey parking structures but their typical cost is $1,000–$2,000 per square meter [26] (in addition to land costs, which can significantly exceed this).

In addition to the advantages of compactness, EN-V's electric propulsion system is more easily integrated with public transport systems than today's petroleum-based vehicles because energy can be provided at a "granular" level through individual power outlets instead of in bulk via a conventional refueling station. If EN-Vs are publicly owned then it is possible to ensure that when they are parked they can be rapidly recharged, perhaps inductively whenever they dock, and that the cabin temperature can be pre-conditioned to a comfortable level. Both of these measures could reduce the size of the battery needed because range anxiety concerns could be reduced and less energy would be needed for heating and cooling the vehicle.

Finally, the connectivity embedded in EN-V facilitates communication with the public transport system so that drivers could be made aware of rapidly changing schedules, for example, or make seamless plans for inter-modal transport while traveling. Connectivity between vehicles and with the infrastructure will make it

Parking garage

Surface parking lot

Fig. 16 Parking (*shaded areas*) accounts for a significant fraction of the land use in the center of Albuquerque

easier to decide in real-time which vehicles should be given priority for recharging and how best to balance the supply and demand for a car sharing fleet.

Expanding the concept of bicycle sharing to a broader user base may require that the types of vehicles offer more range, comfort, utility and security than a bicycle. Hence, one can imagine small EN-Vs being part of a public transport system, available for sharing and parked, with modest cost to the user, at public transport stations dotted throughout the city center. People could ride the public transport system into the city center and enjoy personal mobility without creating the same side effects as with today's automobile usage.

This integration between personal and public transportation could be taken to another level if one draws an analogy between small electric vehicles and bicycles that are sometimes taken on buses and trains or even cars that can be taken on ferries. Imagine a resident of Shanghai, for example, who drives a small electric vehicle to the main train station and docks it onto a high speed inter-city train and enjoys the journey as a train rider. When the train arrives, in Beijing for example, the driver returns to their vehicle and because it was recharged during the long train journey from the electric train it is again able to provide personal mobility inside the city. This idea was demonstrated at 2010 Geneva Auto Show by Rinspeed in their UC? concept vehicle [27]. It could be extended to moving around inside the city and leveraging the buses or trains that run at less than capacity during off-peak hours.

This concept may seem futuristic but it preserves personal urban mobility and leverages high-speed electric public transport. Compared to today's scenario of driving a traditional automobile to the airport, flying in an airplane and then renting another automobile at the other end this scenario would require fewer transport mode shifts and could reduce petroleum consumption and greenhouse gas emissions.

8 Recommendations for Personal Urban Mobility

Transportation is often considered one of the least effective of the urban infrastructures (behind electricity, natural gas, water, telecommunications, security and waste treatment) [28, 29]. This can partly be explained by the simple fact that average travel speeds in city centers are no faster than a hundred years ago [30]! Improving the way we move around in cities in the future will clearly require fresh thinking in terms of urban design, new uses of communications technology, integration of personal and public transportation and rethinking automobile design.

Reinventing the automobile to address the specific challenges associated with urban mobility will require support from governments and from various industry sectors (transportation, energy and communications) because there are significant infrastructure hurdles to overcome and a need for incentives to accelerate the transition towards an electrified and networked mobility system that integrates personal and public transport and may even blur the distinction between the two. Cities will be motivated to make these changes in order to improve quality of life as they compete to attract human talent and new business.

In addition to providing funding and developing appropriate zoning codes for the physical infrastructure, national and city governments can also play a valuable role in supporting the development of test-bed sites that showcase the benefits of the proposed solution in a controlled, but real-world, environment before it is rolled out to existing cities. The creation of new cities in China over the next few decades provides such an opportunity but even within small islands, college campuses, military bases, tourist resorts and assisted living communities there are similar opportunities all around the world today. It is also possible to test components of the complete solution, such as wirelessly networked automobiles that can communicate with each other but still allow the driver to retain control at all times, and autonomous vehicles where the vehicles not only communicate with each other but act on this information to brake, steer and accelerate without driver intervention. Other potential demonstration projects could focus on one-way car-sharing (with conventional or electric vehicles) and lightweight electric vehicles.

Separate initiatives can demonstrate compelling solutions but widespread adoption by governments and by major industry players will require the creation of standards for electric charging and for inter-operable vehicle-to-vehicle and vehicle-to-infrastructure wireless communications protocols. International co-operation is essential if these same standards are to be applied around the world.

Finally, there is a need to rethink the separation of personal and public transport systems. The problems that are created by widespread automobile usage in dense cities sometimes encourage city government officials to consider drastic actions to deter personal mobility. However, it should be recognized that the value proposition offered by the automobile—comfort, convenience, utility, safety, security, flexibility—is likely to remain compelling and that these same benefits can be maintained with a smaller, networked electric vehicle that is designed for urban mobility. The same type of vehicle, designed to address urban needs, is also well suited to being physically and electronically integrated with public transport systems in new and innovative ways. This is important as it gives urban transportation planners more options for considering how to provide people with a better way to move around the city in the future and it allows urban designers to rethink land-use allocation.

The sum total of all these benefits is enormous and far greater than the benefit of simply increasing the energy efficiency of today's automobiles. Although there are challenges to implementing these concepts on a large scale there should be no insurmountable individual technological barriers to realizing this vision. The greatest benefit will be realized when electrification, connectivity and vehicle design are integrated together as a new transportation network, rather than through slow evolution of the existing urban transportation system, since they build on each other to reach their potential of transforming personal urban mobility.

References

1. Mitchell WL, Borroni-Bird CE, Burns LD (January 2010) Reinventing the automobile: personal urban mobility for the 21st century. MIT Press, Cambridge (Source for Figures 9, 12 and 14)
2. Spencer N and Butler D (2010) Cities: The century of the city. Nature 467:900–901
3. http://esa.un.org/unup/p2k0data.asp
4. Forstall RL, Greene RP, Pick JB (2009) Which are the largest? Why lists of major urban areas vary so greatly. Tijdschrift voor economische en sociale geografie 100:277 Source for Figure 2
5. Steven O'Neill (2000) US Census Analysis: New York City Vehicle Ownership (by Household)
6. Parking spaces/community spaces: finding the balance through smart growth solutions, EPA 231-K-06-001 (2006)

7. Mobility 2001: World mobility at the end of the twentieth century and its sustainability, WBSCD
8. http://aspe.hhs.gov/hsp/06/Catalog-AI-AN-NA/NHTS.htm
9. Dinesh Mohan (2008) Mythologies, metros and future urban transport. TRIPP report series, IIT Delhi, p 6
10. Frost and Sullivan (2010) Source for Figure 4
11. Kromer M, Heywood J (2008) A comparative assessment of electric propulsion systems in the 2030 light-duty vehicle fleet (SAE 2008-01-0459)
12. Electric bike use in China and their impacts on the environment, safety, mobility and accessibility, Christopher Cherry, UCB-ITS-VWP-2007-3
13. O'Keefe MP, Vertin K (2002) An analysis of hybrid electric propulsion systems for transit buses milestone completion report, NREL/MP-540-32858 (October)
14. EN-V, GM Press Release, March 24, 2010 www.media.gm.com
15. www.darpa.mil/grandchallenge/index.asp
16. ITS Japan Handbook 2006–2007
17. www.fhwa.dot/gov/congestion_problem.htm
18. www.streetfilms.org/dr-shoup-parking-guru
19. www.kankyo.metro.tokyo.jp/kouhou/english/2008/warming/warming02.html
20. Samuel P (2002) Busway versus rail capacity: separating myth from fact. Policy update 16, p 4
21. Burns LD, McCormick JB, Borroni-Bird CE (2002) Vehicle of change. Sci Am 287:64–73
22. Figure 9, courtesy of Peter Schmitt, MIT CityCar Program
23. Borroni-Bird CE, Verbrugge MW (2012) Fully autonomous vehicles: a far-reaching perspective on potential materials and design implications. In: Ginley DS, Cahen D (eds) Fundamentals of materials for energy and environmental stability, ISBN 978-1-107-00023-0. Cambridge University Press, New York
24. cities.media.mit.edu/pdf/Mobility_on_Demand_Introduction.pdf
25. Mark Childs (1999) Source for Figure 16
26. www.robopark.com
27. www.rinspeed.com/pages/cars/uc/pre-uc.htm
28. Miller D, Hazen G (2007) Megacity challenges: a stakeholder perspective. www.siemens.com
29. Liveable Cities, Economist Intelligence Unit (2010)
30. Automobile Association (2008) (from Transport for London)

Delivering Sustainable Transport in London

Michèle Dix and Elaine Seagriff

Abstract As a leading world city with the oldest metro in the world, London has a well-established transport system which has developed hand-in-hand with its urban growth. Londoners make over 24 million trips on the transport network each weekday; however, this is set to rise to over 27 million trips by 2031, due to increasing population and jobs. The Mayor of London's Transport Strategy, finalised in 2010 after extensive consultation and developed alongside the city's spatial and economic development strategies, sets out the transport agenda for the next 20 years to ensure this growth can be accommodated sustainably. The Strategy builds on the successes since Transport for London was established in 2000 by setting a policy framework for cross-modal sustainable growth and desired outcomes for the future. It addresses the Mayor's goals for supporting economic and population growth, enhancing quality of life, improving safety and security, reducing transport's contribution to climate change and improving its resilience and supporting the London 2012 Olympic and Paralympic Games and its legacy. The Transport Strategy sets out policies and proposals for London to meet these goals, including the further integration of spatial development and transport provision through the land use planning process, efficient use of the transport system, increased capacity and demand management. A cross-modal programme of improvements and policy initiatives to support walking, cycling and public transport use, to better manage road traffic and to reduce emissions from ground-based transport is well underway. With an integrated approach to transport and

M. Dix (✉)
Managing Director Planning, Transport for London, Windsor House, 42-50 Victoria Street, London, SW1H 0TL, UK
e-mail: micheledix@tfl.gov.uk

E. Seagriff
Head of London Wide Policy and Strategy Transport for London, Windsor House, 42-50 Victoria Street, London, SW1H 0TL, UK
e-mail: ElaineSeagriff@tfl.gov.uk

O. Inderwildi and Sir David King (eds.), *Energy, Transport, & the Environment*,
DOI: 10.1007/978-1-4471-2717-8_18, © Springer-Verlag London 2012

land use policy and a clear coordinated implementation plan, the challenges of meeting London's growth in a sustainable manner can be met. The following chapter represents individual reflections of the authors but from the perspective of London's transport authority, Transport for London.

Abbreviations:

CO_2	Carbon dioxide
CSR	UK Government comprehensive spending review
DfT	Department for Transport (UK)
DLR	Docklands Light Railway
EDS	Mayor of London's Economic Development Strategy
EU	European Union
GLA	Greater London Authority
LED	Light emitting diode
LIP	Borough Local Implementation Plan
MTS	Mayor of London's Transport Strategy
NO_x	Oxides of nitrogen
TfL	Transport for London

1 Introduction

London has evolved and changed as its population has grown and transport system has expanded. In coping with increasing pressures on a historic underground system and the capacity of the transport network, Transport for London (TfL) employs a range of policies to meet required targets in, for example carbon dioxide (CO_2) reduction. Continuing this will be important to enable London to accommodate growth in a sustainable way and to continue to succed in its role as the UK's capital and as a world city.

Sustainable development is key theme of national policy and is defined in the Report of the Bruntland Commission, *Our Common Future* (1987) as:

> *development that meets the needs of the present without compromising the ability of future generations to meet their own needs.*

Transport can play a role in delivering sustainable development by encouraging a shift away from private fossil fuelled motorised transport towards public transport, walking, cycling and low and zero-emission vehicles. TfL's formation under the direction of the Mayor of London has provided a huge opportunity to do this, by integrating services, policies, strategies and investment plans effectively. The Mayor's Transport Strategy (MTS) [1] for the next 20 years is integrated with spatial and environmental strategies and contains a common set of targets and goals which organisations involved in the provision of transport in London work towards.

While there is now a broad range of policies, tools and innovative mechanisms available to influence travel demand and behaviour, a strategic approach ensures the correct tools are employed in the best circumstances. It is also important to understand public views as these influence political acceptability of possible transport policy. In this respect the London experience has been illuminating. All major strategic projects and proposals are subject to public consultation—either through the MTS or their own project approval mechanisms. In this way it is possible to gauge local support and political acceptability.

This chapter examines London's development alongside that of its transport system and draws on experiences since TfL was established. It looks to the future and the significant predicted growth, explains the policies to be employed and focuses on the need to reduce CO_2 emissions from transport.

2 London's Urban Development

London's transport system has evolved with the city's urban development and has helped shape it. Greater London covers 1,500 square kilometres and is defined by the administrative boundaries of 33 boroughs. It is generally contained within London's orbital road, the M25. It encompasses central, inner and outer London which has varying employment, population and travel characteristics.

London has grown from the area now broadly defined as central London. This area remains the heart of the UK economy and is home to many corporate headquarters. These represent insurance, finance, leisure and service sectors and central government among others. It occupies a small area—26 km square (just two per cent of the city's land area) but contains 31% of its jobs.

London's spatial planning has reflected market developments and maintained a pattern of concentric growth supported by radial links to the economic centre. Building on economic productivity and agglomeration benefits, employment growth continues to be concentrated in the central area with some growth in metropolitan town centres and business parks.

The surrounding area of Inner London predominantly developed during the period up to the First World War, with large areas developed to expand London's residential capacity. As is seen in Fig. 1, Greater London's peak population of 8.6 million people in the late 1930s was reached prior to the huge growth in use of the private car. In this respect, London has developed without car dependence. This was supported through the extensive public transport network in existence at that time comprising trams, buses, the Underground and rail. As reported in 1934 (London Passenger Transport Board), the average number of public transport trips per person in 1933 was 500 a year, excluding suburban rail. Since most people lived close to their place of work, this is substantial.

Fig. 1 London's population 1800–2011. *Source* ONS data

Outer London grew between the two world wars as older towns and villages merged and expanded with new housing developments, reflecting the emerging preference for lower density suburban living. With the exception of the post-war high-density tower blocks, housing development has dispersed much more than employment. As a result, the largest proportion of the capital's population is contained in Outer London. This is a larger geographical area than Inner London, where densities are greater.

Through changes in legislation the green belt was established around London. In 1938 the *Green Belt (London and the Home Counties) Act* enabled local authorities to purchase land to protect it as open space rather than it being developed. The 1947 *Town and Country Planning Act* ratified this further by allowing local authorities to designate areas for protection in their development plans. The green belt around London remains today and covers an area of around 515,000 hectares, including areas within 19 of London's boroughs and parts of the surrounding counties.

The establishment of the green belt, coupled with population growth and an increasing desire for more suburban homes and space, helped shape the geography of London and its surrounding area. After the Second World War, new towns were established 35–80 kilometres from the centre of London to ease overcrowding. These included Basildon and Harlow in Essex; Bracknell in Berkshire; Crawley in Sussex; and Hemel Hempstead, Stevenage, Welwyn Garden City and Hatfield in Hertfordshire. This pattern of growth was supplemented by more towns in the late 1960s—including two further north of London—Milton Keynes in Buckinghamshire and Peterborough in Cambridgeshire.

Population growth in areas outside the green belt was accompanied by a decline in inner London's population and, to a lesser extent, outer London—for example 23 and 8%, respectively between 1971 and 1991. In total London's population fell by around one million during this time. This has since been reversed and London's population is now at 7.6 million as a result of natural growth and migration. Current population distribution and density is summarised in Table 1, which shows the larger geographical extent and much lower densities of outer London compared with inner London.

Table 1 Population distribution and density

	Inner London	Outer London	Greater London
Total population 2007 ('000s)	3.0	4.6	7.6
Built up land area (sq km)	320	980	1,300
Population density (1,000/sq km)	9.4	4.7	5.8

Source ONS data

3 London's Transport Network

3.1 Development of the Public Transport Network

The public transport system in London comprises two main rail networks—the commuter rail network known as Network Rail and the Underground system which together covers over 1,000 kilometres—an extensive bus network, an automatic light rail system in east London and a tram system in south London.

When the original rail network was developed in the mid-1800s it was not allowed to extend into the centre of the city. As a result, radial lines were constructed to stations including Kings Cross, Euston, Paddington and Waterloo, forming a ring around the edge of central London which still defines the area today. These lines allowed for the development of suburbs in outer London, facilitating movement out of the capital by wealthier residents who could then commute to the central area.

Commuters were forced to make onward connections within the central area by bus, foot or tram until the Underground was developed. It was the first underground railway in the world and remains one of the most extensive metro systems today. The first section, running between Paddington and Farringdon, opened in 1863. The suitability of London's clay subsoil enabled tunnels to be built and electric train services then to be provided. In the 1920s and 1930s the Underground was extended further north to support the growth of new residential areas. The provision of far-reaching rail services and the growth in areas outside of the green belt meant that commuters were travelling ever greater distances.

The demise of trams and trolley buses in London in the 1950s and 1960s was accompanied by the rise in use of the private car and increasing traffic congestion. This resulted in a decline in bus use though in recent years there has been a dramatic turnaround. The Underground system has continued to be expanded in increments–in the 1960s with the Victoria line, and in the 1970s and 1990s with the Jubilee line. More recently the East London line has been extended and upgraded to form a core part of London Overground, an orbital route serving many densely populated areas of the capital. London Underground now has over 400 km of route and 270 stations.

3.2 Governance of Transport in London

Since 2000, London has had a directly elected Mayor with powers devolved from central Government for strategic city-wide matters such as transport and land use planning. TfL was established at this time as the integrated body responsible for the transport system and implementing the MTS in conjunction with partners.

Though many of TfL's constituent bodies were well established before 2000, London had never experienced the level of co-ordination afforded by the new governance arrangements. TfL is one of the most integrated transport authorities in the world, controlling, planning and financing the vast majority of transport for people, goods and services in the capital. Its remit stretches much further than the Underground and buses to a wide range of areas. These include: Docklands Light Railway in east London; Tramlink, a modern tram network in south London; major strategic roads and all traffic signals; many walking and cycling programmes; London Overground; oversight of river services; the licensing of taxis and private hire vehicles; travel information and ticketing; and strategic transport planning. More than 24 million journeys are made on London's transport network every day and the organisation now has around 28,000 employees.

3.3 TfL's Achievements to Date

TfL's first decade of activity is marked by a number of major infrastructural and operational developments. Following years of underfunding, the amount of sustained investment in the transport system has been unparalleled. Through this TfL has been able to introduce a wide variety of improvements for public transport users, pedestrians and cyclists as well as other road users. In addition, a net mode shift from the private car to public transport, walking and cycling of 7% has been achieved—larger than any other major metropolitan city in the world. Some highlights include:

- London Underground now carries more passengers than at any time in its 148-year history—1.1 billion passengers in the past year.
- The Oyster card—an electronic travel smartcard—has been used across TfL's modes, and extended to all suburban rail and river services. More than 1.5 million National Rail journeys are now made each week using Oyster pay as you go.
- The bus network has grown by around 40% to over 8,000 low-floor vehicles and 90% of Londoners now live within 400 metres of a bus stop. Bus usage has increased by 60% and is now at its highest level since 1960 with 2.2 billion passengers a year, equating to around half of all bus journeys in the country.
- There has been a comprehensive improvement in the provision of passenger information with the widespread roll-out of real-time data and increasing use of TfL's 'Journey Planner', TfL's web-based tool to allow people to plan their

journey using up to date information. This is now used by almost eight million people each month.

- On the Overground, a revitalisation of the orbital railway links around London has occurred. By 2012 London will have its first fully orbital rail network.

London's road space is used by almost all freight in and out of the city, and by 80% of all passenger journeys. Total road space (more than 13,000 km in Greater London) is effectively fixed and with the expected growth in population and employment the problem of congestion will continue. Congestion on London's roads is a hindrance to businesses, costing the UK economy around £2bn each year. Therefore, innovative policy development to manage scarce road space has been a major focus. To help address this, a central London congestion-charging scheme was introduced in February 2003 which led to a 20% reduction in traffic. The scheme was extended in February 2007 to cover the area to the west of the original zone (the Western Extension Zone). However, the Mayor recognised that to be beneficial to businesses, road user charging must be accurately targeted to achieve set objectives. Therefore, after a public consultation exercise in 2010, the western extension was removed.

In recent years London has seen a more than doubling of the number of cycling journeys. The take-up of the new cycle hire scheme in central London has been significant—with more than 110,000 people registering within the first few months of operation. This has been accompanied by the opening of the first four of twelve cycle superhighways into central London. Early research on the first two of these indicates a 25% increase in cycling on these routes—with some sections up by more than 90%. In addition, there have been significant improvements to the urban realm—such as Trafalgar Square where, working with Westminster council, a traffic island was transformed into a large open attractive space. This is an example of a scheme on a grand scale but there are many more local examples, for example Windrush Square in Brixton.

3.4 Travel Characteristics

As Fig. 2 shows, the number of trips undertaken in London has grown steadily year by year from 1993 until 2007 when there was a levelling off owing to the economic recession. However trips are now rising again as the economy recovers, with the underground and bus networks both carrying record numbers (1.1 and 2.2 billion passengers per annum, respectively).

The basic trend of a substantial shift away from private vehicles to public transport in London is clear. Mode share for the latter has grown from just under 30% in the early 1990s to around 41% today. Accompanying this there is an established trend of falling levels of road traffic. Total vehicle kilometres fell by 3% between 2008 and 2009, having fallen between 1 and 2% each year since 2000. Within the context of increased overall travel, the net shift in mode share towards public transport in London has contributed to travel becoming more sustainable overall.

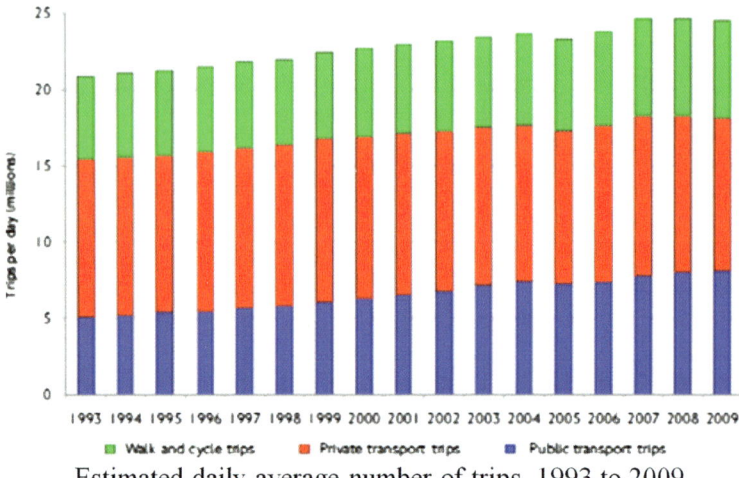

Estimated daily average number of trips, 1993 to 2009

Fig. 2 Travel volumes in Greater London 1993–2009. *Source* TfL Planning

3.5 *The Funded Plan for Transport*

Approximately one-third of TfL's funding comes from a direct grant from the Department for Transport (DfT). Following the Government's Comprehensive Spending Review (CSR) in late 2010, TfL's overall grant funding from the DfT was been reduced by £2.17bn in total, over the four years covered by the CSR, or 21% in real terms in 2014/15, when compared with the previous 2007 settlement.

However, the overall DfT grant is just one element of TfL's funding which includes fares, borrowing and other sources of revenue, such as advertising and commercial partnerships. During the current TfL Business Plan period to 2014/15, the proportion of TfL's income from fares, congestion charge and enforcement income is forecast to rise from just under 50% today to around 60%, whereas the contribution from Government grant will fall from 40% to 30%. Other income and grants will remain broadly level, at 10% of overall income. This package of funding, when coupled with efficiencies savings of around £8bn to 2018, allows a programme of investment in London's transport infrastructure that is the largest for 80 years and is set out in TfL's Business Plan [2]. It includes:

- A comprehensive upgrade of the Underground network—the Underground still has some signalling equipment which is almost 90 years old and infrastructure from the 1860. The upgrade includes new signalling, new and more trains, replacement of track and other expired infrastructure as well as major congestion-relief schemes at Victoria, Bond Street, Tottenham Court Road, Paddington and Bank stations, Thameslink and a massive expansion in the capacity of suburban rail services.
- Construction of Crossrail which will provide up to 24 high-capacity trains an hour between Paddington and Whitechapel during peak periods, and will increase the

Fig. 3 Alignment of Crossrail. *Source* Crossrail

capital's rail capacity by 10%. The route, as shown in Fig. 3, will serve areas in which more than 35% of London's future growth to 2031 is expected. The project itself, with 21 km of twin-bore tunnels and major station construction, is Europe's largest civil engineering project and is expected to provide an economic boost to the UK economy.

- The extension of the cycle hire scheme, providing another 200 docking stations and 2,000 bikes, ahead of the 2012 Games, plus delivery of all 12 cycle superhighways by 2015.
- All of TfL's 2012 Games transport commitments will be delivered.
- A 'New Bus for London' with five prototype vehicles entering service in 2012 offering step-free access to the lower deck. The hybrid vehicles will be 40% more fuel efficient than standard diesel buses, and will reduce nitrogen oxides (NO_x) and particulate emissions from buses.

The funded capacity enhancements described above will increase public transport capacity in the three-hour morning peak by over 30% between 2006 and 2031. They will also ease crowding on some parts of the rail and Underground network and support modal shift in favour of public transport, walking and cycling.

4 Strategic Transport Planning Context

4.1 London's Growth

The city's population is expected to grow by 1.25 million in the next 20 years—from 7.6 million to 8.8 million. Employment is also expected to grow from 4.7 million to 5.5 million by 2031. The concentration of employment growth in central London is likely to persist. Using current forecasts, over the next

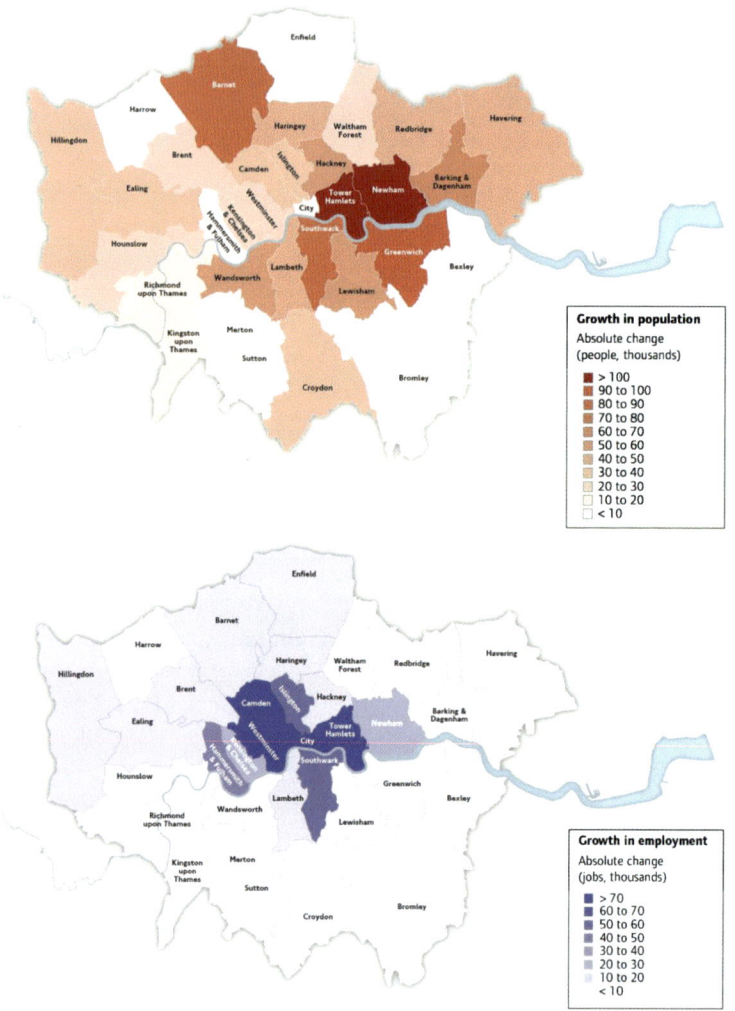

Fig. 4 Spatial distribution of population and employment growth, 2007–2031.*Source*London Plan 2011

20 years, 35% of London's future employment growth is expected to be located within the central area and Canary Wharf Fig. 4. This area will therefore continue to be important not only for London's economy, other areas of London and the surrounding regions, but also for the UK as a whole.

Housing growth will be more dispersed, as in the past, albeit with high rates of growth in Inner London. This pattern of growth has placed, and will continue to place, greater strains on the capacity of the radial transport networks in and out of central London, requiring ever more radial transport capacity. The overall number of trips in the city is forecast to increase from the current level of 24 million a day to more than 27 million by 2031.

Although continuing to support the economic development of the central area will remain a priority, the need to make improvements to support quality of life and the economies of Outer London has been increasingly recognised. The Outer London Commission was set up specifically to investigate and report on how this might be achieved. To support this, the Greater London Authority (GLA) and TfL undertook extensive analysis of various land use, transport and economic development scenarios. These tested differing sets of assumptions about the pattern of future jobs and housing growth, with greater levels of employment assumed in Outer London town centres and proportionately less in the centre of London. In particular, by using high-level transport modelling, the work sought to understand the overall implications of such a shift of emphasis on the key challenges. In transport planning terms, it identified a number of significant implications.

Locating more employment growth in outer London (assuming this is offset by central London employment growing at a slower rate than previously assumed) can reduce trip length and bring about reductions in central London crowding and congestion. However, higher growth in outer London, without any change in transport provision, could lead to more congestion and a small overall rise in London-wide transport-based CO_2 emissions. This is because existing trip patterns in Outer London are more car-dependent and less public transport-focused than trips from Outer London to central London.

On its own, such a change of development focus does not achieve a wholly better transport outcome. Furthermore, the transport improvement required to support such development could not be sustained without significant growth in Outer London which would undermine the 'suburban' character of many areas that the boroughs wish to maintain.

Therefore, with the expected growth of the central area and with increasing densification at key public transport interchanges, increasing mode share in favour of public transport, cycling and walking is considered to be the most sustainable. Accompanying this, development of the capital's specific Opportunity and Intensification Areas, regeneration areas, Strategic Outer London Development Centres and Strategic Industrial Locations, as designated by the London Plan, is being progressed. Large increases in housing and employment are planned for many of these locations on the edges of central London, the Thames Gateway and Outer London. These developments will be supported by transport improvements to ensure they are integrated within London's network, and opportunities for walking and cycling will be maximised.

4.2 The Mayor's Strategies

The Mayor has a legal duty to provide and consult upon a number of strategies setting out London's future direction. The key spatial strategy, The London Plan, sets out the overall visions for London and its geographical development. All other mayoral strategies need to be consistent with the London Plan.

The current London Plan [3] states that:

Over the years to 2031 and beyond London should:
  Excel among global cities—expanding opportunities and enterprises, achieving the highest environmental standards and quality of life, and leading the world in its approach to tacking the urban challenges of the 21st century, particular that of climate change.
  Achieving this vision will mean making sure London makes the most of the benefits of the energy, dynamism and diversity that characterise the city and its people; embraces change while promoting its heritage, neighbourhoods and identity; and values responsibility, compassion and citizenship.

This is a wide-ranging vision for London, one with ambition and inclusivity. It sets the direction for employment and population growth to 2031 and is ultimately approved by Government to ensure conformity with the national planning framework. All of the Mayoral strategies are consistent with the London Plan's planning assumptions, and aim to be supportive of its objectives.

The latest MTS was developed alongside, and takes account of, the policies in the London Plan and the Economic Development Strategy (EDS) [4]. Public consultations for these strategies ran concurrently. Alongside sits the Mayor's environmental strategies including those for climate change mitigation, energy and air quality. Transport provision and policy development are fully integrated with economic and spatial development, while ensuring environmental targets and impacts are predicted, measured and considered. All of these strategies look to the next 20 years forming an integrated strategic policy framework for the medium and long term.

The Mayor's Air Quality Strategy [5], published in December 2010, sets out specific measures to curb emissions of air pollutants from transport—with a focus on particulate matter and NO_x. This includes setting age limits for taxis and private hire vehicles, a pilot of retrofitting older buses to Euro IV standards for NO_x and a trial of dust suppressant machines. In terms of the Low-Emission Zone, from 2012 larger vans and minibuses will be subject to tighter controls if travelling within Greater London, doubling the effect of the earlier phase in cutting pollution. It is important that this and other programmes are progressed to ensure achievement of the EU targets and to deliver health benefits for London. The Mayor's draft Climate Change Mitigation and Energy Strategy [6] also contain substantial policies relevant to transport. These are outlined in more detail in Sect. 6 below.

4.3 The Mayor's Transport Strategy

London has benefitted from a legal framework which enables the Mayor to set direction for all organisations involved in transport provision in the capital. The MTS is the principal policy tool through which the Mayor exercises his responsibilities for the planning, management and development of transport in London, for both the movement of people and goods. The legislative framework for the

MTS was laid down by the *Greater London Authority Act 1999* (and amended by the *Greater London Authority Act 2007*). It specifies that the transport strategy must contain policies for 'the promotion and encouragement of safe, integrated, efficient and economic transport facilities and services to, from and within Greater London', and proposals for securing the transport facilities and services needed to implement the Mayor's policies over the lifetime of the MTS. The MTS is also required to contain the Mayor's proposals for providing transport that is accessible to people with mobility impairments, and may contain any other proposals which the Mayor considers appropriate.

The MTS provides the policy context for TfL and London's boroughs to implement their more detailed plans. TfL's annual Business Plan is effectively its MTS implementation plan. The boroughs prepare Local Implementation Plans (LIPs) to detail how the MTS will be achieved in their area. TfL and London boroughs have also been working together to prepare regional transport plans to set out how the MTS will be implemented across the city's five subregions and to ensure any particular local challenges or opportunities are addressed.

Progress against the MTS goals is reported in TfL's Annual Report [7] and results are monitored and reported each year in TfL's 'Travel in London' report, the latest being Report 4 [8]. This allows regular checks to be taken on the effectiveness of policies against goals to ensure further actions can be taken if needed.

4.4 The Vision for Transport in London

The Mayor's vision for transport in the capital is that:

London's transport system should excel among those of global cities, providing access to opportunities for all its people and enterprises, achieving the highest environmental standards and leading the world in its approach to tackling the urban transport challenges of the 21st century.

To achieve this he has defined six goals for his transport strategy:

- Supporting sustainable population and employment growth—there are 15% more trips in London forecast by 2031 increasing from 24 million each day to more than 27 million. This consists of around 30% more use of the public transport network.
- Enhancing the quality of life for all Londoners—London's transport system is critical for access to employment, education, healthcare and leisure services. There is a need to further improve the quality of passengers' journeys, enhance the built and natural environment and improve Londoners' health through cleaner air.
- Improving the safety and security of all Londoners—The capital's public transport is a safe, low-crime environment and road safety has improved dramatically in recent years. There is a need to continue to reduce road traffic

Table 2 Proposed transport outcomes for London

Goals	Challenges	Outcomes
Support economic development and population growth	Supporting sustainable population and employment growth	Balancing capacity and demand for travel through increasing public transport capacity and/or reducing the need to travel
	Improving transport connectivity	Improving people's access to jobs
		Improving access to commercial markets for freight movements and business travel, supporting the needs of business to grow
	Delivering an efficient and effective transport system for people and goods	Smoothing traffic flow (managing delay, improving journey time reliability and resilience)
		Improving public transport reliability
		Reducing operating costs
		Bringing and maintaining all assets to a state of good repair
		Enhancing the use of the Thames for people and goods
Enhance the quality of life for all Londoners	Improving journey experience	Improving public transport customer satisfaction
		Improving road user satisfaction (drivers, pedestrians, cyclists)
		Reducing public transport crowding
	Enhancing the built and natural environment	Enhancing streetscapes, improving the perception of the urban realm and developing 'betterstreets' initiatives
		Protecting and enhancing the natural environment
	Improving air quality	Reducing air pollutant emissions from ground-based transport, contributing to EU air quality targets
	Improving noise impacts	Improving perceptions and reducing impacts of noise
	Improving health impacts	Facilitating an increase in walking and cycling
Improve the safety and security of all Londoners	Reducing crime, fear of crime and antisocial behaviour	Reducing crime rates (and improving perceptions of personal safety and security)
	Improving road safety	Reducing the numbers of road traffic casualties
	Improving public transport safety	Reducing casualties on public transport networks

(continued)

Table 2 (continued)

Goals	Challenges	Outcomes
Improve transport opportunities for all Londoners	Improving accessibility	Improving the physical accessibility of the transport system
		Improving access to services
	Supporting regeneration and tackling deprivation	Supporting wider regeneration
Reduce transport's contribution to climate change, and improve its resilience	Reducing CO_2 emissions	Reducing CO_2 emissions from ground-based transport, contributing to a London-wide 60% reduction by 2025
	Adapting for climate change	Maintaining the reliability of transport networks
Support delivery of the London 2012 Olympic and Paratympic Games and its legacy	Devetoping and implementinga viable and sustainable legacy for the 2012 games	Supporting regeneration and convergence of social and economic outcomes between the five Olympic boroughs and the rest of London
		Physical transport legacy
		Behavioural transport legacy

Source MTS 2010

casualties and injuries, and improve the perception of personal safety on the transport network.

- Improving transport opportunities for all Londoners—Increasing accessibility across the network will help more people access opportunities locally and across the capital.
- Reducing transport's contribution to climate change and improving its resilience—ground-based transport accounts for about 22% of London's CO_2 emissions leading to strong acceptance of the need to improve the modal share of public transport, walking and cycling further, together with improving the uptake of low-carbon vehicle technology and fuels to meet the Mayor's 60% CO_2 reduction target by 2025 from a 1990 base.
- Supporting the London 2012 Olympic and Paralympic Games and its legacy— TfL is responsible for key infrastructural improvements including demand management, Games route networks and road freight management programme. Up to three million additional trips are expected in London on the busiest day of the Games and a key aim is to ensure that all spectators can walk or cycle or use public transport to get to events. Walking and cycling infrastructure routes into the Olympic Park are being improved to encourage more people to use them during and after the event. Barclays Cycle Hire is being extended eastwards so that by spring 2012 there will be around 8,000 hire bikes and 14,400 docking stations in London. Seven new walking routes will be completed. There will be a significant transport legacy to help east and south east London attain greater socio-economic convergence with the rest of the capital.

The outcomes sought from the MTS, in relation to each of the goals, are set out in Table 2.

5 London's Policy-Led Approach to Transport

Even with the investment outlined above, more is required to meet the Mayor's goals and to support sustainable economic and population growth in London. The policy-led approach entails setting goals, understanding the challenges and determining outcomes, then developing a range of policy tools to deliver these outcomes. The policy tools adopted are described in turn below.

5.1 Integration of Spatial Development and Transport

Through land use policies contained within the London Plan, patterns and forms of development that improve accessibility of services and reduce the need to travel are encouraged. High trip-generating developments will be focussed in areas with good public transport access, or sufficient existing or planned capacity. In east

London, in particular, a priority is to maximise development opportunities around existing or committed transport infrastructure, making the best use of available or planned capacity. Through setting appropriate parking standards, encouraging smarter travel planning and making public transport more attractive, the use of public transport, walking, cycling and car sharing are supported.

While London's boroughs are the local planning authorities for development proposals in the capital, major planning applications that meet certain criteria are referred to the Mayor for his consideration. As part of this process, TfL provides advice on transport impacts and mitigation to ensure that new developments are fully integrated with the network. This includes ensuring that transport accessibility, capacity and connectivity are sufficient to cater for new residential and commercial developments. Where necessary, improvements to the transport network are secured as part of this process. TfL offers a pre-application advice service that enables developers to identify transport issues at an early stage in the planning process. For all planning applications that meet the criteria for referral to the Mayor, comprehensive transport assessments, travel plans, delivery and servicing plans and construction logistics plans need to be submitted in accordance with TfL best practice guidance. These documents aim to demonstrate how the application complies with transport policies in the London Plan and include measures to address likely impacts on the transport network.

East London in general and London's Docklands in particular is a microcosm demonstrating the intrinsic interdependence of land use and transport planning. In the 1950s some of London's highest bus flows were those serving the docks. There was very limited demand for rail transport and connections to central London. With containerisation and the increasing attractiveness of Tilbury and Felixstowe, the future of the Docklands was at risk and plans began to emerge in the early 1970s for their redevelopment. Demand for public transport to the proposed mixed-used redevelopment was initially expected to be very modest—a maximum of 9,000 work trips an hour. With significant Government intervention Docklands has been transformed into a major centre of finance. This has been built in increments as rail capacity has increased.

London's first light rail project was developed in the 1980s. It proved to be inadequate given the level of development in the area and has been upgraded and extended regularly. The substantial office development at Canary Wharf helped to justify the subsequent extension of the Jubilee line, and the area will also be served by Crossrail, giving it excellent rail connections and interchanges allowing fast links across the city. The area is expected to employ 110,000 more people by 2031 and the transport network capacity will be boosted through DLR three-car expansion, Jubilee line upgrade and Crossrail.

5.2 Achieving the Most Efficient Use of the Transport System

A number of policies, many innovative, are being delivered across TfL to improve the efficiency of the transport system, including:

- A package of measures to smooth traffic flow and achieve more reliable journey times. Through the *Traffic Management Act 2004*, TfL and London's boroughs are required to ensure London's road network runs as smoothly and efficiently as possible. TfL's approach comprises: a review of signal timings and removal of unnecessary signals to improve efficiency; the enforcement and coordination of street works through a permit scheme to minimise disruption to traffic flow; and improved incident management to better handle the impact of unplanned events such as collisions. Around 36% of London's traffic delay is due to roadworks, costing the economy around £1bn each year.
- TfL is working with the DfT to implement a lane rental system for roadworks on main roads during peak periods where this would have a significant effect on journey time reliability.
- Simplification of payment methods on public transport including the introduction of payment by touching in and out by contactless-enabled credit and debit cards without a ticket from 2012.

5.3 Providing Further Transport Capacity

Even with the significant investment in the transport system as outlined above, some key areas of the network will continue to be crowded as a result of projected growth in population and employment. Further investment in transport infrastructure will therefore be required.

London's continuing success depends on efficient and effective connectivity at all levels—international, national, inter-regional, London-wide, subregional and local. High-speed rail in the UK and better rail services to Europe are recognised as an alternative to short- and medium-haul air travel. Schemes such as the Chelsea Hackney line (Crossrail 2), an extension of the Northern line to Battersea, new river crossings in east London, a possible extension of the Bakerloo line in southeast London, further capacity enhancements on national rail service and better interchange are required to provide more capacity and improve the connectivity of areas currently less well-served by public transport. Figure 5 summarises these further improvements.

There is now greater recognition of the important role for outer London and the need to enhance its vitality, including improved accessibility to and between metropolitan town centres. Working closely with the boroughs, locally agreed approaches to improving orbital connectivity are being developed through better integration of public transport services, walking and cycling initiatives and improved information provision. Town centres, particularly in outer London, are recognised as priority places for growth and investment in the London Plan and potential measures to enhance accessibility to and within them are illustrated in Fig. 6.

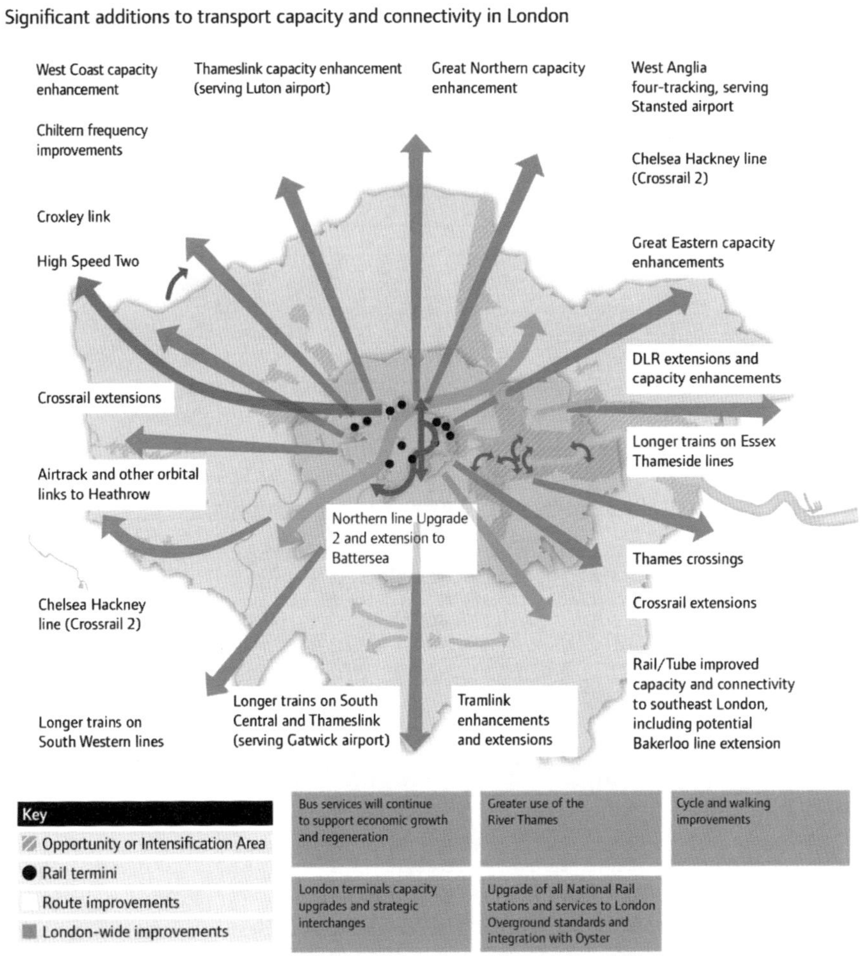

Significant additions to transport capacity and connectivity in London

| West Coast capacity enhancement | Thameslink capacity enhancement (serving Luton airport) | Great Northern capacity enhancement | West Anglia four-tracking, serving Stansted airport |

Chiltern frequency improvements

Croxley link

High Speed Two

Chelsea Hackney line (Crossrail 2)

Great Eastern capacity enhancements

Crossrail extensions

DLR extensions and capacity enhancements

Airtrack and other orbital links to Heathrow

Longer trains on Essex Thameside lines

Northern line Upgrade 2 and extension to Battersea

Thames crossings

Chelsea Hackney line (Crossrail 2)

Crossrail extensions

Rail/Tube improved capacity and connectivity to southeast London, including potential Bakerloo line extension

Longer trains on South Western lines

Longer trains on South Central and Thameslink (serving Gatwick airport)

Tramlink enhancements and extensions

Key

Opportunity or Intensification Area

Rail termini

Route improvements

London-wide improvements

Bus services will continue to support economic growth and regeneration

Greater use of the River Thames

Cycle and walking improvements

London terminals capacity upgrades and strategic interchanges

Upgrade of all National Rail stations and services to London Overground standards and integration with Oyster

Fig. 5 Summary of schemes required to deliver MTS outcomes to 2031. *Source* MTS 2010

5.4 Managing Demand for Transport

To make the best use of London's limited road space, more efficient modes of transport in terms of road space will continue to be encouraged, in particular use of buses, and cycling and walking. Alongside this, TfL is working with the boroughs to deliver smarter travel initiatives to encourage people to increase the share of journeys made sustainable travel modes.

TfL has pioneered the use of smarter travel initiatives. Around 93% of London's schools have a School Travel Plan resulting in the proportion of trips to school by car falling by over 6%. More than 10% of London's workforce also works in locations with travel plans, achieving a 13% reduction in the proportion of car journeys to work at these sites. Smarter travel provides the opportunity to explore flexible working patterns

Illustrative town centre improvements

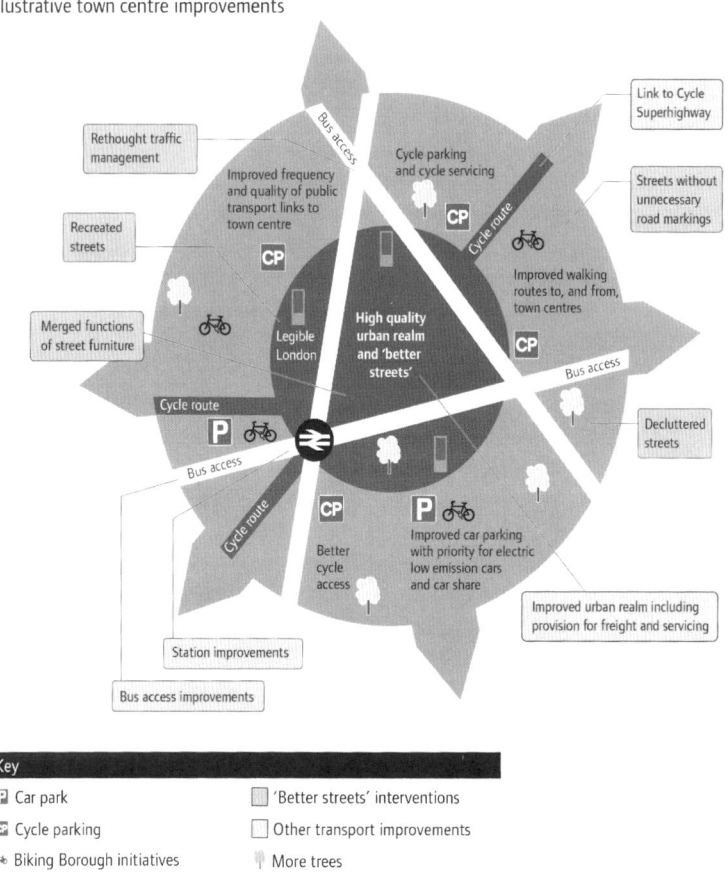

Fig. 6 Potential improvements to town centres. *Source* MTS 2010

and remote working to reduce the need for travel, especially during peak hours. London is a national leader in the development of car clubs with a rapidly growing membership of around 130,000, representing almost 90% of the UK membership, and their expansion and use of low-carbon vehicles in their fleets is encouraged.

In the longer term, road user charging schemes may be required if other measures are insufficient to meet mayoral goals and where there is a reasonable balance between the objectives of any scheme, its costs and other impacts.

6 Reducing Carbon Emissions From Transport in London

TfL is using targeted policies to reduce CO_2 emissions and negative impacts from transport. The Mayor's CO_2 target reduction of 60% emissions by 2025 from 1990 levels, more ambitious that the Government target of 80% reduction by 2050, sets a specific medium to long-term challenge for the capital. Road vehicles currently account for around 72% of ground-based transport CO_2 emissions in London so much effort is needed to reduce them from this source. TfL's approach to reduce emissions from transport is focussed around three main areas:

- Modal shift to public transport, walking and cycling as these are the most sustainable modes.
- Improved operational efficiency of the transport system.
- Enabling the development and deployment of low-carbon vehicles, technology and energy.

CO_2 emissions from transport need to be reduced to between 5.3 m and 4.6 m tonnes to meet the Mayor's 2025 target. The range represents the uncertainty across the distribution of savings between transport, homes and workplaces in London. The greatest step towards achieving CO_2 savings will be from improved vehicle efficiency and the use of low-carbon energy sources. There is also a significant anticipated impact from driver training and enhanced operational efficiency. However, while much can be done at a local level these will be insufficient to achieve the Mayor's target and further action from Government, the Mayor and/or London boroughs will be required. National and international incentives in the areas of industry support, vehicle purchase and scrappage will be crucial to realising the full potential rate of change and scale of impact. All of these actions are subject to regular review in this developing field of expertise.

6.1 Driver Behaviour and Operational Efficiency

Awareness of the environmental impact of travel choices, driving style and vehicle maintenance is growing but needs to increase within the constraints of available technology and infrastructure. Specifically, TfL has run a two-year smarter driving campaign complemented by the nationwide 'Act on CO_2' campaign, to communicate clear, practical methods to improve fuel efficiency.

A key part of TfL's approach in this area is to improve the efficiency of its operations, to reduce energy consumption. The majority of the city's rail-based public transport networks are electrified. Further investment to complete electrification of London's rail network, including the Gospel Oak to Barking line, is needed. On the road network, a trial of light-emitting diode (LED) traffic signals that reduce power consumption by around 60% has been successful; funding is now available to install LED signals at around 300 junctions across London.

6.2 Smoother Traffic Flow

Stop-start traffic conditions and congestion leads to increased CO_2 emissions. Improved management of London's road network (including re-phasing of traffic signals and introduction of a state-of-the-art traffic control centre) and driver information will enable a smoother flow of traffic and ultimately reduce CO_2 emissions given a constant volume of road traffic.

6.3 Improved Vehicle Efficiency

Given the international nature of the climate change challenge, regulations and agreements made by large automobile and aircraft manufacturing companies at an international level have the potential to be particularly effective. EU regulations will enforce average emissions from new cars in Europe of 130 g CO_2/km from 2015 (compared to around 150 g CO_2/km today), with a target of 95 g CO_2/km by 2020. A similar EU directive is under review for vans. This will help improve internal combustion engine efficiency and hybridisation of fleets still further.

6.4 Low- and Zero-Carbon Energy Supply

Biofuels, hydrogen and electric power will have an important role to play in transport's future energy supply. The European Renewable Energy and Fuel Quality Directives require that 10% of transport energy comes from renewable sources by 2020. Electricity is expected to supply a growing proportion of transport energy requirement. Therefore, the CO_2 efficiency of transport and electricity generation will become ever more closely linked. Further decentralised energy production in London will lead to CO_2 savings and improve security of supply as will Government ambitions for decarbonisation of the National Grid supply. Uptake of electronic vehicles will increase demand for electricity. However, it is anticipated that the majority of the additional demand could be met without substantial additional generating capacity if incentives are provided to ensure the vast majority of recharging occurs at night.

Encouraging the shift from petrol to low-carbon vehicles requires incentives that pull people towards low-emission vehicles and disincentives that discourage the use of higher CO_2 emitting vehicles. Low-carbon vehicles do not have to pay the congestion charge, potentially saving drivers around £2,500 a year. Distribution infrastructure networks for alternative fuel sources such as electric charging points, biofuels and hydrogen refuelling facilities, will play a crucial role in the conversion to alternatively fuelled vehicles. Development of the network of electric charge points in London is well underway. TfL is working with a consortium to install at least 1,300 by 2013—by comparison just 104 were in place in 2011. The land use planning process supports the implementation of EV charging points for new developments. It is planned to secure a London EV fleet of 100,000 vehicles by 2020 if not sooner.

In the long term, the combination of electric power and decarbonisation of electricity generation has the potential to go a long way to the decarbonising of car use. This would meet environmental needs, while maintaining the societal and economic benefits realised through the advent of affordable private motorised travel. This approach is in line with the Government's White Paper on Local Transport [9].

6.5 Public Sector Fleet Efficiency

Meeting the emissions challenge necessitates the implementation of strategies that make the public sector fleet more efficient. The implementation of regenerative braking across nearly all rail services in London will provide a CO_2 saving of around 15% as electricity produced while braking is transferred to the power supply network for other trains to use. All new buses are to be low-emission (initially diesel/electric hybrid, before the roll-out of full hybrid vehicles) in addition to a low-emissions taxi programme. A programme also is in place to convert the GLA fleets to low-emission vehicles.

6.6 Smarter Travel and Changing Working Patterns

Continued investment in smarter travel is a key TfL commitment which will have positive impacts on emissions through modal transfer away from private car for short trips. Working with schools, employers and major trip generators will help communicate information regarding individuals' choices and, in doing so, work towards achieving improved modal shares.

As described earlier, one of the approaches in the capital is the encouragement of mode shift from private to public transport. Achieving this necessitates a multifaceted, cross-modal strategic approach which will have a number of positive outcomes, including reducing emissions from transport.

7 Conclusions

The changes experienced in planning and transport policy have been significant over the past decade or so—the effects have been very positive in terms of mode share and delivery of innovative projects and programmes. The level of integration has enabled TfL to make a compelling case for investment in London's transport over the last ten years, planning investment carefully, demonstrating the economic benefits of good transport provision and overcoming years of under-funding. London has benefitted greatly from the strong leadership brought by having a directly elected Mayor. This has helped raise the profile of the role of transport in the city's success. However, it also means that the organisation needs to be very well-informed of local political issues and to work within a strongly political context. TfL is, as an organisation, moving towards an age of maturity and is able to work well within this political context to achieve the right outcomes for the capital.

To achieve desired outcomes in the future, more challenging mode shifts are necessary. The integrated approach adopted by TfL for the capital aims to increase the mode share of public transport, walking and cycling from 57 to 63%. This means that between 3.5 and 4.5 million extra trips each day will need to be made by public transport, walking and cycling to support the growth envisaged in the London Plan.

As is shown by London's historic development, its growth has always been strongly influenced by its transport system. This relates to the spatial provision of transport capacity and services, which was the prevalent factor in the past, but also to the provision of accurate transport information and a higher quality of environment. TfL is trying to encourage people to think differently about travel to encourage a shift away from the private car, to cycle and walk more, to drive differently, to improve road safety, smooth traffic and reduce harmful emissions. The challenge ahead is amplified by the expected rise in population and the need to improve health and quality of life for residents.

The approach adopted for the capital is multi-faceted and fully integrated—across strategies, policy areas and transport modes. This will ensure that land use densities are maintained or increased in areas of high public transport accessibility and, where capacity exists, that an outcome-focussed approach is maintained to ensure goals are monitored and achieved, and that innovative thinking is maintained but in the context of public and political awareness and support. TfL has established an important role in helping to integrate the planning process in London, and in bringing the previously disparate transport modes together. This role will continue to support the capital as it continues to grow and emerge from the recession, helping to balance the practical, political and financial demands of a vibrant world city.

References

1. Mayor's Transport Strategy (2010) Greater London Authority
2. TfL Business Plan 2011/12–2014/15 (2010) Transport for London
3. London Plan (2011) Greater London Authority
4. Economic Development Strategy (2010) Greater London Authority
5. Mayor's Air Quality Strategy (2010) Greater London Authority
6. Mayor's draft Climate Change Mitigation and Energy Strategy (2010) Greater London Authority
7. TfL Annual Report (2011) Transport for London
8. Travel in London Report 4 (2011) Transport for London
9. Creating growth, cutting carbon: making sustainable local transport happen (2011) White Paper Department for Transport

Sustainable Urban Mobility

Debra Lam and Peter Head

Abstract For many cities, traditional transport comprises a sizeable percentage of total carbon emissions. It also contributes to air pollution, poorer health, and resource inefficiencies in the form of higher oil prices, traffic jams, etc. Often city policy-makers do not account for climate change impacts and natural disasters or consider alternative transport options and networks. It does not have to be like this. Cities can continue to develop and grow, attracting industry, high-skilled workers, tourists with sustainable urban design, and mobility. With walking, cycling, green public transport, and shared vehicle use taking the lead, and supported by ICT, cities can become less reliant on traditional and personal transport. Instead, city policy-makers can aim to increase accessibility and convenience to their residents and visitors alike, including rapid and safe mobility in times of emergency. This can be done with good urban design, behaviour change, advance technology, supportive policies, economic incentives, and city engagement and leadership.

1 Introduction

Sustainable urban mobility is about the ease, convenience, affordability, and accessibility of travelling to one's destination with minimal impact on the environment and others. Travel should be safe, at optimal speeds, and by the most direct routes. Options, with real-time information should be readily available for travellers to chose based on time, costs, distance, and other factors. For example, for a short trip, a student may prefer biking, whereas an elderly person might

D. Lam · P. Head (✉)
13 Fitzroy Street, London W1T 4BQ, UK
e-mail: peter.head@arup.com

D. Lam
e-mail: debra.lam@arup.com

O. Inderwildi and Sir David King (eds.), *Energy, Transport, & the Environment*,
DOI: 10.1007/978-1-4471-2717-8_19, © Springer-Verlag London 2012

choose a clean bus. For longer trips, a family with a pet may choose to hire an electric vehicle from a car club, whereas a businesswoman elects riding rail.

Options should not be exclusive. They can be connected to provide the best route. Consumers can choose to walk or bike before riding public transport. Workers can park their vehicles before joining car pools. These options are aided by improved transport and communications technology, higher density living around cities, and changing the business model to provide increased frequency and choices of travel modes.

This chapter will highlight the three main drivers that are pushing sustainable urban mobility away from dominant private vehicles, and describe strategies to help support it in that direction. Sustainable urban mobility cannot be achieved overnight, nor is it a constant state. Rather, it is a process that will help guide stakeholders in improving overall travel and choices.

2 Drivers of Sustainable Urban Mobility

Mobility, economic and social development are very much interlinked. Mobility is essential for development and growth, as we seek comparative advantages in trade and obtain resources. And as we continue to develop, we require and expect greater mobility. In turn, mobility enables more and diverse access to goods and services, wider personal contacts, and greater awareness of the world.

But as we increase the frequency of travel, the traditional means of transport become a burden to our development and quality of life. We can no longer continue to depend on oil consuming, personal vehicles as our main transport means. We need alternative ways to support our growing mobility and development, without the devastating consequences. Mobility will not decrease, but it can change to a more sustainable option.

Personal vehicles have been an important travel means. They have increased our independence, and access. However, their traditional use is now in doubt and driven by three main factors:

- *Climate Change*—rising transport-related carbon driven by oil dependence, and climate change impacts
- *Environment and Health*—resulting in poorer air quality, congestion, and health effects
- *Economic*—rising fuel and congestion costs, wasted time, and resources

3 Climate Change Driver

Transport is currently responsible for 19% of global energy use and 23% of energy-related carbon dioxide emissions. This is expected to rise substantially. Transport-related CO_2 emissions are to increase by nearly 50% by 2030 and over

80% by 2050, more than other sector emission [1]. What are ways to decrease transport-related carbon emissions? How can we reduce our reliance on oil?

Changing transport will be critical if the global community is to decrease carbon emissions, and address climate change impacts. Tackling transport is seen to be more effective as climate change negotiations at the international and national level continue to stall and vary. This is especially true for cities, which are built for high-density and extensive travel plans and networks, but have limited say in the international climate change talks.

The other half of the climate change push from decreasing transport-related carbon emissions is addressing the inevitable climate change impacts with more sustainable transport. Climate change impacts, such as snowstorms, typhoons, and other extreme natural disasters will be more severe and frequent in the coming years [2]. For instance, China's huge snowstorm in 2008 left millions stranded in train stations. Is the transport network prepared to withstand climate change impacts? Is the transport network equipped with emergency transport routes that are publicly known? Is there real-time information accessible to show the latest transport route changes and status?

4 Environment Driver

The second driver for sustainable urban mobility is around the decreasing environmental quality spurred by greater car use. Private car ownership is correlated with rising income, especially in emerging economies. Consumers begin buying cars as necessities when a country's per capita gross domestic product (GDP) reaches USD 1,000 [3]. While on a per capita basis, it is still small, China's car market is the second largest in the world, and is expected to overtake the US within a generation [4].

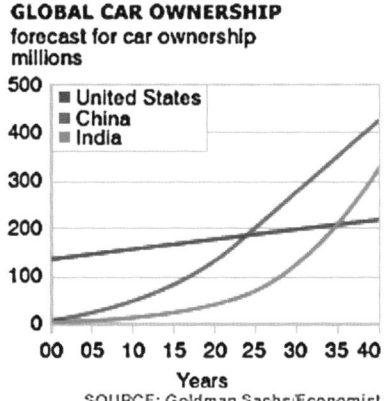

Greater numbers of traditional fuel vehicles result in greater congestion and worsen air quality. Heavy congestion is now a daily part of many commuters' life. According to the US Census Bureau, the average American spends almost 50 min commuting a day [5].

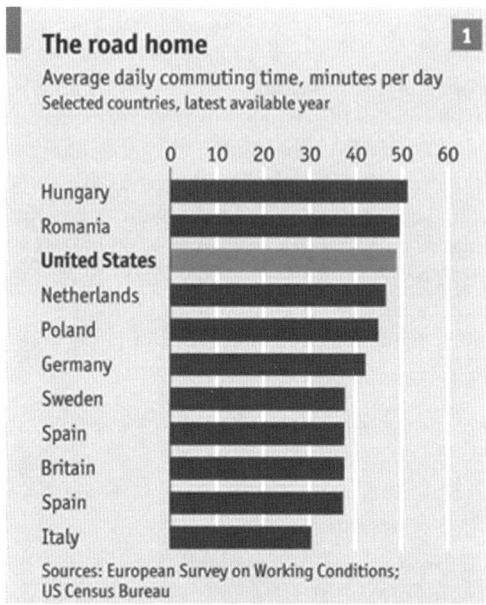

More time in cars and more cars on the road worsen air quality. Many megacities have some of the worse air quality in the world, greatly exceeding World Health Organization recommendations. The numbers of premature deaths and asthma rates, especially in children, have increased as a result. Preliminary assessments indicate that diseases related to the air pollution caused by road transport affect tens of thousands of people in the WHO European Region each year [6]. There should be a better way to travel than sitting in traffic and breathing in poor air.

5 Economic

The final driver is around the economics. Consumers are demanding faster access to goods and services, businesses want to improve ways to communicate and conduct work, and tourists aim to see more places. They all demand that it be cost-efficient. Traditional travel, dependent on personal vehicles, is becoming more expensive especially if you account for the externalities. The 2010 Urban Mobility Report, published by the Texas Transportation Institute at Texas A&M University, paints the most accurate picture yet of traffic congestion in the 439 U.S. urban areas. According to the report, congestion costs have risen from $24 billion in 1982 to $115 billion in 2009. The total amount of wasted fuel in 2009 topped 3.9 billion gallons [7].

Moreover, the supply of the world's main transport fuel, oil, is unlikely to meet increasing demand and is increasing in price. Whether oil has peaked or not, most energy analysts have concluded that the age of oil dominance and cheap prices are over. While technology has improved, it is getting harder to extract unknown and known supplies of oil. Environmental catastrophes as seen in the recent BP spill in the Gulf of Mexico or geopolitical concerns cast further doubt on the stability of supply and prices. The limited and inconsistent oil supply is expected to increase oil prices to over USD 200/barrel by 2035 [8]. The UK Industry Taskforce on peak oil and energy security, to which Arup has contributed, found that oil shortages, insecurity of supply, and price volatility could potentially destabilise economic, political, and social activity by 2015 [9]. It will be increasingly difficult to rely on oil as the main travel fuel.

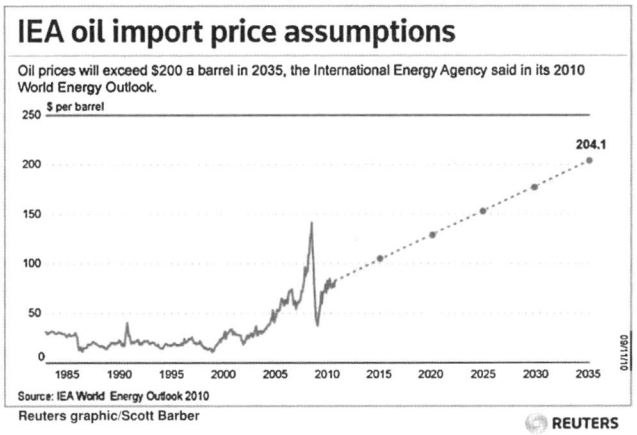

Source: IEA World Energy Outlook 2010
Reuters graphic/Scott Barber

6 Strategies for Sustainable Urban Mobility

With such drivers pushing sustainable urban mobility, we propose the following strategies for cities to lead:

- Integrated urban planning and design
- Expanding eco-vehicle use
- Enact behavioural change
- Low emissions policy

We have seen these strategies implemented at various scales and degrees by other cities, such as London and Hong Kong. They have shown to decrease carbon emissions, improve environmental quality, and serve greater economic and social value. Cities, with their growing populations, economic weight, governance ability, and density, are the places to enact such strategies and pursue sustainable urban mobility. Local governments will play a major role in setting targets and passing legislation but must create the right conditions that allow market forces to operate and business to take advantage of opportunities that arise from improving transport.

7 Integrated Urban Planning and Design

Integrated urban planning and design should promote high density, compact development around major, clean public transportation nodes, green networks, and business links. Travel should be done if necessary, supported by a strong communications system to decrease the need to travel. A strong communications system would allow city policy-makers and residents greater precision in travel choices, speeds, and times. Like travel costs, they can also determine the amount of energy and resources consumed with their choices.

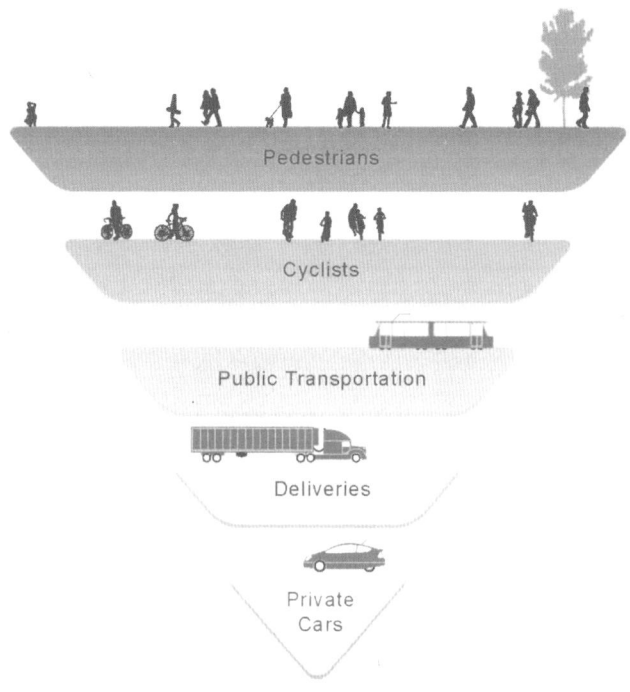

"Tram-stops, bus-stops and train stations are 'informationally-rich' spaces. By offering real-time location-based information they will provide users with a better service, and provide city leadership with data on mobility patterns, occupancy rates and interaction. Users can make informed decisions about journeys, and the city can more effectively adapt the transport network to fit usage patterns" [10].

If travel is needed, it should be tiered, first with walking and cycling, then with eco-public transportation. Private vehicles should be the last option and seldom used. The urban plan would encourage such tiered travel through its integrated design, supporting communications and logistics system, and inter-city connections.

One of the largest differentiators in the ecological footprint of cities is the relationship between urban density and transport energy use. An average urban dweller in the United States consumes about 24 times more energy annually in private transport than a Chinese urban resident [11]. There is a sweet spot of urban density of 75 persons/hectare in which transport energy use is reduced through the economic provision of public transport and there is still ample room for urban parks and gardens [12].

A green corridor with some pedestrian-only streets and dedicated bike lanes should be heavily used and connected to city's overall transport network. Bike locking and parking infrastructure would be widely distributed and convenient. The green corridor would include parks, urban gardens, and ample open space to encourage outdoor activities and community interaction.

Public transport, fuelled by renewable and alternative energy, would also be heavily embedded into the urban design. Public transport investment, aided by increases in urban density, in rail, metro, bus, and tram and better information systems will enable more journeys to be taken by efficient public transport. Relatively low cost bus systems in dedicated lanes have been very successful in places like Curitiba, Brazil [13]. Replacing selected roads with green networks will provide more direct walking and cycling access to work, schools, shops, and public facilities (Fig. 1).

This is a very big challenge in the United States, where transport is extremely energy intensive due to car dependency and the spread of suburban development. Los Angeles is trialling the replacement of low-density single use city blocks with high-density mixed use and this is proving commercially successful. Removing major freeway infrastructure from urban areas would free up valuable land, remove the burden of huge maintenance costs, and provide funding for public transport. Vancouver has demonstrated how well a city works without being dependent on freeways. It comes high up the list of the most liveable cities in the world and has relatively low carbon emissions. High-quality higher density developments are now being built in the city centre rather than extending the suburbs.

- Logistics and communications
  Energy consumed in goods distribution in urban areas can be minimised by the use of consolidation centres around the city perimeter which are accessed by

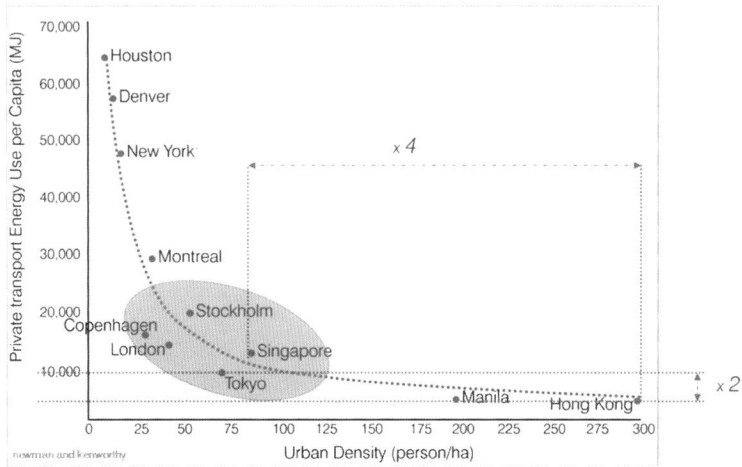

Fig. 1 Transport and urban density [12]

intercity rail and road links. Distribution from these centres can be made using a fleet of zero emissions vehicles on an organised basis to minimise travel distances and congestion. Delivery reliability can also be improved. Use of consolidation centres around the city perimeter for goods delivery will also improve delivery and energy efficiency.

A strong communications infrastructure would enable industries and residents to work and obtain information, services, or goods online. Commuters can easily access real-time information on bus and rail routes, including delays, alternative routes, and its capacity.

• High-speed rail

The energy efficiency of inter-city travel is likely to be achieved through a combination of investment in a high-speed rail passenger networks (eventually running on renewable energy), using bus and car share priority on motorways, improved information and traffic management systems, and improved vehicle and fuel technology. Now that there is a viable high-speed rail network we understand that, where available, high-speed rail is a more attractive option than regional air travel for distances of up to 600 km. When high-speed rail was introduced, rail user numbers doubled and on some routes, such as the 300 km Paris–Brussels route, air travel dropped to a negligible level. This experience has also been confirmed in Japan. High-speed railway investment needs to include the capacity for rail freight movement with links built directly to city edge consolidation centres. This is the area of ecological footprint reduction that will be difficult until renewable fuel supply-powered road vehicles are available at competitive prices for long distance passenger and freight use. This may, however, be the case in the future.

- Air travel

  Growth in air travel from airports is still accelerating because the demand for leisure flights has increased, with decreasing costs (and frills) from budget carriers. If this level of usage continues without any technology changes, then emissions from air travel will become the single biggest source of greenhouse gas emissions by 2050 [14]. Construction of a comprehensive European high-speed rail network connected to all UK regions and airports such as Heathrow and Charles De Gaulle will be important, as will finding a renewable source of aircraft fuel by 2050.

  New airports should be focused on travel over 600 km and they should be located on high-speed rail routes and connected into local urban areas with mass transit systems. Also, high-speed rail investment should have equal priority with roads. The major challenge in USA is to follow these principles and consider putting in place a high-speed rail network that serves urban centres within 600 km of each other and connects to airports.

8 Expanding Eco-Vehicle Use

Personal transportation is one of the areas most in need of change. For all their ill effects, it is unrealistic to abolish private vehicles. Rather, private vehicle use should be the last resort, and connected with other travel modes. Car clubs, hybrid, and/or electric vehicles should be encouraged. Car clubs will enable people to hire vehicles when they need them. Their use is growing quickly in many cities like London. Research has shown that users drive 64% less distance after joining a club and that each club vehicle replaces, on average, 20 privately owned cars [15].

Electric or hybrid private vehicles can be seen as one of the many transport choices, and selected only when walking, cycling, public transport, and rail are not attractive options.

Battery- and hydrogen fuel cell-powered vehicles for private, public, and goods delivery use will be part of the mix with hydrogen sourced from natural gas or other sources. Electric vehicles are seen to have a big advantage in terms of management of intermittent renewable energy supply by increasing the storage capacity of the grid. Vehicles become mobile storage system, controlled with smart network technology.

From the Chevrolet Volt to Nissan Leaf, most major car manufacturers have hybrid or electric cars on the road or in development. Increased research and resources devoted to these types of cars has improved selection, design and prices. While ever-higher demand is expected to drive up the market share of hybrid and electric vehicles. Chinese manufacturers such as Geely have been successful at producing modest, low-priced electric vehicles, while Lexus has dominated the premium market. Expensive and unreliable batteries are becoming less of an issue with improved technology, supporting infrastructure, and changing business models (Fig. 2).

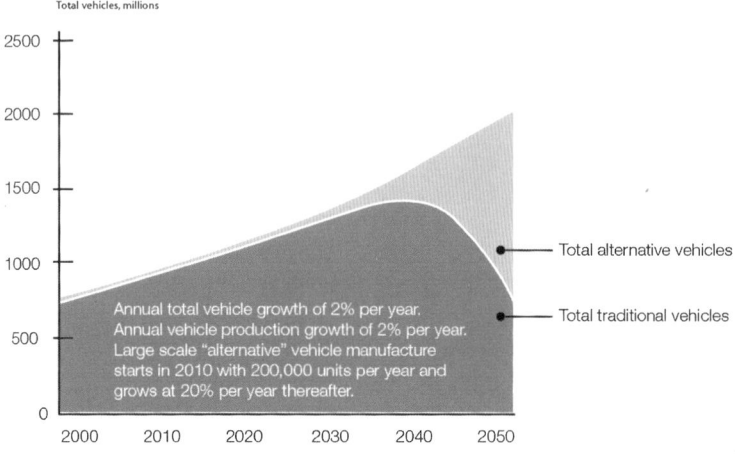

Fig. 2 Increasing alternative vehicles [16]

9 Enact Behavioural Change

Integrated urban design and planning and expanding eco-vehicles will certainly support sustainable urban mobility and steer consumers away from traditional personal vehicles. Fundamentally, however, we need to understand why people gravitate to private vehicles in the first place and how to encourage them to use alternatives. We need to understand the factors that cause people to drive, how they personalise their vehicle, and the status that it carries. We need to enact change management so that consumers can see the benefits of their contribution and change their behaviours.

Owning a car is an important goal for many middle classes in emerging economies. It is a symbol of freedom, independence, and having 'made it'. We need to show that other transport modes- walking, cycling, public transport, and eco-vehicles can provide the same level of convenience, accessibility, and freedom as the personal vehicle, without its negative effects.

Arup has begun to develop a wider transport understanding with its UK Coventry and Birmingham Low Emission Vehicle Demonstrator (CABLED) project. The project aims to increase the use of electric vehicles, through a diverse group of stakeholders–vehicle manufacturers, local authorities, energy providers, and academia—working together. Despite financial incentives, electric car use in the UK is currently low and concentrated in London. It is expected to increase, but for that to happen smoothly, a greater understanding of its drivers and its use of cars needs to be established. The consortium team will trial 110 vehicles on the roads of Birmingham and Coventry to understand driver behaviour. They will then develop the necessary infrastructure to support driver needs and encourage behavioural change.

10 Low Emissions Strategy

A carbon footprint reduction can best be achieved through the formation of a low emission strategy (LES) [17]. The primary aim of a LES is to accelerate the uptake of low emission fuels and technologies in and around a development site. This usually takes the form of an area where some kind of enforcement is carried out to ensure particular types of vehicles are restricted. Some UK authorities are already making effective use of LES. LES are secured through a combination of planning conditions and legal obligations. They may incorporate policy measures and/or require financial investments in and contributions to the delivery of low-emission transport projects and plans, including strategic monitoring and assessment activities.

LES provides a package of measures to help mitigate the transport impacts of a development. Sample LES recommendations include:

- Develop a transportation demand management programme, with short- and long-term strategies, that manages transport demand through limiting car use, increasing alternative travel options, and redistributing travel demand. Strategies include congestion pricing for cars as seen in London or Singapore, or limiting cars in the city to those with certain license plate numbers on particular days as seen in Beijing. They also include better infrastructure for pedestrians, cyclists, and public transport users with improved street signs, designated paths and crossings, parking and waiting areas, and subsidised costs.
- Establish supplier parks to minimise transport costs/times and potentially promote synergies between supply chain members.
- Ensure that a strong education and communications platform is supporting LES so that residents and visitors alike support and are aware and know how to use the transport system.
- *Parking strategies.* Reduce supply by changing requirements for residential areas and employment centres, introduce shared parking facilities (entertainment-office-transit), limit free parking, and provide priority spaces for carpools and low-emission vehicles.

Rather than ad hoc, and isolated sustainable transport initiatives, local governments can adopt a holistic LES, where high transport standards are maintained over a longer period of time, high polluting vehicles are controlled, and all transport trips, including logistics, are made in the most sustainable way possible. The formation of a LES provides a package of measures to help mitigate the transport impacts of a development, and provides further access and travel ease to industries and people. A LES can make a city's regional links more efficient and carbon friendly, while remaining competitive in the transport and delivery of people and goods. Improved transport links might incentivise tourists to visit and residents to stay. Residential living can be improved through better live-work units and improved communications, reducing the need to travel so often.

Fig. 3 Arup's virtuous cycle
[18]

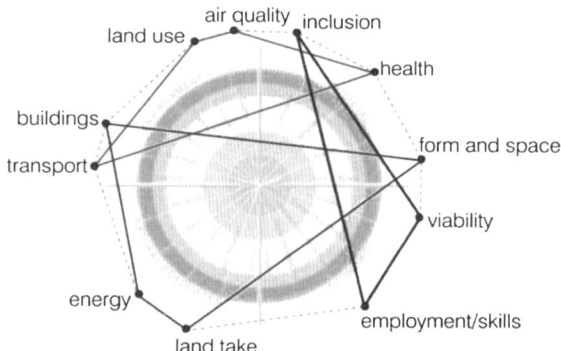

11 Conclusion: Integrated Strategies

Our overreliance on oil, increased carbon emissions, wasted resources, degrading environment, and increasing awareness of climate change is causing us to rethink our transport options. Mobility will not decrease, rather it can become more sustainable with better urban design and planning, expanding use of eco-vehicles, shifting consumer behaviour and low emissions strategy.

While this chapter focuses on sustainable urban mobility, we recommend a more comprehensive approach in pushing for broader sustainable development. It is impossible to implement one technical strategy, in this case, transport, in isolation. But other technical strategies can support the development of sustainable urban mobility and leverage its benefits. We call these linkages and connections "virtuous cycles" to encourage more holistic thinking. A simple example is that with the use of quiet electric vehicles and pedestrian streets, there is less need for noise attenuation, so the facades of buildings can be lighter and therefore consume fewer resources. Similarly, other technical strategies in water, waste, or energy can bring transport benefits for broader sustainable urban mobility. Virtuous Cycles also promote technical specialists or city agencies to work together. The Transport Bureau can work with the energy specialist or the finance expert on what was considered traditionally a 'transport issue'. By connecting the dots or wider spheres of city operations, cities can move beyond singular issues into broader sustainability (Fig 3).

References

1. International Energy Agency (2009) Transport, energy, and CO_2. http://www.iea.org/textbase/nppdf/free/2009/transport2009.pdf
2. The International Bank for Reconstruction and Development/The World (2010) Cities and climate change: an urgent agenda, vol 10, December 2010
3. China's private car ownership tops 10 million. http://www.china.org.cn/english/BAT/67018.htm

4. Schifferes S (2007) Cracking China's market, BBC News, 17 May 2007. http://news.bbc.co.uk/1/hi/business/6658583.stm
5. Economist (2011) Life in the slow lane, 28 April 2011. http://www.economist.com/node/18620944?story_id=18620944
6. Krzyzanowski M et al (eds) (2005) Health effects of transport related air pollution, World Health Organization. http://www.euro.who.int/__data/assets/pdf_file/0006/74715/E86650.pdf
7. 2010 Urban Mobility Report. http://mobility.tamu.edu/ums/media_information/press_release.stm
8. IEA World Energy Outlook (2010)
9. http://www.arup.com/News/2010-02_February/10_Feb_2010_Action_On_Peak_Oil_Crunch_Threat_To_UK_Economy.aspx
10. Buscher V (2011) Urban life: the smart solutions for cities, Arup, London. http://www.arup.com/Publications/UrbanLife_Smart_Solution_For_Cities.aspx
11. Kenworthy JR (2003) Transport energy use and greenhouse gases in urban passenger transport systems: a study of 84 global cities, presented to the international third conference of the Regional Government Network for sustainable development, Notre Dame University, Fremantle. http://cst.uwinnipeg.ca/documents/Transport_Greenhouse.pdf. Accessed 17–19 Sept 2003
12. Newman P, Kenworthy J (2006) Urban design to reduce automobile dependence. Opolis 2:1, Article 3. http://repositories.cdlib.org/cssd/opolis/vol2/iss1/art3
13. Modes for rapid transport, race, poverty & the environment; Curitiba's Bus system (2007), sourced from http://www.urbanhabitat.org/node/344
14. Hickman M (2006) Cheap flights threaten uk target for carbon emissions, the independent. http://www.independent.co.uk/environment/cheap-flights-threaten-uk-targets-for-carbon-emissions-524879.html. Accessed 28 Jan 2006
15. Transport for London information on motor vehicle usage and ownership. http://www.tfl.gov.uk/tfl/search/?keywords=car%20drivers%20research&direction=next&filter=1&restrict=&lastResult=10. Accessed 2008
16. World Business Council for Sustainable Development (2004) Facts and trends to 2050: energy and climate change. http://www.wbcsd.org/DocRoot/FjSOTYajhk3cIRxCbijT/Basic-Facts-Trends-2050.pdf
17. LES differs from a LEZ (Low Emission Zone) which refers to a geographic area where emissions from road transport are mitigated. A LEZ may be one aspect of an LES in an area
18. Arup's integrated urbanism. http://www.arup.com/integratedurbanism/whoweare.cfm?pageid=7937

Road Use Charging and Inter-Modal User Equilibrium: The Downs-Thompson Paradox Revisited

Michael G. H. Bell and Muanmas Wichiensin

Abstract This paper looks at the impact of charging for road use in cities where an intermodal equilibrium prevails, where increased transit use leads to efficiency gains, and where the roads are congested. Demand is assumed to be elastic and transit is assumed not to be directly affected by road congestion, as would be the case where transit uses a reserved track. It is shown that at a stable intermodal user equilibrium a version of the Downs-Thompson paradox applies if the efficiency gains arising from increased transit use are passed on to passengers as reduced generalised costs (reduced fares, increased service frequencies or both). The paradox arises because the imposition of a road user charge not only reduces road congestion but also reduces the generalised cost of travel at the intermodal user equilibrium, including the road user charge for those who choose to drive. The paper then goes on to consider what would be expected if, rather passing on the efficiency gains to transit users, the transit operator(s) as a whole maximise profits, and establishes that the paradox no longer arises. The implications of these findings for the regulation of transit fares are considered.

1 Introduction

The arguments for a congestion charge are widely understood, at least since Hau [4] presented them diagrammatically. If drivers are confronted with the full marginal cost of making a trip by car then the value of driving covers its full cost.

M. G. H. Bell (✉)
Department of Civil and Environmental Engineering,
Imperial College London, London SW7 2BU, UK
e-mail: m.g.h.bell@ic.ac.uk

M. Wichiensin
Department of Civil Engineering, Kasetsart University, Bangkok, Thailand
e-mail: fengmms@ku.ac.th

O. Inderwildi and Sir David King (eds.), *Energy, Transport, & the Environment*,
DOI: 10.1007/978-1-4471-2717-8_20, © Springer-Verlag London 2012

Marginal cost pricing also ensures that rational drivers would make decisions which minimise system cost, appropriately defined. The principal of congestion charging is therefore to add a charge to the cost of driving which equates to the difference between the marginal cost and the cost directly experienced by the driver, usually referred to as the average cost. In an inter-modal context, however, the impact of such a charge may paradoxically be to lower the cost of driving including the charge. This paper explores the circumstances in which this might occur.

Typically trip-makers in urban areas have a number of modes available to them, which can be categorised as *public* and *private*. The public modes (bus, train, metro, tram, etc.) are assumed here to impose a cost per trip directly on the maker of the trip. The attractiveness of the public modes, referred to collectively as *transit*, depend on the fare, the travel time, the waiting time and other factors which combine to form the *generalised cost* of transit travel. The private modes (car, motorcycle, van when used privately, etc.) also impose a cost per trip primarily in the form of travel time, fuel consumption and vehicle depreciation (or rental), which may be combined to form the generalised cost of *car* travel. However, some of the cost per trip falls on the maker of the trip and some on other trip-makers, through the contribution of the trip to congestion. We assume that trip-makers switch between modes so that for any origin, destination and time of travel at the inter-modal equilibrium the generalised cost of travel is the same for all modes that are used. Moreover, any mode that has a higher generalised cost than the minimum prevailing for a given origin, destination and time of travel will not be used.

A feature of transit is that when there is spare capacity in the system, its efficiency improves with increasing patronage because fixed costs are spread across more trips and service frequencies may be increased, reducing passenger waiting times. In this paper we consider the case where the average cost per transit trip falls with increasing transit travel. By contrast, increasing car travel leads to higher traffic densities, lower speeds and therefore longer travel times. Hence we assume that the average cost of car travel rises with increasing car travel.

When the efficiency gains of increased transit travel are passed on to trip-makers, it seems plausible that the imposition of a charge for car use might reduce the generalised cost of travel across all modes. This, if it occurred, would be a manifestation of the Downs-Thompson paradox. According to Wikipedia [11] the "Downs-Thomson paradox, also referred to as the Pigou-Knight-Downs paradox, states that the equilibrium speed of car traffic on the road network is determined by the average door-to-door speed of equivalent journeys by (rail-based or otherwise segregated) public transport". Hence increasing road capacity can worsen road congestion when the shift from public transport causes the operator to reduce frequency of service or raise fares to cover costs. This shifts additional passengers to cars leading to the congestion on the expanded road network that may in certain circumstances be worse than before [3, 8].

The general conclusion, if the paradox applies, is that expanding a road network as a remedy for congestion is not only ineffective but also counterproductive.

This is also known as the Lewis-Mogridge Position [11] and was extensively discussed by Mogridge [6] . It is an extension of induced demand theory and consistent with Downs [3] theory of "triple convergence" formulated to explain the difficulty of removing peak-hour road congestion. Downs suggests that three effects arise from a highway capacity addition: Drivers on alternative routes begin to use the expanded road, those previously travelling at off-peak times shift to the peak,
and transit users shift to cars. Mogridge et al. [7], Holden [5] and Abraham and Hunt [1] consider further the circumstances where the Downs-Thompson paradox may arise.

Rather than expanding the road network, the imposition of a road user charge is considered here. This is equivalent to a reduction of highway capacity since for a given highway flow a reduction in capacity would lead to an increase in density, a reduction in speed, an increase in travel time and therefore an increase in the generalised cost of travel by car. If the displacement of trips to transit causes the intermodal equilibrium cost of travel to fall, this would be a manifestation of the Downs-Thompson paradox but in reverse. We show that when a displacement of trips to transit improves its efficiency and when this gain is passed on to transit users, then under rather general conditions the new intermodal equilibrium cost of travel will be lower than before, confirming the existence of a Downs-Thompson paradox. The elasticity of demand for travel is taken into account in the analysis.

In reality, the transit operator(s) may choose not to pass on the efficiency gains to transit users but rather use the opportunity to increase profits (or reduce losses). This paper extends the analysis to show that if the transit operator(s) as a whole were to maximise profits then none of the efficiency gains would be passed on to passengers and so there will be no Downs-Thompson effect. The introduction of a road user charge under these circumstances will therefore be to increase (or at least not reduce) the inter-modal equilibrium cost of travel. Even for a very high elasticity of demand for travel, transit profits are not increased by reducing fares in response to a road user charge.

The implication of this finding is that the level of competition in transit provision is crucially important for the welfare of trip-makers. Without effective competition, none of the efficiency gain to transit resulting from a road user charge will be passed on passengers. Worse than this, transit operator(s) without effective competition may have an incentive to increase fares. Policy should therefore be focused on ensuring sufficient competition in the provision of transit or, if this is not possible, regulation should ensure that transit fares are reduced. The conclusions of this paper support game theory findings previously presented in Wichiensin et al. [9], Bell and Wichiensin [2] and Wichiensin and Bell [10].

In the following analysis all functions are assumed to be linear. Since we are only interested in the directions of changes at the intermodal equilibrium, there is no loss of generality.

2 Linear Inter-Modal Equilibrium Model

Let

p	Generalised cost of travel, assumed to be equal for public and private transport and therefore not subscripted ($/trip)
AC_t	The average cost of travel by public transport ($/trip)
$q_c \geq 0$	Quantity of travel by private transport (trips/day)
$q_t \geq 0$	Quantity of travel by public transport (trips/day)
$q = q_c + q_t$	Quantity of travel by all modes (trips/day)
$a, b, c, d, e, f > 0$	Parameters

The demand for travel is assumed to be linearly related to the generalised cost of travel:

$$q = a - bp \tag{1}$$

The generalised cost of travel by private transport (by car) is assumed to be linearly related to the quantity of travel by private transport:

$$p = c + dq_c \tag{2}$$

As q_c increases, traffic density increases, traffic speed reduces and consequently p increases. The average cost of travel by transit is related to the quantity of travel by transit:

$$AC_t = e - fq_t \tag{3}$$

We interpret f as the *efficiency gain* when transit patronage increases by one unit. We assume that there is excess capacity in the public transport system, so that as the quantity of transit travel increases, the fixed cost of transit is spread across more trips leading to a reduction in its average cost. This paper looks at two scenarios. In the first, the average cost of travel by transit is passed on to the traveller. In the second, we assume that the transit operator(s) behave in such a way as to maximise profit (or minimise loss). In both cases, we assume that private and public modes are used so that equilibration (the process of switching from more expensive to less expensive options) results in the generalised cost of travel by public and private transport being equal. It is appreciated that this is a gross simplification of the way the urban transport market works, but it allows us to generate some interesting policy hypotheses worthy of further investigation.

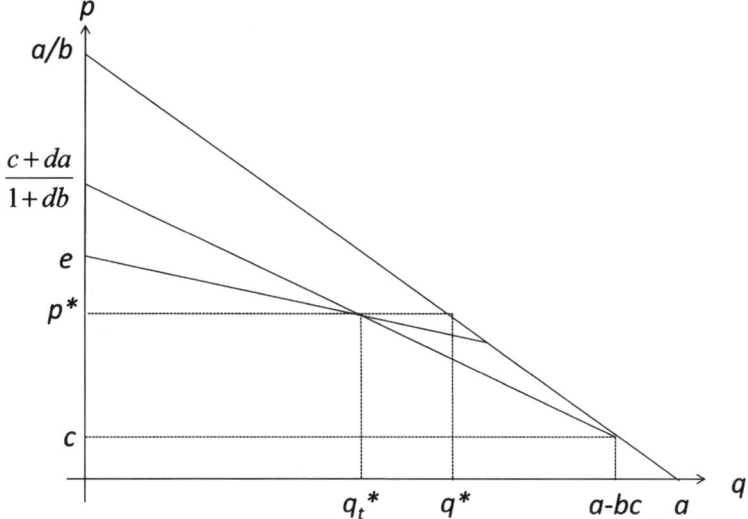

Fig. 1 Stable inter-modal equilibrium for breakeven transit

2.1 Scenario 1: Cost-Covering Transit Operator

The transit operator(s) pass on the average cost to users so:

$$p = e - fq_t \qquad (4)$$

A stable equilibrium is illustrated in Fig. 1. The line from a/b to a represents the demand curve and shows how demand falls with increasing cost per trip. The line from e on the vertical axis represents the average cost of travel by transit as a function of the amount of travel by transit. This falls with increasing transit use due to transit efficiency gains, as these are passed on to the traveller. The line from $\dfrac{c + da}{1 + db}$ on the vertical access represents the average cost of travel by private transport as a function of the amount of travel by private transport, measured from the right and starting at the demand curve, noting that $q = q_c + q_t$. This line starts at $\dfrac{c + da}{1 + db}$ on the vertical axis because when $q = q_c$ ($q_t = 0$) we know that:

$$p = c + dq = c + d(a - bp)$$

so

$$p = \frac{c + da}{1 + db}$$

Conversely, when $q_c = 0$ we know that $p = c$ and $q = a - bc = q_t$.

The inter-modal equilibrium is given by the intersection of the lines from e and $\dfrac{c + da}{1 + db}$ in Fig. 1. At this point we obtain from (1), (2) and (4):

$$p^* = \frac{a}{b} - \frac{q^*}{b} = \frac{a}{b} - \frac{q_c^* + q_t^*}{b} = \frac{a}{b} - \frac{p^* - c}{bd} - \frac{e - p^*}{bf}$$

Collecting terms:

$$p^* \left(1 + \frac{1}{bd} - \frac{1}{bf}\right) = \frac{a}{b} + \frac{c}{bd} - \frac{e}{bf}$$

Hence:

$$p^* = \frac{c + da - \dfrac{de}{f}}{1 + db - \dfrac{d}{f}} \tag{5}$$

and $q^* = a - bp^*$.

Proposition 1 *Stability of the inter-modal equilibrium requires* $\dfrac{d}{f} > 1 + db.$

Proof The slope of the line leading from $\dfrac{c + da}{1 + db}$ on the vertical axis is

$$-\frac{\dfrac{c + da}{1 + db} - c}{a - bc} = -\frac{d}{1 + db} \tag{6}$$

The slope of the line from e on the vertical axis is $-f$. Stability requires that if q_t is below the intermodal equilibrium q_t^* then a generalised cost incentive arises which encourages private transport users to switch to public transport, restoring the inter-modal equilibrium. The converse should occur when q_t is above the inter-modal equilibrium q_t^*, again restoring the inter-modal equilibrium. This incentive arises when:

$$\frac{d}{1 + db} > f$$

namely when:

$$\frac{d}{f} > 1 + db \tag{7}$$

QED

Proposition 2 *At a stable equilibrium an increase in c reduces p^*.*

Proof The impact of a change in the fixed cost per car trip (c) depends on the sign of the denominator of (5). For an increase in c to lead to a reduction in p^* we require (7). Hence stability of the inter-modal equilibrium guarantees that an increase in c reduces p^*.   QED

The implication of Proposition 2 is that if a stable intermodal equilibrium exists in which all modes are used, then a Downs-Thompson effect is inevitable and a road user charge will reduce the intermodal cost of travel. This must be also evident from Fig. 1. When c is increased, the line from $\dfrac{c + da}{1 + db}$ is lifted, shifting the intersection with the line from e to the right and down.

Proposition 3 *If the road network is congested ($d > 0$), then the larger the elasticity of demand the lower p^*.*

Proof This also follows directly from (5). When $d > 0$, p^* falls when b increases, which in turn increases with the elasticity of demand.   QED

The above analysis relates to an inter-modal equilibrium where the transit operator(s) passes on any efficiency gains to passengers. In cities where transit is in the private sector, or behave as if it were in the private sector, efficiency gains may not be passed on to passengers but retained or distributed to shareholders. However, the impact of a road user charge on a profit maximising transit operator(s) is unclear. Will any of the efficiency gain be passed on to passengers?

2.2 Scenario 2: Profit Maximising Transit Operator(s)

Suppose that transit operator(s) behave like a profit maximising monopoly and that $AC_t = e - fq_t$. Profit for the transit operator is therefore:

$$\pi_t = pq_t - (e - fq_t)q_t = \left(\frac{a}{b} - \frac{q_c(p) + q_t}{b}\right)q_t - (e - fq_t)q_t \tag{8}$$

We assume that transit operator(s) as a whole maximise profits. The complimentary slackness conditions at the maximum are:

$$\left.\begin{aligned}\frac{\partial \pi_t}{\partial q_t} = \frac{a - q_c(p)}{b} - 2\frac{q_t}{b} - e + 2fq_t \geq 0 \\ q_t \geq 0\end{aligned}\right\} \tag{9}$$

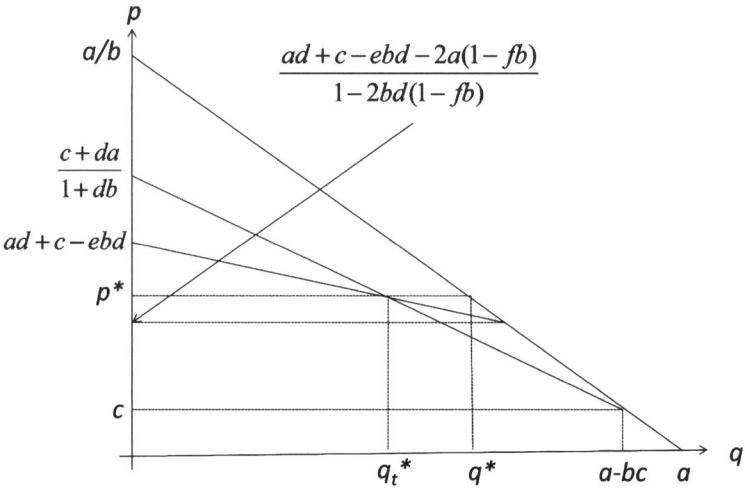

Fig. 2 Stable inter-modal equilibrium for profit maximising transit

Proposition 4 *Conditions (9) imply profit maximisation (loss minimising) if $fb < 1$*

Proof Conditions (9) are necessary and sufficient for a maximum of $\pi_t(q_t)$ if this function is concave. Note that

$$\frac{\partial^2 \pi_t}{\partial q_t^2} = -\frac{2}{b} + 2f \tag{10}$$

If (10) is negative then $\pi_t(q_t)$ is concave. When $fb < 1$ (10) is negative.   QED

Proposition 4 implies that the impact that travel has on the price of travel $\left(\frac{1}{b}\right)$ should be greater than the efficiency gain (f) for (9) to guarantee a profit maximising solution. Where this is not the case, the objective function is convex rather than concave and the profit maximising solution is $q_t^* = \infty$.

Suppose that $fb < 1$ and $q_t^* > 0$. Then at the optimum:

$$\frac{a - q_c(p)}{b} - 2\frac{q_t^*}{b} - e + 2fq_t^* = 0$$

Hence for transit:

$$a - q_c(p) - eb - 2q_t^*(1 - fb) = a - \frac{p - c}{d} - eb - 2q_t^*(1 - fb) = 0$$

which in turn yields the following relationship between p and the profit maximising q_t^*:

$$p = ad + c - ebd - 2d(1 - fb)q_t^* \tag{11}$$

The new inter-modal equilibrium is shown in Fig. 2. The slope of the line from $ad + c - ebd$ on the vertical axis, which traces the relationship between p and q_t^*, is $-2d(1 - fb)$, which is negative because by assumption $fb < 1$. However, this line is no longer independent of c, so when c is increased due to, say, the introduction of a road user charge, both the line from $\dfrac{c + da}{1 + db}$ and the line from $ad + c - ebd$ are lifted and the intersection of the two may no longer be shifted to the right and down. It seems that a Downs-Thompson effect is no longer guaranteed.

Proposition 5 *Stability requires* $fb > \dfrac{1 + 2db}{2 + 2db}$

Proof The slope of the line from $ad + c - edb$ on the vertical axis is $-2d(1 - fb)$. Stability requires that if q_t is below the intermodal equilibrium q_t^* then a generalised cost incentive arises which encourages users of private transport to switch to public transport restoring the inter-modal equilibrium. The converse should occur when q_t is above the intermodal equilibrium q_t^*, again restoring the inter-modal equilibrium. By making use of (6), we see that stability requires:

$$\frac{d}{1 + db} > 2d(1 - fb)$$

This implies:

$$0.5 > (1 + db)(1 - fb)$$

Hence:

$$fb > \frac{1 + 2db}{2 + 2db} \tag{12}$$

QED

Proposition 6 *If* $1 > \dfrac{1 + 2bd}{2 + 2bd} > fb > 0.5$ *an increase in c leads to a reduction in* p^*.

Proof At the inter-modal equilibrium:

$$p^* = \frac{a}{b} - \frac{q_c^* + q_t^*}{b} = \frac{a}{b} - \frac{p^* - c}{bd} - \frac{1}{2bd(1 - fb)}(ad - p^* + c - ebd)$$

$$= \frac{a}{b} - p^*\left(\frac{1}{bd} - \frac{1}{2bd(1 - fb)}\right) + c\left(\frac{1}{bd} - \frac{1}{2bd(1 - fb)}\right) - \frac{ad - ebd}{2bd(1 - fb)}$$

This implies that:

$$p^*\left(1 + \frac{1}{bd} - \frac{1}{2bd(1 - fb)}\right) = c\left(\frac{1}{bd} - \frac{1}{2bd(1 - fb)}\right) - \frac{ad - ebd}{2bd(1 - fb)} + \frac{a}{b} \tag{13}$$

If

$$-1 < \frac{1}{bd} - \frac{1}{2bd(1 - fb)} < 0 \tag{14}$$

then an increase in c leads to a reduction in p^*. This condition reduces to

$$1 > \frac{1 + 2bd}{2 + 2bd} > fb > 0.5 \tag{15}$$

QED

The implications of Propositions 5 and 6 taken together is that if there is a stable inter-modal equilibrium at which both public and private transport are used then there will be no Downs-Thompson effect when transit operator(s) conspire to maximise profits (or minimise losses) because (12) is inconsistent with (15).

3 Conclusions

The foregoing analysis shows that if increased transit patronage leads to efficiency gains for transit and if any such gains are passed on to passengers then the imposition of a (small) road user charge should lead to a reduction in the inter-modal user equilibrium cost of travel. However, if the transit operator(s) as a whole maximise profits then any efficiency gains for transit resulting from a (small) road user charge will not be passed on to the passengers and there will be no reduction in the inter-modal user equilibrium cost of travel. The policy conclusion is that the regulation of transit fares is important when a road user charge is being implemented.

The situation is rather more complicated when there is a publicly owned monopoly provider of transit. While the transit operator may still behave like a monopolist and not pass on any of the efficiency gains to passengers in the form of reduced fares, these gains may be returned to passengers in other ways. Insofar as there is a democratic consensus about such transfer payments, it could be argued that welfare has been increased.

These conclusions are drawn from the analysis of a linear inter-modal equilibrium model. The underlying functions will not in general be linear, although they may be linearised at the equilibrium. The conclusions are still robust for larger road user charges, provided $a, b, c, d, e, f > 0$ remains true for the linearisations of the underlying functions at all the intermediate inter-modal equilibria as the road user charge is incrementally increased.

References

1. Abraham JE, Hunt JD (2001) Transit system management, equilibrium mode split and the Downs-Thomson paradox. TRB conference preprint. University of Calgary, Department of Civil Engineering, Calgary
2. Bell MGH, Wichiensin M (2008) Profit maximising transit in combination with a congestion charge: an inter-modal equilibrium model. In: Richardson HW, Christine Bae C-H (eds) Road congestion pricing in Europe: Implications for the United States. Edward Elgar, Cheltenham, pp 23–38
3. Downs A (1962) The law of peak-hour expressway congestion. Traffic Q 16:393–409
4. Hau TD (1992) The economic fundamentals of road pricing:a diagrammatic analysis. World Bank Policy Research WPS 1071 (reproduced in Transportmetrica 1(2):1–149)
5. Holden, DJ (1989) Wardrop's third principle: urban traffic congestion and traffic policy. J Transp Econ Policy 23:239–262
6. Mogridge MJH (1990) Travel in towns: jam yesterday, jam today and jam tomorrow? Macmillan Press, London
7. Mogridge MJH, Holden DJ, Bird J, Terzis GC (1987) The Downs-Thomson paradox and the transportation planning process. Int J Transp Econ 14:283–311
8. Thomson JM (1977) Great cities and their traffic, Gollancz, London (Published in Peregrine Books, 1978)
9. Wichiensin M, Bell MGH, Yang H (2007) Impact of congestion charging on the transit market:an inter-modal equilibrium model. Transp Res A 41(7):703–713
10. Wichiensin M, Bell MGH (2008) Transit market effects on socially optimal congestion charging. In: Steg L, Verhoef E, Bliemer M, van Wee B (eds) Pricing in road transport: a multi-disciplinary perspective. Edward Elgar, Cheltenham, pp 131–150
11. Wikipedia (2011) Downs-Thompson paradox. Online: http://en.wikipedia.org/wiki/Downs%E2%80%93Thomson_paradox Accessed 14 Feb 2011

Part IV
Aviation

Foreword
Andrew Haines

Aviation's current impact on climate change is relatively small, but as other sectors decarbonise, that impact is going to grow, rapidly. Globally, through ICAO, the aviation industry has the challenging goal of achieving fuel efficiency improvement at 2% a year by 2020. But with forecast demand growth of 5% a year over the next 5 years, emissions are likely to continue to grow over the next decade.

In the UK we face similar challenges, with both the Climate Change Committee and the Government predicting that without additional policy measures, aviation carbon emissions will be greater in 2050 than today. However, sustainable growth is now at the centre of the government's aviation policy. The Secretary of State for Transport has made it clear that there can be no strategy for aviation growth without a convincing plan to manage the environmental impact.

The Government's consultation on a new aviation policy framework is a huge opportunity for the aviation industry to lead the debate, but if the industry is to shape that policy, then it must not only take the environment immensely seriously, but it must be seen, and believed, to be doing so. Many stakeholders remain to be persuaded that the aviation community is doing all that it can to reduce its carbon footprint and there is scant evidence to date that aviation consumers pay real attention to the climate change implications of their flights.

Real progress is being made with improvements in engine and airframe technology, and the development of lower carbon, sustainable fuels. But questions remain around the ability to quickly scale up a sustainable fuel supply and the lead times for new investment are such that these alone probably will not be enough for aviation to meet its targets.

Much more focus will be required on operational and airspace efficiencies. The Future Airspace Strategy developed outline plans for UK airspace up to 2030. A key plank of that strategy is improved environmental performance. The Single European Sky project has a target of a 10% reduction in air journey length with a significant reduction in fuel burn and emissions and this remains fertile territory.

A solid action plan and a real belief in the industry's capacity to meet the environmental challenge, will give the industry a locus in shaping its own future. I am determined that CAA will play a leading and constructive role as the debate moves forward, facilitating the industry's own work where we can add value, and assisting and challenging the Government as they develop their own policies to help create a framework within which a sustainable, successful aviation industry can flourish.

Andrew Haines, CEO
Civil Aviation Authority

Aviation Policy and Governance

Lourdes Q. Maurice and Carl E. Burleson

Abstract Aviation has become a critical part of the global economy in the last 60 years. It has redefined the nature of opportunity and neighbors in our world. Aviation has become a crucial driver of economic development in large parts of the world and supports the world's largest industry—tourism. Yet, the success of aviation growth in opening borders, expanding travel, and integrating economies has produced concerns over its environmental impacts.

1 Introduction

Aviation has become a critical part of the global economy in the last 60 years. It has redefined the nature of opportunity and neighbors in our world. Aviation has become a crucial driver of economic development in large parts of the world and supports the world's largest industry-tourism. Yet, the success of aviation growth in opening borders, expanding travel, and integrating economies has produced concerns over its environmental impacts.

Aviation has made significant strides in lessening its environmental "footprint." In the U.S., over the last 35 years there has been a six-fold increase in the mobility provided by the U.S. air transportation system. At the same time there has been a 60% improvement in aircraft fuel efficiency and a 95% reduction in the number of people impacted by aircraft noise [1].

Despite this success in reducing its environmental footprint, anticipated increases in air transportation demand could generate significant environmental pressures.

L. Q. Maurice · C. E. Burleson (✉)
Federal Aviation Administration, 800 Independence Avenue,
Washington DC, SW 20591, USA
e-mail: carl.burleson@faa.gov

O. Inderwildi and Sir David King (eds.), *Energy, Transport, & the Environment*,   387
DOI: 10.1007/978-1-4471-2717-8_21, © Springer-Verlag London 2012

Current operational trends show that environmental impacts resulting from aircraft noise and aviation emissions are likely to be a principal critical constraint on the growth of capacity in the U.S. in the years ahead. While the future growth of emissions from civil aviation in the U.S. is expected to be relatively modest, worldwide civil aviation emissions are expected to grow at a much faster rate driven by the expected surge in air travel in the other parts of the world.

The U.S. has placed addressing environmental and energy issues at the heart of Next Generation Air Transportation System (NextGen) plan. For sustained mobility, aviation must have a reliable, affordable, and environmentally sound energy supply as well as an effective and balanced approach to simultaneously address aviation noise, air quality, and climate impacts in an integrated and cost-beneficial manner. Aviation's contribution to air quality and climate change-related human health and welfare impacts will likely increase in the future relative to other sources if emissions of air pollutants from non-aviation sources continue to decrease at a faster rate. Independent of its growth, aviation-related environmental impacts for noise and air quality are expected to increase due to an increase in human exposure caused by projected growth in population and urbanization.

Under the auspices of NextGen, the U.S. has adopted and continues to advance and implement a five-pillar strategy to effectively address aviation environmental impacts. This strategy is designed to characterize issues and assess risks, and develop and implement well informed and cost-beneficial solutions to meet environmental targets in a verifiable manner. The elements of this five-pillar strategy are: (1) advance scientific understanding and improve integrated noise, emissions, and fuel efficiency analyses capability; (2) advance and accelerate maturation of clean, quiet, and energy efficient aircraft technologies; (3) develop and implement clean, quiet, and energy efficient operational procedures; (4) develop, qualify, and implement sustainable aviation alternative fuels; and finally (5) use policy, environmental standards, and market-based options.

2 Advance Scientific Understanding and Improved Integrated Environmental Analysis Capability

Even though noise and emissions originate from the same aircraft source, their effects and related environmental and human welfare impacts are distinctly different. There are also various levels of uncertainties associated with each impact. Therefore, characterization of environmental effects and impacts and comprehensive understanding of risks, tradeoffs, and interdependencies as an integrated system is critical so that appropriate targets for environmental goals and well informed cost-beneficial solutions can be developed and implemented.

The Center of Excellence PARTNER [Partnership for AiR Transportation Noise and Emissions Reduction—a center for cooperative aviation research sponsored by the U.S. Federal Aviation Administration (FAA), the National Aeronautics and Space Administration (NASA), Transport Canada, the U.S.

Department of Defense, and the U.S. Environmental Protection Agency (EPA)] continue to address gaps and uncertainties in aviation impacts on noise, air quality, and climate change as outlined in the report convened by the International Civil Aviation Organization Committee on Aviation Environmental Protection (ICAO/ CAEP) to assess the state of knowledge [3]. Aircraft noise continues to be the primary aviation environmental concern to communities in the vicinity of airports [4]. Long-standing noise impact significance criteria and land-use compatibility guidelines may not sufficiently address evolving nature of noise impact and increased public sensitivity. Community response data providing the foundation for determining significant impact is based on reactions to all transportation noise (not aviation-specific), varies across different regions of the world and in the U.S. relies on dated studies. There are gaps in knowledge of noise effects on health and welfare, including sleep disturbance and student learning. The primary metric [day-night-level (DNL)] and the threshold of significance in the U.S. (DNL 65) are not well understood or at times trusted by the public. And ultimately, decisions need common metric/approach by which to evaluate noise relative to other impacts.

The FAA has developed an aviation noise research roadmap with input from domestic and international stakeholders and identified four key research areas: (1) health and welfare impacts in noise compatible areas, (2) analysis of noise impacts on national parks and wilderness, (3) noise propagation modeling enhancements, and (4) social cost of noise impacts. Some activities to support the research roadmap have already been initiated under the Transportation Research Board Airports Cooperative Research Program [5].

Techniques and tools for measuring and modeling the levels of aircraft noise to which people are exposed are generally mature. However, there are gaps in our scientific knowledge base of the extent of impacts of the exposure and our modeling techniques may need to be refined based on the results of focused research.

The FAA is pursuing research efforts with focus on annoyance (including sleep disturbance) and effects on children's ability to learn. Other areas of ongoing focus are investigation of day-night average sound level (DNL) as a metric and its threshold for establishing significant noise impacts for airport communities as well as development of acceptability standards and assess noise impacts of supersonic and future unconventional aircraft.

The persistence of significant levels of aircraft noise in communities around airports is the major impact, but not the only one. There are increasing concerns in areas of moderate noise exposure and public complaints from suburban and rural areas where ambient noise is lower. At noise exposure levels below those involving health and welfare concerns, there are also sensitivities with respect to national resources such as national parks.

Aviation air quality and climate impacts are related to direct engine emissions and their atmospheric evolution. Aviation's impact on air quality, through emissions of specific pollutants, is a growing concern. Emissions of criteria pollutants contribute to surface air quality deterioration, resulting in human health and welfare impacts [6]. At the airport level, about 30% of U.S. commercial service airports are in non-attainment areas that do not meet national air quality standards or maintenance areas in danger of violating these standards. For these airports, emissions issues add to the complexity and uncertainty of expansion proposals. The national air quality standards are expected to become increasingly more stringent in the future, potentially placing more pressure on aviation to reduce emissions despite growth.

In the U.S., efforts to characterize air quality emissions and related impacts are being continually coordinated through the Aviation Emissions Characterization (AEC) Roadmap with participation from national and international stakeholders. In particular, research is ongoing on the extent to which aircraft landing and takeoff cycle (LTO) and cruise emissions impact surface air quality and human health. FAA continues to work to quantify the extent to which future aviation activity growth scenarios would impact human health [7].

The potential effects of aircraft emissions on the climate may be the most serious long-term environmental and related energy issues facing aviation. According to the Intergovernmental Panel on Climate Change (IPCC), aircraft account for about 3% of both national and worldwide carbon dioxide (CO_2) emissions [8]. Aircraft have been projected to contribute a larger portion of greenhouse gas emissions in the future—perhaps 5% by 2050—based on robust aviation growth assumptions and the prospect of easier transition to alternative technologies and fuels for land transport modes. There are additional climate effects concerns specific to aircraft as the majority of emissions from a given flight are directly released into the chemically complex and sensitive region of the upper troposphere and lower stratosphere. While CO_2—accounting for the bulk of aviation greenhouse gas emissions—has the same effects regardless of where it is emitted, certain emissions may have greater effects when released at altitude. In addition, aircraft emissions of water vapor and aerosols lead to the formation of contrails and modification of cirrus cloud distribution—both of which can impact earth's climate. Scientifically, we do not yet know enough about aircraft contrails to determine their impact on climate or adopt measures to deal with them [9]. There are multiple, interrelated impacts due to aircraft emissions with varying degrees of understanding, with CO_2 being the best understood and quantified.

The FAA is continuing research activities under its Aviation Climate Change Research Initiative (ACCRI) program to characterize climate impacts of aviation non-CO_2 emissions for current and future scenarios. There remain a significant number of uncertainties about non-CO_2 emissions. At present, FAA has funded ten research teams comprising domestic and international participants. ACCRI held its first annual science meeting in February, 2011. The ACCRI science meetings will be open to national and international researchers actively working in the field. Aircraft black carbon emissions contribute to both surface air quality and climate impacts. However, there are large uncertainties in these emissions. The FAA has funded PARTNER to refine aircraft black carbon emissions as a function of engine type, thrust, and ambient conditions. PARTNER and DLR (German Aerospace Center) researchers are working jointly on this project.

In the analytical area, the FAA continues to develop and assess a comprehensive suite of models, comprising of the environmental design space (EDS), the aviation environmental design tool (AEDT) and the aviation portfolio management tool (APMT). AEDT is being used to develop integrated inventories of fuel burn, noise, and emissions. EDS is being used for assessment of aircraft technologies that are being developed under FAA and NASA's technology programs [10]. In addition, these aviation environmental analysis tools are presently being used to inform ICAO, the United Nations Framework Convention on Climate Change (UNFCCC), and U.S. NextGen environmental analyses and to support research on emerging aircraft technologies. The FAA will release the first public version of AEDT in early 2012 with regional integrated environmental analysis capability.

3 Advance and Accelerate Clean, Quiet, and Energy Efficient Aircraft Technologies

Advances in engine and airframe technologies have historically produced the largest reductions in noise and emissions The Boeing 777 is 300 times more efficient than the early Convair, Douglas, and Boeing jets [11]. The efficiency of today's aircraft is on par with the primary choice of U.S. mass-market travel—the automobile [12]. A fully loaded Boeing 787 flying from San Francisco to New York's JFK airport offers fuel economy similar to that of a Honda Accord with three passengers at almost 10 times the speed.

All of this translates into fuel savings. Between 1978 and 2008, U.S. airline fuel efficiency improved from 2.92 Revenue Ton Miles/gallon to 6.11 Revenue Ton Miles/gallon, a 110% improvement. This equates to a CO_2 savings of 2.7 billion metric tons, which was equivalent to taking 19.5 million cars off the road each year. Of course total activity has grown. More recently, there have been reductions in absolute fuel burn. Between 2000 and 2010, U.S. airlines reduced their absolute fuel burn and emissions 15%, while increasing passengers and cargo about 15% [13].

New engine/airframe technologies will need to play key roles in achieving aviation environment and energy goals. The U.S. is supporting advances in engine technology and airframe configurations to lay the foundation for the next generation of aircraft. The U.S. technological strategy envisions a fleet of quieter, cleaner aircraft that operate more efficiently with less energy use. The U.S. has an integrated effort to address environmental and energy challenges through aeronautics technology research in the President's National Science and Technology Council's multi-agency National Aeronautics Research and Development Plan [14]. Each agency focuses on different elements but they share the same goals. The FAA's focus is on maturing technologies for near term application through continuous lower energy, emissions, and noise (CLEEN) effort. The FAA has selected five teams: The Boeing Company, General Electric, Pratt & Whitney, Honeywell, and

Rolls Royce to accelerate maturation of promising clean, quiet, and energy efficient aircraft technologies and advance sustainable alternative fuels. Technologies to be developed and demonstrated under CLEEN include lighter and more efficient gas turbine engine components, low NOx combustor, noise reducing engine nozzles, adaptable wing trailing-edges, open rotor and geared turbofan engines, advanced onboard flight management systems for optimized flight trajectories, and sustainable alternative aviation fuels. The CLEEN effort is complemented by NASA's efforts, most notably the recently launched Environmentally Responsible Aviation (ERA) Project [15].

4 Develop and Implement Clean, Quiet, and Energy Efficient Operational Procedures

Improvements in air traffic management and airport infrastructure operational procedures offer near term ways to meet aviation environmental and energy efficiency goals. NextGen will increase the efficiency of aircraft operations, both in the air and on the airport surface. Improving efficiency saves time and fuel. Reducing fuel consumption reduces CO_2 emissions that affect climate and other emissions that contribute to poor air quality. Fuel burn, emissions, and flight times can be cut by Performance Based Navigation (RNAV/RNP) routes and approaches. Optimized Profile Descents can reduce noise, emissions, and fuel consumption. NextGen technology and procedures that optimize gate-to-gate operations are being demonstrated with international partners in Europe and Asia–Pacific to reduce fuel burn, emissions, and noise. The FAA has already implemented a number of procedures that have reduced fuel consumption and emissions. For example, introduction of Reduced Vertical Separation Minimum in 2005 cut fuel consumption by an estimated 300 million gallons annually [16]. Today FAA is producing a significant number of performance based navigation (PBN) routes and procedures, exceeding our fiscal year 2010 goal. Performance-based navigation offers our airline industry better routes, added capacity, improved on-time performance and lower fuel bills. Fuel represents 30–50% of an airline's total expenses, on average. The cost of jet fuel has increased significantly in the last decade, reinforcing airline incentives to reduce fuel consumption (and emissions). In fact, for the first time in over three decades, fuel costs exceed labor costs as the single largest expense for airline operations in the U.S. (see Fig. 1).

Southwest Airlines started using the precision procedures at a dozen airports this year and estimates will save $60 million per year in fuel when NextGen arrival procedures are in place nationwide. Helicopters in the Gulf of Mexico have benefited from ADS-B technology, saving up to 10 min and 96 pounds of fuel each flight. Airlines flying over the Pacific are taking advantage of a combination of improved capabilities to save 200–300 gallons per flight. Use of continuous descent arrivals (CDA) as depicted in Fig. 2 reduces fuel burn,

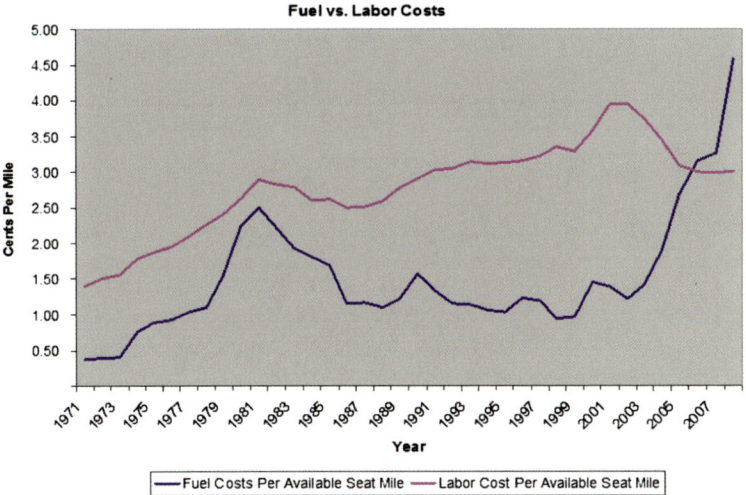

Fig. 1 Fuel versus labor costs

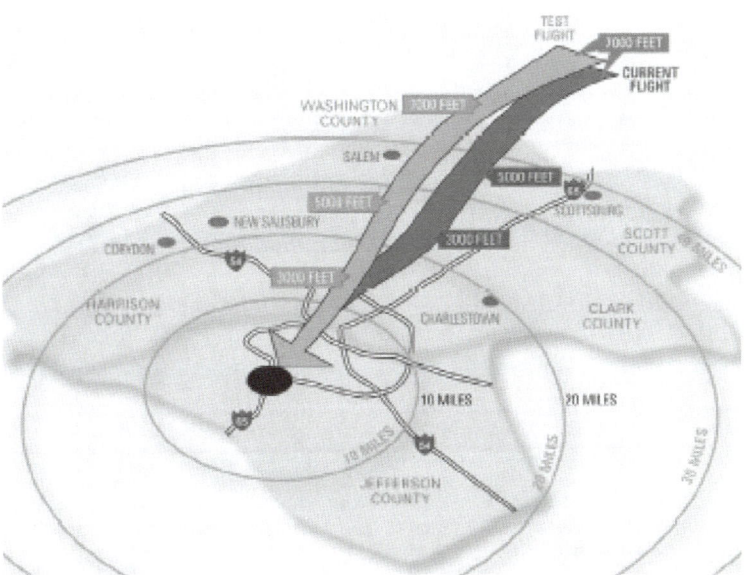

Fig. 2 Schematic of continuous descent approach

emissions, and noise. For example, at Los Angeles International Airport, airlines save over 2 million gallons of fuel each year while reducing the noise footprint.

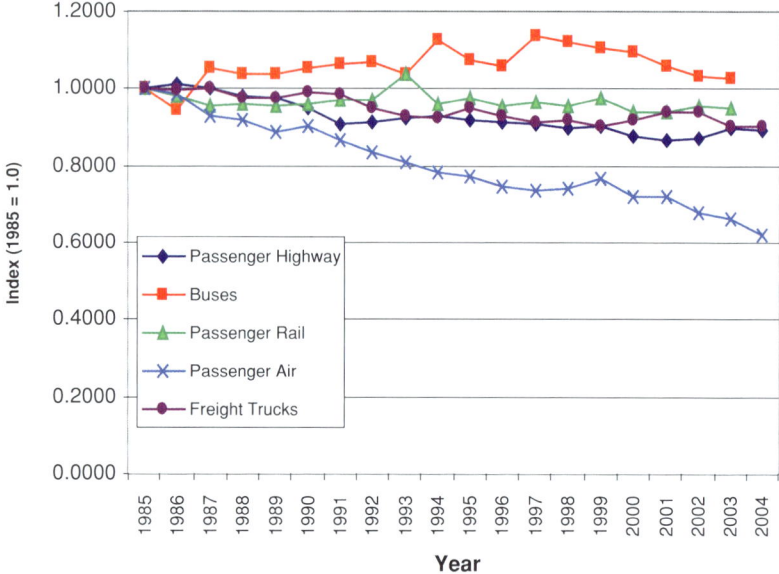

Fig. 3 Trends in energy intensity improvements of modes of U.S. transport [2]

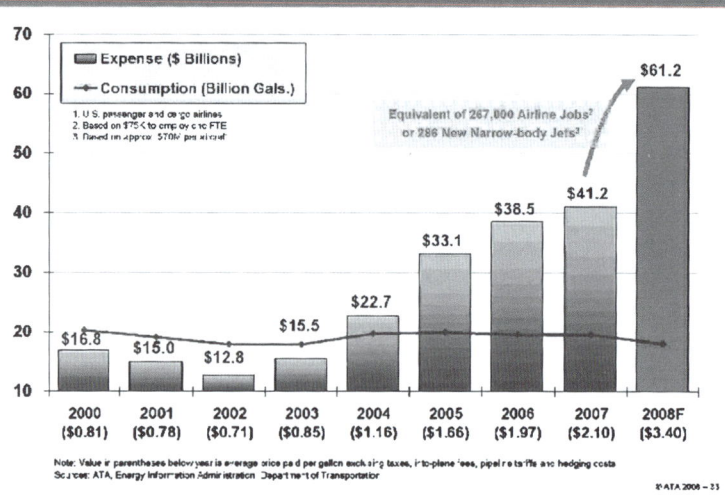

Fig. 4 Jet fuel expenses for 2000–2008

Aircraft engine and airframe advances, together with improved air traffic management and operating procedures, have dramatically improved aircraft fuel efficiency. As noted in Fig. 3, the aircraft energy efficiency improvement over the last twenty years has outpaced other forms of transportation in the U.S. The air traffic modernization planned under NextGen should further improve efficiency by reducing delays and enabling more direct routings. Notwithstanding this success, there is renewed emphasis on improving the fuel efficiency of the aviation system. This is understandable given that fuel currently represents the largest operating cost for U.S. airlines, and this cost category has grown nearly four-fold over that past 5 years (see Fig. 4).

5 Develop and Qualify Sustainable Aviation Alternative Fuels

Nearly 100% of the fuel used in aviation operations is petroleum-based—raising issues of energy supply, energy security, and fossil fuel emissions affecting air quality and climate. Sustainable alternative fuels development and deployment offer prospects for enabling environmental improvements, energy security, and economic stability for aviation. In response to these multiple concerns, the U.S. Government and industry are working cooperatively under the auspices of the Commercial Aviation Alternative Fuels Initiative (CAAFI) to develop and deploy "drop in" alternative fuels that can be blended with or replace petroleum jet fuel with no changes to existing engines and aircraft and ground infrastructure and supply equipment.

CAAFI has crafted a knowledge sharing network with over 600 organizations to address a comprehensive roadmap (see Fig. 5) to develop and deploy sustainable fuels. This includes qualification of new fuels led by the FAA; research and development efforts led by engine and aircraft manufacturers focused on identifying possible fuels for use in aircraft, researching fuel conversion processes and identifying fuel "feedstock" options; an environment effort lead by the FAA to conduct assessments of the environmental impacts of the fuels including emissions measurements and looking at the full life-cycle greenhouse gas footprint of different fuel options to make certain that any fuels used are also environmentally friendly; and a business and economics work led by the Air lines for America to develop a stable market for fuel suppliers by coordinating fuel purchasing efforts of the airlines and by developing long-term purchasing agreement terms.

Near terms efforts include adding new classes of fuels to the two recently approved alternative jet fuel standards by ASTM International, conducting aircraft flight tests using alternative fuels, ascertaining their emissions characteristics, life cycle greenhouse gases, and sustainability. Alternative fuel options that use oils, sugars, or cellulose from plants have the potential to dramatically reduce CO_2 emissions, if produced in a sustainable manner. Generally, all alternative fuel options appear to reduce particulate matter emissions in engine exhausts—a cause of respiratory ailments, although not unique to aviation as a source.

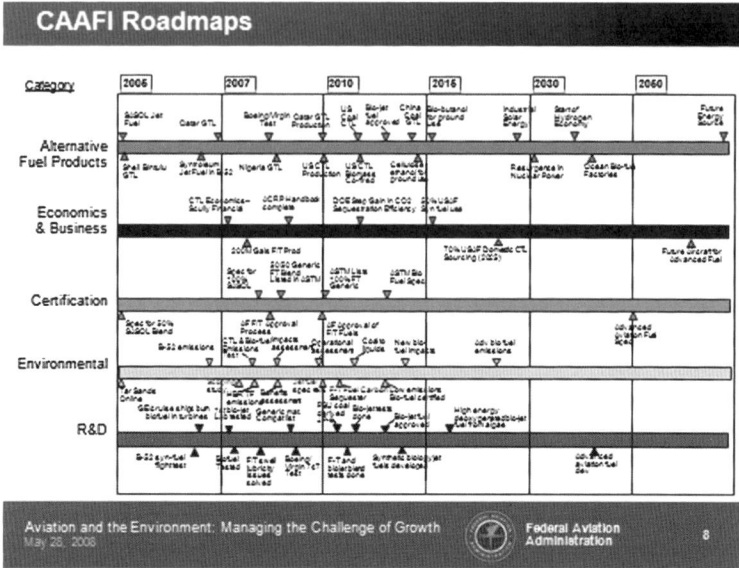

Fig. 5 CAAFI roadmaps (Federal Aviation Administration)

6 Market-Based Measures, Environmental Standards and Regulatory Policies Options

Aviation is an inherently global activity. A flight takes off every five seconds 24 hours a day that will cross an international border. This creates an additional challenge as aviation markets can vary dramatically from country to country. Even for relatively mature markets like the U.S. and Europe, the aviation sectors are very different. The majority of U.S. aviation operates domestically and has only had modest growth in recent years, and has a general aviation sector unseen anywhere else in the world. In Europe, it is difficult not to cross a border in most flights and the later arrival of the low cost carrier revolution has produced double-digit growth in some portions of the commercial aviation sector in recent years.

Development and implementation of appropriate policies, programs, regulations, and market mechanisms are critical to support advantageous technology and operational innovations and accelerate their integration into the commercial fleet, the airport environment, and entire national aviation system. The FAA is developing aviation a NextGen Environmental Management System (EMS) to manage effectiveness of environmental mitigation solutions and to provide guidance on adaptations needed toward meeting NextGen environmental goals and targets. Work is continuing to refine the EMS framework and decision support tools as well as demonstration through pilot studies at U.S. airports.

Internationally, the U.S. is working with the International Civil Aviation Organization (ICAO) to limit and reduce international aviation emissions, including development of a CO_2 standard for aircraft, and a new PM certification requirement for engines. ICAO has additionally agreed to explore more ambitious goals for the aviation sector, including carbon neutral growth in the mid-term and reductions in the long-term. The U.S. is also supporting studies to investigate the technological feasibility of the need, cost, and trade-offs of more stringent noise standards.

Penetration of advanced technologies and sustainable alternative aviation fuels and better management of the airspace system along with implementation of improved environmentally efficient procedures provide an effective basket of measures to meet aviation environmental goals and a secure energy future. How environmental standards and market-based measures may accelerate adoption and integration of such technology is an important dimension that FAA is currently engaged in analyzing.[1]

The FAA continues to quantify the challenges associated with future aviation growth, and assessing operations and fleet evolution scenarios. We are exploring the contributions of the various solution sets and their combination for meeting NextGen environmental and energy goals in an integrated manner. Potential benefits of independently varying infusion rates of mitigation solutions based on penetration of advanced aircraft technologies, alternative fuels, and efficient operational procedures are being investigated. This analysis will reveal gaps in meeting the goals and targets, and provide guidance for additional research and development to produce optimally balanced, cost-beneficial solutions. This analytical framework could also inform CAEP as it strives to meet its environmental and energy efficiency goals.

7 Closing Observations

Aviation flourished in its first 100 years because it constantly redefined the borders of the possible. Innovation has often outstripped even its most knowledgeable participants. Consider that about 100 years ago Orville Wright opined that "No flying machine will ever fly from New York to Paris... [because] no known motor can run at the requisite speed for four days without stopping." Yet today thousands of flights occur each year, and in less than eight hours. So in thinking through the challenges ahead, a few things to keep in mind.

First, aviation has some interesting characteristics in applying environmental strategies. There is an extremely high premium placed on safety that results in only proven and technically sound environmental technologies being incorporated in

---

[1] The FAA is also supporting research through PARTNER to investigate the effects of market-based measures like emissions trading to limit aviation-related greenhouse emissions within the context of greenhouse emissions from all other anthropogenic sources within the APMT.

aircraft. You cannot pull off the side of the road at 30,000 feet. Aircraft involves high capital cost and have a long life span, requiring long lead times for new technologies to be widely incorporated in the fleet and close attention to financial feasibility. Airborne systems must be lightweight and fuel-efficient. Noise, air quality, and climate effects of aviation result from an interdependent set of technologies and operations, so that action to reduce impacts in one area (e.g., aircraft engine noise) can increase impacts in another area (e.g., nitrogen oxides emissions). All these features increase the challenge of achieving ambitious environmental and energy goals.

Second, the strategy being pursued by the U.S. under the NextGen plan—with its portfolio of technology, operational, and policy initiatives—is not unique in any of its elements. In fact, what is apparent is that our portfolio of measures is similar to what you would find if you looked at plans in Europe or the "four pillar" strategy outlined by the International Air Transport Association. The differences perhaps exist in the more integrated approach—dealing with all of aviation's environmental issues—in a unified way.

Third, because of commercial aviation's relatively modest age and the majority of its growth in the past few decades, many developing countries' airlines are major contributors to the sector's greenhouse gas emissions. In fact, over half of the top 20 aviation countries in the world are developing countries under the definition of the UNFCCC. This creates a challenging discussion for international aviation emissions as there are different and competing legal frameworks under the UNFCCC and ICAO in how such emissions might be dealt with. Certainly the discussions over the past few years in both forums, and the vast influx of climate negotiators into ICAO meetings, has highlighted the difficult and conflicting paths in grappling with this problem. Given the moderating growth in developed countries and double-digit increases in air travel in the developing world, the need to find solutions that involve all major aviation countries—regardless of development status—will only grow in importance as we move forward.

Fourth, the inherently global nature of the aviation industry, whose greenhouse gas emissions often occur outside of their own borders or over the high seas, also create some significant challenges. While ICAO has a strong record of performance in setting environmental standards, policy, and guidance,[2] it has struggled to help its members resolve a number of issues involved in applying market-based measures. The problem this creates can be seen in the European Union's attempt to apply its emissions trading system unilaterally to other countries airlines that fly to and from Europe. Not surprisingly, this has generated significant opposition from countries and airlines around the world as this EU action would apply legally binding emission targets and impose billions in costs to other countries' airlines

---

[2] Over the last 10 years, ICAO has fostered adoption by the international aviation community of environmental goals in noise, air quality, and climate change; a new noise standard; two increases in NOx stringency for engines; guidance on operational measures to reduce fuel burn; and guidance for States that wish to adopt emissions trading. Most recently at its 37th Assembly it facilitated an outcome where international aviation became the first sector to adopt goals (even if voluntary) for greenhouse gas emissions performance.

over time. If market-based measures like emissions trading or carbon charges are going to play a constructive role internationally, countries will need to resolve the difficult legal and policy issues involved.

It is imperative that the aviation sector is a good environmental steward as it grows globally. In the U.S. we are implementing this through our NextGen plan to meet the environmental and energy challenges that confront the U.S. aviation system. We will have to find the right balance to manage capacity growth and environmental protection. It will require partnership and shared responsibility with airlines operating quieter and cleaner aircraft, airports providing good planning and local environmental measures, air traffic service providers facilitating environmentally effective procedures, government funding for research and development of new technologies and mitigation, and local governments ensuring proper land-use around airports—if aviation is to continue to provide its benefits to the U.S. and world economics and citizens.

The chapters ahead outline a number of perspectives of how the aviation sector can deliver the innovation and performance to ensure environmentally responsible growth during aviation's second century.

References

1. For Greener Skies (2002) Reducing the environmental impact of aviation. National Research Council
2. http://intensityindicators.pnl.gov/trend_data.stm
3. Assessing Current Scientific Knowledge (2007) Uncertainties and gaps in quantifying climate change, noise and air quality aviation impacts workshop—final report of the ICAO CAEP Workshop, 29 Oct to 2 Nov 2007. http://web.mit.edu/aeroastro/partner/reports/caepimpactreport.pdf
4. For a review of current environmental issues in US system, see GAO report. http://www.gao.gov/new.items/d09554.pdf. This reinforces that aircraft noise remains the primary environmental concern of communities around airports
5. Transportation Research Board Airports Cooperative Research Program. http://www.trb.org/ACRP/Public/ACRP.aspx
6. The focus for commercial aviation is on reducing nitrogen oxides (NOx), particulate matter (PM), sulfur dioxide (SO_2), and hydrocarbons (HC). Lead (Pb) is an issue for general aviation since more than 200,000 piston-engine aircraft rely on leaded AvGas for safe operation and produce about half of all lead emissions in the US
7. Recent PARTNER air quality research indicates that future health impacts due to aviation emissions could be even higher. This is primarily due to interaction of aircraft emissions with changes in background air quality if forecast reductions occur in emissions from non-aviation sources. http://web.mit.edu/aeroastro/partner/reports/caepimpactreport.pdf
8. Intergovernmental Panel on Climate Change Special Report. Aviation and the global atmosphere, 1999. http://www.ipcc.ch/pdf/special-reports/spm/av-en.pdf
9. Assessing Current Scientific Knowledge. Uncertainties and gaps in quantifying climate change, noise and air quality aviation impacts workshop—final report of the ICAO CAEP Workshop, 29 Oct to 2 Nov 2007. http://web.mit.edu/aeroastro/partner/reports/caepimpactreport.pdf
10. Federal Aviation Administration. Environmental tool suite frequently asked questions. http://www.faa.gov/about/office_org/headquarters_offices/apl/research/models/toolsfaq/index.cfm?print=go

11. Colpin J, Altman R. Dependable power reinvented. AIAA 2003–2882, AIAA-ICAS International Air and Space
12. Greene and Schafer, Pew center for global climate change, May 2003. http://www.pewclimate.org/docUploads/ustransp.pdf
13. U.S. Bureau of Transportation Statistics. http://www.bts.gov/programs/airline_information/
14. For the latest version of the plan, please see: national aeronautics research and development plan. http://www.whitehouse.gov/sites/default/files/microsites/ostp/aero-rdplan-2010.pdf
15. For details on the ERA program, please see: environmentally responsible aviation (ERA) Project. http://www.aeronautics.nasa.gov/isrp/era/index.htm
16. PARTNER. Assessment of the impact of reduced vertical separation on aircraft-related fuel burn and emissions for the domestic United States. http://web.mit.edu/aeroastro/partner/reports/rsvm-caep8.pdf

Sustainable Aviation Alternative Fuels: From Afterthought to Cutting Edge

Richard L. Altman

Aviation Alternative Fuels

Acronyms and Terminology

CAAFI	Commercial Aviation Alternative Fuels Initiative—U.S. initiated global coalition that foster the development and deployment of alternative aviation fuels (see www.caafi.org)
FAA	Federal Aviation Administration—Aviation branch of the U.S. government Department of Transportation
R&D	Research and Development
REDAC	Research, Engineering Development Advisory Committee—Advisors from outside the Federal Aviation Administration that guides the aviation Research agenda
AV030	Alpha numeric designator for the United States National Academy of Sciences, Transportation Research Board Committee on Aviation Effect on the Environment
TRB	United States, Transportation Research Board, a part of the National Academy of Sciences
ASTM	American Society of Testing and Materials International—THE internationally recognized independent standard setting body whose document (ASTM 1655) is referenced in Commercial

R. L. Altman (✉)
Commercial Aviation Alternative Fuels Initiative (CAAFI),
402 Hang Dog Lane, Wethersfield, CT 06109, USA
e-mail: rcbaltman@gmail.com

O. Inderwildi and Sir David King (eds.), *Energy, Transport, & the Environment*,   401
DOI: 10.1007/978-1-4471-2717-8_22, © Springer-Verlag London 2012

	jet operating manuals as setting required properties that must be conformed to safe operation of Jet Aircraft using petroleum-based fuels
ASTM 7566	Published September, 2009 the standard for non-petroleum fuel having equivalence to ASTM 1655 for safe operation of Jet Aircraft
JP-4	(Jet Propellant 4" is a Jet fuel specified in 1951 by the U.S. government for use in Military aircraft
JP-8	"Jet Propellant 8") is a jet fuel specified in 1990 by the U.S. government as a replacement for JP4. The U.S. Air Force replaced JP-4 with JP-8 completely by the fall of 1996, to use a less flammable, less hazardous fuel for better safety and combat survivability
ICAO	International Civil Aviation Organization is branch of the United Nations dedicated to Aviation matters including the development of norms for airports and the environment
FRL	Fuel Readiness Level—A systems level-gated risk management process formed by CAAFI and the United States Air Force Research Lab to enable tracking of Research, Development and Deployment status of Alternative fuels. FRL was formed from a merger of the Technology Readiness Level process (TRL) and Manufacturing Readiness Level (MRL) Process. It was approved as a global best practice by the International Civil Aviation Organization of the United Nations at a conference held in Rio De Javier in November, 2009,
NASA	United States National Aeronautics and Space Administration
MIT/PARTNER	Massachusetts Institute of Technology is leader of PARTNER. PARTNER is the Partnership for AiR Transportation Noise and Emissions Reduction, a leading aviation cooperative research organization. An FAA Center of Excellence, PARTNER is sponsored by the FAA, NASA, Transport Canada, the U.S. Department of Defense, and the U.S. Environmental Protection Agency
DLA	Defense Logistics Agency of the United States Government Department of Defense (DOD) is the Fuel Purchasing Agent for the Department
USDA	United States Department of Agriculture—Cabinet level Department reporting to the President of the

	United States on issues of Agriculture. USDA is a co-equal part of the Executive branch of the U.S. Government with Defense, Transportation and Energy (among other Departments)
FSRL	Feedstock Readiness Level is a process formed by the Federal Aviation Administration (FAA) and United States Department of Agriculture (USDA) to allow gated risk management in the maturity of energy crops using systems principals applied to Jet Fuel Research, Development and Deployment identified as Fuel Readiness Level (FRL) above
SWOT	Management sciences term for "Strength, Weakness, Opportunity, Threat" analysis approach to identify and evaluate potential projects
EASA	European Aviation Safety Agency, is the equivalent of the United States, Federal Aviation Administration in the United States in the European Union. EASA has jurisdiction of all safety matters related to Aircraft certified for operation in Europe
Fischer–Tropsch	One process that produces a petroleum substitute jet fuel, typically from coal, natural gas, or biomass http://en.wikipedia.org/wiki/Synthetic_fuel to form synthetic jet or SPK (synthetic paraffinic kerosene). The Fischer–Tropsch process was invented in Germany in the 1920's and named after its founders. Fischer–Tropsch process fuel were the first to be qualified under ASTM 7566 as jet fuel in September of 2009
ASTM D-4054	in October, 2009 ASTM passed International standard D4054, "Guideline for the Qualification and Approval of New Aviation Turbine Fuels and Fuel Additives". D4054 provides candidate alternative fuel producers and jet engine companies a document identifying the procedures required to qualify fuel to the D-7566 specification (above)
DARPA	Defense Advanced Research Projects Agency (DARPA) is a branch of the United States Defense department executing largely classified research on technologies viewed as having significant strategic value. DARPA conducted early research on a number of alternative jet fuels
HRJ	Hydrotreated Renewable Jet—are jet fuels derived from the extraction of lipids from seed bearing

plants such as Camelina, Jatropha for the production of Jet fuel. HRJ sources also include animal fats or tallow. When ultimately certified for Jet usage the fuel acronym was changed to HEFA (Hydrotreated Esters and Fatty Acids)

HEFA   Hydrotreated Esters and Fatty Acids (see HRJ above)

Next Gen   is acronym for Next Generation Air Traffic Management System conceived by the U.S. FAA to automate and increase the efficiency of Air Traffic Management

GHG, GHG footprint   Greenhouse gases and their quantitative signature, A greenhouse gas (abbreviated GHG) is a gas in an atmosphere that absorbs and emits heat and is generally considered to be the cause of long term increases in the earth's temperature or global warming. The primary greenhouse gases are water vapor, carbon dioxide, nitrous oxide and ozone

"Ground to Wake Analysis"   Carbon dioxide from the inception of a biofuels creation (ground) to its exit at the exhaust plain (wake)" analysis generally is viewed as analyzing the complete carbon life cycle of Jet engine fuel production and use. Among green house gases carbon dioxide is viewed as having the longest life in the atmosphere and hence a most critical long term control priority

Clean Air Act   As amended in 1990 the Clean Air Act administered by the U.S. Environmental Protection Agency governs the control of dangerous air pollutants or those classified as having serious health effects if present in the local atmosphere beyond certain critical thresholds

Non-Attainment Area   Areas classified within the U.S. as surpassing the threshold of serious health effects under the Clean Air Act and subsequently subject to regulation by the United States Environmental Protection Agency

P.M. 2.5   Particle Matter of 2.5 microns or less. PM 2.5 has been identified as a carcinogen as solid particles of this size can penetrate into the lungs without being filtered. Consequently any source of PM 2.5 is subject to regulation by the Environmental Protection agency if produced in non-attainment areas under definition above

Section 526	Is a provision of the United States Congresses' Energy Independence and Security Act of 2007. Sec—526 provides a legislated goal and constraint for government purchasers of Alternative fuels including aviation fuels to be lower in Carbon Life Cycle impacts than that of an Oil Refinery in fuel production
IATA	International Air Transport Association—Trade association for the majority of the World's airlines
ATAG	Air Transport Action Group—An association of aviation interests working to ensure sustainable industry growth
ATA	Air Transport Association of North America is a trade association representing 90% of U.S. Airlines and one of four sponsors of the CAAFI coalition
ATA Energy Council	Committee of the Air Transport Association of North America consisting if those airline employees having primary responsibility of fuel purchasing. The CAAFI Business and Economics team is centered among those purchasing agents
ISO 14000	ISO 14000 includes ISO 14001 standard. It represents the core set of international standards used by organizations for designing and implementing an effective environmental management system
ACRP	Airport Cooperative Research Program is a program of the Transportation Research Board (see TRB above). Programs are categorized by objective type in the first two digits of the project designator. A total of four projects from ACRP have targeted alternative fuel for use and/distribution at airports for study using tool suites developed under the programs. There program have number 02-18, -23 and -36)
Crack Spread	Difference between the price paid for a barrel of crude Oil and that paid for a barrel of jet fuel
"Farm to Fly"	Initiative of the USDA, ATA, and The Boeing Company formalized by Memorandum of Understanding signing on July 14, 2010 with the goal of encouraging the development and deployment of sustainable biofuels for aviation in the United States

Section 9000	Energy crop provisions of the USDA Congressional Authorization of 2008. Nearly a dozen sections are highlighted by section 9003 which provides loan guarantees for biofuel production facilities. Aviation fuels are allowable candidates for section 9000 support
FAAC	Future of Aviation Advisory Committee (FAAC) was formed by the Secretary of Transportation to provide recommendations to support the growth and sustainability of aviation in the U.S. Its top recommendation surrounded the benefits of sustainable biofuels
BRDI	Biomass Research and Development Initiative BRDI is a joint activity of the U.S. Departments of Agriculture and Energy that seeks to recommend research pathways for the development of Biomass based energy solutions. Aviation is now actively interfacing with BRDI through the FAA
Pyrolysis	Is candidate production process for jet Fuel involving the thermo chemical decomposition of organic material at elevated temperatures and pressures in the in the absence of oxygen. Pyrolysis oils have high proportion of complex hydrocarbon particles called aromatics that are needed at a minimal level of 8% to ensure seal expansion in aircraft engines and to prevent oil leakage
Alcohol to Jet	Generally thought of as fermentation processes using biological agents (Fermentation Renewable jet) also called synthetic biology or catalytic processes (Catlytic Renewable jet). Alcohol processes can open feedstock availability of biofuels for jet to a much larger ranges of feedstocks that produce sugars form cellulosic sources

1 An Introduction to CAAFI: A Coalition of Informed, focused and connected Industry leaders

From 2006 through 2011 Sustainable Aviation Alternative fuels transitioned from a stagnant research focus project to a multi-dimensional technology development and deployment thrust commanding the attention of airline and defense buyers, advanced biofuels producers and governments worldwide.

To expedite entry and success of aviation into the alternative fuels space the Commercial Aviation Alternative Fuels Initiative was formed by a private/public coalition of Airlines, Manufacturers, Airports and Federal Aviation Administration. Successes in multiple development and deployment focuses has moved Aviation to the forefront of transport sectors seeking to achieve sustainable growth in a scant 5 years.

CAAFI's emergence and successes to date, should be viewed as a "work in progress". Collectively they already form a unique example of how focused, collaborative effort with clear goals with leaders who "own" the processes that they execute can move in a decisive manner and achieve notable results. It is a model from which other efforts within the sustainable biofuels space—both in national and international aviation fuels efforts and biofuels efforts in other sectors may be able to draw.

For purposes of this treatise the effort has been divided into several parts. The parts demonstrate the decentralized nature of the outstanding individual CAAFI teams and build upon the distinctive competencies of team leaders and stakeholder associates that have contributed to CAAFI's success. These leaders, and the organizations they represent, share a willingness to develop and to own CAAFI's collective goals. This ensures the decentralization of execution when communicated aggressively across function areas and optimizes success. How these teams and their leaders have ministered to execution in their areas of unique expertise and what they see as the successes in their disciplines are at the core of the CAAFI story. Team and discipline leaders and the portion of the CAAFI construct that they own are identified in Sects. 5 through 8.

Another CAAFI success model relates to how these leaders have formed needed relationships with government stakeholders beyond the founding FAA sponsor in a most complimentary and aligned manner, from R&D cooperation through acquisition. How some 17 government agencies and their private sector clients have worked together cooperatively, and in a fully complementary manner, is another marker for aviation. Alternative fuels' rise to prominence and is the subject of Sect. 9.

Preceding this central core and providing the "what and how' of execution are snapshot s of the industries position in the sustainable renewable market at CAAFI's outset and today (Sects. 2 and 3). This snapshot is followed by a view of the logic that led to the formation of the CAAFI supply chain pyramid and discipline centric operating model that has facilitated our progress (Sect. 4).

Last but not least are the "next steps" (Sect. 10) that will lead to our continued progress toward a sustainable future in all senses of the word (economic, environmental and socially viable and secure). Clearly the learning obtained in Sects. 1 through 9 of the CAAFI tale has led to the definitions of next steps through the incremental learning which those steps have and continue to produce.

As this piece is being written there is a literal explosion of similar," CAAFI-esque", efforts throughout our great industry. If as you read and contemplate the CAAFI story look at your enterprise and establish how its experience can be applied in a creative manner. In doing so readers will both gain added faith in Aviations ability to continue and in fact accelerate the process CAAFI sponsors and stakeholders have achieved.

2 Aviation Alternative Fuel 2006 Snapshot: "What if Your Family were an Airline"?

The Genesis of what has become the CAAFI coalition started from a simple recommendation from the Research and Development Advisory Committee (REDAC) at the FAA Office of Environment and Energy during the second Quarter of 2005. The committee asked a seemingly simple question. What was the Office doing to expedite developments with the second E... (Energy) in its Charter.

The problem statement was brought to the newly formed National Academy of Sciences, Transportation Research Board Committee on Aviation Effect on the Environment, colorfully named "AV030". A session called "Gas Pains" was organized for the January of the following year at the annual Transportation Research Board annual meeting in Washington.

Entering the TRB meeting the generally accepted premise was that aviation was a poor candidate for Alternatives to petroleum-based fuels. Only a handful of specialists who had toiled for decades since the first Oil shock thought that this approach could work at all.

What we found was an industry that seemingly had

- No Options to Oil.
- Portrayed as Growing Polluter.
- 10 Years to Fuel Qualification discouraging investment.
- Aviation only 10% of Transport Demand.
- Airline balance sheets that left few buyers with investment grade ratings.
- A sector in which was headed toward what became a scant 2 years later fuel costs as its single largest expense—larger than its labor costs.

Putting it in terms that most of us can relate to—what would you do if your family found itself in these desperate straits. The collective wisdom was that Aviation being dependent on liquids would get the last ounce of oil produced from what ultimately would be a dying infrastructure of a fixed resource. Surely the 10% of transport demand represented by aviation could not lead the way to the development of new fuels sources. But if Aviation did not lead the way with fuel delivery infrastructure tailored to address other modes, it would surely be condemned to more of the same.

Such dire circumstances among a close knit family unit can also result in the family coming together, pooling their resources and unique skills and building upon their unique strengths to alter its fate. That is exactly what happened over the course of the 5 years since January 2006 in the aviation family.

3 Aviation Alternative Fuel 2011 Snapshot: On the Cutting Edge of Sustainable Transport Fuels

Entering the second calendar quarter of 2011 Commercial Aviation has moved to a leadership position in the U.S. Aviation. Nothing more fundamentally marked this change than remarks made by the U.S. President Obama in a March 30, 2011 speech on Energy security policy. In the speech the president specifically called our commercial aviation as a user of Advance Biofuels to be developed by the military, Energy Department and Agriculture Department specifically stating that.

> *"I'm directing the Navy and the Departments of Energy and Agriculture to work with the Private sector to create advanced biofuels that can power…not just fighters…but trucks and commercial airliners."*

In reality this statement was as much recognition of what had been accomplished by the CAAFI public–private coalition and its component parts as directive. Specifically major accomplishments in the short span of 5 years include

- The passage of the first all new aviation fuel specification (ASTM 7566) in nearly 20 years in September 2009. The last prior specification change (from JP4 to JP8) addressed aviation safety concerns.
- The creation of gated risk management approach to govern the development and deployment of alternative fuels is called "Fuel Readiness Level" or FRL. In November 2009, the process developed by CAAFI's R&D and Certification team was approved as International best practice by the International Civil Aviation Organization—aviation's branch of the United Nations. This systems engineering methodology has long been used by the Air Force and NASA to conduct complex system development.
- The development of an aviation specific "ground to wake" carbon Life Cycle analysis process as an outcome of CAAFI Environmental Team. FAA team leadership through its MIT led Aviation/Environment Center of Excellence (PARTNER) led this effort that contributed to, and was aligned with, the efforts of the Department of Energy and United States Air Force in Early 2010.
- The execution by CAAFI's Airline Energy Council-led business team of three early Memoranda of Understanding by multiple U.S. airlines along with Canadian, Mexican airlines and Lufthansa of Germany by December of 2009.
- Alliances with the Defense Departments fuel purchasing arm, DLA Energy through a unique partnership of private sector (airline) and public (Defense Department) buyers in March of 2010.
- The creation of the Public/Private Farm to fly initiative in July 2010 between airlines, Boeing and the Agriculture and Energy Departments.
- The signing of a complementary agreement between FAA and the Agriculture Department (USDA) to parallel the aviation sectors FRL process with a Feedstock readiness risk management process (FSRL) that uses Aviation Systems Risk management processes to grow feedstock yields and to increase the ability of energy crops to grow on substandard land.

- March 2011 launch of a Agriculture, Commerce and Energy Department initiative with the private sector to leverage foreign investments in Aviation biofuels.

While these achievements in themselves formed an impressive transformation from Aviation's 2006 status, the recognition in Aviation, Biofuels industry and even financial publications like the Economist revealed a whole new picture of Aviation now on the cutting edge of a Sustainable Fuels future. Recognition came in many forms.

- In February 2010, a leading transport industry publication, Air Transport World recognized CAAFI by awarding it the prestigious Joseph S. Murphy Award for Industry service.
- In October 2010 three of CAAFI's principle leaders, Environmental team co-lead Nancy Young, Business team facilitator John Heimlich and Executive Director Rich Altman were voted among the top 50 leaders in Bioenergy by readers of the Biofuels Digest.
- The respected financial publication "The Economist" declared that for biofuels the future was "looking up" next to a cartoon of scientists fleeing an electric car for the refuge of a passenger jet.
- By the Paris Airshow in June 2011 the participation of Airlines and fuel companies in the newly initiated biofuels showcase had grown four fold in a single year from that displayed in July 2010. What is more, debt investors were now engaged in the industries' exploits.
- Most importantly another biofuels digest readers poll in October 2010 showed that nearly 40% of the publications readers fully believed that one billion gallons of Biojet fuel would be produced annually by 2020. If provided at the certified 50% blend level 10% of all jet fuel consumed annually in the U.S. would be synthotic.

In just 5 short years Commercial Aviation had gone from no more than an afterthought in the Biofuels space to the cutting edge of innovation.

What follows in the story of CAAFI is how the efforts of this small but highly focused and skilled industry and its leaders were able to focus on and to achieve its 2011 status, as told by the men and women who executed this metamorphosis.

4 Leveraging Our Industries Assets in Search of "a New Fuel Dynamic"

Most new endeavors, either consciously or purposefully, begin with a SWOT (strength, weakness, opportunity, threat) assessment. Section 2 explains the weaknesses and threats which the commercial aviation family was dealing with in 2006. Importantly all sponsors and technical and business components of the aviation sector in those early days of CAAFI recognized these liabilities and agreed, via the formation of CAAFI, to respond to those systemic as a matter significantly larger than a price spike in jet fuel.

The more difficult task for those engaged at the time was to recognize sector strengths and more importantly how those strengths could be turned into opportunities.

In the case of commercial aviation strengths, these are realized by examining what is unique to the industry, including constraints, and establishing if and how these limitations can be established as distinctive competencies and hence strengths.

Specifically:

- Aviation is limited to high power density liquid fuel use. Electrification for main propulsion is not an option. Hence investors can be assured that the industry will not shift to alternative liquid fuels.
- The Safety requirements of Aviation set by agencies such as the U.S. FAA (Federal Aviation Administration and its European equivalent EASA and implemented by the standard setting organizations most commonly ASTM International are demanding and create a "barrier to entry" for all but the most serious fuel producers.
- While only 10% of the market, such small size, when consolidated into a limited and informed group of buyers can facilitate group decision making which serious producers can be assured will be data driven.
- Distribution for aviation is confined to relatively few airports. In the U.S. 80% of all traffic flows through 35 destinations.
- Jet fuel producers will execute multiyear off take (purchase) agreements. This is not the case with diesel buyers operating in the similar fuel space.
- Systems integration and gated risk management of product and process development are ingrained in aviation and in fact are requirements of technology and product development by the military, NASA and defense contractors.
- Research in Aviation is well supported by both Commercial and Military sources. Aviation contractors have great expertise in meeting the requirements of government research contracts at the outset of technology development.
- Environmental regulations and rules for safety are governed globally by the United Nations, International Civil Aviation Organization.

What follows in subsequent Sects. 5–8 is an examination of how each of CAAFI's disciplines, Qualification, Environmental, R&D and Business and Economics have sought to utilize these strengths to create the transformation from 2006 to 2011 evidenced in Sects. 2 and 3. Each functional area is explained by illustrating the challenge that the team faced, the solution pathway and the results to date.

Sections 9 and 10 focus on the aspects of integration over some 17 different government agencies in the U.S. that has been instrumental in ensuring that research is executed in the most expeditious manner across the entire supply chain.

Together this text provides the core of CAAFI's record of success.

As is the case in most new technology development successes it is often characterized by a new round of challenges uncovered by initial success. Such developments are road marks on the way to ultimate success and should be embraced as part of the progression of any technology. Section 10 describes the road marks that CAAFI has discovered and is pursuing.

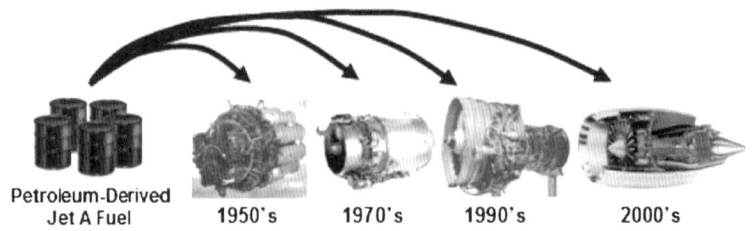

Fig. 1 Legacy jet fuel adoption process

5 Streamlining the Fuel Qualification Process

5.1 The Challenge

At the dawn of jet propulsion in the late 1930s, engines were designed to operate on kerosene. Kerosene was readily available (military aircraft relied on more volatile gasoline) and its properties best suited the Brayton cycle combustion process used by jets. Over time, engine and aircraft designers realized that they needed to more tightly control the properties of kerosene for both commercial and military operations to ensure both safe operation and consistent performance. With the fuel properties known and understood, engineers could then incorporate technological advances into turbine engine designs to achieve significant gains in fuel efficiency and durability. Thus were born the aviation fuel specifications used to control the formulation, manufacture and distribution of jet fuel. These fuel specifications specify "performance-based" properties which are designed to control the known variation of crude oil-derived or petroleum-derived jet fuel.

Early in the twenty-first century, the aviation fuel industry leaders who were strategizing on how to deploy alternatives to petroleum-derived jet fuel, quickly realized that the designs and performance of the many thousands of existing jet engines and aircraft produced and certificated over the ensuing 70 years had all been optimized for this existing, petroleum-derived jet fuel (see Fig. 1).

They faced a seemingly insurmountable challenge to figure out how to design and certify a new alternative fuel for this existing fleet of aircraft (see Fig. 2). The only significant change in fuel specifications that had occurred was a move by military aviation to JP8 fuel from volatile JP4 in the late 1980s. In the commercial world JP8 equivalent Jet A required 10 years and millions of dollars to accommodate a Coal to Liquid alternative fuel from the Fischer–Tropsch process from a single production facility in South Africa.

It was recognized that many different design approaches were taken to accommodate the specified jet fuel properties and this resulted in a myriad of different designs existing on these products. How could each of the many

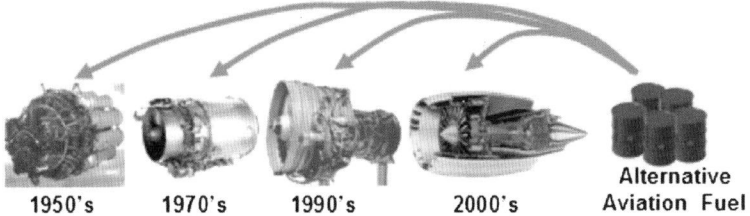

Alternative
1950's   1970's   1990's   2000's   Aviation Fuel

Fig. 2 Adapting existing products to alternative fuels

thousands of engine and aircraft with these many different designs be evaluated and tested to ensure the new fuel was safe and performed in a similar manner to the existing fuel?

5.2 Pathways to Solutions

The CAAFI Certification-Qualification Panel quickly realized that the key to solve this challenge would depend on the ability to prove that the new fuel was essentially identical to the existing, petroleum-based jet fuel. If this could be proven with a thorough technical investigation, then the FAA regulations would not require any certification at all. This was based on the existing, FAA-approved operating limitations for all aircraft and engines which specify the aviation fuel permitted for use. If an aviation fuel qualification process could be established that could prove that the alternative fuel was not a "new fuel", but rather the "same fuel" produced from different raw materials and/or processes, then the alternative fuel would fit under these existing operating limitations. These fuels would be called "drop-in" fuels to reflect their seamless entry into the distribution infrastructure once approved.

The CAAFI Certification-Qualification Panel worked with the key aviation fuel specification-writing organization, ASTM International to expedite the development and approval of a qualification process for new jet fuels. In parallel with that effort, an ASTM Task Force was formed to apply this qualification process to the approval of the initial alternative aviation fuel; Fischer–Tropsch (FT) fuel.

5.3 The Results

On September 1, 2009, ASTM International approved the world's first semi-synthetic aviation fuel specification. This specification, number D7566, entitled "Standard Specification for Aviation Turbine Fuel Containing Synthesized Hydrocarbons", was a significant milestone toward the CAAFI's goal of promoting the deployment of alternative aviation fuels in the commercial aviation world because it allowed the use of D7566 fuels in all existing engines and aircraft. Specification D7566 is considered the "drop-in" fuel specification, because any

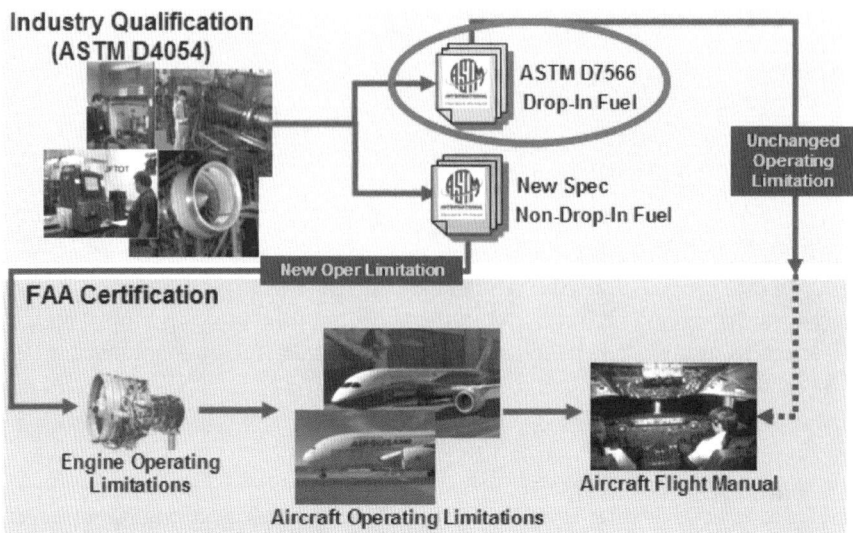

Fig. 3 ASTM advanced fuels process integration

new fuel added to that specification will have been proven to be essentially identical to petroleum-derived jet fuel. The specification is structured to define each new fuel in an annex, with FT fuel included as the first annex at publication. The process from initial tests begun on and Air Force B-52 with Gas to Liquid fuel had taken 3 years and qualified the entire family of fuels from Coal, Gas, and Biomass using the Fischer–Tropsch process.

Also, in October, 2009 ASTM International standard D4054, "Guideline for the Qualification and Approval of New Aviation Turbine Fuels and Fuel Additives", was issued to provide candidate alternative aviation fuel producers with a guide to evaluate their new fuel. Figure 3 shows how ASTM D7566 and D4054 work together to solve the "certification challenge" that at one time seemed insurmountable.

ASTM 4054 documented the experience developed from learning from both the Sasol qualification and from the steps needed to achieve FT fuel qualification under the new ASTM7566 (Fig. 3).

Building on both these standards, CAAFI and the ASTM fuel committee in which its members participate, took on the challenge of fuels from another process, lipids or fats from oil seed plants and tallow from animal fats called most commonly Hydrotreated renewable Jets. It is expected that by the time of publication of this treatise these fuels will form a second annex to ASTM 7566. From the time of initial research via production of small samples were published under the Defense Advanced Research Projects Agency (DARPA) until the mid-2011 approximately 3 years had expired.

Even before the completion of the HRJ fuel qualification, research had been initiated to progress new process types through the qualification protocols (see Sect. 6).

Perhaps the greatest success of CAAFI/ASTM qualification team efforts is that implementation in the case of both FT and HRJ processes had removed the process of qualifying fuel from the critical time path to development and deployment. In so doing a major barrier to investment in Aviation alternative fuels as a "first mover" had been removed while retaining the rigor of the protocols.

6 Implementing Comprehensive Risk Management in Alternative Fuels Research and Development

6.1 The Challenge

Aviation needs as many different sources of alternative fuels as possible to reduce environmental impacts and stabilize both price and energy security. With the success of FT fuels and the template for success that they have provided there has been a proliferation of potential fuel production pathways. While efforts were being exerted by multiple sources within government and biofuels producers, the experience that research agencies (Air Force, NASA, FAA among other had in systems level R&D, defined a need for a risk management tool specific to fuel development. In this manner the status of pathway development and technical suitability of fuels could be communicated. Such tools would also provide a mechanism for tracking research and development (R&D) efforts and identifying gaps.

6.2 Pathways to Solutions

As a result of this challenge, the CAAFI R&D team in partnership with the Air Force Fuels lab sought to develop a "Fuel Readiness Level" (FRL) scale (see www.caafi.org/information/fuelreadinesslevel.html) that adapts the previously existing Technology Readiness Level (of NASA and DOD) to more explicitly cover alternative aviation fuel development. It did so by merging traditional technology measures (TRL) and manufacturing Readiness levels (MRL) used in hardware production to the case of fuel development and production.

The FRL process was sufficiently mature and proven in FT process developments and the early stages of HRJ development that in 2009 it was proposed for global approval to a United Nations, International Civil Aviation Organization Committee (ICAO) Fuels Conference after consultation with European interests who use similar tools.

Like its predecessors the FRL is a gated risk management process. As such it has an associated checklist of pass/fail criteria to be applied first by producers. Subsequent ratification by customer sources, such as the Air Force or Commercial research authorities to move to the next step follows (Fig. 4).

Legend:	R & D	Certification/ Qualification	Business & Economics
FRL	**Description**	**CAAFI Toll Gate**	**Fuel Qty**
1	Basic Principles Observed and Reported	Feedstock / Process *Principles* identified.	
2	Technology Concept Formulated	Feedstock / *Complete* Process identified.	
3	Proof of Concept	Lab Scale Fuel Sample Produced Basic Fuel Properties Validated	500 ml
4.1 4.2	Preliminary Technical Evaluation	System Perf. & Integration Studies Entry Criteria/Specification Properties Evaluated (MSDS/D1655/MIL 83133)	10 gal
5	Process Validation	Sequential Scaling from Laboratory to Pilot plant	80 gal to 225K gal
6	Full-Scale Technical Evaluation	Fitness, Fuel Properties, Rig Testing, and Engine Testing	80 gal to 225K gal
7	Fuel Approval	Fuel Class/Type Listed in Int'l Fuel Standards	
8	Commercialization Validated	Business Model Validated for Production Airline/ Military Purchase Agreements	
9	Production Capability Established	Full Scale Plant Operational	

Legend:	R & D	Certification/Qualification	Business & Economics

Fig. 4 The fuel readiness process as approved by ICAO (*Declaration and Recommendations*, in *Conference on Aviation and Alternative Fuels*2009, ICAO Secretariat: Rio de Janeiro, Brazil. p. B-3, Appendix A: Item 25 (adoption of FRL)

With FRL risk management framework defined the CAAFI R&D team, at its September 2009 meeting held at the United States Department of Agriculture, recommended to CAAFI leadership that its efforts be expanded to encompass feedstock readiness in collaboration with USDA. The thought process was that feedstock yield and tolerance of energy crops to unfriendly environments could be improved. In this manner it would be both clear if the pacing of fuel suitability of aviation use and the ability to produce that fuel could be better matched in schedule. In addition the use of Aviation level system risk management techniques by agricultural researchers for energy crops could lead to rapid development in those areas.

The Fuel Readiness Level-gated risk management approach does include a number of "touch points" to assure that environmental due diligence is progressing. As 2011 dawned and environmental assessments perhaps trailed other technical developments, the expansion of gated risk management to environmental team dealings has been put in place.

6.3 The Results

- In November 2009 the ICAO Conference on Alternative Fuels in Rio De Janievo accepted the CAAFI Fuel Readiness Level process (Fig. 4).

- Since the inception of alternative fuels work the Air Force Research Lab has received hundreds of small fuel samples to execute the initial FRL scale laboratory testing.
- The FRL risk management tool is of particular use in the legitimization of new fuel pathways for which work was initiated following the inception of FRL.
  This scale is being used to help increase the readiness of various other processes, such as "alcohol to jet". ATJ is a process where sugar fermented to alcohols, which is then dehydrated and oligomerized into hydrocarbon jet fuel (Fig. 5).
- The first step in evaluating in alternative aviation fuels is evaluation by the Air Force, this process step is frequently sought by fuel producers and developers when they are introduced by CAAFI to "socialize" their fuel/process to the aviation fuel community.
- In September of 2010 the execution of Feedstock Readiness (FSRL) between FAA and USDA's newly formed research centers was initiated.

7 Structuring and Facilitating Comprehensive Environment Benefits Assessments

7.1 The Challenge

With the advent of climate change/global warming concerns it became clear that the industry needed to find additional ways to reduce its GHG footprint. Sustainable alternative aviation fuels are one of the most promising opportunities. In fact both studies of the multi-government and industry team working on implementing growth in the "Next Gen" Aviation system and the International Air Transport Association identified that equipment improvements offering as much as 1.5% fleet efficiency gains annually, and further gains from air traffic efficiency of new management systems could not achieve the goal of achieving GHG reductions at a rate that would prevent growth from its 2005 base, a metric of environmental success for many.

Policies arbitrarily tied to specific calendar achievements in the case of aviation fail to credit achievements of the industry. Even before greenhouse gas (GHG) emissions' potential contribution to climate change became an environmental concern, the aviation industry was achieving tremendous GHG emissions savings. In fact, the U.S. airlines improved their fuel efficiency by 110% between 1978 and 2009, saving over 2.9 billion metric tons of carbon dioxide (CO_2)—an amount roughly equivalent to taking 19 million cars off the road each of those years. In addition despite its growth the U.S. commercial aviation accounts for only 2% of the nation's man-made CO_2 (and the global commercial aviation sector likewise accounts for 2% of global CO_2).

Further gains in green house gas emissions were to occur over the entire "ground to wake" life cycle for fuels production and use—not only end use itself (Fig. 6).

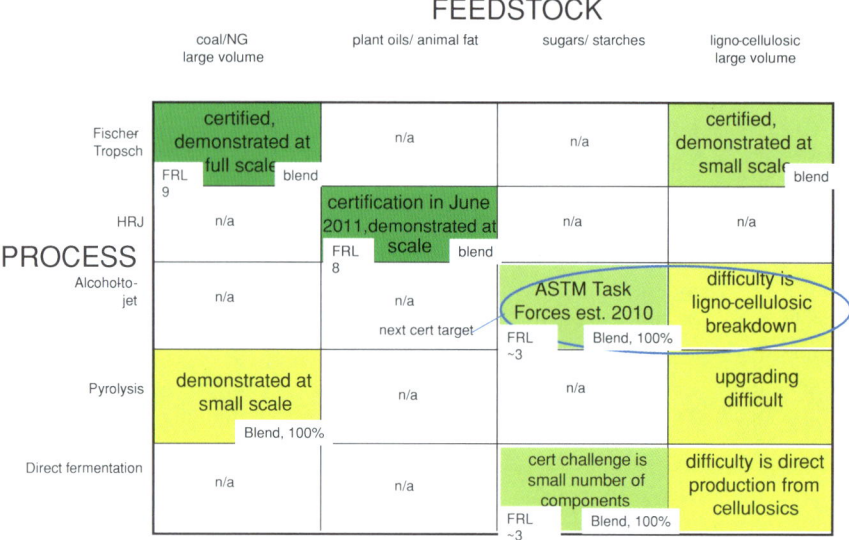

Fig. 5 Feedstock process candidates for jet fuel

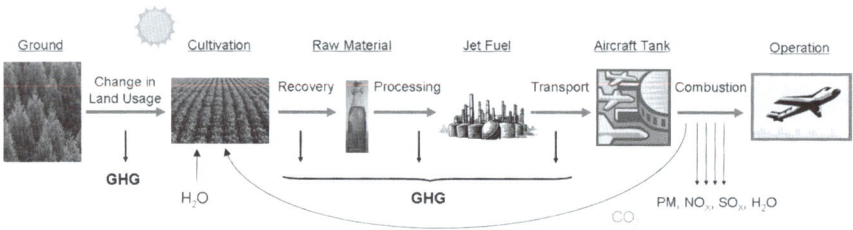

Fig. 6 The "ground to wake" analysis required for green house gas emissions

Quantification of gains in real, discreet and auditable terms was brought into sharp focus in the Energy Independence and Security Act of 2007. There government purchasers were required to demonstrate that alternative fuel purchases would need to demonstrate compliance with the Act's section -526 demonstrating that on a life cycle basis the fuel purchased would be better in GHG Life cycle than fuel purchased from an oil refinery—not a given for any fuel.

Beyond Greenhouse gases the industry was also experiencing added challenges in the area of small particle control called PM2.5 (or particle material of less than 2.5 microns). The concern cited in National Ambient Air Quality Standards is viewed as a precursor to control measures and regulations of that pollutant under the Clean Air Act as amended in 1990. About 60% of all the U.S. Airports are found to be in "non-attainment" areas for this pollutant. The issue to date however has been that globally accepted means of measuring PM2.5 for aviation-specific challenges have yet to be finalized.

Lastly the overall subject of sustainability (combining environmental, economic and social factors to gain acceptance) and the who and how approvals to assure that sustainability criteria are being met will need to be established.

Whether it be GHG life cycle, PM 2.5 or sustainability certification, uncertainty of the outcomes for alternatives in these areas is a true barrier to assuring investors that they can procede with projects knowing that environmental factors will be positive assets for projects rather than barriers to timely decision making.

7.2 Pathways to Solutions

The challenges, while daunting, do allow for clarity in needed steps. These pathways can be categorized as follows.

- Establish universally accepted goals for Greenhouse Gas control by airlines.
- Set means of quantifying real terms carbon and GHG calculation for projects that are specific to aviation.
- Communicate options and certification techniques for sustainability certification.
- Obtain means of quantifying PM 2.5 benefits that are unique to aviation and obtain adequate data to assess benefits.
- Incorporate all algorithms in a comprehensive set of tools that can be utilized to assess project benefits by all stakeholders.

7.3 The Results

7.3.1 Universally Accepted Goals

While Section 526 of the Independence and Security Act of 2007 provided a legislated goal and constraint for government purchasers the airlines were quick to follow with voluntary measures.

On Earth Day 2008, the Air Transport Association in the U.S. put in place a policy for U.S. airlines that paralleled section -526 declaring that "*we believe it is incumbent upon all segments of the transport sector to take voluntary measures to limit their impact on the environment. As combustion of traditional, petroleum-based jet fuel is a source of such emissions, we seek alternative fuel sources having a reduced emissions profile relative to traditional fuels*"

Following these declarations of goals IATA (the International Air Transport Association and the Air Transport Action Group (ATAG) that includes both manufacturers and airlines decided to go further to establish calendar-based goals to achieve Carbon neutral growth. Specifically IATA adopted the goal of

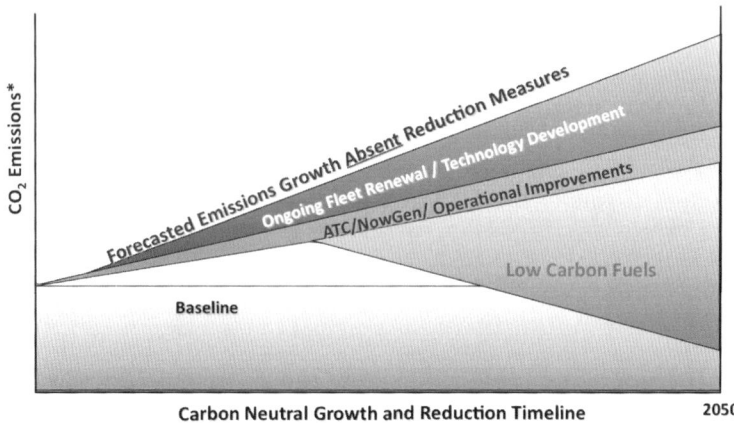

Fig. 7 IATA pathways to carbon neutral aviation growth

achieving Carbon neutral growth for the airline industry by 2020 with a reduction of 50% in GHG emissions by 2050. The approach included assumptions on benefits incorporated from equipment efficiency gains and improved operational efficiency from Air Traffic control gains. The remaining gap defines the need for sustainable alternative fuels insertion (Fig. 7).

7.3.2 Quantifying Real Terms Carbon and GHG Calculation for Projects

Initiated under research executed by the FAA-funded MIT-led PARTNER center of Excellence (Partnership for Air Transport Noise and Emissions Reduction web.mit.edu/aeroastro/partner/index.html) GHG reductions for a variety of processes and feedstocks were embraced by the Air Force and Department of Energy. Further the MIT process bounds projects identifying uncertainties in land use questions for various projects as well as the processes themselves. For example there is considerable issue with the use of algae feedstocks for hydrotreated renewable jet. The energy use in water extraction and retention of temperature in open ponds could add to energy requirements and limit or even eliminate GHG benefits for that process feedstock combination (Fig. 8).

Subsequently the efforts were combined and issued with the Air Force and Energy Department efforts under the report "Framework and Guidance for Estimating Greenhouse Gas Footprints of Aviation Fuels (Final Report) (2009, AFRL-WP-TR-2009-2206) [1]. This document, also colloquially referred to as the "Rules and Tools Document," builds on ISO Standard 14040 and augments and applies the ISO Standard approach to life cycle analysis of aviation fuels. It identifies the steps associated with life cycle analysis for aviation fuels and makes recommendations for dealing with open issues.

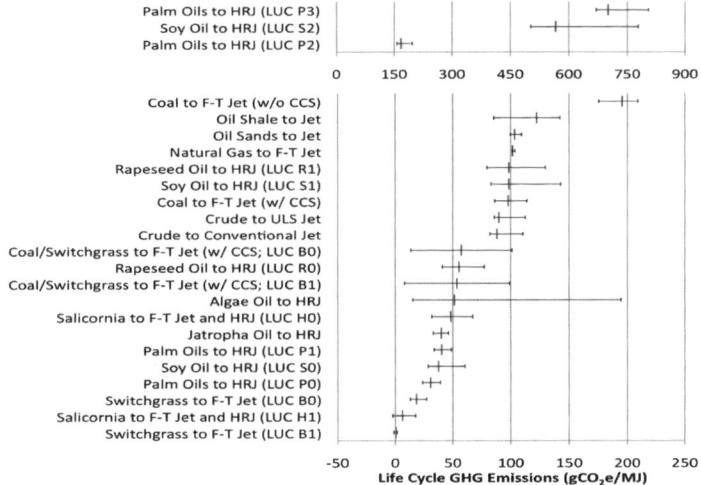

Fig. 8 Summary of PARTNER greenhouse gas life cycle analysis results (as of 03/10)

Work continues on the actual case studies of three different processes, using these rules and tools. In addition PARTNER continues its work expanding feedstocks and processes for review enabling its team members to expand efforts to regional feedstocks and processes.

7.3.3 Communicate Options and Certification Techniques for Sustainability Certification

Green House Gas life cycle quantification from "well to wake" is a subset of the larger issue of full environmental sustainability development and certification. CAAFI efforts to define an aviation-specific path for sustainability certification are not as advanced as GHG quantification but are proceeding. At its August 2010 meeting the CAAFI environmental team put in place a panel of sustainability practitioners to at a minimum communicate options for certification of sustainability and to identify processes that might be used for aviation unique activities.

Of particular interest is to establish.

- The possible use of a product category rule that could be utilized in conjunction with ISO 14000 to provide a process for the certification of Jet fuels.
- The use of elements of processes developed by the Roundtable on Sustainable Biofuels a Swiss-based consortium.
- Consideration of sustainability definition options under consideration by the U.S. EPA and others.

Such sustainability quantification will include factors such as water usage and the use of chemicals in crop propagation. Whether energy crops are harmful to animals or humans as well as multitude of other factors must be considered.

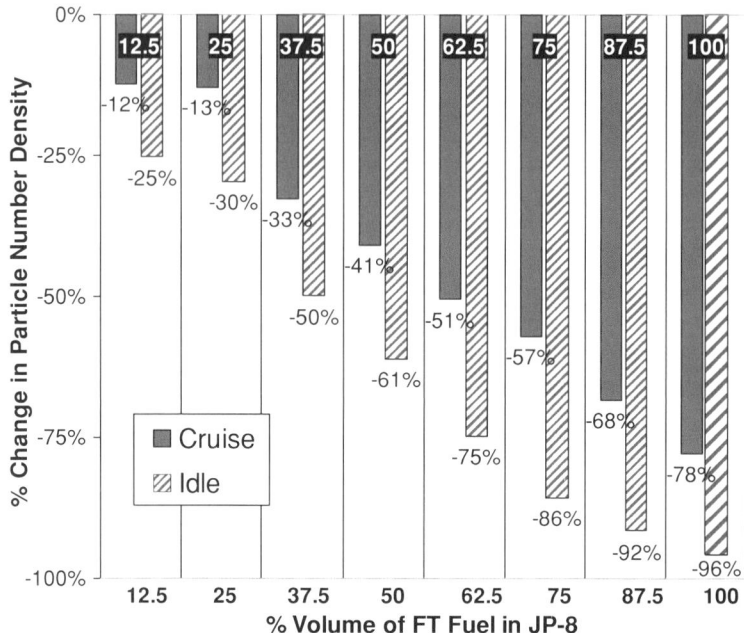

Fig. 9 Measurements of PM 2.5 benefits for fischer tropsch jet fuels (Ref. Corporan, Edwin. Particulate emissions data supplied for an Allison engine using Fischer–Tropsch Fuel. Air Force Research Lab. 2007)

7.3.4 Quantifying PM 2.5 Benefits

As they are dramatically reduced in sulfur content, most alternative fuels are a precursor of small particle formation. Early indications of this progress was obtained in measurements of particle reduction obtained by the Air Force Research Laboratory (Fig. 9).

Execution of further particle measurements and benefit studies is organized under PARTNER Project 20 entitled "Emissions characteristics of alternative fuels". The project seeks to expand the database for such fuels for applications and benefit assessments of both policy and project benefits.

7.3.5 Comprehensive Set of Tools to Assess Project Benefits by all Stakeholders

CAAFI's efforts to provide analysis capability for project use concentrates on mechanisms provided by the Airport Cooperative Research Program of the Transportation Research Board of the U.S. National Academy of Science. ACRP is in the process of executing 3 projects that target alternative fuel benefits assessments for aviation. Specifically these projects are

- ACRP 02-18—Guidelines for Integrating Alternative Jet Fuel into the Airport Setting—led by Metron Aviation.
- ACRP 02-23—Alternative Fuels as a Means to Reduce PM2.5 Emissions at Airports led by AEA of the UK via its U.S. affiliate PPC.
- ACRP 02-36—Assessing Opportunities for multimodal Alternative Fuel Distribution Programs—project contractor to be Announced.

Together these projects provide handbooks that are intended to structure the use of specific analysis defined above into the overall project assessments.

8 Deploying a "A New Fuel Dynamic" Through Public/Private Partnership, and Multiple- Success Models

8.1 The Challenge

Corresponding to a period of consolidation and stress for the U.S. Commercial airlines, the period from 2005 to 2011 displayed an unprecedented roller coaster of unstable cost drivers. While extraordinary measures were taken to control labor and other costs the cost of fuel, in 2006, for the first time exceeded labor as a percent of airline operating costs to become the highest cost element of airline operations—as much as 40%.

The cost run-up from 2005 to 2008 was the most severe the industry had seen. Even with a 13% drop in consumption from in 2007 (20 billion gallons per year) to 2011 (17B plus projected performance) the outcome is eerily similar in industry fuel cost exposure.

Investigation of the underlying cause of this performance reveals an inherent flaw in the current factors leading to Jet fuel production. Owing to the refinery process that produces oil, production of the jet fuel the middle distillate range will not yield more than 10% of a barrel of oil for jet. The result in high demand times has been added increases to Jet fuel price as crack spreads (the difference between crude and jet fuel prices) has expanded to the $25–30 per barrel range an amount higher than the total cost of fuel in the pre-2005 time periods of fuel price stability Fig. 10.

Exogenous factors such as weather shocks (Hurricane Katrina) and dependence on imported oil from politically volatile regions were also factors in the price volatility.

Notable in the period was also the fact that the beginnings of efforts to initiate work on biofuels in the U.S. were exclusively dedicated to ethanol for gasoline—a fuel totally unsuited to use in Jet aircraft. Aviation was not even thought to be capable of use of alternatives to petroleum. Government interest in investing in aviation alternative was limited to a few military sources (notably the Air Force) and the remnants of efforts initiated during earlier oil shocks.

Fig. 10 Airline industry fuel cost (*Left scale*)/ Consumption (Bgals) (*right scale*)

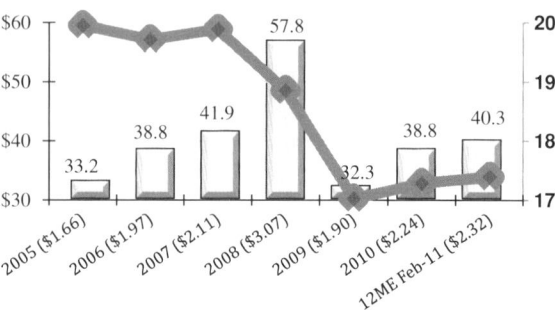

During the middle of the 2005–2011 time period even as credibility for fuel substitutes were being produced through greatly improved qualification success (Sect. 5) and continued disciplined development of multiple alternatives (Sect. 6) the world suffered through a financial crisis of epic proportions. The ability to invest in new alternatives through the private sector was gravely diminished.

8.2 Pathway to Solutions

Given the multiple challenges the construct of a solution required multiple parallel pursuits of solutions building upon and in many cases interacting with the qualification and interacting with the rationale and progress and understanding generated from qualification success, R&D progress and environmental need described in Sects. 5–7. These pathways required simultaneous efforts to

- Convince airline buyers and executives that the current pathway for fuel purchases was structurally flawed, exclusive of the short term price, and that a "new fuel dynamic" involving alternative options is the only way to achieve sustainable industry success from both an economic and environmental perspective.
- Work with government to prove that not only did aviation deserve status as a candidate for alternatives but in fact could be a "first mover" and lead the nation. That outcome would in fact lead to supportive relationships with the energy supply and agricultural industry in government. In fact the term "new fuel dynamic" first appeared in a letter from CAAFI Airline, and manufacturer sponsors and Advance biofuel producers to then President elect Obama on January 16, 2009. The letter declared *"the aviation industry is eager for an entirely new fuel dynamic and will be an enthusiastic purchaser "*. Publication of the letter itself suggested that the job of convincing airline buyers that this is critical prerogative had been largely accomplished.
- Combine efforts of multiple airlines to encourage the development of many success models across the country and around the world. Through the development of these success models and government support for aviation, fuel

private sector investment, including that from major oil companies could be obtained. In so doing the industry could accelerate developments.

- Team with both government buyers and with global efforts to ensure that fuel suppliers viewed aviation supply, though only 10% of their demand, as a single, rationale, global buyer eager for long-term commitments (not the nature of other fuel types) with an assured global market for liquid fuels of the high standard serious suppliers are prepared to provide.

The pursuit of these multiple pathways also involves significant interdependencies. Government declarations embolden private sector researchers and producers to build new pathways knowing that there will be a ready buyer. The qualification authorities respond to "demand pull" by expediting the testing and analysis that leads to qualification. Knowledge that policymakers have been emboldened goes on to indicate to buyers that their efforts will be rewarded. Airline commitments encourage more states and nations to be involved in aviation supply. The visualization of the aviation alternative fuel snowball rolling down a hill gathering mass, and in the ultimate case producing an avalanche of activity, is then possible. The seeds of such growth are now evident in the U.S. and elsewhere.

8.3 The Results

In October, 2010 in the prestigious and conservative UK periodical "The Economist" stated the best summary of Aviation's budding success. In a larger story about the recovering prospects of biofuels rightly noted that, *"There is no realistic prospect for widespread electric air travel: ...if you want low-carbon flying, drop-in biofuels are the only game in town."*[2] It concluded with the apt double-entendre, *"Over the long run, the future for biofuels may be looking up."*

In the same month Biofuels Digest, in a poll of its members, found that nearly 40% now believed that there would be one Billion gallons of biojet in operation by 2020. Taken as a 50% blend and if focused in the U.S. that would mean that over 10% of aviation fuel would be derived from 50% biofuel blends.

Another measure of success is the representation of Biofuels in Aviation global forums. In 2011 alone Commercial biofuels were featured in a three-day forum at the Australian Airshow in March. As this story is being written a dozen biofuels companies and six airlines were committed to join in a biofuels showcase at the June, 2011 Paris Airshow—a fourfold increase from the three fuel companies represented in July 2010 at the Paris Airshow's alternate year counterpart, the Farnborough Airshow in the UK.

These two reports in financial and biofuel, not aviation, publications as well as actions of the fuel companies themselves in becoming paid subscribers to global aviation customer events give apt indication of the success of the measures described previously.

2006

Altair – Washington / HRJ
Rentech – Mississippi / BTL
Rentech – California / BTL
Gevo – Illinois / Alcohols
Multiple airlines

QATAR ✈
AIRWAYS القطرية

TBA – Alcohols
GTL - Shell

SWAFEA /
AlfaBird

Solena – BA - MSW- L
Neste – Lufthansa - HRJ

ABRABA

Amyris - Azul – FRJ
Air BP – TAM - HRJ

2011

Sustainable Aviation Fuel
Users Group

Solena - Qantas -MSW-L
Solazyme - Qantas - HRJ

Fig. 11 Global aviation partnerships producing production arrangements (March 1, 2011)

Discreet progress is best shown in two areas—Project Formation and the concurrent rise of Government focus and support for Aviation in recognizing its move from "afterthought" to the "cutting edge" of Transport alternative fuels.

8.3.1 Project Formation

Now waiting in the wings are more than 20 U.S. biofuel projects in various stages of development, several of which have the potential to produce aviation biofuel. These projects cover a wide range of feedstocks and process.

- Nearly 80% of all aviation biofuels projects are supported by the U.S. biofuel producers in partnership with global entities. Commercial efforts in Brazil, the UK, Australia, Italy and Spain have been launched. Production efforts accompanied by companion private–public partnerships are being formed (Fig. 11).
- Beyond the companies producing projects the investigation of possible projects in some 17 U.S. states from Rural areas to New York City—where the feedstock may be Municipal solid waste, is offering multiple new opportunities. CAAFI stakeholder leaders in many States are seeking to combine the efforts of agriculture, aviation, energy and business development. Efforts in those states are supported by Airline interest from the Air Transport Association Energy Council airline buyer members and CAAFI leadership (Fig. 12).

Arizona
California*
Florida **
Georgia**
Hawaii**
Illinois
Michigan**
Mississippi*
Ohio**
New York**
Pennsylvania**
Oklahoma
Tennessee**
Texas**
Washington*
West Virginia
Wisconsin

* airline/producer MOU's   ** study proposals or Pilot Plants

Fig. 12 Projects under consideration in multiple U.S. States

Obtaining safe, reliable and environmentally preferred aviation biofuels sustains not only the aviation industry but also builds new agricultural and fuel-processing economies as well.

8.3.2 Government Support

In parallel with Commercial developments government support has progressively grown.

On March 30, 2011, in his remarks on America's energy security before a packed audience at Georgetown University, President Obama declared "*...the Air Force is aiming to get half of its domestic jet fuel from alternative sources by 2016. And I'm directing the Navy and the Department of Energy and Agriculture to work with the private sector to create advanced biofuels that can power not just fighter jets, but also trucks and commercial airliners.*"

In fact the President's declaration was not the starting point for Government's focus on alternative aviation fuel, but rather the culmination of a series of events that took aviation from "afterthought" to the "cutting edge" of alternative fuel in transportation from a government perspective.

These efforts actually began in 2008 with congressional action on the Agriculture Department Authorization for its $ 1 Billion dollar Program for Energy crop development and culminated with the Energy Policy statement. Key steps followed:

- In September of 2009 Aviation was the only end-user sector to testify in hearings on the implementation of Section 9000.
- USDA Undersecretary Rajiv Shah in an address to the September, 2009 CAAFI general meeting proposed an Aviation Agriculture R&D partnership.
- On Feb. 3, 2010, The Biofuels Interagency Working Group, consisting of USDA and DOE representatives, released a report spelling out ways to promote the development of the biofuels industry in the United States in connection with the Energy Independence and Security Act target of 36 billion gallons per year of U.S. biofuels production by 2022. The report, "Growing America's Fuel,"[3] laid out the situation and called for "an outcome-driven re-engineered system." Moreover it explicitly recognized the progress and potential in the area of aviation, with a stated goal to *"Secure lead customer purchase commitments to stimulate production of feedstocks and biofuels with a concerted effort directed to our military and airline industry."* The response from DOE was instantaneous. It too would now engage in aviation fuels research—and activity not in its charter until that time.
- In July 2010 USDA Secretary Vilsack, The Airline Industry, via the Air Transport Association, and Boeing launched the "Farm to Fly initiative. The U.S. aviation community had come together with government stakeholders, including USDA, the Department of Energy (DOE), the Department of Transportation (DOT) and the Department of Defense (DOD) to express unified support for the president's goals of environmental stewardship and energy independence. The coalition was formed to ensure that, in the near future, aviation biofuels would become an economical and environmentally preferred alternative to petroleum-based jet fuels through a strong commitment of resources to research, development, demonstration and deployment through public-sector leadership and financial incentives to bring production online. The program was accompanied by regional cooperative feasibility actives in the Pacific Northwest and support for Navy supply efforts in Hawaii. The regional thrusts were in keeping with regional project formation efforts.
- In Oct 2010 the commitment added the creation and implementation of programs and incentives to assist the American farmer in the selection and cultivation of crops that can be converted into affordable and sustainable aviation biofuels. The creation of the Biomass Crop Assistance Program (BCAP) and the liberalization of rules in section 9000 to enable foreign investment in U.S. Projects and to loosen restrictions for Section 9000 for use in more populated areas ensued by year end.
- Lastly, in December 2010, the Department of Transportation, Future of Aviation Advisory Committee (FAAC) endorsed the deployment of sustainable alternative aviation fuels to improve the environmental impact and competitiveness of U.S. aviation, stating that the United States should, *"exercise strong national leadership to promote and showcase U.S. aviation as a first user of sustainable*

alternative fuels. This would involve increased coordination and enhancement of the concerted efforts of government and industry to pool resources, overcome key challenges and take concrete actions to promote deployment of alternative aviation fuels…These actions would affirm a U.S. global leadership position in sustainable alternative fuels for aviation." This clear endorsement of the efforts that USDA and the aviation industry have been undertaking together closed the intergovernmental policy loop meaning that all government agencies are moving in lock step to pursue Aviation alternative fuels.

- Concurrent with the President's announcement the U.S. Department of Commerce in partnership with CAAFI and USDA initiated a series of Webinars leading up to "Investor day" at the Paris Airshow to be held on June 22 of 2011.

9 Interagency Cooperation: Bringing the Government Process Support and their Private Sector Clients Together Effectively

9.1 The Challenge

Unique to the challenge of creating a new source of liquid fuels a primary challenge is how to secure the cooperation

(a) Linking a series of agencies and private sector clients who have never worked together for any purpose.
(b) Leveraging traditional government industry cooperative channels to achieve goals that only a few technical specialists have concentrated upon, to achieve goals well beyond research success.
(c) Learning which agencies have the most critical capabilities needed to execute the Aviation alternative fuels mission
(d) Ensuring that key individuals are both identified and motivated to work toward the collective good of the aviation alternative fuels enterprise even when it is not the primary mission of their agency or organization within the agency.

9.2 Pathways to Solutions

- Recognizing and the CAAFI Coalition Assets as an Effective Enabler, and Focus for Government Interaction with Aviation.

As a coalition of airlines, aircraft and engine manufacturers, energy producers, researchers, international participants and government agencies CAAFI includes all key parties to advance development and deployment in a comprehensive fashion.

As a strong coalition providing direct access to the aerospace industry and particularly to the largest and consolidated fuel buyer group in the U.S. airlines CAAFI has attracted significant interest from federal agencies with missions devoted to energy production (DOE and USDA).

- Build upon Commercial Aviation Strengths among jet fuel purchasers in the Military.

As the largest buyer of jet fuel, nearly 10 times the amount of Jet bought by the Air Force and 40 times that bought by the Navy the attraction to the military services is real. This is particularly important in that nearly 60% of jet fuel purchases by the Air Force are for transports, tankers and other vehicles that use engines that are commercially certified and purchased under commercial terms.

Further Commercial producers have longer term purchase authority which government buyers do not have.

Lastly CAAFI's FAA sponsor, the office of Environment and Energy has capabilities and mechanisms for environmental analysis not present in the military.

- Capitalize on Aviations Strong Assets in Systems Level R&D and Ability to work with Government Contractors.

Beyond purchase collaboration agencies such as NASA and DARPA (the Defense Advance Research Agency) in addition to the services routinely work with Aviation contractors to execute systems level development.

- Recognize Aviation's key role as an a major export asset and desirable vehicle for International collaboration.

The U.S. Departments of Commerce and State recognize the aviation as one of the country's leading exporters and a desired partner in international collaboration. FAA is familiar with and called upon by these departments in addressing aviation matters across the globe.

- Empower CAAFI's Government sponsor, FAA, as the focal point for dialog in the arenas of R&D collaboration, the environment.

By moving FAA Office of Environment and Energy in the position of being government's gateway to aviation CAAFI has helped to focus attention and resources of these agencies with broad energy missions on the needs of the aviation industry as a single entry point to the enterprise. This is particularly important given the traditional alternative fuels focus of these agencies on alternative fuels has been on surface transportation.

FAA and Aviation's efforts have also converged with simple product model the need to pursue high energy density drop-in alternative fuels for singularly focused customer with a consolidated buyer base.

The FAA office of Environment and Energy also has and an asset a culture on a collaborative and global industry consultation model. It is in this manner that it has built its role as a successful industry regulatory authority and serve as well in collaborations with others in Government.

9.3 The Results

9.3.1 Within Government

- In Sept 2010 discussions between FAA and the USDA resulted in the signing of an memorandum of understanding between the USDA's office of Energy Policy and New Uses and the USDA Agricultural Research Service with the FAA's Office of Environment and Energy. The agencies would work together on jet fuel-specific feedstocks leveraging USDA crop expertise and FAA's aviation industry analytical tools expertise to accelerate feedstock readiness (see Sect. 6)
- On a regular basis FAA and the Air Force develop and exchange key data between the U.S. Air Force Certification office, the Air Force research lab and the engine manufacturers. These efforts have led to the development and approval of the first alternative jet fuel standard at ASTM International. This data exchange and collaboration continues as CAAFI's certification and qualification team seeks to approve additional alternative jet fuels. (see Sect. 5). Further it is this grouping that uniquely created a singular Fuel Readiness Scale (see Sect. 6)
- With the 2010 policy focus changed to emphasize aviation, and exchange of key information and technologies with the Department of Energy's Office of the Biomass Program focused on conversion process optimization and techno economic analysis. This is being factored into both R&D and commercialization activities. This is particularly important if the case is to be made that aviation fuel production process costs will be learned out in a reasonable time period to produce affordable fuels.
- FAA now is engaged with the Biomass Research and Development Initiative BRDI as a participant in the team putting together research initiatives.
- In February 2011, FAA formed a cross government bilateral agreement with Brazil. Other discussions are underway with European countries which has included the FAA's hosting of representatives from Germany. That effort has now led to formation of a CAAFI-like process there. Other cooperative arrangements with Australia and Singapore in the Pacific Rim are under discussion.

9.3.2 Between Industry and Government

- In March 2009 ATA entered into a strategic alliance with the Defense Energy Support Center, now known as Defense Logistics Agency Energy, the energy and fuel procurement arm of the U.S. military, to pool resources and buying power to send needed market signals. The ATA/DLA Energy Alliance includes three teams focusing on off take agreement collaboration, deployment strategy and environmental compliance. Most recently the ATA worked with DLA Energy to build regional, bottoms-up, business case Analyses for regions within the U.S.

- In July of 2010, the U.S. Department of Agriculture, the Air Transport Association and Boeing formed the "Farm to Fly" Initiative to focus the deployment of aviation alternative fuels initially through the creation of regional models for deployment in the Pacific Northwest and Hawaii. CAAFI itself has formed collaborations across 17 U.S. states bringing in regional agricultural efforts to rally around Federal and State programs.
- During the first half of 2011 CAAFI in cooperation through the U.S. Department of Commerce, DOE and USDA conducted a series of webinars to attract foreign investment to USDA with the goal of attracting foreign investments to U.S. Fuel projects and feedstock production. Further airline buyers and USDA have met jointly with banks, as potential buyers, and loan guarantee government suppliers to encourage new projects. This effort is culminating with USDA and FAA participation in Fuel showcase activities at the 2011 Paris Airshow.

Most importantly, as successes multiple, the attractiveness of government agencies not centered in Aviation participating in sustainable aviation fuel developments grows at an accelerating rate. Broadened advocacy that has brought aviation from merely an afterthought for all but those closest to it, to an example of being on the "Cutting edge" for all.

10 Next Steps: The Most Critical Prerogatives

Among all else CAAFI continues to be an ongoing learning experience for its sponsors and stakeholders. As in most developments of new technologies success can often be measured not by the number of successes that are achieved but rather by the number of issues that are uncovered on the pathway to those successes. That said here is a simple listing of the next Steps to what appear at the current time to attack the most critical prerogatives.

- *Reliable Consistent Quality from First Time Suppliers.* With the onset of many new sources of sustainable biofuels entering the marketplace issues associated not with the fuels themselves as qualified, but the meeting of quality standards and reliability of delivery for these new fuels. The presence of novice fuel suppliers to the aviation industry creates the need for rigorous quality control systems to ensure delivery.

  Aviation operations personnel…who have not been extensively engaged in the development or purchase of these fuels will then be front and center. Without a doubt initial blame for any turnbacks that occur in reliable delivery or quality will be thrust upon the fuels themselves whether or not it is deserved.

Solution Pathway. FAA is implementing an Advanced Biofuels Quality initiative to attempt to "stay ahead" of any issues on quality product delivery. If issues emerge they will need to be addressed. The speed and effectiveness of those solutions will be critical to how quickly the community accepts the reality of sustainable biofuels in operation.

- *Qualification of Alcohol (Catalytic, Fermentation) and Pyrolysis Processes*. The creation of pathways beyond Fischer–Tropsch and Hyrdotreated Renewable jet (aka, Bio SPK, HEFA) may well be needed to produce adequate supplies to meet aviation goals. Unlike both FT and HRJ fuels these new fuel processes may not have practitioners of essentially the same process to input to the qualification testing for generic types. Further the number of engine and flight platforms available for testing are less available than in times prior to the economic recovery if one is to rely solely on U.S-based platforms. In addition U.S-based funding may be less available in times of government cutbacks.

  Solution Pathway. The FAA does have an advanced biofuels program that will concentrate on new pathways. With growing interest globally and possible test platforms available outside the U.S. global cooperation offers a success pathway.

- *Reducing Sustainable Biofuel cost and increasing supply*. With qualification in hand producing the kind of rapid cost reduction we have become familiar with in such fuels as computer technology needs to be brought into play for biofuels. Collectively we must provide visibility and funding priority to efforts at DOE to reduce product cost, and to increase both crop yields and the ability of crops to grown in adverse environments.

  In keeping with this plan we must resist the temptation to quote the cost of an experimental batch of fuel in limited quantities as the actual cost of fuel as it would be depicted in a long-term production contract.

  Solution Pathway. Provide the highest priority to the reduction in cost of qualified processes, increases in yields and rapid maturity for known viable feedstocks. Incentives such as the USDA BCAP (Biomass crop Assistance Program) are essential to gain farmer acceptance of these new crops even as they are matured through research.

- *Grow Investor Interest and Support Programs Globally*. Private investment in projects is improving but the rate of gain, particularly in debt financing has lagged during the financial contraction process. Growth in investment must spread globally and uniformly to places where it will pay off best.

  Solution Pathway. Airline buyer interest and fuel producer participation and equity investment has grown substantially in the past year. New and improved debt financing support such as section 9003 of the US. Farm Authorization are essential.

- *Move rapidly to reduce environmental uncertainty*. Investments do not happen when the environmental landscape can stop programs arbitrarily and at any time. Further establishing an SAE-accepted particle measurement norm to accurately measure particle reductions is important.

  Solution Pathway. Adopt policies that allow self-certification to a product category rule backed by and ISO standards or their equivalent.

11 Closing Summary

In closing the challenges, solution pathways and results achieved by the CAAFI coalition in taking alternative aviation fuel from an "afterthought to cutting edge" of sustainable fuel development and deployment have been chronicled.

The process has produced extraordinary progress and has exceeded the most optimistic of projections for progress after CAAFI's first five years.

While a good start and a strong methodology to produce this success is evident it must be viewed as just these things. Much remains to be done in what is what will continue to be a major but critical undertaking for our industry.

Aviation's future depends on our remaining focused and dedicated to take on the next set of challenges with the same vigor that we have approached CAAFI's first five years.

References

1. The full title of this report is: Propulsion and Power Rapid Response research and Development (R&D) Support—Delivery Order 0011: Advanced Propulsion Fuels Research and Development Subtask: Framework and Guidance for Estimating Greenhouse Gas Footprints of Aviation Fuels (Final Report)
2. http://www.economist.com/node/17358802, The post-alcohol world: Biofuels are back. This time they might even work (Oct. 28, 2010)
3. http://www.whitehouse.gov/sites/default/files/rss_viewer/growing_americas_fuels.PDF

The Potential Impact of Propulsion Technology on Emissions and Energy Security

Peter Ireland

1 Context

Aircraft engines account for only 2% of the CO_2 emitted to the atmosphere from all sources [20], but the societal impact of these emissions appears to be proportionately much higher. This is partly because of the highly visible nature of engine exhausts under some flight conditions (e.g. contrails at altitude) and partly because air transport continues to grow relentlessly. For instance, Airbus[1] predicts that global passenger traffic will increase by 4.7% per annum over the next 20 years. This impact is mitigated by measures taken by engine companies and aircraft manufacturers to improve the efficiency of their products. In many cases this activity is supported by national and European research programmes. The end result is that specific fuel consumption is reduced on average by over 1.5% per year [17] and there is little doubt that aircraft engines will continue to become more efficient through the twenty-first century. The most important question is the rate at which technology changes will be introduced as some dramatic changes in propulsion technology could lead to dramatic (circa 20%) reduction in fuel burn. The rate at which these disruptive technologies arrive on the wing is tempered by the commercial risks involved in their introduction and the business case for expensive new technologies.

In this chapter, we will discuss some of the work being done by engine and aircraft companies to improve efficiency. The focus of each section is the technology challenges which have led to on-going research aimed at reductions in

---

[1] Airbus 2009–2028 Global Market Forecast.

P. Ireland (✉)
Department of Engineering Science, University of Oxford,
Parks Road, Oxford, OX1 3PJ, UK
e-mail: peter.ireland@eng.ox.ac.uk

O. Inderwildi and Sir David King (eds.), *Energy, Transport, & the Environment*,
DOI: 10.1007/978-1-4471-2717-8_23, © Springer-Verlag London 2012

engine emissions. The emphasis is mostly improved fuel efficiency which directly reduces CO_2 emissions although reduced engine weight (which indirectly reduces CO_2) and reduced NOx emissions are also mentioned.

2 The Modern Aircraft Engine

The majority of air traffic is conveyed by airliners which are propelled by jet engines. The dominant form of jet engine is the turbofan which combines straightforward jet propulsion with thrust from a large fan (Fig. 1). Air is drawn into the front of the engine through the large circular intake and passes either directly to the fan, or into the core of the engine. The fan typically handles significantly more air than the engine core as this improves fuel efficiency. The fan can be considered to be a propeller encased in a duct. The swirling flow from the fan is straightened by a set of stationary vanes which direct the flow along the by-pass duct. The by-pass flow accelerates towards the end of the duct to produce thrust. The compressor is the first component in the core. It typically consists of many stages of alternating stationary vanes (stators) and rotating aerofoils (blades) which pressurise the air. The design of the compressor does not favour heat transfer, so that as the air is compressed its temperature increases to the extent that on a modern engine, the compressor discharge temperature can be several hundreds of degrees. The high-pressure air then enters the combustion chamber. Fuel is added in the combustor and the hot, high-pressure travels through the turbine where its pressure and temperature are reduced. The turbine also comprises a set of alternating stators and blades, but this time, the pressure drops and work is extracted by two or more rotating shafts. Some of this mechanical work is used to drive the compressor, while some turns the fan. The modern turbofans used to power today's aircraft such as the Airbus 380 and the Boeing 787 represent the culmination of over 60 years of continuous technological development. Technologies have been introduced to meet many targets including increased take-off thrust, reduced noise, improved reliability and reduced operating cost. However, one of the main targets has been reduction in fuel burn and hence emissions with the result that a modern airframe and engine uses about 35% (check exact figure) of the fuel per passenger kilometre than the first commercial jets. The fuel efficiency of a gas turbine used for aircraft propulsion is dependent on the performance of many key engine components. It is not possible to discuss all of these here so the author has selected a few of the key components to discuss. These include the compressor, turbine, combustor and fan system.

3 Compressor

The air that enters the core is pressurised by a multi-stage, axial flow compressor. Straightforward cycle analyses show, Oates [16], that the engine efficiency can be increased by increasing the discharge pressure from the compressor, which

Fig. 1 Modern turbofan
engine—from Rolls-Royce

corresponds to the overall pressure ratio (OPR), combined with an increase in the turbine entry temperature (TET). This is achieved by adding stages, higher blade speeds or increased aerofoil loading. It is worth noting that papers by Horlock et al. [5] and Wilcock et al. [26] have reported that the increases in gas turbine temperatures engender other efficiency penalties associated with the real gas behaviour of the combustion products and the requirement for additional turbine cooling.

The compressor extracts power from air and so needs to do this efficiently. The design of the aerofoils has developed to the extent that it is now performed using computer codes which model the full 3D shape of the stators and rotors. Specifically, the computer codes discretise the flow field in an approach referred to as computational fluid dynamics (CFD). The accuracy of these CFD simulations is so high that few rig tests are required for validation. This capability has been achieved by the development of advanced computer codes and the use of high power computers. The design challenge for new engines is to achieve the target pressure ratio at the highest efficiency. The compressor performance is sensitive to flow leakage over the rotor blades (Fig.2), and the compressor casing and rotor must be designed to minimise this leakage. The clearance depends on the cold build clearance, rotor speed and the temperature of the disk, rotor and casing. Since these parameters all change through a flight cycle, the control of the compressor gap over the operation of the engine is a major engineering exercise. There are also inevitabe gaps between the stator and rotor at the hub which, though well sealed, are subject to large changes in pressure as the rotor turns. The pressure changes send flow into and out of the cavity. The outflow can be responsible for losses which reduce the compressor performance. The compressor has to work over a broad range of operating conditions and the operability of the compressor is also a key design feature. The compressor operating range is limited by the onset of an unsteady flow phenomenon called surge. Under surge conditions, the flow separates from the aerofoil surfaces and the flow can reverse through the compressor with dramatic consequences. The secondary flows associated with leakage

Fig. 2 Complex leakage flow over the tip of compressor blade—http://ocw.mit.edu/courses/aeronautics-and-astronautics/16-540-internal-flows-in-turbomachines-spring-2006/

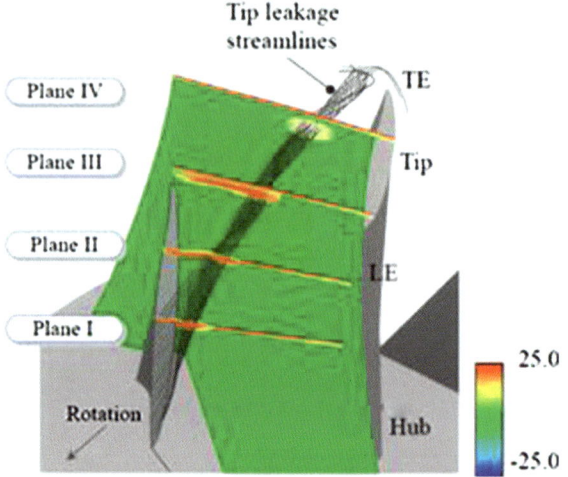

bring forward the onset of stall, [2] and research is being performed to provide understanding of these effects.

The future challenges include accurate prediction of the leakage gaps, leakage mass-flows and evaluation of the impact of these leakage flows on the gas path. Reduction in the core size (to enable increased by-pass ratio) and aerofoil span (for higher OPR) will both tend to increase the impact of these leakage flows on the compressor performance. Ever more powerful computers and larger simulations will lead to progress in compressor design. New technologies are likely to be developed which improve the degree to which component temperatures can be controlled—an early example of axi-symmetric full system modelling for a compressor referenced to engine data is given in Lewis [9]. This understanding presents opportunities to actively control the tip clearance in the compressor using tip clearance sensors and actuators that change the shape of the casing and hence the tip clearance—Garg et al. [4]. There has also been interest in the use of intercoolers to reduce the compressor work and delivery temperature—see for example the reports being generated as part of the NEWAC[2] research programme summarised in Rolt and Baker [23].

Modern civil engine OPRs are so high that the compressor delivery temperature at take-off can exceed the TET of early turbojets. Since some of this compressor air is used to cool the turbine, an increase in OPR makes cooling of the turbine more difficult. In practice, this means that proportionately more air is needed for cooling. Cooling air reduces the turbine efficiency and this phenomenon contributes to the level of OPR used in practice. Optimised cycles with intercooling would be selected to operate at higher OPR—see Fig. 3. Figure 4 shows a sketch of an engine concept which has integrated a heat exchange between the Intermediate-Pressure and High-Pressure compressors.

---

[2] NEWAC: New Aero Engine Core concepts. http://www.newac.eu/12.0.html.

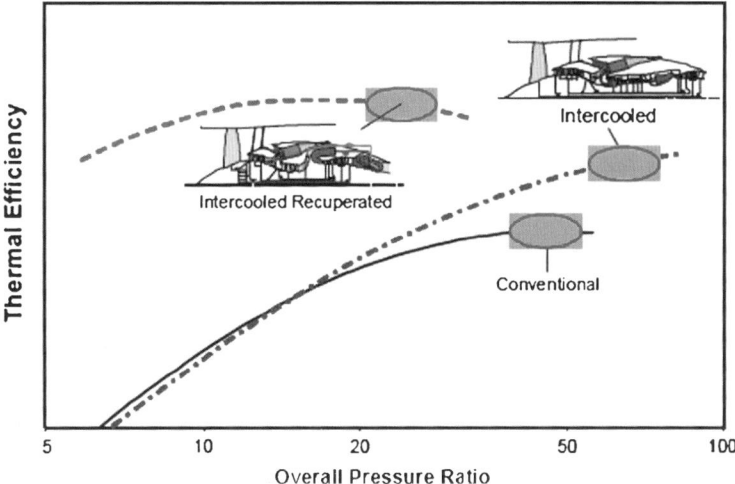

Fig. 3 Efficiency as a function of OPR for various cycles—from [23]

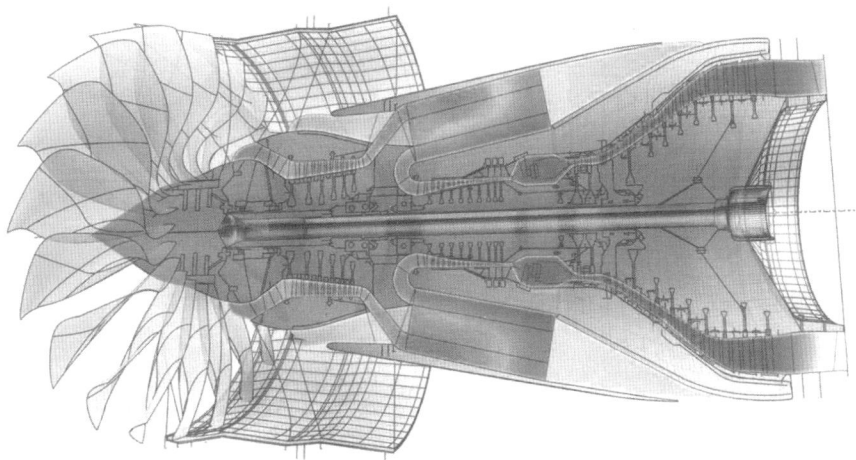

Fig. 4 High OPR intercooled engine—from Rolt and Baker [23]

4 Turbine Stage

One of the most important engine component is the turbine whose efficiency has a large influence on the engine fuel consumption. The high-pressure (HP) turbine stage must operate at high efficiency in the most hostile environment in the engine. The turbine aerofoils and all of the gas swept parts are subject to the engine's most aggressive heat loads as the working fluid supplied to this stage is at the peak cycle temperature and the work generation process in the turbine accelerates the flow to high speeds which results in enormous heat fluxes to component surfaces.

The design of turbine cooling systems remains one of the most challenging processes in engine development—Ireland [7]. Modern high-pressure turbine cooling systems invariably combine internal convection cooling with external film cooling in complex flow systems whose individual features interact in complex ways. The heat transfer and cooling processes active are at the limit of current understanding and engine designers rely heavily on empirical tools and engineering judgement to produce new designs. These designs are developed in the context of continuously increasing turbine entry temperatures as the latter leads to improvements in fuel efficiency, and increases in specific work. Similarly, the stage efficiency of the low-pressure (LP) turbine also has a large impact on the performance of the engine. The cooling flows used in LP blades are subject to complex buoyancy and forced convection interactions which make LP cooling flow prediction and system design very demanding. A more advanced turbine operates at higher temperatures, uses less coolant and has a higher aerodynamic efficiency than a base-line turbine. Progress in turbine technology has been achieved by using better materials and by improving the turbine aerofoil geometries using a combination of better manufacturing methods and better design approaches.

The cooling system is designed to achieve sufficient component life with minimal air flow. In broad terms, a higher turbine entry temperature (TET) allows the cycle efficiency to be increased—but the addition of cooling air reduces efficiency. This balance is indicated in Fig. 5 which shows contours of efficiency (expressed as changes in specific fuel consumption, SFC) for changes in TET and cooling mass-flow rate. For a modern, high by-pass ratio engine, an increase in turbine entry temperature of 50 K improves the SFC by about 1%—if the turbine cooling flow is not altered. The maximum permissible TET is set by technologies including thermal barrier coatings, blade materials and cooling system design. This means that, if the blade technology is not enhanced, an increase in TET is normally accompanied by an increase in cooling flow and the latter acts to reduce the fuel efficiency. This is indicated schematically in the figure. The yellow circle indicates an existing engine and the dark arrows show possible trajectories for successor engines. The horizontal arrow, a, indicates the performance of an enhanced technology engine that takes no more coolant but operates with a TET increased by 50 K. Straightforward arguments can be used to show that increases in convection performance, expressed as increases in the average internal heat transfer coefficient, between 35 and 45% would enable the TET to be increased by the desired 50 K with no increase in cooling flow. The other two trajectories, b and c, show the reduced SFC benefits for engines that require more coolant for the higher TET. Note that the above values should be taken as indicative as they will vary between engines.

The HP turbine rotor blades spin at very high speeds which results in high radial stresses. The blades must be designed to withstand these stresses without appreciable damage for thousands of hours. Nickel-based alloys are normally used to

Fig. 5 Schematic diagram
indicating the benefits of
improved cooling technology
on SFC

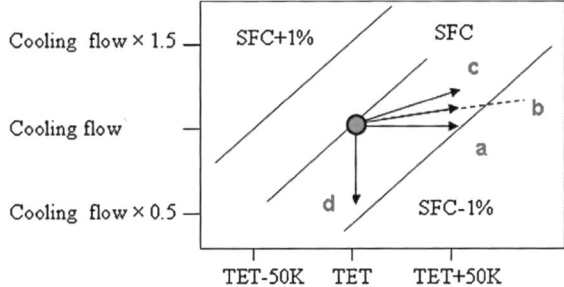

withstand the temperatures and stresses. HP blades experience the highest temperature and are cast in ways that manage the metallic grain structure to achieve acceptable levels of creep life. A casting process was introduced in the 1980s which aligns all of the grains in a single direction. Further resistance to creep was achieved using a method of casting the blade as a single, continuous crystal. Since creep occurs predominantly at grain boundaries, the susceptibility of the blade to creep is dramatically reduced. It is remarkable to note that the blade shown in Fig. 6 is made from a single crystal and has no grain boundaries.

The turbine blade cooling systems are designed by computer but the accuracy to which the blade temperature has to be achieved must be less than 1% of the temperature difference between the hot gas and the coolant. The component temperatures are modelled using Finite Element representation of the blade geometry though the thermal boundary conditions needed to be provided by predictions of the flow using combinations of flow nets and correlations—or directly from computational fluid dynamics. Accurate prediction of the flow field and heat transfer in internal cooling systems presents a considerable challenge, see Iacovides and Launder [6], which means that engine companies often use experiments to validate their designs—before expensive engine testing. Ireland et al. [8] discuss the uses of experiments in engine development programme.

Experiments are also used to improve the accuracy of the computational methods and very sophisticated means of measuring the thermal boundary conditions for the codes have been developed. Some of the most high resolution and accurate use temperature sensitive liquid crystals to measure the distribution of heat transfer coefficient—see Poser and Wolfersdorf [19] for a summary of the state of the art here. A recent development is the use of optimisation codes to develop new cooling system shapes with enhanced performance. Namgoong et al. [12] used this approach to reduce the pressure drop through the 180° bend used in many rotor blade cooling systems. Figure 7 shows the complex 3D shape evolved by the computer which reduced bend pressure loss by 50%. The application of optimisation codes in turbine cooling is a major part of the on-going EC-funded programme ERICKA.[3] Rotation can have a large effect on the cooling passage

---

[3] Engine Representative Internal Cooling Knowledge and Applications. http://www.ericka.eu/.

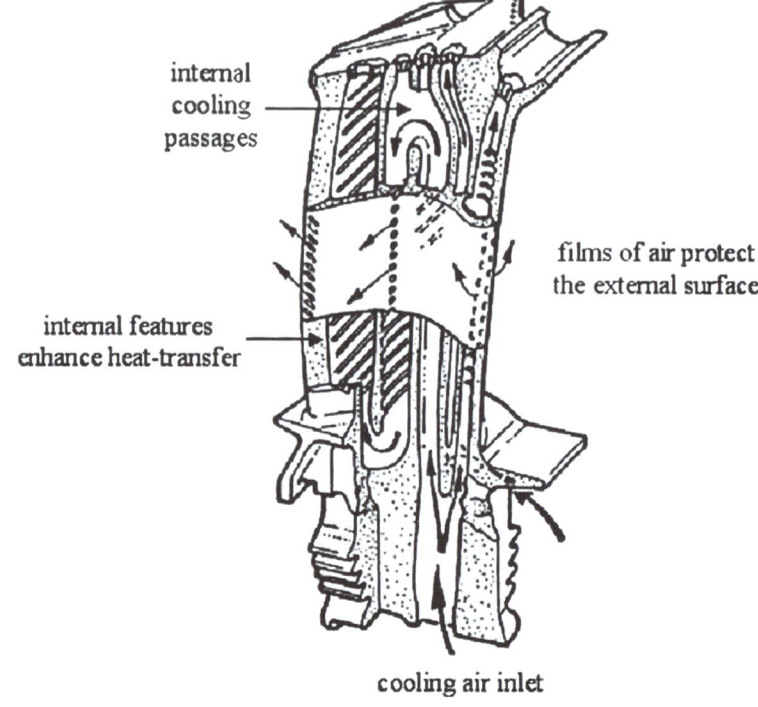

internal
cooling
passages

films of air protect
the external surface

internal features
enhance heat-transfer

cooling air inlet

Fig. 6 Multi-pass turbine cooling system—courtesy of Rolls-Royce

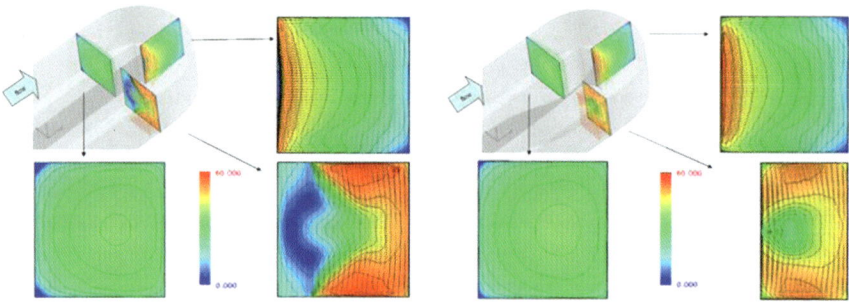

Fig. 7 Contours of velocity for the standard bend (*on the left*) and optimised bend (*on the right*).
A large separation is present at the inner wall for the standard bend but this is eliminated by the
optimised shape. From Namgoong et al. [12]

flow through buoyancy and Coriolis forces which are challenging to predict by
computer. Figure 8 shows one of the two large rotating rigs used in the ERICKA
programme.

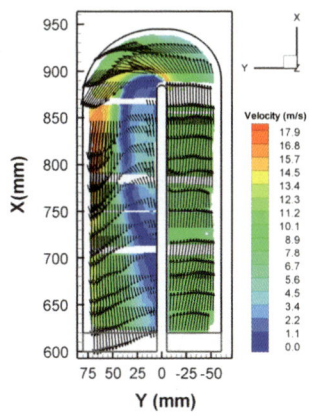

Fig. 8 BATHIRE test rig on the left together with example flow measurements made with rotation—from Ristori et al. [24]

The temperature of the blade needs to be carefully controlled to achieve acceptable life. Some of the air from the compressor by-passes the combustor and is used to cool the turbine stage. All modern HP blades include a complex cooling system such as that shown in Fig. 6. Overheating results can reduce creep life or can cause the metal to oxidise. The blade temperature is also controlled by the use of ceramic coatings. These low conductivity coatings reduce the heat flow to the blade and the load on the cooling system. The coatings are designed to withstand the dramatic changes in blade temperature by the combined use of an interface, or bond, coating and by being applied to produce a columnar structure. These allow the coating to accommodate the expansion and contraction of the blade on heating and cooling. The total heat flow to the blade is reduced as the coating thermal conductivity is reduced and its thickness increased. However, experience has shown that the coatings can be too thick which results in the surface temperature of the coating becoming too hot. If the coating is too hot, it reaches a temperature called the sintering limit at which the columns of ceramic begin to fuse together. This reduces the compliance of the coating which removes the coatings ability to accommodate the mismatch in thermal expansion between the ceramic coating and metal blade. The stiffened coating then cracks and detaches during cyclic heating and cooling. There is also evidence that high coating surface temperatures promote the deposit of complex airborne materials which again act to lock the columns together at the outer surface and hence reduce compliance. One such deposit has been identified as calcium–magnesium alumino-silicate (CMAS)—see Evans and Hutchinson [3].

Considerable research is underway to improve the technology that controls the gap between the tips of the rotor blades and the seals attached to the casing. The leakage flow between the rotor and the inside of the engine casing has a very damaging effect on the turbine performance. The leakage reduces the turbine work, reduces the pressure (and hence its capacity to do work) of the air downstream of the stage and increases the heat load on the blade and the casing. The extra heat load necessitates additional cooling flow which further reduces efficiency. Some of this research involves complex changes to the shape of the blade tips to minimise leakage—see Newton et al. [14]. The flow through the tip gap can be difficult to predict computationally and experiments, such as those reported by Palafox et al. [18] are used to develop modelling approaches. The benefits of the tip geometry on reduced leakage must be weighed up against any increases in required cooling flow—see Tang et al. [25]. Other research is underway to optimise the use of cooling to maintain the casing temperature. This cooling is often implemented using air jets which impinge onto the outside of the casing. The amount of air can be adjusted to control the casing diameter, and hence tip gap, at particular times of the flight.

5 Combustor Technology

Combustion technology has a direct impact on the engine emissions in a number of ways. Firstly, the pressure drop through the combustor corresponds to an irreversibility which directly reduces the cycle efficiency. However, the pressure drop means that the pressure of the air entering the turbine is marginally higher than the turbine cooling air. In practice, this pressure margin can be a major constraint on the minimum combustor pressure drop allowed. Secondly, the temperature distribution from the combustor is far from uniform but will include hot spots associated with the burners and cooler zones towards the casing and hub. The latter are a key part of the turbine cooling technology as the reduced temperatures facilitate cooling of some of the more difficult components to cool. The non-uniform temperature distribution, although desirable for cooling, can have a large effect on turbine efficiency and the subject of the interaction between the combustor and turbine interaction is an important research theme—see Beard et al. [1].

Thirdly, there is considerable on-going effort into reduce the Nitrous Oxide emissions and much of this work has focussed on the technology of lean burn combustion. Rolls-Royce plans to address Lean Burn technology through the ALECSYS programme—Nuttall [15]. The generation of oxides of Nitrogen—collectively referred to as NOx—is highly temperature dependent which means that small zones of high temperature can produce significant emissions. The power generated by the turbine depends, to first order, on the mean combustor exit temperature. Lean burn technology reduces NOx emissions by using enhanced swirl and mixing to reduce peak temperatures. The more uniform temperature then presents challenges to the turbine which shows the importance of designing the combustor and turbine collectively. The increased swirl alters the angle of the flow incident on the first stators [21, 22], which needs to be accounted for in the turbine design.

6 Fan

In a modern wide body jet, the fan produces about 80% of the engine's thrust by pressurising the air in the by-pass which then leaves the nacelle with increased speed. It also supplies compressed air to the core. The aerofoils need to be thin for maximum aerodynamic efficiency, but this is at the expense of increased mechanical flexibility. The fan mass can be considerable and a large amount of effort is expended by engine companies developing technologies which allow fans to be manufactured with low mass. One of the most significant technology steps was the introduction of the hollow, titanium, wide-chord fans pioneered by Rolls-Royce in the 1980s. The hollow fan includes a girder-like internal structure which stiffen the fan. Recently, GE has introduced a composite fan and fan casing on its GEnx engines for the 787 which offer further reductions in weight—Marsh [11]. The fan blades include a titanium leading edge to improve tolerance to foreign object damage.

7 High Bypass Ratio Engines

The propulsive efficiency of a turbofan improves as the by-pass ratio (defined as by-pass airflow divided by core flow) is increased. This has resulted in a progressive increase in fan diameter combined with a reduction in core size. The fan blade tip speed must be limited for reasons of efficiency, stress and noise which means that there is a requirement for larger fans to operate at reduced rotational speed. The fan is driven by a shaft connected to the LP turbine. The turbine aerofoils must travel at high speed for reasons of efficiency and compactness which means that increased fan diameter has a tendency to reduce turbine efficiency. This has resulted in LP turbine shapes with gas path radii which increase with axial distance. The engine weight must be limited for reasons of weight and shaft stability which can result in aggressive divergence, or hade, through the turbine. This divergence can lead to increased losses as the flow field through the turbine becomes increasingly three dimensional.

Pratt and Whitney present a solution to this problem through the introduction of a gearbox between the fan and the LP turbine. This company's Geared Turbofan could offer a dramatic increase in fuel efficiency by allowing the fan size to increase while maintaining desirable LP turbine speeds. The extra mass of the gearbox is offset by the reduced LP turbine mass. The stated reduction in fuel burn for a re-engined Airbus A320 would be 16% reduction relative to current V2500 engines—Lucas [10].

An even more dramatic change in configuration is offered by Un-ducted Fan which uses two counter-rotating rotors to further increase by-pass ratio. This would be mounted high on the air-frame to maximise fan diameter. The Rolls-Royce Open Rotor UDF (Fig. 9) plans to offer a 30% reduction in fuel burn

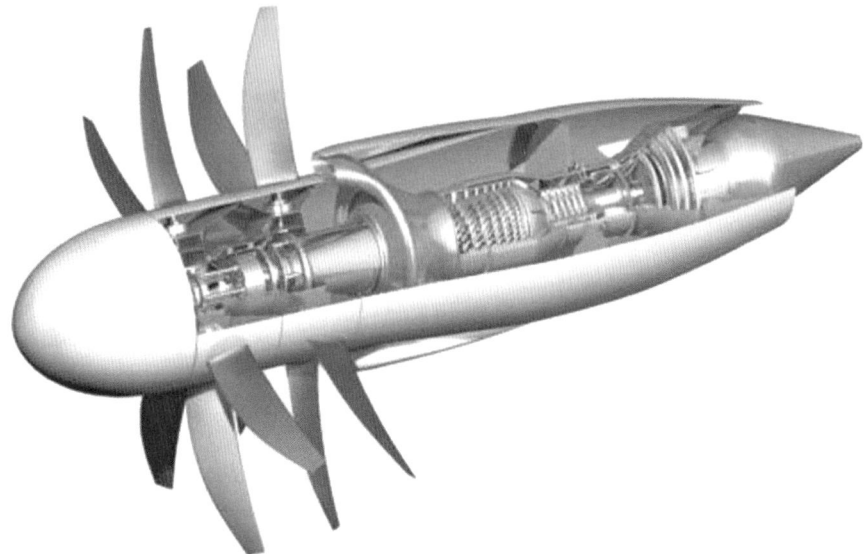

Fig. 9 Open rotor concept—from Rolls-Royce

compared to engines flying today for aircraft in the 100–200 seat market—Nuttall [15]. Implementation of this technology represents a considerable challenge for the designers—not least in the management of the acoustic noise from the rotors.

8 Future Engines

Beyond the above improvements in BPR, there is little consensus on the propulsion systems of the future. Future engines could be powered by Hydrogen to avoid burn fossil fuels. Such propulsion systems could use fuel cells to generate electric power which could then power electric motor-driven fans, Nathan [13]. This offers the possibility of distributed propulsion which allows the air-frame boundary layer to be drawn into the fan—thereby reducing drag.

One of the major technological challenges will be the development of lightweight tanks which can store Hydrogen at high-ressure with acceptable leakage rate. The energy density (J/kg) of gaseous Hydrogen (at 700 bar) is over three times higher than aviation fuel—though the weight of the tanks reduces this advantage. The J/litre figure for Hydrogen is lower than aviation fuel which means that large tanks will need to be integrated into the airframe. The application of a Hydrogen-powered auxillary power unit (APU) is currently being studied by Airbus. Looking further into the future, NASA's Turbo-electric-Distributed Propulsion concept uses cryogenic H2 as both fuel and cooling fluid for superconducting electrical system—Fig. 10. The speed of the power turbine shaft in the

Fig. 10 Hybrid wing-body aircraft with distributed turbo-electric propulsion

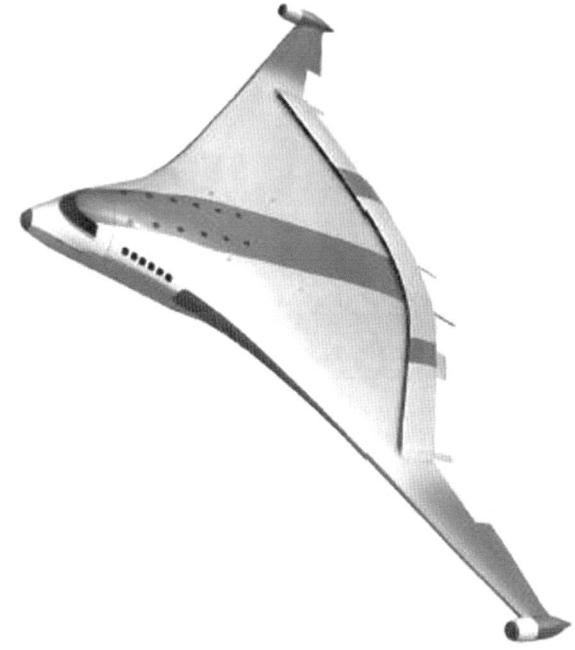

turbine engine is independent of the propulsor shaft speed which effectively means the electrical system acts a variable ratio gearbox. The high effective by-pass ratio provides for a high propulsive efficiency.

References

1. Beard PF, Povey T, Chana KS (2010) Turbine efficiency measurement system for the QinetiQ turbine test facility. ASME J Turbomach 132(1):011002. doi:10.1115/1.3066271 (13 pages)
2. Day IJ, Breuer T, Escuret J, Cherrett M, Wilson A (1999) Stall inception and the prospects for active control in four high-speed compressors. Trans ASME, J Turbomach 121(1):18–27 ISSN 0889-504X
3. Evans A, Hutchinson J (2007) The mechanics of coating delamination in thermal gradients. Surf Coat Technol 201:7905–7916
4. Garg S, Schadow K, Horn W, Pfoertner H, Stihari I (2010) Sensor and actuator needs for more intelligent gas turbine engines. NASA/TM—2010-216746
5. Horlock JH, Watson DE, Jones TV (2000) Limitations on gas turbine performance imposed by large turbine cooling flows. Trans ASME J Eng Gas Turbines Power 123(3):487–494 ISSN 0742-4795
6. Iacovides H, Launder BE (2007) Internal blade cooling: the Cinderella of computational and experimental fluid dynamics research in gas turbines. In: Proceedings of the institution of mechanical engineers, Part A. J Power Energy, 221(3):265–290, 1 May 2007
7. Ireland P (2010) Turbine blade cooling. VKI lecture series 2010−05, Internal cooling in turbomachinery, ISBN 978-2-87516-006-5

8. Ireland P, Mittal V, Jackson D (2010) Heat transfer and flow testing in engine HP turbine cooling system development. Heat Transf Res 41(8):829–847
9. Lewis LV (2002) In-Engine measurements of temperature rises in axial compressor shrouded stator cavities. pp 1349–1360. ASME paper GT-2002-30245
10. Lucas R (2011) The A320 dogfight. Prof Eng 24:6
11. Marsh G (2006) Composites get in deep with new-generation engine. Reinf Plast 50(11): 26–29
12. Namgoong H, Son CM, Ireland PT (2008) U shaped turbine blade cooling passage optimisation, 12th AIAA/ISSMO multidisciplinary analysis and optimization conference, Seattle
13. Nathan S (2011) Hydrogen hopes. The Engineer 296:7819
14. Newton PJ, Krishnababu SK, Lock G, Hodson HP, Dawes WN, Hannis J, Whitney C (2005) Heat transfer and aerodynamics of turbine blade tips in a linear cascade. Trans ASME J Turbomach 128(2):300–309 ISSN 0889-504X
15. Nuttall R (2011) Fly into the future. The Magazine, Rolls-Royce
16. Oates G (1985) Aerothermodynamics of aircraft engine components. American Institute of Aeronautics & Astronautics, Reston
17. Penner JE, Lister DH, Griggs DJ, Dokken DJ, McFarland M (eds) (1999) Aviation and the global atmosphere. IPCC, Geneva
18. Palafox P, Oldfield ML, Ireland P, Jones TV, LaGraff JE (2012) Blade tip heat transfer and aerodynamics in a large scale turbine cascade with moving wall. J Turbomach 134:021020-1–021020-11
19. Poser R, Wolfersdorf J (2010) Liquid crystal thermography for transient heat transfer measurements in complex internal cooling systems. Heat Transf Res 42(2):181–197
20. Quentin F, Szodruch J (2010) Aeronautics and air transport: beyond vision 2020 (towards 2050). Report for ACARE
21. Qureshi I, Beretta A, Chana K, Povey T (2011) Effect of an aggressive inlet swirl on heat transfer and aerodynamics in an unshrouded transonic HP turbine. ASME turbo expo conference 2011 Vancouver, Canada. Paper no. GT2011-46037
22. Qureshi I, Smith AD, Povey T (2011) HP vane aerodynamics and heat transfer in the presence of aggressive swirl. ASME Turbo Expo 2011 Conference Vancouver, Canada. Paper no. GT2011-46038
23. Rolt A, Baker NJ (2009) Intercooled turbofan engine design and technology research in the EU, Framework 6 NEWAC Programme
24. Ristori A, Servouze Y, Barat M, Soulignac F (2007) Characterization of the flow field inside a large scale U-shaped rotating channel using PIV, 7th international symposium on particle image velocimetry PIV, Roma, Italy
25. Tang BMT, Palafox P, Cheong BCY, Oldfield MLG, Gillespie DRH (2010) Computational modeling of tip heat transfer to a superscale model of an unshrouded gas turbine blade. ASME J Turbomach 132(3) Article 031023
26. Wilcock RC, Young JB, Horlock JH (2005) The effect of turbine blade cooling on the cycle efficiency of gas turbine power cycles Transactions of the ASME. J Eng Gas Turbines Power 127(1):109–120 ISSN 0742-4795

Aviation Technology: Aerodynamics, Materials, and Other Options

Christian Carey

Abstract As well as improvements in the propulsive efficiency of aircraft (Chapter 23) there are a number of other technological developments, which enable the reduction in anthropogenic greenhouse gas emissions from aviation. Some are 'tweaks' to current design methodologies; others involve a more radical approach to aviation, all resulting in significant savings in fuel burn and therefore a reduction of emissions from aviation. This chapter will cover developments in materials, aerodynamics and some more radical options for the reduction of the impact of aviation on climate change.

1 A Material Advantage?

Since Orville and Wilbur Wright first decided to power their *Flyer* with a purpose built, cast aluminium engine to meet the specific requirements for power to weight ratio, new materials have been necessary to improve and advance aviation [1]. This improvement in material properties has helped us to travel quickly and inexpensively around the world, by improving the performance and operations of modern aircraft.

With aviation expecting to join the EU-ETS system in 2012 there is now an economic driver to reduce emissions in addition to the social and technical pressures to reduce the environmental impact. With aviation there are a number of methods to increase fuel efficiency, and therefore reducing emissions. If we consider Eq. 1, the formula for aircraft efficiency:

C. Carey (✉)
Smith School of Enterprise and the Environment, University of Oxford,
Hayes House, 75 George Street, Oxford OX1 2BQ, UK
e-mail: christian.carey@smithschool.ox.ac.uk

O. Inderwildi and Sir David King (eds.), *Energy, Transport, & the Environment*,   449
DOI: 10.1007/978-1-4471-2717-8_24, © Springer-Verlag London 2012

$$FE = \eta_0 \left(\frac{\text{Lift}}{\text{Drag}} \right) \left(\frac{W_{\text{PL}}}{W_{\text{PL}} + W_{\text{fuel}} + W_0} \right), \tag{1}$$

where FE is fuel efficiency, η_0 is fuel consumption, W_{PL} is payload weight and W_0 is structural weight. As can be seen, fuel efficiency is mainly the efficiency of the engines, the drag coefficient of the aircraft design and W_0, the structural mass of the aircraft.

While some reductions in structural mass can be achieved through the design and manufacturing phase of aircraft, for example the use of laser welding in place of rivets, the main source of efficiency is in the application of novel, lightweight materials. A weight reduction of 1,000 kg in large turbojet aircraft cuts fuel use by about 1.1–1.5%. Also, as CO_2 emissions equal are in a 1:1 ratio with fuel burn, these reductions relate directly to a reduction in carbon dioxide emissions. Since fuel costs are the largest operating expense for airlines, technologies which reduce fuel use are not only good for the environment, but also have a favourable effect on the bottom line.

1.1 Losing Weight

During the pioneering period of aviation (1903–1930) the minimum weight possible was of utmost importance due to the poor performance of propulsion systems (*the Wright Flyer had about 8 hp*). This led to the use of wood covered with varnished fabric which had limited strength and loading capacities. Aluminium alloys became the baseline for aircraft structures after corrosion issues were overcome by ALCOA Inc. in 1927 [2]. Initial advancement concentrated on the refining of aluminium alloys and the development of new materials, such as composite systems which consists of two or more phases on the macroscopic scale. The mechanical performance and properties of the combined system are superior to those of the constituent materials. These materials were first applied on civil aircraft with the Boeing 707 in 1957, with approximately 20 m^2 of polymeric composites in mainly tertiary roles, such as cabin structures [3]. Increasing use of composite materials was limited, with only a 3% increase observed from A300 to A310. However much larger structural parts, such as the vertical stabiliser (8.3 m by 7.8 m at the base), were now being fabricated entirely from carbon composites. This gives a weight saving of more than 400 kg over an aluminium alloy structure, resulting in approximately 0.5% reduction in fuel burn per hour. Aluminium/lithium alloys, first proposed in the 1950s, were also introduced to reduce the density of components (*1% of lithium reduces the density of aluminium alloys by 3%*) [4]. Production issues initially restricted their use but they are now used in a variety of structural applications.

The latest development in the field of aerospace materials arises from the use of application-specific materials. The Airbus A380, which at 61% has the lowest

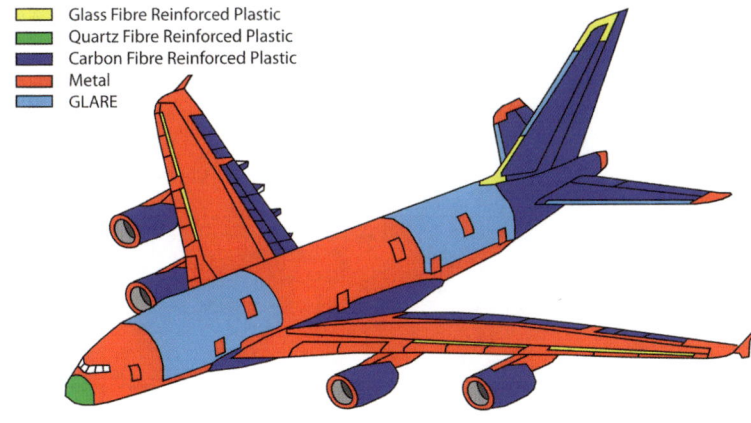

Fig. 1 Composition of Airbus A380

percentage of aluminium by weight of all flying Airbus models, has 20 different alloys and tempers compared to the six utilised on the A320/330 aircraft. The A380 also saw the application of a new material, GLARE, for fuselage skins which shows improved fatigue and impact properties at a lower density than incumbent materials [5]. The composition of the A380 (Fig. 1) illustrates the variation in materials used in modern airliners, in order to ensure that the best material is used for the application, allowing for weight reduction. Significant increases in the amount of composite systems have occurred, with the Boeing 787 and proposed Airbus A350 XWB each having a primarily composite structure (over 50%), with carbon fibre reinforced polymer being used. These material developments have led to the overall reduction of aircraft weight. Whilst advanced materials are not solely responsible for this reduction, they have contributed significantly to this overall improvement in fuel burn.

1.2 Getting Hotter

The introduction of turbines required a development of a new family of materials to cope with the high temperatures and stresses present in the turbine, particularly the so-called 'hot' or combustor/turbine stages, where temperatures can reach over 1,500°C in modern engines. Initial engines such as Sir Frank Whittles W1 used a variety of stainless steels but these were soon replaced with the first super alloy systems, nickel–chromium alloys such as Nimonic and Inconel. The development of high strength materials, resistant to the corrosive environment in the jet turbine, called for improvements in production as well as new materials and alloys. Development of vacuum induction melting technology allowed a much greater control over the composition of superalloys, which increased the component reliability. Commercial production of titanium was also an important development: not only did it find many applications in turbine components, such as the compressor stage, but it also allowed for the development of ducted bypass fans. These work by using excess energy produced during combustion to bypass an amount of air past the core of the engine giving an overall increase in thrust and improvement in specific fuel cost at the cost of top speed and overall engine weight.

The mid-1950s also saw a radical change in the technology of turbine blade production—the use of investment casting. This process allowed the casting of fine channels within the blade, which, with laser drilling, allowed air-cooling of the turbine blade increasing blade-operating temperature. The casting of the blade led to the next leap in turbine technology, the removal of grain boundaries. Standard cast blades contain a large number of grain boundaries, where a number of undesirable events occur. The introduction of directionally solidified (DS) blade (produced by slowly withdrawing the blade from the furnace in one direction) gives no grain boundaries perpendicular to the major stress axis. This improves reliability and maximum temperature by up to 25°C and therefore engine efficiency. This was further developed to single crystal (SX) casting (*first used in Pratt & Whitney's JT9D-7R4 in 1982*), where the use of directional solidification and crystal removal (via *a helix*) led to the production of turbine blades containing no grain boundaries, again increasing maximum operating temperature by 25°C. Thermal barrier coatings (TBC) is another technique used to reduce the relative temperature of engine parts by applying ceramic coatings to hot section parts. The mid 1980s saw the application of polymeric composite materials in engines, in many non-core applications such as fan blades and casings. These have the benefit of reducing the overall mass of an engine and therefore the aircraft, improving efficiency.

The majority of these advancements have led to a vast improvement in engine efficiency by increasing the turbine inlet temperature. Figure 2 shows the turbine inlet temperature of a selection of Rolls-Royce turbines and corresponding material developments, where a significant proportion of the temperature increase can be attributed to advanced materials.

1.3 What Does the Future Hold?

The improvement and development of materials for aviation applications is developing on three main fronts: the development of new materials, the improvement of current material properties by refining composition and novel processing methods for new applications, and the application of current materials in new and novel structures.

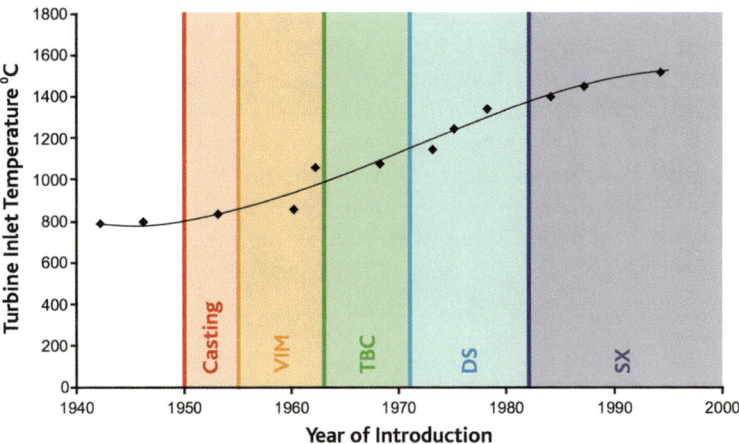

Fig. 2 Turbine inlet temperatures for a selection of Rolls-Royce turbines with major material technology developments indicated

1.3.1 New Materials

New materials can be defined as materials which have yet to be applied in an 'as-designed' application in aviation. Some of materials, particularly Metal matrix composites (MMC) and ceramic matrix composites (CMC) have seen some in-flight testing and are approaching military use but have yet to gain wide ranging acceptance by OEM for various reasons. The following discussion briefly introduces a number of materials that have a potential for applications in next generation aircraft.

CMCs. While consisting of purely ceramic constituents, CMCs utilise a ceramic matrix with reinforcing ceramic fibres and are accepted as a composite system. This creates a material with the excellent thermal properties and with improved mechanical properties, overcoming the limitations of monolithic ceramic (i.e. toughness) and displaying other benefits. The possible applications of CMCs in aviation are generally in the hot section of the aero engines and include turbine disks, combustor linear, turbine aerofoils, transition duct convergent flags and acoustic liners. The use of CMCs would allow an increase in turbine inlet temperature from the current 1,200 to 1,500°C which would lead to a 6–8% increase in fuel efficiency [6].

MMC. These consist of an aluminium or titanium matrix with oxide, nitride or carbide reinforcement and have many advantages over monolithic materials. But toughness is inferior to monolithic materials, are more expensive and difficult to machine. Current research is focused on reducing the cost of production and improving manufacturing processes. Possible applications include highly loaded surfaces such as helicopter rotor blades, turbine fan blades and floor supports [7].

Nanocomposites. As with macro-scale composites, a number of matrix/reinforcement combinations are possible with CMC, MMC and PMC all under investigation. Nanocomposites utilise the huge surface area per mass and high length-to-width ratios of nanoscale objects to improve material properties. Current development issues include producing the necessary quantity of nanoparticles at a commercially attractive price and various production issues such as filler dispersion [8].

Shape Memory Alloys. When shape memory alloys (SMAs) are heated they revert to a pre-deformation shape. They usually consist of copper/nickel based alloys, though other materials can be used. The simplicity of SMA actuators is that they can be used for hybrid applications such as variable jet intake and morphing variable geometry chevrons, where traditional systems are too large and complex when compared with the savings possible [9].

2 Material Improvements

A continuing trend in material development is the improvement to processing and production of incumbent materials to either improve physical properties or to allow their application in new areas and roles.

Aluminium alloys. As the most common of aviation materials, it is unsurprising that a large number of developments are in the pipeline for aluminium alloys. These include further refinement of current alloys to improve specific strength and corrosion resistance as well as developing alloys for specific manufacturing processes such as friction stir welding and laser welding. These advancements will continue the trend for much larger numbers of alloys in aircraft (*A380F has three planed alloys for wing panels*) leading to lighter structures with location specific properties.

Super-alloys. Current research in this area is on fourth generation superalloys containing ruthenium to improve microstructural stability and increased high temperature creep strength.

Titanium. The main area of research in titanium is in improvements in the production process to lower costs. A number of development projects are being carried out with the potential to reduce the cost of final titanium products by very significant amounts in the order of 30% or more.

Steels. Advances in steel alloys have concentrated on improvements in ultra-high strength and toughness. The AerMet family of alloys are a significant development in this area, with similar specific strengths (UTS/density) to common Ti alloys, but with a vastly improved ductility and much higher yield strength. Applications are in safety critical structures, such as transmission gears and parts which require the structural efficiency that steel can offer.

Ceramics. Ceramics exhibit superior thermal properties and major progress has been achieved in improving the mechanical properties so ceramics can now compete with metals in applications for which they were previously unsuitable. Development of ceramic materials has led to the use of these highly thermal stable materials in a variety of applications, such as main shaft bearings, engine seals and thermal barrier coating on turbine blades. The use of ceramics in these applications allows engines to work at a higher temperature, increasing their thermodynamic efficiency.

2.1 New Structure

A number of new structures have been investigated for a variety of materials and are at varying stages of development. Some such as Fibre Metal Laminates have already been applied to aviation whilst others are still at the laboratory stage.

Lattice. One area of particular interest is lattice block, which works on either pyramidal or tetragonal truss arrangements and is produced using investment casting. These structures weigh approximately 15% of a solid plate of the same external dimensions whilst still exhibiting good strength and damage architecture.

Foams. Another major development in the use of aluminium alloys is the production of foam or cellular systems. These are produced by a number of methods such as direct foaming using gas and investment casting but all methods produce a material containing a number of voids. The size, density and structure of the void produced depend on a number of variables and particularly the production method. We think that foam structures will replace honeycomb structures and could lead to higher performance at reduced cost. The use of low density super-alloy foam in noise abatement applications, replacing acoustic liners, would allow for an increase in engine burn efficiency, again reducing fuel burn and emissions.

Laminate structures. A number of laminate systems are under investigation with a variety of constituents. The laminate structure prevents catastrophic failure and exhibits improved impact characteristics. One such material is Fibre Metal Laminate which consists of layers of composite and aluminium and provides high impact strength and directional strength at a low density. A number of different composites have been investigated, such as aramid, glass fibre and carbon fires with a variety of metal layers such as aluminium, titanium and steel. New approaches are investigating asymmetrical lay-up approaches, such as CENTRAL, tailoring the panel properties to the application requirements.

2.2 What Now?

All these developments have led to one of two things, either a lighter overall weight for parts of the same properties in the case of structural materials, or a higher thermodynamic efficiency of the engine with higher temperatures within the engine. If we consider the latest aircraft to be launched, the Airbus A380, a single kilogram of weight saved equates to a 50 ml reduction in fuel burn per hour. This might not sound much, but assuming a 75,000 h life of the aircraft it equates to 3,750 l of fuel. The hypothetical replacement of steel within the A380 (approximately 11,500 kg) with titanium alloy would reduce the overall weight by 5,750 kg, saving 288 l of fuel per hour (22 million litres over the life time) equating to a 2% drop in fuel burn and emissions.

With turbine material improvements an increase in turbine inlet temperature from the current 1,200 to 1,500°C would lead to a 6–8% increase in fuel burn efficiency, equating to a 588 million litre reduction in fuel use over the life of the aircraft, *the equivalent of approximately 300 A380s filled with fuel*. And with Jet-A1 prices exceeding $2.89/gal, these developments offer significant economic as well as environmental benefits in the operation of airliners, even when including the economic and environmental cost of producing advanced materials is taken into account.

2.2.1 The Shape of Things to Come

When asked to describe the conventional aircraft we think of a Boeing 747 and the like, with the classical form of a slender tube with low mounted, swept-back wings, underslung engines and a conventional inverted 'T' empennage (horizontal and vertical tail). In some respects this style has been around almost as long as powered flight itself. Whilst there have been some variations from this design, such as the De Havilland Comet with its embedded engines and the trijets of the 1960s, the majority of commercial aircraft have followed the same conventional layout.

This original design was chosen due for a number of reasons. With the engines in pods below the wing, access for maintenance is relatively straightforward, as well as allowing the use of large bypass ratio fans and the possibility of 'up-engineering' in the future—a restriction with the embedded engines of the Comet. The low-level wings also allow ease of access for maintenance and kept the cabin clear for the wing box and also provided the paying passengers with a clear view, however one key area was to keep costs down to a minimum.

Since the 1980s, the airline industry has gone through a number of changes with a more cut-throat business model in order to attract passengers. Airlines were no longer looking for the very best planes but instead focusing on cost, reliability and size. As a result of this aircraft manufactures concentrated on strenuous cost cutting exercises, concentrating on modifying the current layout rather than research new, more fuel-efficient designs. Gone was the mantra of 'Farther, Faster, Higher', replaced with 'Cheaper, Better, Further'. However, some new developments did occur during this period such as the introduction of digital technology, [Fly-By-Wire (FBW) and Full Authority Digital Engine Control (FADEC)], which produced significant improvements in aircraft efficiency through improved systems control. However, all this occurred while fuel was relatively cheap. Since the oil price increases of the last decade and the emerging importance and concern over environmental impacts, manufactures are starting to look at the benefits of drag reduction more closely and towards the mantra of 'Leaner, Meaner, Greener'.

Drag, otherwise known, as gas or fluid resistance is a constant challenge to the airline industry and an area that continues to be a constant area of research. As a fluid or gas flows past the surface of a body, it exerts a force on it. Lift is the component of this force that is perpendicular to the flow direction and drag is the force parallel to the flow direction. Obviously in aircraft, the lifting force is beneficial, but the parallel drag is not, and requires excess thrust, and therefore requires fuel to overcome it. The total amount of drag on aircraft is a combination of lift-induced drag (approximately 21%), skin friction drag (approximately 50%), interference drag (9%), wave drag (9%), roughness (5%) and miscellaneous other sources (3%) [10].

2.3 Winglets

There are a number of ways to reduce drag, and one of the most common is the application of winglets to the wing tips. Winglets work by increasing the effective aspect ratio of the wing and thus reducing the induced drag produced by wingtip vortices, though they increase skin friction drag (due to the larger surface area of the wing) and add weight. Whilst not a new idea, wing end plates to control vortices were first patented in 1897, development by NASA following the 1970s oil crisis led to the application of winglets, initially by Airbus on the A310-300 and followed by the Boeing 747-400 in 1988. They are now widespread, found on private sail-planes up to the largest of jumbos. Overall, winglets can result in approximately a 4% reduction in fuel use (737-800), though this can vary significantly depending on the type of winglet used and the aircraft it is applied to.

The classic design for a winglet is Witcomb's near-vertical winglet, developed for the KC-135 in the late 1970s and early 1980s. From this base, with the application of good design practise a number of variations on the theme have been developed. Airbus favours the use of wingtip fences. These have surfaces that extend both above and below the wing, giving a more compact unit for the same aerodynamic benefit as traditional winglets. Blended winglets reduce drag by eliminating the sharp join between the wing tip and the winglet that are found on earlier winglet designs. These require more significant alterations and therefore at a greater economic cost when retrofitted to older aircraft. Flight tests on the Boeing Business Jet 737-400 resulted in a 7% drag reduction over a 'clear-wing' 737. For new designs, such as the 787 and A350 raked wingtips are currently preferred. Again they work by effectively increasing the aspect ratio of the wing at a reduced weight over winglets, having the benefit of being an integral part of the wing design.

There are currently a number of promising new designs of winglets. Spiroid wing tips or loop wing tips work by maximising the ratio of trailing edge length to wing span thus reducing induced drag. Although the loop may eliminate the wing tips, it does not eliminate the trailing vortex wake. Flight tests in 2001 showed a 6–10% improvement in fuel burn on short haul flights, though with added weight and complexity. Multiple tip or feathered wingtips are based on observations of birds of prey, whose pin feathers reduce drag. The premise being that multiple small wings are connected at the tip of the main wings at various angles, resulting in a weaker vortex.

There have been other areas or 'tweaks' to this model such as the 'V' or butterfly tail. The butterfly tail can replace the traditional three surface vertical and horizontal stabilizers with two surfaces set in a V formation. This has the advantage of reduced surface area and weight but added complexity due to combined pitch and yaw controls.

It should be acknowledged that there is only so much improvement that can be achieved by altering the current conventional design. To make greater overall reductions on drag, and therefore fuel use, requires significant changes to the current system. One such example is the use of three surfaces. Airbus investigated this layout in the 1990s with the addition of a canard wing immediately behind the cockpit. The main advantage to this layout is the flexibility of the aircraft geometry, which allowed the trimming of the aircraft to the minimum drag at cruise. There are secondary benefits to this layout, such as the lift is provided by both the canard and main wing, allowing a reduction in the main wing area, thus saving weight. The canard also has the advantage that if an increase in lift is required, say if a stretch in cabin size for a larger payload, the canards can be increased without having to alter the main wing. The canard can also supports the operation of the horizontal tail allowing a reduction in mass of that control surface.

2.4 Joined Wings

Another design could be the box or joined wing. This comprises two sets of shorter wings are joined to each other and the fuselage. The design harks back to the biplanes from the early part of the last century and eliminates the wing tip issue by producing a continuous wing. These designs can significantly reduce the wing weight and structure, by each wing providing a brace to the other. The wings can be parallel or a diamond layout, connected either with and horizontal surface in the case of the box wing or encompassing significant levels of anhedral and dihedral angle to join at the tip in the case of the joined wing. Apart from the weight saving the double wings offer high levels of lift, allowing short take off performance, as well as a small reduction in drag of around 5% [11]. Other wing types such as annular and cylindrical wings have been attempted, though development appears to have stalled in recent years. A similar approach was used in the Lockheed Martin Strut-Braced transport, where an external strut was used to brace the wing resulting in an overall lighter aircraft with lower drag. Initial studies at Virginia Tech suggest significant benefit, over traditional cantilever wings, in the order of 6–12% reduction in fuel burn.

2.5 Tail-less Aircraft

The final option could be to explore the tailless aircraft, a design that has been experimented with since the earliest days of aviation with relative success. Tailless aircrafts have the least wetted area, and therefore least skin friction drag, and

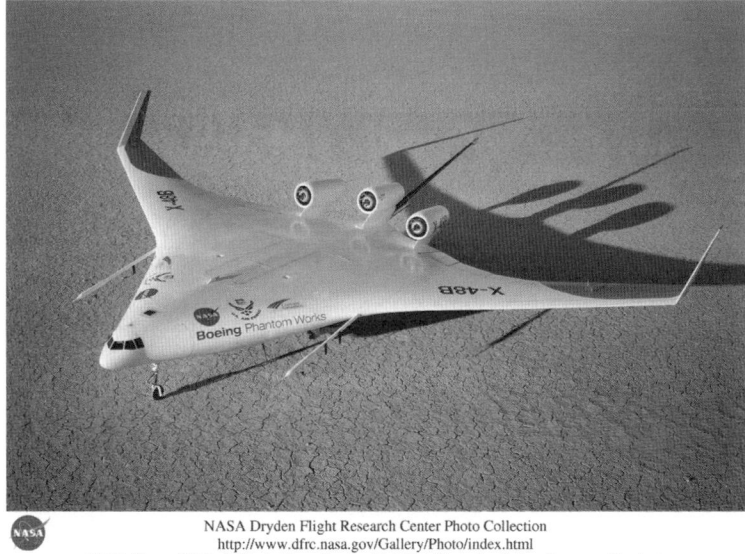

Fig. 3 X-48B blended wing body subscale demonstrator (*Source* NASA)

therefore require the least fuel to travel a given distance. However, the lack of secondary control surfaces can result in stability issues. Tailless designs can be either very stable but with a highly specific centre of gravity, resulting in tight loading restrictions, or they can be inherently unstable, which is not acceptable for civil airliners. Tailless designs can be any one of a number of layouts. Large delta wings, such as found on Concorde are one such example. There are also flying-wings (B-2 Spirit) and a newer type, the blended-wing body (BWB). Flying-wing designs are defined as having no definite fuselage and only a single wing, though there may be structures protruding from the wing. The payload is carried within the thickness of the wing.

The BWB is an evolution of this; it blends the fuselage, wings and engines into a single lifting surface, giving an appearing similar to that of a flying-wing (Fig. 3). This layout produces an aircraft with increased aerodynamic efficiency, as the surface area is reduced, whilst maintaining the required payload space, which can be an issue for pure flying-wing designs. Early Boeing studies showed that for an 800 seat aircraft with a 7000 nautical mile range it is possible to reduce the surface area by up to a third compared with conventional designs. This reduction is mainly from the elimination of the tail and engine interfaces and as it uses less fuel offers an obvious environmental benefit.

Unlike the conventional designs, with their engines hanging in pods beneath the wing, the BWB design has the ability to integrate the engines into the rear of the aircraft which offers a number of benefits: first the overall drag on the aircraft is reduced, thanks to the reduction in surface area; second the embedded engines can ingest the boundary layer improving the propulsive efficiency of the aircraft by reducing ram drag. Third, by having the engine on the upper surface means that the centre body deflects the engine noise upwards, and the exhaust noise is not reflected down by the lower wing surface, as found in conventional aircraft. Finally, BWB also has the potential to remove the need for trailing edge flaps, which are a significant source of noise. This is achieved by the use of distributed propulsion and deflection of the trailing edge jet.

Boeing has been investigating BWB since the late 1980s, initial concentrating on an 800 seat model, but these have since been replaced with a 480 seat model with a 8700 mile range, comparable to an Airbus A380, in both payload and range. The structure consists of seating on the upper level, with cargo on the lower level, as with current designs. Unlike earlier designs, the BWB-450 does not utilise embedded engines with them being placed in pods above the wing, increasing the wetted area by 4% over embedded engines. This was chosen to reduce the 'technology risk' associated with embedded systems. The initial findings are promising; Boeing reports an 18% reduction in maximum takeoff weight and an overall reduction in fuel burn per seat of 32% over the airbus A380. The benefit of lower fuel burn is obvious and the lower maximum take off weight also reduces infrastructural impacts. The BWB-450 is designed to fit within a standard class VI airport 80 m box to minimise infrastructural issues.

The BWB, compared to conventional aircraft, is a relative simple structure. It lacks a number of complex joints and parts found on conventional aircraft, such as the connecting of the fuselage and wing, and fuselage and tail, which produces a 90° joint between two highly loaded structures. Also the BWB lack spoilers and only utilises pivoting control surfaces, removing complex 'track-run' systems found on modern airlines. This equates to an overall reduction in the number of parts required, of the order of 30%, relating to a similar reduction in recurring manufacturing costs.

In the past, it has been common practice for new aircraft designs to be part of a 'family' of aircraft with varying payload capacity and range, achieved by the longitudinal stretching or shrinking of the cabin. This has mainly been done in order to keep 'commonality of parts' and therefore keep costs down. However, the increase in payload usually involves modification of at least the wing, if not other parts. For example, the A340-600 is a 12.27 m stretch over the A340-300 using a wing based on the A330/A340's but is 1.6 m (5.2 ft) longer and has a tapered wing-box insert, increasing wing area and fuel capacity. The BWB offers a much greater level of commonality. The BWB stretches span-wise, by the insertion of a 'central bay' into the centre-body. This automatically increases the wing area and span with increases in passenger capacity. The outer wings and cockpit remain constant with only the aft centre-body and engines changing and a transitional section between the cockpit and centre-body. A common family of nose and outer landing gear could also be utilised, with a centre main gear of varying capacity. This proposed level of commonality comes at a cost of a greater operating empty weight for the smaller aircraft, though this could be overcome with the use of varying gauges of skin dependent on the carrying capacity of the aircraft. This commonality also extends to the interior. In principle the cabin cross-section will be the same for all aircraft in the family and this implies common galleys, lavatories, bag racks, seats etc. offering substantial savings in maintenance and life cycle costs for the operator.

From the point of view of the passengers, the BWB has a number of characteristics that differ from conventional aircraft, apart from the obvious shape change. The vertical walls of the cabin make the BWB feel more like a railway carriage than an aircraft cabin, giving a more spacious environment however apart from the main doors, the cabin will be windowless and while surrogate windows will be provided by external video cameras transmitting to seat back screens there may be some work to be done to convince passengers of the benefits. As with any large aircraft, passenger access becomes a challenge, particularly in an emergency situation. The BWB offers an improvement over conventional large aircraft in two ways. First, they are single deck, removing the issue of emergency slides clearances and second, the internal layout means that passengers are in direct view of an exit, allowing them easily to identify their exit route.

As we move away from the days of 'cheaper, better, further' to 'leaner meaner and greener' it is clear that aircraft designs will have to evolve. The opportunities that tailless aircraft can offer will, in time, make them both economically and environmentally viable, potentially making the days of the 'tube-and-wing' are numbered.

3 Just Hot Air?

The earliest reliable method of manned flight was in lighter-than-air vehicles. Before the development of long-range aircraft, the airship was the choice for discerning travelers. In the 1920s, Zeppelins were regularly crossing the Atlantic and by 1929 circumnavigating the globe in a little over 21 days. However, the development of safer and quicker aircraft reduced airships to tourist rides and mobile advertising boards. Recently, interest has increased in lighter-than-air and hybrid craft, because they are considered environmental-friendly methods of air transport. This article looks at development, both historical and more recent, and analyses the potential uses for modern airships today.

In much the same way that ships float, lighter-than-air craft rely on buoyancy to operate. While small hot air balloons, called sky lanterns, were used in the third century B.C., it was the Montgolfiere brothers who brought lighter-than-air craft into recent history. In 1783, Pilatre de Rozier and Marquis d'Arlandes undertook the first manned flight in a Montgolfiere balloon and started a balloon mania in late eighteenth century France. For the next 60 years, manned flight was undertaken in balloons, using a mixture of lift gases, including hydrogen. These early crafts were not truly navigable due to their shape and lack of propulsion. In 1852, Henri

Giffard attached a small steam powered engine and propeller to an aerodynamically shaped balloon, or envelope, so creating the world's first airship. By replacing steam power with gasoline in 1898, Albert Santos-Dumont produced an airship that was a practical method of flight.

At this point that airship development started to diverge with the production of differing types of envelope, the envelope being the gas containment membrane. These are characterised as rigid, semi-rigid and non-rigid airships. Rigid air ships are able to maintain their shape independent of envelope pressure, by a complex rigid metal framework. They are used for designs exceeding a volume of one million cubic feet, because an unsupported envelope of this size would have problems during pressurisation. Semi-rigid airships have a rigid keel with an aerodynamic shape, which carries the primary loads, while relying on the pressurised envelope to maintain the overall shape of the airship.

Non-rigid airships, or blimps, have their shape maintained solely by the pressure within the envelope. The envelope contains the lifting gas, often helium and ballonets. Ballonets are volumes of pressurised air provided by fans. As air is heavier than the lifting gas, they act like ballast tanks on a submarine, affecting altitude and trim. The lack of any rigid structure means that blimps are easy to design, build and maintain, compared with rigid systems. However, construction of large blimps is difficult due to the long lengths of seaming required of the fabrics, which has an inherent requirement for handling space. Also pressuring large blimps is a delicate procedure because of the relationships between the various structures and the pressurised hull.

Over the next 40 years development continued on all types of airship, with progressively larger models being produced. Of the three types, rigid airships developed most quickly. In 1919 a British airship, R34, crossed the Atlantic and Graf Zeppelin of Germany circumnavigated the globe in 1929. These successes however were overshadowed by several losses, cumulating in the Hindenburg fire in New Jersey in 1937, which ended the golden age of rigid airship construction.

After the Hindenburg, airship research and development languished, but never halted completely. The Goodyear Corporation and US Navy undertook research in the following years and focused on dependable helium-filled non-rigid airships. Various models were considered, though few got past the design stage. The next significant airship to be produced was by Zeppelin Luftschifftechnik in 1997, the NT, a semi-rigid system of which a total of four were completed and used for tourist rides, advertising and special operations such as data transfer and the operation of delicate sensor systems. A further semi-rigid system, also from Germany, was announced in 1996. The Cargolifter CL160 was 260 m long, with a 160 t capacity compared to a 747-400f, which carries 110 t. The design raised interest, particularly in defence circles, where heavy lift airships are seen as a possible replacement for maritime transport, for rapid deployment of heavy forces, such as tanks and other amour, as they could operate from unprepared areas.

In recent years designs have concentrated on the novel or unconventional, particularly in the area of heavy lift systems. The problem with traditional or conventional airships is their low loading capacity, which ranges from -5 to $+8\%$

of the airships mass, because they rely on buoyancy as the sole lifting force. Were additional lifting force available, through either aerodynamic or propulsive methods, a higher load percentage is possible, allowing for larger, more useable loads. On this front, a number of unconventional systems have been investigated and developed, looking at differing shapes, lift mechanisms, lifting gases and propulsion mechanisms.

3.1 Shape

Traditional airship shape evolved from the spherical balloon to an elongated tube as a trade-off between maximum possible lift and minimum air resistance. A number of companies, including 21st Century Airships of Canada and CL Cargolifter of Germany, are developing spherical airships for heavy lifting duties because the spherical shape provides maximum lifting force for a minimum surface area. The reduced size allows operations in confined spaces and much simplified ground handling. They have very high aerodynamic drag, but the proposed applications are for operations which would require minimal horizontal displacement. Lenticular or lens shaped airships have also been developed because their shape can generate aerodynamic lift and improve the craft's manoeuvrability. However, their high surface-to-volume ratio results in high drag, reducing performance. They are also sensitive to payload changes, which make loading and unloading difficult.

The use of multiple hulls, many gas envelopes joined together, is a promising area of investigation as it gives the opportunity to significantly increase gas volume, and therefore load, without increasing the overall length of the craft. The shorter length reduces the sensitivity of the airship to lateral gusts as well as easing construction and storage. Multiple hulls can also be connected by inboard wings, which act as a source of aerodynamic lift. Multiple hulls also reduce the likelihood of catastrophic loss of lift due to damage to the envelope.

3.2 Lift Systems

An obvious choice for improving the lift capabilities of an airship with a given volume is to alter the lifting gas. Hydrogen was originally used, but has fallen out of favour due to its flammability. Hydrogen has largely been replaced by helium, which is inert but more expensive and has a 7.3% smaller lifting capacity. Hot air is also used, mainly in small-scale systems, but has 70% less lifting capability of the same volume of hydrogen. Other gases that have been considered are steam, methane, ammonia and natural gas. These all have various challenges being either flammable, corrosive or, in the case of steam, difficult to maintain at a useable temperature.

A true lighter-than-air craft would be just that, relying solely on buoyancy to provide lift. This has a draw back with limited load carrying capability as well as problems with stability during loading and unloading because rapid changes in weight upset the craft's buoyancy if not counterbalanced. While generating additional hydrostatic lift to offset a load has been contemplated, various technical and logistical challenges have blocked progress. A more successful method would be to use either thrust or aerodynamic force to produce additional lift.

A straight-forward way of producing additional lift would be to attach a high aspect wing to the main vehicle body (Fig. 4a). This would produce substantial aerodynamic lift, improve vehicle stability, decrease drag, as well as increase payload capability. The Ames Megalifter, was one such craft, bearing a significant resemblance to a traditional tube and wing style aircraft. An airship buoyancy envelope had replaced the fuselage. A number of other designs have been analysed and currently, the Dynalifter by Ohio Airships is under development using this kind of assisted lifting.

The next stage of development is to use multiple hulls, linked by an inboard wing. When compared to the previous design of having aerofoils attached to the outside of a single hull, there are a number of advantages. As stated earlier, the craft can be of smaller length for a given volume, and has improved lateral stability. The inboard wing reduces wing bending and twisting under load while the envelopes prevent vortex wing flow and the related loss of lift, making the lifting surface more efficient.

The next stage in producing aerodynamic lift is to use the whole body, producing a 'flying wing'. The load-carrying capability of this kind of hybrid airships depends on the volume of gas for buoyant lift, and on flight speed and altitude for dynamic lift. Aerostatic lift and motor thrust are used for energy-efficient hovering and horizontal landing. The efficiency of hybrid air vehicles is sensitive to size, with large craft being more efficient than small craft. However efficiency of size comes at a cost with increased drag and reduced performance. A number of modern designs, such as the SkyCat by Advanced Technologies Group Ltd in the UK, and the P-791 by Lockheed Martin in the US, use multiple hulls and the resulting increased span to generate lift. SkyCat reports that 40% of lift is provided aerodynamically, with the P-791 receiving 20% from its aerodynamic shape. However, the reliance on aerodynamically provided lift requires a forward velocity in order to take-off, though the distances and speeds required are much smaller than traditional fixed wing aircraft [12].

Another method of increased lift is the utilisation of vectored thrust (Fig. 4b). The Piasecki Aircraft Corporation proposed a heavy lift system, where a traditional streamlined airship was augmented with a number of helicopter rotors to provide additional lift. Goodyear and a number of other companies designed systems using helicopter rotors, with the placement and percentage of overall lift the varying factor. A more unusual design was developed in the 1980s. The Aerocrane featured a spherical, helium-filled centrebody with rotating wings mounted on the equator of the body. The rotation of the center body powered by the wing tip engines produces lift, turning the craft into a lighter-than-air

Fig. 4 Illustrations of possible hybrid lighter-than-air vehicles using **a** dynamic lift and **b** vectored thrust

helicopter. A one-tenth scale dynamic model of a 50-ton payload was built to investigate the stability and control characteristics. The original Piasecki approach has been revived recently, by a Boeing Company and Skyhook of Canada project, the JHL-40. The airship has neutral buoyancy, with four helicopter rotors providing additional lift for payload. The Millennium Airships product, SkyFreighter, uses a similar system with four turbofans as thrust and control systems, providing payload lift and VTOL capabilities. An interesting development is the use of hydrogen as a fuel source. The increased volume that hydrogen requires is not a problem with airships, where it would be on a fixed wing aircraft. Hydrogen provides the same energy level as Jet A1 at 43% of the weight.

It would be unusual for a craft to use just one of these methods to produce lift. A system that combines all these method is the Aeroscraft, which generates lift through a combination of aerodynamics, thrust vectoring, and gas buoyancy generation and management. The Aeroscraft uses a novel approach to buoyancy changes during loading by managing the lifting gas pressure within the hull.

4 Benefits

But why the interest in airships? The potential advantages are; huge load capacity—more than nine times that of a 747-400f; short or vertical take off and landing from unprepared strips or water, and in some cases even no requirement to land to unload; long range capabilities; and high energy efficiency for a reportedly low cost. But there are disadvantages. Airships have a low maximum velocity when compared with fixed wing aircraft, in the region of 140 kph, due to the large gas envelopes and associated drag. There is also the 'Hindenburg effect'—the preconception that airships are dangerous, even though inert helium is mainly used as a lift gas—and concern with stability in high winds, though by utilising aero-dynamic lift, hybrid airships are heavier, and therefore more stable in adverse conditions. Furthermore, in order to be certified for commercial use the craft must meet specific operational requirements in adverse conditions. A further drawback with traditional airships was the manpower required to 'dock' and handle them, with 1920s zeppelins needing over 200 ground staff. This can be overcome by using vectored thrust, managed lifting gases or craft that are heavier than air.

So would there be a market for airships? The slow speed of airships compared to aircraft (140 kph compared to 900 kph) means it is unlikely that airships would encourage passengers to switch away from aircraft, where flight time is a signif-icant factor. However, airfreight is still a large market and according to Airbus, there were approximately 150 billion freight tonne kilometers (FTK) covered in 2008. This is expected to climb by 6% annually until 2028 over which period 72% of freight aircraft will retire. While this represents only 1% of global interconti-nental freight, it equates to 40% of the total value. This is because time-sensitive products, such as IT components, food and cut flowers, can undergo a significant drop in value in the weeks it might take to transport them by ship. Approximately 50–60% of airfreight is transported as 'belly' freight, carried in passenger aircraft, which leaves at least 60 billion FTK which could be carried by dedicated freight airships. The significant load capacity of some designs—Skycat 220 reports loads of up to 220 t compared to 110 t in a 747-400F—suggests that a shift away from fixed wing aircraft would be possible. SkyCat reports a 40% reduction in fuel burn per FTK compared with wide body freight aircraft. In combination with the lower altitude that airships fly at, SkyCat claim a 90% reduction in equivalent CO_2 emissions. To illustrate this, a hypothetical load of 220 t of air freighted straw-berries from Spain to the UK would release around 42 t of CO_2 equivalent. Transporting by airship would release only 4.2 t of CO_2 equivalent.

There are other advantages of using hybrid airships in place of fixed wing aircraft. The V/STOL capabilities of various systems would also allow for changes in supply chain logistics, allowing direct transport from collection centres to distribution centres and removing the road freighting that occurs in between. Airships would not require the use of hard paved runways and produce limited local pollutants meaning they could be operated from sites close to production areas, be they rural or urban. This would free much needed capacity at airports for passenger movements, reducing the need for airport expansion. Also airships operate at much lower altitudes and speeds than fixed wing aircraft, easing their impact on air traffic management systems.

While airships suffer from a comparably low operating speed compared to aircraft, there is a clear advantage over maritime transport whose operating speed is 100 kph slower. The 'as the crow flies capability' of airships also offers inter-esting route flexibility over both road and rail and without the large infrastructure requirements of those modes. In less developed parts of the world, the need for expensive and carbon heavy infrastructure projects could be removed by fleets of 'lorries in the sky' providing multi-drop capability, and connecting distant com-munities with international markets. This lack of infrastructure also appeals to aid agencies, which see airships as a possible method of delivering significant levels supplies to disaster areas quickly and at much lower cost than current multimodal methods.

These advantages look promising, although all the designs discussed here remain under development. A number of projects have stalled through lack of funds though recently Northop Grumman Corp has entered into a $500 million contract with the US Army to develop a system. Airships still need to demonstrate their potential. Airships do have very significant advantages, particularly in the transportation of freight, where they are inherently very competitive in both economic and environmental impact.

References

1. Gayle FW, Goodway M (1994) Precipitation hardening in the first aerospace aluminum alloy: the Wright flyer crankcase. Science 266(5187):1015
2. Dix EH, NACA-TN-259 Alclad: a new corrosion resistant aluminum product N.C.F. aeronautics, Editor 1927, NASA Technical Library, p 17
3. Frankovic I, Rados B, Rados J (2005) Composite materials with fibres. In: DAAAM, Opatija, Croatia, DAAAM International
4. Giummarra C, Thomas B, Rioja RJ (2007) New aluminium lithium alloys for aerospace applications. In: Light metals technology conference, Canada
5. Vlot A (2001) Fibre metal laminates: an introduction. Springer, New York
6. Baldus P, Jansen M, Sporn D (1999) Ceramic fibers for matrix composites in high-temperature engine applications. Science 285(5428):699
7. Hunt WH, Herling DR (2004) Aluminium metal matrix composites. In: Advanced materials and processes. p 39–42
8. Chaiko DJ, Leyva A, Niyogi S (2003) Nanocomposites. In: Advanced materials and processes
9. Chau ETF et al (2006) A technical and economic appraisal of shape memory alloys for aerospace applications. Mater Sci & Eng A 438:589–592
10. Anders SG, Sellers WL, Washburn A (2004) Active flow control activities at NASA Langley. In: 2nd AIAA flow control conference. Portland, OR: AIAA
11. Wolkovitch J (1985) The joined wing: an overview. In: AIAA 23rd aerospace science meeting: Reno, NV
12. Liao L, Pasternak I (2009) A review of airship structural research and development. Prog Aerosp Sci 45:14

The Role of Air Traffic Control in Improving Aviation's Environmental Performance

Ian Jopson

Abstract With aviation under increasing pressure in terms of its environmental performance, more efficient use of airspace can deliver significant benefits in terms of reducing fuel burn and emissions. Air Traffic Control has a major role to play, working with airlines, airports and manufacturers to optimise the whole flight process to identify better ways of operating. NATS was the first Air Navigation Services Provider (ANSP) to benchmark its environmental performance and this chapter will explore how ANSPs can, through benchmarking, identify ways to deliver better flight profiles, procedures and routings to reduce airline fuel burn and emissions. Air traffic controller assistance tools, airspace designs and even corporate culture all have a part to play and this chapter will describe how these can be harnessed to secure a more sustainable future for the global air transportation system.

In 2007 NATS made an environmental commitment through our "Acting Responsibly" programme, one of six Brand Values which we stand for as a business, values which we want to be associated automatically with our company. It was a bold statement of our intent to develop our business sustainably in a world which was becoming ever more aware of aviation's contribution to global warming.

Today, we are the first air traffic service provider in the world to be developing an environmental metric which will be part of our regulatory framework along with metrics on safety, service and value. In other words, environmental considerations underpin our business—from planning to delivery, every day and through every one of our employees.

Let us be clear. It makes perfectly good business sense to improve environmental performance, not only to us but also to our customers. Improving ATC procedures so as to reduce fuel burn saves money—and the more efficient we can make our airspace, the more money our customers will save.

I. Jopson (✉)
NATS, 4000 Parkway, Whiteley, Fareham, Hants PO15 7FL, UK
e-mail: ian.jopson@nats.co.uk

O. Inderwildi and Sir David King (eds.), *Energy, Transport, & the Environment,*   469
DOI: 10.1007/978-1-4471-2717-8_25, © Springer-Verlag London 2012

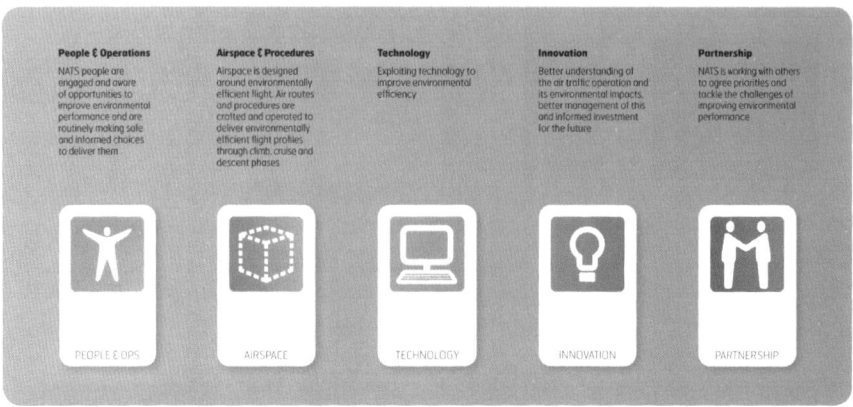

Fig. 1 Five strategic lines of action

Having set ourselves the challenge, we had to make it happen. We were the first air traffic service provider to set challenging targets on our operational CO_2 emissions, and the first to benchmark the airspace system under our control—the area across the UK and extending half way across the north Atlantic, in order to understand the scale of the challenge ahead. It took a year to complete the benchmarking work.

Once we had a better understanding of what it was we needed to change, we were then able to identify the five key areas of our business where we could influence change and improve performance: people and operations, airspace, technology, innovation and partnership (Fig. 1).

Driving better environmental performance in each area, as well as setting out some successes and challenges, we believe we now have a good understanding of how air traffic control can be a catalyst to secure a more sustainable future for the global air transportation system.

1 People and Operations

Most organisations claim that their people are their most important asset; in our case they are among the best in the world, controlling some of the busiest and most complex airspace in the world. Streaming aircraft, offering access to fuel efficient cruise levels, managing speeds, offering direct routes and assisting with smooth continuous climb or descent profiles are all part of the day job and already significantly reduce aircraft fuel use and CO_2 emissions on a daily basis.

Giving controllers the best technology and optimum airspace design means they can build on current good practice and identify further opportunities for improvement. Building a culture of environmental awareness is at the core of delivering environmental performance improvements in our day to day.

Our aim is to help staff identify opportunities to improve environmental performance and support them in making safe and informed choices to deliver them. To help achieve this, we have built an extensive environmental awareness programme including a detailed two-day NATS-specific environmental awareness course available to all staff. Highlighting examples of best practice gives people ideas and permission to explore future environmental solutions.

Obviously the front line of ATC will yield the most opportunity for long-term environmental efficiency. We train all new controller recruits in environment issues as part of their basic training programme and again as they gain validation on operational airspace. We have another tier of awareness training for Watch Managers and individual Watches, which we revisit on a regular basis. It is vital to make the company's environmental aspirations relevant to people across the whole organisation, which is why we have designed an online environmental awareness e-learning package, available to all our employees.

Enhanced awareness across the organisation has triggered a flood of ideas for environmental improvements—some big and some small. In order to act on them, we established an Airspace Efficiency Group at each of our two ATC Centres in Swanwick and Prestwick, marking a significant step in linking operational practice with the longer term environment plan. The groups act as a focal point for acting on opportunities and driving them into the operation to meet our fuel burn and CO_2 reduction targets. At the airports where NATS provides ATC service, we have trained airport environment focal points who play a similar coordinating role.

2 Airspace

AIRSPACE

Large-scale airspace and procedural improvements will deliver the biggest CO_2 savings over the long term. However, there are small changes to airspace, routes and procedures which can generate shorter term, annual fuel savings. This is where we have concentrated the greatest scale of action and seen the greatest rate of progress since we launched our environmental programme.

Supporting this action and tracking progress is the NATS Airspace Efficiency Database, established as a collaboration between our Customer Affairs, Network Management, ATC Procedures and Environment teams. It provides a central database to help identify, prioritise and deliver near-term fuel burn and CO_2 savings and now holds more than 170 potential improvements suggested by airline and airport customers and NATS staff.

Our in-house environmental analysis team assesses each opportunity to estimate its likely fuel and emissions benefits in order to prioritise the best ideas. More than 50 of these fuel and CO_2 savings ideas have been delivered into operation over the past two years. Most of the changes take the form of flight plannable direct routes and/or changes to procedures, for example:

- A new shorter route between Belfast City and Newcastle airports provides savings of 460 tonnes of fuel (1,450 tonnes of CO_2) per annum.
- Removal of a restriction allowing aircraft to stay higher for longer operating inbound to Edinburgh saved 1,250 tonnes of fuel (4,000 tonnes CO_2) per annum. This change affects over 4,000 flights a year, each saving approximately 350 kg of fuel and nearly 1 tonne of CO_2.
- A joint initiative with the Irish Aviation Authority (IAA) through the collaborative UK Irish functional airspace block has enabled a change to the procedures of aircraft between Irish and UK airspace saving 1,250 tonnes of fuel (4,000 tonnes of CO_2) per annum.

We undertook a fundamental review of our airspace to remove some of the restrictions on height profiles and the delivery of point-to-point direct routes. This review is now an ongoing process. We are also working with the Ministry of Defence, another key user of UK airspace, to share airspace more effectively. This has led to more use of military airspace for civil air traffic when it is not being used for military training purposes, which results in more direct routes, less fuel burn and fewer emissions. It sounds simple, but it is only a recent development—delivered through active partnership. We need more of such collaborative working.

The 1970s oil crisis generated an innovation at Heathrow which is now employed at airports all over the world. The 'Continuous Descent Approach' (CDA) was pioneered by NATS with British Airways and other UK airlines, allowing aircraft to approach an airfield at low engine power levels, effectively gliding smoothly with minimum power, reducing noise on the ground and emissions in the air. In the UK, CDA is now built into the procedures at all major airports and achieved on more than 90% of arrivals.

A CDA should start from as high a level as possible given the constraints of the airspace, including the level of the transition altitude (the height at which aircraft change from Flight Level based on pressure above sea level, to height above airfield), and the proximity of other airports, airways or military danger areas. In the London Terminal Manoeuvring Area it is recognised that the highest practicable level a CDA can commence for Heathrow, Gatwick and Stansted is 6,000 ft, but even here a CDA can save up to 0.3 tonnes of fuel (nearly 1 tonne of CO_2) and reduce noise levels by up to 5 dB (decibels). A CDA from top of descent has the potential to offer even bigger fuel and emissions savings. We continue to extend the use of CDA across the UK system, and are continually looking to enable CDA from higher levels and further out when considering airspace redesign.

At the other end of a flight, the opportunity for 'Continuous Climb Departures' (CCD) has exciting possibilities, which our analysis shows has potential to offer transformational fuel and emissions benefits. Modelling of this new departure technique for Heathrow gave results of between 10 and 20% fuel efficiency improvement on a standard profile in use today; that equates to up to 1.5 tonnes of fuel (4.5 tonnes of CO_2) per aircraft compared to a typically held profile. Continuous climb departures are currently a core requirement for our future airspace designs.

Airspace design by nature is long term. NATS has a long-term investment plan to develop and redesign airspace to accommodate improved navigation capabilities, and to implement new operating procedures to deliver smoother and more efficient flight profiles. The airspace in the South East of the UK is a priority area for development. Already the most complex airspace in the world with interacting departure and arrival routes from five major airfields and more than 40 smaller airfields, management of this kind of environment is NATS' specific expertise. The plan we are currently scoping will result in a 10-year programme of change, delivering a once-in-a-generation opportunity to redesign the airspace to optimise the efficiency of today's air traffic control and aircraft technology. This brings with it fresh challenges in terms of the impact on people on the ground; noise is an emotive—and subjective—issue which proves challenging for politicians. Changing routes may mean relief for some areas of the population at the expense of others, and inevitably proves controversial. We are working with the CAA to establish clear guidance for future developments; this clarity will be fundamental to the delivery of our challenging environmental aspirations.

We are factoring environment alongside safety and capacity, from the outset of this large-scale investment programme; the latest designs and operating concepts, driven by NATS' understanding of emissions-optimal profiles are being considered as a core part of the design principles.

Change on this scale can only be delivered with collaboration across the industry, which means working in tandem with airports, airlines and manufacturers to prove the concepts.

In July 2010 NATS, in collaboration with British Airways and BAA, turned a normal Saturday evening service from Heathrow to Edinburgh into the UK's first 'perfect flight' (Fig. 2).

Every step of the journey—from pushback from the stand and taxi, to an optimised flight profile and continuous descent approach—was calibrated to achieve minimal emissions and delay, testing techniques we are seeking to employ

Fig. 2 Representation of perfect flight trajectory

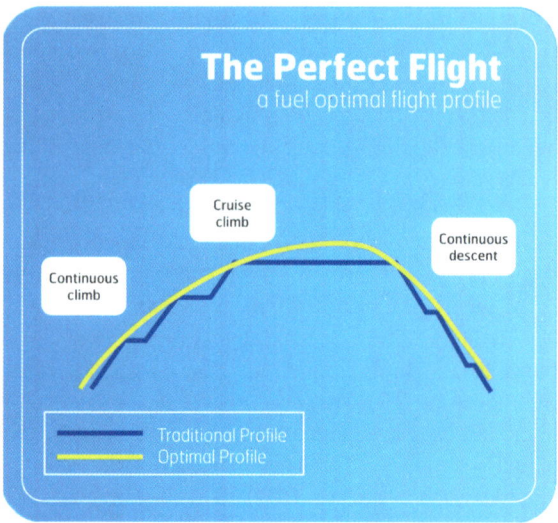

on a regular basis in the future. Within NATS teams from flight planning, operational controllers from Heathrow and Edinburgh control towers and the numerous airspace sectors in between all worked together to deliver the perfect track and profile. Importantly, this also involved ensuring that no other flights were adversely affected. The flight used around 350 kg less fuel generating around one tonne less CO_2 (about 11%) on the norm for the route. The congested airspace in the UK currently limits our ability to achieve perfect flight as the norm, but we have proved what it can deliver and it is a great example of what can be achieved when the industry works together. It forms the blueprint of NATS' vision for future flight.

3 Technology

Air traffic control relies heavily on advanced communication, surveillance and navigation technology to ensure the safe and efficient movement of aircraft. NATS invests heavily in research and development of tools, systems and procedures to improve the service for airspace users. Many of these improvements also help improve environmental efficiency by enabling aircraft to fly closer to their optimum profile, minimise interaction with other traffic flows and regulate speeds and level changes.

Interim Future Area Control Tools Support (iFACTS) will provide our controllers with detailed information about the expected position of aircraft at a future point in time—enabling better planning of aircraft trajectories and with it more efficiency, resulting in reduced fuel burn and CO_2 emissions. iFACTS entered full operational service in November 2011.

Sophisticated tools to smooth arrival and departure flows in UK airspace are increasingly being used in our airspace to support reduced airborne holding, improved flight profiles and lower CO_2 emissions. At major airports Airport Collaborative Decision Making is being installed, which allows operational stakeholders to share accurate information in real-time. Such data flow is vital in ensuring the best use of runways, stands and aircraft movements and reducing emissions in the air and on the ground.

4 Innovation

NATS has a strong track record of innovation in developing bespoke solutions for a range of air traffic-related challenges and now, for environmental improvement. Our environmental work is pioneering in its approach and scope.

Fig. 3 Identifying
environmental hotspots
in the UK

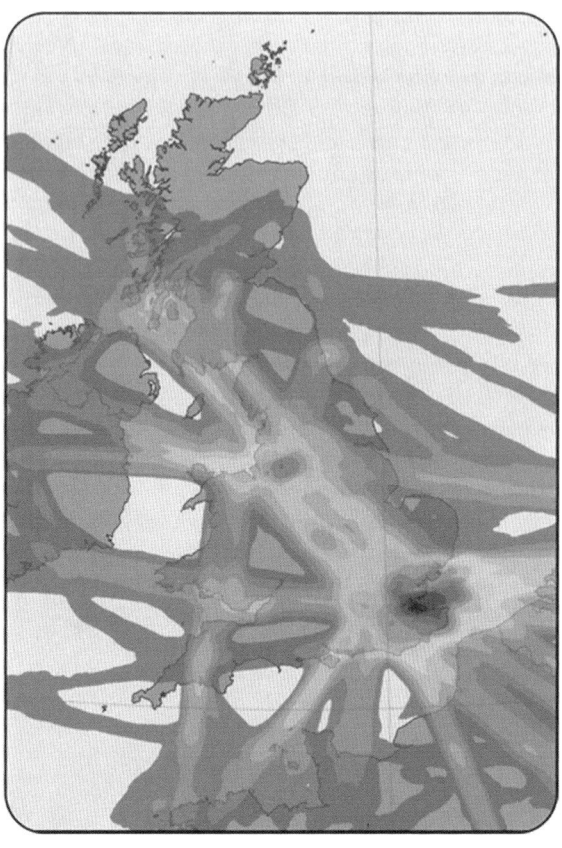

Fig. 3 Identifying environmental hotspots in the UK

Innovation began with the environmental analysis of our operations. All air traffic control organisations understand their systems from a safety or delay perspective. We were the first to consider our system from a holistic CO_2 viewpoint. We now have the capability to track the CO_2 performance of UK airspace using a range of analysis techniques, tools and metrics—giving us a view of the hotspots in our airspace (Fig. 3).

Recent refinements have enabled us to calculate the CO_2 performance of our controlled airspace on a daily basis. This data is helping us understand environmental performance within UK airspace and has enabled us to work throughout 2009 and 2010 with our regulator, the UK Civil Aviation Authority, and our customers to understand whether our environment performance can be financially incentivised—another first for an air traffic control organisation.

Understanding and tracking our environmental performance also enables the most fundamental embedding of environmental performance into our business, through investment requests. All requests for investment must now incorporate emissions analysis and carbon costing to track their contribution to NATS'-targeted CO_2 reduction.

For NATS, innovation sometimes means thinking outside our own airspace, particularly in the development of the European Union's vision for a 'Single European Sky'. NATS is a core member of the R&D programme that has been tasked with delivering this "seamless" sky for Europe. NATS leads the environmental 'transversal' project, seeking to ensure that environmental performance is embedded in all the 300-plus projects in this €2.1 billion programme. The experience we have developed through our own environmental programme at UK level, will offer valuable lessons for the development of a sustainable European airspace network.

5 Partnership

This is important work, but obviously no-one can achieve this alone. Customer airlines and airports are working closely with us to deliver the environmental performance improvement promised. Cutting emissions and saving fuel are vital to the business health of air carriers. Airports, too, are committed to reduce their environmental impact and NATS' plan will be in service of their objectives. Partnership with our Regulator, Government, other air traffic organisations and communities is increasingly important in helping make progress on some of the more challenging areas such as how to manage a changing number of aircraft with a fixed number of runways and within defined scheduling and flight planning arrangements.

We are working with airlines, airports and the CAA to realise opportunities to establish more fuel-efficient routings, holding procedures, arrivals and departures and ground movements. We have regular dialogue with our airline customers leading to a number of subtle changes in the route structure, procedures or tactical handling to enable more efficient flight profiles; and introduced pilot/controller workshops to look for specific fuel and emissions saving opportunities. We capture all proposals in the Airspace Efficiency database, which provides a structured approach to tracking and progressing ideas through to implementation.

At the local level, at Heathrow for example, we have established environmental partnership specifically to generate ideas and actions for reducing CO_2 emissions on the ground and in the air. The first 'Big Green Event' was held at the Heathrow control tower in June 2010 and saw NATS and BAA working in partnership to engage the wider airport community and aviation industry. The event raised awareness of operational changes to reduce emissions and generated further new thinking on airport environmental issues. Our objective is to expand this type of collaboration across all of the airports where we provide air traffic control.

Within the wider UK industry, we were a founding member of the Sustainable Aviation coalition of airlines, airports, engine and airframe manufacturers and ATC which is using cross-industry expertise to safeguard long-term sustainable growth of the industry. Through Sustainable Aviation, the UK aviation industry is committed to target to reduce aviation CO_2 emissions to 2,000 levels by 2,050.

At the international level we work with industry groups, including the International Civil Aviation Organisation (ICAO), the Civil Air Navigation Services Organisation (CANSO) and other air navigation service providers, in order to play our full part in reducing the environmental impact of aviation.

Environmental aspirations must be backed with evidence-based environmental plans and programmes. In NATS we have used our environmental assessments to drive change by engaging our people, through our airspace, procedures and the technology we develop and use, as part of our constant innovation and in our partnership with others. We are early in a programme that will deliver in 2020, and do not yet have all the answers, but we have made impressive progress and proved a good pace-maker for the industry.

We need to continue the momentum and spread awareness and commitment globally if our industry is to prove its ability to grow in a sustainable manner. Saving CO_2 in the UK does not create a virtuous halo above us—aviation is a global industry and it requires effort on every continent.

Strategy and Organisation at Singapore Airlines: Achieving Sustainable Advantage Through Dual Strategy

Loizos Heracleous and Jochen Wirtz

Abstract Singapore Airlines has consistently outperformed its competitors throughout its four-decade-long history, in the context of an unforgiving industry environment. We examine how Singapore Airlines has achieved its outstanding performance and sustained its competitive advantage, through effectively implementing a dual strategy: differentiation through service excellence and innovation, together with simultaneous cost leadership in its peer group. We examine the organisational elements that have allowed the company to do so, illustrate its strategic alignment using a vertical alignment framework and conclude by highlighting the significant challenges ahead.

Keywords Sustainable advantage · Strategic alignment · Dual strategy

This chapter is based on Heracleous and Wirtz 2009, Strategy and organisation at Singapore Airlines, *Journal of Air Transport Management*, 15:274–279. Reprinted with permission.

L. Heracleous (✉)
Warwick Business School, University of Warwick,
Coventry, CV4 7AL, UK
e-mail: loizos.heracleous@wbs.ac.uk

J. Wirtz
NUS Business School, National University of Singapore,
1 Business Link, Singapore, Singapore
e-mail: jochen@nus.edu.sg

O. Inderwildi and Sir David King (eds.), *Energy, Transport, & the Environment*,
DOI: 10.1007/978-1-4471-2717-8_26, © Springer-Verlag London 2012
479

1 Strategy and Organisation at Singapore Airlines: Creating a Global Champion

The airline industry has been plagued by several factors such as overcapacity, commoditization of offerings, cutthroat rivalry exacerbated by the entry of low cost carriers, and intermittent periods of disastrous under-performance [1, pp. 89–100, 4]. Several macro-level socio-economic factors such as rising oil prices, the SARS crisis, frequent concerns about the eruption of bird flu, the Asian tsunami and rising terrorism concerns have further impacted profitability adversely. According to IATA, the global aviation industry suffered $31.7bn cumulative losses during the period 2001–2010.[1] The outlook for the industry onwards remains bleak; not surprisingly, it is regularly rated as one of the worst performing industries in the Fortune Global 500 rankings.

In this challenging industry environment, Singapore Airlines has consistently outperformed its competitors. It has never posted a loss on an annual basis, has achieved substantial and superior returns compared to its industry and has received hundreds of industry awards for its service quality. We suggest that SIA has achieved this outstanding performance by implementing a dual strategy: differentiation through service excellence and innovation, together with simultaneous cost leadership in its peer group (Heracleous and Wirtz 2010).[2] Such a strategy has been deemed unachievable by Michael Porter, who held that differentiation and cost leadership must be mutually exclusive since they require different kinds of investments across the value chain [6]. We examine the elements of this dual strategy, outline SIA's strategic alignment using a vertical alignment framework (alignment among environment, strategy, core competencies and organisation); and conclude by highlighting the significant challenges ahead for Singapore Airlines.

1.1 Case Study Research

Our in-depth case research on Singapore Airlines over the past 9 years examined the company's strategy and competitiveness, in particular its organisational competencies that support the delivery of service excellence in a cost-effective manner. Further, we explored how these competencies are developed and supported by SIA's operational configurations and functional strategies, such as the human resource development strategies and internal innovation processes. We collected both primary and secondary data relating to these issues.

---

[1] IATA Financial Forecast, March 2011, www.iata.org/economics.

[2] In Heracleous and Wirtz (2010) we present an alternative conceptualisation of how SIA has achieved this dual strategy, based on balancing four paradoxes: providing service excellence at low cost, integrating centralised and decentralised innovation, being both a technological leader and follower, and balancing standardisation and personalisation in its processes.

In addition to researching database resources on SIA and the airline industry, gathering materials such as annual reports and press reports, we have conducted 18 in-depth interviews (a list of interviewees is provided in the Appendix) at the SIA headquarters, had several informal conversations with SIA employees and took field notes during our visits. These data have allowed us to gain a deeper appreciation of how SIA has configured its operations and internal processes to develop the core competency of cost-effective service excellence, achieve sustainable competitive advantage and outperform other airlines in its peer group for decades. We have transcribed and analysed the interviews to identify the practices and common themes that we outline in this paper.

1.2 Singapore Airlines' Strategy

Singapore Airlines is positioned as a premium carrier with high levels of innovation and excellent levels of service, and has made a strategic choice of giving priority to profitability over size. The internal organisational practices outlined in this paper, such as continuous people development and rigorous service design are key aspects of operationalising and sustaining this positioning and strategic choice.

At the corporate level, SIA follows a strategy of related diversification. The Singapore Airlines Group has 13 primary subsidiaries and a number of associated companies, with the key subsidiaries being Silk Air, SIA Engineering Company and SIA Cargo [8]. Its airline subsidiaries which comprise 100% ownership of regional carrier Silk Air, budget carrier Tiger Airways (32.9%) and Virgin Atlantic (49%) cover the key customer segments within the industry. This has been a long-standing strategy. According to CEO Chew Choon Seng "we intend to play in all the segments—SIA at the high end, Silk Air on middle ground and Tiger Airways at the low end" [5]. The shareholders in Tiger Airways include Temasek (the Singapore government's investment arm as well as SIA's majority owner) and Irelandia Investments, the private family investment vehicle of Anthony Ryan, the founder of Ryanair, one of the world's leading budget carriers. Temasek owns 55% of Singapore Airlines [8], making SIA one of a very rare breed of companies which although state-owned, act as market-oriented, competitive entities and achieve global leadership positions in their industry.

As part of its international strategy, in April 2000 SIA joined the Star Alliance, one of the three major airline alliances (the other two being Oneworld and Skyteam). In the meantime various divisions of the SIA Group have been investing in China and India through strategic alliances with local organisations (cargo division, airport services, engineering services and catering).

Use of information technology is an essential feature of SIA's strategy both in enhancing customer service as well as increasing efficiency. SIA's web site is one of the most advanced and user-friendly in the industry, where customers can check schedules, buy tickets, check into a flight, manage their Krisflyer (frequent flyer) account, find out about promotions and even choose their meal for their next flight

(Singapore Airlines' "book the cook" service). Given that agents' commissions can be up to 7.5% of the total operating costs (and reservations/ticketing a further 5.4%) [3], effective use of IT can significantly reduce costs and also enhance service levels since transaction costs for the customer are lower for online transactions. When Chew Choon Seng took over as CEO in mid-2003 (until Goh Choon Phong's appointment as CEO from 1st January 2011), cost cutting was on the top of Chew's agenda with particular emphasis on cutting non-fuel costs by 20% within 3 years, and outsourcing IT functions to IBM. The sustained drive for efficiency as well as quality has enabled SIA to increase the spread between breakeven load factor and actual load factor to 6.7% by 2006.[3] This spread was 4.1% in 2010–2011, when actual load factor was 78.5% and breakeven load factor was 74.1%.

With regard to business-level strategy, Singapore Airlines has managed to deliver premium service to very demanding customers (achieving differentiation); at a level of costs that approach those of a budget carrier (as we discuss further below). This achievement challenges Porter's suggestion that differentiation and cost leadership are mutually exclusive strategies [6]. Singapore Airlines supports this dual strategy of differentiation and internal cost leadership through the core competency of cost-effective service excellence, enshrined in a unique, self-reinforcing system of organisational processes and activities.

1.3 SIA's Organisational Activity System: Five Pillars of Cost-Effective Service Excellence

The five pillars of SIA's organisational activity system are rigorous service design and development, total innovation, profit consciousness ingrained in all employees, achieving strategic synergies and developing staff holistically.

1.3.1 Rigorous Service Design and Development

Over two and a half decades ago, services marketing professor Lyn Shostack noted that service design and development was characterised by trial and error rather than by a structured process as was the case in manufacturing [7, pp. 133–139]. Things appear to have changed little since then for most service organisations. SIA however views product design and development as a serious, structured effort. SIA's initial commitment to exceptional levels of service and innovation began in 1972, when, after its separation from Malaysian Airlines, it chose not to be a member of IATA, whose rules SIA considered too constraining.

---

[3] Singapore Airlines analyst presentation, http://www.singaporeair.com/saa/en_UK/content/company_info/investor/analysts.jsp, accessed 19th June 2007.

SIA has a Service Development department that hones and thoroughly tests any change before it is introduced. This department undertakes research, trials, time and motion studies, mock-ups, assessing customer reaction, to ensure that a service innovation is supported by the appropriate procedures. Underpinning the continuous innovation is a corporate culture that accepts change and development as not just inevitable, but as a way of life; a cultural element that is also inculcated at the national level by Singapore's government. A trial that fails or an implemented innovation that is removed after a few months is acceptable, and damages no one's reputation.

At SIA it is expected that any innovation may have a limited shelf life. SIA recognises that to sustain its differentiation, it must maintain continuous improvement, and be able to dispose of programs or services that no longer provide competitive differentiation or that could be offered in a different way. According to SIA's senior management, "It is getting more and more difficult to differentiate ourselves because every airline is doing the same thing ... the crucial fact is that we continue to say that we want to improve. That we have the will to do so. And that every time we reach a goal, we always say that we got to find a new mountain or hill to climb ... you must be able to give up what you love" (Yap Kim Wah, Senior Vice President, Product and Services).

The stakes are raised for SIA, not only by its competitors but also by its customers, who have sky-high expectations: "Customers adjust their expectations according to the brand image. When you fly on a good brand, like SIA, your expectations are already sky-high. And if SIA gives anything that is just OK, it is just not good enough." (Sim Kay Wee, former Senior Vice President, Cabin Crew). Combined with its extensive customer feedback mechanisms, SIA treats its customers' high expectations as a fundamental resource for innovation ideas. Weak signals are amplified; every customer letter, be it complaint or compliment, creates a reaction within the airline. SIA initiated a program called "SIA", for "staff ideas in action", where staff can propose any ideas they have that would improve service or cut costs. Additional sources of intelligence are the IATA GAP (Global Airline Performance) survey, and SIA's "spy flights", where individuals travel with competitors and report detailed intelligence on competitive offerings.

Lastly, SIA recognises that its competition does not just come from within the industry. Instead of aiming to be the best airline its intention is to be the best service organisation. To achieve that, SIA employs broad benchmarking not just against its main competitors, but against the best-in-class service companies.

1.3.2 Total Innovation: Integrating Incremental Development with Un-anticipated, Dis-continuous Innovations

SIA does not aim to be a lot better but just a bit better in every one of its functions and offerings than its competitors. This not only means constant innovation but also total innovation—innovation in everything, all the time. Importantly, this also supports the notion of cost-effectiveness. Continuous incremental development comes at a lower cost than radical innovation, but delivers that necessary margin of value to the customer: "It is the totality that counts. This also means that it does not need to be too expensive. If you want to provide the best food you might decide to serve lobster on short haul flights between Singapore and Bangkok for example, however you might go bankrupt. The point is that, on that route, we just have to be better than our competitors in everything we do. Just a little bit better in everything. This allows us to make a small profit from the flight to enable us to innovate without pricing ourselves out of the market." (Yap Kim Wah, Senior Vice President, Product and Services).

In addition to incremental improvements, SIA also implements frequent major initiatives aiming to sustain service excellence. Organisational initiatives include SIA's "Outstanding Service On The Ground" program, "Transforming Customer Service" and "Soar", for "Service above all the rest". As a way of inspiring discontinuous service innovations, SIA strives to gain a deep understanding of trends in customer lifestyles, and debates their implications for the future of better service in the air. According to the Senior Vice President (Product and Service), "Most new changes that really secure the wow effect are those things that customers never expected … we have our Product Innovation Department that continuously looks at trends and why people behave in a certain manner, why they do certain things. And then we do a projection of 3–5 years of what is going to happen … for the airline, it's not just about having a smoother flight from A to B. That will be taken for granted. It is really about what are the customers' lifestyle needs. Can you meet these lifestyle needs?" Earlier examples of such innovations include the Krisworld on-demand entertainment system for all classes, internet and phone check-in for all classes and the full-size "space-bed". SIA was the first airline to fly the A380 jet (when it was finally delivered after long delays).The offerings on this aircraft include suites, or "a class beyond first" in SIA's words, as well as the widest seats in the sky that have helped to perpetuate its differentiated positioning. Another investment in innovation included a US$1 m simulator that mimics the air pressure and humidity in the air, so that food can be tasted under these conditions, which affect taste buds. One decision was to reduce spices in its food.

SIA has made a clear strategic choice of being a leader and follower at the same time. It is a pioneer on innovations that have high impact on customer service (for example in-flight entertainment, gourmet cuisine that includes fine wines, the ability to order one's choice of dishes in advance by internet, 'beds' in the air). However, it is at the same time a fast follower in areas that are less visible from the customer's point of view, such as revenue management or CRM systems. In doing so, SIA relies on proven technology that can be implemented swiftly and cost-effectively; this reduces the implementation risk while delivering the necessary functionality.

1.3.3 "Profit-consciousness" Ingrained in All Employees

Despite SIA's focus on service excellence and innovation, managers and staff are simultaneously aware of the need for profit and cost-effectiveness. This derives from the company culture: "It's drilled into us from the day we start working for SIA that if we don't make money, we'll be closed down. Singapore doesn't need a national airline. Second, the company has made a very important visionary statement that 'We don't want to be the largest company. We want to be the most profitable.' That's very powerful." (Yap Kim Wah, Senior Vice President, Product and Services). It is due to this policy of pursuing profitability, rather than size, that SIA has one of the highest market capitalisations in the airline industry globally, even though its revenues are relatively modest compared to competitors such as the Air France-KLM Group, British Airways or the Lufthansa Group.

Any proposed innovation is analysed carefully on the balance of expected customer benefits versus costs. A solid business case needs to be made to support all proposed innovations and new service offerings. Station managers and frontline staff know that they should balance passenger satisfaction versus cost- effectiveness in their decisions. The importance of efficiency in the company culture is reinforced by SIA's physical spaces. In contrast to the company's world-class fleet, there are no grand or expensive decorations and furnishings at the company's headquarters for example. The HQ is characterised by a simple, functional design that epitomises the drive for internal efficiency.

Further, SIA has a rewards system that pays bonuses according to the profitability of the company; the same percentage for everyone—the same formula is used throughout the SIA Group. As a result there is a lot of informal peer pressure from individuals within the organisation, and staff and managers can challenge decisions and actions if they see resources being wasted. In 2006, the profit sharing bonus formula has shifted to place more weight on the performance of individual companies (subsidiaries) in the SIA Group in order to increase cost and profit consciousness in these companies and motivate them to increase their business with third parties, so that they will be less dependent on the airline [9].

SIA builds team spirit within its 6,600 crew members through its "team concept", where small teams of 13 crew members are formed and then fly together as far as possible for at least 2 years. This leads to the development of social bonds within the team that reinforce the culture of cost-effective service excellence and the peer pressure to deliver SIA's promise to customers. Based on both cabin crew feedback and efficiency issues, this team concept has recently been under consideration for further refinement. The aims include the improvement of rostering efficiency, enhancement of the cabin crew evaluation system and providing cabin crew the opportunity to meet other colleagues who are not on their team.

Supported by this mindset and organisational practices, the productivity of SIA employees is one of the highest in the global airlines industry (second only to Korean Airlines), at 1,028 thousand available tonne-Kms per employee[4] [3]. For comparison, the figure for budget airlines such as easyJet is 494,000, Jetblue is

---

[4] A measure of airline productivity derived by multiplying the number of tonnes of capacity available for carriage of cargo and passengers (each passenger estimated at 90–95 kg including luggage) on each flight by distance flown. Figures cited in [3], ch. 5.

522,000, and Southwest is 410,000. Calculated per $1,000 of labour cost, SIA is at 20,768 available tonne-Kms as compared to easyJet at 14,629, Jetblue at 12,799 and Southwest at 9,348.

1.3.4 Achieving Strategic Synergies Through Related Diversification and World-class Infrastructure

SIA utilises related diversification to reap cost synergies and at the same time control quality and enable transfer of learning. Subsidiaries serve not only as the development ground for well-rounded management skills, and a corporate rather than a divisional outlook through job rotation, but also as sources of learning. Related operations (such as catering and aircraft maintenance) have healthier profit margins than the airline business itself because the industry structure is more favourable in those sectors.

Singapore's Changi Airport is regularly voted as among the best airports in the world. This excellent airport management and infrastructure entices passengers who are traveling to Australia, New Zealand or other countries in the region, to pass through Changi and to choose SIA as their carrier. Changi Airport is also one of the most cost efficient major airports, with cheaper landing charges compared to Hong Kong or Narita airports for example [2].

SIA's subsidiaries operate under the same management philosophy and culture that emphasises cost-effective service excellence. Although they are part of the group, they are quoted separately and are subject to market discipline with very clear Profit and Loss expectations. In SIA the conventional wisdom of outsourcing (outsource "peripheral" activities and focus on what you do best) does not readily apply. External suppliers might find it difficult to offer the value offered by SIA's own subsidiaries. SIA's related diversification leads to strategic synergy benefits in terms of reliability of key inputs, high quality, transfer of learning and at the same time cost-effectiveness.

A usual metric of airline costs is cents per available seat kilometer.[5] An IATA study[6] found that full service airlines had ASK costs of between US$8 and 16 cents in Europe, 7–8 cents in the US and 5–7 cents in Asia. Budget carriers had costs of between 4 and 8 cents in Europe, 5–6 cents in the US and 2–3 cents in Asia. SIA's average cost during the period 2001–2009 was US$4.57 cents (Singapore $7.47 cents, converted using average exchange rates for this period), which means that SIA has the highest efficiency levels of any full service airline anywhere in the world.

---

[5] Available seats multiplied by distance flown.

[6] From IBM Institute for Business Value, 2006, "Aviation 2010" report, drawing among other sources from IATA Economics Briefing No. 5, 2005, "Airline cost performance".

1.3.5 Developing Staff Holistically

Senior managers at SIA believe that "training in SIA is almost next to Godliness". Everyone, no matter how senior, has a training and development plan with clear goals. The famous "Singapore Girl" undergoes training for 15 weeks, longer than any other airline and almost twice as long as the industry average of 2 months. This training includes not only functional skills such as Food and Beverage serving and safety training, but also soft skills of personal interaction, personal poise, grooming and deportment, and emotional skills of dealing with the consequences of serving very demanding passengers. SIA's training of the Singapore Girl is likened to a "finishing school": "The girls are transformed from coming in, and by the time they come out, they look totally different. Their deportment, the way they carry themselves ... There's a great transformation there" (Sim Kay Wee, former Senior Vice President, Cabin Crew).[7]

In addition to such training, SIA also encourages and supports activities that might, on the surface, be seen as having nothing to do with service in the air. Crew have created groups such as the "Performing Arts Circle", staging full-length plays and musicals, the "Wine Appreciation Group" and the "Gourmet Circle". These activities help to develop camaraderie and team spirit. During their initial training and subsequent career, crew employees also spend time at welfare homes, to get a close-up engagement with the less fortunate, who have to depend on others for their survival. This is aimed to help them develop empathy for others and put themselves in the shoes of the passengers. The contents of the training change to reflect customer expectations. "While our Singapore Girl is our icon, and we're very proud of her and her achievements, we continue to improve her skills; we continue to improve her ability to understand appreciation of wines and cheeses for example, or our Asian heritage ... the enhancement must be continuous" (Yap Kim Wah, Senior Vice President, Product and Service).

Cabin crew can select refresher courses, and on average attend 3–4 days of such courses a year. Popular courses include "transactional analysis" (a counseling-type course), leadership courses and European languages. The company is moving from a system of directing which courses cabin crew should attend, to one of "self-directed learning", where staff take responsibility for their own development.

Even before development starts, there is substantial effort to ensure that the company hires the right staff. For example, entry qualifications for cabin crew applicants are both academic (at least polytechnic diploma, meaning that they have spent 13 years in school), as well as physical attributes. The recruitment process is extensive, involving three rounds of interviews, a "uniform test", a "water confidence" test, psychometric tests and a tea-party. Over 16,000 applications are received every year, and the company hires around 500–600 new cabin crew, to cover attrition rates of around 10%. This includes both voluntary and

---

[7] In addition to the named quotations, this section draws on in-depth interviews with Choo Poh Leong, Senior Manager Crew Services, and Toh Giam Ming, Senior Manager Crew Performance.

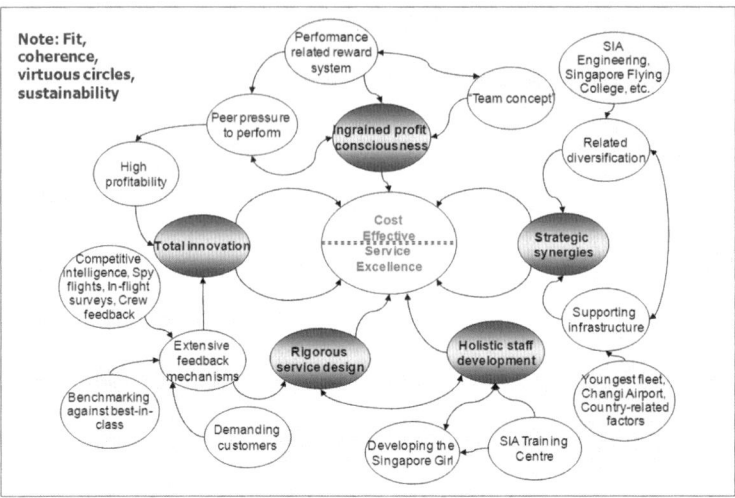

Fig. 1 Singapore Airlines' organisational activity system supporting cost-effective service excellence (*Source* Heracleous et al 7)

directed attrition. After the Singapore Girls start flying, they are carefully monitored for the first 6 months, through a monthly report by the in-flight supervisor. At the end of the probationary period, 75% get confirmed, around 20% get an extension of probation and 5% leave.

As noted above, the five pillars of cost-effective service excellence are interconnected into an organisational activity system characterised by self-reinforcing virtuous circles and high levels of fit. It is this level of fit and mutual reinforcement among the elements that supports the *sustainability* of competitive advantage at SIA. It is relatively easy to copy individual elements of the system, but incredibly difficult to duplicate the whole system, which has evolved historically and is held together not only by formal processes but also by intangible elements such as organisation culture. Figure 1 below illustrates organisational activity system of SIA, where the five pillars support the core competencies of cost-ffective service excellence.

1.4 Achieving a Dual Strategy of Differentiation and Cost Leadership, in the Context of Strategic Alignment

Strategies of differentiation and cost leadership have usually necessitated different and incompatible investments and organisational models. A strategy of differentiation for example implies high quality offerings, and significant investments in innovation, staff development and branding, leading to higher costs than average. SIA achieves a differentiation strategy, but intriguingly, without a cost penalty.

Table 1 Elements of differentiation and cost leadership strategies at SIA

Elements of differentiation and cost leadership strategies at SIA	
Differentiation	*Cost leadership*
Positioning of service excellence and superior quality, brand equity (marketing strategy)	Young fleet (fuel efficiency, lower maintenance costs, effective fuel hedging, paying case for planes)
Developing the Singapore girl (HR development policies)	Labour costs compared to major competitors (16.6% SIA vs 20.1% average of all major airlines); continuous drive for productivity, cost reduction programmes
In-flight experience (young fleet, entertainment system, gourmet cuisine–operations strategy)	Related diversification through efficient subsidiaries that contribute to bottom line
Cultural values and practice of constant innovation and learning	Cultural values: cost consciousness, obsession with reducing wastage
Changi airport one of world's best (related infrastructure)	Innovations not only increase differentiation but also efficiency
Premium pricing in Singapore and in business/ first class, and higher load factor as differentiation indicators	Changi airport one of most efficient (related infrastructure)

In fact, as noted above, SIA has significantly higher efficiency than its peer group, the key feature of a successful cost leadership strategy. Table 1 outlines many of the elements discussed above in relation to the dual strategy of integrating elements of differentiation and cost leadership[8]:

It is worth noting that SIA has one of the youngest fleets in the world, with an average age of 75 months, less than half the industry average which is 163 months [8, p.194]. This is a conscious strategic choice that contributes both to efficiency and service excellence. From a service perspective, younger planes are more comfortable and quiet than older ones. From an efficiency perspective, a young fleet has added benefits in addition to lower repair and maintenance costs and lower fuel consumption. Fewer problems during service and maintenance for example mean lower costs in terms of compensating passengers for flight delays or cancellations, and replacement aircraft costs. Further, a key benefit of a younger fleet is a higher utilisation rate as planes spend less time in the hangar for repairs, checks and maintenance; at SIA, utilisation is 13 h per day, versus an industry average of 11.3 h per day.

Strategic alignment can be represented as consisting of four key elements: First, environmental conditions (macro and micro elements relating to the industry), second, the strategy of the company which should be appropriate for the environmental conditions, third, the core competencies that should effectively support the strategy, and finally the organisational level (including elements such as

---

[8] SIA's labour cost as percentage of total cost was 16.6% in 2007–2008, and 12.5% in 2008–2009. IATA's Economic Briefing, February 2010, notes that the mean figure for all major airlines in 2008 was 20.1%.

Fig. 2 A representation
of strategic alignment
at Singapore Airlines

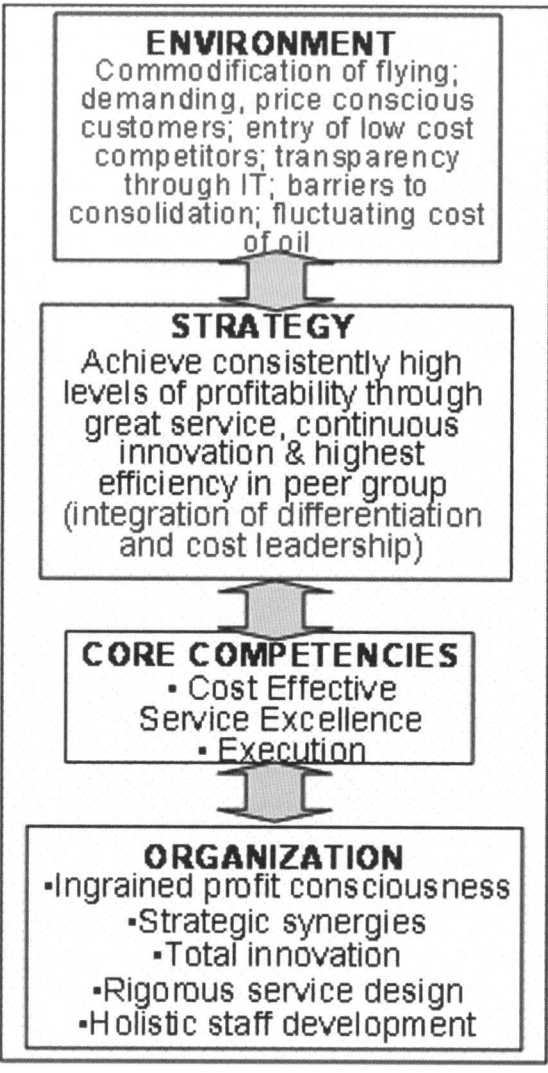

processes, culture and functional strategies) that should deliver the necessary core competencies.

The elements of SIA relating to the five pillars, as discussed above, are principally located at the level of organisation. This is the basic and key level of strategic alignment, which delivers the core competencies of the organisation. SIA's core competence is the ability to achieve a differentiated offering with exceptional levels of efficiency, which we labelled "cost effective service excellence". This capability supports SIA's dual strategy, which in turn is aligned with the environmental (macro and micro-level) market conditions (see Fig. 2).

1.5 Turbulence on the Horizon?

Competitive conditions in the airline industry are becoming more challenging. Apart from wildly fluctuating fuel prices and security concerns, another wildcard for many airlines is the risk of long range aircraft by-passing their hubs. The Boeing 777-200LR launched in 2005, for example, is capable of flying 17,500 km, almost half way around the world. These planes can allow airlines to by-pass hubs like Singapore on flights from Europe to Australia for example. SIA has been seeking rights to fly directly from Australia to Europe, and from Australia to the US as a way of mitigating this risk.

Competitors are hot on SIA's heels, trying to close the gap in both service excellence and efficiency. This is not always easy to do; Malaysian Airlines' service quality is high for example, but its efficiency is nowhere near SIA's (available tonne-kms per employee is 355,000, around just one-third of SIA's—[3]). Other competitors have embarked on aggressive growth while also competing on service quality. For example, Emirates has placed successive orders for 45 A380 aircraft (at a cost of about US$19b),[9] and prices tickets at levels generally lower than its main competitors.

Higher paying jobs elsewhere regularly tempt SIA employees that are sought after in other service organisations, many of whom decide to take up new challenges. Further, with regard to internal conditions at SIA, the need to reduce employee numbers and introduce a variable component to wage packages based on company profitability after the 2003 SARS crisis has been stressful for its industrial relations climate. Singapore's then Senior Minister Lee Kuan Yew intervened to resolve these issues, given the importance of Singapore Airlines and the aviation sector to Singapore's economic prosperity. Further, delays in delivery of the A380 during 2006–2007 have increased SIA's launch costs and delayed the realisation of its capacity plans.

In 2006 the Singapore media expressed concerns regarding service issues at SIA [10]. The Skytrax World Airline Awards ranked SIA 7th in their "Airline of the Year" rankings in 2006, down from 4th in 2005, prompting some to wonder about the effectiveness of sustaining the balance between efficiency and quality at SIA. In both the 2007 and 2008 Skytrax rankings, SIA regained the top position as airline of the year. In 2009 it came in at 2nd place after Cathay Pacific, in 2010 2nd after Asiana Airlines and in 2011 also 2nd after Qatar Airways, sustaining its record of service excellence.

Meanwhile, critics and competitors complain that much of SIA's success is due to environmental factors and the role of government rather than its own capabilities. Analysts note that one benefit of Temasek's 55% stake is lower perceived debt risk by lenders and therefore lower cost of borrowing (even though SIA does not need to borrow significantly). The industrial relations climate in Singapore is

---

[9] See http://www.emirates.com/a380/news_emiratesLargest Operator.asp, accessed on 20 June 2007.

deemed to be less adversarial than elsewhere, enabling SIA to implement policies that would have caused significantly more friction in many other airlines. Critics also suggest that SIA's acquisitions have not fared that well. In 1999, SIA bought 49% of Virgin Atlantic, and wrote off 95% of the investment soon after 9/11. In 2000, it acquired a 25% stake in Air New Zealand, which was seriously impacted by the collapse of its debt-laden Australian arm, Ansett Airlines; this investment was also written off.

Many on the other hand disagree with the suggestion that SIA's success is due to the state. Indeed, high levels of state aid to airlines that have supported many of SIA's competitors [3] have never been awarded in Singapore, where deregulation and encouragement of competition has been the norm. According to SIA's Chairman "We are unlike many of our competitors: we have never had government protection, or underwriting of our business in difficult times. We operate on a commercial basis and our people know that our customers have a choice of airlines" [9, p. 6].

SIA continues its dual focus on the customer experience though service excellence and innovation, as well as continuously striving for efficiency. According to CEO Chew Choon Seng, "the day we stop having visions or objectives to work to, then that is the day we atrophy. I can assure you we have no intention of doing that" [2].

Appendix

List of Interviewees

In this paper we draw from interviews conducted with the SIA executives listed below, between 2001 and 2006. We note the designation of the individual at the time of the interview, and any changes in designation are indicated in parentheses.

- Gladys CHIA Ai Leng, Assistant Manager, Training
- CHOO Poh Leong, Senior Manager, Crew Services
- Timothy CHUA, Project Manager New Service Development (now Senior Manager Inflight Services (Projects))
- Dr GOH Ban Eng, Senior Manager, Cabin Crew Training (now Senior Manager Human Resource Development)
- LAM Seet Mui, Senior Manager for Human Resource Development (now Senior Manager Cabin Crew Training)
- LEONG Choo Poh, Senior Manager, Cabin Crew Performance (now Senior Manager Crew Services)
- LIM Suu Kuan, Commercial Training Manager
- LIM Suet Kwee, Senior Rank Trainer, SIA Training School
- Patrick SEOW Thiam Chai, Inflight Supervisor, Cabin Crew Division
- TOH Giam Ming, Senior Manager, Crew Performance (now General Manager, Taiwan)

- SIM Kay Wee, former Senior Vice President, Cabin Crew
- Betty WONG, Senior Manager, Cabin Crew Service Development (now Acting Vice President Inflight Services)
- YAP Kim Wah, Senior Vice President Product and Services (until 26 December 2010)
- Dr YEOH Teng Kwong, Senior Manager, Product Innovation (currently with another company)

References

1. Costa P, Harned D, Lundquist J (2002) Rethinking the aviation industry. McKinsey Quarterly, Special edition: Risk and Resilience 89–100
2. Doebele J (2005) The Engineer. Forbes Asia 1(9):34–39
3. Doganis R (2006) The Airline Business, 2nd edn. Routledge, Abingdon
4. Heracleous L, Wirtz J, Pangarkar N (2009) Flying high in a competitive industry: secrets of the world's leading airline. McGraw-Hill, Singapore
5. Outlook (2004) November, quoted in Doganis R (2006) The Airline Business, 2nd edn, Abingdon: Routledge, p 263
6. Porter M E (1985) Competitive Advantage: Creating and Sustaining Superior Performance. New York: Free Press; London: Collier Macmillan
7. Shostack GL (1984) Designing services that deliver. Harv Bus Rev 62(1):133–139
8. Singapore Airlines (2011) Annual Report
9. Singapore Airlines (2006) Annual Report
10. The New Paper (2006) Slipping up, SIA? June 18

Uncovering the Real Potential for Air–Rail Substitution: An Exploratory Analysis

Moshe Givoni, Frédéric Dobruszkes and Igor Lugo

Abstract While air to rail substitution is much discussed, in the literature and in the policy debate, there has still been no attempt to quantify the potential for such mode substitution in order to examine the extent to which High Speed Train (HST) might address the main problems faced by the air transport industry: the capacity shortage on the one hand and the environmental problem on the other. This chapter aims to fill this gap and to examine the worldwide potential for mode substitution based on the current air transport network and the supply of air transport services. The potential for reduction in CO_2 emissions as a result of mode substitution is also examined.

1 Introduction

There is an unprecedented interest in high-speed train (HST) around the world. Various elements contribute to that, but perhaps most of all is the image of HST, as opposed to that of conventional trains, as a modern, sophisticated and fast mode of transport. Taking advantage of their high-speed and location at (mainly) city centers, HSTs can now rival and directly compete with the aircraft. They can potentially offer a better service: faster city-centre to city-centre travel time, better travel conditions (more spacious travelling environment and more opportunities to utilize travel time) and faster check-in/out times (attributed to fewer security

M. Givoni (✉)
Transport Studies Unit, School of Geography and the Environment,
University of Oxford, South Parks Road, Oxford, OX1 3QY, UK
e-mail: moshe.givoni@ouce.ox.ac.uk

F. Dobruszkes · I. Lugo
Joint Research Unit No. 8504, Géographie-cités,
French National Centre for Scientific Research (CNRS), Paris, France

O. Inderwildi and Sir David King (eds.), *Energy, Transport, & the Environment*,
DOI: 10.1007/978-1-4471-2717-8_27, © Springer-Verlag London 2012

delays and an easier boarding and alighting process). On top of that, HSTs are considered environmental friendly, or green, at least more than aircraft, and investing in them is expected to contribute to, and even generate, economic growth through employment and agglomeration effects. Even their high cost, especially at times of austerity, does not seem to deter policy makers and planners, although this is likely to be a barrier in practice.

At present, according to the International Union of Railways [25] the world's HST network[1] consists of 14,700 km, most of it in Asia and Europe, and especially in the following countries (in order of network size): China, Japan, Spain, France and Germany, with China only in the last decade joining this club of countries. A further 9,703 km of lines are at present under construction, mainly in China (81%) and Spain (18%) and in addition, 17,594 km of HST lines are planned (mainly in China, France, Spain, Turkey and Portugal). If all these plans will materialize, the world's HST network will spread over 41,997 km, with China's HST network by far the largest (13,134 km), similar in length to the current HST network around the world (Table 1). The potential for the world HST network to grow much further than that is large. In theory, and depending on how high-speed is defined, HST could replace in the distant future most of the current conventional network. To illustrate the scope, the European (EU-27 countries) rail network included in 2008 212,842 km of lines, only 5,745 were considered as high-speed [10].

In parallel to the interest in, and development of the HST, air transport has been facing two major problems or challenges. First, lack of runway capacity, or its direct result—congestion, and second its operation impact on the environment, and especially its contribution to anthropogenic climate change, local air pollution and noise. With demand for air transport forecast to continue growing over the next 20 years, by around 5% annually [1] and with no near term alternative to jet kerosene these issues will likely constrain the operation and development of the air transport industry. Such constrains might come from high oil prices, various regulations, lack of capacity and most likely a mix of these.

One of the often-stated motivations for the construction and expansion of the HST network is to facilitate mode substitution, particularly from air to rail transport.[2] For example, mode substitution has been a priority for the EU for a long time. In its 2001 transport White Paper, the EU calls for cooperation between air and rail modes and states that "network planning should therefore seek to take advantage of

---

[1] The UIC [24] adopts the following broad definition: high-speed is a combination of all the elements which constitute the "system": infrastructure (new lines designed for speeds above 250 km/h and upgraded lines for speeds up to 200 or even 220 km/h, rolling stock and operating conditions. The length of the world's HST network is calculated by the UIC based on lines or sections of lines in which speed is greater than 250 kph (ibid).

[2] The introduction of HST services also results in substitution between car and rail transport. Depending on the route and service characteristics, such mode substitution can be significant and even greater than the substitution from aircraft. Such substitution is important, although it does not get much attention in the literature, but it is outside the scope of this analysis, which focuses on air–rail substitution.

Table 1 Current, planned and under construction HST networks around the world [25]

	Situation Countries	In operation	Under construction	Planned	Total
Europe	Belgium	209			209
	France	1,896	210	2,616	4,722
	Germany	1,285	378	670	2,333
	Italy	923		395	1,318
	The Netherlands	120			120
	Poland			712	712
	Portugal			1,006	1,006
	Russia			650	650
	Spain	2,056	1,767	1,702	5,525
	Sweden			750	750
	Switzerland	35	72		107
	United Kingdom	113		204	317
	Europe Total	6,637	2,427	8,705	17,769
Asia	China	4,175	6,058	2,901	13,134
	Taiwan-China	345			345
	India			495	495
	Iran			475	475
	Japan	2,534	508	583	3,625
	Saudi Arabia			550	550
	South Korea	412			412
	Turkey	235	510	1,679	2,424
	Asia Total	7,701	7,076	6,683	21,460
Other HST systems	Morocco		200	480	680
	Argentina			315	315
	Brazil			511	511
	USA	362		900	1,262
	Total other HSTs	362	200	2,206	2,768
	WORLD TOTAL	14,700	9,703	17,594	41,997

the ability of high-speed trains to replace air transport and encourage rail companies, airlines and airport managers not just to compete, but also to cooperate" [3, p.53]. Furthermore, the White Paper states that "we can no longer think of maintaining air links to destinations for where there is a competitive high-speed rail alternative" (ibid: 38). Since then the EU policy in this respect has not changed. In its 2011 White Paper, the European Commission clearly defines ten "goals for a competitive and resource efficient transport system: benchmarks for achieving the 60% GHG emission reduction target" (European Commission [11], p 9). Goal number four is "by 2050, complete a European high-speed rail network. Triple the length of the existing high-speed rail network by 2030 and maintain a dense railway network in all Member States" (ibid.). The 6th is defined as "by 2050, connect all core network airports to the rail network, preferably high-speed ..." (ibid.).

While mode substitution is much discussed, in the literature and in the policy debate (see for example: [2, 7, 9, 17, 20]), there has still been no attempt to quantify the potential for such mode substitution in order to examine the extent to which HST might address the main problems faced by the air transport industry. The analysis presented here aims to fill this gap and to examine the worldwide potential for mode substitution based on the current air transport network and the supply of air transport services.

Based on various assumptions on the conditions for mode substitution, the analysis presented below is only exploratory in nature, due to various methodological and data limitations, but it provides important results. In addition to an estimate of the share of air transport services that can potentially be shifted to HST, it also provides an estimation of the potential environmental benefits from mode substitution. Based on the results presented, the policy implications are discussed. It is important to make explicit at the outset that if, as a result of mode substitution, the freed runway capacity is used for other flights no congestion or environmental benefits will result, more likely the opposite. Therefore, the potential discussed here is assuming the freed runway capacity remain 'freed', an issue we return to in the last section.

This chapter continues with three sections. The next section presents the methodology and the data used for the analysis, followed by a section that presents the results. First, the potential for mode substitution is presented and second, the corresponding environmental benefits in the form of reduction in CO_2 emission. The last section offers a discussion of the results and the main conclusions that can be drawn from it.

2 Methodology and Data

To derive the worldwide potential for mode substitution, a top-down approach was adopted using the world's scheduled commercial airline services, supplied in January 2010, as the starting point. On this, the 100% potential mark, several criteria were applied to calculate the most reasonable estimation, given the available data and methodology adopted.

Based on the traditional approach to estimate demand for transport services, the potential for air to rail mode substitution was assumed to depend on the route distance, which is used as a proxy for travel time, the main factor likely to determine the choice between aircraft and HST [17]. In addition to overall distance, routes which cross large water bodies (i.e. oceans, seas, lakes) of over 40 km were also assumed to be not suitable for mode substitution.[3] Finally, the level of demand for travel on various routes was adopted as the third criterion to identify on which routes mode substitution might take place. To estimate demand, the level of supply of airline services on the routes, number of monthly seats by city-pair, was used as a proxy. This is the traditional approach as it considers a journey from airport-to-airport (or HST station to HST station) and not the entire journey, door-to-door from origin to destination and it does not account for any other travel attributes of the

---

[3] 40 km was considered as the maximum length a tunnel would be feasible.

journey like convenience, reliability, accessibility (of airport or station), safety/security, cost, etc. that can strongly influence mode choice.

The first step in the analysis was to reduce the worldwide network of airline services to only those routes that are within the distance range where mode substitution might take place. This range was assumed to be between 100 and 1,500 km. Although in the majority of cases the HST is unlikely to effectively compete with the aircraft on a 1,500 km route, as this implies about 6 h train journey with current technology compared with around 2.5 h flight, on some routes and in some circumstances (e.g. much higher fuel prices and/or other restrictions on air transport) passengers might prefer to use the HST over the aircraft.[4]

For the majority of airport/city-pair routes, the distance considered in the analysis is the direct (aircraft) route between them. For lines where the aircraft route passes more than 40 km over water bodies and might still be suitable for mode substitution, an integration of geoprocessing and complex network methods was applied using programming routines to calculate the distance via an alternative route.[5] An example of such a city-pair is London–Brussels. An aircraft flying from Brussels to London flies more than 40 km over water and hence is not suitable for mode substitution based on the criteria applied. However, if the HST route does not follow the aircraft but instead is directed via France and the Chunnel Tunnel (as is the case) such a route is kept in the analysis. The number of city-pairs for which such an analysis was required was relatively small. The distance as measured in this analysis suggests that in the majority of city-pairs no account is made of the fact that rail (and HST) services do not, in most cases, follow the 'direct' great circle route, as the aircraft does. With this approach, the route distance for the HST services, compared to that of the aircraft, might have been underestimated for most of the routes considered. This limitation is important to bear in mind when considering the results and is discussed in the last section.

HST services, and therefore mode substitution, are only feasible on high demand routes. To account for this, two demand criteria were considered. At the lower end, a minimum demand threshold was considered, represented by a level of service equal to 23,250 aircraft seats per month on a route per direction. This seating capacity equals five daily flights from origin to destination by a relatively small (narrow-body) jet aircraft of 150 seats (e.g. an Airbus A320 or a Boeing 737 aircraft), a dense short-haul route in airline terms. The more strict demand threshold considered was 163,618 seats per month. This threshold is based on a

---

[4] In 1999, on the route between Tokyo and Fukuoka in Japan (1,180 km) air transport captured 88% of the market share while the Shinkansen (HST) services captured 12% [4].

[5] In the first step, a Delaunay Triangulation was adopted as a good approximation to represent real transport networks [21]. Based on a set of points which represent airports and additional points along coasts and large water bodies the triangulation connects the points in a manner that none of them intersect each other. This stage creates a set of triangles with empty circumscribed circles without points inside. The distance of these lines is then calculated by the great circle distance formula or, for those lines that 'bypass' large water bodies, by forcing a route which connects two cities/airports via a route that does not cross water bodies for over 40 km.

HST perspective and assuming 14 daily HST services (from origin to destination, on average an hourly service) by a TGV Réseau or Thalys trains with a capacity of 377 seats.[6]

To identify routes where the above demand levels have been met supply of aircraft seats offered by the world airlines in January 2010, as published in the official airlines guide (OAG), was used.

Finally, GIS tools were used to present the routes identified as suitable for mode substitution on maps. The results are presented next.

3 Results

3.1 The Potential for Mode Substitution

In January 2010, the world's network of scheduled airline services comprised of 15,567 routes (city-pairs) on which airlines provided 2,344 thousands flights, 292 million seats or 492 billion seat-km. This is the maximum potential for mode substitution assumed in this analysis. How these services were spread over different route distances is depicted in Fig. 1. It is clear that most of the world's air transport services are short haul in nature (in airlines' terms) [8] suggesting, at the initial stage, large potential for mode substitution. The level of this potential is determined by the assumption on the threshold distance for mode substitution. Assuming a 1,500 km threshold, about 75% of the flights offered by airlines would be suitable to be served instead by HST. These routes account for about 65 and 30% of airlines' seats and seat-km, respectively. With a more realistic threshold of 1,000 km the potential remains relatively large, about 60% of the flights, 50% of airlines' seats provided and about 17% of the seat-km provided. From a policy and airline operation perspectives, the number (or share) of flights suitable for mode substitution represents the potential to free runway capacity at airports, and potentially contributes to reduction in congestion. And, the number (or share) of seat-km that can be transferred from air to rail represents the potential for environmental benefits from mode substitution (when accounting also for aircraft size and number of flights).

Adding to the analysis the demand criteria result in a very different picture (Table 2). Starting with the stricter criterion, which suggests the level of demand shifted from aircraft to HSTs should support at least 14 daily HST services of 377 seats each, the potential for mode substitution worldwide is relatively small. If HST can substitute for the aircraft on routes up to 1,500 km, only 27 routes (out of 15,567) currently operated by airlines match the criteria, and only 13 routes if

---

[6] Such trains might be considered as 'small' HST trains in comparison to many other HSTs. The Eurostar trains, for example, have a capacity of about 750 seats and the Japanese Shinkansen E4 model has a capacity of 1,634 seats.

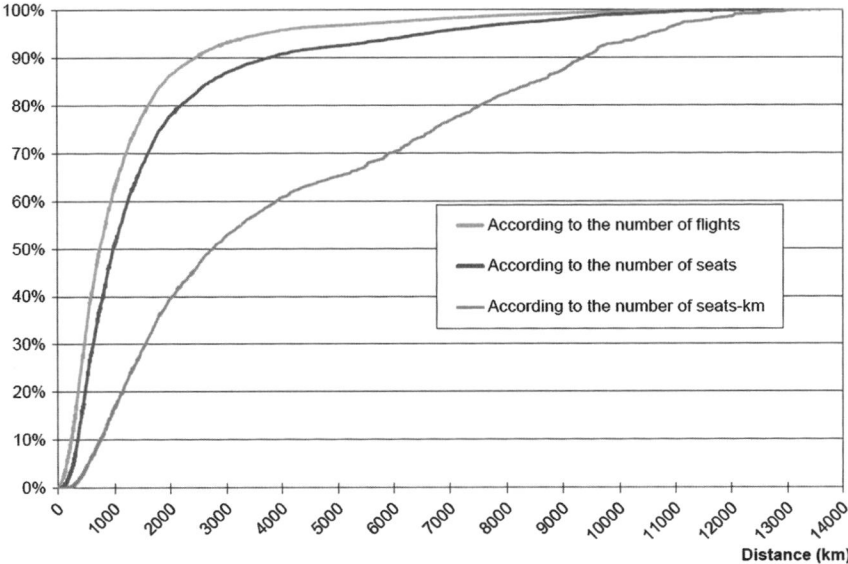

Fig. 1 Cumulative supply of airline services by distance (January, 2010). *Source* OAG data

Table 2 World potential for mode substitution by distance and level of demand

Distance threshold	750 km		1,000 km		1,500 km	
	Demand at least 163,618 seats per month					
City-pairs	13	0.1%	17	0.1%	27	0.2%
Flights (thousands)	41	1.7%	50	2.1%	76	3.2%
Seats (millions)	6.1	2.1%	7.5	2.6%	12.0	4.1%
Seats-km (billions)	2.9	0.6%	4.1	0.8%	9.3	1.9%
	Demand at least 23,250 seats per month					
City-pairs	425	3%	558	4%	742	5%
Flights (thousands)	365	16%	473	20%	621	26%
Seats (millions)	47	16%	61	21%	83	28%
Seats-km (billions)	23	5%	34	7%	60	12%

the threshold distance assumed is 750 km. On routes of up to 750 km, airlines operated about 41,000 flights, 1.7% of worldwide flights they offered. At the same demand level, the share of seat-km that can be shifted to HST services is only 2% of the world airline seat-km when the threshold distances is 1,500 km, and less than 1% if the distance is 750 or 1,000 km.

Given the nature of HST services and as evidence shows, most of the demand for HST services originates from the conventional ('low-speed') rail services. Therefore, a lower level of demand from the airlines might still merit mode substitution. As expected, the potential for substitution is much higher with a demand threshold of only 23,250 seats per month, which represents five daily aircraft services by a

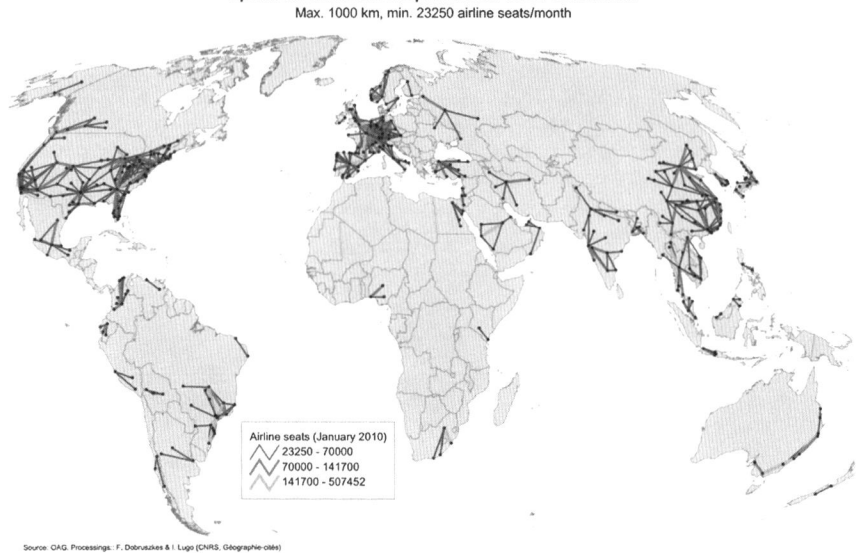

Fig. 2 The spatial distribution of the potential for mode substitution

150-seat jet. With these criteria, the potential is of a different magnitude. With a distance threshold of 1,500 km, 742 routes would be suitable for mode substitution and 425 routes with a threshold of 750 km. On these routes, airlines supplied about 621,000 services (26% of total, 1,500 km threshold) or about 365,000 flights (16% of total, 750 km threshold). In comparison, in February 2011, there were 68,509 flights from and to Atlanta airport, the largest in the world, while at London Heathrow airport there were 36,542 flights.[7] At this demand level, the routes identified account for about 12, 7 and 5% of airlines' seat-km supplied when the thresholds distances are 1,500, 1,000 and 750 km, respectively. Comparing Table 2 with Fig. 1, it is clear that the level of demand, rather than the distance (on routes between 100 and 1500 km), is the main determinant of the potential for mode substitution.

The spatial distribution of routes on which there is potential for mode substitution is shown in Fig. 2, based on a distance threshold of 1,000 km. The regions in which there exists a potential for a HST network, based on demand shifted from aircraft, are clearly visible. These include the US, Europe (mainly western and central) and Asia (mainly China and Japan). Of the 34 billion seat-km that can be transferred from air to rail, assuming distance and demand thresholds of 1,000 km and 23,250 seats, 35% are in North America, 21% in Europe, 20% in East and South-East Asia, while the rest of the world regions account each for less than 10% of the overall potential for mode substitution (the largest of which is South America with 9%).

---

[7] Airport Council International (http://www.airports.org/cda/aci_common/display/main/aci_content07_c.jsp?zn=aci&cp=1-5-212-231_666_2__ Accessed 14 June 2011).

Table 3 The regional potential for mode substitution (share of total seat-km supplied by airlines that can be transferred to HST services)

	Airlines' seat-km (billion)	Minimum demand: 23,250 seats		
		750 km (%)	1,000 km (%)	1,500 km (%)
North America	137	5	9	15
East and South-East Asia	113	4	6	13
Europe	105	5	7	10
Middle East	34	3	5	6
South America	24	8	14	19
Oceania–Pacific	20	6	7	12
India and surroundings	16	3	6	17
Central America and Caribbean	16	3	3	4
Africa	13	3	3	9
CIS	12	1	3	9
Maghreb	3	0	0	0

Ranked by the level of supply of airline services

It is interesting to note in this respect the level of HST development in each of those regions. In the US, HST is very much high on the political agenda but currently there are no HST services, with the exception of the Acela Express services on the Northeast Corridor (between Washington, D.C., and Boston via Baltimore, Philadelphia and New York) that has a maximum operating speed of 150 mph (241 kph). In China, as discussed above, HST lines are rapidly built, with an extensive network of lines already completed and in operation. In Europe, it is clear that there is still potential for mode substitution despite a relatively developed HST network and this is the case in Japan which arguably have the most developed, and dense HST network in the world, which is in operation for many years.

Table 3 shows the regional potential for mode substitution as a share of the total regional air transport services provided (in January 2010). Not surprising, the largest air transport markets, North America and Europe, have the most potential. Based on a demand threshold of 23,250 seats and a distance threshold of 1,000 km, about 9% of seat-km supplied by airlines in North America could be shifted to HST services, 12 billion seat-km out of a total of 137 billion seat-km flown in January 2010. In South America, perhaps surprisingly given the relatively small size of the market (24 billion seat-km), the potential for mode substitution is even larger than that of North America and Europe, nearly fifth of airlines' seat-km with a 1,500 km threshold, and 14% with a 1,000 km threshold. Most of this potential is visibly concentrated in Brazil (Fig. 2). This illustrates how the characteristics of the urban system (namely the size of the cities and the distance between them, which largely determine the interactions between them) influence long-distance travel patterns and in turn the potential for mode substitution.

Considering the potential for mode substitution in North America (Fig. 3) it is evident that there exists only potential for two separate HST networks, with likely

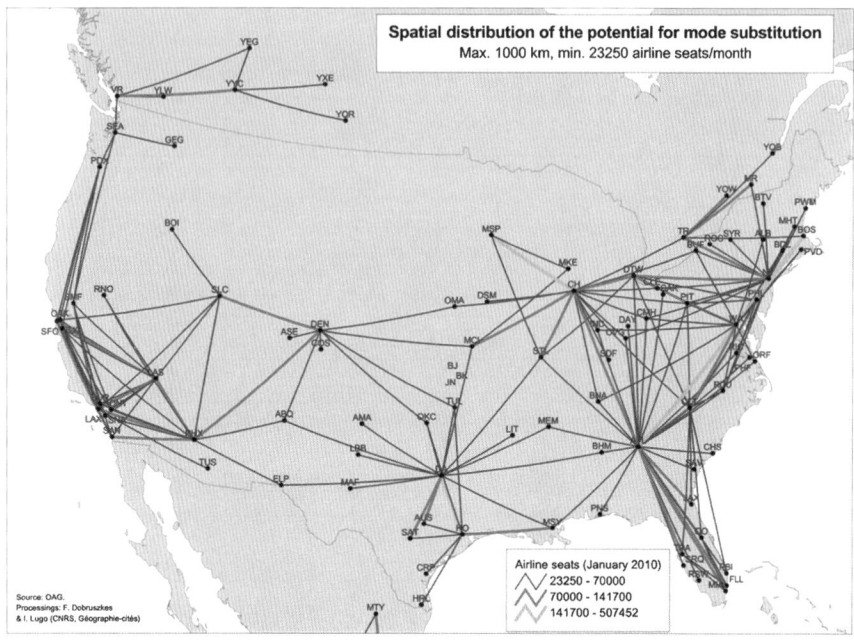

Fig. 3 The spatial distribution of the potential for mode substitution in North America

no connection between them, along the east and west coasts. Figure 3 aptly shows some of the limitations of the current analysis. The potential for mode substitution (as indicated by the number of 'lines' in the figure) in airports (cities) like Atlanta, Dallas or Denver is very much influenced by the adoption of these airports as hubs in airlines' Hub and Spoke networks. This suggests that demand for services on any given spoke (route) from the hub is larger than the actual demand between the hub city and the city on the spoke. With HST services such 'hubbing' would not be feasible and many of the routes between cities that are now connected (by aircraft) via the hub will not have sufficient demand to justify direct HST services between them. In this respect, the current analysis overestimates the potential.

Figure 4 suggests that many European city-pairs with high mode substitution potential remain domestic (in UK, France, Italy, Germany, etc.). The international ones often link small countries (like Switzerland, The Netherlands or Austria) together or with large countries. As is the case in the USA, the potential for mode substitution in Spain is probably overestimated as well, and for the same reason— the function of Madrid airport as a hub. However, the size of the Iberian peninsula, or the shorter distances between cities within the Iberian peninsula suggests that one HST line, for example, from Barcelona to Lisbon via Madrid (which is under construction), can serve at least three airline routes currently served by aircraft. This characteristic of rail operation (the relatively small cost of stopping or 'landing' in several places along the route, when the infrastructure is provided) is one of the advantages of rail, and HST, over aircraft. This would suggest that the

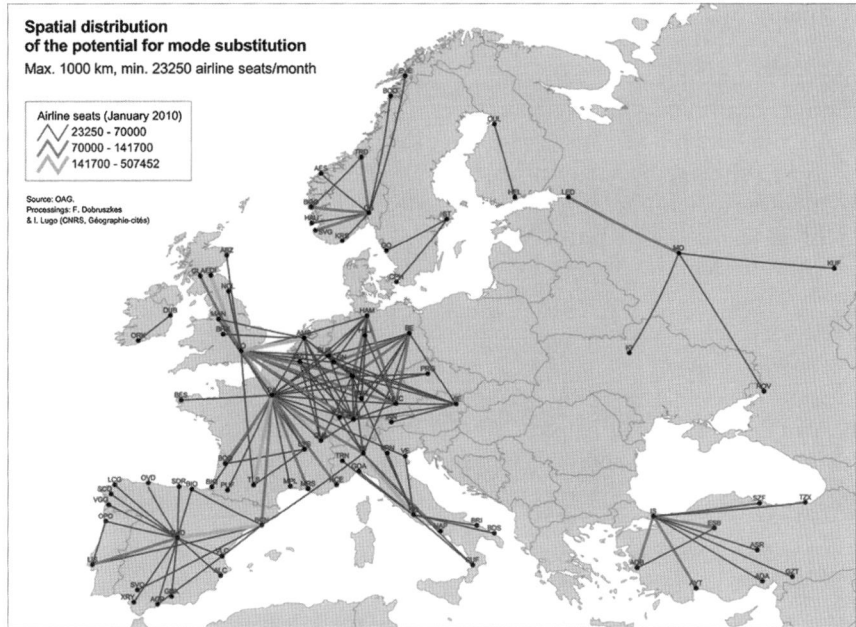

Fig. 4 The spatial distribution of the potential for mode substitution in Europe

demand of several routes currently served by aircraft (e.g. the demand on the routes: Barcelona–Madrid and Madrid–Lisbon) could be summed up to derive the overall demand for mode substitution on one HST line. This, what we term the 'corridor' potential for mode substitution that current analysis is not accounting for, suggests the potential for mode substitution has been underestimated.

As a general principle, the denser is the concentration (or spatial distribution) of routes suitable for mode substitution the more potential there is for a HST line or network to replace several airline services currently served by aircraft. Thus, such potential is visible in Western Europe (Fig. 4) and the East Coast in North America (Fig. 3).

3.2 The Potential Environmental Benefits from Mode Substitution

The extent to which substitution of aircraft services by HST services will result in environmental benefits depends on many factors, but first on whether the flight that is taken off the runway is replaced by another flight. This, a very likely outcome under the current regulation, management and operation of the airline industry, will result in increasing environmental impact from air transport as it can be expected that a long haul flight will replace the short-haul flight that was substituted by HST. The potential for environmental benefits from mode substitution nevertheless exists, and can be estimated assuming freed runway capacity will not be used for other services, a strong assumption as indicated above.

Air and rail transport impact on the environment is mainly in terms of noise, local air pollution and climate change, which are directly linked to the level of operation and its characteristics, and various other impacts related to the provision of the infrastructure. To properly account for those impacts, a life cycle analysis (LCA) is required (see [5, 6]), which is beyond the scope of this chapter. Instead, the approach adopted here to explore the potential environmental benefits from mode substitution only aims to provide a ballpark figure estimate of the reduction in CO_2 emission as a result of mode substitution. Such an estimate, although crude, will provide a good basis to begin the discussion. For this, representative CO_2 emission factors per seat-km for aircraft and HST are used, based on several studies (Table 4). These are then applied to the total seat-km identified above as suitable for mode substitution. Three different representative emission factors are used for aircraft operation depending on the maximum distance threshold chosen. For the HST, the representative value chosen is based on the analysis by Givoni [15] which was carried out for the London–Paris route, and the analysis by Miyoshi and Givoni [22] which was carried out for the proposed London–Manchester route. The estimate for the London–Manchester route given in Table 4 (of 10.98 g/seat-km) is based on demand projections for 2035, forecast energy mix in the UK for that year (less oil and more gas and renewable sources than today, and a new HST model, the Alstom AGV trains). Basing instead the analysis on the current UK energy mix and current Eurostar Class 373 model trains will result in an emission factor of 20.5 g/seat-km. Since the distance of each of the routes represents the aircraft distance while HSTs routes are likely to be longer, by 25% on average—the assumption adopted here,[8] a 25% increase in the emission factor for HST was applied.

Using this approach and the representative emission factors, the potential CO_2 savings for each seat-km transferred from the aircraft to the HST are calculated as: 93.75, 68.75 and 58.75 g/seat-km for route distances of 750, 1,000 and 1,500 km, respectively. Applying these for the potential monthly air traffic that can be substituted by HST results in annual CO_2 savings of 25 million tonnes when the threshold distance is 750 km, 28 million tonne when the threshold distance is 1,000 km and 42 million tonne for distance threshold of 1,500 km.[9]

Based on the same approach, Table 5 shows regional potential for CO_2 reductions from mode substitution. This potential depends on the total volume that

---

[8] For example, the great circle distance between London and Paris is 348 km. The aircraft is expected to fly about 383 km on this route (10% added to account for takeoff and landing procedures) while the HST route is 492 km, 30% higher. Using the same data sources and assumptions, on the Tokyo–Osaka route the HST route is 25% longer. At the same time, on the Rome–Milan route the HST journey is expected to be slightly shorter than the aircraft route [13].

[9] Due to the nature of the analysis there is no real meaning to reporting a more precise estimate.

Table 4 CO_2 emission factors for aircraft and HST operation

Aircraft	g/pkm	LF	Seats	Distance	g/seat-km
Domestic[a]	158	0.65	–	463	102.7
Short haul[a]	175	0.7	150	480	122.5
short-haul Intl.[b]	130.4	0.65	139	1,108	84.76
Medium haul[a]	107	0.7	150	1,402	74.9
500 nm[c]	–	–	150	927.5	89.6
Representative emission factor for aircraft (Route < 750 km)					110 g/seat-km
Representative emission factor for aircraft (Route < 1,000 km)					85 g/seat-km
Representative emission factor for aircraft (Route < 1,500 km)					75 g/seat-km
HST	g/seat		km		g/seat-km
London–Paris[d]	7,200		525		13.7
	g/pkm		LF		g/seat-km
London–Manchester[e]	18.3		0.6		10.98
Representative emission factor for HST					13 g/seat-km
Representative emission factor for HST (25% distance penalty)					16.25 g/seat-km

Note g/pkm—gram per passenger-km. LF—Load Factor. g/seat-km—gram per seat-km
Sources [a] Miyoshi and Mason [23] (Table 2); [b] Miyoshi and Mason [23] (Table 1); [c] Based on Givoni and Rietveld [19]; [d] Givoni [15]; [e] Miyoshi and Givoni [22] (Table 2)

Table 5 Potential yearly regional CO_2 emission savings from mode substitution

	Seat-km transferred (billion)	CO_2 savings (million tonne/year)
North America	12.0	10.0
Europe	7.1	5.9
East and South-East Asia	6.7	5.6
South America	3.2	2.7
Middle East	1.6	1.4
Oceania–Pacific	1.4	1.2
India and surroundings	1.0	0.8
Central America and Caribbean	0.5	0.4
Africa	0.5	0.4
CIS	0.3	0.3
		28.4

can be shifted from air to rail and not on the share of air traffic that can be shifted to HST. The largest CO_2 emission savings from mode substitution are in North America, followed by Europe and amount to 10 and 5.9 million tonnes CO_2 per year, respectively.

This crude estimate of CO_2 emission savings from mode substitution can be put in context by considering the total contribution of transport, and air transport specifically, to CO_2 emission. In 2005, transport was responsible for emission of 8.83 Gt of CO_2, 27% of the world 32.56 Gt CO_2 emission. Of that world total, air transport contributed 0.73 Gt [12]. These estimates imply that if mode substitution takes place on all the routes identified, the reduction in CO_2 from air transport will be of the order of 0.1% of the world total CO_2 emission in 2005. This reduction will be of the order of only 4% of total CO_2 emission from air transport in 2005.[10]

In Europe (EU-27 countries), the transport sector was responsible in 2007 for 1,297.6 million tonne of CO_2 equivalent emission (24.2% of the total), and air transport for 161.2 million tonne of CO_2 equivalent emission, 12.4% of the transport sector contribution, including international bunkers (European Commission [10], Table 4.1.6). Based on these estimates, the potential for mode substitution within Europe[11] can result in CO_2 emission savings of about 0.5% of the CO_2 equivalent emission of the transport sector and less than 4% of emission from air transport.[12]

The results presented above are sensitive to the assumptions made, but the scope for mode substitution and the potential environmental benefits are not expected to vary much. The load factor achieved by HST operators and the carbon intensity of the electricity available to them (the electricity generation mix) are two important variables that the results will be sensitive to.[13] Mode substitution is clearly not a panacea for the environmental problems caused by air transport. However, the simplified analysis presented above suggests it can provide some contribution, but this contribution is probably not large enough to justify, on its own, investments in HST, an issue that will be taken up in the discussion below.

If mode substitution does take place, and the freed runway capacity is not taken up by other flights, other environmental benefits from mode substitution can also be expected, mainly noise and local air pollution reduction. Both could be more substantial than the reduction in CO_2 emission at the individual airport level and will depend on the reduction in the number of flights. In addition, mode substitution could in theory postpone the need to construct new runways, but this suggests that new demand is met which will offset any other environmental benefits from mode substitution.

4 Discussion and Conclusions

The exploratory analysis of the worldwide potential for air to rail mode substitution presented above has numerous limitations, but the results nevertheless provide useful indication and a basis to draw some conclusions.

---

[10] More precisely, 3.5% for a threshold distance of 750 km, 3.9% for a threshold distance of 1,000 km and 5.8% for a threshold distance of 1,500 km.

[11] The EC [10] figures refer to the EU27 countries while Europe in our analysis is larger than that, although most of the potential for mode substitution is within the EU-27 countries.

[12] The EC [10] estimates are for CO_2 equivalent while the mode substitution analysis is based on only CO_2, thus the potential is larger than reported since other emissions from air transport are not accounted for.

[13] The extent to which changing such assumptions will affect the results on a particular route was examined by Miyoshi and Givoni [22].

Overall, and assuming there is no potential for mode substitution on aircraft routes longer than 1,500 km, it appears that mode substitution is sensitive to the assumption on the level of demand (aircraft seats) that is considered sufficient to justify air to rail substitution. When assuming that an airline route with at least 5-daily flights (by a 150-seat jet) has sufficient demand to merit substitution, then the worldwide potential is quite significant, 20% of the flights and 7% of the seat-km for the 1,000 km threshold distance.[14] If this volume of traffic is shifted to the HST, without new flights using the freed capacity, there is a potential reduction of 20% in congestion and 7% in GHG emission from aircraft operation. Alternatively, the 20% reduction in the number of flights can be used to allow other flights, to destinations that cannot be served by HST, to take place, in essence, providing the industry with new capacity but without the need to build additional runways. Which of the two will be the result of mode substitution is a political and a policy decision. It is clear that without some regulation of the number of flights, or a substantial increase in the price of fuel—for example as a result of peak oil, and under the current market conditions and industry practices, mode substitution will be used to meet additional demand. In comparison to a situation where new demand is met through the construction of new runways and additional flights, mode substitution still offers environmental benefits.

From an environmental perspective, and again depending on how the freed runway capacity is used, it is clear that mode substitution is important but will not solve the environmental problems of the airline industry. It is likely to play a greater role in reducing air pollution around airports than contributing to reduction in GHG emission and anthropogenic climate change.

Given the potential for mode substitution presented above and the fact that this potential is only significant under the low-demand criterion, it can be assumed that mode substitution on its own would not justify the construction of HST lines. Development of HST lines depends primarily on the demand for rail services and the lack of capacity on the conventional rail network [14]. This has important planning implications. Since demand between airports is not the main justification for the construction of HST lines, many HST plans bypass the main airports. The assumption is that wherever mode substitution will be advantageous to the passenger it will take place. This assumption has two main caveats. First, many passengers use the large airports to transfer between short haul and long haul flights, and if mode substitution can take place on the short haul flight but the HST service does not include the airport, substitution will not take place. Second, short-haul flights are still important for airlines, and in the face of mode substitution and in response to direct competition from HST operators, airlines are likely to opt to increase their service frequency, not reduce it, canceling any benefits from mode substitution [15, 16]. As advocated by Givoni and Banister

---

[14] Interestingly, an analysis of the scope for mode substitution at London Heathrow arrived at similar estimate of 20% of flights in 2002 [17].

[17, 18] and in the EU White Paper [11], the HST network must include the large airports and in addition, airline and railway services need to be fully integrated. Thus, when a decision is made to build a HST line, the location of the main airports and the potential for mode substitution should be a central planning consideration. Otherwise, much of the potential for mode substitution presented above would be lost, or missed.

The above results call for a more detailed and robust analysis of the potential for mode substitution, an analysis that is probably best to perform at the regional, not global, level. Such an analysis is likely to show considerable higher or lower potential for mode substitution depending on regional characteristics. It can be expected that the analysis presented above overestimated the potential since it relied on aircraft distances, and it did not account for physical barriers such as mountain ranges, the construction cost of HST lines, and the organisation of some airline services in a Hub and Spoke network. At the same time, it can be expected that the analysis presented above underestimated the potential since it did not account for the forecast growth in the demand for air transport; the fact that two or more separate aircraft routes might be served by one HST line (thus allowing to consolidate demand on these routes), and the better quality of service likely to be offered when using HST and not aircraft. Finally, the analysis did not take account of changes in travel destinations HST can generate. For example, with HST in place passengers from Amsterdam might prefer to visit Paris (by HST) than Barcelona (by aircraft). Such factors can be considered in future research. It is also worth noting that technological development in both aircraft and HST will influence the potential, and even the need—from an environmental perspective, for mode substitution. Development of 'clean', quiet aircraft and an alternative for jet kerosene would remove most of the environmental motivation for such substitution and would lift most of the environmental restrictions on expanding airports. 'Clean' electricity and lighter and more efficient HSTs, on the other hand, would further increase the potential environmental benefits from mode substitution. The ability of an increasing number of HSTs models to operate beyond the HST network, on the conventional rail network, and often at relatively high-speed—using titling technology, extends the spatial reach of HST services and therefore the potential for mode substitution. In this respect, it is important to reiterate the need to examine the potential for mode substitution from road to rail.

The potential for mode substitution from air to rail provides yet another strong evidence for the need of a truly integrated transport system. Whether passengers opt to use the plane or train does not only depend on the difference between the flight and the HST journey but also very much on how 'easy' it is to get to/from the airport or the train station. Whether short or long-distance travel, national, regional or international it is the quality of the door-to-door journey that will determine the choice of mode and only strategic consideration of the entire transport network in planning transport infrastructure can facilitate mode choice decisions which are better for passengers and society at the same time.

References

1. Boeing (2010) Current market outlook. Boeing. http://www.boeing.com/commercial/cmo/pdf/Boeing_Current_Market_Outlook_2010_to_2029.pdf
2. Buchanan and Partners (1995) Optimising rail/air intermodality in Europe. European Commission—DG VII, London, November
3. CEC (2001) Commission of the European Communities "European transport policy for 2010: time to decide", Commission of the European Communities, COM(2001)370, Brussels, September
4. Clever R, Hansen MM (2008) Interaction of air and high-speed rail in Japan. Transportation Research Record: Journal of the Transportation Research Board, No. 2043, Transportation Research Board of the National Academies, Washington, D.C., pp 1–12. doi:10.3141/2043-01. http://thinkmetric.com/pubs/japan/airHSRinteraction.pdf
5. Chester MV, Horvath A (2009) Environmental assessment of passenger transportation should include infrastructure and supply chains. Environ Res Lett 4:024008
6. Chester MV, Horvath A (2010) Life-cycle assessment of high-speed rail: the case of California. Environ Res Lett 5(1):014003
7. Dobruszkes F (2011) High-speed rail and air transport competition in Western Europe: a supply-oriented perspective. Transp Policy 18(6):870–879
8. Dobruszkes F, Cattan N (2011) Are worldwide air services really long-range and international? An empirical analysis. NECTAR conference, Antwerp
9. European Commission (EC) (1998) Interaction between high speed and air passenger transport. Final report on the Action COST 318, European Commission, Luxemburg
10. European Commission (EC) (2010) EU energy and transport in figures. European Commission, Brussels
11. European Commission (EC) (2011) White paper: roadmap to a single European transport area—towards a competitive and resource efficient transport system. EC, COM(2011) 144 final, March 23. http://eur-lex.europa.eu/LexUriServ/LexUriServ.do?uri=COM:2011:0144:FIN:EN:PDF
12. European Environment Agency (EEA) (2009) Transport scenarios scoping report. EEA, TRL, and TSU, November
13. Givoni M (2005) Aircraft and high speed train substitution: the case for airline and railway integration. Unpublished Ph.D. thesis, University College London, London
14. Givoni M (2006) The development and impact of the modern high speed train. Transp Rev 26(5):593–612
15. Givoni M (2007) Environmental benefits from mode substitution—comparison of the environmental impact from aircraft and high-speed train operation. Int J Sustain Transp 1(4): 209–230
16. Givoni M (2007) Air–rail intermodality from airlines' perspective. World Rev Intermodal Transp Res 1(3):224–238
17. Givoni M, Banister D (2006) Airline and railway integration. Transp Policy 13:386–397
18. Givoni M, Banister D (2007) The role of the railways in the future of air transport. Transp Plan Technol 30(1):95–112
19. Givoni M, Rietveld P (2010) The environmental implications of airlines' choice of aircraft size. J Air Transp Manag 16(3):159–167
20. Janic M (2003) The potential for modal substitution. In: Upham P, Maughan J, Raper D, Thomas C (eds) Towards sustainable Aviation. Earthscan, London
21. Lugo I (2010) Arteries of Mexico highway system. Cardus-Revista de Estudios Urbanos, Universidad de las Américas-Puebla, Número 3
22. Miyoshi C, Givoni M (2011) The environmental case for high speed train in the UK: examining the London–Manchester route. Int J Sustain Transp (forthcoming)
23. Miyoshi C, Mason KJ (2009) The carbon emission of selected airlines and aircraft types in three geographic markets. J Air Transp Manag 15(3):138–147

24. UIC (2011) High speed principles and advantages. http://www.uic.org/spip.php?article443# outil_sommaire_0 (last accessed 11.04.2011)
25. UIC (2011) High speed lines in the World—UIC High Speed Department. http://www.uic.org/ spip.php?article573 (last accessed on 11.04.2011)

No Third Runway? The Thames Estuary Option

Christian Carey

Abstract In 2008, prospective Tory candidate for London Mayor, Boris Johnson, suggested building a major new international airport in the Thames Estuary to replace the 'planning error' that is Heathrow. Once elected, Johnson commissioned a serious study, headed by civil engineer Douglas Oakervee, into the potential for construction of an airport in the Thames Estuary. This chapter looks at the discussion held around the possible development of a fourth London airport in the Thames Estuary.

1 Why?

But what is the demand for the expansion of aviation, particularly in the south east of England? In 2000 the Department of the Environment released forecasts for the growth of air traffic in the UK for the next 30 years. These indicated a huge growth in passenger numbers—4.25% per annum—and were accompanied by the caveat that past forecasts have often underestimated demand. Were this demand to be met, five new runways—three in the south-east—would be required. As a major aviation hub, Heathrow was chosen for the site of one of these runways, as announced by transport secretary Geoff Hoon in January 2009. The importance of Heathrow to the south-east and to Great Britain as a whole cannot be underestimated. It plays a vital role in the economy and employs at least 100,000 people directly, and is part of a sector contributing £11 billion a year to the economy. Heathrow annually

C. Carey (✉)
Smith School of Enterprise and the Environment, University of Oxford,
Oxford, UK
e-mail: christian.carey@smithschool.ox.ac.uk

O. Inderwildi and Sir David King (eds.), *Energy, Transport, & the Environment*,
DOI: 10.1007/978-1-4471-2717-8_28, © Springer-Verlag London 2012

handles some 68 million passengers, on 477,000 flights, running at 99% of its total capacity [1]. Heathrow's European competitors—Frankfurt am Main, Paris Charles de Gaulle and Amsterdam Schiphol—are running at 75% capacity [1]. This is reflected in the number of routes flown regularly from Heathrow: 133 destinations are served at least once a week, 20% fewer than competitors [2].

There are a number of issues that constrain the expansion of Heathrow. The airports operational capacity is limited by several factors, including airspace, the number and length of runways, the area available for use as taxiways and aprons, the number and size of terminals and landside facilities and access [3, 4]. While infrastructure issues—the need to provide space and access for new facilities to be built—are common for the expansion of any airport, environmental aspects must also be considered.

Of the environmental issues, probably the single most important impact of an airport on its local environment is noise. The frequency of aircraft movements, the sound level of the individual aircraft and the relative proximity of the airport's arrival and departures routes to local communities are the main factors affecting noise levels and therefore operational constraints. A number of technological advances have been made, such as shielding landing gear and faired slats (high lift devices on a wing leading edge), that reduce the noise generated by aircraft. But these have been offset by the increased growth in air travel to such an extent that most major airports have operational constraints or capacity limits based upon aircraft noise. These can be operational restrictions on noisier aircraft, night curfews, limits based on noise budgets or the extent of a noise exposure contour. At Heathrow a quota count (QC) system is used where each aircraft type is classified and awarded a QC value depending on the amount of noise it generates under controlled certification conditions. Heathrow has a fixed QC depending on the season (summer/winter), which is gradually reduced year on year to reduce overall airport noise. Further constrains are placed on night flights, with Heathrow preventing the use of aircraft with a QC greater than 2 (95.9EPNdB) such as the arrival and the departure of Boeing 747-100/200/300 [5, 6]. The UK government uses a 57 dB sound level exposure to determine whether communities are significantly affected by aircraft noise, which equated to 264,000 people at Heathrow in 2003. Any expansion of capacity will increase the number of individuals who fall under this limit, increasing the number of operational constraints on the airport.

The quality of air near an airport is also a major contributor to capacity constraints. Local air pollution is not only generated from aircraft movements, but apron activities and ground transport related to the airport's function are also significant factors in air quality. This has caused several airports to introduce stringent emission controls, particularly in relation to nitrous oxides and volatile organic compounds. To reduce these emissions, a reduction in ground transport by greater use of public transport, together with improvements in aircraft technology and apron activity procedures are required. However, as Heathrow is surrounded on three sides by housing and by the London orbital motorway (M25) on the fourth, the potential for expanding public transport is limited.

Similarly, the risk to surrounding communities also rises with airport expansion as accident rates on approach and departure paths also rise. While aviation accident rates are falling, the increasing growth in air traffic has offset this benefit [7]. In the UK and the Netherlands, a risk contour measuring approach is used which predicts the area within which it is unacceptable to live or work due to the increased level of risk. A 1997 NATS report indicated that 2,222 people lived within an area where there was a 1:100,000 annual risk of death due to an aviation accident due to operations at Heathrow [7]. As a result, airports can be obliged to purchase and demolish properties to remove people from the areas of highest risk.

A lesser issue with Heathrow is the effect airport expansion has on biodiversity. Airports cover large areas of land with either inhospitable areas (built environment) or ecological monocultures (mown grass land) and therefore represent a challenge to the biodiversity of an area. This is particularly so where airports are built on greenbelt land surrounding major conurbations, which can restrict airport expansion.

All these issues could be involved with the expansion of any of the current airports in the south-east. Land is required for construction and it is inevitably communities will be affected by increased air and ground traffic. However one possibility for reducing these impacts is to locate airports away from communities, such as on reclaimed land in the Thames Estuary.

2 Not a New Idea

The possibility of an expansion of airport capacity by construction in or around the Thames Estuary was initially suggested previous to Heathrow accepting civilian traffic. In 1943 a combined airstrip and flying boat port was proposed at Gravesend by a Mr FG Miles. The project was expected to handle eight million passengers a year and 'great quantities of freight' but received no government backing [8]. A Thames site was again proposed during the selection of a secondary site for a new London airport, with a site at Cliffe Marshes being rejected in favour of developing Gatwick [9]. The site was dismissed due to air traffic control (ATC) issues, limited transport infrastructure, construction cost and poor weather. Thames sites were again considered when the site for the third London airport, which eventually became Stansted, was under investigation [10]. A number of onshore and offshore sites were considered including Cliffe Marshes and Gunfleet Sands (8 kilometres offshore from Clacton-on-Sea) [11]. The advantages of the Thames sites were the low value of the land and proximity to London, but again they suffered from poor communication links to London, ATC issues, flood risk and secondary uses such as firing ranges, weather and the high construction costs of reclaimed land when compared to a dry inland site. The choice of Stinted was a controversial one with a government enquiry and commission (the Roskill Commission Inquiry) as well as a number of local government and independent groups producing reports [12]. When the Roskill Commission reported after nearly two years of research in 1970, it recommended Cublington in Buckinghamshire which

Edward Heath's Conservative government then overturned in favour of Foulness (or Maplin Sands) as the site for London's third airport, opposing the earlier choice of Stansted [13, 14]. This finding was backed by an earlier report, published on behalf of the Noise Abatement Society, which concluded that Foulness was preferable to Stansted [12]. The Maplin Sands project was then considered for the next nine years until finally dropped under Margaret Thatcher's Conservative government in 1980, on the grounds of environmental damage, cost and development time. The idea for a Thames Estuary international airport resurfaced as the Marinair proposal. Located five kilometres north east of Minster, on the Isle of Sheppey, the scheme was not the 'son of Maplin' as discussed in newspapers of the day, but a true offshore island, with more in common with the Gunfleet sands proposal from the 1960s. The original scheme was estimated at £20 billion and was based on a two-centre model with terminals at East Tilbury and runways on an artificial island 35 kilometres away. High speed trains running in tunnels would link the passenger terminals to the aircraft waiting on the offshore island. The scheme failed to be shortlisted by the Government following the South-East Regional Air Services Study (SERAS) and was dismissed in the White Paper 'The future of air transport' due to insufficient information and the prohibitive cost of road and rail links (even with support from then Mayor of London Ken Livingstone [15]). The future of air transport White Paper also considered a number of alternative Thames Estuary proposals such as Thames Reach (on the Hoo peninsular) which were also discounted due to cost, environmental impact and an over reliance on rail access [16]. Sahara Group has recently proposed a multi-modal scheme which includes a container port and an airport with four parallel runways on an 18 by 2 kilometre reclaimed island off the Isle of Sheppey. It is linked to the southern shore by overland road and rail and the north shore by a tunnel. It also includes plans for a level of flood protection but not a lower Thames barrier. A more holistic approach is offered by Eleanor Atkinson, who proposes an offshore location linked to both shores by road and rail and incorporating tidal lagoons for energy generation, tidal barrier as well as shipping and leisure facilities [17].

The reason for such a high number proposals is the considerable advantages of locating an airport in the Thames. With few human neighbours, the airport could operate round the clock because noise levels would be less significant. There would be a less pollution because a new airport would be much 'greener', and there would be better public transport than is possible at Heathrow. The third party risk is also removed to a significant extent with approach and departure over water. The possibilities of expansion of the airport at a later date and no requirement for compulsory purchase of property or the demolition of historic buildings, due to the nature of the site, are also significant benefits. The regeneration effects in the Thames area are also of note. Also, possibly surprisingly, an offshore airport would have better visibility than Heathrow.

But there are also disadvantages with this approach, not least the £40–50 billion cost of constructing an airport on reclaimed land, and improving transport links with central London. This compares to £10–13 billion for a third runway at Heathrow. The technical challenge of reclamation has been achieved at a number of locations including Hong Kong's Chek Lap Kok and Haneda Airport in Tokyo Bay, and less successfully at Kansai International airport, Osaka which has suffered from high rates of settlement; it has sank nearly 3 m in the 14 years since it opened, so successful construction is not a forgone conclusion [18–20]. However the geology in the Thames Estuary, clay overlying chalk, is well suited to this sort of scheme. Another challenge is the airlines themselves. When legislation changed in 1991 and 2000, there was a desire for the big international airlines to leave Gatwick for Heathrow so to attract them away from Heathrow will require some thought. Also, while the development of a new airport east of London will financially benefit the people there, there is a corresponding deficit for the west of London. An extra airport would considerably complicate air traffic control in the south-east, requiring wholesale changes to its structure. Also, the proximity of the eastern airspace boundary of the UK's ATC zone and the busiest airspace in Europe increases the difficulties for air traffic management. The location also has a number of physical challenges such as wind turbines at the London Array and Kentish Flats, but these would have served their working life by the time an airport opened in 30 years' time, and could be re-located on renewal. The estuary has features dating back to the Second World War; the wreck of the SS Richard Montgomery and a number of forts collectively known as the Maunsell Sea Forts. The SS Richard Montgomery was a liberty ship which sank carrying ammunition. While most of it has been removed, the equivalent of 1,500 tonnes of TNT remains is a hazard to local shipping and is protected by an exclusion zone [21]. The Maunsell Forts are anti-aircraft and anti-shipping platforms in a poor state of repair, some of which have already collapsed into the sea.

It is not just an airport that has a claim on the Thames Estuary; the site is a busy shipping area with some 53 million tonnes of freight being carried to the various Port of London locations [22]. There is also the new London Gateway Port expected to open in 2011 which is expected to handle 3.5 million containers a year making it the busiest container port in the UK. This will have an impact on any significant construction in the area. Leisure and fishing industries on the North Kent coast and in South Essex will have to be considered as well as there is a significant bird population. The Thames Estuary is an important area for birdlife and the increased noise and pollution, as well as potential habitat loss due to the construction of the airport, would be detrimental [23]. The risk of bird strike and related incidents could also increase, but an offshore site would be several miles from the mudflats where they feed. Other uses for the Thames Estuary are as a flood defence for London and the surrounding area and energy generation. Several power plants both on and offshore are already in operation with others planned.

So while there are a number of benefits that are only possible in a Thames Estuary location, such as 24-h airport operation, there are also issues to be considered. Not least of these is the huge cost and difficulty of construction with offshore sites. Increasing capacity may be possible in other ways and should be considered. Freeing up capacity by shifting domestic flights to a renewed rail system, improving links between central London and Gatwick, Stansted and Luton or reviewing landing charge control to increase competition between airports should all be considered. However if Heathrow is, as claimed by Johnson, a 'planning error of the 60's' is it not time to put this right?

References

1. Adding Capacity at Heathrow: Decisions Following Consultation (2009) Department for Transport, London
2. Boon B, Davidson M, Faber J, Nelissen D, Van de Vreede G (2008) The economics of Heathrow expansion. CE, Delft
3. Upham P, Thomas C, Gillingwater D, Raper D (2003) Environmental capacity and airport operations: current issues and future prospects. J Air Transp Manag 9(3):145–151
4. Atkin JAD, Burke EK, Greenwood JS, Reeson D (2009) An examination of take-off scheduling constraints at London Heathrow airport. In: The 10th international conference on computer-aided scheduling of public transport (CASPT06), Leeds, June 2006
5. Monkman DJ, Deeley J, Beaton D, McMahon J, Edmonds LE (2007) Noise exposure contours for Heathrow airport 2007. Directorate of Airspace Policy, ERCD, Civil Aviation Authority
6. Heathrow London (2007) London Gatwick and London Stansted airports noise restrictions notice 2007. NATS, Civil Aviation Authority
7. Evans AW, Foot PB, Mason SM, Parker IG, Slater K (1997) Third party risk near airports and public safety zone policy, National Air Traffic Services Ltd, R&D Report 9636
8. Land and sea terminals by Thames estuary (1943) The Times, London, 21st July
9. Boyd-Carpenter J (1954) Gatwick Airport (Development), Department of Transport and Civil Aviation
10. Decision for Stansted (1967) Flight International, 18th May
11. Alternatives to Stansted (1967) Flight International, 25th May
12. Foulness: A Feasible Alternative? (1967) Flight International, 9th Nov
13. Report, Commission on the Third London Airport (1971) Roskill Commission
14. Researching for Roskill (1970) Flight International, 5th Mar
15. Waugh P (2002) Livingstone backs Eighties plan for £30bn floating airport on Thames. The Independent, London 3rd Aug
16. The Future of Air Transport (2003) Department for Transport
17. Atkinson E (2009) Thames Estuary Airport, www.thamesestuaryairport.com/
18. Pickles AR, Tosen R (1998) Settlement of reclaimed land for the new Hong Kong international airport, Geotechnical Engineering, Proc Inst Civil Eng, 131
19. Douglas I, Lawson N (2003) Airport construction: materials use and geomorphic change. J Air Transp Manag, 9
20. Nakada H, Akimoto K , Kanazawa H, Tsuji Y, Inada, Haneda M (1997) Airport offshore expansion project, civil engineering international, Proc Inst Civil Eng, 120
21. SS Richard Montgomery Survey Report (2003) Maritime and Coastguard Agency
22. Annual Review 2008 (2008) Port of London Authority
23. Mulholland H (2009) Thames estuary airport would have 'disastrous' environmental impact London mayor told. The Guardian, Manchester 14th Jan

Part V
Sea, Rail and Cargo

Foreword
Brian Collins

World trade depends upon our ability to move goods and raw materials over planetary distances efficiently and economically. In the past optimising these two factors has been sufficient for planning and operational purposes. Now we realise that other factors such as sustainable use of resources, impact on environment and most of all the reduction of GHG emissions have to be taken into account. Analysis of these factors combined with those of efficiency and economy necessitate a system of systems level analysis of the combination of rail and sea transport when moving bulk cargo. Innovative approaches are needed to balance such factors in both the design of solutions and their operation. These essays illustrate how such innovations are occurring, what research is still needed and how policy makers need to embrace such factors, especially when considering large investments such as port modernisation, high speed rail and ocean going ship design.

Governance of such processes of analysis is further complicated by the international nature of the problem, the differing balance between private and public ownership in different countries, and the different levels of maturity of the actors in establishing their long-term programmes of activity, each commencing from differing starting points. But the prize to be won in alleviating food and energy policy, mitigating GHG-induced climate change effects and in raising the living standards of the human race world wide is enormous. Only by tackling the global issue at the top level will success be achieved and this work of scholarship provides a welcome and insightful stimulus to that activity.

Professor Brian Collins CB, FREng
Director of Centre for Engineering Policy
University College London

Reducing Energy Consumption and Emissions in the Logistics Sector

Alan C. McKinnon

Abstract Logistics is a relatively energy-intensity sector which is rapidly expanding mainly as a result of globalisation. This chapter assesses its share of global energy consumption and greenhouse gas emissions and considers how this is likely to change over the next 40 years. It then reviews the numerous ways in which energy consumption by logistical activities and related emissions can be reduced. This is done within a framework built around a series of seven key parameters. By altering these parameters companies and governments should be able to decouple the growth in demand for logistics from the associated energy requirements and externalities. The parameters relate to the freight-intensity of the economy, the division of freight traffic between modes, the utilisation of vehicle capacity, the energy efficiency of logistics operations (comprising transport and warehousing) and finally the ratio of emissions to energy use. While changes in these parameters will offset much of the underlying growth in demand for logistical services, there seems limited prospect of energy use and carbon emissions in this sector dropping sharply in absolute terms over the few decades.

1 Introduction

Logistics comprises the movement, storage and handling of products as they move from raw material source through the production system to point of use. It has become a major growth sector in the global economy. Traditionally, the amount of

A. C. McKinnon (✉)
Department of Business Management, School of Management & Languages,
Heriot-Watt University, Edinburgh EH14 4AS, UK
e-mail: A.C.McKinnon@hw.ac.uk

O. Inderwildi and Sir David King (eds.), *Energy, Transport, & the Environment*,
DOI: 10.1007/978-1-4471-2717-8_29, © Springer-Verlag London 2012

freight movement has expanded broadly in line with economic growth. The globalisation of production, procurement and marketing in recent decades, however, has increased the 'freight transport intensity' of the economy, generating more tonne-kilometres of freight movement per unit of output. Taking account of these trends, the International Energy Agency (IEA) [23] has forecast, on a business-as-usual basis, that tonne-kms moved by trucks and ships will double between 2005 and 2050, while rail freight volumes will rise by 50% over this period. Airfreight volumes are projected to grow at an even faster rate of around 5–6% per annum, at least until 2030 [2]. No attempt has yet been made to forecast increases in the level of related logistical activity in goods handling, warehousing and IT, but this is likely to be substantial over the next few decades.

This rapid growth in demand for logistical services will have important implications for energy use and emissions. To assess these implications, it is necessary to estimate how much energy is used in logistics and the related level of emissions. This is difficult to do on the basis of currently available data. Using statistics compiled by the IEA [23], it is estimated that freight movement by truck, rail and ship accounts for roughly a third of the total energy used by transport and around 6–7% of total global energy consumption. To this figure has to be added energy used in moving goods by air, in vans and by pipeline. This is likely to raise the total to around 10% of total energy use, though this should be regarded as an order of magnitude estimate. No macro-level data are available on energy consumption in warehousing, goods handling and related IT operations.

It has been estimated, on the basis of figures from the Inter-governmental Panel on Climate Change [27], that freight transport is responsible for around 8% of energy-related CO_2 emissions. The World Economic Forum and Accenture [50] have attempted to carbon footprint logistical activity worldwide, calculating its share of total greenhouse gas (GHG) emissions, rather than just energy-related CO_2, and including an allowance for the 'logistics buildings'. This suggests that logistics accounts for around 5.5% of global greenhouse gas (GHG) emissions, roughly 90% of which come from freight transport and the rest from warehouses and freight terminals. This may seem a rather low % considering the vital role that logistics plays in sustaining life and maintaining living standards around the world. It is, however, broadly consistent with the other energy and emission estimates for this activity.

The current energy requirements and emission footprint of logistics may be relatively modest, but the projected growth in demand for this activity over the next few decades will significantly increase its share of both. The proportion of CO_2 emitted by logistics will also rise quite steeply because it will be relatively difficult to 'wean' off fossil fuels. Many other sectors of the economy are powered by electricity and will thus benefit from the decarbonisation of electricity generation. While some logistics operations, such as urban van deliveries, electrified rail freight services, terminal handling and warehousing can run directly or indirectly (via batteries or hydrogen) on low-carbon electricity, many others, such as shipping, long haul trucks and aircraft will have to rely on liquid, carbon-based fuel. Some of this fuel may be produced from plants and waste material and hence, on a life cycle basis, emit less GHG per litre consumed than fossil fuel. But it is likely that large quantities of fossil fuel will still be required to move freight two or three decades from now.

Against a global target of cutting total GHG emissions by 50% by 2050 and a declared objective of cutting by 80–90% in the EU, public policy-makers are unlikely to treat logistics as a special case and grant it the right to enlarge its carbon footprint. On the contrary, they have high expectations of the potential for carbon reduction in this sector. For instance, the recently published white paper on transport from the European Commission [15] declares the goal of achieving '*near zero-emission urban logistics*' by 2030.

How then can the predicted growth of energy use and emissions by the logistics sector be restrained and ultimately reversed over the next few decades without adversely affecting economic development and eroding living standards. The remainder of this chapter examines a series of initiatives that will help to achieve this. The next section outlines a framework within which the relative contribution of these initiatives can be assessed.

2 Assessment Framework

In much of the literature in this field (e.g., [16, 23]) efforts to reduce energy consumption and environmental impacts are grouped under three headings:

Avoid: reduce the demand for freight transport
Shift: transfer freight traffic to more energy-efficient and green modes
Improve: increase the efficiency of freight transport operations

A more detailed framework has been developed which defines the relationship between the growth in the material output of an economy and logistics-related energy use and environmental impact with reference to seven key parameters. The first five relate exclusively to the transport operation while the remaining two apply to the full range of logistical activities:

1. *Average handling factor:* this is the ratio of the weight of goods produced or consumed to the weight of goods lifted onto vehicles, vessels or aircraft at the start of a journey. The movement of products from raw material source to final point of use generally comprises several journeys, representing links in the supply chain. The handling factor is therefore a crude measure of the complexity of a supply chain.
2. *Average length of haul:* this is the mean length of each link in the supply chain.
3. *Modal split* indicates the proportion of freight carried by different transport modes. Following this split subsequent parameters need to be calibrated for particular modes. As road is typically the main mode of freight transport, the rest of Fig. 1 has been defined with respect to this mode.

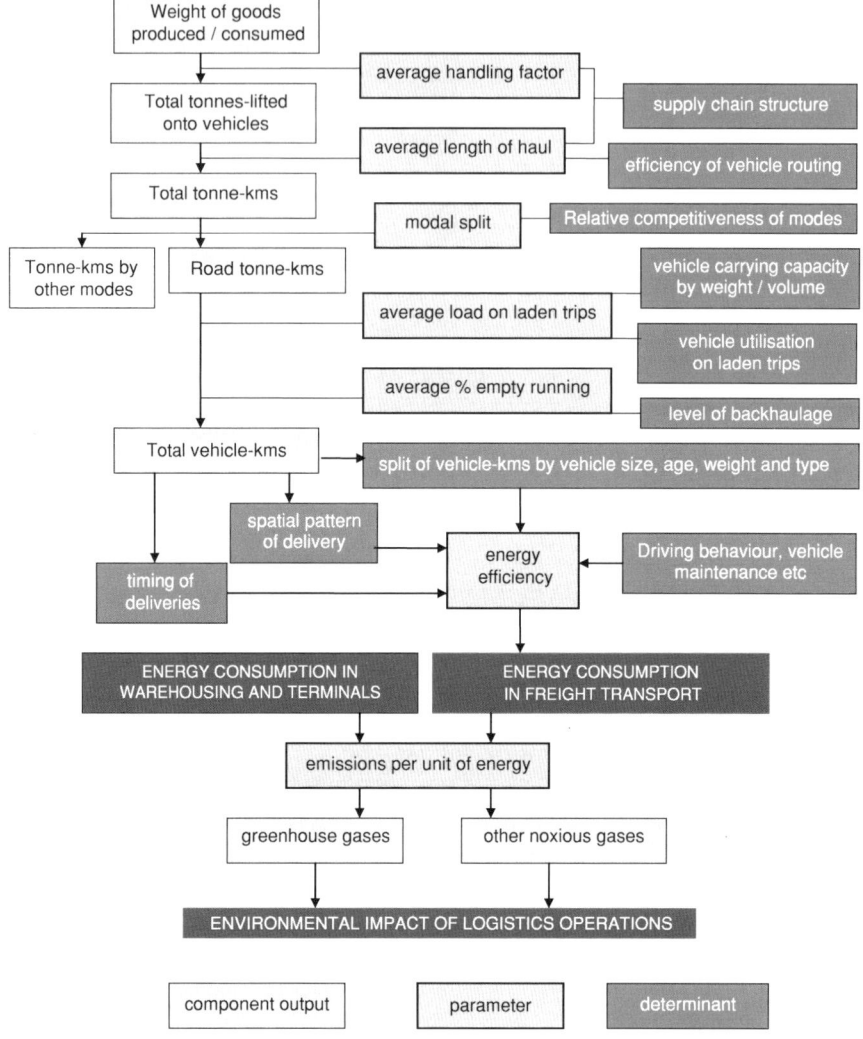

Fig. 1 Assessment framework (adapted from [36])

4. *Average payload on laden trips:* this is generally measured in terms of weight, though it should be noted that the value of this parameter is sensitive to the average density of the load.
5. Average proportion *of vehicle-kms run empty*.
6. *Energy efficiency*: in the case of transport this is defined as the ratio of distance travelled to energy consumed. For other logistical activity, energy use is related to other measures of output such as warehouse throughput or the number of containers handled by a terminal.

7. *Emissions per unit of energy:* the quantity of CO_2 and noxious gases emitted per unit of energy consumed can vary with the type of energy/fuel, the nature of the processes converting this energy into logistical activity and exhaust filtration systems.

Parameters two and three indicate the extent to which, for a given amount of economic output, it might be possible to 'avoid' the need for the freight transport, parameter 1 relates to modal 'shift', while the remaining parameters measure the opportunity for 'improving' the efficiency of the logistics operations. Figure 1 maps the inter-relationships between these critical parameters, the factors affecting them (i.e. 'determinants') and the 'component outputs' that the parameters themselves can influence. These component outputs ultimately determine total energy consumption by logistical activities and their overall environmental impact. By altering the seven parameters companies and governments can substantially reduce both energy use and environmental impact. The following sections examine the opportunities for modifying these parameters in a way that achieves these outcomes. In this discussion some parameters are paired (1 with 2 and 4 with 5) because they are closely associated. The freight-intensity issue is addressed first as it concerns the overall demand for freight movement, prior to the division of freight traffic among the various transport modes.

3 Reducing Freight Transport Intensity

The first and second parameters, relating to the average number and length of links in the supply chain, essentially determine the freight transport intensity of the economy, measured by tonne-kms of freight movement per unit of economic output. Much of the growth in freight movement in recent decades can be attributed to the extension and restructuring of companies' supply chains. It has been due more to each unit of freight being transported over greater distances than to the physical mass of goods in the economy expanding. This trend has been driven primarily by two processes, (i) the wider sourcing of supplies and (ii) the centralisation of economic activity. These processes cannot continue indefinitely. Within several developed countries they are now at an advanced stage [34]. In many others at an earlier stage of development and at the continental and global scales they remain active and show little sign of abating.

Reversing these trends and returning to more localised trading and decentralised logistics would, nevertheless, reduce the demand for freight movement and related energy consumption and emissions. There are several problems with this proposal, however. First, it would be very difficult to achieve. The historic increase in the average length of haul has been a consequence of major global business processes that will be very difficult to reverse. Simulation modelling of logistical systems indicates that the cost trade-offs which companies make between transport, inventory management and warehousing are very robust. Tilting these cost trade-offs sufficiently to induce a return to more localised and decentralised patterns of production and distribution would require steep increases in freight transport costs, particularly as these costs represent only a small percentage of the selling price of most products (e.g. generally less than 5% for consumer products). Second, if a return to more localised and decentralised logistics could be engineered it would conflict with economic development and poverty reduction goals. Third, it might not have the desired effect of reducing overall energy consumption and environmental damage. After all, reducing the average distance that goods travel may cut freight-related energy use and emissions, but, for most products, freight transport represents only a small proportion of total energy consumption and emissions on a full life cycle basis. Production operations are typically responsible for a much greater proportion. The energy- and emission-intensity of production is therefore a much more important determinant of a product's life cycle impacts than the distance it is transported. The environmental benefits of decentralisation can be challenged on similar grounds. Any transport-related savings in energy and emissions accruing from decentralisation would have to be weighed against the loss of scale economies in energy use and the greater environmental footprint of more fragmented production and warehousing systems. So before pursuing a localisation-decentralisation strategy, as many environmentalists now advocate, one needs to model its wider effects on energy and externalities and not simply those directly associated with freight transport.

4 Switching Freight to More Energy-Efficient and Greener Transport Modes

Transport modes vary enormously in the average amount of energy they require to move freight (expressed as energy use per tonne-km). As the vast majority of freight movement is powered by fossil fuels, there is a correspondingly large variation in the carbon intensity of transport modes measured by gCO_2 per tonne-km. Table 1 illustrates the variation in the average carbon intensity of moving chemical products by different transport modes [41]. There is a much weaker relationship between energy use and other externalities as a result of differences in vehicle technology, the nature of the transport operation and the stringency of environmental regulations. For example, government regulation has driven down emission standards for NOx, particulate matter and SO_2 to a much lower level for trucks than for freight trains or ships. Governments around the world are promoting freight modal shift as a means cutting energy consumption and reducing the adverse environmental effects of goods movement. In most cases they are trying to reverse a decline in the share of the freight market held by rail and waterborne transport, i.e. the greener transport modes. Across the EU 25, for instance, road's share of tonne-kms rose from 72% in 1995 to 78% in 2009, while rail's share contracted from 21 to 17% and inland waterway's from 7 to 5% [19].

Table 1 Variations in the average carbon intensity of transport modes used in the movement of european chemicals

Transport mode	gCO_2/tkm
Pipelines	5
Deep-sea tanker	5
Deep-sea container	8
Short sea shipping	16
Intermodal road/short sea	21
Rail transport	22
Intermodal road/rail	26
Barge transport on inland waterway	31
Intermodal road/barge	34
Road transport— heavy truck	62
Airfreight	600

(*Source* [42])

To reverse these trends, governments are investing in rail and inland waterway networks, supporting the development of intermodal terminals, providing subsidies for the purchase of vehicles and equipment and offering revenue-support for rail and water-borne services. At a continental level, the 2011 EU white paper on transport has set a target of shifting 30% of road freight moving over distances of 300 kms or more to rail or water by 2030 and 50% by 2050. One policy instrument which the European Commission wishes to deploy to help attain this goal is full internalisation of the marginal social costs of transport, incorporating a realistic valuation of freight-related externalities. At present the prices charged by different freight modes do not fully reflect differences in their environmental impact [8]. The use of taxation or emissions trading to correct this anomaly would favour the use of greener, more energy-efficient modes. These fiscal measure will be reinforced by future increases in the market price of oil.

The use of rail and waterborne services, however, is inhibited by factors other than cost which government economic incentives alone cannot correct. Intrinsically, these modes have lower flexibility and accessibility and, in the case of short sea shipping and inland waterways, much longer transit times. Their competitiveness can also be impaired by poor service quality and inadequate marketing. There is evidence too that freight purchasing decisions are often biased towards particular modes, as logistics managers do not objectively appraise the various modal options. In recent years, however, many large manufacturers and retailers, such as Procter and Gamble, Baxter Healthcare, ASDA and Tesco, have made a serious effort to expand their use of the rail and/or inland waterway networks, ostensibly for environmental reasons.

Over the next few decades new technology will improve the energy efficiency and emission performance of some transport modes more rapidly, and to a greater extent, than others. Much-quoted figures about the energy- or carbon- intensity of mode X being Y percent lower than that of mode Z need, therefore, to be kept under constant review. Companies and transport policy-makers also need to exercise caution in interpreting much of the comparative energy and emissions data for freight transport modes currently in the public domain. Quoted figures for energy consumption and emissions per tonne-km by a particular mode are very sensitive to the averaging of load factors, fuel efficiency and the carbon content of energy used. Inconsistencies in this averaging process can bias the modal comparison. Published statistics also give only a partial view of a mode's overall carbon impact because they exclude the emissions from the construction, maintenance and dismantling of vehicles and infrastructure. To make a comprehensive assessment of the energy and environmental benefits of altering the freight modal split it is necessary to extend the system boundary around the calculation to include this broader range of activities.

5 Improving Vehicle Utilisation

The fourth and fifth parameters can be combined under a single heading as they determine the utilisation of vehicle capacity. Improving the loading of vehicles reduces the amount of traffic (measured in vehicle-kms) needed to move a given quantity of freight (measured in tonne-kms). There is a corresponding reduction in energy consumption and environmental damage per tonne-km.

On the basis of currently available data, it is very difficult to estimate the potential for improving capacity utilisation. Most EU governments compile statistics on the empty running of trucks and the average weights of payloads they carry, but not on the proportion of available weight-carrying capacity actually used [37]. Weight-based statistics also give only a partial view of vehicle loading as many low-density consignments 'weigh-out' before they 'cube-out'. A series of 'transport KPI' surveys commissioned by the UK government between 1997 and 2009 provided greater insight into the weight and volumetric loading of trucks, but these covered only a few industrial sectors and have no parallels in other countries [35]. The statistical deficiencies that exist in the road freight sector are serious but minor when set against the almost complete dearth of official data on capacity utilisation across other freight transport modes. The available survey evidence, mainly for road, does nevertheless suggest that there is substantial under-utilisation of freight vehicle capacity. In EU countries, for example, an average of 25% of truck-kms are run empty, with national averages ranging from 13 to 38% [18].

The under-utilisation of vehicles does not necessarily mean that companies are mismanaging their logistics. Vehicle loading is constrained by many factors relating to market conditions, government regulation, type of equipment, infrastructure and internal company management [38]. One of the most critical factors affecting vehicle loading is the relationship between transport and other activities such as production, procurement, inventory management, warehousing and sales. Companies often quite rationally assign the optimisation of these other activities priority over transport efficiency. For example, inventory savings from just-in-time replenishment or reductions in handling costs accruing from the use of roll-cages rather than wooden pallets may exceed the additional cost of running a truck only part-loaded. In some cases, the resulting loss in energy efficiency in the freight transport operation will be more than offset by energy efficiency gains in other logistical and production activities. So, once again, there is a clear need for a cross-functional trade-off analysis of energy/emissions impacts—this time relating to vehicle loading. Nevertheless, as pressure mounts to conserve energy and cut environmental impacts there is likely to be a re-ordering of corporate priorities, placing greater emphasis on freight transport optimisation.

A broad range of measures can be applied to improve vehicle loading, and thereby save fuel and cut emissions. These include improved backloading [39], the use of more space-efficient handling systems and packaging [28], the adoption of more transport-efficient order cycles and consolidating freight in larger/heavier vehicles [33, 49]. In the road freight sector, the last of these options usually requires a relaxation of legal limits on truck size. This is currently a very contentious issue in the EU because it can have the 'second order' effect of displacing freight from more energy-efficient, lower carbon rail and water-borne modes [20]. A substantial body of evidence has accumulated, however, in those countries operating longer and heavier trucks to show that their introduction yields net energy, environmental and safety benefits [43].

To achieve a step-increase in vehicle loading it will be necessary for companies to collaborate and share vehicle capacity. This increases opportunities to increase the consolidation of outbound loads and eliminate empty backhauls. Recent bilateral collaborations by companies such as Kelloggs and Kimberly Clark [4] and Nestle and United Biscuits have yielded significant fuel and emission reductions. A major study is currently underway in the UK involving 27 large manufacturers and retailers in the grocery sector to assess the extent to which 'multi-lateral' collaboration can cut truck-kms, fuel consumption and related emissions by an even greater margin.

6 Increasing Energy Efficiency

Improvements in the energy efficiency of freight transport modes over the past few decades have been impressive. Over the past 40 years the average fuel efficiency of new trucks has increased at a rate of around 0.8–1% per annum [14]. The main improvements were made in the 1970s and 1980s. Since 1990 incremental fuel efficiency gains from the refinement of existing vehicle technology have been diminishing partly as a result of tightening controls on emissions of noxious gases, mainly of NOx and particulate matter. Cutting these emissions has, unfortunately, carried fuel and CO_2 penalties, bringing climate change and air quality objectives into conflict. Over the past 40 years the average fuel efficiency of commercial aircraft has risen by 70% [21], while, according to the Container Shipping Information Service [10] 'a container ship now typically emits about a quarter of the CO_2 it did in the 1970s as well as carrying up to ten times as many containers'.

The prevailing view is that opportunities exist to achieve further substantial improvements in the energy efficiency and emissions performance of new freight vehicles, aircraft and vessels over the next 20–30 years. Fuel efficiency improvements of 40% have been forecast for new trucks from a combination of technologies such as turbo-charging, improved aerodynamic profiling, hybridisation and idling control [22], a 20% improvement in the energy efficiency of rail-freight by 2050 [23], 35–45% reductions in the CO_2-intensity of new aircraft from engine and fuselage redesign by 2020 [1] and a new generation of 'super-eco' container ships is envisaged post-2030 emitting 69% less CO_2 per container than the average vessel afloat today [42].

Plans have also been prepared for a new generation of 'eco-warehouses', which will operate with much lower energy use and emissions [32]. The world's largest warehouse property developer Prologis, for example, has designed a prototype distribution centre which 'can achieve a 69 per cent reduction in operational carbon and energy compared with a typical UK warehouse' [11]. This dramatic reduction is achieved mainly by 'reducing the demand for heating through good insulation and airtight construction methods and reducing the need for artificial light by increasing the use of daylight supplemented by energy-efficient lighting systems'.

The diffusion and commercialisation of energy and emissions-reducing technology in the logistics sector can be slow processes, however, particularly where it involves fundamental redesign. The replacement cycles for ships, planes, railway locomotives and warehouses typically extend over 25 years or more—a long time relative to the 40 years over which we have to cut total GHG emissions by 50% to avoid catastrophic global warming. Compressing these cycles would not only be expensive; it would also temporarily increase energy consumption and emissions from the dismantling and construction of vehicles and buildings. Retrofitting energy-saving and emission-reduction devices to existing vehicles and logistics facilities can offer a quicker and more cost-effective pathway to sustainable logistics [5].

For technical improvements to be widely diffused and vehicle manufacturers to be incentivised to make them, operators will have to attach greater importance to energy efficiency and emissions in their vehicle purchasing decisions. The rates of technical advance and adoption can also be accelerated by legislation. The Japanese government, for instance, is currently pioneering fuel economy standards for new trucks as part of its 'top runner' programme. Truck manufacturers serving the Japanese market will have to raise the average fuel efficiency of their new vehicles from 6.30 kms/l in 2002 to 7.09 kms/l in 2015 [30]. The European Commission is planning to set fuel and emission standards for new vans and is considering doing the same for trucks.

Although new vehicle technologies tend to grab the headlines, potentially greater savings can accrue from improvements in the operation, loading and maintenance of freight vehicles. Companies can apply a wide range of fuel conservation measures, which collectively raise energy efficiency by a significant margin [3, 13]. One of the most cost-effective measures in the road freight sector is driver training, combined with telematic monitoring of driving behaviour and incentive schemes to embed 'eco-driving' principles in the workforce [47]. Reports of 6–10% improvements in the average fuel efficiency of drivers are quite common. The adoption of a tighter maintenance regime can ensure that any excess fuel consumption and emissions are minimised. A wide range of technical imperfections, such poor combustion, fuel leaks, under-inflated tyres and axle mis-alignment can prevent a truck from operating at optimum fuel efficiency.

Better management of existing warehouses can also yield substantial energy and emission savings. The Carbon Trust [7] has estimated that the energy consumed by the typical 'storage and distribution property' can be reduced, on average, by 28% (from 228 to 164 kWh/m^2) by applying good practice.

The energy efficiency of freight movement is constrained by the capacity and management of transport infrastructure. Freight trains generally share rail infrastructure with passenger trains and the latter invariably take precedence. Repeatedly stopping and starting freight trains to clear pathways for faster passenger trains is a very energy inefficient practice which erodes some of rail's environmental advantage. Increasing the capacity of the rail network and more effective train scheduling can help to ease this problem. In the air freight sector, air traffic management practices, affecting both the airborne routing of aircraft as well as their taxiing on the ground, significantly increases energy use and emissions. ACARE [1] estimates that 'between 13 and 15% of fuel is consumed through excessive holding either on-ground or in-flight and through indirect routing and non-optimal freight profiles' (p. 84). It has set a target of 5–10% fuel savings from 'radical changes to the air traffic management system'. The energy efficiency of moving freight by road is also seriously impaired by traffic congestion. For example, where a 40 tonne truck, running at an average speed of 50 km per hour, has, on average, to stop twice per kilometre because of congestion, total fuel consumption increases from about 28 to 84 l/100 km (VDA quoted in [31]). Better management of road space and the rescheduling of freight deliveries into off-peak periods can offer significant energy and emission savings. Research in the UK has found that, on the basis of 56 journeys altering the departure times of freight deliveries to avoid congestion on the trunk road network can reduce fuel consumption and CO_2 emissions by around 5–6% [44].

7 Reducing Emissions per Unit of Energy

Modifying parameters 3, 4, 5 and 6 in the ways outlined above will cut energy consumption per tonne-km and at the same time reduce energy-related emissions relative to freight movement. There are additional measures that can be applied, however, to further decouple these emissions from energy consumption. These broadly fall into two categories:

(i) Use of alternative forms of energy: at a national level, many governments have ambitious plans to decarbonise electricity generation by switching from coal,

gas and oil to renewables, nuclear power and the installation of carbon capture and storage systems on the remaining fossil-fuel power plants [9]. Assuming that these plans come fruition, the challenge for the logistics sector will be to access this low-carbon electricity. As noted earlier, it can be directly transmitted to electrically hauled freight trains, giving governments and railway companies an added incentive to electrify more of the rail network. It can also be used to recharge the batteries of electric vans and rigid trucks running on multiple-drop and collection rounds. Electric and diesel-electric hybrid power is likely to be confined to these smaller freight vehicles, leaving the heavier, long haul trucks dependent on internal combustion engines for the foreseeable future. These engines will be increasingly powered by those forms of biofuel that meet stringent environmental criteria and yield a net GHG saving on a well/field to wheel basis. The IEA [23] sees the switch to alternative forms of energy as being the main method of de-carbonising trucking operations by 2050, yielding a 27% reduction in GHG emissions against the baseline trend, by comparison with 24% from improved efficiency and 6% from modal shift.

Airfreight operations will also be denied access to low-carbon electricity as the energy density of batteries will continue to fall well short of that of kerosene. Several airlines have made successful trial flights with one or more engines powered by biofuel/kerosene blends though as Airbus [2] acknowledges 'no game-changing alternative to burning kerosene is foreseen in the short to medium-term' (p. 18).

Opportunities for 'repowering' the shipping fleet are also limited. Greater use can be made of wind and solar power on vessels, though this will be used mainly for ancillary equipment rather than propulsion. In the longer term new ships may derive some power from fuel cells. The '*current assumption*', however, is that '*ships will continue to burn fossil fuels for the foreseeable future*' [24]. The nature of this fossil fuel will have to change, however, for shipping to meet tightening restrictions on sulphur emissions. At around 2.7% the sulphur content of marine bunker fuel is much higher than that of the fuels used by other freight transport modes [25]. Bringing it down 0.5% as the ICCT proposes, would require a huge investment in new refining capacity around the world, significantly inflating the average cost of marine fuel and hence probably restraining the growth in demand for maritime transport. It will also, however, increase carbon emissions from the refining process.

Warehousing and terminal operations run mainly on electricity and so should directly benefit from the decarbonisation of electricity generation. It is also pos-sible to install wind turbines and solar panels to permit 'micro-generation' of renewable energy locally where weather conditions permit and 'feed-in' tariffs offered by operators of the national electricity grids are financially attractive. Ports have become a favoured location for wind-turbine development and two British ports aim to become carbon–neutral partly by tapping local wind power. Temperature-controlled warehouses (and vehicles) present a particular decarbon-isation challenge as the refrigerant gases they use can have a global warming potential thousands of times higher than that of CO_2. HFC123, for example, has a global warming effect 11,700 times greater [12]. Strenuous efforts must be made, therefore, to minimise the leakage of these gases from refrigeration systems across the supply chain.

(ii) Redesign of vehicle engine and exhaust systems: the engines propelling the movement of freight by road, rail, sea and air have been substantially redesigned over the past 20 years to meet tightening controls on the emissions mainly of NOx and particulate matter (PM10). In the road freight sector, where reductions, respectively, of 87 and 95% in NOx and PM10 emissions have been achieved since the early 1990 s for new vehicles (to meet the Euro 5 emission standard), future 'cleaning' of diesel engines is likely to be marginal, particularly as greater priority will be given to the minimisation of fuel consumption and CO_2 emissions. Engine technology in rail, aviation and shipping can be further refined to cut emissions of noxious gases, though as in road transport a difficult environmental balance must be struck between cutting emissions of NOx and CO_2. Significant advances have also been made in the use of traps to filter emissions of particulate matter from diesel exhausts, with trucks again ahead of the other freight modes in the application of this technology. As in the case of the fuel economy innovations discussed earlier, the uptake of these emission-reduction technologies usually occurs in line with the vehicle replacement cycle which in the rail, shipping and aviation tends to be relatively long.

8 Wider Effects of Climate Change Mitigation and Adaptation

Forecasts of the future demand for logistics services appear to under-estimate the effects of climate mitigation measures in other sectors of the economy and the need to adapt our built environment to a climate changed world:

8.1 Decarbonisation of Other Sectors

Logistical activities pervade all aspects of the economy and society. They will therefore be affected by wider efforts by governments, companies and individuals to decarbonise. This makes it extremely difficulty to model both the potential magnitude and cost-effectiveness of GHG-abatement in the logistics sector. As explained earlier, critical to the creation of low-carbon economies will be the decarbonisation of electricity generation. The resulting transformation of the national energy mix is likely to prove freight transport-intensive during the construction phase. For example, wind power requires the installation of much more steel and concrete per megawatt of power generated than coal-fired, natural gas or nuclear power stations [45]. On the other hand, once they are installed wind turbines can remove the need to move large quantities of fossil fuel. If, for example, sufficient wind-turbine capacity was installed in the EU27 by 2020 to generate 180 gigaWatts of power (enough to meet the needs of 107 million households), around 28 million tonnes of oil could be removed from the fuel supply chain annually [17]. The substitution of fossil fuel by biofuel, batteries and, possibly, hydrogen, will also require complete restructuring of the energy supply chain. Bonilla and Whittaker [6] have shown that anticipated expansion of bioenergy production and distribution in the UK will have major implications for the freight transport system. New global supply chains will have to be established to transport large quantities of lithium from the major sources of this mineral in countries such as Bolivia and China to battery manufacturing plants around the world.

The expansion of rail-based public transport systems to divert personal travel from aircraft and private cars will also require a high level of logistical support. So too, for example, will steep increases in the insulation of homes, commercial premises and public buildings and the creation of new value chains for the batteries required to 'electrify' much of the car, van and rigid truck fleet. Ironically the freight transport intensity of national economies may rise temporarily as governments strive to decarbonise the built environment. There is a danger that we could get into a logistical 'vicious circle' with our efforts to cut GHG emissions, particularly on capital projects, actually generating a net increase in emissions, at least in the short to medium term [40].

8.2 Adapting to Climate Change

Given atmospheric and ecological time lags there is a significant amount of climate change 'in the pipeline' which we cannot avert and to which we will have to adapt our built environment. Assessments of the vulnerability of infrastructure [29, 48] and settlements to climate impacts, indicate that over the next few decades huge investment will be required in the 'climate-proofing' of existing buildings, flood protection schemes, the realignment of road and rail networks and the relocation of population and economic activity. Vast quantities of construction materials will have to be moved greatly increasing the demand for logistics services.

Changes in temperature regimes, water availability and disease will cause significant shifts in agricultural zones. Supply chains will have to be reconfigured to this new geography of food production and distribution, in some cases becoming more transport-intensive, while in others permitting more localised sourcing of agricultural produce [26].

Should current efforts to cut GHG emissions drastically over the next few decades fail or prove inadequate, it may be necessary to resort to the much more radical, geo-engineering options, such as fertilising the oceans with nutrients to promote 'algal blooms' which would absorb more CO_2, enhancing the weathering of silicate rocks and dispersing aerosols, mainly of sulphur dioxide, in the stratosphere. These geo-engineering measures would have to be applied on a planetary scale to exert enough influence on global climate. Little research has so far been done on the logistical support required to implement these geo-engineering measures. Clearly they would entail the movement of vast amounts of material by land, sea and air. Order of magnitude estimates have been made of the quantities of material that would have to be distributed for some of the options. For example, the enhanced weathering option would entail mining and grinding huge quantities of rock and dispersing it widely over fields. A study by the Royal Society [46] has estimated that 'a volume of about 7 km^3 per year (approximately twice the current rate of coal mining) of such ground silicate minerals, reacting each year with CO_2, would remove as much CO_2 as we are currently emitting' (p. 14).

Efforts to reduce the overall energy demands and environmental impact of logistics may therefore be frustrated by wider climate change mitigation and adaptation measures. As many of these measures will be pivotal to future climate change policy, logistics may require some dispensation in the setting of sectoral-level targets for future GHG reduction.

9 Conclusions

Given the strong growth in demand for logistics services it is going to be difficult to cut energy consumption and emissions in this sector in absolute terms. There are numerous ways in which the energy- and emissions-intensity of logistics can be reduced and in combination they may be able to offset the overall increase in logistics activity. It is unlikely, however, that the logistics sector can achieve a 50% reduction in CO_2 emissions by 2050, in line with global targets, while accommodating the forecast growth in freight movement over this period. The additional demands imposed on logistics by the decarbonisation of other sectors of the economy and adaptation of the built environment to climate change, which are not factored into current forecasts, will make this even less likely.

This paper has reviewed energy- and emission-reducing measures within a systematic framework built around seven key parameters. Many of these measures will cut energy costs while reducing emissions, generating streams of economic and environment benefit. In energy and emissions terms, there is still a good deal of 'low hanging fruit' in this sector, comprising measures that are relatively easy and quick to implement, require little capital investment and offer short payback periods. Future increases in energy prices which increase the rate of return on these investments. The fundamental redesign of freight vehicles which will be required to achieve a step-reduction in energy use and CO_2 emissions will, however, take much longer, particularly in the rail, maritime and aviation modes where vehicle replacement cycles span decades. It will also be difficult for large parts of the logistics sector to take advantage of the decarbonisation of electricity generation. Given their continued dependence on liquid fuel, much of it from fossil sources, strong emphasis will need to be placed on operational and fuel efficiency, mode shift and the reconfiguration of supply chains. More radical proposals to cut freight-related energy use and emissions by forcing a return to localised sourcing and decentralised logistics need to be appraised on a full life cycle basis. After all, minimising the amount of freight movement will not necessarily minimise energy use and emissions across the global economy. By comparison with geographical variations in the energy- and emission-intensities of other economic activities the energy and emissions footprint of freight transport is small. In absolute terms, however, it is still significant and growing worldwide at a rapid rate.

References

1. ACARE (2008) The challenge of air transport system efficiency. Brussels
2. Airbus (2008) Flying by nature: global market forecast 2007–2026. Paris
3. Ang-Olson J, Schroeer W (2002) Energy efficiency strategies for freight trucking: potential impact on fuel use and greenhouse gas emissions. Paper presented to the 81st annual meeting of the transportation research board, Washington, DC
4. Anon (2008) Collaboration brings savings for Kelloggs and Kimberly-Clark. *Logistics Manager*, 13th Oct
5. Baker H, Cornwell R, Koehler E, Patterson J (2009) Review of low carbon technologies for heavy goods vehicles. Ricardo report for Dept for Transport, London
6. Bonilla D, Whittaker C (2009) Freight transport and deployment of bioenergy in the UK. Working paper 1043, Transport Studies Unit. Oxford University, Oxford
7. Carbon Trust (2000) The designer's guide to energy-efficient buildings for industry. GPG 303. HMSO, London
8. CE Delft (2008) Internalisation measures and policy for the external cost of transport. Delft
9. Committee on Climate Change (2008) Building a low carbon economy. London
10. Container Ship Information Service (2009) Environment.http://www.shipsandboxes.com/eng/keytopics/environment/
11. Dalton, M (2009) Take the green route out of the red. Supply Chain Standard, May
12. DEFRA (2007) Guidelines to Defra's GHG conversion factors for company reporting. London
13. Department for Transport (2006) Fuel management guide. London
14. Duleep KG (2007) Fuel economy of heavy duty trucks in the USA. Presentation to the IEA workshop on 'Fuel efficiency for HDV standards and other policy instruments' Paris
15. European Commission (2011) Roadmap to a single European transport area—towards a competitive and resource efficient transport system. White Paper, Brussels
16. European Environment Agency, Transport Research Laboratory and University of Oxford (2009) Transport scenarios scoping report. Copenhagen
17. European Wind Energy Association (2008) Wind energy: the facts. Brussels
18. Eurostat (2009) Panorama of transport. Luxembourg
19. Eurostat (2011) http://epp.eurostat.ec.europa.eu/portal/page/portal/transport/data/main_tables
20. German Environment Ministry (2007) Longer and heavier on German roads: do megatrucks contribute towards sustainable transport? Berlin
21. ICAO (2007) ICAO environmental report 2007. Montreal
22. IEA (2008) Energy technology perspectives. Paris
23. IEA (2009) Transport, energy and CO_2: moving towards sustainability. Paris
24. International Chamber of Shipping (2010) Shipping, world trade and the reduction of CO_2 emissions. London
25. International Council on Clean Transport (2007) Air pollution and greenhouse gas emissions from ocean-going ships: impacts, mitigation options and opportunities for managing growth. Washington

26. International Food Policy Research Institute (2009) Climate change: impact on agriculture and costs of adaptation. Washington
27. Inter-governmental Panel on Climate Change (2007) Climate change 2007: mitigation of climate change. Cambridge University Press, Cambridge
28. Kearney AT (1997) The efficient unit loads report. ECR Europe, Brussels
29. Koetse MJ, Rietveld P (2009) The impact of climate change and weather on transport: an overview of empirical findings. Transp Res Part D 14:205–221
30. Konuma N (2007) Japanese current activities for fuel efficiency improvement: top-runner standard. IEA workshop on 'Fuel efficiency for HDV standards and other policy instruments', Paris
31. Larsson S (2008) Data and indicators for road freight transport. Presentation to the IEA workshop on 'New energy indicators for transport', 28–29 January, Paris
32. Marchant C (2010) Green warehousing. In: McKinnon AC, Cullinane S, Browne M, Whiteing A (eds) Green logistics: improving the environmental sustainability of logistics. Kogan Page, London
33. McKinnon AC (2005) The economic and environmental benefits of increasing maximum truck weight: the British experience. Transp Res Part D 10(1):77–95
34. McKinnon AC (2007) The decoupling of road freight transport and economic growth trends in the UK: an exploratory analysis. Transp Rev 27(1):37–64
35. McKinnon AC (2009) Benchmarking road freight transport: review of a government-sponsored programme. Benchmarking: an International Journal 16:5
36. McKinnon AC (2010a) Environmental sustainability: a new priority for logistics managers. In: McKinnon AC, Cullinane S, Browne M, Whiteing A (eds) Green logistics: improving the environmental sustainability of logistics. Kogan Page, London
37. McKinnon AC (2010b) European freight statistics: limitations, misinterpretations and aspirations. ACEA, Brussels
38. McKinnon AC (2010) Optimising the road freight transport system. In: Waters D (ed) Global logistics, 6th edn. Kogan Page, London
39. McKinnon AC, Ge Y (2006) The potential for reducing empty running by trucks: a retrospective analysis. Int J Phys Distrib Logist Manag 36:5
40. McKinnon AC, Kreie A (2010) Adaptive logistics: preparing logistical systems for climate change. In: Whiteing A (ed) Proceedings of the logistics research network annual conference. University of Leeds, Leeds
41. McKinnon AC, Piecyk MI (2010) Measuring and managing CO_2 emissions of European chemical transport. Cefic, Brussels. www.cefic.org
42. NYK (2011) NYK Super Eco Ship 2030. http://www.nyk.com/english/csr/envi/ecoship.htm
43. OECD/International Transport Forum (2010) Moving freight with better trucks. OECD, Paris
44. Palmer A, Piecyk M (2010) Time, cost and CO_2 effects of rescheduling freight deliveries. In: Whiteing A (ed) Proceedings of the logistics research network annual conference. University of Leeds, Leeds
45. Petersen PF (2006) Current and future activities for nuclear energy in the United States. Presentation to CITRIS Research Exchange, University of Berkeley, Oakland
46. Royal Society (2009) Geoengineering the climate: science, governance and uncertainty. London
47. SAFED (2008) www.safed.org.uk/Presentations/SAFED_summary%20document.doc
48. Transportation Research Board (2008) Potential impacts of climate change on us transportation. TRB special report 290, TRB, Washington, DC
49. Vierth et al (2008) The effects of long and heavy trucks on the transport system VTI.McKinnon AC, Kreie A (2010) Swedish National Road and Transport Research Institute, Linkoping
50. World Economic Forum and Accenture (2009) Supply chain decarbonisation: the role of logistics and transport in reducing supply chain carbon emissions. Geneva

Low Carbon Ships and Shipping

T. W. P. Smith

Abstract This chapter discusses the scope for increasing the energy efficiency and decreasing the carbon (and other GHG) emissions of ships and shipping. An overview of the fundamentals of the shipping industry is presented (why does shipping exist, how does demand develop, where do ships go, how is shipping structured, what are its impacts) to provide context for further detail on the energy efficiency, regulatory options and technology options that could be employed to help transition shipping to a low carbon future. Two specific examples of trends which could be important to this transition (an increase in ship size and a decrease in speed) are then analysed in greater detail to reveal their potential and also some of the practical implementation issues that will need to be considered.

1 Context

1.1 Why Does Shipping Exist?

International trade has existed for centuries, including the transport of specialist goods and commodities for long distances at sea (e.g. cotton, wool, spices and other agricultural products). In some instances, that trade was driven by the demand for a good that cannot be sourced locally (e.g. spices due to global climate and agricultural productivity variations). Whilst this remains an explanation for the existing patterns of trade (particularly for extracted commodities which for geological reasons are distributed unevenly around the world and not necessarily

T. W. P. Smith (✉)
UCL Energy Institute, Central House, 14 Upper Woburn Place, London, WC1H 0NN, UK
e-mail: tristan.smith@ucl.ac.uk

O. Inderwildi and Sir David King (eds.), *Energy, Transport, & the Environment*,
DOI: 10.1007/978-1-4471-2717-8_30, © Springer-Verlag London 2012

collocated with their consumption location), a more recent trend in demand growth has been one of the last 50 years' most notable economic phenomena— Globalisation.

Globalisation refers to the increasingly global nature of the market in which goods and services are produced and consumed. The principle can be summarised, for a good consumed in country "b", by Eq. (1):

$$CP_a + CT_{ab} \leq CP_b, \tag{1}$$

where CP is the cost of production and CT is the cost of transport. Providing that the combined cost of production and cost of transport from country "a" to "b" is less than the cost of local production (in "b") then it is economically viable to locate production in country "a". Wage and skill differentials between countries can be one explanation for a geographical differential in cost of production, however these have always existed. What has not always existed is a supply of safe, reliable and cheap transport connecting the world's consumers with the world's raw materials and skilled, low cost labour markets.

That connection for the majority of the world's trade is the shipping industry, and as the industry and global transport infrastructure has developed and increased its efficiency, distance to market as a parameter that identifies the competitiveness of many goods and commodities, has reduced in significance.

1.2 How Does Demand Develop?

Demand, T_d, can be considered as the product of the quantity transported, Q_d, and the distance that it is transported, D_d.

$$T_d = Q_d.D_d \tag{2}$$

In the case of maritime freight demand, mass is commonly used as the quantity parameter, in the case of passenger demand number of passengers is commonly used as the quantity parameter. Therefore the development of demand can be thought about as the development of each of those parameters in isolation or their development in combination.

Figure 1 shows the historical trend in the transport demand (in units of billion tonne miles: 1e9 × mass in tonnes × distance in nautical miles) for some of the main commodity groups of the shipping industry. The graph shows that, at least for global aggregations of flows of commodities, there is a high level of correlation between GDP and transport demand. However, since trade growth is a contributor to GDP growth this should not be construed as GDP growth causing trade and therefore transport demand. Many of the commodities that are moved by ship (bulky, low value commodities) are those associated with heavy manufacturing and infrastructure development (e.g. coal, iron ore) and so if global maritime transport demand is disaggregated between the developed and developing world,

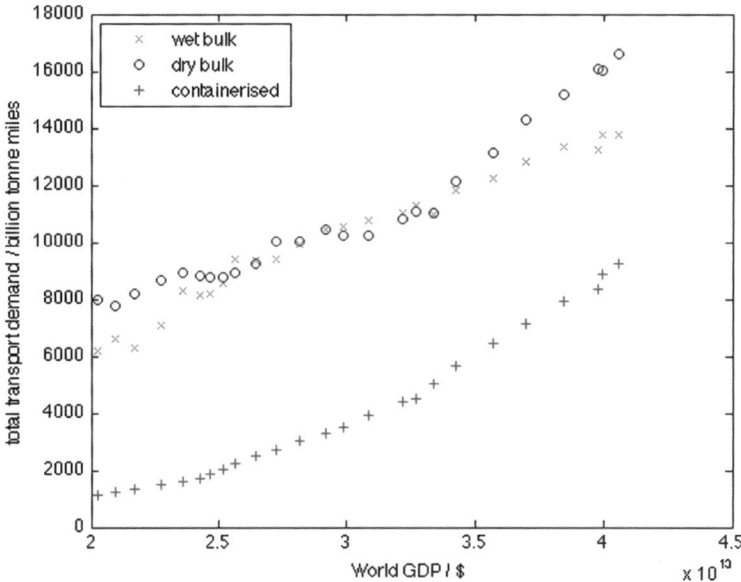

Fig. 1 Relationship between world GDP (1985–2009) and transport demand for different marine transport commodities. *Source* UNCTAD

Table 1 Transport demand in 2050

SRES scenario	A1B	A2	B1	B2
Ocean-going shipping	320	240	220	180
Coastwise shipping	320	270	220	220
Container	1230	960	850	690
Average (all ships)	540	421	372	302

these correlations may not be observed for developed nations—particularly those with economies that have diversified away from significant contribution to GDP from their manufacturing industry.

Despite this philosophical complication and the 'shakiness' of making extrapolations from historical trends to estimate what might happen in the future, some transport demand scenarios are developed using correlation with GDP forecasts (e.g. those available for IPCC scenarios, SRES). This method can produce dramatic growth rates and so others have tried to apply resource constraints or wider mitigation scenarios (e.g. finite oil), for example [22, 2]. Using the latter approach, Buhaug et al. estimate the transport demand in 2050, as a function of different SRES scenarios to be represented by the values in Table 1. The values are indexed to the 2007 transport demand (100). The range of values shows an expectation that total maritime transport demand will increase by between a factor of 3 and 5.4 over the next 40 years.

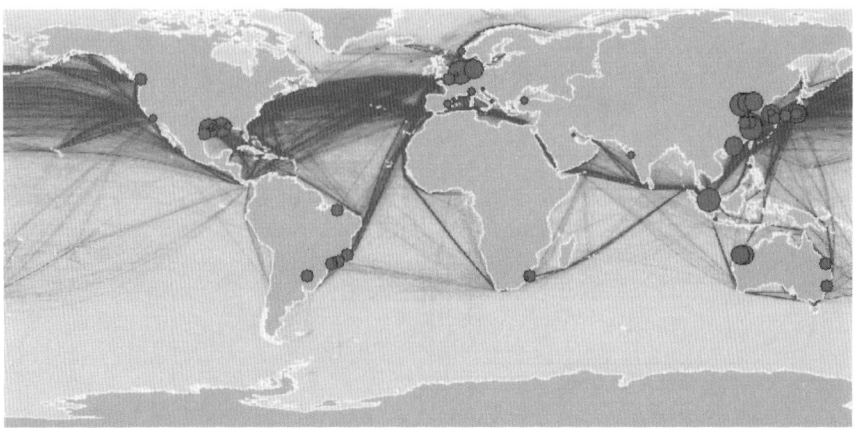

Fig. 2 Trade routes and port volumes

1.3 Where Do Ships Go?

In order to understand maritime transport demand at a finer granularity than
global trends for individual commodities requires an understanding of trade flows
and trade routes. Trade routes for ships are an ever-evolving pattern depicting
the geographical distribution of origins and destinations attributable to global
production and consumption. Wherever a new route appears to be profitable a
service, and possibly competition, will establish itself. Wherever a flow of trade
dries up or reconfigures, services are abandoned and ships redeployed. Figure 2
displays two sources of data:

- Ship locations (source: AAPA), and
- Port Volumes (source NCEAS).

The ship location data (longitude and latitude) is reported from a sample of
the global fleet of ships and the volume data for the flow of trade through a port
(with the magnitude of the flow in tonnes represented by the radius of the circle).
The image therefore displays, qualitatively, both the global routes that are of most
significance (through the intensity of the ship location data) and the dominant
origins and destinations of global trade flows.

Two particular routes have specific constraints on the size of ships:

- Panama Canal (constained to Panamax size ships, connecting the Atlantic and
  Pacific).
- Suez Canal (constrained to Suezmax ships, connecting the Persian Gulf and
  Indian Ocean to the Mediterranean).

Therefore some trade flows have the choice between movement on longer
unconstrained routes or shorter constrained routes.

As conveyed in Eq. 2, the distance that freight and passengers travel is a key parameter in transport demand and two potential developments could create reductions in this distance, sometimes referred to as the average haul:

- Expansion of the Panama Canal (expected 2014)—This will increase the capacity and maximum ship size potentially leading to trade being routed through the canal that had previously been carried greater distances on larger ships.
- Opening of the NE and NW passage (unknown)—Dramatic reduction in ice cover in the Arctic in recent years has led to speculation that the two major routes across the arctic may become economically viable and significant trade routes. The routes offer distance savings of approximately 25–50% for certain trade flows [5].

But beyond this conceptualisation of shipping as a process of moving freight and passengers on a series of origin–destination routes around the globe, it is also important to transport demand that a number of "styles" of shipping operation can exist, which explain why for some commodities the route travelled to get from origin to destination is not necessarily the shortest:

Hub and spoke—this characterises parts of the container shipping industry where the large variation in the destinations of a ship's cargo and the ease of handling of the units of freight (containers) facilitates multi-ship movement from origin to ultimate destination port—also called trans-shipment. This in turn enables larger more cost-efficient ships to carry out the long distance movement and smaller more versatile ships to carry out collection and delivery from/to the origin and destination. It should also be noted that transshipment can occur in the movement of bulk goods too, either facilitated by a port or terminal, or through ship to ship transfer whilst at sea.

Multi pick-up, multi-drop off—this characterises parts of the bulk industry, where the comparatively low value and low urgency (in terms of time to market) associated with a commodity, justifies the need to minimise transport cost through use of a large ship that calls at a number of origin ports in one region and a number of destination ports in another region. This fulfils a number of origin to destination flows at approximately the cost of one large ship's long distance voyage.

For some freight, the destination is unknown at the start of the journey or is changed en route. This can happen due to arbitrage after departure from the origin—the instantaneous buying and selling of part or all of a ship's cargo, because of cargo value volatility and geographical variability. The extent and impact of this phenomenon on transport demand is not known, but for some types of freight (e.g. crude oil and oil products) it is believed to be significant.

Just as in the maritime transport hub and spoke concept described above, ports themselves are transport hubs with intermodal links (road, rail etc.) connecting the maritime transport mode to ultimate destinations. For this reason, routing decisions for ships are as susceptible to the economics, reliability and capacity of those other transport modes as they are to the trade-offs of these variables between different ships and carriers. This has led to research that considers shipping in a supply chain perspective, inclusive of other transport modes and the inter-operation between modes [25].

1.4 How is Shipping Structured?

The shipping industry is made up of a diverse mix of sectors that differ in respect to: market structure, type of ship used and its requirements, port and infrastructure requirements, national and international scope and the nature of the competition (other ships or other transport modes) etc. Some level of disaggregation and categorisation of the shipping industry is provided by Table 2.

Each sector in Table 2 can also be decomposed to different subsectors, which may each have their independent markets or may behave in substitution with each other. In many cases those subsectors are organised according to vessel size (for example the wet bulk tanker sector, with approximate deadweight (payload capacity) ranges in tonnes):

- Handysize—30,000–60,000
- Panamax—60,000–80,000
- Aframax—80,000–120,000
- Suezmax—120,000–200,000
- Very Large Crude Carrier (VLCC)—200,000–320,000
- Ultra Large Crude Carrier (ULCC)—320,000–550,000.

As will be shown in Fig. 4, ship size has an important influence on energy efficiency and so understanding the demand and growth of each of the subsectors is important for understanding the potential growth of the industry's carbon emissions.

There are two dominant styles of operation of shipping, liner and spot. Within each of those styles of operation, a number of market structures exist which characterise the responsibilities of ownership, operation responsibility and components of cost carried. Liner shipping is dominated by owner operated and time-chartered vessels and the spot market is dominated by voyage charter and time-chartered vessels [27].

A particular relevance to carbon emissions and energy efficiency is that the party that carries the cost of fuel and who therefore takes the risk on fuel price and the benefit of any intervention on energy efficiency differs depending on the charter/owner arrangement. As discussed in Rehmatulla [23], this has the potential to create a split incentive, and act as a barrier to emissions reduction in the shipping industry.

1.5 What are its Impacts?

The environmental impacts of the shipping industry can be divided between the impacts due to ships and those due to the infrastructure (e.g. ports). For the purposes of this discussion the "system boundary" for impacts is defined around the ships themselves, as ports will normally obtain their energy from the

Table 2 Breakdown of the different sectors and subsectors of the shipping industry (*source* NEA)

Personal travel/ tourism	Trade										
Passenger flows	Freight flows										
	Short sea					Deep sea					
	Unitised		Nonunitised			Unitised	Nonunitised			Other	
Cruise	Ferry	Ferry	Container	Dry	Wet	General cargo	Container	Dry	Wet	General cargo	Car carriers	Reefer ships

Table 3 Major impacts of shipping industry

Impact type	Origins	Mitigation options
Exhaust emissions (climate)—GHG, SOx, NOx and PM (black carbon)	Fuel and lubrication products	Scrubber/capture technology, energy efficiency, fuel substitution
Exhaust emissions (health)—SOx, NOx, PM	Fuel and lubrication products	
Ozone and VOC emissions	From equipment installed on board and cargo venting	Technology substitution and VOC capture
Leaching	Chemicals in anti-fouling underwater coatings	Coating substitution
Waste (sewage/grey, oily water)	Ejected	On board storage or treatment
Nonindigenous species invasion	Transportation of species from one ecosystem to another either attached to hull or in ballast water	Ballast water treatment, hull scrubbing and anti-fouling

infrastructure of their host nation so cannot easily be considered in isolation. For ships, some of the main environmental impacts are listed in Table 3.

Many of ship's environmental impacts can be mitigated at comparatively low cost, and are the subject of command and control regulation (or ongoing attempts to introduce regulation) to ensure that this is done. Whilst shipping has experience in controlling environmental impact, the focus of this discussion (exhaust GHG emissions) has not yet been brought under international, regional or local regulation. As in many other transport modes, this is because of a lack of economically viable preventions (substitution fuels and technologies) and cures (capture or exhaust treatments).

The climate impacts of shipping are complicated by the fact that some of the exhaust products have a warming impact (e.g. GHG) and some have a cooling impact (e.g. SOx). This has led some to describe the net impact of shipping as a global cooling effect [13], however regulations on the sulphur content of fuels and other health impacting pollutants is expected to alter this balance in future decades to net warming. The climate impacts of shipping's emissions have different impacts over time—the climate cooling impacts (due to sulphur) are more short-lived than for example the CO_2 impacts, and so some [3] have argued that taking a longer (and more relevant to climate change) view, shipping is net warming.

A counter argument, sometimes deployed by the industry, to appeal against punitive regulation of the climate impacts is that it is the 'least bad' transport mode. That is to say that for a unit of transport supply (for example movement of 1 tonne of goods 1 km), a ship requires less energy and has lower emissions than any alternative mode (aviation, road, rail). However, it is possible that over time, some of those other transport modes (e.g. rail) may, due to their technology and infrastructure characteristics, be able to decarbonise at low cost, to the extent that shipping is no longer the "least bad" transport mode. Indeed, this may already be the case for some short sea and coastal voyages.

Seen from a wider perspective than a competition between transport modes, transport (and therefore by association shipping) is socially beneficial (e.g. because it enables the provision of goods at least cost, facilitates employment and provides opportunity for economic development). These benefits, can be thought of as positive externalities that are not currently internalised. If we assume a relationship between transport cost and transport demand (whereby higher costs due to environmental cost internalisation reduce demand) then this raises the question:

"Is the social disbenefit of curtailment of transport demand warranted by the social benefit obtained by internalising shipping's negative externality?"

The answer to this question is important and nontrivial as there is uncertainty associated with many of the sub-questions:

1. What is the extent of the social benefit that transport provides?
2. How are transport cost and demand related?
3. What is the size and impact of the externality?
4. What is the cost of internalising the externality?

Unfortunately, the urgency implicit in the trajectory of global emissions mitigation that has been proposed by the UNFCCC, IPCC and others, for the avoidance of dangerous climate change, requires that decisions and at least the establishment of a process, be initiated in spite of this uncertainty. Many (industry stakeholders, commentators, NGOs and regulators) have already formed their opinion on what shipping should do—it must put in place measures to control its GHG emissions.

1.6 The Existing Fleet and Fundamentals of Ship's Energy Efficiency

Ships have evolved their efficiency (both energy and economic efficiency) over time. Today's VLCCs can carry 300,000 tonnes of crude oil at a speed of 15 knots (travelling approximately 25,000 nm on the 7,000 tonnes Heavy Fuel Oil capacity). Nevertheless, globally, shipping was estimated to be responsible for 3.3% of anthropogenic CO_2 emissions in 2007 [2]. Figure 3 displays the breakdown of

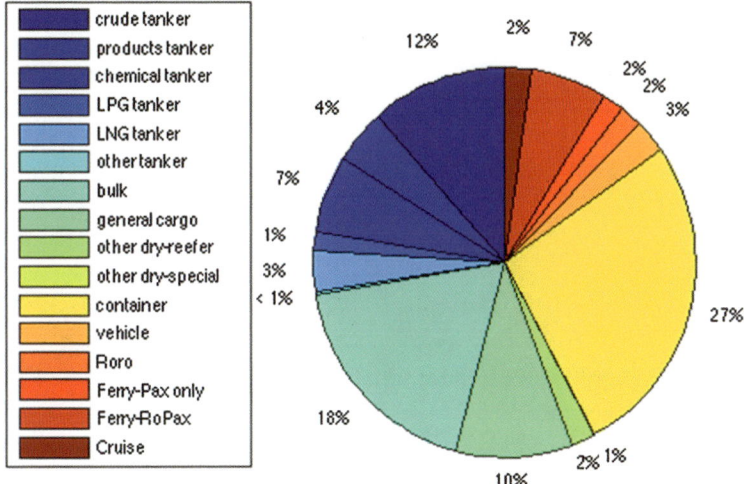

Fig. 3 estimated shares of emissions (totalling 977 million tonnes of CO$_2$) of the highest emitting sectors of the shipping industry, data from IMO [17]

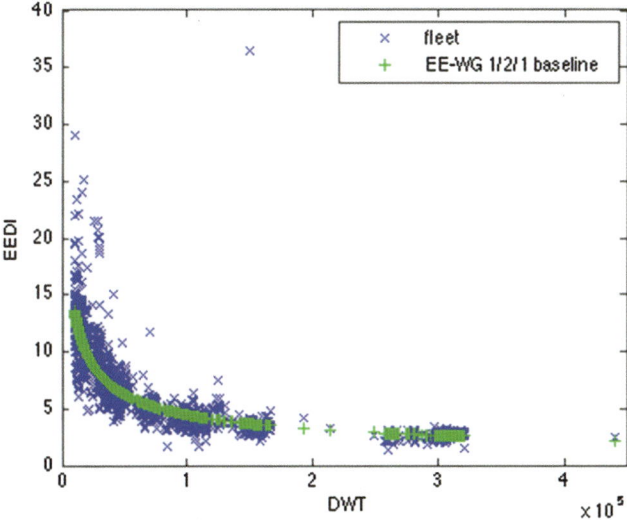

Fig. 4 Energy efficiency design index calculations for the tanker sector. *Source* Clarksons World Fleet Register

emissions for the dominant sectors of the shipping industry, estimated in 2009 to be responsible for a total of 977 million tonnes of CO$_2$.

Ships of a certain size and type are heterogenous in their energy efficiency, due to variations in design speed, hullform and machinery. A depiction of that heterogeneity is given in Fig. 4 which shows the scatter of energy efficiency (calculated according to Eq. 6) for 10,948 ships in the tanker or wetbulk sector. Individual ship's energy efficiencies are indicated with an 'x'. This can be compared with data points that correspond to a regression fit through the data, calculated according to the method detailed in IMO [18].

It is useful to understand some of the fundamental relationships important to a ship's energy efficiency. The Admiralty Formula, (3), can be used in ship design to obtain the first estimate of the propulsive power required to move a vessel of displacement Δ te at speed V kts.

$$P = \frac{\Delta^{2/3} V^3}{C} \tag{3}$$

C is a constant specific to ship type (a reflection of hull form and dimensions). From this formula, it can be seen that for a given size and type of ship, speed and power are non-linearly related—a small reduction in speed corresponds to a greater (in proportion) reduction in propulsive power. The output or 'transport supply' S that a ship provides in a specific time period t is commonly quantified in tnm or tkm—the product of payload carried and distance carried (Eq. 2). In a time period of 1 h, this transport supply can be expressed as (4).

$$S = dwtVU, \tag{4}$$

where dwt is the deadweight of the ship (approximately the payload capacity) and U is a factor expressing the ships utilisation.

A quantification of energy consumption pertinent to ships is their fuel consumption. This is commonly derived as the product of the power of the vessel and the specific fuel consumption sfc which is in units of kg/kWhr. If the energy efficiency, E (units of kg (fuel)/tnm supply), of a ship is the amount of energy required to produce a unit of output, then taking the definitions obtained from Eqs. (3) and (4) above, this can be expressed as:

$$E = \frac{sfc\Delta^{2/3} V^2}{CdwtU}. \tag{5}$$

This relationship is of a structure common to the IMO's EEDI and EEOI formulae, and behind the common assumption that slower ships are more energy efficient than faster ones.

In practice, however, ship design is the considered trade off between a number of objectives, one of which is energy efficiency at design speed and so a pure efficiency/speed trade off is rarely executed. However, as voyage costs (fuel costs) are a significant portion of total operating costs of most ship types, optimisation for a design condition (clean hull and still water) is commonplace. This optimization means that:

- Ships of the same size/type (e.g. Panamax containers ships) that are designed for different speeds will not have the same propulsion, machinery or hullform.
- The same ship operated away from its design point, will to some extent (through propulsion, machinery and hullform) be operating in an off-design condition.

The consequence of being in an off-design condition is a deterioration of performance, which can be thought of as a failure to achieve the energy efficiency that is theoretically attainable from Eq. (5). The further away from the design point, the greater the compromise in efficiency. Consequently, Eq. (5) should only be considered as an *upper bound* on energy efficiency.

1.7 The IMO, Regional and National Regulation

The shipping industry is controlled at a number of levels. At the highest level, the International Maritime Organisation (IMO), a UN agency, has the responsibility of developing and maintaining regulation, according to the consensus of its member states, that is implemented through those member states' jurisdictions. At a more regional level, the EU is also able, with the support of its member states, to develop regulations that will impinge on ships, and any state (or region within a state), is able to develop and apply rules that impinge on a ship.

A number of limitations constrain unilateral action by member states from being particularly effective, including:

- Transport and trade flows are versatile and can re-route to avoid significant cost-imposing regulation, losing business for local ports and services
- If a state feels that a regulation is causing distortion in the competitiveness of trade with another state, it can bring a challenge under the World Trade Organisation (WTO)
- Shipping and regulation of shipping must not contravene the United Nations Law of the Seas (UNCLOS), which prevents individual territories from making regulation that extends extraterritorially.

For some of these reasons, no significant unilateral action has yet been taken for the control of shipping's GHG emissions (although it is under discussion within the EU) and the focus here is on measures under development at the IMO, which could apply global regulation to the shipping industry. Two concepts are attracting attention at present and deserve further discussion:

- Energy efficiency design index (EEDI),
- Market-based measures (MBM).

EEDI is a quantification of a ship's CO_2 intensity (the amount of CO_2 emitted by a unit of transport work). It is not an energy efficiency index, despite its name, as the measurement is of CO_2 emitted not work performed per unit of energy consumed, however for existing technology on ships, this is a reasonable proxy for energy intensity. A simplified version of the formula is given by:

$$EEDI = \frac{C_f sfc P_d}{dwt V_d},$$
(6)

where C_f is the carbon factor of the fuel, P_d is the power output of the main propulsion machinery at the design speed and V_d is the design speed. Further terms exist to include the auxiliary machinery and specific design details of a vessel. The power law regression fit through data calculated with the formula (shown in Fig. 4) can be thought of as a "fleet average efficiency baseline", and the objective of an IMO regulation is to impose a minimum efficiency requirement (defined relative to the baseline) for newbuild ships. Over time, the stringency of the minimum efficiency could be increased to enable a gradual phasing in of a new generation of increasingly energy efficient ships. The proposal at the time of publication is to have stringency increments every 5 years (2015, 2020, 2025), with a stringency, by 2025 that all newbuild ships must be 30% more efficient than the baseline.

The majority of ships are in service for 30 years before it becomes economically more attractive for them to be scrapped [19]. Therefore, the global fleet or 'stock' has significant inertia in terms of its characteristics, performance and energy efficiency, that due to a comparatively slow turnover will take many years to change. This is one reason why many see an EEDI regulation as an important first step in the shipping industry. However, there is concern that the regulation will result in little measurable reduction of shipping's actual CO_2 emissions because there is nothing in the currently proposed regulation to:

- Influence the way a ship is operated and
- Control the condition of the ship after it has been built.

Some commentators [9] have suggested that EEDI is "…absurd…" because they suggest that the theory and corresponding regulation do not equate to a rational scientific basis for energy efficiency. Others [16] have shown how existing designs can be modified without changing fundamentals such as design speed or technology, to achieve significant gains in EEDI over a baseline ship.

In parallel to the discussions on EEDI, there has also been a proposal to consider the development of Market Based Measures (MBMs) at IMO that could create economic incentives by pricing the negative externality (in this instance CO_2) into the operation of a ship. The three main groupings of the proposals are:

- Emissions trading
- Fuel levy
- Command and control with credit trading or fine.

The latter category does not conform to a standard internalisation of a negative externality, as there is no representation of carbon price. Regardless, it was chosen by the IMO to remain under consideration in the MBM debate. The emphasis in the command and control MBM is on the reduction of emissions from the global fleet itself rather than the purchase of offsets outside of the shipping industry, both of which can occur in the former two groupings.

As in many other sectors of the economy, the benefits of emission trading (a cap on emissions and a market to efficiently find the least cost carbon price) are being pitted against the implementation simplicity of the fuel levy. However, the details of the implementation, the management and use of any revenue and any conflict between the IMO's principles of no more favourable treatment against UNFCCC's common but differentiated responsibility are also key to the debate. All three MBM groupings are undergoing detailed analysis and comparison, an example of which is IMO [20].

1.8 Technological and Operational Abatement and Energy Efficiency Options

To meet any future incentive (economic or regulatory), a number of options exist for either the increase of energy efficiency or the abatement of CO_2 and GHG emissions. These options can be applied to newbuild ships and in some cases also for retrofit to existing ships. Examples of attempts to compile lists of these options, their characteristics and economic attractiveness can be found in IMO [19]. To summarise briefly here, the options can be broken down to incremental changes and more substantial changes.

Incremental changes:

- Drag reduction

  - Total drag reduction e.g. devices for wave-making drag reduction
  - Friction drag reduction e.g. hull coatings and air lubrication
  - Appendage drag reduction e.g. high lift rudders
  - Parasitic drag reduction e.g. fairing of hull openings
  - Aerodynamic drag reduction e.g. fairing of superstructure

- Propulsion energy efficiency increase

  - Propulsor efficiency increase e.g. alternative propulsors
  - Propeller inflow and outflow efficiency increase e.g. pre-swirl ducts and vane wheels

- Engine and machinery efficiency increase e.g. engine tuning or control.

  - Auxiliary energy efficiency increase

  - Heat recovery systems
  - Electrification and control of auxiliary systems

- Operational changes

  - Increased frequency of hull and propeller cleaning
  - Increased frequency of engine maintenance

In each case, specifying changes from standard technology (those associated with low cost, volume ship building) incurs a cost, but also provides a benefit. If the benefit also increases energy efficiency then that can be a reduction in fuel costs per unit of transport work. A common method of presenting analysis of the order in which options might be adopted and the likelihood of investment, particularly for policy work, is the Marginal Abatement Cost Curve (MACC), examples of which for shipping can be found in Faber et al. [12], Buhaug et al. [2], IMO [19], IMO [20], Det Norske Veritas [7].

Besides the inherent shortcomings in MACC analysis [15], for shipping it is commonly undertaken with an incomplete representation of costs and little representation of risk (beyond the investment rate of return). The result from several analyses has so far been the identification of substantial (e.g. up to 30%) abatement potential using options that often appear to be cost-negative at current fuel prices, and yet this contradicts the logic that a competitive industry with a dominant energy cost should be overlooking opportunities to increase efficiency at a profit. This has led to an interest in identification of market failures and barriers that could be obstructing the efficiency of the fleet and causing this non sequitur.

But beyond the challenge of understanding the sequence and rate of energy efficiency and abatement uptake, the majority of the options available will only provide incremental improvements, and for a ship of a given size, at present the only options for substantial abatement appear to be:

- Significant speed reduction—speed change, whilst offering substantial reduction in emissions (20% speed reduction is approximately a 50% efficiency increase for a given ship) impinges on revenue and for this reason will be discussed in greater detail below.
- Adoption of wind power—sails and kites can provide significant propulsive power, even for large ships. Both however, are technologically immature and possibly impractical for application close to the power outputs that would be required to propel larger ships at existing speeds. This power requirement would reduce and the availability of wind power would increase (due to the reduction of the apparent wind created when motor-sailing) if ships were designed to operate at lower speeds. Only then, and certainly not for all ship sizes, might wind provide significant emission reduction.
- Adoption of renewable fuel—further to wind power, ships could be propelled using synthetic fuels (such as ammonia or hydrogen) or bio liquid/gas fuels either in an internal combustion engine or a fuel cell. Amonia has the advantage that it can be co-fired in internal combustion engines with conventional fossil fuels with comparatively little technological intervention. However, the prices and availabilities of these fuels will need to be viable for widespread adoption
- Nuclear power—marinised fission reactors are already in operation in ice-breakers, aircraft carriers and submarines and so the technological challenge of deployment in international commercial shipping can be anticipated [4], although the risk, insurance and cost issues are a challenge to analyse.
- Exhaust scrubbing—carbon capture and storage (CCS) technology is being developed for land-based fossil fuel combustion emissions and scrubber technology has also been trialed for ships that can remove CO_2 from exhaust emissions. Attention must be paid to the life cycle of that CO_2, particularly if it is not stored for disposal.

2 Considerations Relevant to the Energy Efficiency of Ships

Increasing the energy efficiency of shipping and decreasing its CO_2 intensity requires some compromise between the potential and the practical (or rather economically viable). Before steam engines arrived, there was a long history of maritime trade being propelled with zero operational CO_2 emission—by the wind. Since then, shipping has changed, both in terms of scale (volumes of goods transported) as well as expectations of maritime transport's role in supply chains (2 weeks for a container movement between Asia and Europe), so sudden change is not possible without significant cost and disruption. What then are likely to be the key challenges that will need to be dealt with in order to manifest a substantial decarbonisation? Beyond the technical challenges associated with commercialising any of the options identified above, two pathways which seem inevitable in their contribution—increase in ship size and decrease in ship speed—will be examined in detail here, because they expose some of the implementation challenges that will be faced.

2.1 Ship Size

Large ships are more energy efficient than smaller ships. This statement is justified both by Eq. 5 and also by the calculations and plots of EEDI characteristics (Fig. 4). Therefore one trend that could create an increase in energy efficiency across the fleet is an increase in the average size of ships—both by increasing the largest ships and shifting transport supply from small ships to large ships. Constraints to the design and operation of larger ships can be divided into three categories: technology, market and infrastructure.

2.1.1 Technology

Constraints on ship size could come from processes associated with designing and building the hull. Historically, new ship designs have been evolutionary steps rather than revolutionary. Designing a ship that survives the harsh marine environment, particularly the wave and operational structural loading, is challenging and the knowledge that is used even today (e.g. classification society rules) still enshrines much empirical knowledge, due to the complexity of the fundamental physics and the financial risk associated with a newbuild ship. However, significant progress was made during recent decades with the arrival of low cost computational power and the analysis capability needed to construct more revolutionary designs.

Propulsion machinery has also evolved and the largest size of two stroke diesel engines grown (the largest currently available has in excess of 80 MW power).

Conventional merchant ships have been single shaft (single main engine, shaft and propeller), as this has proved to be the most economical. There are exceptions for higher speed and specialist vessels, where greater propulsion power density, safety or layout issues have favoured multiple engines and shafts. Therefore, increasing the number of engines or shafts could, if necessary be a method of increasing propulsion power availability for larger ships, should the continued growth in power outputs of two stroke diesel prove unsustainable or unsuitable for use on ships.

Therefore, it is possible that the inertia in the global fleet (due to the long life of ships and slow turnover of the global fleets), and the inertia in the ship building infrastructure (large ship yards and dry docks are required to build and maintain large ships) act as brakes to increase the average size of ships at the rate that demand can support. But it is unlikely that technology and knowledge alone are a significant obstruction in increasing the size of ships.

2.1.2 Market

The costs associated with operating a ship do not vary linearly with ship size. One of those costs is the fuel cost and so as energy efficiency increases with ship size, economies of scale should be apparent. Other costs, such as capital costs and operational costs (e.g. manpower and maintenance) can also be envisioned to produce economies of scale. Evidence of those economies can be found in Fig. 5, which displays the unit mass freight rates (transport price per tonne per day) for different sizes of tanker. The specifications of the ships and their routes are in Table 4. Whilst there are fluctuations over time, and the price data is for different trade routes, a clear trend that the largest ships offer the lowest prices for a unit of transport can be seen.

The wetbulk sector is useful to illustrate this point because the dominance of spot market activity leads to a level of price and route competition that should minimise market distortions or localisms in the data. However, as the technology and operation of ships is broadly similar across the merchant fleet it can be expected that similar economies of scale exist, at least in terms of costs, if not price, elsewhere too.

A customer needing a good to be moved from A to B (or shipper) does not just see the cost of the transport. There are inventory costs whilst goods are in transit or storage, risks on price (both of the good transported and transport itself), and assuming that the production and consumption of the good are continuous processes and the transport of it a discrete process there are some costs associated with storage (potentially both at the origin and the destination). For these and other reasons, some [14] have suggested that a shipper's selection of ship size is not determined solely by the economy of scale. Kendall expressed formulae for the components of cost seen by the shipper including:

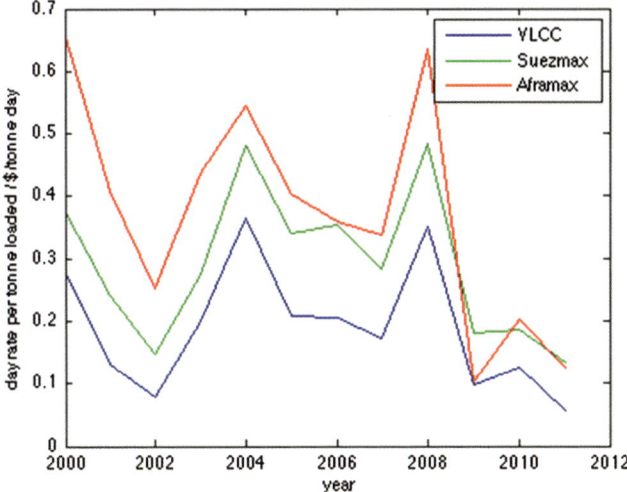

Fig. 5 Freight rates for three sizes of ships for the period 2000–2009. *Source* Intertanko

Table 4 Ship characteristics and voyage specification for freight rate data	Size	Capacity/tonnes	Route
	VLCC	250,000	Arabian Gulf to Japan
	Suezmax	130,000	West Africa to US Atlantic coast
	Aframax	80,000	North Sea to UK continent

- The cost of the sea voyage, *F*
- The handling cost, *H*
- The storage cost, *S*

Taking Kendall's equations for the cost components, an expression for the 'total' cost can be obtained, which when differentiated with respect to ship size (Qs) reveals the relationship:

$$\frac{\partial T}{\partial Q_s} = \frac{VI}{Q_y 200} + \frac{\partial F}{\partial Q_s}, \tag{7}$$

where *V* is the value of the commodity, *I* is the percentage rate of return on the capital invested in it (during transport and storage) and Q_y is the total annualised flow of the commodity on a specific trade route. Besides the importance of the relationship between freight rate and ship size, this relationship suggests that as *V* or *I* increase, or as Q_y decreases, the 'optimal' ship size decreases.

Kendall undertook validation of the equations against ship size selection for the movement of some bulk commodities (iron ore, grain) and found a reasonable agreement against historical data, although whether the assumptions still hold for the present or future scenarios is untested. Whilst the model is only applicable to commodities that are moved in bulk (container movement could not be modelled in this way), it is potentially a useful model to create indicative trends for demand and utilisation as a function of ship size, and to understand some of the shipping markets.

Anecdotally, it is interesting that the largest wetbulk tanker ever built was built in 1979, *Seawise Giant*, with a capacity of 565,000 dwt tonnes, and has not been superseded in the decades since (despite substantial growth in demand for wetbulk transport). The largest wetbulk tankers currently have over 20% less capacity than that vessel. This clearly illustrates that technology or infrastructure are not ultimate limits for ship size in this sector of the shipping industry and that the economics could provide a ceiling to ship size.

Without further work, it is difficult to conclude whether there is an economic constraint on ship size, or whether the economics are influenced by the infra-structure constraint which in turn reduces the flexibility and therefore productivity of larger ships. A similar ceiling to the anecdote of the *Seawise Giant* does not appear to have been reached in the container shipping industry, which has seen steady increases in maximum ship size with time.

2.1.3 Infrastructure

A ship can only be deployed on a trade route if it can; safely navigate at either end of its voyage, pass through any constrictions during its voyage (e.g. the canals in Panama and Suez) and the port or terminal can support the vessel when alongside and load or unload its cargo.

Further to safe navigation, the route must be navigable, and a common constraint is that the draught of the vessel (or its air draught) must be less than the depth of the water or the air gap beneath the lowest bridge. The cost of dredging and development of port infrastructure mean that port access is still a significant obstacle in many countries, particularly in the developing world, to the operation of large ships. This also presents an interesting trade off between a ship's design for minimal resistance (which occurs at a certain combination of the main dimension parameters length, beam and draught) and design for operational flexibility (the ability to operate on the largest range of trade routes and therefore selection of values of beam and draught to accommodate physical restrictions).

Port infrastructure can also impose limits and for the container shipping industry, the maximum outreach of cranes (used to load and unload containers) is an important parameter. The maximum beam of the largest container ships (and the number of containers that can be carried widthways) cannot be greater than the crane width, and with the arrival of Maersk's Triple E class ships [21], this limit will be reached.

Therefore, infrastructure is a significant constraint on both the average and the maximum size of ships.

2.2 Ship Speed

Ship speed is another parameter that, as illustrated in Eq. 5 can create a significant increase in energy efficiency. However, the extent to which it can have an impact is as with ship size, effected by a number of practical considerations.

2.2.1 Economics

Some have used observation of reports of slow steaming (operational speed reduction) to try and elucidate trends and patterns in operational speed [10]. One of the hypotheses presented by Dinwoodie et al. is that the recent trend of slow steaming is a response to a depressed freight rate resulting from the global financial crisis of 2007/2008. Should the shipping markets return to profitability, then the speed of ships will rise again.

This hypothesis appears to be consistent with theoretical analyses that attempt to understand the economic justification for a certain speed. Many of these theoretical studies focus on the ship owner as the agent involved in selecting an optimum operational speed. That optimum can be calculated, given energy costs and freight rates, through profit maximisation [8, 24, 6]. Eefsen [11], adds in the preference of the shipper, an important consideration as the shipper is also an agent that influences the economics of ship operation. However, Eefsen only seeks to calculate operational speed selection as a response to an agent cost-minimisation assumption, which is unlikely to reveal the more rational profit maximising behaviour. A model that could consider both the shipowner and charterer's preferences and their relation to shipping's negative externality is proposed in Smith et al. [26].

Both these pieces of evidence suggest that if increase in fuel price or carbon pricing arise in the shipping industry, it is not unexpected for there to be an incentive to reduce speed or sustain lower speeds. However, that this will only occur as long as the freight rate environment make this rational profit maximising behaviour. If the dynamics of the shipping market lead to a sustained period of high freight rates, and voyage costs do not themselves increase the incentive for energy efficiency, it will be expected that operational (and potentially design speeds) will increase, with negative consequences for shipping's energy efficiency and carbon emissions.

2.2.2 Safety

One of the key requirements of a ship is its safety. Ships operate in a harsh environment and accidents resulting in the loss of ships often result in fatalities and significant damage to maritime ecosystems, even before the value of lost cargo is considered. One cause of these accidents is a ship's inability to manoeuvre to safety in a hazardous seaway. If a ship has insufficient propulsive power, tides, wind and waves can conspire to overcome a ship's manoeuvrability which leave

		Handymax bulk	VLCC	Panamax container
Table 5 Representative specifications for the analysis of total CO_2 emissions	Cfsteel (t CO_2/t)	2	2	2
	Lightship (t)	8,000	45,000	25,000
	Cffuel (t CO_2/t)	3	3	3
	k	0.013	0.03	0.012
	Das (days)	240	280	250
	l (years)	30	30	30
	Vopt (kt) ~	3	4	4.5
	Vdes (kt)	13.5	15	24.5

the ship, its cargo and its crew vulnerable to navigational hazards. Speed, or more appropriately, installed power, can often therefore be seen as a safety measure and this needs to be taken into consideration when considering speed from an energy efficiency perspective.

2.2.3 Life Cycle Emissions

The energy consumed and carbon emitted during the operation of a ship is only a part of the totals associated with its life cycle (other portions of which are attributable to construction, maintenance and disposal). This is a consideration in the debate about ship speed because, caeteris paribus, a reduction in ship speed will mean an increase in the number of ships in the global fleet (and therefore at a total fleet level an increase in non-operational life cycle emissions). Failure to pay attention to the non-operational components of a ship's life cycle emissions could result in a significant offsetting against the operational savings obtained through speed reduction. Smith et al. [26] develops a theoretical estimate of the total life cycle emissions (operational and non-operational). Differentiating the expression provides a formula for the speed V_{opt} at which the life cycle emissions are minimised. That expression is given in Eq. 8:

$$V_{opt} = \sqrt[3]{\frac{12 C_{fsteel} \text{lightship}}{C_{ffuel} k T_{aS} l}} \tag{8}$$

where, C_{fsteel} is the carbon factor (kg CO_2/kg steel) associated with the non-operational emissions, lightship is the mass of the steel used in the ship's construction, C_{ffuel} is the carbon factor of the fuel, k is a representation of the fuel consumption of the ship, T_{aS} is the time spent at sea and l is the life of the ship. For any speed above the calculated value of V_{opt}, the operational emissions (dominated by fuel consumption) will exceed the non-operational emissions and vice versa.

Table 5 describes characteristics for some representative ships from different sectors and size categories, and the values of V_{opt} which result from the application of 8.

Clearly, for all the ships considered here and the assumptions of the input variables to 8, the speed of ships has a significant scope to reduce before the emissions savings will be negated by non-operational emissions. But it should be noted that depending on the decarbonisation of the energy used in ship building, maintenance and disposal and the evolution of the energy efficiency and carbon intensity of ship fuels, this threshold speed could change in the future.

2.2.4 Charterparties

A charterparty is a legal agreement between the owner of a ship and a charterer, and in many instances specify the speed at which at voyage should be undertaken. Standard charterparty contracts (e.g. BIMCO Gentime) stipulate that a chartered vessel must sail at 'utmost despatch' without consideration of berth availability at destination ports [1]. This provides the charterer an incentive to instruct the master to sail at full speed to the ports which admit vessels on a first come first serve basis (FCFS). FCFS (and the inability to pre-book berths) is one of the key drivers for charterers to adopt full speed clauses on charterparties.

3 Futures of the Shipping Industry and Concluding Remarks

Without regulation, carbon emission abatement will only be incentivised by rising energy prices and the social conscience of shipping's owners and operators (or pressure from shipping's customers) leading to voluntary internalisation of shipping's negative externality.

If the trajectory associated with a lack of regulation is untenable (and the Business As Usual Scenario in Buhaug et al. [2] implies that this is the case if shipping is to contribute in any way to the avoidance of dangerous climate change), then robust and effective regulation will need to be developed. The slow rate of global fleet turnover requires that for regulation to create impacts in decades, it needs to be substantial and in place in years from now.

Some of the analysis that has been developed to enable consideration of the opportunities available to reduce shipping's CO_2 emissions, and the impacts that may result, is presented in this chapter. However, the shipping industry does not have a substantial body of literature describing its future evolution and there is still significant analysis work to be done to provide rigorous support to the policy making progress. This chapter has hopefully identified some of the areas for that further work and explored the breadth and depth of perspective that will be required.

Acknowledgments This chapter is based on work undertaken for the ongoing project "Low Carbon Shipping—a Systems Approach", the authors would like to thank RCUK Energy, Rolls Royce and Lloyd's Register who have funded the research, as well as the academic, industry, NGO and government members of the consortium that support the research with in-kind effort and data. The chapter uses many ideas that have been developed with colleagues at the UCL Energy Institute, including Mark Barrett, Sophie Parker, Eoin O'Keeffe and Nishatabbas Rehmatulla.

References

1. Alvarez F, Longva T, Engebrethsen E (2010) A methodology to assess vessel berthing and speed optimization policies. Marit Econ Logist 12(4):327–346
2. Buhaug Ø, Corbett JJ, Endresen Ø, Eyring V, Faber J, Hanayama S, Lee DS, Lee D, Lindstad H, Markowska AZ, Mjelde A, Nelissen D, Nilsen J, Pålsson C, Winebrake JJ, Wu W-Q, Yoshida K (2009) Second IMO GHG study 2009, IMO London, UK
3. Borken-Kleefeld J, Berntsen T, Fuglestvedt J (2010) Specific climate impact of passenger and freight transport. Environ Sci Technol 44(15):5700–5706
4. Carlton JS, Smart R, Jenkins V (2011) The nuclear propulsion of merchant ships: aspects of engineering, science and technology. Proc IMarEST Part A J Marine Eng Technol 10(2):47–59
5. Corbett JJ, Lack DA, Winebrake JJ, Harder S, Silberman JA, Gold M (2010) Arctic shipping emissions inventories and future scenarios. Atmos Chem Phys 10:9689–9704
6. Corbett JJ, Wang H, Winebrake JJ (2009) The effectiveness and costs of speed reductions on emissions from international shipping. Transp Res D 14:593–598
7. Det Norske Veritas (2009) 'Pathways to low carbon shipping'
8. Devanney J (2010) The impact of bunker price on VLCC spot rates. The Center for Tankship Excellence, USA
9. Devanney J (2011) EEDI Absurdities. Center for Tankship Excellence, USA
10. Dinwoodie J, Tuck S, Landamore M, Mangan J (2010) Slow steaming and low carbon shipping: revolution or recession. Logistics research network 2010 conference, Harrogate
11. Eefsen T (2010) Speed, carbon emissions and supply chain in container shipping. IAME 2010. Lisbon, Portugal
12. Faber et al (2009) Technical support for European action to reducing greenhouse gas emissions from international maritime transport. CE Delft, Delft
13. Fuglestvedt J, Berntsen T, Eyring V, Isaksen I, Lee D, Sausen R (2009) Shipping emissions: from cooling to warming of climate and reducing impacts on health. Environ Sci Technol 42(24):9057–9062
14. Kendall PMH (1972) A theory of optimum ship size. J Transp Econ Policy 1(2):128–146
15. Kesicki F (2010) Marginal abatement cost curves for policy making—Expert-based vs. model-derived curves. Energy Institute, University College London, London
16. Kristensen HOH (2010) Model for the environmental assessment of container ship transport. Trans Soc Nav Archit Mar Eng
17. IMO (2010a) Prevention of air pollution from ships. Information to facilitate discussion on GHG emissions from ships MEPC60/WP.5
18. IMO (2010b) Further improvement of the draft text for mandatory requirements of EEDI and SEEMP. EE-WG 1/2/1
19. IMO (2010c) Reduction of GHG emissions from ships MEPC61/Inf.18
20. IMO (2010d) Reduction of GHG emissions from ships—full report of the work undertaken by the Expert Group on Feasibility Study and Impact Assessment of possible Market-based Measures. MEPC 61/Inf.2
21. Maersk (2011) www.worldslargestship.com. Accessed 19 June 2011
22. Ocean Policy Research Foundation (2009) The World's Changing Maritime Industry and a Vision for Japan, OPRF Tokyo
23. Rehmatulla N (2011) Applying systems thinking approach for qualitative analysis of GHG emissions regulations in shipping. In: 1st International Conference on Maritime and Maritime Affairs, Plymouth, UK
24. Ronen D (1982) The effect of oil price on the optimal speed of ships. J Oper Res 33(11): 1035–1040
25. Saldanha JP, Tyworth JE, Swan P, Russel D (2009) Cutting logistics costs with ocean carrier selection. J Bus Logist 30(2):172–195
26. Smith TWP, Parker S, Rehmatulla N (2011) On the speed of ships
27. Stopford M (2009) Maritime economics, 3rd edn. Routledge, London

Energy for Railways

Roderick A. Smith

Abstract Within the context of overall energy use for transport, the railways offer the easiest path to low carbon through electrification and its generation through low- or zero-carbon sources. The railways also generally claim to be a "greener" form of transport, although this claim depends critically on the passenger occupancy achieved. For freight movements there are clear energy advantages, but issues exist related to the transfer and final delivery of goods. The freight areas in which rail excels, in its traditional markets of coal and heavy goods, have declined in developed economies, the most common cargo now being the freight container. This chapter will flesh out the above issues and discuss some suggested routes to low energy, low emissions such as lightweighting, hybrid trains, fuel cells and bio-diesel and driving style.

1 Background

The writer of a letter to The Times, 4 April 1912, said he was between 1845 and 1850 a junior partner in a Newcastle Glass Manufacturing firm, in which R Stephenson and G Hudson were also partners. G Stephenson came to see the firm in 1847, and said, "I have credit of being the inventor of the locomotive, and it is true I have done something to improve the action of steam for that purpose. But I tell you, young man, I shall not live to see it, but you may, when electricity will be the great motive power of the world."

R. A. Smith (✉)
Future Rail Research Centre, The Department of Mechanical Engineering,
Imperial College, London Exhibition Road, South Kensington,
London SW7 2AZ, UK
e-mail: roderick.smith@imperial.ac.uk

O. Inderwildi and Sir David King (eds.), *Energy, Transport, & the Environment*,
DOI: 10.1007/978-1-4471-2717-8_31, © Springer-Verlag London 2012

Fig. 1 Length of railway
(thousands of km) in major
regions of the world

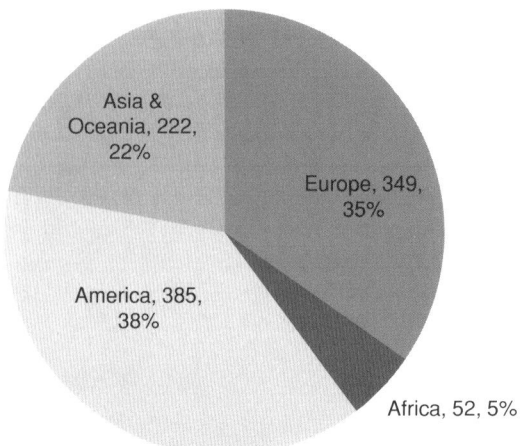

Fig. 1 Length of railway (thousands of km) in major regions of the world

Railways have been with us since the early nineteenth century. Their introduction was primarily to move freight, mostly coal, but passengers rapidly became an important market. The railways caused a step change in the distance over which it was possible to make a return journey in a day. They enabled travel to take place more cheaply, in greater comfort and with greater reliability than the horse drawn coaches they replaced. Our cities rapidly expanded as the industrial revolution blossomed, aided by our mastery of the energy obtained from coal and contained in steam. Energy and coal consumption grew in an exponential manner. That railways were a major contributor to this revolution is not in doubt. They were responsible for, inter alia, the standardisation of time, the expansion of the electric telegraph system, the definition of a nation state (based on a days journey) and were a huge contributor to the quality of life experienced by the common man. They spread rapidly, through Europe, west across America and east to Asia and the Far East. Their supremacy as a mode of transport was unchallenged up to their Edwardian zenith, but the depression of the 1930s, the rise of the motor car and later, in the 1950s, the aeroplane, all contributed to a decline and contraction of rail networks throughout the world. But recent years have seen a reversal of that decline. A catalyst for longer distance travel was the introduction of high-speed travel on a dedicated new line, initiated by the Japanese shinkansen (Bullet train) in 1964 [5]. Congestion caused by urbanisation has seen a raise in rail borne commuter traffic and the building of new subway, tram and metro systems in many cities worldwide.

Figure 1 illustrates the distribution in 2009 of railways in the major regions of the world. The map of Fig. 2, in which countries are drawn in proportion to the percentage of the world's total railway length contained in the country, neatly emphasises the predominantly north hemisphere nature of the railway. Figure 3 shows the passenger freight split in the major regions, and the tiny role of the railway in Africa, the almost complete dominance of freight in America and

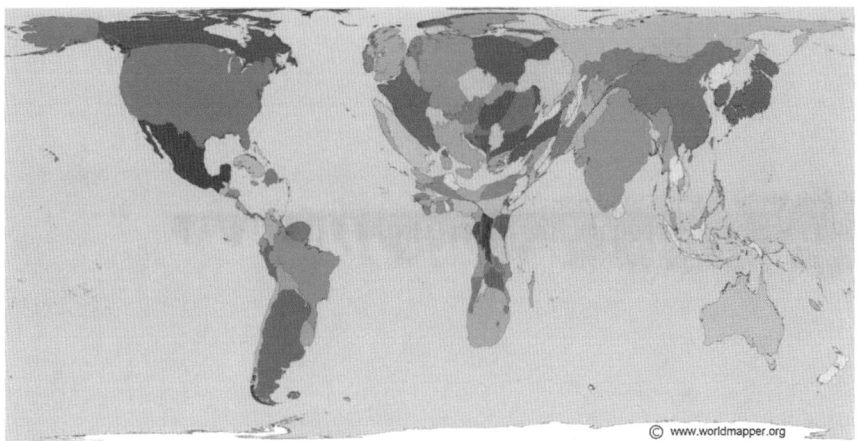

Fig. 2 A world map distorted to show *territory size in proportion to all railway lines in the world found there.* © Copyright SASI Group (University of Sheffield) and Mark Newman (University of Michigan)

Fig. 3 Railway utilisation, passenger and freight in major regions (millions of passenger km + millions of freight km per annum divided by line length)

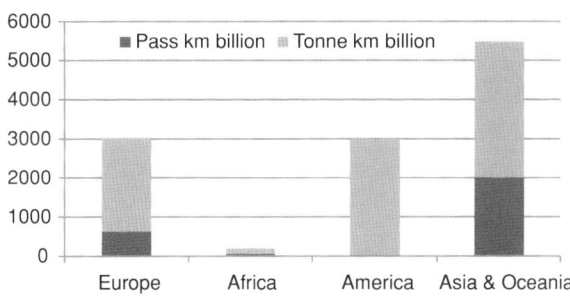

the large reliance on the railway for both passengers and freight in the Asian and Oceania region. A crude calculation can be made for a figure of merit for the carrying efficiency of the railway by dividing the sum of the millions of passenger km and freight tonne-km per annum by the length of line. It is, of course, recognised that there is no one-to-one equivalence between a passenger km and a tonne-km. This yields an approximate value of 4 for Africa, 8 for America, 9 for Europe and a considerably larger value of 25 for the railways of Asia and Oceania. This macro overview illustrates one of the difficulties in calculating an environmental benefit of the railways: their performance varies hugely in major regions of the world and they perform different tasks. These differences can become even more marked if we examine performance by country, or by region or even particular route within a country: but nevertheless, the macro figures are a good starting point.

At their inception (mechanical) railways were hauled by steam engines. Since they often carried coal, their fuel was at hand and cheap. Energy efficiency was relatively unimportant: until their demise in the 1960s and 1970s steam engines very rarely achieved a thermal efficiency of more than 10%, added to which they needed considerable maintenance and were unavailable for a large proportion of the working day. Very few steam engines doing real work exist today, but there are a considerable number of preserved locomotives which live on to serve the nostalgia and tourist markets. Electric traction was established before the end of the nineteenth century, particularly for street trams in cities, and was very soon adopted for underground systems, but rather slowly for main lines. Diesel traction developed strongly in the 1930s, for example the Flying Hamburger in Germany achieved high speeds while the steam speed record of 126 mph (202 kmph) was being established by the streamlined Mallard locomotive in 1938. After World War II, steam was rapidly replaced by either diesel or electric traction. The power supply for electric trains was largely high voltage AC and the concept of the locomotive-less multiple units found favour. Counties electrified at varying rates. Clearly those with cheap hydropower, like Switzerland, were able to do so faster than the coal-reliant countries. Today, Switzerland has had 100% electric traction for many years, while the ratio of electrified lines in the UK is 33%, in Ireland only 3%, more typically 58% in Germany and 52% in France. Worldwide in round figures 25% of the rail network is electrified, but that 25% carries 50% of the traffic.

Given that there are many advantages of electrification (to be explored later), we might ask, Why have the railways of the world been so slow to convert?

The answer is, of course, economic. Railways need expensive infrastructure, expensive both to build and to maintain, and electrification is perceived to be expensive. Unless a railway carries considerable traffic, its income falls short of its expenses. As far as passenger traffic is concerned, we may usefully define a usage parameter as thousands of passenger kilometres per route kilometre per day. This yields overall country KPIs of 40 for Japan, 6 for Germany and 9 for the UK. If individual routes are examined then for example, in Japan, the highest riderships are achieved by the Tokyu railway which on its tiny 103 km route generates about a fifth of the passenger km of the UK's 16,000 km route at a density of 270. The Tokaido shinkansen, linking Tokyo and Osaka, is the busiest intercity route in the world with a density of 212. On the other hand, in the geographical extremities of Japan, the railways play a much less significant role and have generally low riderships.

These usage figures also vary with time of day. The morning commuter peak (and to a lesser extent the evening) produces maximum demand for which a railway must be equipped. Vehicles are often idle for much of the day which follows and ridership drops significantly.

Given the huge variations in the figures discussed above, it should be no surprise that most railways in the world are subsidised (or invested in, depending on viewpoint) by the public purse. The question of profitability is governed by the choice of the system boundary through which costs and benefits pass. The tighter the boundary is drawn, the less likely the balance will be positive. The more generous the boundary, the more likely it is for a railway to be profitable or worthwhile. This generous calculation must include external costs, and one of the key drivers in such a calculation is the environmental costs, and in this energy is a major factor.

2 Energy Efficiency

If we define energy efficiency of transport as the energy per unit distance needed to transport a passenger, then most important single factor determining the energy efficiency of all vehicles is the passenger load factor: that is the proportion of carrying capacity filled by passengers. For trains this factor varies hugely. Commuter trains are often crush filled, with all seats occupied and standing passengers filling all available space. For passenger comfort this is undesirable, but for energy efficiency, it produces spectacularly good results. But commuter flows are strongly unidirectional, and the trains travelling at the same time but in the opposite direction have low load factors. Some typical overall energy data is collected in Table 1.

Tables 1 and 2 illustrate the "green" claim of the railways. In terms of mode share the railways punch well above their weight in terms of emissions: but this latter claim depends critically on the origin of the electricity use. These figures are from Japan which has an extremely high mode share for passengers of nearly 30%. Much lower figures of 5–7% are typical. Note too, the figures in this table are the energy inputs to the vehicle: in the case of electric propulsion the greatest loss in its production from a primary energy source stems from the thermodynamic efficiency loss in the steam cycle. For diesel trains, the major loss occurs within the engine itself as the fuel is combusted internally. It is therefore important to define exactly where in the chain energy measurements are being made and what is being compared. Typical energy chains for electric and diesel traction are shown in Fig. 4.

There are well-known operational advantages of electric traction which include: lower running and maintenance costs and instant availability of rolling stock, higher power-to-weight ratio which results in fewer locomotives, faster acceleration, higher limits of power, higher speeds and quieter operation with less vibration transmission into the passenger space. The removal of heavy engines and the requirement to carry fuel leads to lower mass, leading not only to improved performance, but a lighter footprint on the track and large potential savings on track maintenance. The regeneration of energy on breaking and its feedback is facilitated by electric traction. The environmental advantages include reduction of local pollution in the vicinity of the train, and, particularly increasingly important in the future, the lack of dependence on crude oil fuel. To this list can be added the fact that electric traction has the potential to be as carbon free as the generation route allows. If the supply comes from renewables or nuclear, then largely carbon-free traction is available, without recourse to increasingly expensive and, in the future, scarce fossil fuels. There is pressure to limit the particulate emissions of

Table 1 Approximate energy use figures for average passenger loadings

	MJ/pkm	kWh/100 pkm
Car	2.4	68
Bus	0.68	19
Rail	0.22	6
Air	1.8	51
Sea	2.1	57

Table 2 Passenger Mode share, energy and carbon emissions (Japan 2007 and 2009)

	%p. km	% energy	%CO$_2$
Railway	29	7	5
Plane	6	5	3
Car	51	76	83

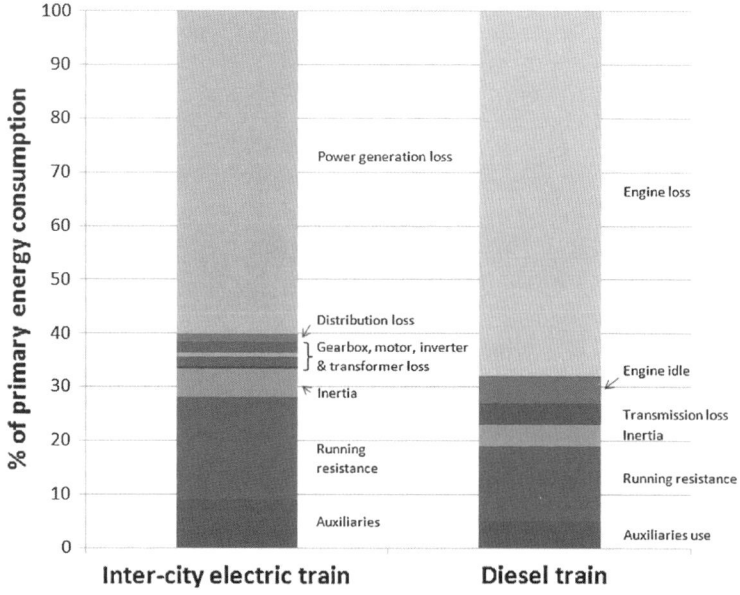

Fig. 4 Primary energy consumption of typical inter-city electric (Class 390 Alstom Pendolino) and diesel (Class 221 Super-Voyager) trains (Peckham C. *Improving the efficiency of traction energy use.* Rail Safety and Standards Board (RSSB). Report number T618, 2007)

diesel trains, with a tightening European requirement which must be met in the near future, which adds to the benefits of electrification. There are, of course, downsides. Expense has been mentioned, although the other considerations are reducing the significance of this factor. On older infrastructure, tunnels may need to be deepened to allow sufficient clearance for the overhead equipment, and low bridges may have to be heightened. Unless route coverage is near complete, a change of traction mode may be required on many longer journeys, the overhead supply needs to be constructed robustly and the visual intrusion of the supply needs to be minimised. The overhead supply line adds a further interface between the vehicle and the infrastructure, for example, electromagnetic radiation can interfere with the signalling system, but these downsides can be engineered out and rarely outweigh the advantages. It is assumed that any new electrification will consist of high voltage 25 kv overhead supply. Experience over the years has shown the advantages of this system, particularly over the low voltage dc third rail system. It is worth noting that the extensive third rail main line system south of London in the UK, is the only such system remaining in the world, and was declared obsolete[1] more than 100 years ago!

3 Improving Rail Vehicle Energy Performance

In general, and for all forms of transport, increasing speed uses more energy. As air resistance increases with the square of the speed, so does the energy requirement for the same vehicle as its speed increases. If we do not want to use any energy we must not move! Similarly, an empty vehicle or a stationary vehicle with a running engine has an infinite energy consumption per passenger kilometre travelled. There are clearly trade-offs here. We have lived through an era of relatively cheap energy but are entering a time of relative scarcity and higher cost, as well as recognising that the pollutants caused by transport fuel use, carbon dioxide being one of several causing concern, need to be much reduced.

Although passenger load factor is so important in determining energy use, it is clearly useful to reduce the energy requirements of vehicles. The three main ways of achieving is by improving the aerodynamic performance of the train consist, by reducing its weight and by improving the efficiency of on-board converting and transmitting equipment. It is often said that basic physics dictates that for a given vehicle the resistance to motion, the aerodynamic drag, increases with the square of the speed, thus doubling the speed requires a fourfold increase in energy. What is necessary as speeds increase is to develop new vehicles which can overcome the limitation of this simple physics extrapolation. Figure 5 illustrates this effect for the Toakaido shinkansen family showing the improvements made from the original Series 0 bullet train introduced in 1964. Although the speed of operation has increased from 220 to 270 km/h, the energy consumption is only 68% of the original. The principal technological development which has made this possible is the design of the aerodynamic profile of the train to minimise drag. Figure 6 shows the development of the front end shape which has allowed a 75% reduction from the original drag factor, a reduction achieved by a combination of real and model wind tunnel testing and computation using the techniques of Computation fluid Dynamics (CFD). Figures 7 and 8 detail the mass reductions achieved for the

---

[1] Letter to the Times, 8 October 1904, from Silvanus P Thompson: *The "live rail" is itself already an obsolete device, discarded in the latest types of electric railway. In ten years time there will probably be no "live rail" left.....It is an engineering blunder.*

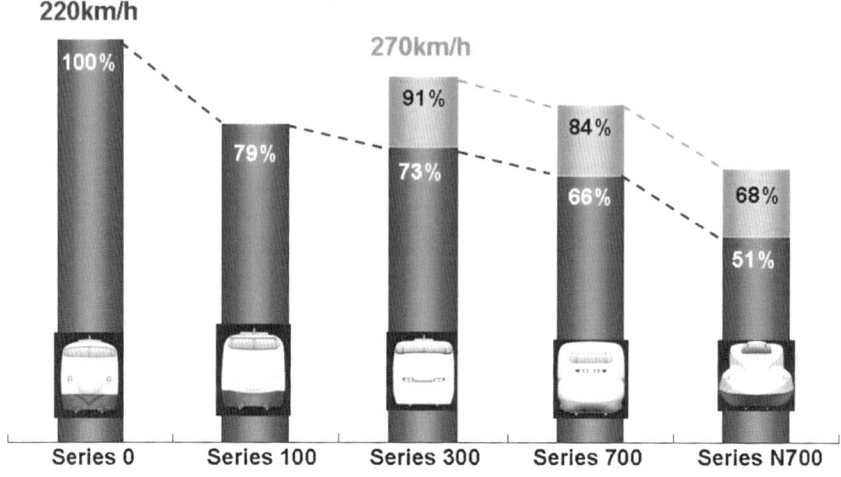

Fig. 5 Evolving energy consumption in the Shinkansen series (JR Tokai)

Fig. 6 Reduction of drag of the Shinkansen, principally by nose shape development. AR is the aspect ratio of nose length to the car body cross-sectional hydraulic radius. (JR Tokai)

Type		Cd ratio	AR
0 		100	4.4/1.77
100 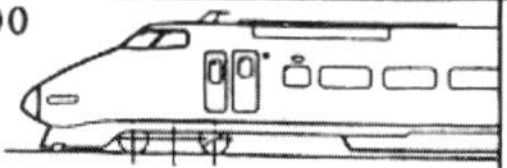		84	5.5/1.77
300 		58	6.0/1.68
700 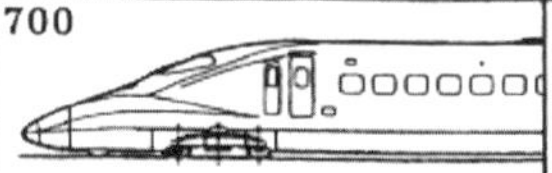		25	9.2/1.65

traction motors and bogies. The shinkansen trains are not hauled by a power car, but have distributed traction meaning that nearly all axles are individually motored. This has the advantages of removing the peak axle loads which occurred under the power car of achieving better traction and having built-in redundancy and reliability should a motor fail. The huge improvements of power density and volumetric efficiency have been possible because of the developments made in magnetic material used in motor construction.

Item	Series 0	Series 100	Series 300	Series 500
Output (kw)	185 (100)	230 (124)	300 (162)	300 (162)
Weight (kg)	876 (100)	825 (94)	390 (45)	375 (43)
Weight/Output (kg/kW)	4.74 (100)	3.59 (76)	1.30 (27)	1.25 (26)

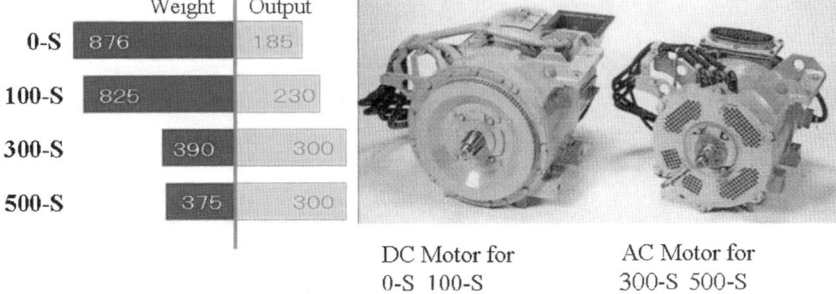

DC Motor for
0-S 100-S

AC Motor for
300-S 500-S

Fig. 7 Developments in electric motor technology (JR Tokai)

Lightweight Bogie

Bogie for Series 100
9.7 t
(Bolstered type)

Bogie for Series N700
6.8 t
(Bolsterless type)

Fig. 8 Light weighting of bogies (JR Tokai)

3.1 Lightweighting

The reduction of mass in the bogies is worth a special mention because the bogies and wheelset comprise what is known as the unsprung mass, that is, the mass below the main suspension of the vehicle. This mass plays a major role in determining the magnitude of the dynamic forces generated at the stiff wheel rail interface, which in turn determines the rate of deterioration of the track and the maintenance costs. The reduction of weight of the unsprung mass presents a particularly difficult technical challenge as these components are particularly vunerable to fatigue loadings and failures are critical and likely to have serious consequences.

The energy and cost values of reducing mass can be quantified. For example, we have modelled on a simulator the route and timetable of a so-called High-Speed Train[2] (HST) between London Paddington and Bristol. The calculations are route specific and depend on, inter alia, the overall mass of the train, its speed, acceleration and braking capability and regeneration capacity (zero in the case considered), the traction against speed relationship, gradients, curves, number of stops and speed limits and, of course, the timetable. Numerous simulations allowed us to derive a value for the gradient of the energy consumption against mass curve of 3.6 kwh/tonne/journey. The effect of various masses is illustrated in Fig. 9. Based around the structural mass of the train, tare mass 446 tonnes, the 15 tonnes of fuel for example adds 54 kwh to the energy consumed for a single Paddington Bristol journey (190 km), a figure which becomes very significant if the total distance travelled by the whole fleet of similar trains is considered. In a similar manner we can investigate the effect of passengers and the lightening of the train from a diesel to an electric version. These kinds of calculation can be extended over the lifetime of the vehicle and applied to the population of similar vehicles in the world.

An example of the former is in [4], in which for the particular train and mission of their study, a gradient was calculated of 0.0259 kW h/km/tonne. They then considered various carriage body construction materials, a steel version at 11.5 tonnes, Aluminium 9, a steel composite hybrid 8.5 and a full composite of 7.6 tonnes. Taking an estimated life of 25 years and 7.5 million km (say 300 days per year at 1,000 km/day) it follows that the energy saving moving from the heaviest to the lightest construction is given by. This figure is then used to justify the addition monetary and energy cost of the composite construction.

Weights saving calculations for a range of vehicle types and their world populations have been used in [1] to produce the results shown in Fig. 10. A saving of 100 kg per vehicle is assumed and the energy saving per vehicle is multiplied by the population of such vehicles in various regions of the world. It is of no surprise that because private cars are so ubiquitous they show the greatest potential for energy saving. Likewise the total population of trains is so low, energy saving efforts here have little effect on the world energy and emissions problem.

---

[2] The UK HST (Class 43), does not travel on dedicated track, was designed more than 40 years ago and has a top speed of 125 miles per hour (200 kph).

Fig. 9 The energy/mass relationship for a HST on the London Bristol route

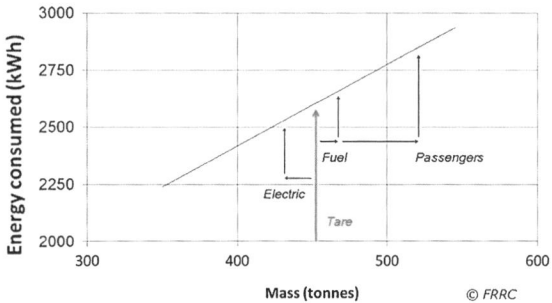

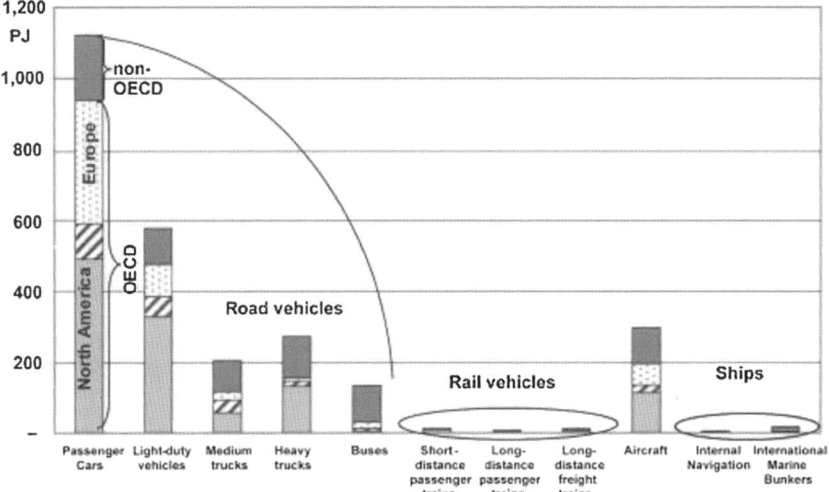

Fig. 10 Potential global annual primary energy savings by light-weighting [1]

This figure serves to emphasise that if overall transport energy use and emissions are to be reduced, then the railways can only play a part if there is a large mode switch from road to air. Nevertheless, the incentive remains for rail operators to reduce operational costs by both the energy saving potential of mass reduction and to benefit from lower maintenance costs by having a lighter track footprint.

4 Alternative Traction

Although it is clear that electrification is an energy efficient method of railway propulsion and furthermore offers a route to low carbon operation depending on how electricity is generated, alternatives have been investigated either as short-term solutions or for operational convenience. These alternatives can be grouped as hybrids, fuel cells and bio-fuels.

In a hybrid train, an onboard rechargeable energy storage system (RESS) is placed between the power source and the traction transmission system connected to the wheels. Surplus energy from the power source, or energy derived from regenerative braking, charges the storage system. During acceleration, stored energy is directed to the transmission system, boosting that available from the main power source. In existing designs, the storage system can be electric traction batteries, or a flywheel. There have been several such trains produced to trial the technology. However, the additional mass of the RESS offsets the advantages of mass saving previously discussed. Flywheels offer mechanical simplicity and the ability to charge and recharge indefinitely [2]. Battery technology on the other hand is limited by the speed of charge and discharge, the loss of capacity of the battery with number of changing cycles and the low energy density of existing batteries. With current battery technology it is hard to see more than niche application of hybrids to rail: a situation which might change with developments in battery capabilities. However, flywheel trains have been introduced into service on the Stourbridge Branch line in the West Midlands of the UK. Single car, twin axle cars, carrying 30 up to seats and 30 standing passengers, are powered by a small 2 litre ic engine burning LPG, with energy storage in a 500 kg steel laminate flywheel operating at up to 2,600 rpm. The manufactures claim that 3.4 times less CO_2 is being produced that would be generated by a convention diesel powered railcar.[3]

Fuel cells have been suggested as an on-board source of electricity for rail, either independently or in a hybrid configuration. The limitations of current fuel cell technology as well known and include the low proportion of mass of hydrogen which can in a fuel tank made to withstand the high storage pressures required, the cost of the platinum catalyst of the fuel cell and the carbon produced in the manufacture of the hydrogen. Potentially these problems might be overcome by research financed by and directed towards the automobile industry: eventually niche application may be found in rail, but this is not a route of huge practical promise, despite successful demonstrations of the technology. For example, in May 2003, JR East started test runs using a *NE* (*New Energy*) train. The design had two 65 kilowatt fuel cells and six hydrogen tanks under the floor, with a lithium-ion battery on the roof. The test train was capable of 100 kilometres per hour with a range of 50–100 kilometres between hydrogen refills. Research is underway into the use of regenerative braking to recharge the test train's batteries, intending to increase the range further, a production version entered limited local service in 2007.

Biofuels may supplement liquid fossil fuels, and indeed there is currently a European Directive on the promotion of the use of biofuels and other renewable fuels for EU transport. The directive entered into force in May 2003, and stipulates that national measures must be taken by countries across the EU aiming at

---

[3] http://www.parrypeoplemovers.com/index.htm, accessed 20 August 2011.

replacing 5.75% of all transport fossil fuels (petrol and diesel) with biofuels by 2010: a target which has been missed comprehensively. However, at the start of 2008 it was announced that the EU was rethinking its biofuel programme due to environmental and social concerns. The EU was particularly concerned about the impact of biofuels on rising food prices, rainforest destruction, notably from palm oil production and concern for rich firms driving poor people off their land to convert it to fuel crops. These concerns are well founded if the required scale of crop-based biofuels is considered. There is about 64,000 km^2 of arable land in the UK. The yield of rapeseed oil is in the order of 1000 kg/hectacre, so all our arable land might produce 6.4 m tonnes which would make 6.2 m tonnes biodiesel. This is approximately just 12% of current demand.

5 Operational Factors

In a well-designed and operated railway, if the timetable is properly developed, then unplanned signal stops would be unnecessary. In practice, such perfection cannot be achieved for a huge number of reasons, but nevertheless, improved punctuality and reliability can combine to produce energy savings by reducing unwanted stops. Just as we are aware (but do not always act upon) the idea that our own driving style in cars can produce fuel consumption saving, so too on the railway driving style can generate economies. It is common practice to achieve line speed as quickly as possible and to hold it for as long as possible before braking as rapidly as possible to a station stop. This style often generates dwell times at stops before a timetable departure can take place. We have investigated such patterns of driving for a UK train operator by using our in-house modelling system, TREN,[4] in for example [3]. By employing techniques such as coasting to a halt over a much longer distance than the minimum breaking distance but still arriving to a stop on time, we have shown theoretically how as much as 10–15% of energy can be saved on specific journeys. Experimental measurements on the live railway have supported the accuracy of these predictions. These trails are replicated in several such similar experiments both in the UK and elsewhere, as the interest of train operating companies to reduce their fuel consumption has been highlighted by the recent rapid rise in fuel prices. Although currently, such techniques can only be empirically employed by the real-time intervention of the driver, if fully automated in-cab signalling systems are employed in the future, thus obviating the need for signalling blocks in the track, then algorithms based on any optimisation desired based on journey time, fuel economy, capacity of the line, etc., can be implemented.

---

[4] Train Energy modelling system, Future Rail Research Centre, Imperial College London.

6 Freight Operations

In a sense, we now return to the beginnings of the railways by briefly mentioning the use of energy in the transport of freight. In round figures rail freight is around 4 times more energy efficient per tonne-km moved than road transport. Recent typical figures (quoted with all the caveats mentioned previously) suggest that for 1 unit of CO_2 produced per tonne-km for an electric train, the figure might be 8 for a diesel train, 21 for a large (>32t lorry) and 140 for a small (3.4–7.5t) lorry. Thus the historical advantage of rail for moving freight is still maintained and any ways of "getting more freight onto rail" would be environmentally advantageous. However, there are issues. Despite some bright spots in the world such as the US and Australia, it has proved difficult in many countries to maintain freight traffic levels as the traditional bulk goods of rails markets, principally coal and steel products, have declined. Limited capacity of existing networks is a major issue with passengers being seen as a more important market. Some networks, for example the UK's, are generally physically too small in terms of loading gauge size to take full size fright containers which have revolutionised freight shipping. Transhipment for final delivery causes delays and added costs. But clear advantages to reducing road congestion can be seen if "lorries can be taken from the roads". One significant advantage of a substantial switch of passenger traffic onto new dedicated high-speed lines would be the freeing of capacity for slower freight trains on existing congested conventional lines.

7 Concluding Remarks

Railways use either diesel fuel on-board or externally supplied electricity to generate traction power. The latter method has many practical advantages and, importantly, offers the possibility of reducing emissions if the generating method is decarbonised. The cost of providing electrical supply currently limits its use to the more intensively used routes, but this cost will become less significant as the price of fossil fuel rises.

The best way of improving the energy efficiency of railways is to increase the carrying load factor of trains. Currently this is generally low, partly because of the distorting effects of commuter peaks.

Lightweighting of rail vehicles not only saves energy, but also gives the train a lighter footprint which reduces the deterioration of the track and thus reduces maintenance costs.

New high-speed lines offer opportunities for intensive intercity use and can be energy efficient if the load factor is high. Further, the building of such lines can release capacity of existing networks for freight traffic. But the high construction costs can only be justified if these new lines succeed in attracting traffic from the roads, and for journeys less than about 800 km, from air.

Freight movements by rail are potentially much more energy efficient than by alternative modes. But rail is best suited to heavy bulk traffic which has generally declined as economies have developed. However, the container revolution in sea traffic offers opportunities for rail.

Although the rail industry is vociferous in promoting its "green" credentials, in terms of worldwide issues, the impact of rail is small. The industry must be prepared to shed lightly used lines and to concentrate its efforts to increase traffic on more populated routes. Furthermore, it needs to win greater mode shares by encouraging passengers and fright to transfer to rail. The small mode shares of most railways, in the order of 5–7%, indicate what a huge ask this is, a task that cannot be achieved without significant increase in the carrying capacity of existing railways.

References

1. Helms H, Lambrecht U (2007) The potential contribution of light-weighting to reduce transport energy consumption. Int J Life Cycle Assess 12(1 Special issue):58–64
2. Read MG, Smith RA, Pullen KR (2009) Are flywheels right for rail? Int J Railw 2(4):139–146
3. Read MG, Griffiths C, Smith RA (2011) The effect of driving strategy on hybrid regional diesel trains. Proc Inst Mech Eng Part F: J Rail Rapid Transit 225(2):236–244
4. Schwab Castella P, Blanc I, Gomez Ferrer M, Ecabert B, Wakeman M, Manson J, Emery D, Han S, Hong J, Jolliet O (2009) Integrating life cycle costs and environmental impacts of composite rail car-bodies for a Korean train. Int J Life Cycle Assess 14(5):429–442
5. Smith RA (2003) The Japanese Shinkansen: catalyst for the renaissance of rail. J Transp Hist 24(2):222–237

Does Britain Need High Speed Rail?

Chris Stokes

Abstract Until recently, British governments have been sceptical about the case for construction of high speed rail lines, but by 2009 all three major parties supported the concept, and in March 2010 the government published details of the proposed route for a high speed line from London to Birmingham ("HS2"), with connections to the existing network to allow faster journey times to Manchester, Liverpool and Glasgow. Following the general election in May 2010, the new coalition government committed to a larger project, with a network linking London with Birmingham, Manchester and Leeds, together with links to Heathrow Airport and the existing high speed line to the channel tunnel. The supporters of high speed rail argue that the project has a number of major benefits for Britain: (1) Conventional transport economic benefits, as a result of time savings and congestion relief; (2) Environmental benefits, particularly as a result of substitution of rail for air travel; (3) The provision of additional capacity to meet rapidly growing rail demand; (4) Regeneration of the North of England, reducing the North–South divide in Britain. The evidence for each of these claimed benefits is reviewed, together with experience from high speed rail projects elsewhere in the world. The review concludes that there is no strong case for construction of the HS2 project.

1 Background

Until about 10 years ago, Britain had at best been an interested observer of the development of high speed rail, with commentators admiring the Japanese Shinkansen and the French TGV routes, but only in a detached way—there was no

C. Stokes (✉)
3 Rothschild Road, Leighton Buzzard,
Bedfordshire LU7 2SY, UK
e-mail: chrisjstokes@btopenworld.com

O. Inderwildi and Sir David King (eds.), *Energy, Transport, & the Environment*, 577
DOI: 10.1007/978-1-4471-2717-8_32, © Springer-Verlag London 2012

clamour for building such lines here. The rail industry was seen as being in structural decline, even though the nationalised British Rail had been successful and innovative in its development of fast InterCity services. Nevertheless, overall passenger volumes had slowly declined over the period 1955–1995, reflecting progressively increasing car ownership and the development of the motorway network.

The industry was broken up and privatised in 1993–1996. Since then there has been a rapid, sustained growth in passenger numbers, in part due to better services and improved marketing by the new private operators, but also because of exogenous factors: increased road congestion, near saturation of car ownership, sharply increased fuel costs and strong economic growth. The growth in rail use changed attitudes towards the industry, and the provision of extra rail capacity moved rapidly up the political agenda.

Nevertheless, prior to the appointment of the strongly pro-rail Lord Adonis as Secretary of State for Transport in 2009, the Labour Government had not supported high speed rail. The 2006 Eddington Transport Study [1] argued against constructing new high speed routes, as journey times between major UK cities compared favourably with those in other European countries. The 2007 White Paper *Delivering a Sustainable Railway* [2] also rejected new high speed routes as inappropriate and unnecessary.

That began to change over time, with pressure from lobbying groups like Greengauge21. The Liberal Democrats were the first party to support the idea of HSR. The Conservatives were next, committed to a policy of building a high speed rail line linking London to the Midlands and the North. They saw this as a crucial step in developing environmentally friendly transport and providing the additional capacity needed to meet ever growing transport demand. They argued it would eliminate the need for domestic air travel, and with links to the existing high speed line from St Pancras to the channel tunnel, it would reduce short haul flights to Europe too, avoiding the need for a third runway at Heathrow.

The former Secretary of State for Transport Lord Adonis set up a Government owned company, HS2 Ltd, completing the consensus between the three main parties. It had a remit to prepare plans for a high speed route from London to the West Midlands, to report by the end of 2009. By the election in May 2010, all three main parties were committed to construction of a new high speed line, and this was formalised by the new government in the coalition agreement:

> We will establish a high speed network as part of our programme of measures for creating a low carbon economy. Our vision is of a truly national high speed rail network for the whole of Britain. Given financial constraints, we will have to achieve this in phases.

Despite the political consensus, it is not clear that the country as a whole is seized with the need for major rail investment, and road remains overwhelmingly the dominant mode of transport. Research for the RAC Foundation by Ipsos MORI into the transport priorities of ordinary people found fixing pot holes was their highest priority; high speed rail barely registered as a concern: 70% of respondents prioritised spending on road and pavement maintenance, while only 8% wanted to protect expenditure on high speed rail [3]. But politicians have so far pressed ahead with HS2 regardless.

The HS2 Ltd report—and the accompanying white paper—was published shortly before the election [4]. The new Secretary of State, Philip Hammond, then asked HS2 Ltd for further work, to evaluate the best options for going further North to Manchester and Leeds. And on 20 December 2011, he confirmed that the project had expanded. It is now a £32 billion scheme for a new route from Euston via a West London interchange station at Old Oak Common to an interchange station near the existing Birmingham International station and Birmingham airport. The line divides North of the interchange station, with a spur to the centre of Birmingham, a route to Manchester, and a second route via the East Midlands and Sheffield to Leeds.

The plans now also include a link between HS2 and HS1 (the Channel Tunnel Rail Link), and, in the longer term, a direct link to Heathrow. Both these two additions are very surprising, as they fly in the face of the analysis carried out by HS2 Ltd for the previous Government. In its evaluation of linking with HS1, HS2 Ltd stated:

> Running direct services to Paris or Brussels….would bring Birmingham within three hours and attract a significant market share, but the market would not be big enough to fill a 400 metre train a day in 2033. Direct services to destinations North of Birmingham would attract a smaller market share but are competing in a bigger market and might fill another train a day (Para 3.8.12)

In other words, the Government is proposing to build a long, expensive tunnel in West London for two trains a day each way. The economics of though services to Europe would be improved if security restrictions were eased and the services were able to carry domestic as well as international passengers, but this appears unlikely for the foreseeable future.

HS2 Ltd was no more enthusiastic about a direct link to Heathrow:

> ….the total market for accessing Heathrow from the West Midlands, North West, North and Scotland is currently around 3.7 million trips. Our modelling suggests relatively little of this would shift to HS2, with the rail share increasing by less than 1 percentage point (about 2000 passengers per day, or just over one train load each way) (Para 3.3.10)

In taking these links forward, the Government is making a political and strategic statement which flies in the face of the analysis carried out by HS2 Ltd, its own development company.

2 Business Case

Table 1 sets out the summary business case published by the Department for Transport (DfT) in February 2011 [5].

This updates the business case published just for the London–Birmingham section in 2010. Even in that period, the case for the project has dramatically worsened: the BCR for the London Birmingham section has declined from 2.7 in 2010 to 2.0 in 2011, only just at the level which is normally taken as a cut off point below which schemes are not regarded as being value for money.

Table 1 Quantified benefits and costs (£ billions) of the Y network (2009 PV/prices) and the resulting BCR

(1)	Transport User Benefits	Business	£25.2 bn
		Other	£13.1 bn
(2)	Other quantifiable benefits (excl. Carbon)		£0.4 bn
(3)	Loss to Government of indirect taxes		-£2.7 bn
(4)	Estimate of additional released capacity		£1.3 bn
	benefits facilitated by the Y network		(£0 – £2.6 bn)
(5)	**Net Transport Benefits**		**£37.3 bn**
	= (1) + (2) + (3) + (4)		(£36.0 bn – £38.7 bn)
(6)	Wider Economic Impacts (WEIs) (London – West Midlands only)		£4.0 bn
(7)	Estimate of Additional WEIs from the Y network		£2.3 bn
			(£0 – £4.7 bn)
(8)	**Net Benefits including WEIs**		**£43.7 bn**
	= (5) + (6) +(7)		(£40.0 bn – £47.4 bn)
(9)	Capital costs		£30.4 bn
(10)	Operating costs		£17.0 bn
(11)	Estimate of additional classic line cost		-£3.1 bn
	savings facilitated by the Y network		(£0 – -£6.1 bn)
(12)	**Total Costs**		**£44.3 bn**
	= (9) + (10) + (11)		(£47.4 bn – £41.3 bn)
(13)	Revenues		£27.2 bn
(14)	**Net Costs to Government**		**£17.1 bn**
	= (12) – (13)		(£20.2 bn – £14.1 bn)
(15)	BCR without WEIs		2.2
	= (5) / (14)		(1.8 – 2.7)
(16)	**BCR with WEIs**		**2.6**
	= (8) / (14)		(2.0 – 3.4)

Source: HS2 Ltd

N.B. the numbers in brackets represent a range around the central numbers presented above them.

There are major flaws in key aspects of the business case. But even if this were not the case, the overall business case is extremely weak. HS2 claim a benefit cost ratio of 2.6—not good for transport projects, and way below the cut-off point for road schemes, even though 80% of travel in Britain is by road. Moreover, the project comes nowhere near to paying for itself, as only 38% of the total claimed benefits (net transport benefits + revenues) are captured through fares. The value of the net revenues once it has been built—fares (£27.2 billion) less operating costs (£17.0 billion) estimated over a 60 year project life—only cover 34% of the capital costs. And that assumes the revenue forecasts are realistic: there is compelling evidence that they are overstated.

3 Review of Claimed Benefits

3.1 Demand Growth

The case is based on extraordinary estimates of demand growth. HS2 is forecast to deliver growth of 216% by 2043. That is more than three times the number of passengers carried on the existing InterCity services out of Euston. However, the

Fig. 1 Comparison of forecasted and actual passenger number 1996–2008

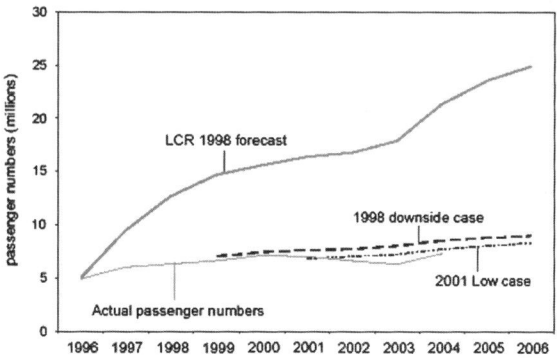

Department for Transport (DfT)'s own National Transport Survey shows that overall transport demand is no longer growing with GDP; the average distance travelled per person slightly declined over the period 1995–1997 to 2007–2008, before the start of the current recession

Many major rail projects in the past have been based on forecasts that prove to be wildly optimistic. Eurostar's passenger numbers in 2009 only reached 37% of the level forecast for 2006 when the project to build the existing high speed route to the Channel Tunnel began.

The Public Accounts Committee reviewed the results and produced the following graph (Fig. 1), showing the extent that demand projections were inaccurate [6].

They reported that:

> The Department told us that it has now learned from all this experience, and that the next time it considered undertaking a major transport project, it would factor more severe downside assumptions into its business case analysis.

Unfortunately, the HS2 business case shows that these lessons have not been learned.

Over-optimistic demand forecasts are by no means just a British phenomenon. Research at Aalborg University in Denmark [7] concluded that:

> For nine out of ten rail projects, passenger forecasts are overestimated; average overestimation is 106%

Recent examples of projects elsewhere in the world include the new Amsterdam–Rotterdam–Antwerp high speed line, and the Taiwan high speed line; in both cases passenger volumes are much lower than forecast. This has also been the case for the domestic services between London and destinations in Kent on the British High Speed 1 route. Many passengers on the routes affected have continued to use the cheaper, slower services on the "classic" routes, despite these trains having been slowed down as the result of additional stops, and low loadings on the high speed services have led to the number of train sets in daily use being significantly reduced.

So are HS2's growth forecasts are credible? They are certainly at the top of the range of serious forecasts, as shown in Table 2, based on the March 2010

Table 2 Forecasts of long distance rail travel demand

Source	Date	Period	Increase (%)	Annual rate (%)
DfT all [8]	2007 (July)	2006–2027	65	2.4 (1.8 from 2017)
DfT all [9]	2007 (July)	2006–2030	73	2.3
Network Rail	2010 (August)	2008–2034	70	2.1
Network Rail (WCML)	2009 (June)	2007–2036	47–89	
Prof J Dargay (for Independent Transport Commission)	2010 (January)	2005–2030	35	1.2
HS2 Ltd (Atkins)—WCML	2010 (February)	2008–2033	133	3.4
HS2 Ltd (Atkins)—all	2010 (February)	2008–2033	62	1.9

business case. While HS2 expected background growth of 133% by 2033, Network Rail, for example, forecast growth in the West Coast corridor of a much lower range of 47–89% by 2036, 3 years later. And Network Rail has no incentive to produce low forecasts.

The demand forecasts have subsequently been reduced in DfT's latest iteration of the business case (February 2011), to reflect both the impact of the recession and the Government's decision to allow regulated rail fares to rise by 3% above the Retail Price Index for 3 years. But the new forecasts are still higher than the alternatives shown above; and demand growth has now been capped at 2043, instead of 2033 in the 2010 Business Case.

It is probable that there will be significant growth in rail demand, at least for the next 5–10 years, but there must be real doubts about whether growth will continue indefinitely. The models used to estimate demand are essentially short term; 15 years ago, rail demand had a negative time trend, so at times of moderate economic growth, rail volumes remained broadly static. Indeed, overall passenger volumes were largely static over a 40 year period, until about 1995, as shown in the chart (Fig. 2).

It is also appropriate to look at how demand might develop on specific flows, to assess whether there is a plausible hypothesis to support a view that "background" demand will double by 2043, with a further equivalent growth as a direct result of the HS2 journey time reductions.

3.1.1 London–Paris/Brussels

Eurostar is the market leader for travel between London and Paris and Brussels, with a mode share of 80–85%. The quality of the rail product is very high, with journey times of around 2 h, and high speed operation on both sides of the Channel Tunnel. St. Pancras, the London terminus, is now an iconic location. But the evidence is that the market is saturated; the improved rail service has captured mode share but has not led to rapid growth in the total transport market. Given this experience, why should HS2 produce a further step change in total travel demand on domestic routes in Britain?

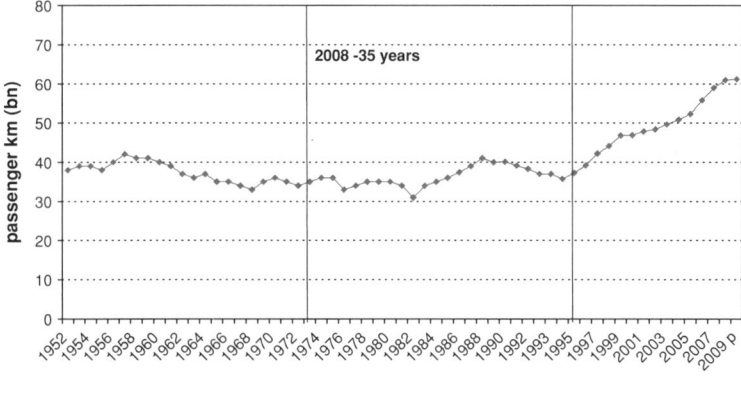

Source: Transport Statistics of GB, 2010 release, Table TSGB0101

Fig. 2 Long term rail growth

3.1.2 London–Manchester

Here again, rail is the market leader for travel from Greater Manchester to central London. The improved rail services introduced after completion of the West Coast Main Line upgrade at the end of 2008 have led to a significant modal shift from air to rail, with rail now carrying c80% of the combined market. Rail frequency is now every 20 min, with end to end journey times of just under 2 h 10 min—this is by any standards a high quality link. But is it credible to argue that a reduction in journey times to 1 h 15 min will produce a further step change in volumes on this flow? Existing rail demand models may predict this, but these are based on empirical data from situations in which rail could grow its business through increased mode share. Mancunians are not likely to triple their travel to and from London just for the pleasure of riding on HS2.

Other key HS2 flows such as Birmingham and Leeds to London have similar characteristics, except that there are no existing flights to London from which to capture mode share.

If future demand is at half the level forecast by HS2—still more than double present levels—the business case for HS2 would collapse, as the scale of the benefits delivered directly relates to passenger volumes.

3.2 Forecast Revenue Growth

The revenue forecasts for HS2 are not only a function of the demand forecasts, but are also based on two significant assumptions: continued real pricing at 1% above the Retail Price Index until 2033, and lack of competition from operators on the existing network.

The pricing assumption is aggressive, but the competition assumption is heroic. There will be parallel services on the existing network for many flows, for example London–Birmingham and London–Manchester, as these are vital to serve significant intermediate flows to both ends of the route. And the routes themselves will have spare capacity, with parallel services able to operate at an attractive frequency, probably half-hourly. So HS2 is unlikely to offer a significant frequency benefit, and the economics of rail operation are such that operators will have an enormous incentive to price to fill spare capacity, which in turn will reduce yields on HS2.

It is interesting that even Philip Hammond, the Secretary of State for Transport, has recently offered this prospect. He told the Daily Telegraph (1st April) that:

> Our proposed new high speed rail network would free up a huge amount of space on the current railways for more trains to operate. Building a whole new line would create scope for people who live on the current lines to have more frequent services that are less crowded—I would also hope that this additional competition could mean cheaper fares as well

The overall additional revenue included in the HS2 business case [10] is £27.2 billion (net present value over 60 years). Competition with existing routes will undoubtedly halve this, probably more, effectively destroying the Government's business case.

3.3 Transport User Benefits

The majority of the transport user benefits relate to the time saved by passengers using the new service. In principle, this is conventional transport economic analysis, but the values used are an order of magnitude too high. The value of time for business travellers assume both an implied average salary of £70,000 pa, and that all time spent on the train is wasted: are all the passengers who read reports or work on laptops wholly unproductive? This is a key issue—the time saving benefits make up a major part of the £37.3 billion net transport benefits claimed for the project. The transport user benefits claimed for business travellers alone are £25.2 billion, only just below the claimed total revenue improvement of £27.2 billion.

So there are clearly major problems with the current assumptions, which significantly overstate the value of the time savings and need to be seriously revised in order to reflect reality. It appears that the DfT themselves recognise this, but refuse to reduce the claimed benefits, asserting that HS2 will have reduced overcrowding, and there will be benefits for business travellers who transfer from air and road to rail and will now be able to work during the journey. This analysis is simply not credible.

4 Environmental Impact

HS2 Ltd themselves say that the project is at best neutral in terms of carbon emissions. Most of the passengers it is forecast to carry are either transferred from existing rail services—where faster trains inevitably increase carbon emissions—or are additional journeys as a result of the faster trains, which also increases emissions. There will also be an environmental impact from the construction work. HS2 Ltd forecast only modest transfers from road and air, with passengers transferred representing only 7 and 6% respectively of HS2's passengers. The project also has only minimal benefits in terms of reduced road congestion—HS2 Ltd evaluate that the new railway will reduce congestion on the M1 motorway by only 2%.

HS2's supporters claim that the environmental benefits of transfer from air are understated, as domestic air traffic at Heathrow has declined because of slot constraints. This is certainly true, but overall domestic air volumes to *all* London's airports declined by 26% between 2004 and 2010—yet HS2 Ltd's 2010 business case assumed continued high growth in domestic air traffic.

Similarly, supporters argue that new high speed trains will be more efficient than the existing InterCity rolling stock. But the laws of physics are quite clear: the same train travelling at 300 km/h will consume much more energy than if it was only travelling at 200 km/h.

If the Government wanted to reduce the carbon impact of transport, Ministers would be working out how to reduce travel by all modes or make more environmentally friendly modes cheaper, not encouraging growth and expensive alternatives. In this context, it is ironic that Government has two consultations running in parallel, for the HS2 project, and for "Alternatives to Travel" [11].

5 Provision of Additional Capacity

Ministers have consistently argued that it is vital to provide additional rail capacity to meet future demand growth, in effect espousing a doctrine of "Predict and Provide" for rail which was abandoned years ago for road, and has de facto been abandoned for air also, given the lack of any strategy for dealing with future potential growth in air demand in London and the South East.

The capacity required is of course a function of future demand, for which the forecasts for HS2 are almost certainly greatly overstated. But analysis of the technical specification for HS2 shows that the project is not in reality capable of delivering the claimed capacity.

5.1 HS2 Technical Specification

The technical specification for HS2 reflects unbounded ambition for the project, the engineering equivalent of the blind faith of its political sponsors. Despite the relatively short distances in Britain, the route is to be designed for speeds of 400 km/h, and to have initial operating speeds of 360 km/h, faster than anywhere else in Europe. The only country to have operated its high speed services at close to this speed so is China, at 350 km/h, where the distances are such that there may be useful potential journey time savings. But China has recently reduced speeds to 300 km/h, to reduce unsustainable maintenance and energy costs. In contrast, the time saving as a result of operation at 360 km/h between London and Birmingham is of the order of 4 min against what could be achieved at 300 km/h; the latter would save construction, energy and maintenance costs.

The planned utilisation of the route is 18 trains/h in peak periods, 15 trains/h off peak. This is higher than that achieved on any other high speed line in the world, and most industry experts do not believe this level of utilisation is achievable, particularly as 6 out of 18 trains are shown as originating on the existing network. The Tokaido Shinkansen operates at the highest capacity, with up to 14 trains an hour at peak periods, despite the constraints of varied stopping patterns—slow trains are overtaken several times on-route. French high speed lines operate at up to 12 trains/h at present. German, and Spanish routes operate at lower levels of capacity, in the case of Spain typically at no more than four or five trains an hour.

And in a "Why we need HS2" supplement (April 2011, page 56), Modern Railways reports that Jacques Robouël of Systra stated at a recent HS2 conference that:

> the present signalling on high speed lines allows a dozen trains an hour in each direction— the European Rail Traffic Management System is probably not going to increase this number

Systra is SNCF's consultancy arm, so the company has an enormous knowledge of high speed rail, and a clear interest in promoting it. Yet its staff believe that 12 trains an hour, not 18, is the practical maximum for a high speed line.

Greengauge21, the pro-HS2 lobby group, published a useful and comprehensive technical note on its website, as Appendix B to its report "Fast Forward: a high-speed rail strategy for Britain" [12]. This gives considerable detail on the technical capacity of high speed lines.

The report also includes a table setting out the "technical headway", the absolute minimum time between two trains at various maximum speeds. This is shown in Table 3.

The best technical headway quoted in the Greengauge21 report is 2.27 min, at 350 km/h, close to the 360 km/h claimed by HS2 Ltd for their operation. However, no European high speed line in fact operates at this speed at present.

The table indicates that the design capacity varies between 16.8 and 22.1 trains/h. However, this is a purely theoretical capacity, as it makes no allowance for *any* delay whatsoever, even of a few seconds. The more realistic figure is the "trains/h at 75% of design capacity" column, which represents the maximum realistic level of operation, ranging from 12.6 to 16.6 trains/h.

It is therefore clear that the claimed 18 trains an hour for HS2 is not achievable. The key constraint is not signalling technology but the braking distance for trains

Table 3 Techincal Headway

Maximum speed limit (km/h)	Speed with 5% punctuality margin (km/h)	Number of blocks	Block lengths (m)	Headway (m)	Speed (m/minute)	Technical headway (min)	Trains/hr, design capacity	Trains/hr 75% of design capacity
300	285	7	1600	11600	4750	2.78	21.6	16.2
300	285	7	2000	14400	4750	3.36	17.8	13.4
320	304	8	1600	13200	5067	2.94	20.4	15.3
320	304	8	2000	16400	5067	3.57	16.8	12.6
350	332.5	8	1600	13200	5542	2.72	22.1	16.6
350	332.5	8	2000	16400	5542	3.29	18.2	13.7

from full speed to a stop, which increases in relation to the square of the speed—if a train comes to a sudden halt for any reason, it is essential that the following train can stop safely without running into the train in front.

In summary, operation of the planned 18 trains/h is almost certainly impractical. Based on experience in other countries, the maximum realistic capacity is 12–15 trains/h. A reduction in planned use to 12–15 trains/h will significantly reduce the available range and frequency of HS2 services to London, with a major adverse impact on the business case for the project.

5.2 HS2 Service Plans

As with the technical specification, the detail of the service pattern to be operated has not been thought through. This would not be particularly important if the planned level of utilisation was conservative, but needs to be rigorously developed at the very levels of capacity utilisation planned for the route.

As an example, the specification does not include any trains to Heathrow or mainland Europe via the HS1 connection. The document states:

> further work is being done to determine which of the above services might serve Heathrow….and which might run to mainland Europe [13].

No information is available on the proposed frequency of Heathrow and HS1 services, other than it is stated that Heathrow trains may join and split on-route, presumably at Birmingham Interchange. This operating pattern represents a major constraint on timetable planning, and may in practice not be possible.

Furthermore, operation of trains to Heathrow and mainland Europe will further reduce the range and frequency of HS2 services to London, reducing capacity and further weakening the business case. Detailed analysis of the case for these links confirms that their financial performance would be dreadful, with services on both links requiring operating subsidies. There is no obvious policy reason for subsidising regional rail services to Europe, given that flexible, low cost air services are already provided by airlines such as Flybe without subsidy, using low capacity planes to offer frequent departures on routes such as Birmingham to Paris and Manchester to Brussels.

5.3 HS2 Reliability

It is also clear that operation at or near the claimed level of utilisation would be extraordinarily problematic. There are a number of factors which will impact on HS2's reliability

- The requirement for absolute precision in all aspects of operation. Operation at the claimed level of 18 trains an hour requires trains to operate at an average interval of 200 s, little more than the absolute technical minimum.
- Presentation of trains from the existing network. Southbound, six out of the proposed 18 trains an hour will have started their journeys on the existing network, in some cases having travelled significant distances over busy main lines, for example from Glasgow and Newcastle. It is not realistic to expect that these trains will always be precisely on time—but if they miss their "path", there will inevitably be significant consequential delays, as there is no resilience or spare capacity with 18 trains/h operation.
- Presentation of through services from mainland Europe. These trains will inevitably be subject to risk of delay, having travelled on TGV Nord, through the Channel Tunnel, HS1, the busy North London Line and the single track tunnel between Camden and Old Oak Common—again, if they miss their "path" from Old Oak Common, there is a real risk of consequential delays to other HS2 services.
- The pattern of operation proposed for Heathrow trains, with joining and splitting on-route, adds significant complexity and risk to the planned operation. Without a detailed timetabling exercise, which it is clear has not yet been carried out, the ability even to plan the proposed Heathrow services, with trains splitting and joining at Birmingham Interchange, is unproven, and indeed may well be impracticable.

As with timetable planning, it is clear that no work has been done to simulate the reliability of the planned use of HS2. It is also clear that there is no prospect of HS2 delivering reliable operation at the claimed level of utilisation.

5.4 Alternative Provision for Growth

Rigorous evaluation of proposals to construct a £32 billion rail project should properly include consideration of all alternative options, with the project itself evaluated against the best alternative, rather than an artificial "do minimum" case, as has been the case with HS2. Options should be considered incrementally, starting with proposals which prime facie offer the best value for money. The options would include:

- More effective demand management, including use when appropriate of obligatory reservations.
- Rolling stock reconfiguration, particularly conversion of some first class vehicles to standard class.

- Operation of longer trains, to the extent that this is possible without major infrastructure expenditure.
- Targeted infrastructure investment to clear selected bottlenecks to enable frequencies to be increased.
- Construction of new infrastructure.

It should be noted that the Department for Transport (DfT) and HS2 Ltd have given *no* consideration to improved demand management and rolling stock reconfiguration, and have not optimised their evaluation either of train lengthening, or of incremental infrastructure investment.

HS2's supporters argue that the work carried out on alternatives has been comprehensive and shows that the better investment is in a new line, not more upgrades, and that the new line has most value if it is built for high speed. However, a detailed review of the work published by the Government doesn't substantiate this.

Rail Package 2 [14], prepared by DfT in its review of alternatives, would cost less and can be delivered faster, has a much better benefit cost ratio and provides for 15–16 InterCity trains an hour from Euston. This is clearly effectively the limit on what can be achieved on the existing route, but at present there are only 9–10 trains an hour from Euston so it does represent a major increase.

At the same time, additional vehicles are already under construction to lengthen most trains from 9 to 11 cars, giving 150 extra standard class seats—an increase of 50%. There are of course some trains with standing passengers now, before the extra vehicles already on order are introduced, but the scale of this shouldn't be overstated. Network Rail recently published its "Route Utilisation Strategy" for the West Coast Main Line [15], as part of the key rail industry planning process. This shows there are currently two Virgin West Coast train a day which have standing passengers, out of a total of 287; there are 10 on Fridays. A combination of the already committed extra capacity, and smarter yield management to reduce the artificial peaks caused by time restrictions for regulated "saver" fares, could resolve this problem for quite a few years. And the existing load factor for the route—the percentage of seats occupied—is probably only about 50%.

The economics of rail would be transformed if the industry achieved airline load factors, but this may be considered too ambitious. It's reasonable to argue that load factors could be increased to 60% through improved IT and more effective yield management. In addition, trains can be lengthened to 12 cars, except for Liverpool services, because of the physical constraints at Liverpool Lime Street. One first class vehicle in each train could be converted to standard class and, with incremental infrastructure investment to relieve a small number of specific bottlenecks, frequency could be increased by one or two trains an hour. Taken together, this package of actions can deliver a total standard class capacity uplift of 215%, far in excess of the background growth forecast for HS2. Capacity increase on this basis can also be delivered incrementally, as and when it is needed, in contrast to HS2, which is an "all or nothing" solution.

In summary, it is clear that the alternatives to HS2 have not been optimised, and the economic case for the project has been assessed against an artificially constrained "do minimum" base, not, as should be the case, against the best alternative option.

6 Economic Regeneration

Regenerating the Midlands and the North has become one of the Government's central arguments for high speed rail—it sounds good, and it's very difficult to quantify. To understand the issue in quite a straightforward way: Manchester to London already has a train every 20 min, taking just less than 2 h 10 min. Will improving this to, for example, a train every 15 min taking an hour and fifteen minutes transform Manchester's economy?

It's not clear that everyone in Manchester really believes it will. The Manchester Independent Economic Review (2008) states [16]:

> For additional investments within the North of England as a whole, including Leeds–Manchester, the case is stronger than for additional investments on the route to London

This may reflect the current poor services in the North West: for example, train services between Manchester and Liverpool are slightly slower and no more frequent than a 100 years ago [17].

Some economists have argued that HS2 is likely to exacerbate the North–South divide rather than reduce it. It would be even quicker for those living outside of London to make the journey into the prosperous capital. Professor Henry Overman from the LSE gave cautious evidence to the Transport Select Committee:

> […] claims about the transformational nature of transport investments for particular areas should be greatly discounted…..because they have no convincing evidence to support them.

His fellow witnesses in the same session echoed these sentiments. One young PhD put it to me rather more brutally: "the best way to reduce the North South divide would be to lengthen the journey times to London".

HS2 Ltd's own analysis supports their conclusions. Their estimate of the present value of "wider economic impacts", which includes the regeneration effects, is £6.3 billion, much lower than the £30.4 billion capital costs.

It is, of course, likely that the immediate area close to each of the HS2 stations will become a development honey pot, creating an illusion of more general economic growth and dynamism. But in reality, this is much more likely to represent a concentration of development within a city or region. So the southern part of Greater Manchester is likely to see development focused on the proposed South Manchester parkway station, but this will be at the expense of the northern side of the city region, and at a wider level, other cities in the region such as Liverpool.

Some of the economic benefits arise from use of the spare capacity created on the existing network. HS2's supporters point out that these benefits are a direct result of HS2. This is largely true, but it is clear that most of the extra services that generate these will require further subsidy, and it is unclear whether the cost of this is included in the HS2 business case. Thus Centro (the Passenger Transport Authority for the West Midlands) argue the case for additional services, but are silent on the need for additional long term subsidies.

In this respect, HS2 represents a trap for the taxpayer: every time a new high speed station is built, there will be demands for major investment, and ongoing subsidy, to deliver the promised regeneration or provide the transport links needed for passengers to access the new service. Euston is the most dramatic case: the Victoria Line is already full, and the Mayor of London has already publicly stated that the additional passengers forecast for HS2 mean that Crossrail 2 has to be built, at a cost of up to another £9 billion. "A billion here, a billion there, pretty soon it adds up to real money" [18].

7 Comparison with Other Countries

Ministers have argued that Britain must invest in high speed rail to keep pace with other countries. These comparisons are not valid.

However, most countries that have invested in high speed rail have done so for journeys which are much longer than in Britain: Tokyo to Osaka is 553 kms, much further than London–Manchester (296 kms); In contrast with Britain, Madrid to Barcelona took seven hours by rail before the high speed line was built, so almost everyone flew; now it takes less than three hours, and most people go by train. Distances in Britain are simply not long enough for high speed rail to deliver an equivalent step change, except for journeys between Southern England and Scotland, which make up a relatively small proportion of total long distance travel.

Capital costs are also much lower in other European countries. The cost per mile for HS2 is much higher than in mainland Europe—typically twice as much or more—reflecting a combination of easier topography and lower unit costs, an issue which Infrastructure UK is seeking to address.

Other countries generally also started off with rail services which were significantly slower. The existing British network already offers fast, frequent and increasingly reliable rail services from London to major British cities. In contrast, in Germany, for example, there is a new high speed line from Frankfurt to Cologne, the same distance as London–Birmingham. But the old route from Frankfurt to Cologne takes 140 min, compared with 84 min for Birmingham–London today.

Rail has also been gaining market share. The Association of Train Operating Companies has recently published details of rail's growth at the expense of air for key long distance flows where rail and air compete [19]. Rail's share of the combined rail/air market grew significantly between 2008 and 2010, for example from 69 to 79% between London and Manchester. Rail has even done well on the much longer distance Anglo-Scottish routes, growing from 12 to 20% between London and Glasgow.

The financial results of many high speed lines have also been poor, with risk of bankruptcy for schemes taken forward as PPP projects such as in Taiwan and the Netherlands. And at a national level, Spain has invested proportionately more than anyone else; it's certainly increased the country's debt, but less clear it's helped to drive economic growth. Also, President Obama's ambitions for high speed rail in the United States have essentially ground to a halt.

8 The Opportunity Cost

Britain is going through a period of stringent public expenditure cuts, unprecedented in recent years. Yet expenditure on HS2 is protected, with £750 million earmarked for planning and development of the project in the Government's 2010 spending Review. Actual construction expenditure on HS2 will not start for at least 5 years, so the development expenditure itself will undoubtedly represent an opportunity cost for the industry. It is almost certain that smaller scale improvements will be squeezed out. Some cuts have already taken place, like the £50 million 'Better Stations' fund and the uncommitted HLOS1 [20] additional rolling stock, and HLOS2 is likely to be steady state at best.

The Midland Main Line represents a substantive example. Network Rail's "Network RUS—Electrification" [21], published in October 2009 claims that there is a financial case for electrification of the Midland Main Line, the only route evaluated for which this is the case. However no decision has been taken by DfT on whether to proceed with electrification of the route, and the Secretary of State has already indicated his approach in answer to questions in the House of Commons from Members with constituencies served by the Midland Main Line following his statement on electrification of the Great Western Main Line:

> The announcement today does not include provision for the midland main line. The hon. Gentleman mentioned bi-mode trains, and I am sure that he has also been lobbying for the electrification of the line, as have many other midlands Members. The debate about the line's future also has to take account of the implications of High Speed 2, however. Once the High Speed 2 consultation, which began yesterday, has been completed and the Government have announced their definitive plans later this year, it will be much easier to plan for the long-term future of the midland main line.

The clear implication is that there will be no substantive improvement to routes which will eventually be affected by HS2 until its completion, in this case when Phase 2 is built in 2032/3 at the earliest.

9 Impact on Rail Industry Finances

The last government set up the Rail Value for Money Study, led by Sir Roy McNulty. His report, "Realising the Potential of GB Rail" [22], published on 19th May raises a number of new and important issues in relation to the case for High Speed Rail.

The study was set up to investigate why the costs of the rail industry in Britain are significantly higher than for comparable European networks. The report confirms that the efficiency gap could be as high as 40%. One of the key conclusions was that an important factor is the lower level of train utilisation in this country, with on average fewer passengers using each train [23]. The report therefore recommends that there should be much better use of existing capacity:

> There should be a move away from 'predict and provide' to 'predict, manage and provide', with a much greater focus on making better use of existing system capacity [24]

The study also identified a bias in the planning system towards capital expenditure [25]. This is illustrated by the Network Rail "Route Utilisation Strategy" process which captures the plans and aspirations for using key parts of the rail network from existing and potential users. However, rather than prioritising these on the basis of economic value, the process tends to look for physical solutions which enable all the aspirations to be met, in effect, "predict and provide", an approach which was dropped for the road network some years ago, and has now also been implicitly dropped for airport capacity in the South East.

These issues are brought together in recommendation 6.3.7:

> The Study considers that industry, together with the ORR and the DfT, should review incentives and responsibilities for the efficient management of capacity. There needs to be at least as much focus on train utilisation (the number of passenger km per train km) as there is on track utilisation (the number of train km per main track km). Existing approaches appear to focus much more on track utilisation and the provision of train paths, but whilst that is important, the unit costs of carrying passengers are influenced heavily by train utilisation, which does not appear to be a primary focus for any organisation within the present system.

This report's recommendations strongly point to incremental improvements to the rail network. In contrast, the proposed HS2 project is a clear and dramatic example of the failures that he has identified. Even on DfT's own optimistic evaluation, the project would have a net cost to the taxpayer of £17bn over 60 years.

10 Passenger Profile

Research shows that nearly half (47%) of long distance rail journeys in Britain are made by people from households in the top income quintile (Fig. 3).

Expenditure on HS2 therefore represents regressive taxation, as the people likely to use it are typically much wealthier than the average for the whole population.

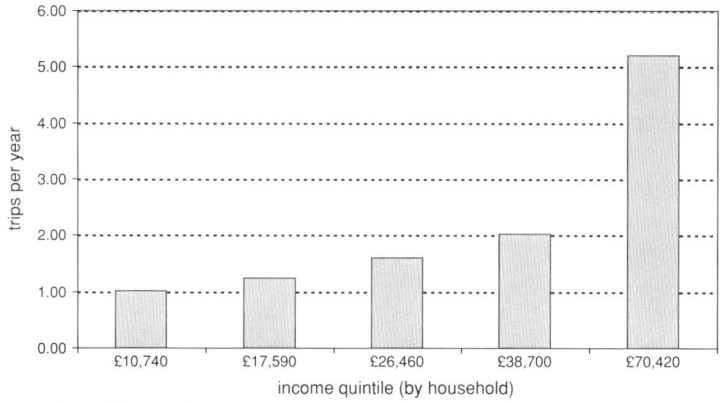

Source: 'Modelling Long-Distance Travel in the UK', Charlene Rohr, James Fox, Andrew Daly, Bhanu Patruni, Sunil Patil, Flavia Tsang. RAND Europe, NTS 2002/5, income data 2005/6 ONS

Fig. 3 Lond distance rail trips by income (*Source* NTS)

11 Conclusion

Fitch Ratings published a report on high speed rail projects in April 2010 [26]. This identified over-optimistic revenue forecasts as a major reason for the failure of a large number of public–private partnerships (PPPs) for high speed rail projects. One of the reasons cited is the political environment:

> Rail projects are often high profile. This exposes them to "political entrepreneur syndrome" where the public authorities overestimate the benefits of the project to get it approved for the purposes of political gain

There is mounting evidence that HS2 is now in this position; Government Ministers are increasingly making subjective, emotional arguments to support the project. The case for HS2 is based on flawed demand and revenue forecasts and an inflated evaluation of transport user benefits. The technical specification for the project is also flawed, as it is almost certain that the scheme will not be able to deliver the promised capacity, so some regions which now expect to have high speed services in the future are unlikely to see these in reality.

Britain has relatively short distances between its major centres of population and already has fast, frequent InterCity rail services. High speed rail is not needed in this country.

References

1. http://www.thepep.org/ClearingHouse/docfiles/Eddington.Transport.Study%20-%20Rod.pdf
2. http://www.dft.gov.uk/pgr/rail/whitepapercm7176/
3. http://www.racfoundation.org/media-centre/holes-over-high-speed-rail

4. http://www.dft.gov.uk/pgr/rail/pi/highspeedrail/hs2ltd/hs2report/
5. The economic case for HS2. http://highspeedrail.dft.gov.uk/sites/highspeedrail.dft.gov.uk/files/hs2-economic-case.pdf, p 12
6. http://www.publications.parliament.uk/pa/cm200506/cmselect/cmpubacc/727/72705.htm
7. Flyvbjerg B, Holm MKS, Buhl SL (1999) Inaccuracy in traffic forecasts. Department of Development of Planning, Aalborg University
8. Delivering a Sustainable Railway: Summary of key research and analysis July 2007, slide 27
9. Delivering a Sustainable Railway, Cm 7176, Dft, July 2007, paragraph 6.6, p 60
10. Economic Case for HS2 February 2011, p 12
11. http://www.dft.gov.uk/consultations
12. Greengauge21 (2009) www.greengauge21.net/publications/fast-forward-a-high-speed-rail-strategy-for-britain. Appendix B, §2.4–2.6
13. http://highspeedrail.dft.gov.uk/sites/highspeedrail.dft.gov.uk/files/hs2-economic-case.pdf, p 61
14. http://www.dft.gov.uk/pgr/rail/pi/highspeedrail/alternativestudy/pdf/railintervention.pdf
15. http://www.networkrail.co.uk/browse%20documents/rus%20documents/route%20utilisation%20strategies/west%20coast%20main%20line/westcoastmainlinerus.pdf. Published July 2011
16. http://www.red.mmu.ac.uk/documents/ent_files/strategies/mier_Review.pdf
17. Bradshaw's April 1910 Railway Guide
18. Senator Dirksen 1896–1969
19. http://www.atoc.org/media-centre/latest-press-releases/shift-from-air-to-rail-heralds-turning-point-in-how-people-travel-between-uks-main-cities-100571
20. High Level Output Statement—HLOS1 is the statement by DfT of the outputs it wants the rail industry to deliver in the current Network Rail regulatory control period
21. http://www.networkrail.co.uk/browse%20documents/rus%20documents/route%20utilisation%20strategies/network/working%20group%204%20-%20electrification%20strategy/networkrus_electrification.pdf
22. http://www.dft.gov.uk/pgr/rail/strategyfinance/valueformoney/realising-the-potential-of-gb-rail/pdf/realising-the-potential-of-gb-rail-summary.pdf
23. Realising the Potential of GB Rail. Executive Summary, paragraph 4; also section 2.3.4, figure 2.12
24. Realising the Potential of GB Rail. Executive Summary, paragraph 23
25. Realising the Potential of GB Rail. Section 4.4
26. Fitch Ratings: High Speed Rail Projects, 6th April 2010

Part VI
Finance and Economics

Foreword
James Cameron

We are close to the point of giving up on finance and economics as disciplines for resolving highly complex systemic problems. We have so much information to absorb that we crave simplicity—in ways that might be dangerous in the political arena—and yet we are left unsatisfied with simple economic positions that we intuitively feel fail to explain how things are, let alone how they should be. It gets worse when these propositions are attached to a theory of action. In all this complexity what is the right thing to do and for what reasons, for we must have reasons. To have a vision for how life might get better and a means for ordering and directing human behaviours in pursuit of that position requires reasoned argument in a language that can form and inspire a societal response. It requires authoritative and real understanding of real-world facts—good data about things that matter. It requires a carrier of ideas to be free enough to connect with the many real-world actors that make a difference. Ideas must be tested, proved and experienced in many transactions, both intellectual and material, and in markets of many forms. In addition, we require a more heightened awareness of the emotional, imaginative and physiological response to the complexity. That will encourage attention on human relations as much as data and the rational response to economic stimuli.

The financial world has been in a kind of man–world tumult. Hysterical swings in sentiment, vast herd instincts decision making, concentration risk, shocking unjustifiable rent seeking, public money bailing out private errors of judgement with almost no sanction placed on those who caused loss or created harm to the common good. The flight from risk and hoarding of credit continues to damage growth prospects; where are the long-term investors? As risk capital dries up who will fund the transformation to a safer and more secure economy?

'*Energy, Transport and the Environment*' offers insight as to how it might think and ultimately cut our way out of this malaise.

James Cameron
Vice-Chairman and Founder of Climate Change Capital
Vice-Chairman of the World Economic Forum's Agenda
Council on Climate Change
Member of the Prime Minister's Business
Advisory Group

I Want to Ride My Bicycle! Financing Sustainable Transport

Julie Hudson

1 Financing Sustainable Transport Defined

A sustainable economy 'meets the needs of the present without compromising the ability of future generations to meet their own needs' [3] and a sustainable transport system should support this overall goal. The definition of sustainable finance applied in this chapter thus goes beyond the basic idea that financial institutions need to be robust enough to survive turbulent economic conditions. It also goes beyond the idea that government and private sector entities (represented in financial markets by instruments such as government bonds, corporate bonds and equities) must be sound enough to survive volatile markets. Sustainable finance is a broader concept, denoting financial institutions and structures in which being financially sound is necessary but ancillary to their broader goal of facilitating and supporting a sustainable economy, and within that, a sustainable transport system.

The way in which economic activity, and, within it, the transport system, are financed is likely to have an important influence on how sustainable or otherwise

Although Julie Hudson is an employee of UBS, her contribution to this work is independent of her employment with the company. The views and opinion expressed herein are her own. *Disclaimer.* This article draws on material from UBS and Julie Hudson. The views and opinions expressed in this article are those of the author and are not necessarily those of UBS. UBS accepts no liability over the content of the article. It is published solely for informational purposes and is not to be construed as a solicitation or an offer to buy or sell securities or related financial instruments.

J. Hudson (✉)
UBS, Finsbury Avenue, London, EC2 M 2PP, UK
e-mail: julie.hudson@ubs.com; julie.hudson@smithschool.ox.ac.uk

J. Hudson
UBS, London, UK

O. Inderwildi and Sir David King (eds.), *Energy, Transport, & the Environment*, DOI: 10.1007/978-1-4471-2717-8_33, © Springer-Verlag London 2012

they turn out to be, in the broadest sense of the word. In practical terms, transport is no different to any other sector in being funded by a mix of investment stocks (capital on the relevant private—or public-sector balance sheets) and investment or spending flows from government, corporate and consumer sectors. However, unless we understand what 'needs' we are dealing with, thinking in terms of a taxonomy of funding mechanisms is unlikely to be very informative in respect of sustainability. Before funding mechanisms can be identified, the question is what is needed for a sustainable economy.

The first key point is that *transport is needed for all sustainable economic activity*; this becomes apparent if we give transport a more generic description—such as networking, or connecting. Given the several negative environmental and social externalities[1] inherent in travel as currently configured, engaging in less of this fossil-fuel intensive economic activity without further ado might help the environment and save money. It might also have unwanted economic and social side effects that would leave everyone including future generations worse off. Maslow's well-known hierarchy of needs suggests that once physiological needs such as air, water, food, shelter and warmth are met when human beings urgently need to connect to other human beings for belonging, esteem and personal growth. Networks both tangible and intangible can be identified as fundamental to the success of the civilizations of yesteryear that succeeded in creating significant wealth. They satisfy what appears to be a psychological need, and this point cannot and should not be separated from the fact that they also support and sustain human-welfare-improving economic activity as well as financial systems. Insufficiently networked countries tend to develop more slowly; in warfare armies seek to destroy networks as an effective way of incapacitating the population and damaging the economy; and in the modern era so-called cyber-attacks are feared because of the potential financial and social damage they can do.

As one of many possible examples demonstrating the importance of networks to civilization and the role money plays in creating them, the well-known Silk Road was not just a 'road'. Defined broadly it was a dynamic many-stage multi-modal, multi-nodal transmission network facilitating the movement of a wide range of goods, technology, culture and human beings. As its name—the *Silk* Road—suggests, a nexus of financial dealings played an important part in driving its development. The possibility of getting rich by trading lucrative goods such as silk, spices, precious stones and slaves drove the early movers and the generations that followed to invest capital and effort in forging the first caravan and the connecting shipping routes.

A glance at the growth of almost any of the well-known transmission networks of history—canals, railway systems, shipping lines, road networks or the Internet—finds the same potent mix of behavioural forces at work, with markets and finance in the broader mix as a facilitator. The arrival of new networks tends to foster a whole host of quality-of-life-improving, welfare- and wealth-generating

---

[1] Environmental pollution, congestion, noise, impacts on human health, to name a few.

human activity, often stimulating a significant volume of financial and other market transactions at the same time. It should be pointed out that this is not a one-way street. The arrival of new networks can render old ones obsolete. In the case of the Silk Road, pure economics dictated the eventual switch from land to sea made possible by a shipping route round the Cape of Good Hope. There were without doubt financial as well as economic, and social winners and losers. As the new infrastructure arrived caravanserai were scrapped.[2]

2 Sustainable Transport Investment: A Financial Hybrid

As the above sections suggest, any transport system or project is supported by a hybrid mix of environmental, social and financial costs and benefits which, between them, contribute to the development of sound economies, and sound transport systems. In an ideal world *all* transport finance would be sustainable. Sustainable finance would play a facilitating, co-ordinating role in the delivery of an optimal balance among the economic, social and environmental needs of society in general. This presupposes balance in financial markets themselves, but, as history suggests, money is capable of stimulating growth in networks at the same time as laying the seeds of their future destruction, as in the well-known nineteenth century American boom and bust in railroads. A failure to embed broader sustainability goals throughout the relevant market can also be responsible for destroying financial and economic as well as environmental and social value. It must never be forgotten that finance is a two-edged sword, capable of value creation or value destruction, depending on how it is wielded.

The profit motive, sharpened by an increasingly competitive market[3] and facilitated by finance, drove the technology innovations that delivered high speed transatlantic passenger ships in the early twentieth century. Market conditions and 'hopelessly outdated' regulation [4, p. 72] may also have been a factor in decisions relating to the speed of travel and number of lifeboats on the Titanic. According to one writer ship owners' decisions were 'largely based upon the economical factor' [15] Kindle locations 1955–1969), rather than upon safety and security practices that could (with hindsight) have rendered this particular transport business more sustainable.

The so-called dot-com bubble at the end of the twentieth century left a tide of wealth destruction in its wake. Yet it also left behind the foundations for an increasingly networked world opening up opportunities in economic and cultural exchange and perhaps even future transport-related solutions. This bubble may one day, from a longer historical perspective, be seen as a moment in which the social value created far outweighed the financial wealth that went down the drain in

---

[2] With thanks to Paul Donovan for this suggestion.

[3] 'If Cunard wanted to build big, the White Star would build bigger…' p. 10

2000–2001. In the longer-run, indeed, financial gains may also have outweighed the shorter-run financial losses, although probably not always for the same individuals or organisations at the same time.

The general point implicit in all of the above is the significant potential for risk and timing mismatches in the context of networks in general, and transport specifically.[4] This problem is compounded by the need to deal with the intangible and tangible costs and benefits typically found in the context of value created on the basis of networks. On a positive note, dealing with timing mismatches and risk by facilitating exchange is the stock-in-trade of financial markets. This suggests that it should be possible to leverage finance with the aim of delivering sustainable transport choices. On the other hand, to function properly markets need clear, relevant, accurate and timely information. Unfortunately this is very often unavailable in the context of timing mismatches and intangibles, which can turn structured project assessment exercises like cost-benefit analysis into little more than educated back-of-the-envelope guesses. This is not as much of an obstacle as it sounds: all it means is that markets and the analytical tools used in their context cannot be expected to deliver sustainable transport in isolation. For financial practitioners seeking to participate in the delivery of sustainable transport it must be kept in mind that transport is a socio-economic hybrid therefore requires a range of stakeholders in the decision-making mix. This point needs to be able to influence the shape of financial markets as well as the price of risk.

Blending social, environmental, economic and financial value means accepting that financial markets inevitably deal with asynchronous and sometimes quite uncertain financial flows. As social organisations they blend the visible and the invisible, the quantifiable and the unquantifiable, the tangible and the intangible. Dealing with the consequences of such complexity is precisely what sustainable transport finance should be about. As Michael Porter [20, 21] said of a concept he has named 'shared value': '[The] principle of shared value [...] involves creating economic value in a way that also creates value for society by addressing its needs and challenges,' the overall goal being 'economic success'.

3 Sustainability and Finance: Key Concepts

At the most basic level, markets help to address society's needs and challenges by facilitating exchanges between two or more counterparties who each want something they do not have and each have something else to offer in exchange for it. Financial markets can seem esoteric but they also simply facilitate exchange. In so doing, they play a valuable social function—one that may be potentially very

---

[4] The dominance of the fossil-fuel driven combustion engine in twentieth century transport can be seen as an accident of timing at the point of inception: among other things there was little information on the environmental impact of fossil fuels when widely used and therefore no real opportunity to embed this then future externality in the relevant markets.

Fig. 1 Basic swap transaction

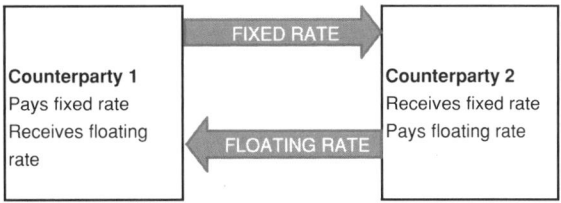

helpful in the context of sustainability. For anyone new to finance the swap transaction is a good place to start, for it is an important building block for many other financial structures. Figure 1 depicts a generic interest-rate swap—a 'fixed to floating' swap transaction in which Counterparty 1 resolves a timing problem thanks to the desire of Counterparty 2 for greater certainty over income inflows over a given period of time.[5] Almost anything in financial markets can be 'swapped': interest, credit, equity returns, insurance, mortgage fund flows, pension fund liabilities and so on. This point is likely to be relevant in the context of sustainable transport where timing issues abound. Indeed, time itself is often a driver of travel decisions and many journeys could be envisaged as time-for-energy or indeed time-for-money swaps.

The success of any financial instrument—swap, bond, corporate credit, equity or derivative—rests on a clear specification of the transaction conditions and good disclosure. Healthy underlying markets and institutions are also needed to support sound pricing. With these prior conditions in place financial instruments can potentially be structured to deal with some of the timing problems faced by organisations seeking to deliver sustainable transport.

In a different context, environmental permitting markets such as the European Emissions Trading System also facilitate an exchange. Hypothetical Firm 1 on one side of the table is suffering from a timing or capital availability problem and therefore is unable to reduce greenhouse gas emissions (GHG) on time. Hypothetical Firm 2 has a different timing problem, being ahead of target with its own GHG emissions reduction programme. Firm 1 can decide whether it makes better financial sense to rush new carbon abatement investment through, or to buy permits and defer investment until it is more affordable. Firm 2 can look to cover abatement costs by selling permits. Assuming governments would like to see more cyclists on the roads in the context of the sustainability agenda, the question is what financial markets could do to facilitate an exchange between drivers and cyclists. More broadly, whether and when it may make sense to leverage financial tools to facilitate exchange in the context of transport is one of the concepts explored in this chapter.

The building blocks of finance and valuation can also be used to conceptualise the potentially complex relationships embedded in the Michael Porter concept of shared value. As an example, in the so-called residual claims approach to finance

---

[5] This figure is generic. Similar figures can be found in Gastineau et al. [9].

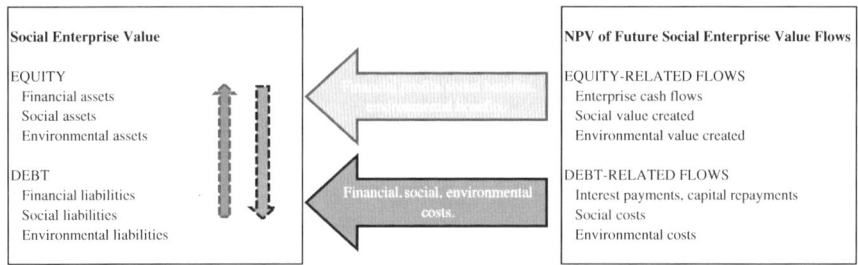

Fig. 2 A hypothetical firm, reflecting financial, environmental and social value

(explored in [12, 13]) the financial value of any given enterprise can be calculated as the market value of all claims on the business—equity, debt, pension fund liabilities and any other potential liabilities.[6] It can also be calculated as the net present value of enterprise cash flows. However, this is not the whole story when the goal is to deliver sustainable businesses in the broadest sense of the word. Any hypothetical firm funded by the conventional mix of equity and debt in the transport (or any other) business is also supported by far more than the mix of equity and debt shown on a conventional balance sheet; it could be represented as shown in Fig. 2.

On the left, we see the total value of the enterprise. To the usual picture of enterprise value from the balance sheet perspective has been added intangible environmental and social assets and liabilities. In conventional accounting frameworks these would not usually be recognised unless a catalyst event made it highly probable that they would crystallise in financial terms. Similarly, environmental and social value that might be created on a flow basis over the life of the firm or project has been added to the right-hand side of the figure. These, too, would not usually appear in conventional accounts unless likely to be converted to cash. If, at any point, hidden social or environmental assets or liabilities should look likely to become financially material, then financial markets can and do respond ahead of company accounts. As soon as those environmental or social intangibles become tangible, then, other things equal, the market value of the conventional firm would change, rising with the value of newly recognised assets, or falling with the in proportion to the size of any newly recognised liabilities. This concept captures the potentially significant social and environmental value that flows between the enterprise and the broader economy.

This brings us to Sect. 2, where we consider the connectedness of transport systems to life, the universe and everything.

---

[6] UBS defines 'enterprise value' as the sum of the *market value* of claims on the business: equity, debt, off balance sheet liabilities and any other relevant claims on the business. The so-called residual claims approach to the valuation of cash flows to the firm is also described in Damodaran [6, p. 107].

4 Transport Systems Interconnectedness Explored

Access to affordable housing for the population at large is the main determinant of access to jobs, education and healthcare, but the state of the transport system plays an important indirect role. The price of residential accommodation is heavily influenced by its position in relation to transport networks. Access to transport is therefore arguably an important determinant of the 'standard of living people enjoy across the economy' [16, p. 180]. In developing countries, in particular, inadequate transport systems are known to result in a structural divide between rich and poor, potentially slowing growth for the broader economy. However this problem is not confined to developing countries. With sufficient space, it would be worth exploring the extent to which the ability to afford housing close to the relevant transport networks has contributed to the widely reported fall in social mobility in the UK by shaping who has access to important social enablers like education. One of the potential opportunity costs of reduced social mobility within the schooling system could be a less productive workforce in the medium term. Issues of social mobility in the current generation will affect the social welfare of future generations. Therefore, in the context of sustainable transport, there is a strong argument for allocating transport from a holistic perspective and according to need, rather simply than willingness to pay. This is often regarded as a job for governments rather than private sector firms, but this may be a short-sighted view. Given the importance of the transport system to any sustainable economy this point surely must impact corporate strategy for the private sector firms who gain from the same transport structure. However, it should be noted that expecting either of these two sectors—government, or private—to be a prime mover for change in this direction may be optimistic. Both tend to be constrained by a relatively short-term outlook dictated by the electoral cycle in one case and profit as a performance driver in the other. Nevertheless the point remains that benefits delivered through adequate access to transport are important to both.

A summary of potential benefits arising from access to transport is shown in Table 1. This is not intended to be exhaustive but focuses on benefits that should, other things equal, become visible somewhere in the system in financial or economic terms. Hence, transport can facilitate political aims and assuming these to be benign this should read through to a lower cost of capital in both government and private sectors. When transport permits efficient access to raw materials, this should be supportive of manufacturing profit margins in the private sector, other things equal. The labour specialisation facilitated by transport should also be good for profit margins for firms in the private sector, who should gain access to a more diverse workforce. Moreover, as economies become wealthier, transport can become a value-added consumer good in its own right: 'The proportion of expenditure that is devoted to transport tends to rise overall with household income, reflecting the "superior good" nature of the activity.' [5, p. 32]. The choice for the consumer in matters of where to live and work can also influence land values. Although culture is hard to monetize it surely connects in some way to

Table 1 Summary of potential gains from transport systems

	Politics	Raw material access	Industrial production	Consumption
Economic and social benefits brought by travel	Facilitates regional integration; efficient defence of the realm; opportunities for regional economic development	Allows labour force to move to where raw materials are, locally, or internationally. Allows firms to move raw materials for processing	Specialisation in industrial processes	Permits choice of living and working in different locations. Allows firms to produce and/or offer consumer choice. Travel is in itself a discretionary good
Economic or financial value likely to be generated	Other things equal: lower risk premia in financial markets	Other things equal improved labour market efficiency. Cost savings in manufacture	Improved labour productivity. Allows economies of scale therefore cost savings in manufacture	Facilitates labour market flexibility. Can boost land values

Cultural Exchange is a further possible column.
Source UBS

social stability without which none of the above is likely to happen. So-called economic externalities are often thought of as environmental and potentially negative in a financial context. However, as the above suggests, in the context of transport social externalities can turn out to be potentially economically or financially positive, something that may not always be taken into account in considering how transport is financed.

This table, and the paragraph that precedes it, introduce a critical point from the perspective of finance. The presence of a functioning transport network helps private sector firms in a range of sectors make money by facilitating exchange. Such exchanges do not always entail cash-based transactions. The hypothetical stimulus to revenues or profit margins facilitated by access to transport therefore does not appear in conventional financial statements as an 'inflow' from outside. However, when the system stops working (as for instance in the heavy snowfalls of the winter of 2010) the financial damage done acts as a forceful reminder of the dependency of financial value on the often invisible contribution of the transport system. There is no such thing as a free lunch. The question is what the costs are, how the system should be paid for, by what means and by whom. The private sector firm might argue that the boost to profit margins and other benefits described above are paid for through taxation, but this is clearly not the whole story.

At this point, it is important to emphasize three points in the context of the discussion of the above paragraphs and other points made in Table 1:

1. Transport is a social good. Whether or not the transport system works on an affordable, equitable basis can have far-reaching, intra-generational social and economic effects. The transport system must therefore be seen as a socio-economic hybrid, when viewed from a financial perspective. In practical terms this may mean a shift in the risk profile of a financial instrument. Somehow this needs to be priced in.
2. Access to transport is financially valuable. Whether transport functions efficiently (or not) can have a range of direct financial impacts. This is not the only reason for shaping finance to take the social and environmental aspects of transport into account. However, it reinforces the argument from the perspective of financial practitioners because point 1 hints at the presence of so-called 'tail risk'.[7]
3. Transport is a consumer good in its own right. For most of the 'sectors' shown in Table 1 transport is ancillary to the sector but for the consumer transport is both facilitator and consumer good. In the context of free market economies constraints on consumer choice can potentially influence election results. This may make coherent policy formation, implementation and financing challenging.

---

[7] 'Tail risk' is an insurance term, and usually refers to the tails under the statistical bell-curve in a normal distribution. Used in the vernacular it refers to so-called 'one hundred year' events—extreme accidents that should appear only rarely. Sustainable economics can be about reducing economic, environmental, social or financial tail risk for future generations.

The next step is to consider the several food chain inputs required to deliver most modes of transport. As the above points suggest, each may require a different approach, from a financing perspective.

5 Transport Markets: An Interconnected Complex Food Chain

Transport is delivered by a number of different markets. The transport food chain includes infrastructure, fuel, equipment, service provision, users and financial services. Service provision breaks down further into repair and maintenance, refuelling, traffic management, safety, information provision. One perspective on this food chain that makes up the transport sector is shown in Table 2, which summarises the food chain inputs to five travel contexts—virtual travel, urban land transport, rural land transport, air and sea.

Although the table presents the transport 'food chain' components as a discrete market, in reality they are intimately connected with each other. This must inevitably affect the financing mechanisms used to deliver them in the context of the overall aim of rendering transport systems more sustainable. In fact—a key point for this chapter—some financial innovation may be needed to deal with cross-sectoral effects.

Let us consider the private car transport market, for instance. The relationship among the infrastructure, fuel and transport equipment markets is very simply described for the private car market in Fig. 3. What this depicts is widely known. The three markets shown—infrastructure (roads), energy (liquid fuel and delivery infrastructure) and equipment (the private car)—are interdependent. Changes in one of these markets will have some sort of impact on the other two, and the speed of change in any one of these markets may be dependent on the other two. Thinking back to the hapless traveller in the title of this note ('I want to ride my bicycle!'), making it possible for him or her to travel sustainably by switching from the car to the bicycle might require significant changes to infrastructure. Making car exhaust cleaner might require significant changes to fuel, or to car engine technology. Financing such a change will require more than one input.

6 Market Structures, Pricing, Regulation and Finance

In some of the 'food chain' markets price signals determine the balance between supply and demand—for instance, in the private transport equipment market. (Sometimes this is the case for fuel, too, but not always). This suggests price could potentially be a driver of behaviour in a positive way with sustainability goals in mind. In others—infrastructure for example—price signals cannot work effectively because of the presence of sunk costs and ratchet effects. Where the price point

Table 2 Inputs to transport modes in summary

Input	Virtual	Urban land	Rural land	Air	Sea
Infrastructure	Eco-system services (land use). Landline, wireless, local loop etc	Eco-system services (land, air). Roads, fuel delivery, rail, tramlines, monorail, underground ('tube'), garaging/parking. Potentially—charging infrastructure for electric vehicles	Eco-system services (land, air). Roads, fuel delivery, rail, garaging/parking	Eco-system services (air and land). Air traffic control, airport landing slots, parking/station/port for connecting modes	Eco-system services (sea and land). Sea lanes, ports. Parking/station for connecting modes
Fuel	Electricity	Liquid fuels, electricity?	Liquid fuels, electricity?	Specialist liquid fuel	Liquid or solid fuel. Wind power. Currents and tides
Equipment and components (transport user perspective)	A range of electronic equipment: PC, laptop, phone, remote data storage (server farm)	Trains, trams, buses, cars, motorized two-wheelers, bicycles, footwear, navigation equipment	Trains, trams, buses, cars, bicycles, foot, navigation equipment	Planes, helicopters, airships	Ships/boats/yachts and so on
Ancillary services, user perspective	Repair and maintenance, connectivity	Infrastructure repair and maintenance, vehicle repair and maintenance, fuelling, safety, traffic control	Infrastructure repair and maintenance, vehicle repair and maintenance, fuelling, safety, traffic control	Airport services (transit, car parking, restaurant, hotel, luggage handling). Choice of modal connections	Port services (transit, car parking, restaurant, hotel, luggage handling). Choice of modal connections

Source UBS

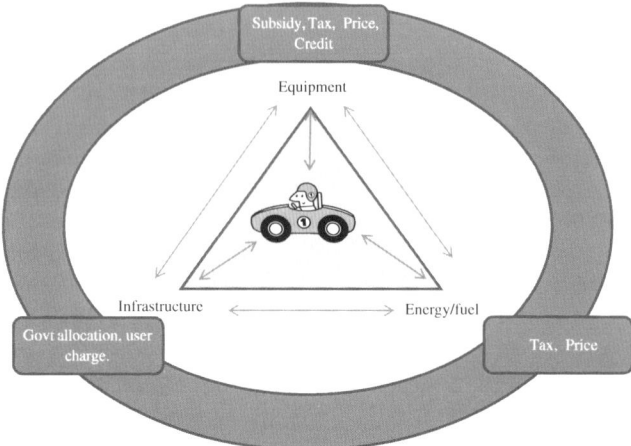

Fig. 3 Interdependent markets supporting car transport

falls from the perspective of the consumer varies and may not be ideally situated in the structure from the perspective of sustainability. An obvious example is the pricing of road journeys by private car, where once car and fuel have been bought and annual road tax paid access to the road network is very often free of any financial charge. This inevitably encourages overuse. In a nutshell, when viewed as a 'market' transport is a complex hybrid.

The hybrid nature of the transport market can make it more difficult to implement sustainable transport strategies. This compounds the issue that environmental and social externalities are not 'in the price' and therefore not in any signalling system to which the consumer can readily respond. This may sound somewhat discouraging from the perspective of the sustainable transport goal under discussion here. The point to retain is that the pricing mechanism can nevertheless be enormously powerful as a force for change, just as long as someone has a firm grip of the sustainable steering wheel and puts the pricing engine in gear. The question of who this might be is a rather more vexed question.

In the next few sections, we take a slightly closer look at selected transport subsectors as set out in Table 3.

6.1 Infrastructure and Transport Demand

Sometimes infrastructure regulates usage only in the sense that once the network is full, usage can grow no further. Congestion is the only brake on many roads. In other cases (e.g. airport landing slots) capacity is set by a regulator. The decision maker facing severe congestion on an important trunk road has a choice, from the perspective of sustainable transport. He or she could decide to leverage

Table 3 Markets unlikely to deliver sustainable transport in isolation

Market/Subsector	Price determination	Market structure
Raw materials	Open market	Basic materials—competitive market Oil—tends towards oligopoly
Infrastructure, and its repair and maintenance	Regulated pricing (e.g. rail) or open access (e.g. roads)	Monopolistic. High exit barriers. Risk of sunk costs associated with obsolescence can inflate costs
Fuel delivery/distribution	Complex: market and regulation (tax can be a significant %)	Oligopolistic
Equipment and components	Open market	High (social) barriers to exit (e.g. cars). Barriers to entry can arise in the context of large contracts (e.g. rail and road)
Ancillary services—equipment repair and maintenance; navigation; insurance; financial services (credit)	Open market	Competitive market
Transport modal choice facing the user	A mix of regulated pricing and market-driven pricing	Mixed, and only rarely competitive

Source UBS

the existence of the network capacity limit in favour of the environment, avoiding further changes to land use and greenhouse gas emission growth, by calling a halt to volume growth, but with significant financial costs.[8] Or he or she could focus purely on the economic costs of congestion. In this case, the answer would be to extend the road network to accommodate increased traffic. The first of these routes might be better for the environment, by accident rather than design. However, without the provision of viable transport alternatives this approach might have social costs. The second entails an allocation of finance to resolve the congestion issue but at a number of significant potential economic, environmental and social costs.

The interesting point about such costs in this context is the possibility of capturing them in relevant markets. As an example housing situated close to traffic noise tends to be lower in price, suggesting that, for new roads, a financial exchange could be arranged between the beneficiaries (road users) and the noise sufferers based on the sensitivity of house prices to noise. In practice this would be enormously difficult to implement. The question is thus whether it is practical to consider addressing purely environmental or social costs through financial

---

[8] The Eddington Report [8] suggested the time costs of congestion (UK) would amount to £22 billion by 2025.

markets. The above sections suggest that this is likely to depend on specific circumstances.

6.2 Fuel and Transport Demand

The price of fuel is known to drive transport volumes directly. The difference between fuel tax structures in Europe and the USA, respectively, has given rise to a significant difference between their car fleets, where the European car fleet is more fuel efficient on average. Higher fuel prices can cause transport usage to adjust in the short run. In the UK for instance a *YouGov* poll of 2682 adults reported in the *Sunday Times* on 20 March 2011 (in the context of high petrol prices) found a number of changes. Reportedly, 31% of respondents were driving more slowly, 27% using the car less for leisure activities and 23% using the car less for shopping. Smaller numbers reported taking other measures such as removing excess weight from the car, or using the car less for commutes.

Such changes in behaviour might only be temporary. What happens next will depend on what transport users expect to happen to prices. If the expectation is for fuel prices to remain extremely high for a significant period of time this will provide a financial incentive for capital investment aimed at removing structural impediments to change in transport usage. This could come from one or more or of several directions. Consumers might take decisions to reduce car ownership; consumer firms might look for ways to prevent the fall off in trade that might arise from changes in car use. Alternatively people might relocate, or governments might increase investment in public transport on the basis that high transport costs give rise to social inequity. In short structurally high fuel prices can act as a financial incentive to change the transport system and the sectors that rely on it, and this could result in an improved environmental performance of the transport system. However, unless this is carefully done, the social costs could be considerable.

6.3 Transport Equipment and Transport Demand

In the context of public transport, transport equipment can have a significant impact on the efficiency of the system, on its environmental performance, and also on the desirability or otherwise of travel by train, light railway, tram, underground train or bus. Well designed equipment is more likely to be capable of delivering a clean, safe, comfortable, timely journey that might be able to compete with the car, whereas simply reducing ticket prices is unlikely to attract customers, suggesting this gambit has no value as a competitive tactic in the land transport market. As Button [5, p. 93] comments: 'One of the more interesting points is the almost total insensitivity of the demand for urban car use to the fare levels of both bus and rail public transport modes.' The short case studies below hint at the key role technology (such as GPS and related navigation equipment) may also be able to play in delivering intelligent transport systems that can adjust according to passenger volumes. The question is where the incentive to the providers of transport equipment and related technology to invest in the R&D needed for product development may come from.

Given the hybrid picture of the sector painted above, both government and private sectors will need to come into play if the system is to change. One possible model is for government regulation to act as the initial catalyst, triggering R&D in market-driven transport subsectors. Examples would include new transport equipment technology or fuel that would then compete in the market on their own merits. An alternative route to change might be through technology innovation in private or academic sectors. Real dilemmas, however, arise when such shifts have social costs, such as job losses. Elsewhere, in global markets for highly traded goods, a great deal can be achieved by co-ordinating performance standards. They are only indirectly connected to finance but once in place, they can act as a catalyst, triggering competition that, in turn, moves finance to where it is needed. (A good example is found in the field of transport itself. Over 90% of world petrol is now unleaded [22, p. 599]). However on the flip side, attempts to put global standards in place can have adverse impacts. For instance they can be interpreted as protectionist, and in that case they could have the opposite effect to that originally intended.

7 Case Studies: What Has Worked and What Has Not

7.1 Road Pricing

Several countries have tried road pricing as a means of reducing road transport demand. The outcome seems to depend on what other measures are taken in the same context. As an example, the effectiveness of Singapore's electronic road pricing (implemented in 1998) in operational terms is described by Moavenzadeh and Markow [17] as 'remarkable'. (Kindle Locations 2254–2257). Capital costs were US$140 million, however the revenue picture was clouded by the fact that this scheme replaced another variant of road pricing, the Area Licensing Scheme, making it impossible to compare before and after revenues and profitability. All may not have agreed that this was a 'remarkable' system, for, as a regressive tax, this or similar road pricing systems may put pressure on consumers when the economy is in a downturn. A further development of interest is the 2007 agreement by the Dutch government to introduce national charging from 2011 for freight vehicles to 2012 for private cars, with the full scheme implemented by 2016 [18]. Time will tell how well this works. Toll roads may have been around for hundreds of years but, as this brief section suggests, the merits of road-pricing are unclear from economic or environmental perspectives.

7.2 Using Infrastructure and Equipment Investment to Change Travel Patterns

Car transport demand could be controlled by managing car ownership directly. Ownership quotas (as tried in Singapore) are unlikely to work in contexts in which doing without a car is currently impossible because there is no transport substitute. In any case, strategies leveraging more than one mode, involving a mix of infrastructure and equipment investment, are more likely to be effective than quotas in isolation.

As described by Moavenzadeh and Markow [17] Bogota had an extreme congestion problem. It could take an hour to drive five miles even though 80% of the population used public transport while at the same time private vehicles were taking up 95% of road space. (Kindle locations 1994–1998). The policy response worked upon several fronts, both regulatory and financial. Measures taken included a 300 km bicycle network; some 'non-car' days in which the roads were open to cyclists and pedestrians; bans on car travel at peak times according to registration number; a petrol tax; a steep increase in parking fees; land use and zoning regulations. In addition there was a reorganisation of the existing (dysfunctional) bus system to turn it into a Bus Rapid Transit ('TransMilenio') system facilitated by GPS, and a system of trunk and feeder routes; and plans for a long, shaded walkway. (In hot countries, outside temperatures can make non-car transport uncomfortable and hazardous). The financing of TransMilenio was almost entirely public sector (Locations 2128–2132), and Guerrero [10] estimated an internal rate of return (IRR) of over 60%. (Locations 2140–2144). This is not something that can be done on a one-off basis as the Bogota experience also illustrates. *The Economist* (March 12th 2011) commented: 'The bright red articulated buses of Bogota's Transmilenio, with their dedicated lanes and station-style stops, were once a symbol of a city that had been transformed from chaos and corruption in the 1980s to a model of enlightened management admired and imitated across Latin America. Today [the] Transmilenio buses are horribly overcrowded […] attracting muggers and pickpockets.' (p. 59).

7.3 London UK: Infrastructure, Equipment and Market Signals

The urban cyclist is dependent on transport infrastructure that was designed for the car and is seen as owned by the car. In order to allow the hapless commuter who spoke the words of the title to ride a bicycle, action is needed on several fronts. These could include a reduction in the numbers of cars on the roads and their speed; a designation of more road space for cycles (less space for cars) and provision of secure cycle parking at important transport nodes. Space might be made to accommodate bicycles on tubes, buses and trains. Finally work-places might be incentivized to provide sufficient facilities. Some of these changes would require a significant up-front cost in infrastructure, with no guarantee that people would actually abandon the car in favour of the bicycle. To make the case for up front financial investment, it would help to have an indication of latent demand. The arrival of Boris's Bikes could be seen as an actual demand indicator for non-car transport in London. According to the *Guardian* (11 October 2010) it took just ten weeks to clock up a million rides. The scheme also contains some useful pointers from the perspective of the (financially sustainable) design of price and payment mechanisms. Journeys shorter than 30 min cost nothing once the upfront access charge has been paid, while journeys longer than 30 min are charged on a time basis reaching £50 for the 24 h maximum.[9] This, together with the fact that docking stations are near public transport means the bikes are more likely to be docked than chained to the railings. For the provider, this is likely to keep costs associated with stolen bikes low. The only missing piece in the equation from the perspective of the user wearing sustainable transport spectacles is the lack of a technology-driven feedback loop. There is thus an identifiable information gap for the provider suggesting that docking station provision may not always evolve in an optimal fashion as time passes. Finally, it should be noted that a scheme such as this could be far more powerful than mere finance in another way, by triggering a cultural shift among transport users.

7.4 Brazil, Proalcool: Changes to Fuel and Car Technology

The first and second oil shocks of the 1970s changed the way energy is used in many ways, globally speaking. In Brazil these shocks can be said to have triggered a significant change in approach to private transport provision. This took place in several stages, in the programme known as Proalcool and is fully described by Afionis [1]. Cutting a long story short, after the second (1979) oil shock Brazil moved towards ethanol as a full substitute for petrol. This was an astonishingly bold move that can be usefully viewed from the food-chain perspective of this chapter, to think about why it was possible, and what went wrong (and was then corrected), as an object-lesson for other countries. Drawing on Afionis [1, pp. 13–19]:

Raw material procurement: the agricultural and industrial sectors were supported with cheap credit, purchase commitments and price guarantees.
Infrastructure: specialist distilleries were built, but these were ring-fenced for the transport market, reducing the capacity to respond flexibly to price changes.
Fuel: government subsidy was required to render ethanol competitive, in price terms, with petrol. Existing fuel distribution systems could continue to be used.
Transport equipment: public research centres invested heavily in car engine R&D, passing the technology on to private sector manufacturers.

---

[9] See Barclays Cycle Hire/costs, at www.tfl.gov.uk/roadusers/cycling/14811.aspx. Website accessed in March 2011.

Consumer: encountered sub-par vehicle performance and difficulty finding retrofitters.

Leaving aside niggles such as car performance during the transition, a protracted fall in oil prices led to this scheme being regarded as a 'total failure.' [1, p. 26]. After this point, however, two important developments arrived:

Raw Material Procurement and fuel: the government stopped intervening in the two relevant commodity markets.

Technology: in 2003–2004, thanks to the efforts of the auto manufacturing sector, the flex-fuel vehicle arrived.

For the purposes of this chapter, the Proalcool case study is important because it is the most comprehensive attempt seen to date to shift the private car away from dependency on fossil fuels. This important experience indicates that finance at times both helped and hurt. What it suggests is that:

1. Significant staying power is required, going beyond normal government terms of office, and based on a vision that goes beyond short-run financial impacts.
2. Even though the focus was on fuel, the change affected all levels of the food chain—raw materials, infrastructure, fuel, equipment and the consumer.
3. A significant (and longer than expected) period of subsidy may be needed.
4. The corporate sector needs to be able to respond flexibly to pricing signals.
5. Technology innovation may be needed, and can take time to arrive.
6. Early ideas may not work perfectly therefore working on a portfolio of ideas is likely to be ideal. This would permit withdrawal from failing ideas without killing the overall direction of travel in the direction of sustainable transport.[10]

7.5 Multi-Strategy Approaches to Land Transport

As the above suggests, it is likely that the most successful transport policies will work on several fronts. Munich is cited by Rye [24] as an example of a city that has seen high economic growth and the rapid formation of new, smaller households, while holding the car as a share of transport constant over no less than two decades. City size (1.3 million) may have helped. Nevertheless how this was achieved is significant: investment in light rail and tram to the tune of £2.5bn (2006 prices); speed limits enforced on all minor roads; high quality segregated cycle routes on all main roads; regional integration of public transport systems and (affordable) fares; controlled development; and integrated parking strategies. (pp. 207–208). A mix of urban planning, investment in infrastructure, pricing of public transport and regulatory

---

[10] This chapter is too short to permit a foray into so-called 'real options', conceptually a useful idea when considering investments that are not the most financially attractive in the short run but may open up very significant opportunities in the medium term.

controls are combined in such a way as to address environmental, social and economic needs. There is no information available in regard to the obvious follow-through from the above. This is the likely boost to private sector activity—infrastructure building firms and local sports equipment providers in the private sector for instance—as the scheme was implemented. Not only did this policy work on different fronts but it rested on a mix of government sector funding in the context of regulation. At no stage did it depend on free markets or the private sector, rather, private sector effects were left to follow through under their own steam. That is not to say government-led initiatives will always be optimal in all situations. The invisible ingredient for success in the above could well be cultural.

7.6 In Summary, What the Case-Studies Say

A number of important points drop out of the inevitably slightly random nature of the above section. In several cases the idea did not work properly first time around. Alternatively, the idea worked at first but then worked less well as time passed. This has several general implications for implementation. It suggests that a trial and error approach needs to be designed into new schemes, enabling tweaks to be made that improve their performance, up to and including replacement with something better. It indicates that staying-power and consistency over the longer term is a must, implying ownership and commitment beyond time horizons usually at work in political or private sector circles. It indicates that maintenance costs must not be underestimated, and contingency plans should be made for the nice problem to have—take-up with higher volumes of use than expected. Secondly, piecemeal approaches to sustainable transport are risky on the basis that perfectly good ideas may be undermined by impediments elsewhere in the transport system, the built environment, or the broader economy. Thirdly, sustainable transport schemes are likely to be more robust over the longer run if they are designed to flex constructively with the economic cycle.

Some of the above may be easier said than done, requiring as it does quite sweeping cultural and political change across public, private and consumer sectors. This brings us to Sect. 3, where we address the question of when and how finance should be leveraged in the delivery of sustainable transport projects.

8 Finance is a Constrained Resource: How and When Should it be Leveraged?

As Porritt [19, p. 279 citing, 14] puts it: The 'apparently innocuous and perfectly rational choice by individuals—to buy and drive cars more since other transport options appear less able to provide access to amenities wanted—have led cumulatively to disbenefits nobody consciously sought[…]. [Each] perfectly sensible choice to make a journey by car instead of bus will slightly: reduce the fare income to buses while increasing the congestion they face, this making the service a bit less effective; reduce the safety and attractiveness of cycling; encourage shops and other amenities to move to car-accessible rather than bus-accessible locations; encourage better-off people to move out of more heavily trafficked central areas to suburbs with better car access to new amenities.'

Financial decisions taken by the better off in the context of open market competition can be inequitable from a social perspective. As an example the train and the private car tend to be used by higher income groups, the bus by poorer sectors of the community—cf [5], p. 32. The broader issue is that they can ulti-mately lead to an unsustainable transport system from both social and environ-mental perspectives and an economy suffering from structural flaws, such as a lack of equality of opportunity. Once such structural decisions have been made financially driven decisions tend to be powerless to change behaviour. As Button (p. 51) observes: 'The decision whether to use public transport for the journey to work, besides depending upon [previous decisions] regarding location and car ownership, is thought to be influenced by the quality of local public transport, and the [needs] of non-working members of the household to use a car.'

In a nutshell, there are times when simply letting markets get on with the allocation of resources on a competitive basis works, but, equally, there are times when it will not. Table 4 above summarises some of the things that can be done to align incentives in such a way as to arrive at a sustainable transport system. This table is clearly not exhaustive, and, returning to the idea presented in Sect. 2, there may be times when financial market instruments could facilitate cross-sectoral exchange in ways that have not yet been envisaged.

The question not answered by Table 4 above is how to decide when money is the answer to the transport problem, or when it would be more efficient to apply non-financial means such as quotas or regulations. In short, what is the most efficient route to behavioural change in the transport sector. At this juncture, it may be useful to bring Michael Porter into the picture again, and think about different transport markets in the context of his well-known five forces of competition.

Figure 4 is an attempt to put five transport-related 'markets' side by side, from the perspective of the incumbent provider or system: infrastructure (roads,); energy (usually liquid fuel' or electricity); public sector equipment (buses, train rolling stock); private sector equipment; and 'modes'. This last item denotes the choice of travel modes available to the consumer deciding which mode of travel to use. The asterisked markets are markets that could be changed by an external catalyst such as new regulation (R), financial subsidy from the government sector (G) or the arrival of new technology (T). Hence, the competitive pressure cooker in the auto sector which is arguably one reason for high ownership could be changed by the imposition of government quotas on ownership (as in Singapore). A hypothetical 'new entrant' to the transport modes market (virtual travel) could arrive as a result of technology developments. The power of the provider of infrastructure to control its availability ('supply') could be enhanced by regulation or pricing. Substitute

Table 4 Regulation markets and the transport food chain

Sector	Financial and regulatory mechanisms—immediate transport-related sector	Cross-Sectoral financial and regulatory mechanisms[a]	Relevant investment markets
Raw materials (for use in equipment manufacture; development of alternative fuels)	Regulatory constraints. Carbon taxes, carbon trading. Fiscal Incentives—such as reduced taxes/investment tax credits on biofuels	Equipment makers' need for safe, energy efficient equipment potentially provides a significant market-driven incentive for raw materials providers to invest in R&D	Public sector: funds or grants for technology start-ups. Private sector firms—R&D specialists
The built environment and its infrastructure	Government procurement policies e.g. for rail, light rail. Subsidy e.g. for rural public transport. Regulatory targets or restrictions can connect to finance by triggering innovation (e.g. virtual travel)	Land use planning: car parking quotas, association of public transport provision with new build. Subsidy of broadband provision to rural areas. Subsidy of GPS/other technology to improve transport reliability	Public sector: 'Sustainable Transport' govt bonds. Private sector: real estate—potential for sustainable credentials to result in a price premium. Green REITS as an asset class
Fuel delivery/distribution	Oil prices. Fuel taxes, carbon taxes, carbon trading. Tradable quotas—such as the Renewables Obligation and the Renewable Fuels Transport Obligation in the UK. Can affect volume usage as well as triggering R&D	Government subsidy of alternative transport technology e.g. electric car and charging infrastructure. Decarbonisation of power generation—e.g. shift to Renewable Portfolio Standard contingent on new transport infra build	Government sector: 'Sustainable Infrastructure bonds'. Seed capital for new fuel R&D/start-ups. Private sector R&D
Equipment and components	Targets and quotas (e.g. CAFE standards) become financial by triggering innovation in technology	Car ownership linked to place of abode. Cross-subsidy—car ownership to public transport	Government sector: provide retraining for auto fab employees
Ancillary services—equipment repair and maintenance; navigation; insurance; financial services	Incentives or targets to embed sustainability in product design	Insurance products that cost less for green driving, leveraging navigation technology. Credit costs cheaper for low carbon vehicles	None direct—many indirect, hard to break out

[a] C.f. Stern Review [22, p. 416]. Box 16.6 gives examples of existing deployment incentives.

Source UBS

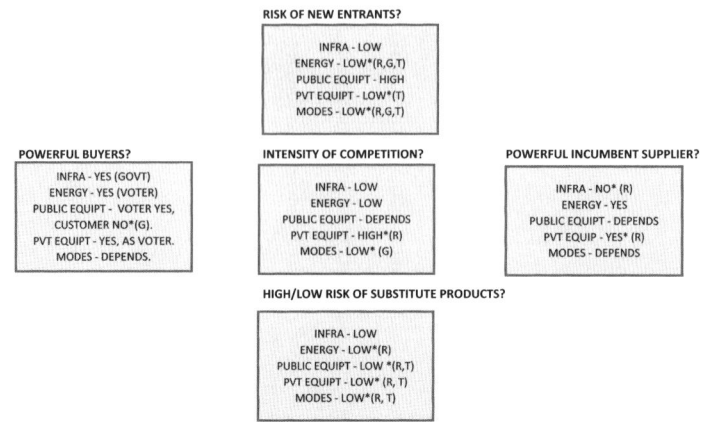

RISK OF NEW ENTRANTS?

INFRA - LOW
ENERGY - LOW*(R,G,T)
PUBLIC EQUIPT - HIGH
PVT EQUIPT - LOW*(T)
MODES - LOW*(R,G,T)

POWERFUL BUYERS?

INFRA - YES (GOVT)
ENERGY - YES (VOTER)
PUBLIC EQUIPT - VOTER YES,
CUSTOMER NO*(G).
PVT EQUIPT - YES, AS VOTER.
MODES - DEPENDS.

INTENSITY OF COMPETITION?

INFRA - LOW
ENERGY - LOW
PUBLIC EQUIPT - DEPENDS
PVT EQUIPT - HIGH*(R)
MODES - LOW* (G)

POWERFUL INCUMBENT SUPPLIER?

INFRA - NO* (R)
ENERGY - YES
PUBLIC EQUIPT - DEPENDS
PVT EQUIP - YES* (R)
MODES - DEPENDS

HIGH/LOW RISK OF SUBSTITUTE PRODUCTS?

INFRA - LOW
ENERGY - LOW*(R)
PUBLIC EQUIPT - LOW *(R,T)
PVT EQUIPT - LOW* (R, T)
MODES - LOW*(R, T)

*The situation could change in response to regulation (R), government support/intervention (G), or technology change (T).
The significance of competitive conditions will vary with context. For example, buyers/users of private transport equipment are also
voters with strength in numbers; in this sense they are powerful buyers (left). They have the power to block attempted shifts towards
sustainable transport, or indeed to propel sustainable transport forwards, depending on the context.

Fig. 4 A Porter perspective on transport. (This figure is based on the well-known and widely used Porter theory of competitor analysis. This may or may not fully reflect the model as intended by Michael Porter. All errors are the author's. See Porter [20, p. 4] for the well-known five forces figure on which the above is based.) *Source*: Porter, UBS

products could arrive as a consequence of the price of journeys (rising because of the oil price) or as a consequence of regulatory regimes designed to increase the cost of private travel.

The key point to emerge from the figure is that competitive conditions in the markets that make up the transport food chain markets depicted here are not consistent. They are constrained from being so in different ways, as discussed in an earlier section. Infrastructure changes slowly because of the time it takes to enter or exit the market. The relevant fuelling system is not easily changed because of the (sunk costs of) the existing (oligopolistic) delivery infrastructure and the existing technology model in equipment such as the private car. The public equipment market is held back by the fact that most of the time the procurement markets cannot be fully competitive for reasons of size (or sometimes politics). None of this is new. However, what does give pause for thought is the contrast between two segments of the market in particular. On the one hand is the highly competitive nature of segments of the private equipment (particularly the auto) market, where the provider firms compete fiercely for market share on the basis of product, price and financing packages. On the other is the uncompetitive nature of the 'modal' market where the consumer rarely has a menu of transport modes to choose from. It is likely to be this relationship that is holding back the intending sustainable traveller who spoke the words of the title to this chapter.

The question is what could be done to change the relationship between the pieces of the transport food chain described above (such as the car, and the 'modal menu') in such a way as to move towards a sustainable transport system. Ideas suggested by the above framework might include the following:

1. Low risk of new entrants: Increase the risk of new entrants in the private equipment market (for instance by subsidising new technologies). Increase the choice of transport modes—for instance, make it possible for the traveller in the title to choose to travel by a combination of public transport and bicycle or car and bicycle. (This change could require cultural change on the part of employers in general, and the real estate sector practices).
2. Powerful incumbent suppliers of private equipment: Changing the market structure in the auto industry would be controversial for it might mean reducing the size of incumbent players. This would require some long term social investment (c.f. Fig. 2) because the immediate costs would be social (e.g. unemployment).
3. Consumers are powerful buyers of private auto equipment because they are voters as well as consumers. However the consumers of public transport tend to be non-powerful buyers in the 'modal' market (as anyone who has attempted to travel by train on a weekend in the UK can testify). Regulation (for instance quotas) could be used to redress the balance in the car market and subsidy could be used to redress the balance in the modal market (but only if both done at the same time).
4. Low risk of substitute products in all transport food-chain markets. Here, technology could have a powerful impact. It could for instance put virtual travel on the table as a meaningful substitute to air or land travel, or open up the choice of fuels to facilitate markets in low carbon fuels. Alternatively a change to transport ownership models might be needed, and, perhaps surprisingly, this could be private-sector led: see Streetcar (below).

This brings us to the final section of this chapter, which is to consider what could be done with financial tools or mechanisms to bring about change. This section is emphatically exploratory and therefore tends to be hypothetical in nature, the key point being that financial innovation has potentially a great deal to bring to bear on the issue.

9 Financial Tools, and How They Might be Leveraged

Finance serves many purposes, one of which, in the modern economy, is to facilitate consumer choice. As a consumer of transport services, I want to ride my bicycle yet I hardly ever do. Most regular journeys are long enough to require Olympian fitness. Cycling helmets feel as useless as the proverbial chocolate tea-pot in the context of fast-moving glass and metal. Even the most robust rider may be haunted by fears that the stress-inducing lack of road-space and the possibly lung-damaging stink of exhaust fumes will more than wipe out the health benefits of exercise. Finance—specifically, the financing of growth in unsustainable transport—is precisely what has left us with so little choice in the matter of how, and indeed, whether or not, we travel. The rapid growth in fossil-fuel-driven transport in recent decades is the outcome of a myriad of financial decisions taken by a range of economic agents including consumers. This is a clear demonstration of the power of finance to drive change albeit in the wrong direction from a sustainability perspective. The question addressed here is whether and how finance could be leveraged to allow consumers to make sustainable transport choices. By collectively changing the way we travel, we could change the world: *ukuhamba kukubona*.[11]

Returning to Fig. 2, one of the problems inherent in using it is that it is very often impossible to put a value on social or environmental value potentially created or destroyed because there is no benchmark. This may not matter all of the time. Many financial decisions are made on the basis of incomplete information. However, the larger the investment, the more analysis is generally required, and the more demand there tends to be for numbers to make the financial case. The challenges are legion. Even in the field of carbon regulation, which has been around for long enough for financial models to be somewhat developed, the practical difficulties must be recognized. As Tol [23] observes in the context of greenhouse gas mitigation: 'In theory, the marginal damage costs should be equal to the tax on carbon, to the price of tradable permits, or to the marginal costs of emission reduction. In practice, the carbon tax (etc.) should not be too far from the estimated marginal damage costs because otherwise environmental and economic objectives would be too far out of step with each other [...]. At the same time, it allows for the coordination of climate, energy, transport and other policies. This is important for economic efficiency but also for accountability. *Why would* CO_2 *emissions from cars be treated differently from power plant emissions?*' (Tol, in Helm, Kindle Locations 2065–2071). It is hard to argue with this, however, Tol also notes fiscal policy measures such as carbon taxes and/or trading based on such cost estimates *will' affect behaviour only to the extent that economic players believe them*' (Kindle Locations 202–207).

In the following sections, the financial instruments and mechanisms relevant to the different levels of the food chain discussed in this chapter—infrastructure, fuel, transport equipment and modal choice—are reviewed. The food chain segments are taken in turn, but financial innovation in one area could affect another, as discussed above. Before proceeding it should be noted that financial innovation is best undertaken in the context of well regulated markets by a diverse group of finance professionals in response to a wide range of financial and economic needs. What follows in the next paragraphs is thus inevitably incomplete and none of the precise ideas described below may prove workable in practice. The aim of the exercise is to suggest possible directions for development. The question explored here (and not necessarily answered) is whether financial instruments could be designed to align incentives optimally in the context of the holistic perspective of Fig. 2, in Sect. 1, on a case-by-case basis.

---

[11] *Travel opens a window on the world* (translated from Xhosa). Source: on the paper cups on South African Airways.

An important health warning is also in order. Financial market history presents more than one example of perverse financial, economic and social consequences arising from financial engineering. In this complex sector there is huge scope for such accidents, a strong argument for keeping financial engineering as simple as possible. So, it is with caution that we proceed to hypothetical examples.

10 Financial Instruments Relevant to Infrastructure

Section 2 suggested that financial innovation may have potential when it comes to dealing with cross-sectoral effects. This is likely to be particularly the case in the context of the infrastructure sector, which affects the economic, environmental and social performance of activity in sectors directly dependent upon it such as the real estate sector, or the communications sector. The main financial instrument in the field of transport infrastructure is project finance, which can be funded by an appropriate mix of debt and equity. Here, the return on a specific financial instrument such as a bond, or the dividend yield in the case of equity, can be directly connected to the return on capital invested in the project, or some other financial metric. Financial innovation here could include financial assets and liabilities in which regulatory environmental or social targets determine the yield to maturity.[12] A small number of hypothetical examples can illustrate the general idea.

Contingent Infrastructure Bonds: The annual yield on new infrastructure (such as a railway) funded by bonds could be contingent on an annual gradually reducing target for greenhouse gas per passenger mile. The payout could rise to reflect income from carbon trading if the issuer was on or ahead of the target, or fall in the reverse case. For this to be possible, the issuer would need to be able to monetize carbon mitigation success (for instance, by selling CO_2 emissions permits) in order to be able to deliver on the coupon. To make this easier to implement, the coupon could vary with the CO_2 price. Similar ideas could be applied to convertible bonds. These can be converted by the holder from debt into equity on the basis of specified conditions which could also be 'green' in this context.

Congestion-Reduction-Contingent Municipal (Local Authority) Bonds: A local authority seeking to build a new ring-road project could link the interest costs of the related bond to a congestion-related target, such as a reduction in the number of car journeys over a given period of time in the relevant city centre, reducing congestion and greenhouse gases at the same time. To align incentives in the right

---

[12] This idea is analogous to the so-called Social Impact Bonds in which social outcomes (such as a reduction in hospital admissions are a result of preventative measures, or a reduction in criminal reoffending) determine the return on the bonds. In this way funds are raised for a social purpose and the cost savings generated by a successful social outcome accrues to the bond holders (suppliers of capital). See 'Towards a new social economy' (March 2010), Social Finance Limited. www.socialfinance.org.uk

direction, the payout to investors would be higher if targets were not met. Note that some bond investors might not find this very attractive, but, read on.

Such an authority might take several steps (unrelated to the specific project) to help meet the car-use reduction target. This could include rethinking the design of built environment due to be rebuilt or refurbished in the normal course of events, by tweaking aspects of the mass-transit system. Alternatively local firms could be encouraged to 'telecommute' through a build-out of the broadband infrastructure. If bond holders were also residents of the city, then, in effect, they would potentially receive financial compensation for the social costs of failure to deliver on a policy target. (This would be consistent with the concept underpinning Fig. 2).

Green Yield Equities: This class of shares could pay out a lower dividend yield in the event that green targets were met, leaving the firm with more to reinvest in the business (especially if carbon markets could be used to monetize GHG reductions). A higher dividend yield in the event that environmental targets are not met would potentially exert financial pressure on the firm, especially if permits also had to be bought. This may sound counterintuitive but, hypothetically speaking the investor should be indifferent between receiving a dividend (offset by the lower ex-dividend share price) or holding shares with no payout. The cash not paid out by the successful firm would hypothetically be reinvested in the business therefore theoretically 'received' as capital. Carbon trading gains for this hypothetical firm should (other things equal) be supportive of the share price. [13] Over the medium term, the firm failing to meet green targets would hand more cash back to shareholders, thereby shrinking the firm. The pressure on cash-flow resulting from environmental or social underperformance might have an impact on the entity's credit rating. In the extreme this might lead to a gradual withdrawal from the market, leaving the greener firms in the market. The above instruments would clearly work in an optimal fashion in the context of well regulated financial markets with a healthy carbon trading sector in place.

Green Lottery Bonds: These could work in the same way as the UK's National Savings Premium Bonds. The average return to holders is in line with other National Savings Certificates, and in any given month most of the holders receive nothing. Consequently sometimes very large sums can be paid on a lottery basis to the numbers that come up. As relatively low cost finance for the issuer they could be permitted for the funding of projects that find it hard to get off the ground such as rural bus services. The government might need to stand behind them in much the same way as they do for other National Savings products.

Season Ticket Holders' Equity: Savers in UK building societies received an allocation of shares (or could opt to buy shares in) the company, many moons ago. In the here and now (2011) Spanish football season ticket holders automatically become a 'socio' in the club.[14] In effect, they own the club, and have the right to

---

[13] For this to work optimally the government might need to make the returns on such green investments tax exempt, or, alternatively, to equalize taxes on income and capital gains.

[14] With thanks to Paul Donovan for this suggestion.

elect the club's President.[15] In the same way, transport season ticket holders could (in selected public transport facilities) automatically become shareholders in the company at no further charge. This idea could also potentially apply to some cycle rental schemes. Whether becoming a shareholder should come automatically, or should entail a further financial charge, would need to be decided on a case-by-case basis, the aim being to align owners and users as far as possible. When a financial transaction is involved (and the shares become tradable) this often results in a separation of ownership and control. Whether this is a good thing or a bad thing, can only be decided on a case-by-case basis, but, a rule of thumb is that if the underlying asset is an important social or environmental asset to the user, separation is best avoided.

Some of the hypothetical financial instruments described above have the advantage of (literally) investing those who most need the relevant service in the equity. This would help overcome the problem that government policy is beset by a time-inconsistency problem (Tol)—as everyone knows governments have a tendency to change policy goal posts from one government to the next. In the context of climate change policy, Helm et al. [11] suggest delegation to an independent agency (Locations 3917–3922).[16] For some transport projects incentives might be best aligned by having that 'agency' strongly connected to the transport user, who often has a vested interest in consistency. In the above case the so-called agency could be a board of directors elected by regular transport users, following the Spanish football model.

11 Financial Instruments Relevant to Fuel

Perhaps the most powerful financial instrument in the context of transport is the oil price. In response to the 1970s oil shocks, the world became less dependent on oil, some countries significantly so. When liquid fuel becomes sufficiently expensive, this has both short run and long run effects, as discussed in Sect. 2. In the short run, people will economize on journeys, choosing to travel less. If prices are expected to stay high for long, then structural changes in behaviour are likely to follow. These may include changes in infrastructure such as a build-out of the infrastructure needed to support the electric car to facilitate some diversification of the fuel source in the transport industry. Alternatively the relationship of firms and individuals with existing infrastructure might change as they may move to locations that would require less travel for freight or people, switch transport modes, or where possible put more emphasis on 'virtual' travel. High fuel prices might also

---

[15] See www.spanishfootball.info/guide/the-press/ Accessed on 17th April 2011.

[16] Dieter Helm, Cameron Hepburn and Richard Mash, 7 July 2005, 'Credible Carbon Policy', in *Climate Change Policy*, by Dieter Helm (Oxford University Press USA, via Kindle).

spark increased R&D in alternative fuels; and increased efforts to develop more efficient engines.

Meanwhile many countries have fuel taxes in place. Over the long run their presence can be seen to have had significant structural effects on the fuel market from one country to the next. As discussed above this is reflected in the transport equipment market—hence, European cars tend to be more fuel efficient than their US counterparts. Fuel taxes are not politically popular, however. Last but not least, carbon trading regulation may, in the long run, also change the behavior of liquid fuel users in permanent ways. 'The conventional economic literature focuses on the choice between carbon taxes and tradable permits. The former sets the price of carbon, letting the quantity adjust. Permits start the other way around' setting the targets in quantities. Which is preferable 'depends in theory on the shape of the marginal damage- and marginal abatement-cost functions', and, since climate change (other things equal) 'tends to have steeper, shorter-run marginal cost functions than marginal damage functions', theory suggests carbon taxes are preferable. (Helm, Location 165–170). In a cap and trade scheme (such as the European Emissions Trading System) a limit is set on the number of tonnes of greenhouse gases that can be emitted by 'factories, power plants and other installations in the system' (EU website) and the price varies with supply and demand. In the current year (2011) the scheme covers power stations, combustion plants, oil refineries, iron and steel works, and manufacturers of cement, glass, lime, bricks, ceramics, pulp, paper and board. On 1st January 2012, airlines joined the scheme[17] and the petrochemical, ammonia and aluminium industries will follow in 2013. Profit margins for the airline companies will depend on whether carbon costs can be passed on to the consumer either within the ticket price or as a surcharge. Unless the change in air travel costs is significant, whether it will have much direct impact on travel demand is a moot point. It may turn out to have a more powerful influence as a signaling mechanism eliciting cultural change. In that case it might become the 'done thing' for firms or consumers to keep their carbon footprint down.

The Renewables Obligation (RO) is the financial mechanism by means of which electricity providers are incentivized to increase the proportion of energy derived from renewable resources The scheme is indirectly relevant to transport in the general sense that it is a system of exchange between high emitters and low emitters. The question is whether a similar system of exchange between (for example) frequent travelers and less frequent travelers could be feasible in the context of travel for organizations who might, through a system of 'travel exchange' be incented to change employee business travel habits, or commuter patterns for the cohort of company employees. In the context of the transport sector a 'Travel Avoidance Obligation' could work as follows: In return for a given number of miles of fossil fuel driven travel avoided the relevant organization

---

[17] See the UBS Q-Series® Global Aviation report written by Jarrod Castle which cites financial costs to the airlines industry of €300 mn in 2012, €600 mn by 2014 and more than €1 bn by 2014, depending on the CO_2 price prevailing at the time. (See pp. 1–4).

would receive one 'Travel Avoidance Obligation Certificate'. If the relevant organization failed to accumulate enough TAOs to meet the relevant target in any given year, they could be required to pay into a fund, which would be disbursed to the organizations that managed to exceed the target.[18] As in the above instances, this is no more than a hypothetical example, intended to suggest that some of the ideas implemented in other sectors could be adapted for the travel sector.

12 Financial Instruments Relevant to Transport Equipment

Transport equipment tends to be produced in the context of competitive product markets, undertaken by major industrial equipment firms who look, for funding, to bond and equity markets. In the case of equipment used in public transport such as rolling stock for train companies, and buses sustainability tends to be increasingly embedded within the business model. Such firms increasingly find that their product suite is responding to demands for a more energy efficient product, or for products that can be properly recycled at the end of life. This has the effect of producing innovation, driven by the need to compete to reach the customer.

For the car, it would probably be fair to say that, currently, there is no available financial instrument to leverage in the direction of sustainability in such a way as to escape from the volume driven car sales model currently in place. Taxes, congestion charges, parking costs and so on are in place but do not appear to have much impact on demand. Indirect effects appear to have greater impact. Should alternative convenient, well connected, pleasant to use transport modes be readily available, and fuel prices *also* stay high enough for long enough to facilitate changes to infrastructure, then perhaps the enthusiasm for car ownership and individual car travel might fall off, in the long run. Encouragement can be drawn from the success of Munich described above, and also from an idea in the next section, both of which suggest there could be hope for the hapless cyclist in the title.

13 Financial Instruments Relevant to Modal Choice and Other Aspects of Driver Behaviour

A recent innovation in the insurance market—which has so far failed to take off—is known as 'pay as you drive'. Data is collected by the insurance company through mobile phone or satnav or similar technology. Insurance premia can be set according to frequency of car use, typical speed and typical journey length. The driver can reduce his or her premium by taking fewer, shorter trips and by not

---

[18] Information Note, 4 February 2011, 'The Renewables Obligation Buy-Out Price and Mutualisation Ceiling 2010–2011', Ofgem

breaking the speed limit at any time. From the perspective of the insurance company, this reduces asymmetry of information, better aligning insurance premia with risk. From an environmental perspective, greenhouse gas emissions per passenger mile are reduced. From a social perspective, safer young drivers (and safer female drivers) could enjoy evidence-based fair rates, cash constrained individuals could actively move to reduce their transport bills, and last but not least cyclists would be safer. The disadvantages of such a scheme lie primarily in its potential infringement on privacy, as well as higher data security risks, together with potential inequity arising from inaccurate data. However, a further advantage of such data collection is that it could facilitate systems of exchange hinted at above. Examples could hypothetically include:

- Car Sharing Certificates: This mechanism could engineer payment from individuals choosing to drive a car as a single passenger to individuals sharing a car, limited to specific roads with the aim of avoiding congestion.
- Congestion Reduction Certificates: Payment from individual car users on short journeys to individuals who choose to switch some journeys from the car to other modes such as the bus.
- Green Lottery Tickets: Passengers taking the bus could be automatically have their ticket entered into a lottery.

Such schemes might also prompt a response from the private sector. For instance, the potential to credit 'points cards' with the credit earned by switching from the car to the bus might prompt supermarkets to provide bus services. These could be based on clean technology thereby reducing the environmental impact of the average shopping trip.

As it happens, an example of 'pay as you drive' in action is up and running in London and other UK cities, but, in this case, it works by changing the ownership structure in relation to the car. This is Streetcar.[19] Instead of owning a car, club members can call and reserve a car for anything between half an hour or 6 months. Thanks to a Smartcard, charging stops as soon as the car is returned. This scheme, which was the brainchild of two individuals (Andrew Valentine and Brett Akker) is particularly interesting in the context of the Porter analysis in Fig. 4 above, because by changing the basis on which the decision to drive is made, it opens up the possibility of choosing from a transport menu rather than taking the car because the car is there. In *From Red to Green?* (penned by UBS colleagues Donovan and Hudson 2011), one of the writers is not infrequently referred to as a 'dismal scientist' [7]. Dismal or not, the economics of changing the private car ownership structure are potentially quite significant. Citing a conversation with Donovan (2011) who read this chapter ahead of publication:'Of course what is happening here is that the pricing mechanism is shifting incentives through an altered marginal cost pricing structure. For an urban user, Street Car may actually *lower* the cost of driving relative to owning a car but the visibility of the marginal

---

[19] See www.streetcar.co.uk/whatisstreetcar.aspx Accessed on April 17th 2011.

pricing mechanism shifts behaviour'. The result: a potential shift *away* from the car towards other modes for some journeys. It should be emphasized, however, that such behavioural shifts may depend on a rational consumer. As we also point out in *From Red to Green?*, consumers tend not to be rational.

14 The Potential for the Financial Services Sector to Deliver a Market in Green Transport Investments

The report of the Green Investment Bank Commission[20] states that the scale of investments required to meet UK climate change and renewable energy targets could reach £550bn between now and 2020. (p. xiii) [26]. It is not yet known what portion of this might be transport-related, but in a sense this is academic. Deep financial markets in this area would make it easier to crystallize (when appropriate) some of the environmental and social factors shown in Fig. 2. An investment requirement on this scale will require capital from a wide range of sources, and a 'Green Investment Bank' could facilitate capital-raising at all stages. The report lists a range of potential financing vehicles including early stage grants, equity co-investment, wholesale capital, mezzanine debt, the purchase of completed renewables assets, purchase and securitization of project finance loans, insurance products and long-term carbon underwriting (p. iv). As the report points out, the flip side of financial provision is the presence of long-term investors. The report mentions green bonds, and green ISAs as well as debt funding, and energy bill levies as other potential sources of funds.

The arrival of a significant market in green assets could be a boon to pension funds, and individuals looking after their own pension fund provision. The field of transport is likely to be particularly helped by the arrival of such an institution, for what is needed in the context of transport funding is, above all, the ability to respond flexibly to different conditions and to experiment. For, in finance, innovation often arrives by trial and error. Infrastructure, fuel and equipment are, as has been pointed out, quite different markets, but there may be considerable currently unrecognized scope to cross between them. As an example, an electric car system (infrastructure and equipment) is likely to have to be an integral part of the so-called 'smart grid'. As things stand currently, from an investment perspective, it is possible to invest piecemeal in some parts of the potential electric grid, battery technology being the obvious example. What is needed is a seed-bed of transitional investments ranging from early stage projects to fully fledged tradable assets across the entire smart grid food chain in the longer run. What is also needed is careful experimentation in financial innovation.

---

[20] Bob Wigley et al., Unlocking Investment to Deliver Britain's Low Carbon Future. Green Investment Bank Commission, June [26].

15 Conclusion: 'When I see an Adult on a Bicycle, I have Hope for the Human Race'

The author of this chapter was asked to focus on financing sustainable transport projects.[21] In this chapter, private car transport and its potential substitutes have tended to be the main focus, on the basis that potentially infinite demand for private car transport is one of the most salient problems in the sector, given potential resource constraints. It was impossible (in one chapter) to deal comprehensively with the entire system, but the hope is that ideas here will apply to other transport modes.

The credit crunch of the early twenty-first century provides a salutary reminder of what can go wrong when financial institutions and markets are not working as they should. In the context of this chapter, the message is that, for finance to do its job, the right conditions need to be in place, and, by implication, for a green investment bank to do its job, a strong supervisory body is required. Experience in the climate change field to date suggests that, for projects with a timing mismatch problem, an independent agency may be needed. The discussion here also suggests that current transport ownership structures may need to evolve.

These important points having been made, perhaps the clearest conclusion to drop out of this chapter is that finance is a powerful facilitator of change. Moreover individual financial instruments are capable of delivering on the minutiae of specific conditions often required to facilitate the delivery of capital to where it is needed. However as this chapter has also made clear, financial markets will be unable to re-engineer the transport system alone. The engine-room of change for transport may ultimately lie in a much wider cultural shift.

Acknowledgments Thanks are due to the following UBS people who kindly read this chapter and offered some helpful suggestions: Paul Donovan, Patrick O'Bryan, Jarrod Castle and Hubert Jeaneau. All errors are the author's.

References

1. Afionis S (2009) Brazil's ethanol fuel programme: a history and analysis. Lambert Academic Publishing, Koln
2. Booz & Co and Temple (2011) HS2 London to the West Midlands appraisal of sustainability. HS2
3. Brundtland (1987) Report of the world commission on environment and development, general assembley resolution 42/187, 11 December
4. Butler DA (1998) Unsinkable. The full story. Stackpole Books, Mechanicsburg
5. Button K (2010) Transport economics, 3rd edn. Edward Elgar, Cheltenham
6. Damodaran A (1996) Investment valuation. Wiley, New York
7. Donovan P, Hudson J (2011) From Red to Green? Earthscan, London and New York

---

[21] Attributed to H. G.Wells

8. Eddington R (2006) 'Key Findings and Recommendations' in The Eddington Transport Study. UK Department for Transport
9. Gastineau GL, Donald J, Smith DJ, Todd R (2001) Risk management, derivatives and financial analysis under SFAS No. 133. Research Foundation of the Association for Investment Management and Research, Charlottesville
10. Guerrero DH (2001) Transmilenio: the Mass Transport System of Bogota. In: Prepared for the Latin American Urban Public Transport Congress, Havana
11. Helm D, Cameron Hepburn C and Richard Mash R (7 July 2005) 'Credible carbon policy'. In: Helm D (ed) Climate change policy, Oxford University Press USA accessed through Kindle
12. Hudson J (2006) The social responsibility of the investment profession. Research Foundation of CFA Institute, Charlottesville
13. Hudson J (2005) Why Try to quantify the unquantifiable? UBS Investment Research, London
14. Levett R, Christie I, Jacobs M, Therivel R (2003) A better choice of choice. Fabian Society, London
15. Majors D (2010) The Titanic Book: Facts, details and the story behind the sinking of the titanic. Kindle. 20 December 2010.
16. Mallard G and Glaister S (2008) Transport economics: theory, application, and policy. Palgrave Macmillan, Basingstoke
17. Moavenzadeh F, Markow MJ (2007) Moving millions: transport strategies for sustainable development in megacities. Springer accessed through Kindle, downloaded in February 2011
18. Parkhurst G and Dudley G (2008) 'Roads and traffic'. In: Docherty I, Shaw J (eds) Traffic jam, The Policy Press, Bristol
19. Porritt J (2005) Capitalism as if the World Matters. Earthscan, London
20. Porter ME (1980) Competitive strategy techniques for analyzing industries and competitors. Free Press, New York
21. Porter M, Kramer MR (2011) 'The big idea: Creating shared value'. Harvard Business Review, January-February. (Downloaded 8 Mar 2011 at http://hbr.org/2011/the-big-idea-creating-shared-value/ar/1)
22. Stern N (2007) The economics of climate change. Cambridge University Press, Cambridge
23. Tol RSJ (2005) 'The marginal damage costs of carbon-dioxide emissions'. In: Helm D (ed) Climate change policy, (Oxford University Press USA, accessed through Kindle in March 2011)
24. Rye T (2008) 'Mind the gap! The UK's record in European perspective'. In: Docherty I, Shaw J (eds) Traffic jam, The Policy Press, Bristol
25. United Nations. 5 November 2010. The Report of the Secretary-General's High-Level Advisory Group on Climate Change Financing, United Nations
26. Wigley B et al (2010) Unlocking Investment to Deliver Britain's Low Carbon Future. Green Investment Bank Commission, London

Economics and the Future of Transport

Phil Goodwin

Abstract Economic analysis applied to transport is faced with four crises. They are (a) evidence of a profound (and favorable) shift in long term trends, which should change forecasts of future traffic levels; (b) doubts in the axiom that transport investment necessarily supports economic growth; (c) rethinking the relevance and reliability of long established formal methods of assessing the benefits and costs of transport projects; (d) unfavorable pressures to reduce spending on small, widely spread, local policy initiatives which assist behaviour change and sustainability, in favour of larger 'flagship' investments whose impacts may be the opposite of intentions. It is argued that a broad evidence base now supports sustainable transport policies which can reduce car dependence, improve health, and give both environmental and economic benefits.

1 Introduction and Overview

The single most important theoretical contribution of economic theory to transport is the concept of road pricing, demonstrating that in certain conditions increasing the price of transport leads to net increases in economic welfare. It has been repeatedly proposed, studied, modelled, approached and, mostly, rejected for over half a century. On the other hand, the most successful practical innovations in transport policy over the same period have been the pedestrianisation of city centres, which has had great economic benefits in reviving the commercial success of central areas, and the development of 'smarter choices' in which travel choices

P. Goodwin (✉)
Faculty of Environment and Technology, Centre for Transport and Society,
University of the West of England, Frenchay Campus Coldharbour Lane,
Bristol BS16 1QY, UK
e-mail: philinelh@yahoo.com

O. Inderwildi and Sir David King (eds.), *Energy, Transport, & the Environment*,
DOI: 10.1007/978-1-4471-2717-8_34, © Springer-Verlag London 2012

are modified by a combination of mostly rather cheap and uncontroversial measures. In neither of these has formal economic analysis played any decisive role, and indeed economists have mostly ignored both.

'Economics' is a broad church, and transport economics has included studies ranging from the descriptive tradition of economic history to the detailed econometric modelling of traffic flows on a network. This chapter discusses four of its contemporary strands of interest, mostly focussed on the official guidelines for determining which projects, and sometimes which policies, should be supported. These relate to (a) a strong view about how economic growth has driven traffic growth, and may continue to do so; (b) some analyses, or hopes, about how transport improvements will contribute to economic growth in the future; (c) interpretation of the impacts of transport in economic terms even where no market exists to express their cash value, by attributing values to effects such as travel time, pollution, accidents and carbon emissions; and (d) a renewed concentration on the rather complex financial implications of transport taxes, subsidies and spending on public expenditure. These come together in cost-benefit analysis (CBA) and the ratio of benefits to costs (BCR), in recent years organised under the general heading of the Department for Transport's New Approach to Transport Appraisal (NATA) and its web-based advisory guidelines (WEBTAG), summarised in DfT [8]. A fifth strand, the analysis of regulation and competition under public and private ownership, has been important but is not discussed in this chapter.

There is a very substantial literature of empirical studies, some more convincing than others, such that the collection of statistical data on traffic and travel patterns is an industry in its own right, and textbook concepts like 'demand elasticity' are measured, often to several decimal places. An odd feature of these studies is that far more is spent on analyses to predict the future needs and effects of transport initiatives than on factual analyses of what the effects have been in reality. Some (but not all) analysts call the forecasting assessment 'appraisal' and the ex post assessment 'evaluation'.

For many years, there was a tradition among civil servants attending international conferences, for example, of claiming that as a result the economic analysis of transport impacts in the UK is 'the best in the world', defining best practice for other countries and other sectors. This filters through to Ministers. In November 2010 for example Philip Hammond, UK Secretary of State for Transport, said in an interview with the transport planners' journal 'Local Transport Today' (Issue 558 November 12 2010).

"We have the best appraisal system in Whitehall, there's no question about that. The Department for Transport's appraisal system is more objective, more quantitative than anything else across Whitehall".

On the other hand, he continued, expressing a rather different view "But we've got to make it more reflective of the current priorities, in particular decarbonisation, to make sure that things such as walking and cycling initiatives score the real value they deliver in terms of environmental improvement". This tension between the claim to have a superb systematic approach and awkwardness about its failure to cope with core issues of transport policy has been long-standing. A year before

Professor Peter Mackie [23] (who is generally among the most supportive of academics towards the successive DfT approaches) expressed it in a rather stark way:

> ...it has probably been reasonable to claim that the UK has been one of the leading European countries in transport modelling and appraisal...And yet it would be difficult to assert that the transport sector is the crowning glory of the British economy

He observes

> ...transport appraisal in the UK is a hotly contested space, and the idea that transport CBA in the UK is somehow a stable, agreed concept is not true.

Another active participant in this argument has been Professor Alan Wenban-Smith, formerly responsible for land use and transport policy in Birmingham, who engaged in an illuminating discussion with former Secretary of State Lord Adonis about some 'unfinished business' from a time when as minister he had seemed to be less doubtful about the official appraisals than he subsequently became. Wenban-Smith (2011 personal communication) summarised the concerns that had been raised about the economic underpinning of transport decisions as follows:

- "...equilibrium models routinely used have a bias towards serving the continuation of past trends, and towards large-scale schemes;
- Time savings are valued over a 60 year period, even though it is well-known that most are dissipated by changed locational choices in a much shorter time. Although generally accepted (since the 1999 SACTRA report) that the loss of time savings is essential to the realisation of economic benefit, not only is it generally assumed that the conversion is 'perfect', but that the remaining time savings also have an intrinsic value.
- The value of work time savings (both converted and unconverted) is taken as equal to employment costs to businesses, even though this is not 'real' time at all, but the difference from a theoretical 'do nothing' case. In many real-life situations, the effect of investment is not that time is saved, but that 'delays get worse more slowly'. Valuing this as though it is 'clock time' involves a heroic leap. (Where independent direct estimates of economic impact are made—e.g. in regeneration areas—they cannot not normally identify more than 10% of those inferred from time savings).
- The low discount rates currently used mean enormous weight is given to unknown (and unknowable) future events (up to 60 years off), well beyond the range of any of the transport models in use (more like 10–15 years).
- This has grossly distorted the balance between different forms of transport, most particularly that between inter-urban roads and urban public transport (outside London, which has always been given special treatment—as I know from leading the 1990 'London's Transport' project for the Corporation of London that resulted in the first (of many) go-aheads for CrossRail—from Mrs Thatcher)".

And finally in this brief overview, one of the most persistent critics of UK transport appraisal practice has been Professor David Metz [24], tellingly the former Chief Scientist in the Department for Transport, who proposes that limits of the rather stable amount of time that people are prepared to spend travelling have resulted more or less in a saturation level for the amount of daily travel per capita. Future growth in the volume of traffic would therefore be driven by population growth. Because of the travel time constraint, congestion is self-limiting in that slower traffic speeds lead to shorter trips, and therefore preventing future growth in congestion would not be so important a rationale for large infrastructure investments.

So the question is, how can a field which is often described as the jewel in the crown of economic cost-benefit analysis be at the same time one where there is such widespread discontent about results? Prima facie, it seems almost axiomatic that the application of better tools of appraisal and good quality quantitative data should lead to improved decision making.

My proposition is that in each of the four areas listed above, there is a form of intellectual or policy crisis which has not yet been fully confronted, in all of which conventional thinking is under very considerable stress.

First, the underlying trends which have characterised transport since about 1950 have been dominated by an increase in private car use and a decline in the use of public transport, walking and cycling, driven by a critical economic relationship with disposable income, and reinforced by various feedback processes with land-use and the location of economic activities. Since about 1992 those relationships have weakened, and may have undergone a structural shift connected with both social and economic relationships. The received wisdom however is that the traditional relationships are still the long term reality, and policy thinking has not kept pace with the empirical evidence. The axiom 'economic growth leads inexorably to increasing car traffic' needs rethinking.

Second, there has been a postulated link—very powerful in economic history—such that investment in new transport infrastructures is a decisive driver of economic growth, often thought to be a necessary condition (though never a sufficient one). In recent years, claims for economic regeneration or improvements in economic efficiency have been made in respect of almost every large-scale infrastructure investment, for road, rail and air. But the empirical evidence is much weaker than the influence of the hype of promoters and hopes of governments, and the real effect may be small, outweighed by other much bigger factors, and sometimes even counter-intentional. The axiom 'transport investment supports economic growth' is not self-evidently true.

Third, the transport sector has been the location of perhaps the most elaborate formal social cost-benefit appraisal methods of any sector, applying economic principles of welfare maximisation, willingness-to-pay, economic values of time, life, reliability and carbon, discounting over a long time period (currently 60 years into the future) to the calculation of detailed, specific, quantified benefit-cost ratios. This is embedded in national advice and much local practice, and indeed supports an industry of specialist expertise. But methods based on linear relationships and achieved equilibrium in a rapidly changing nonlinear world look increasingly implausible, and there is little consensus that the methods used are correct other than a temporary and contingent tactical one among those who are required to apply them. The axiom 'British transport appraisal methods are the best in the world and support selection of the best projects and policies' is widely questioned.

And finally, the exigencies of financial cuts and pressures, while described in terms of 'economics', leads to many economically sub-optimal decisions, as policies with very high rates of return (such as behaviour change measures applied at a local level) are sacrificed to other pressures. The axiom 'cuts should focus on the least beneficial activities' is not systematically being applied.

These four crises have led some commentators to suggest that the choice is between two incompatible alternatives—'economics' or 'the environment'. Such a divergence can be seen both among those who then would support the 'economics' choice and those who favour the 'environment' choice. The chapter argues that this is a false dichotomy, and the concepts of economics can enrich and assist the choice of policies which are simultaneously environmentally friendly and support economic growth. But it is not clear that this potential is currently being achieved.

2 Does Economic Growth Always Lead to Traffic Growth? The Argument About Unimpeded Growth, Saturation, and 'Peak Car'

All the infrastructure schemes appraised in the last two decades have been on the basis of assumptions about future traffic using, or based on, Department for Transport forecasts. These have typically assumed high and continuing rates of growth of car use, which increases expected future congestion and hence raises the estimated benefits of expanding road infrastructure, as well as putting downward pressure on the demand for non-car transport. However, since 1989 the real growth in traffic has been very much less than forecast, and currently is actually reducing. Fig. 1 shows a distinct flattening, and recent fall, in trips made and distance travelled by car, per person.

There has not yet been a consensus of explanation of the various causes of these shifts, much of the discussion simply focussing on the narrower question about whether it is a temporary shift due to recessionary pressures (after which growth would continue as formerly) or some as yet undefined structural shift due to other causes. Metz [24] offers an interpretation based on NTS time series data showing that average distance travelled increased over time until the mid-90s, while travel time, trip frequency and journey purpose remained broadly unchanged. He suggests that this increased distance permitted increased access and choice of jobs, homes, shops, schools etc.: since access and choice increase with the square of the speed (determining the area of a circle of opportunities) while choice is characterised by diminishing marginal utility, saturation is to be expected.

The Department for Transport has not (at the time of writing) engaged directly in this argument, but its latest formal traffic forecasts, used for appraisal of major transport projects, are those published in March 2010, and are essentially driven

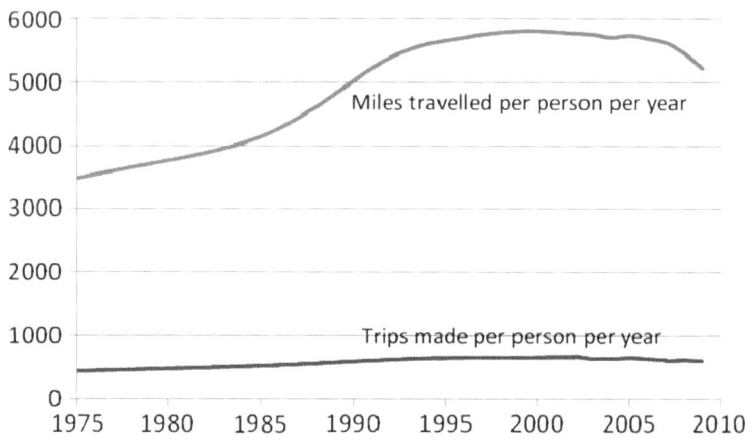

Fig. 1 Car trips and car distance travelled, per person, 1975–2009 *Source* Chained from DfT National Travel Surveys, [4, 7]

Fig. 2 DfT central, high and low 2035 traffic forecasts, england *Source* adapted from DfT [7]

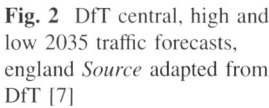

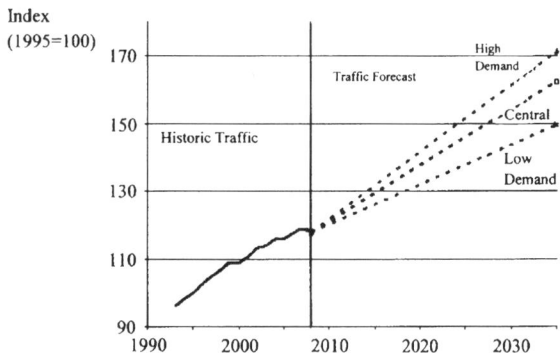

mainly by expectations of future trends in incomes and oil prices. These envisage that even under a combination of low economic growth, high fuel prices, and little improvement in fuel economy (all of which would be expected to depress demand), traffic would grow by 31% from 2003 to 2035, and by up to 50% under more favourable economic assumptions. This is shown in Fig. 2.

Under the central scenario, traffic would grow by 43%, congestion (measured as time lost per kilometre) by 54%, and journey time per kilometre by 9%. Carbon dioxide emissions however are expected to decline by 22%, even with this traffic growth.

Thus by implication, the official interpretation is that the levelling off and decline in car use is a 'temporary blip due to recession', with growth expected to restart strongly alongside improved economic circumstances. Some commentators have agreed with this interpretation, notably Professor Steven Glaister [11], director of the RAC Foundation, who argues "total traffic has grown in a quite remarkable way since the 1950s, I would suggest, more or less a straight line, with deviations from a straight line depending on the current economic circumstances.

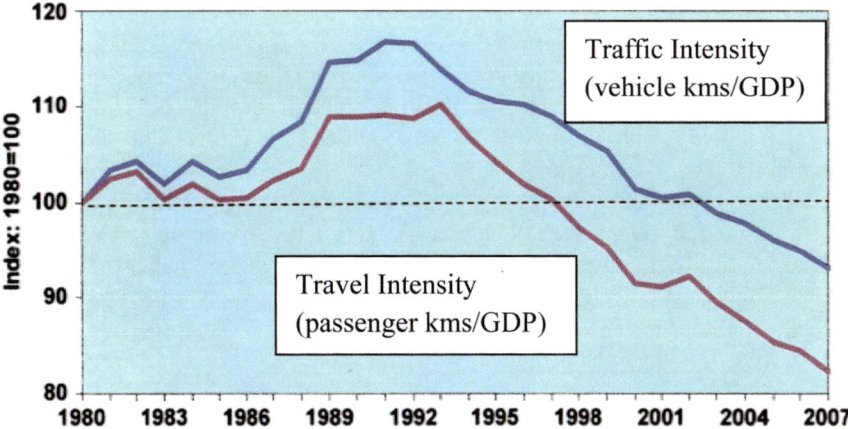

Fig. 3 'Decoupling': a shift in the relationship between traffic and economic growth *Source* adapted from Fig. 1.2b, Department for Transport [6]

In the last two or three years total traffic has indeed fallen a bit. It's what you would expect to happen in view of the history and the fact we have quite a severe economic recession. What that says to me is that you must expect that, when the economy recovers, the demand for the road network will recover as well".

The main argument against this is simply that the dates do not match up. It is obvious by inspection of Fig. 1 that that the shift in slope predated the current recession by a considerable period. By eye, one would point to somewhere around the early 1990s. A more precise indication can be found in the indicator 'transport intensity', which measures the association between traffic growth and economic growth. Generally speaking they tend to rise and fall together, to some extent (which does not of itself prove a cause-and-effect relationship) but it is notable that this relationship has changed substantially over the last 20 years, as shown in Fig. 3. This indicates very clearly that a shift was taking place in the period 1989–1992, after which intensity fell more or less continually.

Thus until the early 1990s, economic growth was associated with high and increasing traffic growth. Since then, it has been associated with lower and decreasing traffic growth. This is precisely what policy would have intended, giving very favourable possibilities of economic growth without excessive congestion or environmental damage: declared policies of both Conservative and subsequently Labour administrations sought to achieve this. Indeed it is interesting that the time of a policy shift away from road building and giving more attention to environmental impacts, by 'reducing unnecessary car use' occurred in the Conservative period 1990–1994 (notably associated with Gummer, Portillo and Mawhinney) with a consistent development in the Labour period 1997–1998 (notably associated with Prescott).

At the time, it must be said that most observers (including the author) did not think that their policies could have been powerful enough to result in such a

marked shift on their own. That judgement may need revision: there is now sufficient evidence to make a prima facie case that a social and cultural shift in
attitudes to car use is taking place which goes beyond the classic economic forces,
as tentatively discussed by the author [12] under the heading of 'peak car', and in
an international context by Millard-Ball and Schipper [25] and the OECD
International Transport Forum [20].

However, whatever the reason, it is arguable that that the scale of car traffic
growth implied by the earlier DfT forecasts, and used by successive governments
in their appraisal of transport projects, is no longer plausible. Whether the future is
of continuing falls in car use, or a return to some stable or slightly increasing level,
there is no evidence for such large growth as previously. The effect will unam-
biguously be to reduce the estimated value for money of road schemes compared
with the background library of apparently justified schemes which the present
Coalition government has inherited.

The scale of this impact may be judged from an associated policy appraisal,
made during 2006, of the interaction between road building programmes and road
pricing. The issue is what would happen if some future government were to
implement road pricing at some point in the future during the 60 year appraisal
period of road schemes. This would cause some reduction in overall traffic
volumes, and a more efficient allocation of them on the network, hence reducing
the problem of congestion that had been the core justification for building the new
roads. In that case, it would be found that at least some of the roads had been built
unnecessarily, and the question is how big the effect would be. Table 1 shows
the DfT [5] calculations, seen in a published but little remarked Annex to the
Eddington [10] report (discussed further below).

The results show that the incremental benefit of extra road construction declines
the more one builds, both with and without road pricing. The reduced overall
traffic level, and its relocation in less congested conditions, resulting from road
pricing substantially reduces the estimated benefits obtainable from road building:
the modelling suggested there would be an economic case for building 3,250 lane
kilometres by 2,025 if road pricing is *not* implemented, but only 700 lane kilometres if it is, a reduction in the warranted road programme of nearly 80%.
Comparing like with like, road pricing reduces the BCR of road spending substantially, e.g. by 70% in the roughly overlapping category 1,450–2,250 lane
kilometres in the table (from a good BCR of 3 to an unacceptable BCR of 0.9,
a reduction of two-thirds. The reason for this is mainly that the problems of future
congestion which the road building had been intended to prevent, would be largely
already solved by more rational pricing, so that the extra benefit of building the
roads is small compared with their cost.

Thus one of the indirect financial consequences of road pricing would be the
saving of funds on unnecessary road building. (Since it also generates net revenue
itself, there is a double whammy effect on public finances, a feature which may
lead to a regrowth in policy interest by Government in the future. Indeed it may be
a triple whammy, since it would also be expected to reduce the need for at least
some public transport revenue support, because of the more buoyant market

Table 1 Marginal benefit-cost ratios for with and without road pricing, according to the DfT's National Transport Model

Additional lane (km)	Marginal BCR without road pricing	Additional road (km)	Marginal BCR with road pricing
		350–550	1.5
		550–700	1.5
		700–850	1.1
		850–1,150	1.0
		1,150–1,500	1
1,450–2,250	3.0	1,500–2,450	0.9
2,250–2,750	2.3	2,450–3,700	0.7
2,750–3,250	1.2		
3,250–3,350	0.7		
3,350–4,450	1.0	3,700–4,600	0.7
4,450–5,200	-0.1		
5,200–6,150	0.2		

conditions that would apply, and therefore increase the proportion of public transport infrastructure spending that would be profitably funded internally).

The mechanisms by which road pricing brings about a reduction in congestion are not the same as those by which a reduction in traffic growth for other reasons brings about a reduction in congestion, but the orders of magnitude of effect seem prima facie comparable. It may be reasonable to assume that the scale of impact on congestion which would be available from road pricing is broadly similar to the scale that would be caused by a good combination of other transport policies (e.g. a combination of smarter choices, public transport improvements, etc., though these would certainly be a more expensive way of achieving them), together with the results of the partly spontaneous shift in the traffic growth trend already noted. This would be a conclusion of great importance for the role of the transport sector in carbon reduction targets, since the policies are much more favourable to carbon reduction than the currently assumed case of continued traffic growth which could only be changed by expensive, politically unpopular and growth-threatening restraint. An interim suggestion would be that this is an unnecessarily pessimistic view and transport can make a significantly greater contribution without causing economic difficulty.

3 Do Transport Improvements Lead to Economic Growth? SACTRA, Eddington and the 'Two-way Road'

3.1 SACTRA

Whether this interim conclusion can be supported depends on a closer look at the rather complex (and disputed) theories about how transport problems may inhibit economic growth, and transport improvements contribute to it. There have been

two rather different major exercises to address 'transport and the economy' initiated by the DfT in recent years. These were the study by the Government's
Standing Advisory Committee on Trunk Road Appraisal, SACTRA [27], and then
the review carried out by Eddington in 2006, with a great reliance on empirical
work done for the Department for Transport by Graham et al. [16–18].

The 1999 SACTRA report on Transport and the Economy was well-received in
its time—not the easiest to read, but thorough and thoughtful. It was the bearer of
two new messages, one of which has lasted and the other of which is invisible:
there is an indrawn breath, almost a Victorian 'oh no we never mention her' sense
of bad taste and tactlessness, if the issue is raised. The welcome message was that
there could be additional economic effects of transport improvements over and
above those, such as time savings, accounted for in conventional cost-benefit
analysis. The unwelcome message was that these additional effects might be bad
for the economy, not good for it. These conclusions depended on a formal chain of
argument that had eight links.

1. Traditional cost-benefit analysis does not need provision for any extra economic benefits over and above what is embodied in the calculations, primarily
   time savings. Wider benefits can certainly exist, but they arise by conversion:
   for example time savings to travellers get converted into profit increases for
   companies, or wider employment opportunities from bigger catchment areas, or
   land value increases for landowners, but they are not extra.
2. It follows arithmetically from that, that if there are real economic benefits in
   such other forms, but no increase in their overall value, the initial value of the
   time savings to the travellers must logically be taken away from them, for
   example by the price they have to pay for land increasing, transferring the value
   of the benefit from them to the landowner. Then the political prospectus
   offering time savings to motorists and extra jobs and better standards of living
   and higher property values would be false, consisting of double counting. This
   would clearly make it more difficult to build political support, if understood, but
   it did not matter because the implication was buried in the argument, and
   received virtually no attention at the time, or since.
3. The theory that the value of the time savings directly expected from the
   improvement measures the total benefits which later arise from those savings
   depends on the economists' model of "perfect competition". This includes
   perfect information, contestable markets, all prices aligned with costs, no
   external costs, no monopoly power over prices, no subsidies, etc.
4. So if the economy is not perfect, then it may be that the value of the time
   savings does not add up to the total economic benefits.
5. So it is necessary to consider the *specific* nature of the imperfections in the real
   world which departed from the ideal of perfect competition. SACTRA classified these according to whether or not market prices were well aligned to cost,
   as a key short cut to identify the visible effects of such imperfections as faulty
   information, uncharged external costs, monopoly, subsidies, certain types of
   taxation, etc. The prices were considered in two groups—prices in the transport

sector, and prices in the 'rest of the economy' (or, strictly, the 'transport-using' sector). In each of these there are three cases: prices may be higher than, equal to, or lower than the properly measured full social marginal costs.

6. So that makes nine cases altogether, a three-by-three matrix which logically includes all the possible combinations of circumstances. The centre cell is perfect competition (prices equal to costs both in the transport sector and in the transport-using rest of the economy). Here, the conditions for conventional cost-benefit analysis apply, and any wider economic effects are well valued by the initial transport effects. Any conversion of immediate benefits into wider benefits then means the initial beneficiaries are only transitory. The other eight cells represent some sort of imperfection in the transport sector or the transport-using sector, or both. These imperfections might be, for example, uncharged congestion and pollution costs in transport (which would make transport prices less than transport marginal costs) or monopoly or subsidised industries (which would make prices more than or less than costs respectively in the nontransport sector).

7. In each of these eight cases, conventional cost-benefit calculations, dominated by time savings, would not add up to full economic value. There could be additional benefits. This was, as it were, where the analysis was expected to finish.

8. But by simple symmetry, the same argument led to the identification of conditions under which transport investment could actually have negative wider economic impacts. For this reason, one of the SACTRA recommendations was not to use the term 'wider economic benefits', but 'wider economic impacts', to avoid overlooking the negative case.

This was where the SACTRA approach suddenly went from being 'helpful' in uncovering a nice class of extra benefit to make transport projects seem more worthwhile, to being a double-edged sword which might make transport projects less worthwhile. It all depended on the specific nature of the imperfections that were being considered. In retrospect, that is where the SACTRA approach became sidelined. In subsequent DfT guidelines and the Eddington study, that possibility was accepted in principle, in a very low key way. Their main thrust was different: the question was what are the extra classes of *benefit*, and not what are the extra classes of impact. Quite rightly, the DfT's guidance, in the form of Webtag and various notes, has stated that a 'wider economic' case for a project must be supported by careful evidence. From time to time, this is actually done, though more often the case is asserted rather than proved, and is sometimes asserted by application of very insubstantial or inconsistent evidence or (quite often) voluminous compilations of statistics and surveys which demonstrate anything except the most important proposition that the proposed investment will be followed by its desired effect. But there is a huge and ominous gap in the literature. As far as I know there has not been a single case—not one—where any scheme has been assessed using conventional cost-benefit criteria, and then a calculation has been made that it would have a negative wider economic effect and therefore the true value is less than calculated, not more.

Table 2 SACTRA's argument that 'extra economic impacts' could go either way

Transport using sectors of the economy

	Prices greater than marginal costs	Prices equal marginal costs	Prices less than marginal costs
Transport prices *less than* marginal social costs	Transport prices and general prices pull in opposite directions: indeterminate effect on CBA	Ignore general price effects, but reduce traffic levels by increasing user charges	General subsidies and unchanged external costs: CBA will overestimate economic benefits of transport improvements Better to reduce traffic levels
Transport prices *equal to* marginal social costs	External costs can be ignored, but benefits are underestimated	Perfect competition: CBA results unbiased	Ignore external costs, but benefits overestimated
Transport prices *greater than* marginal social costs	Goods overpriced because of monopoly, transport also overpriced: CBA will understimate benefits of transport improvements Should reduce transport prices	Ignore general prices effects, but should increase transport usage, reduce user charges	Transport prices and general prices pull in opposite directions: indeterminate effect on CBA

So the emerging practice became in effect that the promoters of schemes which may have a positive wider economic case were allowed, if they can, to make that case. But schemes which may have a negative wider economic impact were allowed to ignore the whole issue (Table 2).

3.2 Revisiting the Economic Impact of Transport: The Eddington Report

The Eddington [10] Report was also a long and complex study, and like SACTRA its main messages have not all been remembered. It focussed on reducing congestion in urban areas, key inter-urban corridors, and key international gateways, these being the areas where it concluded that economic impacts could be most helpful. It did not, however, simply conclude that road building was the best way of doing so. For a full picture it is necessary not only to look at its summary, but also the technical work, notably in the main report, but also its important Volume 3 [19], and a technical annex with modelling results provided by the Department for Transport [5].

Table 3 DfT modelling results showing congestion getting worse, in the absence of road pricing, even with a very large road building programme

England, road lane (Km)	2015	2025		
Scenario	'Baseline' scenario	'Baseline' scenario	Economically justified, no road pricing	Economically justified, with road pricing
HA Road Lane kms-additional to 2003	1,590	3,500	4,850	2,300
HA Road Lane kms-change from 2,015 Baseline		1,900	3,250	700
HA Road Lane kms-change from 2,025 baseline	–	–	1,350	−1,200
Traffic(Change from 2003)	22%	31%	32%	22%
Average Delay per vehicle km (Change from 2003)	25%	30%	28%	−37%

Source adapted from department for transport [5]

These showed that the attempt to meet these objectives by expanding infrastructure investment but without road pricing would not in fact lead to an improvement in congestion, but steadily worsening travel conditions. It 'made things worse more slowly' rather than 'making things better'.

At a time when road pricing is off the political agenda, at least for a while, this message becomes very salient. There is an argument about whether the effects of transport on the economy are in addition to the conventional assessed transport benefits especially time savings, or simply a re-expression of them, but in any case it is clear that if the transport benefits are weak, there is little potential source of wider economic benefits. People do not make investment and increase employment because they are convinced that things could be even worse, but because they see evidence that things will get better. The evidence for such estimated worsening conditions is seen in the DfT model results in the Annex (Table 3).

In the last row there is the preferred measure of the severity of congestion. It is seen that with road pricing (and with the substantially reduced road programme that would then be implied) there is a significant net *improvement* in the level of congestion, measured as a 37% fall in the average delay: most of this effect is due to the pricing itself. However, in the case without road pricing, even with the very much larger road construction programme that would then (under the assumptions) be warranted, congestion would actually get worse, giving an increase of 28% in average delay.

This reinforces a widely observed phenomenon, namely that most or all road proposals, appraised assuming no road pricing, provide their benefits in the form of 'slowing down the pace at which congestion gets worse', as measured not against an observable starting point, or any actual experience of road user, but against a 60 year forecast sometimes called the 'do-nothing' or 'base-line' option, being a forecast of what would happen in the absence of the proposed scheme. Thus the appraisal will interpret this difference between the two forecasts as the benefit from the scheme, but the road user will *experience* a progressive worsening of travel conditions. It is not sensible to expect that this will lead to wider economic benefits, except in a peculiarly negative sense. It is notable that descriptions of the potential benefits of transport investments almost never clarify that these are benefits by comparison with a different, worse future, and do not represent improvements from current experience. Thus there is a dissonance between the language of project promoters and the common language of road users, who generally expect that an improvement will make things better, and are allowed to think that that is what they are being promised. In some cases the promoters of the scheme themselves seem also not to have understood this distinction, because of the nearly universal use of the language of 'benefit'.

3.3 The 'Two-Way Road' Problem

A further issue is what is usually called the 'two-way road' problem. Suppose road improvements are used with the intention of improving the balance of international trade (usually called 'to help exports') or to increase the economic productivity of an area (usually discussed in terms of 'attracting inward investment'), or giving access to jobs over a wider area (usually described as 'helping the unemployed to get to jobs'). The problem is that such projects may instead have the opposite effect, i.e. increasing imports rather than exports, or encouraging outward invest-ment, or giving easier access to the few local jobs to people living outside the area, etc. The SACTRA report 'Transport and the Economy' emphasised the importance of considering such unintended effects as well as the intended ones, and formal advice from the DfT reminds scheme promoters that they need to include such effects in their analyses, but in practice it is virtually never done. All the effects are assumed either to be positive, or negligible, never negative. (Hence the use of phrases like 'wider economic benefits' rather than 'wider economic impacts).

The issue is well captured in the following example (set as a student exercise, but also the cause of some interesting recent conference discussions among professionals).

Consider a long but rather simple country, with uniform density, which has a single rather poor quality road, running East–West (Fig. 4). There is a single distribution company, considering where to locate. It may be proved (and is intuitively rather obvious) that the best place for it to locate is at the half-way point. Here it will get maximum access to the whole country with minimum transport costs. Now this half way point happens to be the boundary between two regions, who are competing for tax revenue, employment opportunities and the signs of economic progress. They are controlled by different political parties, and the party in the East decides to make a substantial improvement in its half of the road, straightening and widening it, to enable faster travel. This is intended to attract the distribution company over the border into its region. The adminstration in the West does nothing. The directors of the company are in fact wondering about relocating because their lease is up for renewal. They now consider—with

Fig. 4 Do road
improvements attract
company relocation?

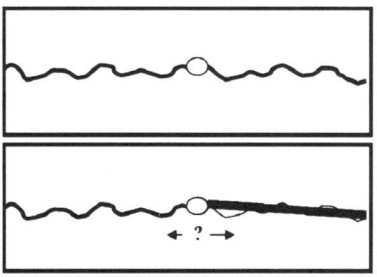

the new transport arrangements, where is the best place for them to locate? Should
they move? And if they do, in which direction?

4 Indirect Tax

In the period 2003–2009, the DfT used a rather non-standard method of social
cost-benefit analysis, such that the indirect tax consequences of a project were
included as changes to the cost of a scheme. For example, if a road scheme
induced traffic which generated more fuel tax revenue, the extra revenue was
treated as a reduction of the cost of the scheme. This approach became increasingly criticised as it appeared to be biased towards roads schemes and against
public transport, and especially against those where traffic reduction was actually
intended as a policy objective. It appeared to build in an incentive to public
stakeholders to adopt policies which were in direct conflict with objectives of
efficiency and environmental protection. The quotation from Philip Hammond
above refers indirectly to the concern about this feature of the appraisals. As part
of the preparation for such a change, DfT officials [26] retrospectively reworked 10
Highways Agency schemes, from which Fig. 5 is adapted.

Each block consists of the appraisal of one, anonymised but real, road scheme.
The criticised standard approach, the left-hand column in each of the blocks, is
labelled 'NATA'. The other columns are an average of three other contending
methods (which actually gave rather similar results to each other). In eight of the
10 cases the BCR given under the former NATA approach was higher than any
other criterion, and in three of these cases the difference is very substantial indeed,
a BCR of the order of 12–15 in cases where the three other methods all give BCRs
of the order of 3–5. This result is reinforced by work by Buchan [1], who carried
out an analysis of a small number of specific schemes, including some whose BCR
would be better under an alternative approach. His results are shown in Table 4.

With some caveats spelled out by Buchan, the direction and order of magnitude
of change for road schemes is broadly consistent with the DfT schemes reported
above. In some other cases the change is in the opposite direction, notably the
busway, rail freight scheme, to a lesser extent Merseytram, and very substantially
for the Cycle scheme where an already very large BCR (calculated by the DfT) is

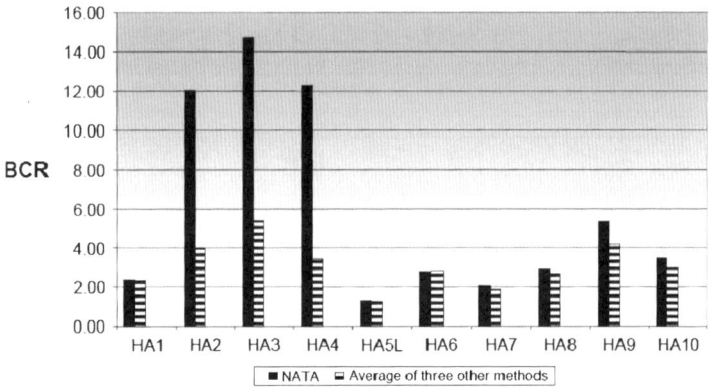

Fig. 5 DfT Calculations showing difference between the NATA method of calculating Benefit-Cost Ratios, and other methods

Table 4 Buchan's [1] calculations of comparative benefit-cost ratios under different assumptions

Project	BCR original appraisal model	BCR revised	BCR revised with further reforms
Tram (Merseytram)	1.97	1.07	2.85
Cycle path (Grand union canal)	38.4	75.0	75.0
Road improvement (A14 ellington-fen ditton)	10.83	6.69	1.3–3.25
Guided Busway (Cambridge-St Ives)	4.8	6.4	7.9
Rail Freight (Felixstowe-Nuneaton)	5.25	10.4	10.4

made substantially greater. This is entirely in accordance with what one would expect. The biggest class of schemes affected are those whose indirect effect is to increase tax revenue, primarily by inducing more traffic. These are mainly the bigger road schemes. The new rules, it was expected, would reduce their BCRs, compared with the 2003–2009 rules which produced the data used by Eddington.

Other schemes which would be affected in the opposite direction are those whose indirect effect is to reduce tax revenue, primarily by reducing traffic, but potentially also by increasing fuel efficiency. This would include smarter choices, cycling and public transport improvements. The new rules would be expected to *increase* their BCRs compared with the 2003–2009 rules.

In 2009 DfT decided (rightly, in my view) that the former approach was not going to be continued, and new rules were worked out during 2010, being eventually published in May 2011. However, at the time of writing it is not at all clear that the change has had the expected effect, as described above. In one well reported case, the effect seemed to be exactly the opposite of the argument above, namely a single carriageway road project of 4.3 km from Thornton to Switch Island in Merseyside. Sefton Council's earlier appraisal of the scheme had claimed a benefit-cost ratio of 12: but on redoing the calculations using the DfT's revised appraisal method, the new BCR had increased to a massive 35:1 (i.e. £35 of benefit for each £1 of cost) and the Council reported [28] that the main reason for this was, or appeared to be, the change in the treatment of indirect tax, i.e. fuel tax, in the economic appraisal of the scheme. The change is both very large, and in the opposite direction from the expectations from the earlier analysis described above. In this case, the counter-intuitive result seems to have been driven by the arithmetic behaviour of the benefit-cost ratio when the indirect tax element is moved from the denominator to the numerator—the ratio falls if the project increase fuel consumption, and rises if the project reduces fuel consumption. The project was deemed to reduce fuel consumption because of a reduction in distance travelled and congestion forecast to result from the scheme. Leaving aside the question of whether the forecasts are well-judged, there is cause for concern that apparently technical issues of appraisal methodology can have such a large effect on the apparent benefit of schemes, even where the underlying data remains little changed. If the results are so volatile, how can they be well-founded? A bigger strategic policy question however is highlighted by these results. At a time of great pressure on public finances, to what extent should projects which would not be justified on grounds of resource efficiency or environmental impact, be supported because they generate tax revenue? The appraisal techniques used only mirror this question, they do not answer it.

5 Criticisms, or Critique, of NATA

Putting these points together, the sort of comments made by Wenban-Smith above may be extended to a sort of checklist of areas in which NATA, the New Approach to Appraisal, has been criticised. It was not clear whether these were criticisms of detail which fitted into the DfT call for advice on how to 'refresh' NATA, or went more deeply to challenge its underlying assumptions. The author's summary of these gives 20 suggestions for change necessary to give a better fit between the appraisal methods and the policy context.

1. NATA rightly already advocates a broad multimodal approach including demand management. But in practice it is too often confined to narrower calculation of benefits of road projects providing small time savings or increased car travel.
2. NATA is rather weak on demand management, walking, public transport quality, cycling, land use planning especially favouring settlement patterns which reduce car dependence, pricing systems reflecting full external costs, smarter choices, redefinition of the styles of street management including traffic calming and the reallocation of scarce road capacity, better information, recognition of reliability and variability both in conditions and choices, emphasis on neighbourhood access rather than long distance journeys. Yet these are precisely the most critically important issues for sustainable transport systems.

3. Therefore making NATA fit for purpose will involve substantial strengthening of focus and content, not marginal tidying-up of current practice.

4. Options must be full and comprehensive, including at least one realistic well-conceived option which constitutes a genuine alternative to major infrastructure provision, carried right through the appraisal process.

5. Appraisal of all policies must include the forecast option against an experienced reality (a base year, or the present time). When an appraisal depends on describing a forecast future which is worse than the present, but 'not as bad as it might be otherwise', this should be said explicitly and with emphasis, to avoid the misunderstandings when worse conditions are described as a benefit.

6. The NATA benefit-cost ratio is not robust in its treatment of tax, especially for very cheap, revenue-increasing or demand management measures. Tax effects should be considered transparently in their own right, not mixed with the resource calculations in a benefit-cost calculation.

7. People with higher incomes spend more money than poor people to secure time savings, life, comfort and quality of life, but that does not mean their time, lives, comfort and quality of life are worth more.

8. Time spent on good quality public transport can be spent on productive work, relaxation and thought. This 'reduces the value of time' which should be treated as a benefit, not a loss. Very small time savings should always be reported separately and transparently, and gainers and losers should be distinguished.

9. Currently carbon values have much less effect on appraisals than time values. This needs to be reconciled with commitments to make very large reductions in carbon emissions. The value of carbon used in appraisal should not be derived from assumed success in meeting carbon reduction targets, but be an instrument in achieving them.

10. When appraising wider economic benefits, unintended negative effects must be scrutinised at least as carefully as intended positive ones. 'Agglomeration' calculations tend to suggest improving transport facilities for the richest areas, and 'regeneration' for the poorest areas, but the evidence base for both is still very weak.

11. Where transport policies or projects are claimed to produce different patterns of employment, the traffic effects of those differences must be included.

12. Discounting is all very well, but the impact of air pollution, climate change or deaths due to traffic accident may not be less important to future generations than to the present one. And an hour saved from travel will not be of greater usefulness to future generations than to us, just because they will be richer.

13. With 60 year appraisals and around 20 year forecasting ability, many schemes claim that a majority of the estimated benefits will happen later than the most distant forecast made. Where this applies it should be reported transparently and with emphatic caveats on the results.

14. Long term considerations are vital, but long term forecasts are misconceived. The longer the period, the greater degree of possible adaptation of behaviour must be considered, with diverging scenarios of what may happen, also to

policy, economic growth, traffic policy, energy prices, demographic structure, and car dependence.

15. Health benefits from walking and cycling, together with health costs of 'lazy' travel behaviour, should be included in appraisal of policies and projects.

16. Access to social, cultural, leisure, jobs, education and social services brings benefits. Relocation of facilities should be included in appraisal, reporting the distribution of costs and benefits of each major impact for different social groups.

17. An assessment summary table can never provide a full description of all the outcomes, by definition, but one test of the clarity and indeed truth of a summary is that the main supporters and opponents of an option should both be able to recognise that aspects have not been hidden, even when they disagree about the assessment of them.

18. All aspects of NATA which depend on research for their verification should be open to scrutiny and challenge. Matters of theory and evidence must not be treated as 'policy and therefore unchallengeable'. The DfT should take care to eliminate ambiguity or loopholes which enable scheme promoters to claim that they were 'following Government advice' in not considering aspects of core importance to policy goals.

19. Appraisals should recognise that policies and projects are contested, and research and evidence are debated. Peer review and professional audits should reflect the full range of professional thinking, not be confined to rather similar consultants approving each others' reports. Objectors are not lesser beings than promoters: on some occasions, they have been a truer voice of the future than has been offered by received wisdom.

20. It should be normal to carry out ex post appraisal of forecasts and assessments, say one and five years after all big schemes; on the other hand it is important not to make the costs of appraisal a barrier for small, cheap initiatives which can be implemented speedily and smoothly. Why not require the same amount of money to be spent on monitoring as on forecasting?

6 'Smarter Choices': Rethinking the Nature of Economic Analysis

While all this was going on, changes of a different sort were coming from increasing interest in a new, rather pragmatic, policy approach which stemmed from a need to do *something* if road pricing was for ever going to be too controversial and infrastructure expansion, of the scale discussed, unaffordable. Smarter Choices became the term for a combination of school and workplace travel plans, personalised travel guidance, better marketing, internet-based activities such as tele-shopping and tele-commuting, car-sharing and similar measures. They used to be called 'soft measures' until a Minister refused to put his name to being 'soft'. They have been some of the most important developments in transport policy and planning in recent years, taking their place alongside reallocation of road capacity as demonstrations that travel behaviour does not change in the way the old models forecast.

The background was that until 2003 the DfT was estimating that the combined effect of such measures might eventually reduce travel overall by around 5%. Since at the same time traffic growth much greater than this was being forecast for other reasons, the expected effect was hardly noticeable. However, the 5% figure was quite controversial—it was just about equal to the lowest figure that any of six different empirical and review studies had found. The DfT commissioned a new study, Cairns et al. [2], a two-volume report of some 750 pages. It concluded that with a serious, committed, reasonably resourced build-up of implementation over a 10 year period, smarter choices could stimulate a behaviour change equivalent to a nationwide reduction in all traffic of around 11%, and a reduction in peak period urban traffic of over 20%. A later study of the partial application of such a programme in three designated sustainable travel towns, Worcester, Doncaster and Peterborough, gave results of about half this level, which was consistent with the observation that they were carried out at about half the rate of spending, and for half the period, as had been proposed [29]. The spending gave very good value for money. The political balance was also very favourable, since smarter choices in general have no natural opposition, and are cheap. There are caveats, but they are not impossible to solve, mostly by suitable supporting policies in service levels and allocation of road capacity.

The DfT took all these results very seriously, and undertook a programme of 'mainstreaming' smarter choices which ran well, albeit with barriers in staffing, policy consistency and centrality at both local and national level. There were still many professionals in transport, including some voices within the DfT itself, who did not feel that this was as 'real' as building a bypass or a new train service.

They were reinforced in this by a great inertia when it came to incorporating such measures in the formal procedure of policy appraisal, cost-benefit analyses and traffic forecasts. This is partly because the essential feature of the smart measures is that you can stimulate a change in travel behaviour by altering the form and content of information and administrative opportunities offered to the traveller, but without first changing the speed or money cost of transport, by expensive subsidies, controversial charges or expensive and controversial infrastructure. That does not fit well into the framework of conventional economic analysis, since it implies that people are currently making some of their choices with imperfect information, or by not optimising their use of it. But the units of benefit in the benefit-cost calculations depend absolutely on the idea that choices are well-informed and optimal. If that is not true, the adjustments that have to be made to calculate changes in consumer benefit are complex and opaque.

So instead of testing a range of different packages of smart measures and forecasting their effects, the forecasts typically just 'make an allowance' for their effect, either taking off a given percentage of the traffic, or ex post rationalisation suggesting that somehow this had already been done automatically. This deduction is then applied as the same percentage or amount, to the 'with' and 'without' forecasts of a traditional infrastructure scheme. This certainly does affect the results, but it denies to smart measures any possibility of consistency or independence or constituting an alternative viable strategy. The point is not whether the allowance is the right level or not, but that even if it is right, it is not embedded in the appraisal of alternatives.

Thus it was not at all uncommon for the argument at a public inquiry to be between 'the project', whatever it might be, and 'the alternative', (not studied), which consists of stronger demand management including much greater emphasis on smarter measures. There may be many who think such a choice is oversimplified, but it does form one of the recurrent threads in transport policy in the last 30 years, and will continue doing so. It then hinders fair decision making for the tools used to prevent any formal comparison of those two, or different ways of making a balance between them.

The implication is that policy appraisal will have to do four things. First, it should provide for the centrality of smarter measures to be included in appraisals. Second, it must provide some guidance on how to make forecasts, or good judgements, of the effects of different levels of implementation instead of the assumption of some sort of average. Third, it should enable the level and type of smart measures to be implemented in combination with 'the project' to be different from those without. Occasionally promoters will want to argue that they can expand these measures with the project, but can do little without. This is in principle a legitimate argument (though in practical terms sometimes dishonest or specious) but if it is valid then the effect should be included in the appraisal—the project plus the measures compared against no project and no measures. More realistically, however, exactly the opposite will be true: the real alternative if a project rejected is a *more* intensive programme of smarter measures and other techniques of demand management, and that must then be included in the 'do-something else' option, a recurrent theme in the refreshment discussions.

And finally, the behavioural changes produced must be accorded a user welfare benefit just as other projects claim. This can be seen by considering what the equivalent changes in price or travel time would have had to be to produce similar changes in behaviour.

In spite of this, current experience is that the smarter choice policies are being seen as fringe areas of spending and first to be cut in many local authorities. They are only very rarely seen as providing a competition which increases the value for money from transport spending as a whole.

7 NATA is Dead. Long Live NATA

In the event, some but not all of these changes were announced by the DfT in May 201 1. The journal Local Transport Today headed its editorial on the announcement 'NATA is dead. Long live NATA', as a commentary on the Secretary of

State's statement 'This means there will no longer be a separate process called NATA", clarified by 'a DfT spokeswoman' that the processes underpinning NATA would be retained and used. In summary, the dominance of the formal economic appraisal is now reduced, at least in theory, as it will be carried out with some changes within a wider 'business case' which requires parallel strategic, economic, commercial, financial and management cases to be given prominence. An environmental case is not prescribed as a separate exercise, though at least part of it will be embedded within the economic case by application of a higher notional value attributed to carbon. Other technical changes include the treatment of indirect tax, unpacking the claimed value of time savings so that large and small are distinct (albeit still valued at the same rate), attention to social and distributional consequences, and changes in some parameters. Although much of this relates to technical and theoretical arguments, the core—and still unresolved— question is how the technical changes relate to implemented policy where it touches on both environmental and economic objectives. The comments of the Campaign for Better Transport (formerly called Transport 2000) seemed to speak for many professionals in an uneasy welcome of the changes subject to what it would mean in practice. Its chief executive, Stephen Joseph [21] said: "The review and associated changes to transport appraisal include many changes we've been arguing for, including looking at a wider range of options, testing proposals to see how well they fit with transport policies, and changing some of the technical detail in the appraisal. This is very welcome, but the real test will be how this plays out in practice. If it results in transport decisions that reduce traffic and carbon emissions and improve alternatives to cars and lorries, then it will be worthwhile. However, the review shows up the lack of a national transport strategy against which projects can be tested."

8 What Type of Transport Spending Should be Prioritised, in the Context of an Overall Spending Reduction, in Order Best to Support Regional and National Economic Growth?

The last report of the Commission for Integrated Transport [3], whose publication was, the next day, followed by the announcement of its abolition, concluded that

> Even with reduced spending limits, a good deal more net benefit could be generated by rebalancing the residual spend away from road capacity, to be focused instead on lower cost, high return schemes. These include road safety, and travel behaviour change through "smarter choices" measures, like school and workplace travel plans, car clubs, cycling, teleworking and internet shopping.

The work was informed by a series of 'think-pieces' commissioned from academics and specialists, included with CfIT [3]. The author's contribution [13–15] made an attempt to build on the Eddington results by making adjustments for traffic growth, tax, and the inclusion of a large number of empirical studies of the

Sector	Number of projects	Average BCR
Highways agency road schemes	93	4.66
Local road schemes	48	4.23
Local public transport schemes	25	1.71
Rail schemes	11	2.83
Light rail schemes	5	2.14
Walking and cycling	2	13.55
Total	184	

Table 5 Dodgson's [9] summary of schemes considered by Eddington

effects of other policies. The resulting figures would no doubt be no less subject to detailed criticism as the DfT's own previous assessments, but are given here as an indication of a line of argument which has been material within local authorities in the current discussions.

A good summary of the key results of the Eddington Report itself, expressed in terms of Benefit-Cost Ratios (BCRs), is provided by Dodgson [9] as shown in Table 5.

At face value the results seemed to suggest that walking and cycling schemes gave the best value for money, then highway schemes, then public transport schemes. Policies such as smarter choices were not included as there was insufficient data available (or at least, known to the Eddington team) at that time. These results expressed as average BCRs do include, within the constraints of the assumptions made, comparison of the projects' real resource costs with a wide range of economically-valued benefits including effects on some or all of congestion, accidents, carbon, health, local environment, travel time, consumer satisfaction and wider effects on the economy which might be generated by these benefits. It is not a perfect measure, not coping well with strategic interactions of policies and projects or considerations of fairness and political acceptability, and in practice very many assumptions are built in which can have the effect of giving answers which are biased for or against certain types of projects.

The approach I used was similar in underpinning to Eddington's, but with a somewhat different presentation. It is based on the idea that for each area of expenditure (road building, public transport improvements, smarter choices, etc.) the benefits of properly judged spending increase as spending increases, but usually at a declining rate, so that the best projects at the top of the list will have a bigger benefit and higher value for money than the marginal projects at the bottom of the list. This can be depicted as in Fig. 6, for the expenditure classes A, B, C etc. Each of these are then adapted to allow for the effects of reducing the assumed future rate of traffic growth, and what at that time was anticipated to be the impact of changing the treatment of indirect tax.

It follows generally that the more is spent on a particular area, the lower the benefits. Although there is no presumption that decisions already made in the past will have been optimised, in general it is likely that the more mature the field of application, the more of the very best projects will already have been identified and

Fig. 6 Incremental benefits
from successive increases in
spending

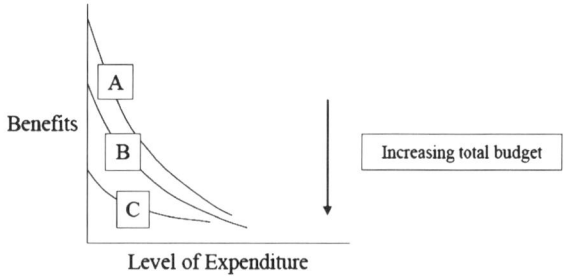

Benefits

Increasing total budget

Level of Expenditure

carried out, so a mature class of expenditure may well have lower BCRs than it used to have, and lower also than new emerging fields. This does not mean that the new is always better than the old, but does mean that it is more likely to have unexplored potential.

8.1 Results

Table 6 and Fig. 7 summarises the results, explained in more detail in Goodwin [12–14].

A strong pattern emerged of which types of transport expenditure have the greatest value for money in terms of speeds, travel times, safety and other economic costs such as health. In summary, by far the best returns came from smarter choices, local safety schemes, cycling schemes and the best of local bus and some rail quality and reliability enhancements, together with new light rail systems in some places. Traditional road capacity schemes gave much lower estimated value for money than cited in Eddington.

In current circumstances, it followed that by far the best value for money was being gained from a group of policies and projects consisting of low budget items, namely local safety schemes, smarter choices and cycling schemes. The next in order of value for money is a second group, including some cheap and some more expensive public transport improvements, namely the best local bus schemes, and the best new light rail and conventional rail schemes. However, a third group, consisting mainly of Highways Agency and Local Roads schemes, gave much poorer results (even for the best schemes). It should be acknowledged that the case of very large 'flagship' projects such as high speed rail did not fit comfortably into this sort of marginal analysis, and they were not included, and nor was the spending on concessionary fares for the elderly, on which no study was available of their effect on overall travel patterns. Road maintenance, pedestrianisation of town centres, traffic calming (other than safety schemes) and traffic management were not yet included in the analysis, though it would be useful (and with some care possible) to do so. Road pricing is not itself included in the analysis: since it produces both revenue and net social benefit, it will inevitably count better as 'value for money' than any of the spending policies included. However, it is

Table 6 BCRs by quartile of expenditure in nine areas of spending

Exp £b	Local safety	Smarter choices	Cycling	Conc bus fares	Local bus	Local roads	New light rail	HA roads	Rail
0.125	50								
0.2		30	20						
0.25	30			6					
0.375	20								
0.4		15	10						
0.5	10			6	10	1.3			
0.6		10	5						
0.625	0								
0.75				6					
0.8		6	4						
1		0	0	12	4	1.1			
1.25							7		
1.5					3	0.6			
2					1.5	0.5			
2.5					0	0	4	1.5	6
3.75							2		
5							1.5	1.25	3
6.25							0		
7.5								0.7	1.5
10								0.6	1.2
12.5								0	0

Results after adjustment for traffic growth, indirect tax, and omitted elements

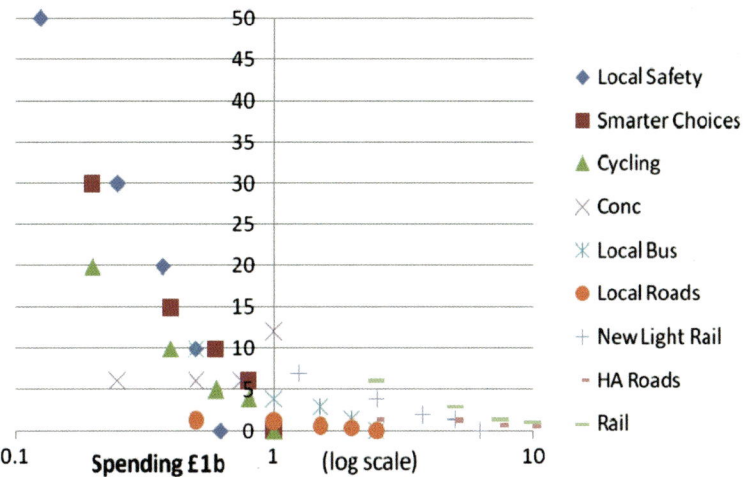

Fig. 7 Value for money related to expenditure

allowed for in testing the robustness of other policies: road pricing would (like reduced traffic growth for other reasons) reduce the value for money of road building. It could also increase the ability for public transport to fund its own improvements.

Overall, of the areas studied, the best value for money will be gained from *increasing* the expenditure on the first group, *protecting* the best projects in the second group, and *making savings* mostly from the worst projects in the second group and all except the very best in the third. Although carbon considerations have played only a very small part in the calculations, the resulting pattern of recommended expenditures is very supportive of carbon objectives.

The results are provisional, to test the feasibility of the method, availability of data and robustness of the conclusions. Qualitatively they seem in line with common sense and strategic priorities, but the exact numbers would be influenced by detailed data which could be published by the DfT, and further more substantial analysis and remodelling. The results are broadly in line with those which have been argued by various green and campaigning organisations, and in an interesting convergence also some business interests, such as the industry lobby group [22] who suggested that the wider economic benefit was the highest proportion of total benefit for urban public transport schemes, with interurban road schemes having a very much smaller effect.

The unresolved question in all this remains the scale and nature of behaviour change, both as embodied in the empirical evidence about the effects of past initiatives, whose interpretation remains contested, and the ambivalence towards future policies depending on such change. The House of Lords Science and Technology Committee (2011) summed up its inquiry into behaviour change as follows:

> We are not clear about the extent to which Government intend to reduce carbon emissions by reducing car use but, if they hope to achieve a significant reduction, the evidence suggests that regulatory and fiscal disincentives to car use will be required. We recommend that the Government (a) establish and publish targets for a reduction in carbon emissions as a result of a reduction in car use; (b) pubish an estimate of the percentage reductions in emissions which will be achieved through reducing car use and the timescale for its achievement; and (c) set out details of the steps they will take if this percentage reduction is not achieved by this time.

This analysis in this chapter would support that approach, and suggests that it might not be as difficult as is sometimes assumed.

9 Conclusion

There have been some quite complex layers of discontent about the methods of economic appraisal, building up over the last 10 years, and with some uncertainty about whether the latest round of changes have resolved these issues or not. There is a rather querulous astonishment among the veteran professional transport appraisal theorists and practitioners at how little progress has been made in resolving some core outstanding issues from decades ago—small time savings, indeed the value of time in general, for example. New developments in behavioural economics, regeneration, wider economic impacts and the dynamics of behavioural change are ignored, until they are suddenly treated with a boldness which is not nearly as well-founded in evidence as is claimed. A widespread mood of professionally grounded consensus has been replaced by a resigned 'jumping through the hoops' by consultants and local authorities dependent on central funding, a lack of confidence about how much effect formal appraisal has on actual decisions anyway, and a degree of cynicism about the quality of evidence on which all the important parameters are based.

This has recently increased as it becomes apparent how arbitrary changes in the values of these parameters have disproportionately large effects on results, especially the estimated values of benefit-cost ratios purporting to be the core quantifiable, evidence-based, scientific part of the appraisal process. While only 5 years ago 90% of estimated project benefit-cost ratios were tending to convergence either just above, or just below the acceptable passmark for funding—say in the region of 1.5 to about 4—now we see values of 20, 30 and 40 being claimed, by many different stakeholders (including, I should say, me), but with contested patterns of which types of project give the best returns. If the results are so unstable, how can they at the same time be well-founded?

Understanding this may be easier if we look internationally at the experience of cities where the greatest changes have taken place in transport projects and strategies. The great tide of rethinking about cars, traffic, pedestrians, public transport, walking and cycling is surely one of the great transport changes of our time. There is a high—though far from complete—degree of consensus about what sort of projects should be carried out. But this rethinking has manifestly not been led by formal appraisal methodologies; indeed, the methodologies have lagged about a decade behind, struggling to keep up with the policies. While the policy thinking may indeed be described as a 'tide', there seems to be little sign of a similar tide of eager international application of a common set of appraisal tools and formal methodologies. Multinational EU projects have had very limited success in producing a consensus among cities about what sort of appraisal will lead to successful policies.

A hypothesis consistent with these observations is that existing tools, ignoring the culture which produced them, have quite failed to understand that there are several different 'models' of how cities and nations approach sustainable transport policies, depending on their cultural, historical, ideological and political background. My suggestion would be that there are three, or perhaps four, quite different transport policy appraisal cultures.

Model A is a very elaborate nationally recommended economic benefit-cost core, detailed modelling and forecasting, generally assuming economic principles of behaviour (hence a main focus on costs, travel times etc., somewhat moderated by consideration of environmental and social aspects). There is some discretion for cities about how to apply this, but generally access to funds is dependent on satisfying these rather formal guidelines. The defining question is 'which projects give the best value for money?' This is a model often advocated in the UK.

Model B is driven by strong local political leadership (sometimes with directly elected mayors) based on strategic conception of the city and its future, funded by local sources and based on political processes. Technical considerations are acknowledged but have second place to strategic determination, and the level of confidence in the technical 'truth' of forecasts and hence appraisals is notably less. It is led by a strong, explicit, past and future vision of the city, owned by the city itself. The defining questions are 'what sort of city are we? What sort of city do we want to be?' There are some French and German examples, and at a local level it seems often to be true in Italian and Spanish cities.

Model C is a marketing approach informed by psychological and sociological insights, and using the methods of political science to seek consensus among complex stakeholders. The 'battle for hearts and minds' is not seen as a battle, but as an ever-changing shift in balance and influence. The defining question is 'What are the areas of agreement among powerful stakeholders?'. This approach was seen in the early stages of discussion about road pricing in Stockholm, with a senior lawyer appointed to find a consensus rather than to find a good transport policy, implemented as a process of negotiation between different interest groups. Intriguingly, after this exercise faltered to a halt, it was replaced by a different, but equally politically-savvy, approach based on the now well-appreciated process of timing of implementation of trials and referendum.

There may be a Model D, whose cities feel they are constrained by the overwhelming pressure of geographical or historical necessity with only marginal opportunity for appraisal to influence anything. The defining question is 'what infrastructure do we have to build?'. Cities in far peripheral regions, and those usually called 'transitional' in Eastern Europe, often use these arguments.

Our own tradition sees it as almost axiomatic that application of a formalised appraisal system must lead to selection of better projects, and therefore better strategies. I am sure that this is an arguable hypothesis, but the evidence simply does not allow us to treat it as an axiom. If it were true, by now we would be leading the world in the efficiency, attractiveness and contributions to environmentally benign economic growth of our transport systems. Rather, it is a proposition that needs to be tested empirically, by close comparative examination of the outcome of transport decisions, and monitoring real impacts in real time, not by the size, level of detail and theoretical claims of the appraisal guidelines used.

References

1. Buchan K (2009), cited in Cary R, Phillips R, Harwood J (2009) The right route: improving transport decision-making, Green Alliance, November. Available at http://www.green-alliance.org.uk/grea_p.aspx?id=4619
2. Cairns S, Sloman L, Newson C, Anable J, Kirkbride A, Goodwin P (2004) Smarter choices: changing the way we travel, Department for Transport, ISBN 1-904763-46-4 (2 volumes), London. (Available at www.dft.gov.uk sustainable travel)
3. Commission for Integrated Transport (2010) Transport challenges and opportunities: getting more for less, CfIT, London dated May, published September. Available at http://cfit.independent.gov.uk/pubs/2010/tco/report/pdf/tco-report.pdf

4. Department for Transport (2001, 2004, 2010) National Travel Surveys, DfT London. Available at http://webarchive.nationalarchives.gov.uk/+/dft.gov.uk/pgr/statistics/datatablespublications/nts/
5. Department for Transport (2006) Transport Demand to 2025 and the Economic Case for Road Pricing and Investment, DfT London December
6. Department for Transport (2009) Transport Trends, 2008 Edition
7. Department for Transport (2010) Road transport forecasts 2009: results from the Department for Transport's National Transport Model, London, DfT. Available at http://www2.dft.gov.uk/pgr/economics/ntm/forecasts2009/pdf/forecasts2009.pdf
8. Department for Transport (2011) Transport analysis guidance—WebTAG, DfT London. Available at http://www.dft.gov.uk/webtag/index.php
9. Dodgson J (2009) Rates of return on public spending on transport, RAC Foundation, London. Available at http://www.racfoundation.org/assets/rac_foundation/content/downloadables/rates%20of%20return%20-%20dodgson%20-%20190609%20-%20report.pdf
10. Eddington R (2006) Transport's role in sustaining UK's productivity and competitiveness: The Case for Action. Department for Transport, London
11. Glaister S (2011) Evidence to the transport select committee inquiry on transport and the economy, questions 430-460, house of commons 2 Mar 2011. Available at http://www.publications.parliament.uk/pa/cm201011/cmselect/cmtran/473/10120703.htm
12. Goodwin P (2011) Three views on 'Peak Car', special issue on 'A future beyond the car', guest editor S. Melia. World Transp Policy Pract 17(4)
13. Goodwin P (2010a) Opportunities for improving transport and getting better value for money, by changing the allocation of public expenditures to transport, Commission for Integrated Transport, London
14. Goodwin P (2010b) Improving value for money in the context of transport expenditure cuts: feasibility study, Centre for Transport and Society UWE Bristol
15. Goodwin P (2010c) Transport and the Economy: submission to the House of Commons Transport Select Committee. Available at http://www.publications.parliament.uk/pa/cm201011/cmselect/cmtran/writev/economy/te43.pdf
16. Graham DJ (2007) Agglomeration, productivity and transport investment. J Transp Econ Policy 41:317–343
17. Graham DJ, Gibbons S, Martin, R (2009a) Transport investments and the distance decay of agglomeration benefits. Report to the Department of transport
18. Graham D, van Dender K (2009b) Estimating the agglomeration benefits of transport investments: some tests for stability. Discussion paper 2009–2032, ITF, Paris
19. HM Treasury (2006) Volume 3—Meeting the challenge: prioritising the most effective policies. Available at http://collections.europarchive.org/tna/20070129122531/http://www.hm-treasury.gov.uk/media/39E/F8/eddingtonreview_vol3.0_011206.pdf
20. International Transport Forum (2011) Peak Car travel in advanced Economies? Chapter 3, Transport outlook: meeting the needs of 9 billion people, OECD/ITF, Paris
21. Joseph S (2011) Real test of government's decision making review will be how it is applied, Press Norice 27.4.2011, Campaign for Better Transport, London. Available at http://www.bettertransport.org.uk/media/apr-27-decision-making-review
22. London First (2010) Greater Returns Transport Priorities for Economic Growth, London
23. Mackie P (2010) Cost-benefit analysis in transport—a UK perspective, OECD/ITF Round Table 21–22 October, OECD, Paris
24. Metz D (2010) Saturation of demand for daily travel. Trans Rev 30(5):659
25. Millard-Ball A, Schipper L. (2010) Are we reaching peak travel? Trends in passenger transport in eight industrialized countries, Transport Reviews, pp 1–22
26. O'Sullivan P, Smith S (2009) So you thought you understood value for money? GES Conference July, DfT
27. SACTRA (1999) Transport and the Economy, standing advisory committee on trunk road appraisal, Department for Transport, London

28. Sefton Council (2010) announcement reported in Local Transport Today, 24 Dec 2010
29. Sloman L, Cairns S, Newson C, Anable J, Pridmore A, Goodwin P (2010) Effects of Smarter Choices Programmes in the Sustainable Travel Towns, UK Department for Transport, 2010. Available at http://www.dft.gov.uk/pgr/sustainable/smarterchoices/smarterchoiceprogrammes/pdf/effects.pdf

Deciding What Transport is for: Connectivity and the Economy

Bridget Rosewell

Abstract The evaluation of transport projects rests on comparing the costs of investment with their benefits. How we describe these benefits therefore has a strong impact on whether investments are made. One approach is to use time savings, but this abstracts from trip generation and economic impacts, and leaves it hard to incorporate environmental constraints. This however is still a dominant methodology amongst transport analysts. This chapter will critically evaluate these methodologies and the impact they have had on the ability to consider transport projects, with particular reference to the UK.

1 Does Transport Matter?

It is really rather odd how hard it is to answer this question. At one level it is clearly obvious that without transport there is no trade and without trade there is no economy. The discovery of wheeled transport, the training of animals to pull this transport created networks of towns and markets which were the backbone of mediaeval Europe. Moreover, transport is even more the backbone of military might. Roman roads are to this day referred to and even used—they were built to move troops, military material and all important provisions. Napoleon Bonaparte is credited with saying that an army marches on its stomach. Romans knew this just as clearly.

Harnessing steam power made railroads possible and it is hard to imagine the Industrial Revolution taking place at the pace and extent that it did without the

B. Rosewell (✉)
Volterra Consulting and Greater London Authority, 56-58 Putney High St,
London SW15 1SF, UK
e-mail: brosewell@volterra.co.uk

O. Inderwildi and Sir David King (eds.), *Energy, Transport, & the Environment*,
DOI: 10.1007/978-1-4471-2717-8_35, © Springer-Verlag London 2012

power of railways to move goods, people and food across continents. It is perhaps not surprising that it is the railways which feature in such descriptions of American capitalism as Kipling's Captains Courageous, and Ayn Rand's Atlas Shrugged.[1]

Containerisation is the further transportation improvement which dramatically reduced the cost of moving goods and has helped make possible the current wave of globalisation that we are experiencing.[2] The effort and investment which this required and the fortunes made and lost in the process are just as compelling as the stories of the railway barons of the early nineteenth century, though not nearly as well known. If moving goods is one axis of transportation, then moving people is the other. People movements are equally essential to trade as merchants, but the twentieth century has seen most extensively the rise of personal mobility, not only in cars, but also by air. Such movements have created entirely different labour markets as well as a whole new leisure industry. The OECD [2] has concluded that the outcome is that infrastructure needs are increasing across the globe and more funding is required.

But if transport is important, it seems remarkably hard to prove it, and economists have struggled with various methodologies. These have been well summarised by Crafts and Leunig [3] in a background paper for the Eddington Report into Transport and the Economy for the UK government. It is apparent from this study that creating a general framework is fraught with difficulty. A growth accounting framework is likely to lead a reliance on unmeasured spillover assumptions, while the analysis of gains from trade has no allowance for the benefits of variety and new products becoming available. Crafts and Leunig rely on case studies to illustrate the benefits of transport systems for the economy and the way in which the returns to transport systems have been measured. It is clear from their account how different forms of communication systems, from canals to roads, have had strong positive impacts which have then faded away, to be replaced by new structures. It is this phenomenon which needs to be at the heart of any understanding of transport and the economy.

For example, an investigation of the relationship between the growth of road traffic and that of output in the UK from 1950 to 2009 shows a strong short-term relationship but distinct phases over the medium term (Fig. 1). Up to the first oil crisis in 1973 there was strong growth in road traffic moving cyclically with output. In the period from then to the early 1990s both output growth and road traffic growth slows down while thereafter the cyclical pattern is much less obvious. At the same time, the relationship between traffic growth and output growth is distinctly different, with higher output growth compared to traffic growth than in the 1950s and 1960s.

This illustrates how the relationships can evolve over time for any given technology, and that an understanding of such relationships and the impact of investment will not be straightforward.

---

[1] In both stories, the efficiency and speed of communication are used as metaphors for power and commitment.

[2] The story of containerisation is dramatically described in The Box, by Marc Levinson [1].

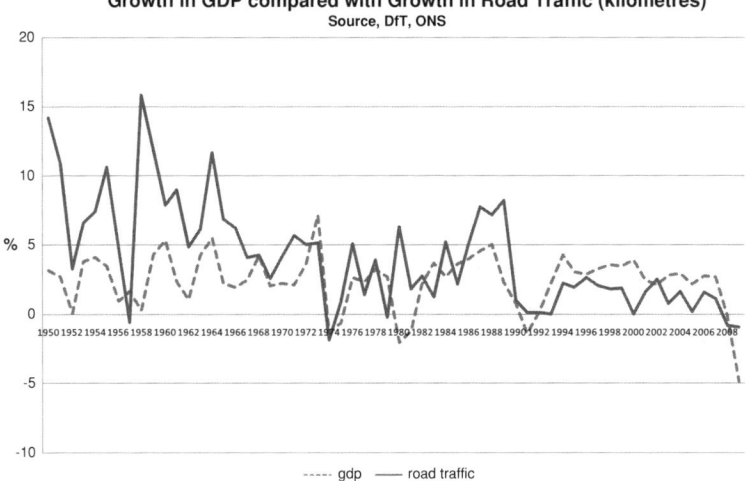

Fig. 1 Growth in GDP and Road Traffic

Moreover, it is noteworthy that, as far as people are concerned, the amount of time spent in travel remains fairly static over a lengthy time period. Table 1 illustrates this. While the survey shows an increase of about 50% in average trip length, the average trip time has barely changed. This suggests that persons engage in some form of time budgeting which will only allow for a certain proportion of time spent travelling.

Improvements in transport availability allow longer trips rather than saving time as Metz [4] has shown and it is this which creates greater market access for both people and goods. So the basic relationship between output and transport continues to be powerful as Fig. 2, taken from an analysis by the UK Department of Transport, shows. This data indexes all forms of mobility including both passengers and freight. Although road traffic growth has slowed down, the relationship with output as a whole has remained strong—suggesting that we are a long way from ending the role of transport in growing our economy.

However, the uncertainties surrounding these trends and of measuring the impact of transport have been compounded by the proposition that new technologies will make physical transport systems redundant as we move into the digital age.

2 The Digital Economy and Transport

It is not yet entirely clear what we mean by this term. It could be argued that the digital economy is anything that happens using a computer (based in other words on bits). However, most commentators concentrate on aspects of the economy which are mediated in some way through the internet.

Table 1 Passenger trips, distance travelled and time taken: Great Britain 972/1973 to 2009, National Travel Survey

| Year | Per person per year | | | | | | Unweighted sample size (individuals) |
	All trips[a]	Trips of 1 mile or more	Distance travelled (miles)	Time taken (h)	Average trip length (miles)	Average trip time (minutes)	
1972/1973	956	594	4,476	353	4.7	22.2	15,879
1975/1976	935	659	4,740	330	5.1	21.2	24,692
1978/1979	1,097	736	4,791	377	4.4	20.6	18,433
1985/1986	1,024	689	5,317	337	5.2	19.8	25,785
1989/1991	1,091	771	6,475	370	5.9	20.4	26,285
1992/1994	1,053	742	6,439	359	6.1	20.5	24,671
1995/1997[b]	1,086	794	6,981	369	6.4	20.4	22,861
1998/2000	1,071	810	7,164	376	6.7	21.1	21,868
2002	1,047	819	7,135	380	6.8	21.8	16,886
2003	1,034	812	7,192	381	7.0	22.1	19,467
2004	1,026	806	7,103	382	6.9	22.3	19,199
2005	1,044	818	7,208	385	6.9	22.1	19,904
2006	1,037	812	7,133	383	6.9	22.2	19,490
2007	972	786	7,103	377	7.3	23.3	19,735
2008	992	800	6,923	376	7.0	22.7	18,983
2009	973	774	6,775	372	7.0	22.9	19,914

Great Britain comprises England, Wales and Scotland

[a] There is an apparent under-recording of short walks in 2002 and 2003 and short trips in 2007 and 2008 compared to other years

[b] Data from 1995 onwards has been weighted, causing a one-off uplift in trip numbers, distance travelled and time taken between 1992/1994 and 1995/1997

Fig. 2 Taken from: Foresight Land Use Futures Project: Final Project Report (2010), The Government Office for Science, London (page195)

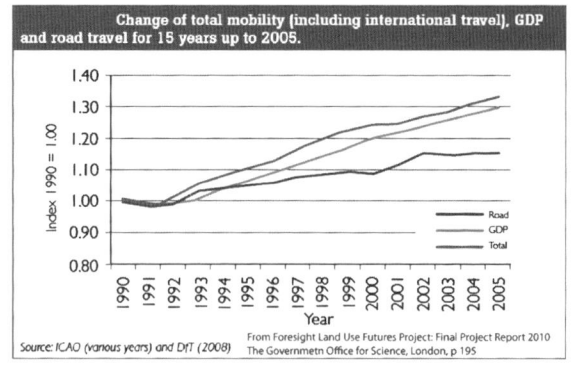

However, this is not necessarily the most useful way of thinking about the digital economy. Conceptually, we are most interested in those aspects of activity which are facilitated by Internet activity or which take place entirely online. If I buy my music online, I may never have a physical object. However, the production of the music in the first place is not (yet) an entirely digital event. My grocery shopping equally relies on a cyberspace catalogue but certainly requires physical activity at many other levels.

In principle, therefore, we are most interested in the added value that such activity represents and whether the margins on internet sales are higher than those on bricks and mortar shopping. Since many retailers now effectively offer both, this is not a straightforward question. It is noteworthy that HMV has recently announced the closure of many of their stores to cut costs, precisely because customers are now buying entertainment products entirely online.

Businesses also purchase over the internet; the UK Office for National Statistics estimated that the value of this totalled £360bn—in GDP terms this counts as intermediate purchases and does not add to output. However, if these purchases are better value for money then the cost reduction can be passed on in lower prices or better profits for shareholders.

There are also products which are produced entirely online. An example is internet advertising, on which companies are estimated to have spent £4bn in 2010, now representing 30% of the total advertising spends.[3] In the US, spend reached a record $26bn.[4]

Software is not necessarily a digital product. A piece of software delivered to me on a physical medium and used on a standalone system is a more traditional product. Software downloaded and with internet capability in its operation is more clearly digital. At present, we do not have statistics which distinguish these categories. Counting the firms engaged in this type of business might be the best that can be done.

Various EU studies have looked at the impact of digital trading and broadband access. The biggest impact has been procurement savings by manufacturers. Broadband access also adds to service sector productivity.

The measurement of GDP was originally intended to serve as a measure of how well off a set of consumers was—this was the value of their total product. As economies have developed, it has become harder to make such measurements. Service sector products are harder to value, and there is more focus on consumer surplus and welfare—a source of the new interest in 'happiness' measures.

Consumer surplus is conceptually the difference between what I am willing to pay for something and what I actually paid for it. An innovative economy and information constraints mean that this is an uncertain concept. What I would have paid for a product that did not exist is a moot point. Estimates of price elasticity have been used to get a handle on this, and research for BT showed the value of broadband access in this way.

---

[3] Internet Advertising Bureau, http://www.bancmedia.com/news-online-advertising-value-passes-4billion/

[4] http://www.iab.net/about_the_iab/recent_press_releases/press_release_archive/press_release/pr-041311

One of the most significant aspects of internet economy products is their impact on information availability. Standard analysis suggests we consume information up to the point where the cost of acquiring additional knowledge is less than its benefit. However, this implies that we have knowledge of the benefit of this additional information. But to know this is the same thing as saying that we actually have the information in the first place. Under these circumstances, traditional measures of value become meaningless.

More useful therefore are the measures of internet use, connections and sales and measures of intensity. UK estimates[5] show that:

- 19 million households have an internet connection.
- 25 million UK residents are members of Facebook.
- 31 million adults made online purchases during 2010, spending c£50bn in 2009.
- More than a quarter of mobile phones were smartphones by the second quarter of 2010.

BCG [5] has calculated that the value of the internet economy in GDP terms was around £100bn in 2009, of which the largest share is consumer transactions. This is using an expenditure-based estimate, valuing all products and services purchased and it represents 7.2% of GDP.

This description of the digital economy shows how much physical activity takes place in it as well. Goods need to be delivered and indeed manufactured. Even a completely digital product will be hard to create without human interaction.

Of course there are those who think that such interaction can take place in the digital domain itself. However, all the evidence suggests that this is likely to be a minority taste. Edward Glaeser's recent book 'The Triumph of the City [6] describes how cities provide the essential underpinnings to civilisation and the face-to-face interaction that economies and in particular innovation require.

He points out that in countries which are more urban, people report being happier. As the share of the country's population that is urban rises by 10%, the country's per capita output increases by 30%.[6]

It is noteworthy that the more we are able to communicate the more we congregate rather than spread out. From road to rail to the internet, cities have become more important rather than less, and we have now reached the point where it is estimated by the UN that more than half the world's population live in cities.

This all leads to the conclusion that the digital economy will not be one where transport does not matter and where physical proximity somehow becomes irrelevant. However, it certainly implies that the growth of digital products will change the production methodologies and spatial organisation of activity. As a result, it is more than ever important to think about how transport infrastructure investments are decided upon and how their contribution to a sustainable economy should be evaluated.

---

[5] BCG [5].

[6] Glaeser p. 7.

3 The Current Methodology

The preferred methodology for an evaluation is to compare costs and benefits. Indeed, one might say this is pretty obvious at first sight and bears useful comparison with the methods used by business to decide on investments.

The challenge is to decide however, what the relevant costs and benefits actually are. In practice, they can become so complicated that the relationship with simpler business concepts becomes misleading. The evaluation of transport projects in the UK in particular is an arcane art conducted with large and complicated models by a priesthood of experts skilled in their use. Outsiders can find it hard to judge the results and unable to penetrate the conclusions.

Let me try and set out the parameters. In order to judge a transport project it is of course necessary to assess how many will use it. Over the last 30 years London has developed a suite of models to produce these results. The model takes as inputs pre-existing forecasts of employment and population changes which have been allocated to a large number of transport zones (1,700 at the latest count). These zones are very small in the centre and tend to get larger as you move outwards and indeed include commuter zones outside London. These forecasts are used to predict the use of the network of roads, rail, bus and underground, given the costs of this use. Costs include the time taken as well as money spent and include adjustments for difficulties such as having to make a change of train. The model is solved on the assumption that people will minimise the cost of making the trips they need given the distribution of where they work and where they live.

In order then to understand the effects of a new transport investment, we create a new solution of the model having made the changes to the cost of the network that new linkages imply. We still have the same number of trips, but now they are being made more easily as a result of the new investment, which means crowding can go down and trips take less time. This is the key measured benefit of any transport scheme.

Of course time savings are made in minutes, while the cost of the scheme is in money. So time savings need to be turned into money in order to establish the value of the scheme. The establishment of a value for time is a non-trivial question, as some introspection should quickly show. It will vary by person, by income, by kind of trip and so on. The most recent review of this question was undertaken in 2003 [7] and its conclusions are the basis for the current levels which have been set by the Department for Transport.

An hour of leisure time is currently valued at £4.46 per hour and an hour of work at an average of £26.73[8]. The studies on which these values are based are not uncontroversial. Much of it used experimental studies which give people options of various trips. Answers can be inconsistent but these will often be ignored because rational economics suggests that they 'ought' to be. These values are part of the whole edifice of rational, expert, cost benefit analysis which rests on an intellectual construct which is increasingly under challenge. The value of working time is perhaps less controversial as it essentially rests on wage rates—these are at least directly observable whatever actually drives them. However, for most studies they provide a relatively small proportion of benefits. Even in the Crossrail[7] case, where a value of work time of £55 per hour was applied for trips to Heathrow, the contribution was only 30% of the total time savings.

These basic calculations are those used to compare all UK road and rail projects where there is public investment. The approach is based on wanting to capture all the benefits of improving people's welfare when considering the costs.

Saving time can be considered the welfare benefit of an investment. Since time can also be thought of as money, then it implies somewhere a willingness to pay for a commodity (travel) which is not being charged for, or where the charges do not actually cover the costs. The approach is described in the official guidance (WEBTAG) as follows:

> The basic strategy of the willingness-to-pay (WTP) calculus is to arrive at a money measure of the net welfare change for each individual that is brought about by the project under consideration, and then to sum these. The welfare change for any individual is measured by the *compensating variation*, i.e. the individual's WTP for benefits or the negative of his/her willingness to accept compensation for disbenefits. The principle behind this calculus is the Kaldor-Hicks *compensation test*: a move from one state of affairs to another passes this test if, in principle, those who benefit from the move could fully compensate those who lose (without themselves becoming losers). When the cost-benefit accounts are presented in this way, there often are items which appear as benefits for one person and equally valued costs for someone else: such items are *transfer payments or pecuniary externalities*. Items which do not cancel out in this way are *social* costs or benefits (sometimes called *resource or real resource* costs or benefits). The word 'social' is used to signify that these are costs or benefits which fall on 'society as a whole', understood as the aggregate of all individuals.
> The calculus of social costs and benefits seeks to measure the value of the 'resources' used by, and the benefits created by, a project.[8]

These dense jargon ridden paragraphs can essentially be translated into English in a single sentence. Namely, the benefit of an investment can be defined as the willingness to pay of an individual—in this case for an extra minute of time. This willingness to pay is clearly an individual matter based on the enjoyment of individual benefits. It takes no account of broader economic benefits or indeed the impact of one individual's enjoyment on another.

This approach to transport projects was reinforced by a major study of the analysis of road schemes which reported in 1999. The report showed that economic theory suggested that time savings and economic benefits were two sides of the same coin. In principle, and in a competitive economy, time savings could be converted into economic activity as trips increased and time savings were competed away. As SACTRA concluded, "If these conditions hold, we concur that the value of the estimated costs and benefits to transport users (notably time savings, operating costs

---

[7] This railway, involving a new tunnel under Central London, is now under construction. It is a large project costing in the region of £16bn and will add around 80–90,000 to London's commuter capacity.

[8] WEBTAG, Unit 3.5.4, Box 2 [8].

and accident reduction), and to nonusers (notably environmental impacts—provided that they have all been identified and a money value attributed to them) would give a full and unbiased estimate of the value of the overall economic impact. This is equivalent to the statement that no 'additional' economic value exists."[9]

This means that there are circumstances in which an analysis of costs and benefits based on time savings will give us the transport investments which are needed to drive economic growth, since these can be considered to be one and the same thing. It is therefore crucial to take a look at the likelihood this will actually be the case.

In fact, there are many reasons why these conditions will not hold. First, there is the question of whether all costs and benefits have been correctly valued. Second, there is the question of whether the assumptions of perfect competition hold. And third there is a challenge to the time period over which any such changes will take place.

4 Valuing Costs and Benefits

It is easy to gloss over estimating costs. All projects are however bedevilled with this question. In large projects there is great scope to get things wrong and since they also take a long time to come to fruition, many costs can change in this period too. The UK government insists on 'optimism bias' in costing which adds 50% to costs on the basis of past over-runs. Giving firm deadlines can concentrate minds on delivery but also escalates costs as it approaches. It is well known that the Millennium Bug generated eye-watering fees for computer consultants as the end of 1999 approached—and it is still not obvious whether the bug was anything more than a minor irritant, rather than a deadly sting.

Even so, getting the costs right is a minor problem compared to getting the benefits right.

The process of time evaluation described above has a lot of embedded assumptions which go far beyond whether we can measure the value of time correctly for any set of individuals. At the heart of this is how we view the process of trip generation.

The benefits of time saving accrue, fairly obviously, to those who make the trips. For any investment, these are future trips and therefore there is a forecasting process involved.

Forecasts are notoriously wrong. They are nonetheless central to any investment evaluation—indeed this applies to costs as well as benefits. The trouble with the benefits is that they have to take place over a longer horizon than the costs and thus become still more uncertain.

The truth of forecasts. A forecast is likely to be right when the variable in question is not too random and where its causation is likely to be stable and direct. Merely stating this shows how unlikely this outcome is. Where there is any circularity of causation—your forecast leads me to take a different decision—there is already a

---

[9] SACTRA Final Report, para 24, [9].

problem as with any structural changes which are going on. In the short run, output of the economy has a large random element and the noise overwhelms the signal.

There is far too little active consideration of what we might call 'forecasta-bility' and far too much reliance on the need to simply have a set of numbers. Figure 3 shows the percentage change in UK GDP from 1956 with each quarter as the percentage growth on the same quarter the previous year. Two things jump out—the volatility in growth, and perhaps the unusual stability of the period between 1994 and 2008. If we want to examine 'the business cycle' it is not really obvious where we should look for it.

Looking at a chart like this, it is hardly surprising that forecasters struggle to get the following year right and often disagree. Precisely why each turning point occurs can be established in hindsight but their timing looks pretty random. Signal is dominated by noise.

On the other hand there is some underlying stability exhibited by the same data when we look at levels of GDP, as Fig. 4 shows. This certainly illustrates how unusual the last couple of years have been in historical terms compared to previous slow downs, but also how the general path of growth has continued to march upward. Thus it becomes an important judgement to consider whether recent and unprecedentedly sharp falls herald a completely different path for the economy—a break with the whole of post war history—or rather that growth will eventually return. In planning it is important to rise above the short term and the mood of the moment to consider the whole sweep of relevant history.

It is equally important to consider the level of aggregation. Employment or output may be fairly stable over the long term at a regional or national level. They may be quite variable and unpredictable further down. There is often a demand for consistent forecasts, where views taken at one level are the same as those taken at another. But the actuality may not be consistent at all. Individual areas may buck a trend. For example, funding in Hull for housing renewal was being put at risk by forecasts which showed employment and output plummeting. Yet this was inconsistent with what was actually being observed in the city where the trajectory seemed more stable. The forecast was based on a national and regional prediction of falls in manufacturing, which indeed was the case more generally. But Hull had been hanging on to its manufacturing despite the trend elsewhere. Should the forecast therefore capture Hull's own history or should it show a break with the past and capture the patterns of the rest of the country. Either position is equally viable.

Another example is Hounslow, which lost 40,000 jobs between the late 1980s and the mid 1990s and then gained 30,000 back again by the turn of the century. Should this strong variability be part of a future projection, or should it try and abstract from this over the longer term? There is no right answer—planning for variability is extremely hard to do in spatial terms, but on the other hand its existence would lead you to foreground a need for flexibility.

The forecastability question obscures a more important issue. This is the extent to which the outturn is independent of the investment. This is one of the ways in which confusion is generated by the evaluation process.

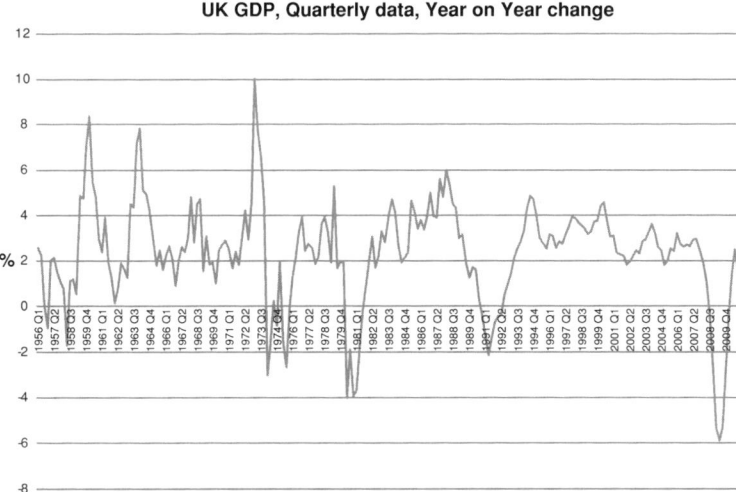

Fig. 3 Quarterly growth, year on year, UK GDP, Source ONS

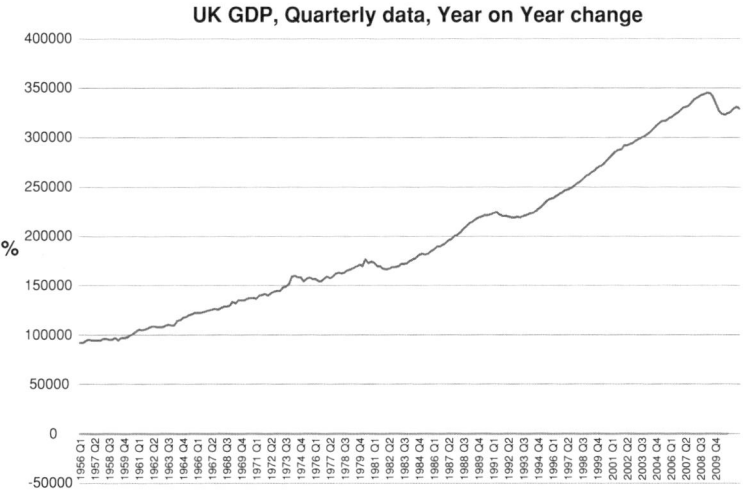

Fig. 4 UK GDP quarterly data- year on year change

It is like considering an investment in extending your house. You get a quotation for the building work, and imagine the benefits of living in the new kitchen, dining room and conservatory. But you fail to imagine the process which is essential to getting from the original state of the house to the final state and the other spending (from kitchen equipment to curtains) and the time involved in the process. You also (generally) fail to realise how the project will need to change as you go along, as it turns out you really want the windows somewhere else, or the building regulations change.

A similar process is going on with any transport infrastructure investment. The benefits are thought of as time savings for people who make trips using it. But how they use it and why they will change as a result of the new opportunities—and their choices will involve their own investments (the kitchen equipment). All of this will happen in real time as the feedbacks emerge between the ideas and the reality. The trips that will use the new infrastructure do not emerge independently of the investment, much as the new kitchen and curtains do not emerge independently of the new extension. They are complementary and one only exists because of the other.

Thus it is the trips and their associated output that comes first—the welfare benefits are a by-product. It is putting the cart before the horse to start with welfare and to consider the economy as an afterthought.

Moreover, this is a process that happens in time. It may take many years to generate the full consequences of a major investment. Even in the case of the house extension, construction takes time, fitting out takes time, and deciding on furniture may take still more time. Each stage involves decisions and changing decisions.

This kind of world, in which we must consider the result of an investment as something that happens in time and should be assessed that way is quite against the assumptions of perfect competition. Indeed it is the assumptions of perfect competition which lead to a comparison of one outcome with another and no consideration of the path involved. It also leads to the ability to equate welfare with economy.

In perfect competition there is perfect information. No need to change your mind about the colour of the curtains—you will have got it right the first time. Moreover, you only need information about yourself, your tastes and preferences. Fashion plays no role—since fashion implies that you are affected by what others do. There is never any need to look round and think that the colour scheme looks dated.

In other words, perfect competition is a theoretical construct established to enable economists to generate neat and tidy results. It should be used with the utmost care, as it has a tendency to behave like a weed and swamp more careful analysis.

Once it has been dug out, we can clear the way for thinking more clearly about the links and feedbacks between transport and its effects without assumptions that mean no one cares about another's choices, or that the future is completely known to us.

5 What Next?

Whether evaluating transport investment with welfare-based tools, or assessing an 'approved' cost of capital with a regulated asset base, we have allowed economic models to get in the way of reality.

This means that costs and benefits are assessed by different standards and with insufficient attention paid to payback, as distinct from benefit cost ratios. Private investment needs to get a payback which can cover interest and principal and over which the investor has some control. Where the returns are under the control of the public sector (fares, charges, prices) a lender must have faith that the monies will be sufficient and robust to that regulation. This is a difficult judgement and made more difficult every time the rules are changed.

In the case of Tax Increment Funding, there is a further judgement that tax revenues will flow through to a project to make this payback. Creating this ring fence at a sufficient scale and size to cover the costs is still untested outside the US.

The Treasury will only be prepared to hypothecate taxes which they view as additional to those they might otherwise collect without the investment. This is quite a stringent view of how the economy works and depends on strong economic modelling assumptions of the kind that are extremely hard to test—another example of how economists have imposed a particular view of reality.

Another implication of the 'modelled' approach that we take to infrastructure is that the real economic benefits are seen as the add-on and the less real are taken to be the basic benefit. In the case of a transport investment, for example, it is time savings, valued by some limited techniques, which is the core method of thinking about rail and road infrastructure.

The weakness of this approach is self-evident:

- It relies on forecasts of economic activity and population and assumes this growth will happen anyway.
- It relies on assumptions of the value of time which are in turn dependent on survey evidence.

The larger and longer term an investment the less likely it is that growth will be independent of the investment. The value of time (even if measured correctly) will only be a good measure of welfare and economic benefits if the stringent and highly unrealistic assumptions of perfect competition apply.

Nonetheless this approach is used in generating the cost benefit ratios which are a key public sector decision rule.

More recently, some variations on this theme have been permitted—a bit like the addition of further epicycles to Ptolemaic astronomy rather than accepting that the earth goes round the sun. These variations permit additions to benefits if it can be shown that:

- Investment increases activity in highly productive centres, increasing the output of the UK as a whole.
- Investment improves land use and regenerates locations which are the subject of policy.

Both of these add further degrees of arcane analysis and are difficult, if not impossible, to prove. It is undoubtedly true that increasing activity in productive centres by relieving constraints on transport systems, such as Crossrail does, increases the output of the UK economy as a whole. Indeed it is possible to show that the taxes on such activity are sufficient to pay back the investment, while the fares are capable of covering the interest charges. In this context, it would seem that the whole paraphernalia of traditional cost benefit analysis is unnecessary—the investment will pay for itself.

In the case of the second effect, this has been defined in the guidance in such a way as to provide an almost impossible test, especially for any investment on a large scale. Requirements include the need to provide a model of how transport interacts with land use known as LUTI models. Such models are even larger black boxes than the transport models, but like them the scale of investment that is made in them by both researchers and clients mean that there is a tendency for faith to replace judgement. These models are opaque and almost impossible to calibrate to real data. Relying on them for any policy decision is entirely irresponsible.

The case for the Thames Gateway Bridge in London had to rest on proving the potential for the bridge to create economic activity. Models in social science do not meet the standards of proof of engineering models and this proved a serious difficulty.

If we are to be able to invest in transport infrastructure it is essential to find a better way to rank investments and especially to create a more transparent method which can both be better understood and challenged. Public sector priority setting is a key element in the investment process, whether there is public sector funding or not. Even where there is a private sector funding, planning processes will still require analysis to meet the standards set by transport policy makers.

The tests that assessments need to meet are:

- Transparent in process and assumptions.
- Clear about the split between financial and non-financial benefits.
- Clear about the purpose of investment and what form of benefits are expected.
- Clear about who controls the benefits.

One way it has been proposed to do this is to create a future asset base valuation (RAB) on which an agreed rate of return can be made. Unfortunately this will fail the transparency test. The way in which a future asset base is estimated has a lot of similarities with the existing process of modelling. Again, it relies on forecasts and an assessment of benefits in the same framework as before.

The easiest way to create transparency is to start with an analysis as if a project is a private sector investment. Benefits would clearly be financial and it would be clear to what extent these could actually be captured by the investors. This would also create a framework for analysing the extent to which private investors would have an interest in investing—if returns can only be captured by institutions under the control of the public sector, this will clearly limit private sector funding.

Subsequent to this initial analysis, which can be provided in a form in which it will be possible for third parties to understand and challenge the analysis, it is then perfectly possible to consider external effects which might make the investment either more or less attractive, such as:
Negatives:

- It reduces the value of other activity (cannibalisation).
- There are environmental negatives.
- There are distributional negatives.

Positives:

- There are growth impacts.
- There are environmental benefits.
- There are distributional benefits.

It would need to be obvious that these were not captured in a monetary analysis before they could be accepted—but if they were, this would be a clear signal for government investment.

It must be accepted that such an approach to investment appraisal for transport would go against the development of large-scale models and associated guidance which has been built up over 30 years or more. There is much vested interest associated with the current system and intellectual capital that has been built up around understanding and presenting this complex analysis.

However, it is clear that this system militates against principles of good governance, even if the analysis could be done perfectly. It lacks transparency of both process and analysis. Moreover, the embedded assumptions in the analysis are risky and lack calibration to the real world. The current decision making process has therefore failed at a very basic level.

A simpler system must have a better chance of getting a sensible set of decisions which can generate the infrastructure we need.

6 Institutions and Finance

In infrastructure, institutions come in three flavours—planning, delivery and finance. While the skills required in each are different, they need also to be related to each other and creating new institutions is rarely an answer to a problem. A willingness to create a long-term vision and act on it in a flexible way is much more important. It requires a willingness to create proper financial vehicles which can control costs and manage revenues, whether from charges or from taxes.

At a recent conference on High Speed 2, the speaker from France stressed finance, profitability and return on investment. The UK speakers did not mention any of these concepts. Instead they talked about systems, general economic benefits and management. This was noteworthy from a French system which is nonetheless heavily subsidised by the public sector.

The attack on investment in High Speed 2 in the UK has focused on the damage to properties in a rich part of England in return for '20 min off the trip to Birmingham'. If this were indeed the only return on such an investment, it is hard to see why it would be work spending up to £40bn on it. This is the consequence of an evaluation framework based on time savings. In fact such new connectivity is focused on economic regeneration, and exploiting opportunities for growth in cities in the North of the country. This case is not well made at present.

The stress on model-based benefits and evaluation tends to lead to larger projects and a focus on cost benefit ratios, rather than a focus on deliverability and revenues. The best becomes the enemy of the good. Large projects obscure the identification of benefits and who could pay.

I have recently worked with a small group looking at finance for High Speed 2. We concluded that the public sector should pay only for the basic infrastructure of tracks and signalling. Trains, stations and associated development should be the subject of separate appraisal and separate finance. Indeed the spurs from the main track to city centres are not necessarily a general taxpayer benefit, although the organisation of local government finance means that it is hard to see how a city could raise the necessary funds.

References

1. Levinson M (2006) The box: how the shipping container made the world smaller and the world economy bigger. Princeton University Press, Princeton
2. OECD (2006) Infrastructure needs to 2030, OECD futures project
3. Crafts N, Leunig T (2005) The historical significance of transport for economic growth and productivity. Background paper for the Eddington report
4. Metz D (2008) The myth of travel time savings. Transp Rev 28(30):321–336
5. Boston Consulting Group (2010) The connected kingdom: how the internet is transforming the UK economy, Google
6. Glaeser E (2011) The triumph of the city
7. Institute for Transport Studies (2003) Values of travel time saving in the UK, Leeds University
8. WEBTAG Unit 3.5.6 (2011) http://www.dft.gov.uk/webtag/documents/expert/pdf/unit3.5.6.pdf
9. Standing Advisory Committee on Trunk Road Assessment (1991) Final Report, Department for Transport

Global Consequences of the Bioenergy Greenhouse Gas Accounting Error

Tim Searchinger

Abstract Like the global financial crisis, which resulted in part from misguided accounting of mortgages, global policies to expand transportation biofuels and bioelectricity reflect an accounting error. Although the carbon accounting in these policies assumes that plant growth offsets all carbon released by burning biofuels, only "additional" plant growth can provide an offset. Because they double count biomass and land already used by people or sequestering carbon, many policy proposals aim for bioenergy to supply 20% or more of the world's energy by 2050. That would require almost doubling the present global harvest of plants for all uses, which would likely lead to extensive deforestation and increase greenhouse gases. Fixing the accounting would focus policies on the more limited potential for truly low carbon biofuels.

1 Introduction

Can an accounting error that has crept into world efforts to curb global warming threaten much of the Earth's natural forests and savannas? Can that error be large enough to cause efforts to reduce global warming to increase it? Most rational people would probably react to such suggestions with grave doubts. Anything to

Thanks to Ralph Heimlich for substantial data analysis and to Mary Booth for many contributions to the understanding of emissions from bioelectricity.

T. Searchinger (✉)
Woodrow Wilson School of Public and International Affairs, Princeton University,
Robertson Hall, Princeton, NJ, USA
e-mail: tsearchi@princeton.edu

do with accounting seems too obscure to have such large effects, and if the problem were that serious, surely they would already know about it.

Yet, the world economic crisis from which we are still recovering results to a great extent from flawed accounting. The gross overvaluation of risky U.S. mortgages once combined into securities played a critical role in the housing bubble, whose burst touched off the 2008 financial collapse. A look back at previous financial disasters, such as the U.S. Savings and Loan crisis of the 1980s and the Enron bankruptcy, would show that flawed accounting played a similarly dominant role.

At least three features of the financial crisis explain the potential of accounting errors to have such vast consequences. First, although justified by layers of complex computer modeling or other analyzes, the mortgage financial error had such a simple origin that it could be repeated broadly. It was the belief that housing values never fall at the same time in cities all across the U.S., which meant securities that combined mortgages from all over the country would hold their value regardless of how little income people had to pay them back [31, 33]. This accounting concept proved false because it led to a flood of easy money across the country that itself created a nationwide bubble and a nationwide collapse.

Second, the error distorted incentives for vast numbers of people. Because the error led financial markets to reward mortgage agents for even the most inappropriate mortgages, agents had the incentive to enroll as many mortgages as possible regardless of their integrity. If accounting rules reward thousands or millions of people for doing things that lose money, people can lose a lot of money.

Third, public officials had little incentive to fix the errors because they could do nothing and claim credit for apparent economic gains while fixing the errors would require battles with a range of powerful financial and housing interests. Who would force adequate reserves on banks and the quasi-government lenders that backed U.S. mortgages when their loans and loan guaranties helped keep the economy humming at no immediate and therefore apparent cost to taxpayers. Flawed accounting allows people temporarily to have their cake and eat it too. Over time, as more people and companies come to rely on the flawed rules, fixing those rules become even harder.

Because of the power and difficulty of changing accounting rules, governments need to count greenhouse gas emissions accurately from the start as they start to curb global warming. Most policies do a good job, but in one critical respect to date, they do not. The error involves bioenergy. In particular, it involves a convention that any accounting of emissions from bioenergy should always ignore the very real carbon dioxide emitted through exhaust pipes and smokestacks from burning the biomass. Although the accounting rule does count the carbon dioxide released by using fossil fuels to generate liquid and solid biofuels, the actual carbon in the biomass itself, turned into carbon dioxide through burning, does not count regardless of how or where the biomass is produced. In that sense, the rule treats bioenergy like solar and wind energy as inherently "carbon neutral." The error makes forests worth more dead than alive, can influence the behavior of millions of people and vast areas of land, and can do to the planet what the mortgage error did to the world economy.

2 The Assumption that Biomass is Carbon Free: The Basic Accounting Error

Roughly three quarters of the World's use of biomass for energy still remains the burning of wood for traditional home cooking and heat or the conversion of that wood to charcoal for the same purposes [9]. But bioenergy is expanding now through the modern effort to turn crops into ethanol or biodiesel for use in cars and trucks, and to replace coal and natural gas in power plants that generate electricity. Bioenergy is also heating factories and homes in developed countries, and in some developing countries, charcoal is powering steel production. Projections of bioenergy for the future rely on increasing the harvest of world forests, and turning vast tracts of the world's land over to specially bred, fast-growing grasses and trees.

Much of this new interest in bioenergy results from the belief that bioenergy is inherently carbon neutral but for the fossil fuels needed to make it, and to its proponents, the reason is obvious. Bioenergy is just carbon recycling. The carbon emitted to the air from burning biomass is the same carbon absorbed by plants when they grew and generated the biomass. Unfortunately, this reasoning makes a baseline error. It takes land to grow plants for bioenergy, and that land almost always grows plants and absorbs whether used for bioenergy or not. Bioenergy can only reduce global warming by growing plants if it results in the growth of additional plants.

To understand the accounting better, it is useful to recognize that bioenergy is at best a way of offsetting emissions from energy use, not reducing the emissions from energy combustion. When cars or power plants burn biomass, they emit carbon dioxide, just as if they burn coal, oil or gas. No matter how much bioenergy is used, no less carbon dioxide is going to spew out of tailpipes and smokestacks— and, in fact, bioenergy releases more carbon in this way for the same amount of energy if only because biomass contains less energy than fossil fuels per each atom of carbon [21].[1] Although bioenergy reduces emissions from burning fossil fuels, it directly just substitutes emissions from burning biomass. Bioenergy can only reduce carbon in the atmosphere if in some other location, the production of biomass for energy either is absorbing more carbon from the air, or leading to less release of carbon to the air from another source.

---

[1] A ton of carbon biomass released by processing and burning biomass only displaces a fraction of a ton replaced by processing and burning fossil fuels for a variety of reasons, including greater energy in processing (even if some of that energy comes from the biomass itself), a higher carbon to energy ratio for biomass than fossil fuels, the higher water content of biomass, and in many situations, a lower generation efficiency.

The key word is "more." As in any form of compensation, it is double-counting to count plant growth as an offset if the carbon would be absorbed anyway. If an employee works overtime, the company may offset her time by providing more vacation days but it cannot count the vacation days she would earn anyway. That is also why a power plant cannot claim carbon offset credits by pointing to an existing forest that would grow or remain intact anyway; it can only claim credit for planting an additional forest.

That too is why bioenergy crops can reduce emissions if grown on land that would otherwise remain barren. For example, if a farmer plants bioenergy crops in the desert—and putting aside questions of water use—the growth of the crop absorbs additional carbon, which offsets the carbon released when the crop's biomass is used for energy. For the same reason, if bioenergy uses methane captured from a landfill, or crop residues that would otherwise be burned or decomposed in the field, it reduces emissions from the landfill or field. While it is common to think of the greenhouse gas reduction occurring in the power plant or car, the reduction truly occurs in the field or landfill.

On the other hand, if power plants harvest the world's forests and turn them into parking lots, they have not absorbed any additional carbon. Worldwide, trees, grasses and soils store at least four times as much carbon in the terrestrial ecosystem as the atmosphere holds, and burning up those ecosystems transfers carbon to the air just as much as burning up stores of carbon in coal, oil and gas underground. What happens if bioenergy is generated by clearing forests to grow bioenergy crops? While the bioenergy crop absorbs carbon, the initial outflow of carbon will exceed the annual inflow for decades, increasing global warming in that time. In these examples, there is no offset to the carbon emitted by the bioenergy use, and while few would challenge these examples, they by themselves show that the accounting rule is wrong.

The same accounting problem occurs if biofuels use crops that would grow anyway, such as the maize generated by Iowa farmers that is now diverted to ethanol plants. Because the farmers would grow this maize anyway, using the crops for biofuels does not result directly in any additional plant growth or absorption of carbon and therefore does not result in a legitimate offset. (Whether the diversion of these crops indirectly spurs reductions or increases in carbon emissions through market forces is another matter, discussed below.)

In short, whether bioenergy can reduce greenhouse gas emissions depends on the source. For bioenergy to offset tailpipe and smokestack emissions and thereby reduce greenhouse gases in the air, it must capture additional biomass and carbon either by increasing plant growth or by reducing its decomposition. Just using existing levels of plant growth differently, in and of itself, does not help the climate at all.

3 The False Intuitions

Many minds hold such an fixed view that plants are inherently carbon free sources of energy that it is worth addressing these intuitions directly.

The Wrong Baseline. Some people hold in their minds a fixed two-picture story. In the first frame, the picture shows land growing crops for energy, and in the second frame, it shows them being burned. If the story included only these two frames, bioenergy would inherently absorb and give off the same carbon and so would always be "carbon neutral." But land is not normally empty, and the true baseline picture frame should normally be of land growing plants, and absorbing carbon, for other uses.

Recycling Versus Long-Sequestered Carbon. Because carbon cycles relatively rapidly between plants and the air, many people believe plant carbon must be more environmentally friendly than carbon taken from fossil fuels stored underground for millions of years. That would only be true if there were some force that automatically reestablished a balance quickly between carbon in air and plants. Yet no such force exists. Carbon in the air has the same warming effect regardless of its source. If biomass carbon did not matter, the present focus on curbing deforestation to address global warming would be fundamentally misplaced.

Over long time frames, clearing carbon from the land may create additional potential to absorb carbon into vegetation. But there is no guarantee that will happen, and the timing is important.

Carbon Gains from Reduced Food Consumption. Annual crops do not typically store carbon beyond their year of growth. Diverting crops to bioenergy is carbon free by this way of thinking because the crops would be broken down and give up their carbon to the air anyway. But crops do not typically store carbon for long only because they are consumed by people or their livestock, which breathe out much of the carbon and expel the remainder as wastes, which microorganisms then consume and respire. To the extent bioenergy diverts crops or other biomass to energy, and they are not replaced elsewhere, bioenergy does reduce greenhouse gas emissions, but it does so by reducing the consumption of food or other plant products and the resulting respiration. In reality, at least some of the displaced crops will be replaced elsewhere, which leads to a cycle of additional carbon releases. But even to the extent crops are not replaced, bioenergy and biofuel advocate do not generally seek to reduce global warming by reducing consumption of food.

Both Carbon Stocks and Flows. Some observers note that if annual bioenergy crops replace mature forests, which no longer sequester additional carbon each year, those crops will increase the flow of carbon from the atmosphere into vegetation, which they intuit should reduce global warming. But even if bioenergy would increase the absorption of carbon per year, the conversion of the forest creates a massive initial outflow to the air. That large initial outflow must be balanced against the increase in inflow, and it will often take many years to pay off what is now known as the "carbon debt."[2]

Sustainable Does Not Mean Carbon Free. To some people, the word "sustainable" is so positive it must also mean carbon free. "Sustainable" has many meanings, but a strong version of sustainable forestry would mean that harvests do not exceed annual forest growth. Because such a rate of harvest would not reduce carbon stocks, many people believe bioenergy is carbon neutral up to such a harvest level. But forests all over the World are gaining carbon, particularly the forests of the northern hemisphere that are re-growing from earlier harvests or from abandoned cropland, with a total gross sink of 14.5 Gt of carbon dioxide [41]. Sacrificing this carbon "sink" contributes to global warming in the same way as increasing a source.

Renewable Does Not Mean Carbon Free. Many people have a firm intuition that "renewable" must mean carbon free, but bank interest provides a useful analogy to appreciate why not. Just as a piece of land produces new plant growth each year, so a savings account gives off new interest, and in that sense both plants and bank interest are renewable. But that annual bank interest may help to pay the rent, or it may remain in the bank account to be used later, so if it is diverted to pay for a fancy car, it is not available for these other purposes. People similarly use each year's plant growth for necessities like food and wood products, and leave other plant growth to accumulate carbon in ecosystems, which helps to keep down global warming. Using that plant growth for energy also sacrifices other uses. To change a bank account to make you richer—without giving up existing needs like paying the rent—it must provide more interest, and for land to reduce global warming through bioenergy—without giving up plants for food, and timber products—that land has to produce more plants.

Fortunately, in one important respect bank accounts and plant growth differ. Unlike typical bank interest, some plant growth is "use it or lose it" because it will burn or otherwise decompose. This category includes some residues and wastes. Using this biomass for bioenergy can therefore help to hold down global warming, and provides one category of potentially "additional biomass."

---

[2] In reality, even when bioenergy results from harvesting mature forests, it is in reality sacrificing an alternative use of ongoing plant growth although that growth may no longer result in additional net sequestration. A mature forest that not no longer adds carbon storage each year actually continues to absorb carbon from the air through plant growth in great amounts. Its net primary productivity—its degree of annual carbon absorption—may be even higher than that of growing forests. It no longer accumulates carbon or accumulates carbon slowly only because the rate of consumption of its carbon by heterotrophic organisms—particularly microorganisms—breaks down as much biomass as the forest annually accumulates. In this case, the annual carbon uptake is in reality serving to replenish the stored carbon that would otherwise decline. For this reason, the use of plant growth for bioenergy comes at the cost of using that plant growth to replenish existing stores of carbon in the existing forest.

4 The Source of the Error

The source of the accounting error is surprisingly easy to trace, and it did not derive from a scientific error but from a misinterpretation of scientific guidance [48].

In the 1992 UN Framework Convention on Climate Change (UNFCCC), governments agreed to start reporting their emissions of greenhouse gases at a national level and sought guidance from scientists about how to do so. The scientists advised them to develop different sets of accounts, including one set for emissions from land use change, such as the carbon released by cutting down forests, and a second set for emissions from energy use, primarily the carbon released by burning coal, gas and oil. But the scientists recognized that this system had the potential to double-count emissions from bioenergy. If the carbon in a tree is counted as an emission in the land use account as soon as the tree is cut, that carbon should not be counted again as it goes up the smokestack if the tree is burned. To prevent double-counting, the scientists suggested that such carbon should only be counted in the land use account, which meant that power plants did not need to count emissions from burning biomass themselves.

Governments adopted the recommendation, and the rule works for the original UNFCCC in principle. Because that treaty requires all countries in the world to report their emissions from both land use and energy, ultimately all bioenergy emissions are properly counted once. For example, if a tree is cut in Europe and burned in Europe, it is counted in the land use account. If a tree is cut in Canada or Latin America and burned in Europe, Canada or Latin America report same carbon as an emission in their land use accounts, so Europe does not need to count it.

The accounting problem arose in 1998 when thirty-one developed countries approved the Kyoto Protocol. The Protocol set actual limits on the carbon dioxide countries could emit from energy use. But its limits on land use are riddled with loopholes, so land use emissions even for developed countries do not fully count. In addition, because developing countries face no Kyoto limits for any emissions, biomass imported from them cannot possibly count toward legally enforceable limits on land use emissions. Under the Kyoto Protocol, therefore, if bioenergy emissions are to be counted at all, they need to be counted in the smokestacks and tailpipes where the carbon is actually emitted. The Protocol therefore needed a different accounting rule, but that need appears to have gone unnoticed. The accounting rules adopted for the Protocol for bioenergy do not differ for bioenergy from the underlying UNFCCC. The erroneous result means that shifting from coal, oil or natural gas to bioenergy is treated as a 100% reduction in carbon dioxide in all cases—even in such extreme cases as burning up the world's forests. In effect, under the Protocol, forests can now be worth more dead than alive.

The same error applies to any law that exempts carbon from bioenergy limits if it puts limits on energy use without applying comparable limits to emissions from land use. In fact, exempting bioenergy emissions can only work if limits on land use apply worldwide. Otherwise, a country or company can import biomass and burn it regardless of the source. Otherwise, a country can also claim emissions reductions for burning up its own biomass, such as crops, and reducing exports. That by itself does not generate additional carbon and instead causes foreign countries to plow up more land to replace the food.

Most researchers and government officials misunderstood the exemption: They thought the rule against counting carbon from burning biomass flowed from the idea that biomass carbon is inherently free. The error became deeply set, and it is now reflected in hundreds of papers and many laws that wrongly treat bioenergy as inherently carbon neutral.

5 The Role of Time

Whether biomass diverted to bioenergy is additional and increases or reduces greenhouse gases can only be judged within a specified time frame. The conversion of a mature forest that is no longer growing much to an annual bioenergy crop causes a large initial loss of carbon but has the potential to increase the annual uptake of carbon, which can eventually pay off the "carbon debt".[3] Whether biomass is "additional" can therefore only be judged within a certain time frame.

Required to choose a time frame for legal purposes, the European Commission has so far chosen 20 years, and the U.S. Environmental Protection Agency and the California Air Resources Board have chosen 30 years. The academic treatment has so far been inadequate. One paper, for example, argues that trading off early emissions to avoid late emissions is likely to be beneficial because the ocean and land absorbs much of the carbon emitted over time, and eventually the result should be a reduction in peak atmospheric warming [29]. But this thinking ignores the harm from ocean acidification as the ocean absorbs carbon, implicitly focuses only on peak temperatures rather than cumulative warming, and ignores a variety of other important economic factors. Part of the challenge in thinking quantitatively about the appropriate time frame is the great variety of factors and scientific uncertainties involved in making this judgment. These factors include:

- The likelihood and degree of adverse feedback effects from enhanced earlier warming.
- Likely reductions over time in the costs of mitigation as technology improves. Because we need both reductions now and down the road to reach the goal of

---

[3] Net ecosystem productivity is the absorption of carbon after accounting for decomposition of plants by animals and microorganisms.

stabilizing the planet, early reductions are more economically valuable than later reductions for the same level of reductions.[4]

- Committing to bioenergy strategies in the near term that only reduce emissions in the long-term requires a bet that alternative technologies will not make bioenergy less competitive or obsolete. The risk that this bet will fail adds to the costs of bioenergy.

A thirty-year payback period also does not mean that the planet is necessarily cooler in the thirty-first year if biofuels are to be phased in over time. After 30 years, the use of bioenergy has paid back the carbon debt from the first year's land use change, but the additional bioenergy generated in year 30 still requires another 30 years to pay off its carbon debt. If bioenergy is phased in a constant basis over 30 years, cumulative emissions do not decrease until year 45.

In addition, although total net carbon emissions may decrease after a break-in point, in the above example year 45, the net cumulative warming will remain higher for many years more because the period and degree of higher atmospheric carbon exceeds the period and degree of lower carbon. That is true even while accounting for the re-absorption of atmospheric carbon through ocean and land uptake [38]. If the economic consequences of each degree of increased temperature are the same, the time is therefore further extended before the damages from global warming are equal between fossil fuels and biofuels.

Alternative mitigation strategies are also among the factors that should influence the thinking about time. Many biofuel advocates implicitly treat the only alternative to bioenergy as fossil fuels because of the poor alternative ways of fueling cars. By this thinking, the biofuel question then becomes whether increases in greenhouse gases over the short-term, even decades, are nonetheless preferable to increases in greenhouse gas emissions over hundreds of years. But the true greenhouse gas alternatives to evaluate during this time include ways of reducing emissions other than by changing the fuel we use for transportation. Those include all other ways of reducing emissions immediately. In fact, reviews have typically found that alternative mitigation strategies are far cheaper than biofuels even using old lifecycle analyses that assume biomass is carbon neutral in the way this paper discusses [39]. Regardless of biofuel benefits and alternatives down the road, it is difficult to justify biofuels now on carbon grounds so long as other, cheaper, mitigation measures that reduce emissions immediately remain to be fully exploited.

---

[4] This point is confused by some analysts. Policymakers are likely to require greater emissions reductions in the future than at present, and the more reductions are required, the higher the cost, which means the market value of emissions reductions are likely to increase in the future if any policy phases in reductions. But the market value in such a case reflects the policy not the value of the reductions. Such a policy in fact probably would reflect the judgment that costs of emissions reductions will decline over time and so should be phased in rather than required immediately. Such a policy would not mean that emissions reductions in the future are more valuable than achieving *those same levels of reductions* today.

Because all of these factors are uncertain, any mathematical effort to fit them together seems like an exercise in false precision. Yet, these factors can be synthesized with a thought experiment. Imagine a twenty-year payback for biofuels, and an emissions-trading system in which private investors receive credits for greenhouse gas reductions due to biofuels over time but also have to pay up-front for other emissions reductions to offset the emissions from associated land use change. In such a system, the investor must pay for 20 years of emissions credits in year one (or close to it), and while the biofuels generate credits thereafter, the investor must also pay for the added financial cost of producing biofuels rather than fossil fuels. Only after 20 years, the investor is able to sell net emissions credits. Risks include the possibility that even after 20 years the costs of using biofuels would be greater than the costs of alternative carbon-reduction strategies. Would private investors be willing to make such an investment? The answer is probably no, and that helps to explain why 20 years should be an outside payback period for biofuels.

6 The Double-Counting in Most Global Bioenergy Potential Studies

One way in which the error plays out is in estimates of global bioenergy potential. Nearly all of the large estimates of bioenergy potential result from double-counting biomass that is already storing carbon or serving other useful purposes. Recent bioenergy potential analyses both by the IPCC [9] and by the International Energy Agency [5] are based on this double-counting.

Use of Potential Cropland. The IPCC made perhaps the crudest double-counting error in its 2001 report on climate change mitigation [36, Table 3.31]. That report estimated bioenergy potential at 441 exajoules in 2050, a little less than total world primary energy demand today, even while projecting that world cropland would need to expand by more than 400 million hectares. To obtain this estimate, the authors assumed that all potential world cropland not otherwise required, 1.3 billion hectares in 2050, could be devoted to bioenergy at high yields. Unfortunately, the world's unused potential cropland consists mostly of forests, wetter savannas or wetter pasture lands. Using this land for bioenergy would release immense stores of carbon and sacrifice a fair amount of food.

"Excess" Forest Growth. Excess forest growth provides another large source of estimated bioenergy potential according to the International Energy Agency [5] and others [49]. This estimate assumes that all forest growth in excess of wood harvests can be devoted to bioenergy as a carbon free fuel. Unfortunately, as discussed above, this excess forest growth is already reducing atmospheric carbon by storing the carbon in forests [41]. Burning up this biomass, instead of allowing it to accumulate, cannot provide any additional benefit.

Crop Residues. Although unused crop residues can supply additional carbon, large estimates of their potential also double-count carbon. The IEA estimates residue potential at 100 EJ [5], one fifth of the world's primary energy use in 2007. But most of the world's residues are used for feed, bedding or energy already [51]. These estimates assume a percentage harvesting of all the world's crop residues not merely those otherwise unused [7]. In addition, most "unused" residues help maintain soil fertility. The extent to which additional residues can be removed without serious fertility impacts remains mostly uncertain.

Abandoned Agricultural Land. The largest estimates of bioenergy potential rely on abandoned agricultural land, which comes in several flavors [5, 25]. Some estimates assume that bioenergy can freely use abandoned land that is already occurring because of the shifting of cropland from one location to another even as overall cropland expands. For that reason, some studies estimate that bioenergy potential will even grow as climate change itself causes more cropland shifts even while cropland continues to expand. However, abandoned cropland overwhelmingly reverts to forest or grassland and sequesters carbon. This fact explains why net deforestation is far lower than gross deforestation, and why abandoned cropland contributes a major part of the terrestrial carbon sink, so it is already counted as offsetting emissions from agricultural expansion. Counting this land for bioenergy is again double-counting.

Other estimates hypothesize that potential agricultural intensification can reduce the need for agricultural land and free up land for bioenergy. For that to be true, the typical projections of growing agricultural land to feed a growing population will have to turn out false [53]. Regardless, even this estimate is at least partially double-counting. Unless bioenergy is the cause of this great expansion in agricultural intensity, other forces would lead to this land abandonment, and these lands would revert to some other use and sequester carbon if not used for bioenergy. Any advantage from bioenergy would occur only to the extent that using this land for bioenergy produces greater greenhouse gas benefits than those generated by allowing forests and grasslands to re-grow.

Using Savannas. Some papers exclude existing forests, agricultural lands, and protected areas and then estimate bioenergy potential on the remaining lands. In reality, these estimates focus on the World's savannas and sparser woodlands with limited livestock, most of which are concentrated in Africa [11, 25]. Although the method is designed to exclude the highest carbon lands, once these exclusions occur, the authors typically count all the biomass that can be produced on these savannas as carbon free while ignoring the lost carbon storage even of these savannas. A paper that excluded the same kinds of lands up-front, but estimated the losses of carbon storage from converting savannas, found that producing bioenergy on most of them would not pay back the carbon debts within ten years, even assuming high bioenergy yields [6].[5] These types of analysis also incorrectly assume that biodiversity is protected simply by retaining protected areas, or even modestly adding to those protected areas. Many savannas contain extensive wildlife populations and are highly diverse. Although some savannas may be capable of generating additional biomass over a couple of decades, depending on the yields bioenergy crops can obtain, their potential has not yet been properly estimated nor their implications on biodiversity and water properly appreciated.

7 Proper Accounting and Estimates of Bioenergy Potential

Despite the proliferation of improper accounting and flawed estimates of bioenergy potential, bioenergy has some potential to reduce greenhouse gas emissions by producing or capturing "additional biomass." The most likely sources include forest harvest residues, whose estimated potential varies considerably, but appears to have a mean estimate of 27 exajoules per year (around 5% of world energy use today) [24]. Crop residue estimates are more suspect because of limited data on their use, because the estimates tend to arbitrarily estimate the percentage of residues that can be sustainably harvested, and because none of the estimates truly confront the fertility challenge. It is particularly hard to believe that low-yielding parts of the world could spare additional removal of residues as their lands produce few residues already, which are already in high demand for animal feed, and as many of their soils are already degraded by the inadequacy of residues left behind. High-yielding lands in developed countries, where residues are already less used, should have greater potential to spare residues, but a prudent approach would carefully test such harvests on fertility.

The even bigger unknown involves relatively unproductive land. To generate "additional biomass" in general, it is not enough for bioenergy crops to be produced with low yields on otherwise low plant-producing lands, or to produce high yields on highly productive lands already; bioenergy must generate high yields on lands that otherwise generate limited carbon storage, food, forage or timber products. Where can that occur? One example involves deforested lands in Indonesia covered with an invasive grass that burns frequently and has little economic use. Analyses have shown net greenhouse gas savings for the use of these lands for biodiesel from palm oil [10, 19].

Whether wetter but badly managed grazing lands can also supply "additional biomass" is a more difficult and subtle question. Some such lands in Brazil, for example, will generate far more biomass if turned into sugarcane fields for

---

[5] In this study, most of the bioenergy potential existed in shrublands in Asia and temperate zones. By visual inspection, this analysis is likely to be capturing many re-growing forests, and the study did not estimate carbon losses from the foregone sequestration of such lands.

ethanol—in that case generating directly additional carbon—but the ultimate effects depend on whether and how the livestock products are replaced, which is highly disputed. As discussed more below, if livestock production moves into denser forest, the net effect will increase greenhouse gas emissions; if livestock production only intensifies on existing grazing land, the net effects will reduce emissions. Some scholars have argued for combined efforts to intensify livestock and ethanol production, which could guarantee additional biomass if they are causally linked.

Agave provides an example of the type of plant that could help generate additional biomass if grown on dry land. Agave is reportedly capable of generating yields of 10 tons of dry matter per hectare per year even on land that receives only 200–400 mm of rainfall, and it has a high sugar content relatively easy to convert into ethanol [57]. This land is mostly otherwise used for low intensity grazing, whose food production in absolute terms could probably be reasonably replaced on significantly less land. Success from a pure carbon standpoint requires that agave production actually occur on dry lands, and that displaced livestock products not lead to further forest conversion. These lands are also typically used by poor pastoralists, so a shift to bioenergy would often present many social challenges. Today, agave is expensive to grow because it must be grown by harvesting and planting rhizomes from roots, and it has not been adapted to many world areas. These are challenges worth addressing.

Regardless of the true potential, true opportunities for additional biomass are unlikely to be exploited unless the accounting rules are fixed because the laws would otherwise provide no incentive to generate biofuels that use it.

8 The Role of the Accounting Error in Lifecycle Analysis of Liquid Biofuels: The Proper Understanding of "Indirect Land Use Change"

The bioenergy accounting error first reached policy consciousness through the lifecycle greenhouse gas accounting of ethanol and biodiesel. Such an analysis attempts to estimate all the greenhouse gas emissions that occur because of the production and use of biofuels and compare those emissions to the emissions associated with conventional gasoline or diesel. In February of 2008, two papers published in Science magazine, including one from this author, pointed out that these typical lifecycle estimates did not account for the emissions associated either directly with converting a forest or grassland to grow the biofuel crop, or indirectly if the biofuel uses crops from existing cropland and forests and grasslands are plowed up to replace the food.

Since that time, some government agencies have tried to incorporate these land use change estimates, economists have generated more estimates of the emissions from indirect land use change, and critics have charged that the whole enterprise is inappropriate. To them, the indirect land use change (ILUC) that may occur in Latin America, Africa or Asia to replace crops diverted to biofuels in the U.S. or Europe should be the responsibility of those converting the land, not those making the biofuel. They also argue in any event that those emissions are too speculative to reflect in policy. These critics fundamentally misunderstand the biofuel accounting error and the role of ILUC in correcting it.

ILUC is the method of determining how much biomass carbon is additional and can be ignored when burned

What the critics assume is that they can ignore carbon emitted from the exhaust pipes of cars that burn biomass, as well as the carbon emitted in the ethanol refinery from the fermentation of crops into ethanol. Table 1 shows a typical lifecycle analysis comparing the greenhouse gas emissions for biofuels and fossil fuels, in this case the default analysis established by the European Commission's Joint Research Center. As it shows, this assumption implicitly assigns an offset credit to the biofuels for all of the carbon in the maize (excluding the carbon in by-products), and amounts to 107 g of carbon dioxide per mega joule of ethanol. That by itself is larger than total emissions from gasoline. Without this assumption, biofuels from crops always increase emissions substantially compared to gasoline and diesel because of the fertilizer and fossil energy also required.

Unfortunately, this automatic assumption of carbon neutrality is wrong for all the reasons discussed in this paper. If biofuel crops directly displace forests and grasslands, the foregone carbon storage and sequestration must be subtracted from the carbon absorbed by the biofuel crop. Alternatively, if biofuels use crops that would be grown anyway then they cannot absorb any additional carbon as they would by definition have grown anyway.

The ILUC analysis is the means of determining whether additional carbon results from the indirect, market responses of diverting crops to biofuels:

1. The crops are not replaced.
2. The crops are replaced on existing agricultural land through yield increases or additional double-cropping.
3. The crops are replaced by plowing up new land.

The first two effects result in forms of offsets and additional carbon. Not replacing the crops means less respiration by people and livestock (and less of their wastes and its decomposition), which are the source of greenhouse gas benefits. Boosting yields or double-cropping existing cropland means additional plant growth that absorbs additional carbon and provides an offset in that way. If these two were the only effects, then the net indirect effects would be positive and amount to 107 grams per mega joules of ethanol. That is mathematically equivalent to ignoring tailpipe and fermentation emissions as in Table 1, and assuming an ILUC of zero.

The third response, however, not only generates carbon by growing more crops but emits carbon by plowing up vegetation and soils. Depending on the land converted, the net effect may be beneficial or adverse, but the carbon costs will not be zero. Mathematically, when a lifecycle analysis combines the assumption that

Table 1 Lifecycle greenhouse gas comparison for biofuels and conventional fuels: role of implicit carbon credit for plant growth

Source of fuel	Producing feedstock (crude oil or crop)	Refining	Tailpipe emissions	Fermentation emissions	Total GHGs (and % Change) counting tailpipe and fermentation emissions	Implicit credit for plant growth to offset and cancel out tailpipe and fermentation emissions	Total GHGs (and % savings) for biofuels
Gasoline	+4.5	+8	+73.3	–	+85.8	–	+85.8
EU wheat ethanol	+40	+21.2	+71.4	+35.7	+168.3 (+96%)	−107.1	+61.2 (−29%)
Diesel	+4.6	+9.6	+73.2	–	+87.4	–	+87.4
EU rape bio diesel	+35.5	+11.1	+76.2	–	+122.8 (+41%)	−76.2	+46.6 (−47%)

Source European Commission Joint Research Center. Figures in red are implicit but not explicit in the analysis. Figures are emissions of greenhouse gases in CO_2 equivalents per mega joule of fuel

Table 2 Role of reduced crop consumption in ILUC calculations

Model and type of ethanol	Crop consumption reduction (exclusive of by-products) and CO_2 savings from reduced respiration and waste by people and livestock	Total emissions and emissions as percentage of emissions from gasoline if food reduction not credited
GTAP U.S. maize	52% = 56 g/MJ	135.5 g/MJ (+157%)
IMPACT model U.S. maize	36% = 39 g/MJ	123.7 g/MJ (+144%)
IMPACT EU wheat	47% = 50 g/MJ	166.5 g/MJ (+194%)
FAPRI CARD EU wheat	34% = 36 g/MJ	177.6 g/MJ (+206%)
GTAP EU wheat	46% = 49 g/MJ	241.7 g/MJ (+280%)

Source Marelli et al. [34, p. 89, Table 39]. IMPACT was run by the International Food and Policy Research Institute, part of the CGIAR network, and the FAPRI model was run at Iowa State University

biomass is carbon neutral with an ILUC analysis, it does get the accounting correct but in a backwards way, for it is effectively counting whether the indirect responses generate additional carbon. For example, the lifecycle analysis with ILUC for the California Air Resources Board, which estimated an ILUC of 27 g/MJ over 20 years, in effect estimated that for every 107 g of carbon in maize used for ethanol, the market indirectly generated 80 g of a ton of additional carbon. That is a pretty favorable scenario even though most maize ethanol still in the end did not reduce greenhouse gas emissions when production emissions are also counted. LCAs that ignore exhaust and fermentation emissions without factoring in ILUC are simply wrong.

8.1 Is the Good Side Good?

A closer look at the California model also shows that half of that grain carbon is "additional carbon" because the grain is not replaced and people and livestock would eat, and respire (or excrete) that much less carbon. That generates a greenhouse gas reduction of 56 g per mega joule of ethanol. By comparison, gasoline emissions in total are roughly 85 g per mega joule. An analysis by the European Commission's Joint Research Center showed that other models that predict relatively low ILUC from ethanol also predict large food reductions [34] (Table 2).

Some of this reduced crop consumption may occur for relatively benign reasons, such as improved livestock feeding efficiency, or reduced meat consumption by people who eat too much meat. But at least a significant fraction of this reduced crop consumption due to higher world crop prices is likely to occur from the world's hungry and poor. Unlike a tax on beef consumption in Europe or the U.S., world price increases in response to biofuels would indiscriminately affect both the well-fed and undernourished. In reality, the California model is predicting an even worse public policy result than higher greenhouse gas emissions.

How valid are these estimates of reduced food consumption? The strengths and weaknesses of these models present a subject too long for this chapter, but a more direct economic analysis of recent years, using probably more robust economic methods, has mid-range estimates that roughly a third of the grain diverted to biofuels is not replaced because of reduced food consumption [43]. Because these methods inherently focus on short-term responses, and increased supply is likely to play a larger role over time, food consumption is likely to decline less in the long-term. But short-term hunger is still a serious problem. And if biofuels cause less hunger in the long-term, that means they also cause more greenhouse gas emissions from land use change than these models estimate.

Yield increases provide the primary other "beneficial" indirect response from a climate perspective. In this category, ongoing yield gains, or yield gains due to any factor other than biofuels, do not count because they form part of the proper baseline. Only to the extent biofuels trigger higher prices and farmers respond by boosting yields even more do biofuels indirectly lead to additional absorption of carbon that can count as an offset against the emissions from burning those biofuels.

Unfortunately, although preferable to other responses, yield gains come with their own costs. They may occur through increased use of nitrogen fertilizer and irrigation water, with attendant environmental costs: Whole world conferences and libraries of articles devote themselves to the environmental challenges presented by the rising use of these inputs.

More broadly, even without biofuels, the world faces the great challenge of producing more food to feed a growing world population while avoiding land use change to hold down greenhouse gas emissions. The more that farmers exploit the easier opportunities for yield gains to offset biofuels, the fewer easy yield growth opportunities remain available to meet food demands without land use change. The potential effects of biofuels to crowd out other yield gains are large at even modest levels of biofuels. Figure 1 shows the rate of world yield gain from 1996 to 2006, the rate of yield gain necessary to meet world food needs in 2020 without land use change, and the rate of yield gain needed to meet both food needs and a biofuel target of 10.3% of world transportation fuels in 2020.[6] As this chart shows, meeting world food demands alone without land use change would require a much higher rate of yield gain than achieved recently, and adding the level of biofuel demand would make such a goal implausible.

---

[6] The biofuel mix estimate for 2020 was produced by a consulting firm for the British Renewable Fuels Agency, but mostly mirrors the mix of world biofuels in place in 2007 scaled up to reflect government biofuel targets for 2020.

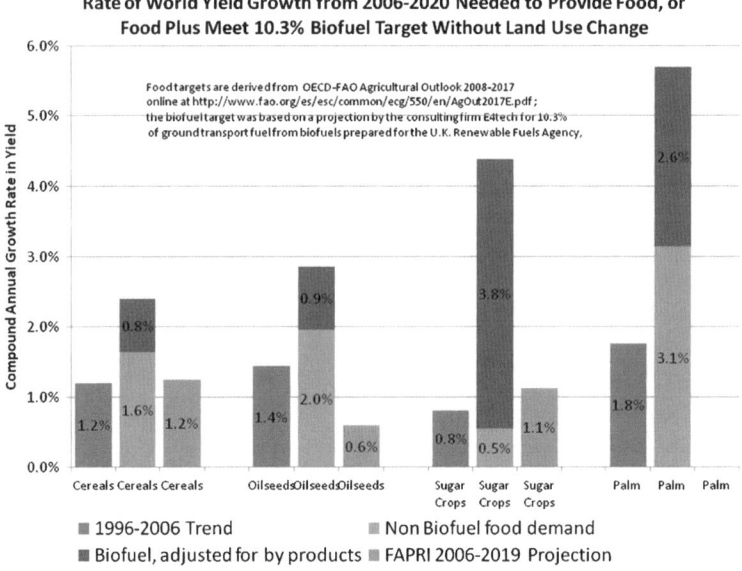

Fig. 1 These middle bars show that supplying both biofuels and food without expanding cropland would require at least doubling the rate of yield growth for each major biofuel commodity

8.2 *Alternative to ILUC Accounting*

Estimating the greenhouse gas consequences of biofuels using an ILUC approach is theoretically correct but inherently uncertain. Any strategy to encourage biofuels that divert existing crops or productive agricultural lands to biofuels effectively amounts to a bet that increasing crop prices will lead farmers to generate additional biomass while avoiding harmful land use change and reduced food consumption. There is little if any evidence to support such a bet. The alternative is to encourage only biofuels that are created by directly generating additional biomass. That would put the onus on biofuel producers to turn unproductive land into productive land, or to use appropriate wastes and residues. Had bioenergy accounting been right from the start, this author suspects that this direct approach would have emerged as the more likely strategy.

9 Biofuels from Sugarcane and Dedicated Energy Crops

To many commentators, Brazilian sugarcane ethanol stands out as an exception to the poor greenhouse gas performance of "first generation" biofuels from crops and provides a model for the world. Although certainly better than biodiesel and ethanol from maize, the benefits of sugarcane ethanol are still ambiguous and any view that it provides a model for other countries misunderstands the argument even of its advocates.

Sugarcane is a high-yielding, perennial crop with relatively modest nitrogen use and whose non-sugar by-products are burned to generate the energy to ferment the sugars and often some electricity on top. Even so, if sugarcane for ethanol directly converts Cerrado or Amazonian forest, it will likely release enough carbon to cancel out its greenhouse gas benefits for 17 and more than 100 years, respectively [18, 47]. The defense of sugarcane starts with the observation that Brazil has already deforested on the order of 175 million hectares for pasture, and that most sugarcane expansion results in the conversion of this pasture, resulting in a likely "payback period" for sugarcane ethanol of roughly 4 years [30]. If that were the full story, sugarcane ethanol could be considered an unqualified carbon success. However, in at least some years, roughly one-third of sugarcane expansion displaces other crops [40], and they must be replaced. In addition, the big question is whether the conversion of pasture lands encourages new conversion of the Amazon to replace the beef and dairy products.

On this point, there are competing claims.

- Goldemberg argues that as sugarcane has expanded in the State of Sao Paulo, pasture management there has intensified sufficiently to make up the lost dairy or beef production, which he views as evidence of a lack of indirect land conversion [20]. This view depends on the assumption that sugarcane expansion drives the intensification in Sao Paulo State and has no effect on the continuing large-scale conversion of Amazon for pasture elsewhere. Put another way, it is based on the theory that the intensification of pasture would not have occurred without sugarcane expansion. This argument assumes that economic signals walk slowly across the State rather than occur simultaneously in remote regions. That cannot be completely true. There are market stories that one could tell to suggest that local intensification is a larger response, but competing stories can also be told.

- Modeling by Andrew Nasser at ICONE in effect embodies the Goldemberg assumption in a more complicated mathematical framework, and while it predicts some indirect land use conversion of forest, that conversion is modest [37]. Like many other economic models, this one incorporates large numbers of elasticities and functions, but the consulting firm Ecofys has pointed out that the model incorporates a variety of assumptions into its structure that essentially assure its result [12]. In particular, the model assumes that displaced crop or pasture production within a region of Brazil is replaced within that region if it can accommodate that increased production, a variation of the Goldemberg assumption. As a result, any intensification of any crop within a region that would occur even without biofuels works in the model to offset the displaced production of that region due to biofuels. That is improper because the model in effect assumes that crop intensification is the result of biofuels without proving it to be true.

- On the other side, a model developed by Lapola et al. predicts that sugarcane expansion will result in widespread Amazonian deforestation that would result in a net increase in greenhouse gas emissions for long periods [30]. But this model incorporated the assumption that the pasture products are replaced by land use change. In essence, this model analyzed where land use change would occur to supply pasture and cropping demand, not whether it would occur.
- One recent paper has estimated that for every hectare of soybean expansion in settled agricultural areas of Brazil between 2003 and 2008, one to eight hectares of additional agricultural conversion of the Amazon occurred [3]. Although the paper analyzed soybean expansion, it implies large effects for sugarcane expansion as well. The paper's method was to correlate rates of expansion of soybean in the Cerrado with rates of expansion in nearer versus more remote areas of the Amazon. It provides some suggestive evidence of a causal link, but there may be other factors that simultaneously influence expansion in the Cerrado and closer Amazon regions and explain the correlation.

At this time, my own view is that the issue remains unsettled. Unless basic economic theory is wrong, intensification of pasture should provide some of the response to displacement of pasture by sugarcane, and area expansion another part of the response, with the balance at this point hard to predict. The dependence of the land response in Brazil on government policies also explains why any prediction is hard. In view of that ambiguity, a political deal might be reasonable—one in which the world agreed to buy biofuels from Brazil in return for successful national efforts to protect forests and intensify grazing land.

Regardless, the key argument by supporters of Brazilian sugarcane is in essence a variation of the marginal land argument. The reason Brazilian sugarcane may reduce emissions is that Brazil has already deforested such vast areas of land and is using so much of it poorly that sugarcane can replace these unproductive pastures with productive sugarcane. In other words, advocates of sugarcane are arguing in effect that sugarcane is generating additional biomass by better using otherwise underutilized land. If true, this argument still does not necessarily apply to biofuels from sugarcane or other plants in other regions except if they also displace underutilized land. For biofuels to reduce global warming the critical need remains a large increase in plant production.

10 Cellulosic Biofuels from Dedicated Energy Crops

Many articles about biofuels, both popular and academic, assume a broad divide between biofuels from food crops and energy crops [17]. But the biofuel challenge does not arise because biofuels use food crops; it occurs because biofuels use land. Energy crops may in some cases use land other than cropland, but if they displace other natural habitats, they are likely to release carbon directly. And if they replace cropland, they lead to reduced food consumption or land use change just like the energy use of food crops.

In general, lifecycle analyses project greenhouse gas savings of 90% or more from cellulosic ethanol on the typical assumption that carbon emissions from tailpipes and fermentation do not count. These analyses also assume that cellulosic ethanol will produce a carbon free electricity by-product that should also be credited to the biofuel. But the land use effects depend on the type of land converted, the yields, and the conversion efficiency of biomass to ethanol. For example, a hectare of U.S. maize ethanol at average yields will generate roughly 4,070 l of ethanol, but it also generates feed by-products. If those by-products are valued at a feed/food value equal to 35% of the diverted grain, that means 0.65 hectares are producing this volume of ethanol, which means that each effective hectare is generating 6,262 l. Cellulosic ethanol would only reach the same level per hectare if cellulosic ethanol can be fermented at 340 liters per metric tonne (90 gallons per metric tonne) and each hectare then yields 18 tonnes of dry biomass per year. That may occur, and advocates of miscanthus claim it will generate 30 tonnes or more per hectare. But the crop modeling analysis for switchgrass used for the EPA's rulemaking estimated less than 8 tonnes per hectare per year for cropland in the Upper Mississippi River basin, and much lower yields in other regions.[7] Meanwhile, miscanthus continues to require expensive, time-consuming production and transplanting of its rhizomes.

Cellulosic ethanol breeders hold out hope for good yields of some biofuel crops such as agave even on dry lands [54]. If so, they may be able to generate additional carbon. But it would be a mistake to place overconfidence in any such estimates as technologies sometimes evolve as hoped and sometimes not. Even with good yields on poor land, proper greenhouse gas accounting will require an analysis of any changes in carbon storage. The amount and type of land needed to replace any lost food or timber products will all have to enter into the estimates of the greenhouse gas consequences of dedicated energy crops.

11 Proper Accounting of Bioelectricity

Electricity generation could ultimately use more biomass than the transportation sector. Paper mills have long burned wood wastes and paper by-products to make their own power and sometimes electricity for the grid. The growth of bioelectricity now results primarily from burning wood harvested from forests. If the wood is truly only slash left behind after a timber harvest, and sufficient residues are left to maintain long-term productivity, its use for electricity may come close to being carbon neutral over twenty-five years (as nearly all the slash would likely decompose in that time).

---

[7] Plevin [42] contains an excellent discussion of these estimates of switchgrass yields.

Table 3 Illustrative emissions from bioelectricity versus fossil fuel electricity

	Tons of carbon (harvesting one hectare)
55 year stand of Douglas Fir Forest (Pacific Northwest)	
Smokestack emissions biomass	102
Net reductions in carbon stocks after 25 years (excluding carbon emitted through smokestacks)	43
Total bioelectricity emissions	*145*
Emissions from fossil fuel electricity for the same electricity	37 (natural gas)** 51 (coal)
Bioelectricity emissions as a percentage of fossil electricity emissions	*∼390% (natural gas) ∼280% (coal)*
95 Year Stand of Douglas Fir Forest (Pacific Northwest)	
Smokestack emissions biomass	+149 tons
Net Changes in Carbon Stocks After 25 Years (excluding carbon from smokestacks)	+41 tons
Total bioelectricity emissions	*190*
Fossil fuel emissions	−54 tons (natural gas) −75 tons (coal)
Bioelectricity emissions as a percentage of fossil electricity emissions	*∼350% (natural gas) ∼250% (coal)*

Smith [52, Table A.17]

[a] The saved emissions from natural gas also include the avoided fugitive emissions from natural gas use based on a U.S. Department of Energy lifecycle analysis cited in Walker [55]

Harvesting whole trees that would otherwise remain standing is another matter. The standard claim for greenhouse bioenergy benefits relies on the fact that the trees will grow back. That is also the most straightforward way to assess the consequences of bioelectricity over any particular time period. The basic calculation is the net effect of the following:

1. The emissions from the smokestack as a result of burning the trees for energy, which is equal to the carbon in the biomass actually burned.
2. The changes in carbon stocks in a forest as a result of the harvest of the live trees for energy. Critically, the comparison is not between the carbon stock in the forest at the time of harvest, and the carbon stock after 20 or 30 years of re-growth. The comparison is between that re-growth level and the carbon the unharvested forest would store after another 20–30 years of growth. When mid-aged forests are harvested, which is common, the proper comparison by itself will often mean bioelectricity increases greenhouse gas emissions because the mid-age forest, if left undisturbed, would continue to accumulate carbon at a higher rate for 20 or 30 years than the regrowing forest postharvest. For this reason, bioenergy greenhouse gas calculations sometimes show better results if the harvest comes from a highly mature forest, with little potential for additional growth. In that situation, however, the up-front carbon debt is larger.

3. The avoided emissions from not burning natural gas or coal for the same amount of electricity. Unfortunately for bioenergy, burning a ton of carbon in wood in a dedicated biomass power plant generally only saves a little more than one third of a ton of carbon from using natural gas. That disparity occurs in part because wood has less energy for each atom of carbon than fossil fuels, in part because wood contains more water that must be burned off, and in part because wood is hard to burn efficiently. Counting in the fugitive emissions from the production and use of natural gas improves the calculation a little and means a ton of wood carbon saves the equivalent of 0.36 tons of carbon from use of natural gas [55]. Using a ton of carbon in wood saves roughly half a ton of carbon from using coal.[8]

Significantly, these savings are for each ton of carbon in biomass actually fed into the boiler. Much of a tree or forest does not make it into the boiler, including roots, some tops and branches left as residue, and some dead standing wood. Most of this wood, will decompose and contribute to carbon emissions. They are reflected in the decline in forest in carbon stocks, but this carbon does not lead to savings by avoiding fossil fuels.

Table 3 provides one scenario for harvests of Douglas fir stands from the Pacific Northwestern U.S. using data from a U.S. Department of Agriculture estimate of carbon stocks in different ages of forests, including the stocks after a clear cut [52]. Douglas fir is a major commercial timber species, often harvested after 30 years, and 55 years is a long rotation. Because this longer rotation generates better results for bioenergy, Table 3 focuses on a 55-year-old plantation and indicates that bioelectricity emissions from this source are roughly four times those of natural gas and almost three-times those of coal over 25 years. Harvesting a 95-year plantation provides modestly better results at 350% more than natural gas and 250% more than coal.[9] One reason the analysis generates such a poor result for bioenergy is that a 55 year old Douglas fir stand would actually sequester carbon at a higher rate over the next 25 years if left alone than if harvested. Harvesting a 95-year-old stand leads to better results because its ongoing growth rate is lower (20 tons of carbon over 25 years compared to 38 tons for the regrowing forest). However, that is to some extent balanced out by a higher up-front carbon debt.

A study for Massachusetts led by the Manomet Center for Conservation Science came out with somewhat better results for bioelectricity [55]. This study followed the same conceptual accounting approach, but with assumptions that biomass would selectively thin mature forests "from above," thereby maximizing light

---

[8] These emissions have been documented from actual plants by the Partnership for Policy Integrity http://www.pfpi.net/carbon-emissions and are the figures used in Walker [55].

[9] Table 17A provides changes in carbon stocks at different ages for live timber, standing dead timber, understory, dead down wood and forest litter. In the analysis here, the dead down wood and forest litter increase postharvest in an amount equal to the loss of standing dead timber, which implicitly assumes that all above-ground vegetation is removed without any amount left for residues. The calculations shown here assume that.

penetration to speed forest regrowth and giving up little ongoing forest growth. The study also credited the harvest of trees in this way with substantial "carbon-free" biomass from the residues of harvests for conventional timber products, which the authors believed would otherwise not be harvested. Even with this nearly ideal harvesting scenario from a carbon perspective, bioelectricity did not lead to greenhouse gas reductions over 30 years.

These calculations highlight that the carbon results depend on how much faster the forest will grow if harvested for energy than it would continue to grow if unharvested, its net re-growth rate. It is possible to calculate the net regrowth needed for bioelectricity to match the carbon emissions from natural gas and coal.[10] Net regrowth must recover 75% of the tree carbon harvested for biomass use to match the emissions from natural gas and 66% of the carbon to match the emissions from coal. That is most likely to occur when harvesting mature trees, and explains why bioelectricity is unlikely to break even over 25 years because few forests generate so much of their growth in that limited time.

Some commentators argue that a stand-specific analysis is overly simplistic and that a proper analysis should focus on the whole forest. It is true that harvesting regimes over a whole forest dictate what harvesting regimes will apply to an individual stand. A stand level analysis must therefore be based on harvest regimes that make sense over the whole forest. However, the impact on forest carbon as a whole is only the sum of the impact on the individual stands, and if a single stand cannot generate greenhouse gas benefits even under a favorable harvest regime, neither can the whole forest. Although a proper whole forest analysis should reach the same result as a proper stand analysis, the risk is that it hides ways in which it double counts carbon [8], or incorporates so many different forest uses into one analysis that it becomes difficult to discern the incremental consequences of bioenergy [44].[11]

---

[10] The math is best illustrated by an example. Start with 100 tons of carbon in live trees in the forest. Harvest them and remove 68 tons to burn, leaving 32 tons of roots and residues to decompose. (That assumes 12% of the carbon is left as unharvested tree tops and branches for long-term soil fertility and wildlife, and that 20% of live tree carbon is unharvested roots [28]. The burning and decomposition of the wood generate 100 tons of carbon emissions, and that saves 25 tons of carbon if the biomass replaces gas, and 34 tons if it replaces coal. That results in an initial increase in emissions of 75 and 66 tons of carbon, respectively, to burn biomass. Before bioelectricity emissions can equal emissions from fossil fuel use, net regrowth must sequester precisely these levels of carbon, which is in turn equal to 75 and 66% of the carbon in the harvested trees.

[11] For example, the Buckholz paper [8] carefully analyzed the availability of biomass through increased forest uptake in the northeastern U.S., but all that biomass would be part of the terrestrial carbon sink and therefore cannot reduce emissions. (The authors were apparently in disagreement about proper bioenergy accounting and noted the counter argument in some caveats although the whole analysis was based on the assumption that forest growth is carbon free.) Schlamadinger and Marland [44] developed one of the first models to look at bioenergy harvests in conjunction with timber harvests and use of timber in long-lived wood products using idealized forest growth figures. A careful review suggests that the bioenergy portion of the harvest appears not to pay off from a greenhouse gas perspective for many decades, but the incremental consequences of the bioenergy are not directly presented.

This form of accounting implicitly assumes that the alternative to harvesting a forest is to leave it standing and growing. In reality, the alternative to harvesting many plantation forests may be harvesting them for conventional timber products, and the alternatives to fast-growing bioenergy trees may be the use of the land for grazing or crops. The carbon consequences then depend on what happens indirectly to make up for the loss of timber supply forage or crops. That could call essentially for an indirect land use change analysis similar to that of liquid for liquid biofuels[12] and presents all the same challenges. There is also the risk of partial market models than estimate only the good market responses and not the bad [1].[13] An alternative might be some kind of opportunity cost analysis that compares the gains to using the land for bioenergy to the gains from allowing the land to reforest naturally.

The direct analysis here shows why simply harvesting trees and letting them grow back does not make bioelectricity carbon neutral over reasonable time periods, and in fact, will often lead to multifold increases in emissions over 20–30 years. As in the case of liquid of biofuels, there remains an alternative, more direct approach to bioelectricity policy. Focus on sources of biomass clearly and directly additional, such as wastes or bioenergy crops produced directly on land that otherwise generates little biomass.

12 Laws Incorporating the Bioenergy Accounting Errors

The modern push for biofuels results from three kinds of government policies, in all of which the assumption of carbon neutrality plays an important role.

Renewable Energy Standards. One push results from renewable energy standards for power plants, and associated tax credits or other incentives. One recent review identified thirty-three U.S. States that had established standards for electric utilities to obtain 10–30% of their electricity from renewable sources. The European Commission has adopted a broader policy to require its member countries to

---

[12] Because of higher prices, some timber products will not be replaced, generating greenhouse gas benefits from reduced consumption but also generating some other greenhouse gas emissions from replacement products. Some forest products will be replaced by harvesting other forests, triggering those emissions and different re-growth consequences. Ultimately, land managers may plant some additional plantation forests either on grasslands or forests, which sequesters more carbon directly but typically triggers land use change (or more reduced consumption) to replace the food products. The main problem with this kind of analysis is that it is extremely difficult and inherently uncertain, with world forest models and data even less developed and trustworthy than world agricultural models.

[13] The Abt paper [1] examined the regional effects of harvesting Southern U.S. forests for electricity and estimated that much of the bioenergy demand would displace other uses of such harvests and also trigger additional forest plantings. Even so, the paper was not particularly optimistic about bioenergy, but this kind of regional analysis ignores the adverse land use changes elsewhere necessary to replace the lost timber products and food.

obtain 20% of all their energy from such sources.[14] Members of Congress in the U.S. have also been considering a national standard. Climate change is a major rationale for these laws, yet in nearly all of them, bioenergy from any or nearly any source of biomass counts as a renewable fuel.

Carbon Emissions Limits. A second push comes from laws that impose explicit emissions limits. These laws include Europe's Emissions Trading System, which limits emissions from large European factories and power plants, and the Regional Greenhouse Gas Initiative (RGGI), a cap and trade system for electric utilities agreed upon by states in the Northeastern U.S. Like the accounting rules for the Kyoto Protocol itself, these systems create incentives for bioenergy because emissions of carbon dioxide from biomass combustion do not count toward the limits. As a result, shifting from fossil fuels to bioenergy is a way of reducing emissions 100%. Many European countries [22] and RGGI impose some kind of sustainability requirement on the biomass,[15] but as discussed above, sustainable does not mean carbon neutral.

Biofuel Mandates and Subsidies. The third push comes from mandates or tax subsidies for liquid biofuels for transportation that have combined to create a rough worldwide goal of 10% of ground transportation fuels by 2020 [46]. That would equal roughly 2% of world energy use. While the tax subsidies usually have no greenhouse gas limitations, some of the mandates do have more sophisticated greenhouse gas accounting rules than other bioenergy policies. For example, the U.S. national renewable fuel standard, as implemented, requires reductions of greenhouse gas emissions over 30 years by differing percentages for different kinds of biofuels, taking into account indirect land use change. The low carbon fuel standard implemented by California has similar greenhouse gas accounting and in effect gives incentives to biofuels based on their level of greenhouse gas reduction. The federal standard also limits the types of lands that can be directly planted for bioenergy [48]. Europe's renewable energy standards require greenhouse gas accounting of emissions from direct land use change, but as of the editorial deadline for this cheaper, is only considering whether to include indirect land use change.

In principle, the U.S. is doing correct carbon accounting for liquid biofuels, but as implemented, the federal law is unlikely to make much difference because

---

[14] DIRECTIVE 2009/28/EC OF THE EUROPEAN PARLIAMENT AND OF THE COUNCIL of 23 April 2009 on the promotion of the use of energy from renewable sources and amending and subsequently repealing Directives 2001/77/EC and 2003/30/EC (http://eur-lex.europa.eu/LexUriServ/LexUriServ.do?uri=OJ:L:2009:140:0016:0062:en:PDF.

[15] RGGI has model rules set forth at http://www.rggi.org/docs/Model%20Rule%20Revised%2012012.31.08.pdf. Emissions of carbon dioxide are not counted if they result from the use of "eligible biomass," which is defined as follows:

Eligible biomass. Eligible biomass includes biofuels harvested woody and herbaceous fuel sources that are available on a renewable or recurring basis (excluding old-growth timber), including dedicated energy crops and trees, agricultural food and feed crop residues, aquatic plants, unadulterated wood and wood residues, animal wastes, other clean organic wastes not mixed with other solid wastes, biogas and other neat liquid biofuels derived from such fuel sources. Biofuels harvested will be determined by the regulatory agency.

according to the regulations issued by the U.S. Environmental Protection Agency, virtually every form of biofuel now in use passes its required greenhouse gas reduction threshold. The rule also provides that any future production of cellulosic ethanol from perennial grasses will also comply with standards [14]. The rule is based on assumptions of where and how the biomass feedstocks are produced, but once the rule is in place, those assumptions do not have to be correct and the crops can be grown anywhere. Regardless of what actually happens, for example, any soybean or rapeseed vegetable oil, and any switchgrass or miscanthus, grown on any existing agricultural land in the world will qualify under the manadate. (That is true even though switchgrass and miscanthus were analyzed solely on the assumption that they were grown in Oklahoma.)

A full breakdown of the EPA model results has yet to occur—and penetrating the model results to understand where the "additional carbon" comes from is a major challenge—but a few features of the analysis are remarkably strange. For example, the EPA found that ILUC emissions from maize ethanol are 81 g per mega joule today (almost alone equal to greenhouse gas emissions from gasoline even without counting production emissions), but will drop to 34 g per mega joules in 2022 [42].[16] That passes legal thresholds for greenhouse gas reductions because EPA decided to judge them based on the emissions predicted for them in 2022 rather than the emissions the biofuels generate today. Increasing yield can explain a modest drop in emissions, but there is no plausible real world explanation for such a large drop. Among the strange contributing factors, switching from soybean production to maize in the U.S. increases emissions from nitrous oxide today but decreases emissions in 2022. Although the full reductions in food consumption are not clear, the final EPA rule estimated that significant greenhouse reductions would occur because of reduced livestock and rice consumption, and both were necessary for biodiesel to pass greenhouse gas thresholds.

An earlier section outlines the environmental and welfare costs of virtually all the possible responses to diverting highly productive cropland to biofuels. The EPA rule in effect blesses U.S. biofuel consumption to rise to the mandated level of 36 billion gallons by 2022 roughly three-times present levels.

13 Scope of the Potential Consequences

The accounting error has remarkable potential to alter the Earth's land surface for two reasons. It can distort the behavior of any of the world's energy users subject to climate laws. Even more importantly, generating a modest percentage of the world's energy requires vast areas of productive land.

The world today uses roughly 500 exajoules of primary energy for all purposes, and depending on the scope of energy efficiency efforts, that use will

---

[16] Perhaps the best analysis of the final EPA rule was by Richard Plevin of Berkeley, and this paper relies on his analysis [42].

probably grow to 750–1,000 exajoules by 2050.[17] Assuming that bioenergy is carbon neutral, and comparing its costs and practicalities to other energy sources that reduce greenhouse gas emissions, the International Energy Agency has estimated that to cut global energy emissions in half bioenergy use should rise to 150 exajoules by 2050, 23% of world energy use in a highly conserving world [27].

The prospect of such large growth in bioenergy is more than abstract, as government policies are already in place to drive well in this direction. According to national action plans submitted as of 2010 by each European country, bioenergy overall will account for 54.5% of the renewable energy target of 20% set by the European Commission for 2020, which means bioenergy will provide 12% of European total energy consumption [4]. Two thirds of this bioenergy is to come from forest and other solid biomass, and biomass is to rise to 18% of electricity production.

The 20% figure for bioenergy is actually modest by comparison with other estimates of global bioenergy. One recent paper put the medium assessments of sustainable bioenergy potential at 200–500 exajoules [13]. Another paper has estimated that bioenergy could provide 1,500 exajoules while still protecting forests and biodiversity [50].

The best way to appreciate the land and water resources required for this production is to compare it to the existing harvest of world plants by people for all purposes. According to the best estimates available today, all the world's harvested crops in the year 2000, if used for their intrinsic energy, could provide 64 exajoules of energy. Along with all the harvested crop residues, all the grass and other forage materials consumed by livestock, and all the wood removed from forests for timber products, paper and traditional energy use, total human plant harvest contains 217–230 exajoules [24]. In other words, many of the paths contemplated would require that the world double, triple or increase even more the total harvest of plants. And these bioenergy needs come on top of needs to generate 70–80% more food by 2050, and growth in demand for timber products of 1.4% per year [45, 53].

Yet, the existing harvests for human needs already involves the manipulation of roughly 75% of the world's ice—and desert-free land [16, 24].[18] Agriculture uses more than 70% of the water diverted from rivers and aquifers [2], and originates most of the nitrogen that pollutes great swaths of the world's coastal waters [23]. As of 2000, land use change had contributed roughly one-third of the human-caused increase in global warming gases [32, 26]. Even so, and despite the

---

[17] The U.S. Energy Information Agency estimates world energy use will grow by 50% just between 2007 and 2035 in their baseline scenario. U.S. Energy Information Agency, International Energy Outlook 2010, World Economic Demand and Energy Outlook (DOE/EIA-0484(2010), Washington) http://www.eia.gov/oiaf/ieo/world.html.

[18] It is impossible to set this number too precisely because it depends on the degree of forest manipulation and grassland use that should be considered human use.

additions of this nitrogen and vast quantities of irrigation water, the net effect of human action to date has been to reduce modestly the annual production of world biomass [24]. Compared to food production, growing plants solely for biomass opens up new possibilities for maximizing carbon uptake because the effort is not constrained by the need to generate biomass in edible form. But any notion that such additional bioenergy demands could be met without great environmental damage is unsupportable—and indeed, as discussed above, is based on double-counting existing carbon.

As important, the accounting error makes the true technical potential of "sustainable" bioenergy production mostly irrelevant because once embodied in legal rules it sanctions the use of biomass from most or any sources. It therefore relieves bioenergy of the need to avoid competition with other uses of water and land even to the extent that is technically possible. The problem remains even if sustainability standards prohibit the direct conversion of high value ecosystems to bioenergy production. The diversion of existing agricultural and timber lands to bioenergy leaves farming and commercial forestry free to replace the lost food and timber by using whatever water they can get, and by expanding into whatever high carbon or ecologically valuable lands are most economical.

Two modeling studies have estimated that the world's natural forests will mostly disappear if policymakers continue to try to reduce greenhouse gas emissions but assume that biomass is carbon neutral [35, 56].

The World has some, but limited, capacity to generate additional biomass, and that biomass needs to be preserved for the most pressing needs, such as aviation, that cannot run on alternative, low carbon fuels.

14 Hope for the Future

There are signs of hope, and one reason is that the accounting flaw seems obvious to many general citizens, who cannot understand why policymakers should ignore the carbon released by power plants that burn trees.

An early sign of protest has occurred in western Massachusetts, where several companies proposed in recent years to develop wood-burning power plants in response to state incentives for renewable electricity. The forests of western Massachusetts are prized heavily by the people who live and vacation there, and the risk to its forests motivated Meg Sheehan, a tenacious local lawyer and philanthropist, to start a grass roots effort to oppose these wood-burning power plants. Fortunately, Mary Booth, a gifted ecologist had just returned to the area from work in Washington where she had acquired strong gifts in using science to analyze policy, and she almost singlehandedly provided the technical analysis to support this burgeoning movement. When scientific articles began to appear questioning the bioenergy accounting, the State ordered up the Manomet study discussed above of the greenhouse gas consequences. In turn, Sue Reid, who led state political efforts for the State's most prominent environmental group, rather brilliantly helped to forge a broader coalition for reform. The State is now in the final stages of adopting regulations that will limit its support for bioelectricity for the most part to additional biomass.

While encouraging, the Massachusetts effort also suggests the intensity and talented leadership needed for reforms. But for the grass roots movement and three remarkable women, the effort would probably have come short. Environmental groups worldwide have by now mostly grasped the accounting flaw and are pushing for various reforms, and some talented individuals are staffing their efforts, but the intensity of this push in the environmental movement overall has not yet matched that achieved in Massachusetts.

At this time, the U.S. EPA and the State of California are struggling with bioenergy accounting. The EPA has recently announced a 3 year hiatus in its decision on how to regulate bioenergy emissions as part of new air pollution regulations for large greenhouse gas emitters [15]. But some of its written statements make it clear that EPA understands the basic accounting flaw. EPA's delay rather openly responds to pressure from the U.S. Congress, which is in response to the expectations of forestry and agricultural interests that their generation of biomass for energy would be rewarded by climate change efforts.

If the problem is not fixed now, a major reason may be the underestimation of the inherent power of accounting errors. The bioenergy error stems from its capacity to distort the incentives of energy users everywhere, and as it spawns more of the wrong kind of bioenergy, the rules become harder and harder to fix. Fixing the accounting is the first step to avoid great harm and to encourage the limited development of truly low carbon biofuels to meet some of our transportation needs.

References

1. Abt R, Galik CS, Henderson JD (2010) The near-term market and greenhouse gas implications of forest biomass utilization in the Southeastern United States. Climate Change Policy Partnership, Duke University, Durham
2. Agriculture CAoWMi (2007) Water for food, water for life: a comprehensive assessment of water management in agriculture. Earthscan, London
3. Arima EY, Walker RT, Caldas MM (2011) Statistical confirmation of indirect land use change in the Brazilian Amazon. Environ Res Lett 6:024010
4. Atanasiu B (2010) The role of bioenergy in the national renewable energy action plans: a first identification of issues and uncertainties. Institute for European Environmental Policy, London
5. Bauen M (2009) Bioenergy—A sustainable and reliable energy source. A review of status and prospects. International Energy Agency, Paris
6. Beringer T, Lucht W, Schaphoff S (2011) Bioenergy production potential of global biomass plantations under environmental and agricultural constraints. Glob Chang Biol Bioenergy 3:299–312
7. Berndes G, Hoogwijk M, van den Broek R (2003) The contribution of biomass in the future global energy supply: a review of 17 studies. Biomass Bioenergy 25:1–28

8. Buckholz T, Canham C (2011) Forest biomass and bioenergy: opportunities and constraints in the Northeastern United States. Cary Institute of Ecosystem Studies, Millbrook
9. Chum H, Faaij A, Moreira J (2011) Bioenergy. In: Special report renewable energy sources: international panel on climate change
10. Danielsen F et al (2009) Biofuel plantations on forested Lands: double jeopardy for biodiversity and climate. Conserv Biol 23:348–358
11. de Vries BJM, van Vuuren DP, Hoogwijk MM (2007) Renewable energy sources: their global potential for the first-half of the 21st century at a global level: an integrated approach. Energ Policy 35:2590–2610
12. Dehue B, Cornelissen S (2011) Indirect effects of biofuel production: overview prepared for GBEP. Ecofys, Utrecht
13. Dornburg V et al (2010) Bioenergy revisited: key factors in global potentials of bioenergy. Energy Environ Sci 3:258–267
14. EPA US (2010) Regulation of fuels and fuel addiatives: changes to renewable fuel standard program; final rule. In: 70–Agency USEP (ed). Federal Register 14670–14904
15. EPA US (2011) Deferral for CO_2 emissions from bioenergy and other biogenic sources under the prevention of significant deterioration (PSD) and Title V Programs. Federal Register 76:43490–43508
16. Erb K, Gaube V, Krausmann F, Plutzar C, Bondeau A, Haberl H (2007) A comphreneisve global 5 min resolution land-use data set for the year 2000 consistent with national census data. J Land Use Sci 2:191–224
17. Fairley P (2011) Introduction: next generation biofuels. Nature 474:S2–S5
18. Fargione J, Hill J, Tilman D, Polasky S, Hawthorne P (2008) Land clearing and the biofuel carbon debt. Science 319:1235–1238
19. Garrity DP et al (1996) The Imperata grasslands of tropical Asia: area, distribution, and typology. Agroforest Syst 36:3–29
20. Goldemberg J (2010) Are biofuels ruining the environment? Biofuel Bioprod Bior 4:109–110
21. Gómez DR et al (2006) Energy: stationary combustion. In: Eggleston S, Buendia L, Miwa K, Ngara T, Tanabe K (eds) 2006 IPCC guidelines for national greenhouse gas inventories. Institute for Global Environmental Strategies, Hayama
22. Group BT (2008) Sustainability criteria and certification systems for biomass production final report: prepared for DG Tren
23. Gruber N, Galloway JN (2008) An Earth-system perspective of the global nitrogen cycle. Nature 451:293–296
24. Haberl H, Beringer T, Bhattacharya SC, Erb KH, Hoogwijk M (2010) The global technical potential of bio-energy in 2050 considering sustainability constraints. Curr Opin Sust 2:394–403
25. Hoogwijk M, Faaij A, x00e, Eickhout B, de Vries B, Turkenburg W (2005) Potential of biomass energy out to 2100, for four IPCC SRES land-use scenarios. Biomass Bioenergy 29:225–257
26. Houghton RE (2008) Carbon flux to the atmosphere from land-use changes: 1850–2005. In: TRENDS: A compendium of data on global change oak ridge. Carbon Dioxide Information Analysis Center, Oak Ridge National Laboratory, TN
27. IEA (2008) Energy technology perspectives: scenarios and strategies to 2050. International Energy Agency, Paris
28. Jackson RB, Canadell J, Ehleringer JR, Mooney HA, Sala OE, Schulze ED (1996) A global analysis of root distributions for terrestrial biomes. Oecologia 108:389–411
29. Kirschbaum MUF (2003) To sink or burn? A discussion of the potential contributions of forests to greenhouse gas balances through storing carbon or providing biofuels. Biomass Bioenergy 24:297–310
30. Lapola DM et al (2010) Indirect land-use changes can overcome carbon savings from biofuels in Brazil. Proc Natl Acad Sci USA 107:3388–3393
31. Lewis M (2010) The big short: inside the doomsday machine. W.W. Norton & Company, New York

32. Malhi Y, Meir P, Brown S (2002) Forests, carbon and global climate. Philos T R Soc A 360:1567–1591
33. Mallaby S (2010) More money than god: hedge funds and the making of a New Elite. Penguin Press, New York
34. Marelli L, Mulligan D, Edwards R (2011) Critical issues in estimating ILUC emissions: outcomes of an expert consultation, 9–10 Nov 2010. European Commission Joint Research Center, Ispra (Italy)
35. Melillo JM et al (2009) Indirect emissions from biofuels: how important? Science 326:1397–1399
36. Moomaw WR, Moreira JR (2001) Technical and economic potential of greenhouse gas emissions reduction. In: Melz B (ed) Climae change 2001: mitigation (climate change 2001 IPCC third assessment. Cambridge University Press, New York
37. Nasser A (2010) An allocation methodology to assess GHG emissions associated with land use change, final report. Icone, Sao Paulo
38. O'Hare M, Plevin RJ, Martin JI, Jones AD, Kendall A, Hopson E (2009) Proper accounting for time increases crop-based biofuels' greenhouse gas deficit versus petroleum. Environ Res Lett 4:024001
39. OECD (2008) Biofuel support policies: an economic assessment. Organization for economic co-operation and development
40. Pacca S, Moreira JR (2009) Historical carbon budget of the brazilian ethanol program. Energy Policy 6:182017
41. Pan Y et al (2011) A large and persistent carbon sink in the world's forests. Science 333:988–993
42. Plevin RJ (2010) Comments on US EPA's final rulemaking for the renewable fuel standard. UC Berkeley
43. Roberts MJ, Schlenker W (2009) World supply and demand of food commodity calories. Am J Agr Econ 91:1235–1242
44. Schlamadinger B, Marland G (1996) Full fuel cycle carbon balances of bioenergy and forestry options. Energy Convers Manag 37:813–818
45. Science UGOf (2010) Foresight project on global food and farming futures: synthesis report CI: trends in food demand and production. U.K. Government Office for Science, London
46. Searchinger T (2009) Government polices and drivers of world biofuels, sustainability criteria, certification proposals and their limitations. In: Howarth RW, Bringezu S (eds) Biofuels, environmental consequences and interactions with changing land use. Cornell University, Ithaca
47. Searchinger T et al (2008) Use of U.S. croplands for biofuels increases greenhouse gases through emissions from land use change. Science 319:1238–1240
48. Searchinger TD et al (2009) Fixing a critical climate accounting error. Science 326:527–528
49. Smeets E, Faaij A (2007) Bioenergy potentials from forestry in 2050. Climatic Change 81:353–390
50. Smeets EMW, Faaij APC, Lewandowski IM, Turkenburg WC (2007) A bottom-up assessment and review of global bio-energy potentials to 2050. Prog Energy Combust Sci 33:56–106
51. Smil V (1999) Crop residues: agriculture's largest harvest. Bioscience 49:299–308
52. Smith JE, Heath LS, Skog KE, Birdsey RA (2005) Methods for calculating forested ecosystem and harvested carbon with standard estimates for forest types of the United States. In: General Technical Report. United States Department of Agriculture Forest Service, Northeastern Research Station
53. Smith P, Gregory PJ, van Vuuren D, Obersteiner M, Havlik P, Rounsevell M, Woods J, Stehfest E, Bellarby J (2010) Competition for land. Philos Trans R Soc Biol Sci 365:2941–2957
54. Somerville C, Youngs H, Taylor C, Davis SC, Long SP (2010) Feedstocks for lignocellulosic biofuels. Science 329:790–792

55. Walker T et al (2010) Biomass sustainability and carbon policy study. Manomet Center for Conservation Sciences, Brunswick, p 182
56. Wise M et al (2009) Implications of limiting CO_2 concentrations for land use and energy. Science 324:1183–1186
57. Yan X, Tan D, Inderwildi O, Smith JAC, King D (2011) Life cycle energy and greenhouse gas analysis for agave-derived bioethanol. Energ Environ Sci 4:3110–3121

Conclusion

Oliver Inderwildi and Sir David King

This treatise is the most comprehensive publication to date on the current challenges facing the transport sector, particularly energy needs and environmental impact. The contributors are leading thinkers from industry, academia, and the public sector. The mobility challenge is so complex that the problems have to be assessed from a variety of viewpoints by industrialists, engineers, scientists, economists, and political scientists working on transport related issues. All aspects of transport—people and freight; air, road, rail and ocean—form an overlapping series of challenges and need to be approached from an integrated standpoint.

Summarizing the insights gained from this unique publication, we were amazed to see that people with very different backgrounds and disciplines have highlighted very similar issues and challenges and, moreover, located the same quick wins.

Here, we provide our summary of the contributions and the conclusions drawn from the wide range of chapters and then map out critical issues that reoccur throughout this book. This assessment is intended to provide a clear roadmap for action.

Energy

Growing global demand for energy, as well as recent geopolitical upheavals, have prioritized energy security concerns in many developed and developing countries. We therefore begin our discussion of the complexity of transport problems with the energy demands originating from transport, as well as resource scarcity and distribution and GHG emission issues.

In the lead article to this section Iain Conn of BP argues that it is time to stop "polishing the 2050 diamond" and take action now. BP's chief executive for refining and marketing urges governments to take practical steps toward a lower carbon, secure energy future as there are many material methods available to them to do this.

O. Inderwildi and Sir David King (eds.), *Energy, Transport, & the Environment*,
DOI: 10.1007/978-1-4471-2717-8, © Springer-Verlag London 2012

Conn calls for energy policy alignment and coherence in the key economic blocs, which should be led through the strong transatlantic relations between the EU and the US, which have very similar challenges with respect to energy security and emissions reduction.

According to Conn we have to face critical choices if we are to achieve progress in decarbonizing our energy systems while also delivering greater energy security. This analysis of the geopolitical and energy security issues is followed by Colin Campbell's assessment of the world oil inventory. According to Campbell, global production of regular conventional oil passed its peak in 2005 and, by 2050, oil supply will have fallen to a level able to support less than half the current world population in its present way of life. The assessment presented by the figurehead of peak oil theory therefore stresses Conn's call for urgent and bold action. There are more optimistic assessments on peak production, but a number of high-level government reports and academic studies agree that the peak of conventional oil production is on the horizon, and this could have drastic consequences both for the environment and the world economy.

Could we cope with a peak oil scenario? Joerg Friedrichs suggests that to mitigate the impact of peak oil, significant investments into alternative technologies and surrogate energy sources is needed. The most appealing energy resource to mitigate the impact of peak oil is, according to Friedrichs, natural gas since each year there is more discovered than consumed at the moment. Natural gas corridors in the United States planned under the American Recovery and Reinvestment Act would be the first example of natural gas as conventional transport fuel and might lead the way. Friedrichs also evaluates the cost and adverse environmental impacts of alternatives such as shale gas, unconventional oil, biofuels, and electricity, while more detailed assessments are presented in the following chapters.

Weber Antonio Neves do Amaral and Adam Liska discuss the critical biofuels issue from the viewpoint of Brazil, the country that hosts the world's most advanced biofuel economy. They see the challenge lying in responsibly establishing sustainable production systems and biofuel supplies in sufficient volume to make an impact on transport fuel supply. According to Amaral and Liska, and best practices are found in Brazilian sugarcane ethanol, which provides a good framework for baseline sustainability compared with other current and future biofuels. The proper planning of sugarcane expansion into new areas will be the next challenge for Brazil and will mark another important step toward the sustainable production of ethanol. Brazil is the prime example for large-scale bioenergy production and is clearly a role model for other countries. However, not every country is endowed, as Brazil is, with sufficient sunlight, water, and fertile soil. Solutions are clearly geographically specific.

Tara Shirvani of Oxford University discusses the potential of alternative fuels to mitigate conventional oil supply shocks and reduce GHG emissions. Unconventional oil resources, synthetic fuels from alternative fossil resources as well as various biofuels are covered in her assessment. Unconventional oil resources are mainly concentrated in developed countries, which is beneficial from an energy security and national economy viewpoint. However these resources provide major environmental challenges. Synthetic fuels from alternative fossil resources such as gas and coal produced using the Fischer-Tropsch process, as developed by Sasol in South Africa, are useful for diversifying the fuel mix, but this benefit comes with the environmental penalty of a substantial increase in GHG emissions. Shirvani shows a useful comparison of the market entrance crude oil prices for liquid fuel production from different resources. This establishes the level of oil price that makes the corresponding alternative resource profitable. The conclusion is that at current market prices, fuels from unconventional resources are already profitable. It must be expected that more and more of these environmentally polluting resources will enter the fuel mix. This trend will have to be mitigated in order to reduce emissions. The Oxford researcher also addresses sustainable biofuel production and concludes that land-use as well as water-availability issues will significantly limit the fuel supply provided by current biofuels. Biofuels that do not impact water and arable land, such as algae and agave derived fuels, provide a viable way forward according to Shirvani.

Alexander Williams assesses the indirect emissions caused by electric vehicles. The drawback is that electric mobility only reduces emission when the electricity grid is low-carbon, as for instance in France. This analysis clearly demonstrates that we have to think holistically when aiming for a decarbonization of the transport sector, as energy and mobility are intrinsically linked. This is a result of obvious importance: all sectors are linked and in order to significantly reduce overall emissions, there is an over-riding priority to de-fossilise regional and national power grids.

It is not only energy resources that are becoming scarce. Jacometti Associates assess general resource scarcity, and conclude that societal changes and our perception of consumption have to change in order to reach sustainability. Information and communication technology will play a major role in dematerialising the economy, but to get this done, we will have to change our mind-set.

Finally, Peter Edwards and co-workers visualize a hydrogen economy based not purely on hydrogen fuel cell vehicles but on the large-scale renewable production of hydrogen. Renewable hydrogen with its intrinsically low carbon footprint could not only be used as fuel in ICE vehicles and fuel cell vehicles but would also be an excellent chemical feedstock for the synthesis of low-carbon liquid fuels and fertilizers. However, the known problems of hydrogen storage and distribution still stand in the way of implementing hydrogen as a large-scale transport fuel.

Road

Road transport is the biggest consumer of oil-derived fuels and within the transport sector consequently the biggest emitter. With increasing demand for individual transport from emerging markets such as India and China, there is no end of growth in sight, and studies predict that soon we will have 2 billion cars on our planet . These vehicles need decentralized provisions of transport fuel, and Mark Gainsborough of Royal Dutch Shell points out that no solution will work if it does not work in the market. People make choices when they buy a vehicle, but also when they choose a fuel, step on the accelerator, or select a route, and all these choices impact our energy security and our environment. So the mobility challenge, especially in road transport, demands solutions that customers can accept and afford and this is most likely going to be a mosaic of various fuels. Shell's executive stresses the role businesses play in the solution of the mobility challenge, but also emphasizes that policy makers have to help to create markets and reduce uncertainties so that private sector players can establish themselves in these new areas. One important fuel in the mosaic described by Gainsborough will be electricity, as it can be generated from various fossil and renewable sources. Electricity in road transport is a recurrent theme in this volume.

Julia King and Eric Ling show in their analysis that in 2030 a third of the car and van fleet of the UK could be powered by electricity without additional generating capacity on the British electricity grid! Key to this is demand-side management by the use of smart meters. In order to achieve emission reductions, however, we have to defossilise the electricity sector in parallel. But that is not the whole story, according to King and Ling. On the one hand, manufacturers will need time to deploy sufficient electric vehicle production capacity to supply a significant share of electric vehicles to the global market. On the other hand, consumers have to be provided with confidence in electric mobility, e.g., through the introduction of incentives for electric vehicles, and a visible widely available charging infrastructure, otherwise consumers will be too reluctant and the uptake will be delayed. The optimal policy for reducing transport emissions using electric mobility is one that delivers the most cost-effective low-carbon technologies, catalyzes manufacturing, and consequently provides a cost reduction while incentivizing consumer uptake in parallel.

Hence, we can conclude that electric, hybrid, and plug-in hybrid powertrains will play a role in decarbonizing transport, but there is still a need for significant research and development to mature the technology. Malcolm McCulloch and co-workers at Oxford University present a study on the emission-efficiency of the three different powertrain setups and, moreover, look beyond the battery and assess the potential of energy storage in ultra-capacitors and high-speed flywheels. The storage in capacitors and flywheels is not quite as efficient as storage in batteries, but could drastically reduce the cost of electric vehicles and hence make them more competitive on the mass market, according to the Oxford team. Making electric individual transport more accessible and affordable will be vital for its success.

Nevertheless, battery technology is an area in which significant progress is under way. The requirements for transport and grid applications are severe in terms of cost, safety, and efficiency, according to Clare Grey of Cambridge University. Despite these stringent requirements, a wide range of lithium-ion battery technologies exist now that are currently ready for use in transport and grid applications. New materials with different chemistries are being developed that have the potential to increase energy density and safety at a reduced cost. Economics of scale will set in when planned large-scale battery production plants come online and drive down costs of state-of-the-art batteries. It would therefore be premature to write rechargeable battery electric vehicles off based on Grey's assessment.

But not only batteries, flywheels and capacitors can also be used to power highly efficient electric engines. Electrochemical combustion in fuel cells can be used to generated electricity on-board and hence circumvent the storage issue. There remain a set of challenges, however, which means that fuel cell vehicles are not yet ready for wide commercial application. Platinum-based cells fuelled with hydrogen efficiently produce electricity. However, due to the continually increasing price of noble metals the cost of these fuel cells restricts applications. Other mobile fuel cells suffer from disadvantages such as long start-up times or inefficiencies due to pre-combustion fuel reforming, according to Holdway.

But it is not only novel technology that can help to reduce emissions, smart use of well-established technology can assist in this transformation as well. Sonia Yeh and Daniel Sperling at the University of California address life-cycle emissions of fuels in their contribution. The academics conclude that low-carbon fuels are a necessity to reduce emissions from road transport in the short run, but that the current ad hoc policy approach to alternative fuels has largely failed to provide these. They propose a low-carbon fuel standard to stimulate innovations in alternative fuels. Such a fuel standard would help to reduce the GHG emissions from road transport by applying life cycle carbon intensity standard and, in addition, incorporate market mechanisms as for instance credit trading, which would spur innovation. However, the authors also state that such a policy has to be combined with other measures such as fuel efficiency standards in order to be most effective.

David Bonilla at Oxford University argues that the fuel economy of road vehicles is directly influenced by fuel price increases. According to the academic's assessment, higher fuel taxes have led consumers to be much more aware of the fuel economy of their cars and, consequently, influence consumer behavior. Taxes on fossil fuels are therefore one way to save transport fuels. European Union fuel taxes have led to improved fuel economy and have consequently saved emissions and increased energy security according to the Oxford economist.

Urban Mobility

More than 50% of the global population lives in cities or agglomerations, making transport in urban environments an issue of paramount importance. The average travel speeds in cities are similar to those 100 years ago. Improving the way we move around in cities in the future will clearly not only require novel transport technologies but also improvements in urban design, communications technology, integration of personal and public transportation, and enhanced transport management. Moreover, we need to rethink the automobile itself, according to Chris Borroni-Bird of General Motors. The separation of personal and public transport systems is an attractive idea; in urban environments the comfort, convenience, utility, safety, security, and flexibility of individual mobility is likely to remain compelling, so instead of penalizing it, governments should incentivize smarter individual mobility in urban areas. Small electric vehicles that are physically and electronically integrated into modern mass transit systems can form a completely new transport network and make cities more liveable, according to this industrialist.

Michele Dix and Elaine Seagriff of Transport for London report on the approach adopted for the British capital; a multi-faceted and fully integrated approach across strategies, policy areas, and transport modes is employed here to optimize transport in Europe's second largest conurbation. The so-called London plan connects previously disparate transport modes and functions. The new integration will become more important as London emerges from the recession and continues to grow, according to Dix.

They point out that, in order to continue providing enhanced urban mobility in the future, more challenging mode shifts are required. Shifting transport from road to mass transit as well as walking and cycling, however, comes with the additional challenge of providing enough capacity on the public transport system and increasing the number of cycle lanes and pedestrian areas.

Debra Lam and Peter Head of Arup argue that it is impossible to implement one technical strategy, as for instance transport, in isolation from a range of other considerations. This is very much in agreement with the authors of the previous papers. Only the combination of several technical strategies can really support the development of sustainable urban mobility and fully leverage its benefits. The author calls for Low Emission Strategies (LES), which are a smart combination of various measures such as demand management, modal substitutions, and demand redistribution. Key to success is that the strategy is tailor-made for a particular city by assessing the needs holistically. Hence, again, this contribution calls for comprehensive, integrated solutions.

One of the main transport problems in agglomerations is congestion, and this can effectively be helped by charging for road use as London and Singapore have shown. Michael Bell and Muanmas Wichiensin show in their contribution that the Downs-Thompson paradox applies if the efficiency gains arising from increased transit use are passed on to passengers as reduced generalized costs, e.g., in the form of reduced fares for public transport. The Downs-Thompson paradox implies that the equilibrium speed of car traffic on a road network is determined by the average speed of equivalent journeys using public transport. The authors conclude that if transit operators maximize profit rather than passing the benefits of road pricing on to consumers, road pricing is ineffective as a means to support modal substitution. Therefore, road pricing has to be regulated in order to achieve a significant modal substitution; Bell and Wichiensin emphasize the importance of including such a regulation in the wider transport policy.

Aviation

When mapping out ways to reduce emissions from transport, aviation is certainly a special case with several critical issues, according to Carl Burleson and Lourdes Maurice of the U.S. Federal Aviation Administration (FAA): First, we cannot compromise on safety and consequently only proven and technically sound applications are being incorporated into aircraft. Second, commercial aircraft fleets have very long lifetimes and therefore the impact of any new technology is delayed by fleet turnover. Third, other prerequisites for aviation, as for instance reduced noise or pollutant emissions, have tradeoffs with the overall efficiency of the propulsions system: more stringent noise or emission requirements might lead to slightly less efficient turbines. Last but not least, aviation is a global industry and a significant amount of emissions occur where no country has jurisdiction, over the high seas. This poses an inherent political problem at the intergovernmental level and is clearly the reason why inclusion of aviation into the EU-ETS has met with significant opposition. This political and judicial problem has to be solved in order to apply market-based measures to this truly global industry, according to these leading American civil servants.

The authors conclude that it will be of paramount importance to balance capacity growth and environmental protection and this can only be achieved if forces are combined. Governments have to incentivize research and development of efficient aircraft and propulsion technologies, low-carbon aviation fuels have to be pushed, and air traffic as well as airport management have to be optimized. Only if all these measures are combined can emissions from aviation be cut while demand increases.

After this high-level introduction by the American regulators, several following chapters shed light on the specific issues.

One of the most promising joint efforts is the Commercial Aviation Alternative Fuels Initiative (CAAFI), a public private partnership of airlines, manufacturers, airports, and the U.S. FAA. Richard Altman, CAAFI's CEO, reports on how a unique example of such a focused, collaborative initiative can promote a particular solution and achieve notable results. Before the initiative was founded, aviation fuel research was essentially a niche area, but this has changed quite dramatically since CAAFI was founded. The initiative has fostered collaborations between the

US Government and their Forces as well as the American industry, and has been critically important in the passage of the new aviation fuel specification. Further major advances fostered by CAAFI include the establishment of an aviation-specific ground-to-wake life cycle analysis for jet fuels and the creation of a gated risk management approach to govern the development and deployment of alternative fuels. Meanwhile, other countries and economic areas have set up very similar initiatives, another proof of CAAFI's success. Richard Altman also maps out what has yet to be done. The next steps will include making sure that a reliable quality of jet fuel can be supplied, mapping out alternatives beyond synthetic fuels made via Fischer-Tropsch or hydrogenation, increasing supply sustainably while reducing cost and spurring more interest from investors. Again, investments are critical for the success of sustainable transport in general and uncertainty has to be reduced in order to attract large-scale investments. In addition to the qualities and sourcing of the fuel, the efficient use of the fuel is crucial. That is why it is important to improve propulsion technology in parallel with reducing the carbon footprint of the fuel. Peter Ireland of Rolls Royce and Oxford University argues in his chapter that there is little doubt that aircraft engines will continue to become more efficient through the twenty-first century. The impact of this, however, depends on how rapidly changes to current technology can be introduced, particularly radical changes which could lead to reduction in fuel burn of up to 20%. He discusses the research and development carried out to reduce emissions and the engine weight of conventional turbines and moreover, discusses more challenging hydrogen powered aircraft. For the latter storage is a major remaining challenge just as with road vehicles. Nevertheless, all this sounds very promising; there is clearly much to be gained from improving propulsion technology.

Another promising option is to reduce aircraft weight and to improve aerodymanics. Christian Carey of Oxford University discusses these options in his chapter. The researcher discusses how the aerodynamics of current aircraft can be improved using, for instance, winglets, which are small wings located horizontally at the tip of the main wing. These winglets are already in use, for instance in the newest model of Boeing's 737 and, although small in size, are found to reduce fuel consumption and hence emissions by up to 4 %. But there is more in the pipeline; more radical approaches to reduced-drag aircraft such as joined or blended wing body planes can reduce drag much more significantly and consequently impact on the environmental performance of aircraft even further. However, these setups are not quite at a commercial stage yet and there is clearly a need for R&D stimulus by governments as the commercial risks are significant.

Carey also points out that novel materials such as alloys or ceramics that decrease weight or allow turbines to operate at higher temperature can have a significant impact. While the savings seem small at first sight, calculated over the lifetime of an aircraft these can be quite significant. Last but not least, Carey assesses the potential of the long forgotten airship; at its peak the airship, invented by the German General and Engineer Ferdinand Graf von Zeppelin, made regular Atlantic crossings, but quickly disappeared after the Hindenburg disaster and was superceded by the arrival of safer, faster airplanes. Carey in his assessment concludes that there could indeed be a revival in lighter-than-air craft, as more modern hybrid airships do not suffer from the disadvantages of old Zeppelins. These modern airships still need to demonstrate their potential but they do have many significant advantages, particularly in the transportation of freight where they are inherently very competitive both economically and environmentally.

But improvements in propulsion technology as well as aircraft weight and design are unlikely to impact emissions from aviation in the short run due to long fleet lifetimes. Improvements in air traffic and airline management are more likely to deliver these reductions in the near future. Ian Jopson of NATS argues that more efficient use of airspace can deliver significant reductions in fuel burn and consequently in emissions. He demonstrates unequivocally that fuel-efficient routing in combination with optimal flight profiles, (the way an aircraft climbs and descends) can significantly curb emissions from commercial aviation. Moreover, NATS is working with airlines and airports to improve holding procedures, ground movements, and passenger handling to ensure smooth operations and minimal delays which further reduces the emissions from aviation. Key to these endeavors is NATS' Airspace Efficiency database, which provides a structured approach to tracking and progressing ideas through to implementation, according to Jopson. All this, however, is contingent on to political will; the European Union would particularly benefit from centalized air traffic governance. With the right will from our political leaders, in the EU and elsewhere, massive savings in cost and emissions would be matched by reduced passenger frustration. This is a triple win awaiting political leadership.

Optimizing air traffic services and airport operations alone is not enough. Airlines themselves play a crucial role as well. Loizos Heracleous and Jochen Wirtz examine how Singapore Airlines have outperformed their competitors in an unforgiving industry. Singapore Airlines has one of the youngest fleets in the business. New aircraft mean more comfort and less noise to passengers and, moreover, less fuel burn and higher occupancy due to reduced maintenance times. Combine this with excellent service and busy schedules, full aircraft, and you are the airline with the lowest passenger–kilometer emissions in the business. This clearly shows the need for optimal management in order to reduce the environmental impact of aviation in the short run.

In order to manage airports and airlines, sufficient space and runways have to be available. Estimates by the British Government forecast a significant growth for air travel in the United Kingdom requiring five more runways, three of those in the South East of England, where land-use issues pose a significant hurdle.

In the second chapter Christian Carey of Oxford University assesses the feasibility of a new airport in the South East of England located offshore in the Thames Estuary. Carey concludes that there are certain advantages of such an airport, such as 24-h operation and a significant increase in airport capacity without impacting landuse. Nevertheless, there are disadvantages as well. Air traffic management will be more difficult and bird strikes are more likely in this area. Lastly, the high estimated cost of £40-50 billion is a significant hurdle. The latter point again emphasizes that financing a renewal of the global transport system is a significant issue that has to be resolved. Carey urges a consideration of other options, such as shifting demand to a renewed railway system. Modal substitution is of paramount importance when trying to mitigate the adverse effect of transport on the environment.

Moshe Givoni and co-workers analyze the worldwide potential for air-to-rail mode substitution. They conclude that where there is sufficient demand (>750 passengers/day) for a route shorter than a 1000-km threshold, modal substitution into high-speed rail is indeed an attractive option. Given this, the worldwide potential for air-to-rail substitutions is significant, since 20% of all flights are within this threshold distance. This translates into a GHG mitigation potential from aircraft operation of up to 7%. Clearly other considerations, particularly passenger convenience and pricing, play a major role: only if the train is the more convenient mode will it be accepted by the consumer. The authors therefore see a need for a truly integrated transport system, the issue that reoccurs throughout this book.

Railways are, however, important in other respects, namely for local and cargo transport, and these were addressed in a separate section.

Sea, Rail and Cargo

Alan McKinnon of Heriot-Watts University examines the highly carbon-intensive freight transport logistics sector, a sector with ever increasing importance in the light of globalization. The academic assesses numerous ways in which the energyand emissions-intensity of logistics can be reduced and concludes that in combination they may be able to offset the growth of the sector. However, he concludes that it is unlikely that the logistics sector can achieve a 50% reduction in CO_2 emissions by 2050, unless other sectors are decarbonized in parallel. The chapter maps out areas with quick wins that require only modest capital investments, such as improved operations. This chapter also stresses that fleet lifetimes are a critical issue: we have to start renewing our vehicle fleets now in order to impact emissions and energy use in the decades to come. This again is an issue that reoccurs throughout this book.

More than 80% of cargo is transported by ships, making it a crucial sector in the logistics arena. Tristan Smith of University College London discusses the potential for increasing the energy efficiency and decreasing the GHG emissions of ships and shipping. The researcher concludes that there are many measures to achieve this, ranging from reducing drag, increasing propulsion efficiency (analogous to the aviation sector), increasing ship size, and decreasing speed. As in other sectors, improvements in operations and investments in infrastructure will be critical, according to Smith. Port access is still a significant obstacle in many countries, particularly in the developing world, and increasing port access will require largescale investments giving returns only in the long run as operations become optimized, a classic time mismatch problem.

According to Rod Smith of Imperial College railways are the easiest route to lowcarbon transport. In contrast to road transport, electrification is relatively straightforward in the case of railways. Countries such as Switzerland have already fully electrified rail networks.. In the case of cargo transport, railways have clear environmental advantages, but for passenger transport the critical factor is occupancy. Smith sees electricity as very useful for highly frequented routes and is in favor of a renewed lightweight train fleet. He concludes that high-speed rail for inter-city use could help to reduce emission for busy routes because load factors would be high.

Chris Stokes, a veteran of the railways business, sees things differently: according to this railway consultant, major conurbations in the United Kingdom are relatively close together compared to those in countries with highly successful high-speed rail networks such as Japan, France, or Germany. Therefore, Stokes concludes that distances in Britain are plainly not long enough to fully utilize the potential that high-speed rail offers. Moreover, the economic case made for HSII in the UK is based on flawed forecasts, according to Stokes, bringing into question the costbenefit ratios of HSII and hence, shedding doubt on its financial feasibility. Financial issues are crucial when planning to revive and renew a transport system.

Finance and Economics

Julie Hudson of UBS states that finance is a power facilitator of change. Individual financial instruments are capable of delivering on the minutiae of specific conditions often required to facilitate the delivery of capital to where it is needed. However, the conditions have to be right in order to attract support from financial markets and it is ultimately down to policy makers to create these conditions. Nevertheless, it is not going to be just finance that revolutionises the transport system. All areas of society, all walks of life have to play their part in the transition.

Phil Goodwin discusses existing tools for economic appraisal that help to assess transport policies. He concludes that these tools are not sufficient since they ignore the different models for urban sustainable transport. He suggests three to four models that could enhance the appraisal mechanism. A more diverse appraisal mechanism would lead to the selection of better projects, and therefore better strategies which would result in the improvement of the sustainability of transport in cities. The assessment of this academic is very much in line with assessments by practitioners from both the public and private sectors.

Bridget Rosewell of Volterra Consulting follows this with an assessment of costbenefit analysis for transport infrastructure. The description of benefits is critical for investment decisions, but this description can have many objectives, such as time savings, trip generation, or economic impacts and often neglects the environmental impact. As an example Rosewell analyzes the British HSII project and comes to the conclusion that this project was assessed on the basis of time savings, neglecting the economic boost such a connection could provide to the less well-off parts of the United Kingdom and also neglecting the environmental benefits an improved railway system could provide. This call for more holistic appraisal methodologies is another reoccurring issue within this volume.

However, holistic appraisal methodologies are not only important when talking about infrastructure investment. Tim Searchinger of Princeton University assesses an accounting error that has crept into world efforts to curb global warming. This mistake might threaten our natural habitat as it incentivizes the use of biomass as an energy source because it overestimates the emissions savings. Burning biofuels, such as alcohol from sugarcane, does of course create GHGs. The misconception is that this is balanced by the growth of the biomass (in this case sugarcane) which consumes CO_2 from the atmosphere. But if green matter—forests, shrubs—were displaced to grow the sugarcane which would themselves have consumed CO_2 from the atmosphere, this would lead to double counting and hence an overestimation of GHG savings. If this accounting error is not corrected we will continue to incentivize the use of biomass too strongly and hence will trigger the destruction of natural habitat and consequently biodiversity. This is yet another example of a flawed appraisal mechanism that could hurt us, while set in place with the best of intentions.

Conclusions

Transforming transport systems over the coming decades is both a necessity and a challenge, and not just a technological challenge: Issues ranging from geopolitics to human psychology affect transport. Energy demands, security of supply, and environmental impact make this problem multifaceted and highly complex.

Leading thinkers from government, academia, and industry demonstrate how, at every level we can transform and improve the transport sector to reduce its environmental impact and to increase energy security. The experts look at transport from very different viewpoints and emphasize different issues. Nevertheless, there are many common threads.

The energy challenge is at the very heart of the transport challenge. The global economy is powered by oil and yet conventional oil reserves, the readily accessible liquid oil we have left to meet the growing demand, is limited. There are indeed copious amounts of fossil resources remaining, particularly gas, and we can convert those into transport fuels; however, the environmental impact of using these resources is higher and would exacerbate the overall impact of transport instead of mitigating it. Biofuels will indeed be part of the solution, but we have to make sure that we incentivize low-carbon fuels and not in the process promote the destruction of our natural habitat, with their ecosystem services. That is why throughout this book authors call for bold, decisive, and practical actions in order to provide future energy security while at the same time reducing emissions from transport.

Critical to the success of emissions reduction will be efficiency, in all areas of transport according to the contributing experts; cars, trains, aircraft as well as ships have to become more energy efficient in order to satisfy increasing demand at competitive prices, and reduce emissions in parallel. Especially in aviation and shipping long fleet lifetimes, and consequently relative slow turnover, require immediate action in order to make an impact over the coming decades. These issues again call for immediate decisive action by policy makers.

But it is not only vehicles which have to be more efficient. Air traffic management, logistics, and public transport have to be optimized in order to achieve emission reductions while satisfying an ever increasing demand. Reorganization of air traffic management will be subject to political will, especially in the EU with its fragmented air traffic service. In the case of mass transit, several authors call for comprehensive solutions; only if public transport is affordable, reliable, and convenient, will it be accepted by consumers as an alternative. This clearly asks for thought-through integrated solutions that will be costly to establish, but there are excellent examples where high-speed rail networks which are connected to integrated urban mass transit and new airports. Efficient integration of individual and public transportation will be the next challenge for engineers.

Another critical issue is the need for a re-appraisal of transport projects. Up to now projects have been assessed based on their time saving, trip generation, or economic impact neglecting environmental or wider economic impacts. This has to change, and this book clearly emphasizes the need to think holistically when assessing the benefits of transport projects. The same is true for assessing the benefits of bioenergy. Bioenergy will only provide benefits if the appraisal process is correct and costs and benefits are viewed holistically. The importance of full Life Cycle Assessments cannot be overestimated.

Renewing the efficiency of the transport system requires two things in particular: political will and capital. That is why throughout this book authors call for policy makers to incentivize energy efficiency and low-carbon developments, create markets, and reduce uncertainty so that consumers and producers both start adapting alternative, low-carbon options. As soon as markets are in place and uncertainty is reduced, the financial sector will act as the facilitator of change and provide capital where it is needed most. But the political decision makers have to boldly take the lead. Clarity of intent,leadership, political will, and decisive action are needed. Here again, there are best practise examples. The lifestyle of the citizens of Bogota, the Colombian capital of 11 million people, was dramatically improved over a period of just 3 years through the inspired vision and leadership of its mayor, Enrique Penelosa.

But there are problems right on our doorstep. Population growth combined with rapid urbanization in developing countries poses severe threats but also enormous opportunities if carefully managed. Many cities in the developing world are imitating the low density growth of cities like Huston in the USA, with transport systems almost totally reliant on the car and hence on oil. Alternative models are Hong Kong, Singapore, and London. The Smith School at Oxford University is advising the Government of Rwanda on how to foster low-carbon sustainable growth and one of the key areas the Smith School team assessed was indeed transport and its oil dependence, and urban development.

It is of paramount importance to help developing countries to leapfrog beyond the mistakes of the developed world and avoid the economically crippling addiction to oil which will otherwise ensue.

Both the developed and the developing worlds have to make informed decisions to avoid lock ins and resource scarcity issues. For this reason the Smith School has initiated a comprehensive futures programme on low-carbon mobility. This programme will generate and translate insights through the development of a more effective futures toolkit and its application in real-world situations, presenting the diversity of different mobility systems in differing regional contexts.

Printed by Printforce, the Netherlands